Modified and Controlled Atmospheres for the Storage, Transportation, and Packaging of Horticultural Commodities

Modified and Controlled Atmospheres for the Storage, Transportation, and Packaging of Horticultural Commodities

Edited by
Elhadi M. Yahia

CRC Press
Taylor & Francis Group
Boca Raton London New York

CRC Press is an imprint of the
Taylor & Francis Group, an **informa** business

CRC Press
Taylor & Francis Group
6000 Broken Sound Parkway NW, Suite 300
Boca Raton, FL 33487-2742

First issued in paperback 2019

© 2009 by Taylor & Francis Group, LLC
CRC Press is an imprint of Taylor & Francis Group, an Informa business

No claim to original U.S. Government works

ISBN-13: 978-1-4200-6957-0 (hbk)
ISBN-13: 978-0-367-38589-7 (pbk)

Library of Congress Cataloging-in-Publication Data

Modified and controlled atmospheres for the storage, transportation, and packaging of horticultural commodities / editor, Elhadi M. Yahia T.
    p. cm.
   Includes bibliographical references and index.
   ISBN 978-1-4200-6957-0 (alk. paper)
   1. Food--Storage. 2. Food--Transportation. 3. Food--Packaging. 4. Food--Preservation. 5. Food--Safety measures. I. Yahia, Elhadi M.

TP373.3.M63 2009
664'.028--dc22

2008036120

Visit the Taylor & Francis Web site at
http://www.taylorandfrancis.com

and the CRC Press Web site at
http://www.crcpress.com

Dedication

To Alicia, Mariam, Naolia-Amina and Tarek

Dedication

Contents

Preface

In the last two to three decades there have been major changes in food consumption habits, including a significant increase in the consumption of fresh fruits and vegetables due to health concerns. These changes have created the need for the development and application of adequate technologies to preserve these perishable food items.

The concept of modified atmospheres (MAs) and controlled atmospheres (CAs) for horticultural commodities originated nearly two centuries ago when Jack Berard in France observed that harvested fruits utilized oxygen and produced carbon dioxide and when fruits were deprived of oxygen they did not ripen, or ripened very slowly. Nearly a century passed before the first principles underlying the scientific basis for the effects of MA/CA were elucidated. It became clear that one of the primary benefits of MA and CA was the effect on the synthesis and action of the plant hormone ethylene needed to initiate fruit ripening and plant organ senescence.

MA and CA have been used for storage, transport, and packaging of foods for the last seven to eight decades. In the 1930s, meat was transported in MA from Australia to Europe. Meat packed and transported in dry ice (carbon dioxide) had better quality than meat packed in conventional ice. In the 1920s, Franklin Kidd and Cyril West in England, and later (1930s and 1940s) several researchers in the United States, including Robert Smock, started the storage of apples in CA as we know it today. The first CA storage room for apples was built near Canterbury, England in 1929. Today more than 10 million tons of apples are stored in CA in many developed and developing countries. The 1980s witnessed major improvements in the technology and a significant increase in its utilization in storage, transport, and packaging of different types of foods. In more recent years there have been more developments in several aspects of the technology, such as modified atmosphere packaging (MAP), transport of fresh commodities in CA, and the use of the technology for insect control (quarantine systems), among others.

MA and CA technologies utilize a process or system in which fresh, perishable commodities (fruits, vegetables, cut flowers, seeds, nuts, and feedstocks) can be either stored, transported, or packaged under narrowly defined environmental conditions (temperature, humidity, and gaseous composition) to extend their useful marketing period after harvest. The absolute or desirable levels of these environmental variables differ according to commodity and stage of development. Moreover, tissue response may vary because of interactions among these variables as influenced by variety and preharvest conditions and climatic factors.

Commercial interest in the development and application of MA and CA for packaging, transportation, and storage of fresh horticultural commodities (fruits, vegetables, and flowers) triggered research in these areas. Commercial application of CA storage is used for apples, pears, kiwifruits, sweet onions, cabbage, and can be used for bananas. Commercial application of MA and CA for long-distance marine transport is used for apples, avocados, bananas, blueberries, cherries, figs, kiwifruits, mangoes, nectarines, peaches, pears, plums, raspberries, melons, grapefruit, strawberries, and some flowers, among other commodities. Commercial application of MA for packaging is used in several intact and fresh-cut commodities. Commercial application of MA and CA for insect control in horticultural commodities has recently been authorized by the United States Department of Agriculture (USDA).

MA and CA involve altering and maintaining (in the case of CA) an atmospheric composition that is different from air composition (about 78%–79% N_2, 20%–21% O_2, and 0.03% CO_2, and trace quantities of other gases); generally, O_2 below 8% and CO_2 above 1% are used and considered as a supplement to maintenance of optimum temperature and relative humidity (RH) for each commodity in preserving quality and safety of fresh fruits, ornamentals, vegetables, and their products throughout postharvest handling. Exposure of fresh horticultural crops to low O_2 and/or elevated CO_2 atmospheres within the range tolerated by each commodity reduces their respiration and ethylene production rates, and therefore results in several beneficial effects such as retardation of senescence (including ripening) and associated biochemical and physiological changes (slowing down rates of softening and compositional changes), reduction of sensitivity to ethylene action, alleviation of certain physiological disorders such as chilling injury, direct and indirect control of pathogens (bacteria and fungi) and consequently decay incidence and severity, and can be a useful tool for insect control. However, outside the range of tolerance, it can lead to incidence of physiological disorders and increased susceptibility to decay, and even fermentation. The objective of applying MA and CA technologies is to extend the useful marketing period for the commodity during storage, transport, and distribution to maintain quality, nutritive value, or market value of the product for eventual consumption over that achievable by the use of controlled temperature only.

The fact that this technology utilizes natural gases and would not cause harmful effects for human health and for the environment made it very appealing as an alternative to chemicals commonly used in foods for the control of diseases and insects. Several MA and CA systems have been developed for several crops as an alternative for agrochemicals, including some insect control protocols that have been accepted a few months ago by USDA as quarantine systems.

Several improvements have been made in recent years on MA and CA technologies, such as better construction of sealed storage rooms and transport containers; better gas monitoring and control systems; new packaging systems; creating nitrogen by separation from compressed air using molecular sieve beds or membrane systems; low O_2 (≤ 1 kPa) storage; low ethylene ($<1\,\mu L\,L^{-1}$) CA storage; rapid CA (rapid establishment of optimal levels of O_2 and CO_2); programmed (or sequential) CA storage (storage in 1 kPa O_2 for 2–6 weeks followed by storage in 2–3 kPa O_2 for the remainder of the storage period); improved technologies of establishing, monitoring, and maintaining MA and CA; polymeric films with appropriate gas permeability to create a desired atmosphere during packaging; quarantine technologies for insect control; among others. Development of processes to artificially establish and maintain controlled levels of temperature, humidity, oxygen, carbon dioxide, and ethylene has, in some instances, gone beyond the research base, which defines the limits within which these parameters may be applied with predictable success in distributing fresh and wholesome produce to the consumer.

The packaging, transportation, and storage of fresh fruit and vegetables are undertaken on a global scale, which are highly dependent on technology. MA and CA are widely applied techniques and important alternatives to chemical preservatives and pesticides. They have great potential for reducing postharvest losses and for maintaining both nutritional and market value. However, the CA storage technology is only widely used for a few commodities (apples, pears, and to a certain extent kiwifruits, sweet onions, and cabbage), and there is still no agreement in the food industry and among researchers as to its usefulness for many other commodities. However, MA and CA for packaging and transport are applied for several types of foods.

Very few books have been written on MA and CA of perishable foods, and none has covered all aspects of the technology. The only instances when CA/MA has been addressed for a very long time have been the proceedings of the "international CA research conferences" held every four years since 1969, which are not directed or comprehensive (just a compilation of submitted articles), and Thompson's 1998 book, *Controlled Atmosphere Storage of Fruits and Vegetables*; although this is an excellent book and very useful as an extensive review of the CA application literature, it does not cover the latest advances in both the application and the basic aspects like modeling, physiology, and biochemistry. The book by S. Burg (*Postharvest Physiology and Hypobaric Storage of Fresh Produce*, published by CABI in 2004), is excellent, but only covers hypobaric storage, which is not commercially used. There have been some other books that have dealt with many other types of foods (usually dedicating one or two chapters to perishables), and almost all of these books have dealt only with MAP, covering very little related to transport and storage of perishables in MA and CA.

Following more than 70 years of research and development on MA and CA, this book will trace the historical developments of this technology, providing information on ideal conditions to be used for many horticultural commodities and outlining the effects of MA and CA on physiology and biochemistry of these commodities and on their flavor and quality.

This book is divided into 23 chapters written by 44 experts in research and industry from 13 countries. All the authors are very well-known in different disciplines, and are engaged in research on the basic concepts as well as application of this very important technology. This book is comprehensive; covering the entire subject for the first time in a single book (storage, transport, packaging, physiological and biochemical aspects, engineering aspects, effects on all commodities, etc.). The chapters cover all aspects of science and application including technological development and applications of MA and CA (storage, transport, and packaging) for all fruits, vegetables, and ornamentals of temperate, subtropical, and tropical origin. A large amount of literature is reviewed on this topic in the book, providing the most comprehensive reference on all basic and applied aspects of MA and CA. The authors are to be appreciated for providing an excellent base of knowledge upon which researchers, students, educators, and industry personnel can advance this important technology. This book is an essential reading material for horticultural researchers and educators, and food industry personnel concerned with transportation, storage, and packaging of perishable foods. The book should also be of interest to regulatory bodies, consumer groups, and students of horticulture, agriculture, food science and technology, and food marketing. It can serve as an excellent textbook for several graduate and undergraduate courses on postharvest technology, food science and technology, food safety, food processing, food handling, and food engineering.

I hope that this book contributes to the improvement of this important technology, which is vital for the preservation of foods, resulting in increased food availability, improved food safety, and reduced losses.

I would like to express my sincere appreciation for the efforts of all the authors and for writing excellent chapters presented in the book. I am grateful to the staff at CRC Press/Taylor & Francis who have been very helpful throughout the review and production processes.

If you have any suggestions or comments on the content of the book, please contact the publisher or the editor at yahia@uaq.mx or elhadiyahia@hotmail.com.

Elhadi M. Yahia, PhD
Queretaro, Mexico

Editor

Elhadi M. Yahia is currently professor of human nutrition at the Facultad de Ciencia Naturales of the Universidad Autonoma de Queretaro in Mexico. Dr. Yahia received his BSc from the University of Tripoli in Libya, MSc from the University of California, Davis and his PhD from Cornell University, Ithaca, New York. At his present position at Universidad Autonoma de Queretaro, he is engaged in teaching, conducting research, and supervising graduate students in postharvest technology and human nutrition. Dr. Yahia is also a courtesy professor at the Department of Horticultural Sciences at the University of Florida and a visiting professor at other universities in Mexico and other countries.

Dr. Yahia taught courses at undergraduate and graduate levels, and presented more than 200 invited lectures at several institutions in more than 35 countries in North, Central and South America, Europe, Asia, and Africa. He has also conducted intensive courses and training programs for different food associations: producers, packers, distributors, exporters, and government personnel in several countries across four continents. Dr. Yahia is the author of 6 books, 40 book chapters, and 150 technical articles on postharvest physiology, technology, and handling of perishable foods.

Dr. Yahia also serves as an advisor/consultant to several organizations and companies such as the Food and Agriculture Organization of the United Nations, the World Bank, USAID, USDA, ICARDA, World Food Logistics Organization, Winrock International, Arcadis Euroconsult, Ronco Consulting, Nitec, CIMO, Mayan Fresh Produce, and several governmental organizations. Dr. Yahia is a member of Sigma Xi, the Mexican Academy of Sciences, the Scientific Advisory Council of the World Food Logistics Organization, the Cold Chain Experts, the Mexican National Research System, the Mexican Federal Commission for the Protection Against Health Risks, the Advisory Committee for Science and Technology for the Governor of the State of Queretaro, the Steering Committee of the National Research and Technology Council of Mexico, and several other international organizations.

Contributors

Jinhe Bai
Citrus and Subtropical Products Laboratory
United States Department of
  Agriculture-Agricultural
  Research Service
Winter Haven, Florida

Elizabeth A. Baldwin
Citrus and Subtropical Products Laboratory
United States Department of
  Agriculture-Agricultural
  Research Service
Winter Haven, Florida

Wayne Benson
Environment Management Systems
Ingersoll Rand Climate Control
  Technologies
Minneapolis, Minnesota

Jeffrey S. Brandenburg
The JSB Group, LLC
Greenfield, Massachusetts

David G. Brandl
San Joaquin Valley Agricultural
  Sciences Center
United States Department of
  Agriculture-Agricultural
  Research Service
Parlier, California

Jeffrey K. Brecht
Center for Food Distribution and Retailing
Horticultural Sciences Department
Institute of Food and Agricultural Sciences
University of Florida
Gainesville, Florida

Patrick E. Brecht
PEB Commodities, Inc.
Petaluma, California

Carlos H. Crisosto
Department of Plant Sciences
Kearney Agricultural Center
University of California
Davis, California

John M. DeLong
Atlantic Food and Horticulture Research
  Centre
Agriculture and Agri-Food Canada
Kentville, Nova Scotia, Canada

Shawn Dohring
Montship Inc.
Montreal, Quebec, Canada

Charles F. Forney
Atlantic Food and Horticulture Research
  Centre
Agriculture and Agri-Food Canada
Kentville, Nova Scotia, Canada

Guy J. Hallman
United States Department of
  Agriculture-Agricultural
  Research Service
Weslaco, Texas

Maarten L. A. T. M. Hertog
BIOSYST-MeBioS
Katholieke Universiteit Leuven
Leuven, Belgium

Q. Tri Ho
BIOSYST-MeBioS
Katholieke Universiteit Leuven
Leuven, Belgium

Ernst Hoehn
Agroscope Changins-Wädenswil
  Research Station
Wädenswil, Switzerland

Judy A. Johnson
San Joaquin Valley Agricultural
  Sciences Center
United States Department of
  Agriculture-Agricultural
  Research Service
Parlier, California

Daryl C. Joyce
Centre for Native Floriculture
School of Land, Crop and Food Sciences
The University of Queensland
Gatton, Queensland, Australia

Adel A. Kader
Department of Plant Sciences
University of California
Davis, California

Angelos K. Kanellis
Group of Biotechnology of
  Pharmaceutical Plants
Laboratory of Pharmacognosy
Department of Pharmaceutical Sciences
Aristotle University of Thessaloniki
Thessaloniki, Greece

Christian Larrigaudière
Postharvest Unit
Institut de Recerca i Tecnologia
  Agroalimentaries
Lleida, Spain

Yaguang Luo
Produce Quality and Safety Laboratory
United States Department of
  Agriculture-Agricultural
  Research Service
Beltsville, Maryland

Susan Lurie
Volcani Center
Department of Postharvest Science
Agricultural Research Organization
Bet Dagan, Israel

Andrew J. Macnish
Department of Plant Sciences
University of California
Davis, California

James P. Mattheis
Tree Fruit Research Laboratory
United States Department of
  Agriculture-Agricultural
  Research Service
Wenatchee, Washington

Edmund K. Mupondwa
Saskatoon Research Centre
Agriculture and Agri-Food Canada
Saskatoon, Saskatchewan, Canada

Lisa G. Neven
Yakima Agricultural Research Laboratory
United States Department of
  Agriculture-Agricultural
  Research Service
Wapato, Washington

Bart M. Nicolaï
Flanders Centre of Postharvest Technology
BIOSYST-MeBioS
Katholieke Universiteit Leuven
Leuven, Belgium

David O'Beirne
Department of Life Sciences
University of Limerick
Limerick, Ireland

Pierdomenico Perata
Scuola Superiore Sant'Anna
Pisa, Italy

Robert K. Prange
Atlantic Food and Horticulture
  Research Centre
Agriculture and Agri-Food Canada
Kentville, Nova Scotia, Canada

Michael S. Reid
Department of Plant Sciences
University of California
Davis, California

Julio Retamales
Valent BioSciences Corporation
University of Chile
Las Condes, Santiago, Chile

Wendy C. Schotsmans
Postharvest Unit
Institut de Recerca i Tecnologia
  Agroalimentaries
Lleida, Spain

Peter L. Sholberg
Pacific Agri-Food Research Centre
Agriculture and Agri-Food Canada
Summerland, British Columbia, Canada

S. P. Singh
Curtin Horticulture Research Laboratory
School of Agriculture and Environment
Curtin University of Technology
Perth, Western Australia, Australia

Zora Singh
Curtin Horticulture Research Laboratory
School of Agriculture and Environment
Curtin University of Technology
Perth, Western Australia, Australia

Leon A. Terry
Plant Science Laboratory
Cranfield Health
Cranfield University
Bedfordshire, United Kingdom

Peter M. A. Toivonen
Pacific Agri-Food Research Centre
Agriculture and Agri-Food Canada
Summerland, British Columbia, Canada

Pietro Tonutti
Scuola Superiore Sant'Anna
Pisa, Italy

Pieter Verboven
BIOSYST-MeBioS
Katholieke Universiteit Leuven
Leuven, Belgium

Bert E. Verlinden
Flanders Centre of Postharvest Technology
BIOSYST-MeBioS
Katholieke Universiteit Leuven
Leuven, Belgium

Clément Vigneault
Agriculture and Agri-Food Canada
Saint-Jean-sur-Richelieu, Quebec, Canada

Elhadi M. Yahia
Faculty of Natural Sciences
Autonomous University of Queretaro
Queretaro, México

Devon Zagory
Food Safety and Quality Programs
NSF Davis Fresh
Watsonville, California

1

Introduction

Elhadi M. Yahia

CONTENTS

1.1 Introduction

Modified atmosphere (MA) refers to any atmosphere different from the normal air (20%–21% O_2, about 0.03% CO_2, about 78%–79% N_2, and trace quantities of other gases), while controlled atmosphere (CA) refers to atmospheres different than normal air and strictly controlled during all time. MA and CA usually involve atmospheres with reduced O_2 and/or elevated CO_2 levels. Hypobaric (low pressure) storage is a CA system involving the use of vacuum to reduce the partial pressure of the gas component of air (Burg, 2004; Yahia, 2004).

The technologies of MA and CA are widely used for the storage, transport, and packaging of several types of foods. They offer several advantages such as delay of ripening and senescence of horticultural commodities, control of some biological processes such as rancidity, insects, bacteria and decay, among others. MA and CA combined with precision temperature management allow nonchemical insect control in some commodities for markets that have restrictions against pests endemic to exporting countries and for markets that prefer organic produce (see Chapter 11).

Major developments have been accomplished in the last two to three decades, such as better construction of sealed storage rooms (see Chapter 2) and transport containers (see Chapter 3), better gas monitoring and control systems, and new packaging systems (see Chapters 4 and 5). Optimum atmosphere for different types of foods is very variable, and depends on many factors such as type of product, purpose of use, physiological age, holding temperature, and duration of treatment (see Chapters 12 through 20). Exposure of different types of foods (especially horticultural products) to O_2 levels below, and/or CO_2 levels above their optimum tolerable range can cause the initiation and/or aggravation of certain physiological disorders, irregular ripening, increased susceptibility to decay, development of off-flavors, and could eventually cause the loss of the product (see Chapters 8 and 9). Most horticultural crops can tolerate extreme levels of gases when stored for only short periods (Yahia, 1998a,b, 2004).

MA and CA should always be used as a compliment and never as a substitute for proper handling techniques, especially optimum temperature and relative humidity (RH). CA storage has been restricted mainly to apples, pears, kiwifruit, and cabbage. MA and CA for transport are used for many horticultural commodities. Modified atmosphere packaging (MAP) is used for some intact and minimally processed fruits and vegetables, in addition to some other types of foods such as meat, poultry, fish and sea foods, cheese and some other milk products, prepared foods, dry and dehydrated foods, coffees, among others.

Commercial interest in the development and implementation of MA/CA for transportation, storage, and packaging of fresh horticultural commodities (Table 1.1) will stimulate research in these areas. Economic principles of supply and demand for the commodity in relation to energy use, cost, benefit, and practicality will ultimately determine the use of MA/CA for postharvest maintenance of horticultural commodities (see Chapter 21). Continued technological developments in the future to provide MA/CA during storage, transport, and packaging at a reasonable cost are essential to greater applications on fresh horticultural commodities and their products (see Chapter 23).

There are basic biologically defined limits to environment that can be employed for the postharvest preservation of fresh horticultural commodities. The basic processes of transpiration, respiration, and biochemical transformations may continue in much the same manner in plant tissues after harvest as before harvest. However, this is done without replenishment of reserves or metabolic control from the parent plant. The nature of these cellular processes can be altered, and the chemical reaction rate can be attenuated or stimulated by manipulating the temperature and concentration of the biologically active gases such as water vapor, O_2, CO_2, and ethylene among others. Beyond certain limits of temperature and gas atmosphere composition and according to the nature and stage of development of the commodity, physiological disorders may develop, which terminate the useful life of the commodity by destroying appearance, flavor, nutritive value, or wholesomeness. Development and application of successful postharvest preservation techniques

TABLE 1.1

Classification of Fruits and Vegetables According to Their CA Storage Potential at Optimum Temperatures and Relative Humidities

Range of Storage Duration (Months)	Commodities
>12	Almond, Brazil nut, cashew, filbert, macadamia, pecan, pistachio, walnut, dried fruits, and vegetables
6–12	Some cultivars of apples and European pears
3–6	Cabbage, Chinese cabbage, kiwifruit, persimmon, pomegranate, and some cultivars of Asian pears
1–3	Avocado, banana, cherry, grape (no SO_2), mango, olive, onion (sweet cultivars), some cultivars of nectarine, peach and plum, and tomato (mature-green)
1	Asparagus, broccoli, cane berries, fig, lettuce, muskmelons, papaya, pineapple, strawberry, sweet corn, fresh-cut fruits, and vegetables

Source: From Kader, A.A. 1986. *Food Technology*, 40, 99, 102.

for fresh horticultural commodities depends largely on understanding certain fundamental aspects of biology, engineering, and economics that are important in the maintenance and distribution of these perishables (see Chapter 22).

1.2 History

MA and CA as techniques for postharvest preservation of foods originated several centuries ago. This was long before the role of O_2 and CO_2 in the basic respiration of plants and plant organs was understood. The centuries-old tradition of burying certain fruits and vegetables in the ground after enclosing them in various wrappings and protecting them from freezing by covering them with soil and insulating materials is an example of a modified environment for extending the life of some types of foods. This practice is still used to a limited extent and most probably will continue in the future, especially in regions with very cold winters. Other examples of subterranean storage to benefit from reduced and more stable temperatures, and modified gas atmosphere include the practice of cave storage of fruits and vegetables as done in China and Turkey, and trench storage of grains, root crops, and silage. Cave storage of apples and pears is still practiced in China, and citrus in the region of Cappadocia in Turkey, and the results of this relatively unsophisticated storage technology rival those obtained by modern means and at only a fraction of the cost (Dilley, 2006). Several countries (especially developing countries) have long practiced the storage of grains and root crops in huts and pits to protect them from extreme, unstable weather conditions and from infestation of insects and rodents. The use of polymeric films and synthetic rubber and metal liners is being adopted as a means to provide atmospheric modification to further extend the useful storage life for some of the staple food and feedstocks (Dilley, 2006).

According to Dalrymple (1969), the first recorded scientific study of CA storage was conducted in the early 1800s in France by Jacques Etenne Berard, a chemist in the School of Pharmacy at Montpellier. His studies revealed that harvested fruits utilized O_2 and produced CO_2, and fruits kept in an atmosphere devoid of oxygen did not ripen. Peaches, prunes, and apricots were found to be stored for up to 1 month, and apples and pears for up to about 3 months at room temperature. Moreover, he found that the fruit would ripen after returning them to air, providing that they were not kept too long in storage without oxygen. Berard's studies published in 1821 won him acclaim and the Grand Prix de Physique from the French Academy of Science. These studies apparently stimulated other scientists to conduct similar investigations, some of which are reviewed in a publication by the U.S. Department of Agriculture (Bigelow et al., 1905).

The first recorded studies on MA/CA of fruits in the United States were made in Northern Ohio by Benjamin Nyce in about 1865 (Dilley, 1990). He constructed an airtight sheet metal-lined storage room of about 4000 bushel capacity that was cooled with ice to maintain a storage temperature of just above the freezing point of water (Dilley, 2006). Apples were stored commercially at the Cleveland storage in saleable condition for periods up to 9 months following harvest, and Nyce obtained several U.S. patents on his storage technology. A subsequent CA storage facility built by Nyce in New York did not prove to be commercially acceptable. The San Jose Fruit Company in California conducted tests in the 1890s with railcar shipments of peaches, nectarines, pears, quinces, persimmons, and grapes in which nonrefrigerated railcars were sealed and gassed with carbon dioxide during shipment of produce from California to Chicago. The tests had partial success and may be considered the forerunner of what subsequently has become known as MA transport of perishables. Shipments of peaches, plums, and pears were made for several years, although some success was seen only in pears. The practice of carbon dioxide enrichment of the transport of perishables was reinstated as a supplement to refrigeration in the 1960s. Strawberries and sweet cherries show beneficial effects from carbon dioxide enrichment of the atmosphere during transport. These early trials with CA or MA storage and transport of perishables, although encouraging, indicated a high level of risk of produce loss, which probably was a factor in slow commercial acceptance of these practices (Dilley, 2006). Major scientific research is still needed to adequately establish MA and CA for transport.

Scientists at Washington State University conducted some of the earliest reported studies on the effects of the gaseous environment on ripening of apples (Dilley, 2006). The results of these investigations, which were conducted in 1903 and 1904, were eventually published by Thatcher (1915). They made the important observation that carbon dioxide was an inhibitor of the ripening process, and 60 years would pass before Burg and Burg (1965) elucidated the mechanism of the carbon dioxide effect as competitive inhibition of ethylene action on ripening. Hill's studies (described by Thatcher) also implicated the possible role of enzymes on the basic respiration of apples during ripening. Again, nearly 60 years would pass before studies at Michigan State University showed that protein synthesis and enzyme activity were closely linked to the ripening process of apples (Frenkel et al., 1968). This was confirmed by the investigations of Hulme and coworkers at the Ditton Laboratory in England (Hulme et al., 1968). In the early 1900s, Hill (1913) at Cornell University made the observation that ripening of peaches could be slowed by storing them in carbon dioxide and, to a lesser extent, in other inert gases. The beneficial effects were ostensibly achieved by restricting the availability of oxygen. This confirmed, to some extent, the observations made at San Jose Fruit Company noted earlier. Hill also made the interesting observation that the respiration rate of peaches exposed to carbon dioxide did not return to the prestorage level. The U.S. Department of Agriculture renewed investigations on the effects of carbon dioxide on storage life and respiration of apples in conjunction with studies on superficial scald of apples (Brooks et al., 1919).

The work that led to the commercial application of CA storage was laid by Franklin Kidd and Cyril West, working at the Low Temperature Research Station at Cambridge, England. They began a systematic study of fruit respiration and ripening as influenced by temperature, CO_2, and O_2 under the auspices of the Food Investigation Board of the U.K. Department of Scientific and Industrial Research. These studies, published over a period spanning nearly 40 years (Kidd and West, 1927a,b, 1930, 1937, 1950), provided the basis for much of the CA storage technology in use today (Dilley, 2006). Their initial studies done on pome fruits and berries have been reviewed by Fidler et al. (1973), where emphasis was directed on apples, but later also included work on bananas, pears, and oranges. Their research was continued at the Ditton Laboratory, which was constructed in 1930 on the grounds of the East Malling Research Station in Kent, England, where a portion of the main building was constructed as a replica of the hold of a ship, presumably to address the problems of temperature and gas atmosphere maintenance encountered during maritime transport of fruits from the Commonwealth countries to England (Dilley, 2006). Fidler and colleagues also conducted CA storage research at the Ditton and Covent Gardens Laboratories, and contributed toward current-day recommendations for the CA storage of various varieties of apples and pears grown in the United Kingdom. The first commercial apple storage to employ the "gas storage" technique in England was constructed in 1929 near Canterbury in Kent County. The commercial success of this venture was realized in 1930 and led to rapid expansion in the adoption of gas storage technology for apples and pears in the decade of the 1930s (Dilley, 2006).

The first commercial CA storage in Michigan was established in Kent County, Michigan, in the early 1950s (Dilley, 2006). This was done by providing a gas-tight sheet metal liner in an existing mechanically refrigerated apple storage, which provided most of the early expansion in CA storage volume in the United States and Canada. It was not until the mid-1950s, when the practice of using bulk bins and lift trucks was introduced, that new construction of CA storages with high ceilings began rapidly replacing the practice of storing apples and pears in 1 bushel field crates. The increased efficiency in fruit handling and labor savings afforded by the use of bulk bins for harvesting, handling, and storing of fruit were important factors in the growth of CA storage capacity worldwide since 1950 (Dilley, 2006).

Overholser (1928) and Biale (1942), investigated CA storage of avocados. In 1935, Allen and McKinnon found that CA storage of Yellow Newton apples at 3°C–4°C allowed this variety to be stored successfully, avoiding low-temperature disorders found when the fruits were stored in air at 0°C–1°C (Dilley, 2006). This confirmed the earlier observations of Kidd and West (1927a) with several important English apple varieties where storage life was found to be limited by low-temperature or chilling injury disorders and these could be largely eliminated by gas storage at 3°C–4°C. This was an important revelation in the 1930s and was responsible for much of the early growth in the adoption of CA storage technology (Dilley, 2006). Robert Smock joined Allen in California and extended the CA storage investigations on apples and broadened the studies with peaches, pears, and plums (Allen and Smock, 1937), and continued his investigations upon moving to Cornell University in the late 1930s. Nearly 20 years would pass before CA storage of apples was an accepted technology in the Pacific Northwest and Archie Van Doren was largely responsible for its commercialization (Dilley, 2006). The commercial development of CA storage of apples and pears in Canada can be largely attributed to the research of Charles Eaves of the Canada Department of Agriculture, and Agriculture Canada Research Station at Kentville, Nova Scotia, where he made numerous significant and innovative contributions (Eaves, 1959, 1960, 1963).

MA and CA research have extended later in different regions and included different commodities. CA studies on citrus were conducted in Florida and California and on cranberries in Massachusetts. The first commercial CA storages in Elgin, South Africa was established in 1935 and 1936, but only operated successfully for a short period of time before being converted into conventional refrigerated air storage (Dilley, 2006). The first commercial CA storage in modem times in the Western Hemisphere was established in Nova Scotia, Canada in 1939, followed in 1940 by equipping existing refrigerated fruit storage rooms for operation as CA storages at Lockport and at Sodus, New York, followed by construction of new CA storage enterprises in the Hudson Valley in New York State (Dilley, 2006). Gradual expansion of CA storage capacity continued in New York and eastern Canada during the 1940s, and in New England in the 1950s, where CA storage was used largely for McIntosh, the major dessert apple variety in these regions (Dilley, 2006). An active period of expansion in CA storage in the United States and Canada began in the 1950s, and the most significant and accelerated expansion began in the 1960s when Washington state apple storage operators and shippers realized the improvements to be gained in dessert quality and marketability of Red and Golden Delicious apples by storing them under CA (Dilley, 2006). By 1960, the total CA holdings of apples and pears in the United States amounted to over 4 million bushels, and by the late 1980s, CA storage capacity for apples and pears exceeded 100 million bushels (Dilley, 2006). CA storage rooms have been established in several countries in all the continents such as several European countries, South Africa, Egypt, Jordan, India, Mexico (since 1980), Chile, Argentina, among others. MA for transport was used already in the 1930s, when ships transporting fruits had high levels of CO_2 in their holding rooms, thus increasing the shelf life of the product. In Mexico, trials of using MA during marine shipping of mango started in 1974.

1.3 Gases Used

MA and CA usually consist of N_2, O_2, and CO_2. It is the altered ratio of O_2 and CO_2 that makes a difference in the preservation of food commodities. By reducing the O_2 level and increasing the CO_2 level, ripening of fruits and vegetables can be delayed, respiration and ethylene production rates can be reduced, softening can be retarded, and various compositional changes associated with ripening can be slowed down. Oxygen is essential for the respiration of fresh horticultural commodities. The removed O_2 can be replaced with N_2, commonly acknowledged as an inert gas, or CO_2, which is a competitive inhibitor of ethylene action and can lower the pH or inhibit the growth of some fungi and bacteria. Respiration rate starts to decrease when O_2 level is decreased to below 12 kPa, and levels commonly used for most fresh horticultural commodities is about 3–5 kPa. The absence of O_2 can lead to anaerobic respiration, accelerating deterioration, and spoilage. High CO_2 levels are effective bacterial and fungal growth inhibitors; however, levels ≥ 10 kPa are needed to suppress fungal and bacterial growth significantly. Atmospheres >10 kPa CO_2 can be phytotoxic to many fresh horticultural commodities. Nitrogen is used as a filler gas since it has no direct biological effects on horticultural commodities, and therefore N_2 is commonly used as the inert component of MA/CA. Replacing N_2 with argon or helium may increase diffusivity of O_2, CO_2, and C_2H_4, but they have no direct effect on plant tissues and are more expensive than N_2 as an MA/CA component.

1.4 Biological Basis of MA/CA Effects

There are basic biologically defined limits to the environment that can be employed for the postharvest preservation of fresh horticultural commodities. Basic biological processes in these commodities such as transpiration, respiration, and biochemical transformations continue after harvest. However, these processes can be altered by manipulating different factors such as temperature, RH, and the concentration of biologically active gases such as water vapor, O_2, CO_2, and ethylene, among others. Beyond certain limits of these factors, especially gas atmosphere composition, and according to stage of development of the commodity, physiological disorders may develop, which can terminate the useful life of the commodity by negatively affecting appearance, flavor, nutritive value, or wholesomeness.

Optimum ranges of O_2 and CO_2 levels can result in several advantages to the food commodity, but unfavorable atmospheres can induce physiological disorders and enhance susceptibility to decay, among other possible problems. Elevated CO_2-induced stresses are additive to and sometimes synergistic with stresses caused by low O_2; physical or chemical injuries; and exposure to temperatures, RH, and/or C_2H_4 concentrations outside the optimum range for the commodity. The shift from aerobic to anaerobic respiration due to low O_2 and/or high CO_2 atmospheres depends on fruit maturity and ripeness stage (gas diffusion characteristics), temperature, and duration of exposure to stress-inducing conditions. Up to a certain limit, fresh horticultural commodities are able to recover from the detrimental effects of low O_2 and/or high CO_2 stresses and resume normal respiratory metabolism upon transfer to normal air. Plant tissues have the capacity for recovery from the stresses caused by extreme gas atmospheres, and postclimacteric tissues are less tolerant and have lower capacity for recovery than preclimacteric fruits. The speed and extent of recovery from reduced O_2 and elevated CO_2 stress depend upon gas level, physiological stage of the tissue, temperature, and duration of exposure.

Elevated CO_2 atmospheres inhibit activity of ACC synthase, a key regulatory site of ethylene biosynthesis, while ACC oxidase activity is stimulated at low CO_2 and inhibited at high CO_2 and/or low O_2 levels. Ethylene action is inhibited by elevated CO_2 atmospheres. Optimum atmospheric compositions retard chlorophyll loss (green color), biosynthesis of carotenoids (yellow, orange, and red colors) and anthocyanins (red and blue colors), and biosynthesis and oxidation of phenolic compounds (brown color). MA and CA slow down the activity of the cell wall, degrading the enzymes involved in softening and enzymes involved in lignification, leading to toughening of vegetables. Low O_2 and/or high CO_2 atmospheres influence flavor by reducing loss of acidity, starch to sugar conversion, sugar interconversions, and biosynthesis of aromatic volatiles (see Chapter 7) (Yahia, 1994). Optimum atmospheres can help the retention of ascorbic acid and other vitamins, resulting in better nutritional quality (see Chapter 6). Severe O_2 and CO_2 stress, especially very high CO_2 atmospheres, can decrease cytoplasmic pH and ATP levels, and reduce pyruvate dehydrogenase activity while pyruvate decarboxylase, alcohol dehydrogenase, and lactate dehydrogenase are induced or activated (Ke et al., 1994, 1995). This causes accumulation of acetaldehyde, ethanol, ethyl acetate, and/or lactate, which may be detrimental to the commodities if they are exposed to stress MA/CA conditions beyond their tolerance (see Chapter 11). However, relative tolerance to severe atmospheres varies among different products, cultivars, ripening stages, storage temperatures and duration of exposure, and in some cases, ethylene concentrations. For example, mango fruit is much more tolerant to extreme atmospheric stress compared to avocado and guava (Yahia, 1998a,b).

MA/CA can have a direct or indirect effect on postharvest pathogens and consequently on decay incidence and severity (see Chapter 9). For example, CO_2 at 10–15 kPa significantly inhibits development of *Botrytis* rot on strawberries, cherries, and other perishables. However, other atmospheres may reduce decay by indirectly delaying ripening and tissue softening. Low O_2 (≤ 1 kPa) and/or elevated CO_2 (≥ 50 kPa) can be a useful tool for insect control in some fresh and dried fruits, flowers, vegetables, and dried nuts and grains, and this effect is more evident at higher temperatures (Yahia, 1998a,b; see Chapter 11).

Superatmospheric levels of O_2 of up to about 80 kPa may accelerate ethylene-induced degreening of nonclimacteric commodities and ripening of climacteric fruits, respiration, and ethylene production rates, and incidence of some physiological disorders such as scald on apples and russet spotting on lettuce. At levels above 80 kPa O_2, some commodities and postharvest pathogens can suffer from O_2 toxicity. The use of superatmospheric O_2 levels in MA/CA will likely be limited to situations in which they reduce the negative effects of fungistatic, elevated CO_2 atmospheres on commodities that are sensitive to CO_2 injury (see Chapter 9).

Development and application of successful postharvest preservation techniques for fresh horticultural commodities depends largely on the understanding of certain fundamental aspects of biology, engineering, and economics that are important in the maintenance and distribution of these perishables.

1.5 Potential Detrimental Effects

Some detrimental effects can result, especially due to the use of O_2 levels lower and/or CO_2 levels higher than the ideal for each commodity (see Chapter 8). The incidence and intensities of these potential detrimental effects can be influenced by several other factors such as temperature, RH, other gases such as ethylene, type of product, cultivar, stage of development of the commodity, storage duration, among others. Some of the detrimental effects can be shown as initiation and/or aggravation of certain physiological disorders such as internal browning in apples and pears, brown stain of lettuce, and chilling injury of some commodities. Irregular ripening of some fruits, such as banana, mango, pear, and tomato, can result from exposure to O_2 levels below 2 kPa and/or CO_2 levels above 5 kPa for more than 1 month. Development of off-flavors and off-odors can occur at very low O_2 levels and/or very high CO_2 levels as a result of anaerobic respiration and fermentative metabolism. Susceptibility to decay can increase due to exposure to very low O_2 and/or very high CO_2 atmospheres. Therefore, inadequate atmospheres can cause the aggravation or the initiation of physiological disorders (see Chapter 8), and can increase the hazard of microbial contamination of minimally processed products (Chapter 10). MA and CA can be deadly to humans getting inside a room or a container without proper security equipments or before the room or the container is properly ventilated (see Chapters 2 and 3). MA and CA can cause structural damage to rooms and containers that lack proper pressure relief systems, or due to inadequate use of some gases such as propane.

1.6   Technological Applications

The most important applications of MA and CA for fresh horticultural commodities are storage, long distance marine transport, and packaging.

1.6.1   Storage

CA storage is not universally adaptable to all crops. It is a technology where fresh perishable commodities are stored under narrowly defined environmental conditions to extend their postharvest life (see Chapter 2 for more details). The ideal levels of these environmental variables differ according to commodity and stage of development. Moreover, tissue response may vary because of interactions among these variables as influenced by variety and preharvest conditions and climatic factors. Commercial use of CA storage is common for apples and pears, less common for cabbages, sweet onions, kiwifruits, avocados, persimmons, pomegranates, nuts and dried fruits, and vegetables. Applications of CA to ornamentals and cut flowers (see Chapter 19) are very limited because decay caused by *Botrytis cinerea* is often a limiting factor to postharvest life, and fungistatic CO_2 levels damage flower petals and/or associated stem and leaves. Also, it is less expensive to treat flowers with antiethylene chemicals than to use CA to minimize ethylene action. CA storage technology does not seem to be as promising for many tropical fruits (Yahia, 1998a; Yahia and Paull, 1997; see Chapter 16) when compared to temperate fruits. This is due to several reasons including those related to crop availability and quantity, preharvest and postharvest handling, and availability of technology. Except for bananas, tropical crops cannot be stored for prolonged periods that would justify the use of CA. Many factors should be considered when evaluating the potential application of MA/CA such as fruit quantity and value, reason for the use of MA/CA (control of metabolism, control of pathogens, control of insects, etc.), availability of alternative treatments, competition with other production regions, type of market (local, distant, export, etc.), and type of preharvest and postharvest technology available in the region. Apple, a fruit very compatible to the use of CA (see Chapter 12), is a high value fruit, produced in large quantities, and characterized by a climacteric respiration and a long postharvest life. In addition, the production and action of ethylene is controlled by CA, a large variation exists in the tolerance levels for O_2 and CO_2, CA permits the use of lower storage temperatures in some cultivars, the fruit is relatively less infected by pathogens and insects compared to other fruits especially of tropical origin, some physiological disorders can be alleviated by CA, fruit can be harvested and stored in bulk, and a great deal of MA/CA research has been carried out. No tropical fruit, except banana, meets these qualifications. Low pressure (hypobaric or LP storage) refers to holding the commodity in an atmosphere under a reduced pressure, generally, <200 mmHg. In this system, the O_2 concentration is reduced depending on the atmospheric pressure. LP has been reported to extend the storage and shelf life of several crops (Gemma et al., 1989) including mangoes (Burg and Burg, 1966; Burg, 1975; Spalding, 1977a; Spalding and Reeder, 1977; Ilangantileke and Salokhe, 1989; Yahia, 1997b), avocados (Burg and Burg, 1966; Spalding and Reeder, 1976; Apelbaum et al., 1977b; Spalding, 1977a), bananas (Burg and Burg, 1966; Apelbaum et al., 1977a; Yahia, 1997d), papayas (Alvarez, 1980; Yahia, 1997c), and cherimoyas (Plata et al., 1987). LP was used for a short period in the 1970s during transport of some food products including meats, flowers, and fruits (Byers, 1977). However, this system is more expensive than the traditional MA/CA systems. Furthermore, some fruits require other gases that cannot be administered during low-pressure storage. For example, CO_2 is an important component in an adequate atmosphere for avocado (Spalding and Reeder, 1976; Spalding, 1977a; Yahia, 1997a). Currently, this system is not used commercially. Several developments in CA storage have been made in recent years including rapid CA (rapid establishment of optimal levels of O_2 and CO_2), ultra low O_2 (1.0–1.5 kPa) CA storage, high CO_2 CA, low ethylene ($\leq$1 μL L-1) CA storage, and programmed (or sequential) CA storage, for example, storage in 1 kPa O_2 for 2–6 weeks followed by storage in 2–3 kPa O_2 for the remainder of the storage period. Other developments included creating nitrogen by separation from compressed air using molecular sieve beds or membrane systems, improved technologies of establishing, monitoring, and maintaining CA.

1.6.2 Transport

MA and CA for long-distance marine transport is used on many commodities, such as apples, avocados, bananas, blueberries, cherries, figs, kiwifruits, mangoes, nectarines, peaches, pears, plums, raspberries, melons, and strawberries (see Chapter 3 for more details). Transport of meat in MA has started since about seven decades ago, and that of fruits and vegetables since about three to four decades ago. The use of MA/CA can encourage the use of sea transport, since it is cheaper than air transport. Atmospheres for transport can be developed passively, semiactively, or actively. Passive systems are MA regimes where the atmosphere is modified by fruit respiration. In the semiactive systems, one or more gases is/are added or withdrawn, most commonly at the beginning, but no strict control is carried out. Active systems imply a strict control of the atmosphere during all transport period. The most common systems for transport in the last 30 years have been developed on a semiactive basis, and used for transport of bananas and strawberries, though usually less efficient, they are less expensive than active systems. The use of CA for transport has been contemplated for several decades; however, several problems have hindered its success, including the unavailability of adequate gastight containers, suitable systems for gas control and analysis, and adequate CA-generating systems. Existing systems and companies before the 1980s were unable to deliver on promised applications and benefits of MA/CA transport systems. Liquid N_2 and pressure swing adsorption (PSA) systems were tried first to create and maintain CA systems in sea containers but were unsuccessful. The disadvantage of the PSA systems is the susceptibility of the carbon sieve to deterioration in high-vibration environments. Liquid N_2 has high boil-off rate and can run out in the middle of the ocean, making it difficult to refill, and thus the CA would be lost. In the late 1980s, the concept of CA transport became more practical due to the availability of gastight containers, adequate gas control systems, and the ability to establish and maintain controlled gas mixes. The use of air separation technologies in the 1980s, especially the introduction of membrane technology in 1987, made CA transport practical and feasible. The use of CA for food transport commodities is now a practical reality, although several systems are still misleadingly called CA when they are in reality MA systems with no capacity to control most of the gases. A CA container has the same features as that of a refrigerated container, in addition to a higher level of gastightness, O_2 and CO_2 control systems, and perhaps systems for control of ethylene and RH. CA systems for transport should be used when transport periods are long and/or food is very perishable. The "Oxytrol system," which was the first commercially available MA system, was developed by Occidental Petroleum Corporation, California. It is a self-contained system designed to be used as an adjunct in refrigerated transport vehicles. The main components of the system are an oxygen sensor, electronic analyzer-controller, liquid nitrogen storage tank, liquid nitrogen vaporizer, gas discharge nozzle, and peripheral equipment. The liquid nitrogen tank can be filled before, during, or after loading. Liquid N_2 is vaporized and warmed prior to injection into the transport vehicle. CO_2 is controlled using hydrated lime. This system has been used for highway and sea shipments of lettuce, celery, papaya, and pineapple. "Tectrol" (total environment control) was developed and first used by Transfresh Corporation, California, in 1969. For transport, tight refrigerated transport units (railroad, highway, or sea) are flushed with the desired premixed gas blend and then sealed. The "Tectrol" unit consists of nitrogen tanks, which are controlled to correct for the deviation of oxygen. This system was only satisfactory when oxygen was the only gas to be controlled, and the trip was short. Lime and magnesium sulfate are used to lower the concentrations of CO_2 and C_2H_4, respectively. Some of the crops transported in this system include lettuce, strawberry, mango, and avocado. Newer Tectrol systems included a controller that monitors, controls, and records O_2 and CO_2. An interface system allows the controller to manage the container environment without external interference. The "CONAIR-PLUS" system (G + H Montage GmbH, Hamburg, Germany) can create a CA system through the introduction of N_2 (generated under pressure) by means of a membrane from the ambient air to the container. This system can also control CO_2 and ethylene, and has been used to transport apple, avocado, melon, and mango. The maritime protection division of Permea, Inc. (St. Louis, Missouri) developed the first membrane-based CA system (PRISM CA) in 1987. This system contains gas analyzers and computer microprocessors for establishing and maintaining the CA system. It was designed to be installed on the weather deck of a refrigerated ship, and to establish, monitor, and control CA conditions. After completion of the voyage, the system can be flown back to the port of origin in the cargo hold of a 747 aircraft (it is designed to occupy the same space as an LD-29 aircraft cargo container). The system was first used by the New Zealand Apple and Pear Board, then by Cool Carriers to transport apples, and lately was used in some other ships. Other systems have been developed in the last few years such as Freshtainer (Maidstone, England) and NITEC (Spokane, Washington), among others (see Chapter 3). Important developments in the transport of foods in MA/CA have been accomplished in the last years, among them are the development of better sealed containers, air separation techniques, better gas monitoring and control systems. However, major problems are still facing the industry to select the adequate MA/CA transport system. MA/CA transport companies need to work much more closely with the industry. They need to investigate in more detail the ideal conditions of each specific MA/CA system for each commodity, especially in relation to stage of maturity, cultivar differences, distances of transport, etc.

1.6.3 Packaging

Modified atmosphere packing (MAP), a technique used to prolong the shelf life of fresh or minimally processed foods, refers to the development of an MA around the product through the use of permeable polymeric films. MAP is used with various types of products, where the mixture of gases in the package depends on the type of product; packaging materials and storage temperature (see Chapters 4, 5, and 18 for more details). Fresh horticultural commodities are respiring products where the interaction of the packaging material with the product is important. If the permeability (for O_2 and CO_2) of the packaging film is adapted to the products, respiration, an equilibrium MA will be established in the package and the shelf life of the product can be improved (Yahia and Gonzalez, 1998; Yahia and Rivera-Dominguez, 1992). In reality, the concept of MAP is not new with early research being reported in the late 1800s. The use of MAs for shelf life extension was conceived as a new food preservation method with the discovery of the preservative action of carbon dioxide over a century ago. In the 1970s, MA packages reached the stores when bacon and fish were sold in retail packs in the United Kingdom. Since then the development has been stable and the interest into MAP has grown due to consumer demand. This has led to many advances, such as in the design and manufacturing of polymeric films. New techniques are designed, like the use of antifogging layer to improve product visibility. Equilibrium-modified atmosphere packaging (EMAP) is a commonly used packaging technology for fresh-cut produce.

MAP is a technique that has gained tremendous popularity in the last years. It can be broadly defined as any process that significantly changes the environment around the product from the normal composition of air in a package. The generic term also includes applications such as vacuum packaging. MAP can be used in combination with a number of other food processing, preservation systems, and packaging techniques to maintain product quality and extend shelf life. MAP benefits to the food product are extended shelf life and maintenance of flavor, taste, texture, color, and nutritive value. Nonetheless, MAP, as it is the case in MA and CA, cannot arrest all degradation processes such as staling, or improve a bad product. Packaging cannot be held solely responsible for upholding the quality and shelf life of MAP products.

Selection of a MAP system depends on a variety of factors, including shelf life, distribution requirements, importance of bloom, product dimensions, and marketing objective. The selection of gas mixtures used in a MAP system is influenced by the microbiological flora on the food product, CO_2 and O_2 sensitivity of the product, and pigment stabilization requirements, among other factors. Although many gas formulations have been investigated to optimize the MAP process, the main gases remain to be O_2, N_2, and CO_2, which can be used separately or mixed together. The best results are obtained when gas mixtures and processes are tailored to specific applications.

The growth in MAP fresh-cut (minimally processed) produce is one of the most dynamic in the last decade. Produce packaging involves a delicate balance of the product's respiration rate, product temperature, and the permeability of packaging material. The selection of the correct packaging material is crucial to maintain the desired balance of O_2 and CO_2 which, in turn, slows the product's metabolism, and thus product senescence.

Although gas atmosphere is important, other criteria must be considered when contemplating MAP technology, which include temperature and RH control, packaging equipment, packaging materials, and food safety issues. High-performance packaging and packaging equipment are continually evolving. Technological advancements in packaging machinery have focused on higher speeds, lower residual oxygen levels, and washdown sanitation construction. High-performance films can be defined as films that truly meet the customer's expectations in various roles as both the communicator and a protector.

Packaging is increasingly important as part of the communicator role to reach the consumer through printing and package appearance. In addition, new packaging technologies are creating advancements in barrier polymers, improved sealant polymers, and abuse properties.

Emerging packaging technologies are continually evolving to prolong the quality of fresh, refrigerated, and minimally processed foods. These emerging technologies include active packaging including oxygen-scavenging films, high oxygen barrier polymers and high oxygen permeable films, and polymers with refined properties. Active packaging involves technologies in which the package interacts with the internal package atmosphere and/or the food in the package. The active system can be an integral part of the package or be a separate component placed inside the package. Substances that can either absorb (scavenge) or release (emit) a specific gas can control the internal atmosphere of the package. Commercial examples of "scavenger" and "emitter" technologies include oxygen scavengers to create low O_2 environment and slow metabolism; ethylene scavengers to slow senescence of produce; CO_2 scavengers to remove CO_2 from coffee; CO_2 emitters to retard microbial growth; and edible/biodegradable barriers to moisture and/or O_2. Other active systems slowly release the active substances onto the food surface including antimicrobial agents, antioxidants, color transfer films; flavor/odor absorbers or releasers, enzyme inhibitors, and moisture scavengers.

Some important MAP developments include, among others, creating nitrogen by separation from compressed air using molecular sieve beds or membrane systems, improved technologies of establishing, monitoring, and maintaining the atmosphere, using of edible coatings or polymeric films with appropriate gas permeability to create a desired atmospheric composition around and within the commodity.

Some of the disadvantages of MAP include added cost, capital cost of packaging, and quality control machinery; added cost of gases and high barrier films, and added transport cost and display space due to increased pack volume. However, the rapid growth of MAP products in the marketplace indicates that the benefits of the system to the retailer, manufacturer, and consumer clearly outweigh the disadvantages.

Food safety must continue to be of prime concern of MAP food. Good manufacturing practices along with the implementation of a hazard analysis of critical control points (HACCP) system are the ideal way to ensure a safe product. In addition, hurdle technology will be imperative to the success of refrigerated, minimally processed foods. The effective and safe use of MAP requires effective control of temperature. Poor temperature control eliminates the beneficial effects of any MAP system and may cause health hazards.

1.7 Some Additional Treatments

The use of carbon monoxide in addition to MA and CA can provide several advantages such as the control of pathogenic diseases and insects (Woodruff, 1977; El-Goorani and Sommer, 1981), and tissue browning and discoloration (Kader, 1986). However, the development of safe methods of application such as the addition of odors is required before any commercial use can be considered (Yahia et al., 1983). The use of CO has to be combined with low (≤ 4 kPa) O_2 atmospheres (Kader, 1986). CO can increase the production of C_2H_4; however, at high concentrations this did not appear to happen (Woodruff, 1977). It has been reported that for fruits that tolerate elevated CO_2, CO appear to be more effective. Ethylene removal during transport and storage in CA can be beneficial in delaying ripening. Atmospheres with very low O_2 content (≤ 1.0 kPa) and/or very high CO_2 levels (≥ 50 kPa) have insecticidal effects (Yahia, 2006). Insect control with MA and CA depends on the O_2 and CO_2 concentration, temperature, RH, insect species and development stage, and duration of the treatment (Yahia, 1997e; Yahia et al., 1997). The lower the O_2 concentration, the higher the CO_2, the higher the temperature, and the lower the RH, the shorter the time necessary for insect control (Yahia, 1998a,b, 2006). MA and CA have several advantages in comparison with other means for insect control. These are physical treatments that do not leave toxic residues on the fruit, and are competitive in costs with chemical fumigants. MA and CA do not accelerate fruit ripening and senescence as compared to the use of high temperatures, and have better consumer acceptance as compared to the use of irradiation. In order to establish quarantine insect treatments for short periods of times, extreme gas atmospheres (≤ 0.5 kPa O_2 and/or ≥ 50 kPa CO_2) at high temperatures ($\geq 20°C$) should be implemented (Yahia, 2006). Some fruits do not tolerate these treatments. For example, 'Hass' avocados are injured when exposed to these atmospheres for more than 1 day at 20°C (Carrillo-Lopez and Yahia, 1990; Yahia and Carrillo-Lopez, 1993; Ke et al., 1995). The application of insecticidal atmospheres at high temperatures can accelerate the mortality of insects (Yahia, 2006).

1.8 Some Future Research Needs

Most chapters in this book contain suggestions for future research needs of each of the topics. Some of the general research needs are specified here. There is a continuing research need to better understand the biological bases of O_2 and CO_2 effects on postharvest quality of fresh whole and fresh-cut horticultural commodities. There is still a strong need for continuing investigations of the physiological and biochemical basis of CA-induced physiological disorders, and to determine the reasons behind genotypic differences in tolerance of a given commodity to reduced O_2 and/or elevated CO_2 concentrations. Research is still needed to elucidate the mode of action of elevated O_2, elevated CO_2, ethylene, and carbon monoxide on postharvest pathogens and insects. Bases for wide variability in tolerance to extreme atmospheres among different commodities still need to be understood. There are still needs for continued development of optimum MAP technologies for various commodities including evaluation of various types of O_2, CO_2, C_2H_4, and water vapor absorbers for their effectiveness in helping to maintain the desired microenvironment within MAP. Investigation is needed to determine the potential of using superatmospheric O_2 levels, especially in addition to high levels of CO_2 for decay control without detrimental effects on fruit quality. Significant research is still needed, especially by MA/CA transport companies, to determine the optimum use of the diverse systems used, the optimum conditions with each type of product, cultivar, stage of maturity, duration of shipping trip, etc.

1.9 Conclusions

MA and CA are important technologies for the preservation of food commodities. They can be used during storage, transport, and packaging. Major research and developments advances have been accomplished in the last few decades, which allowed for the diversity of use of the technology (for more food products, for different types of applications, and in diverse regions in the world). However, major R&D is still needed in all applications in order to increase the use of the technology for more commodities and applications, but especially there is much more need for more research and development on MA/CA transport.

References

Allen, F.W. and L.R. McKinnon. 1935. Storage of Yellow Newton apples in chambers supplied with artificial atmospheres. *Proceedings of the American Society Horticultural Science 1935* 32:146–152.

Allen, F.W. and R.M. Smock. 1937. Carbon dioxide storage of apples, pears, plums and peaches. *Proceedings of the American Society Horticultural Science* 35:193–199.

Alvarez, A.M. 1980. Improved marketability of fresh papaya by shipment in hypobaric containers. *HortScience* 15:517–518.

Apelbaum, A., Y. Aharoni, and N. Temkin-Gorodeiski. 1977a. Effects of subatmospheric pressure on the ripening processes of banana fruit. *Tropical Agriculture (Trinidad)* 54:39–46.

Apelbaum, A., C. Zauberman, and Y. Fuchs. 1977b. Prolonging storage life of avocado fruits by subatmospheric pressure. *HortScience* 12:115–117.

Biale, J. 1942. Preliminary studies on modified air storage of Fuerte avocado fruit. *Proceedings of the American Society Horticultural Science* 41:113–118.

Bigelow, W.D., H.C. Gore, and B.J. Howard. 1905. Studies on apples. *United States Department Agriculture Bureau Chemical Bulletin* 94:9–10.

Brooks, C., J.C. Cooley, and D.F. Fisher. 1919. Nature and control of apple scald. *Journal of Agricultural Research* 18:211–240.

Burg, S.P. 2004. *Postharvest Physiology and Hypobaric Storage of Fresh Produce.* CABI Publishing, Wallingford, U.K., 654 pp.

Burg, S.P. and E.A. Burg. 1965. Ethylene action and the ripening of fruits. *Science* 148:1190–1196.

Burg, S.P. and E.A. Burg. 1966. Fruit storage at subatmospheric pressures. *Science* 153:314–315.

Byers, B. 1977. The Grumman Dormavac system. In: Dewey, D.H. (Ed.), *Controlled Atmospheres for the Storage and Transport of Perishable Agricultural Commodities. Proceedings of the 2nd National CA Research Conference Horticulture Report 28.* Department of Horticulture, Michigan State University, East Lansing, MI, pp. 82–85.

Carrillo-Lopez, A. and E.M. Yahia. 1990. Tolerancia del aguacate var. Hass a niveles insecticidas de O_2 y CO_2 y el efecto sobre la respiración anaeróbica. *Tecnología de Alimentos (México)* 25(6):13–18.

Dalrymple, G.D. 1969. The development of an agricultural technology: Controlled atmosphere storage of fruits. *Technology Culture* 10(1):35–48.

Dilley, D.R. 1990. Historical aspects and perspectives of controlled atmosphere storage. In: Caldron, M. and Barki-Golan, R. (Eds.), *Food Preservation by Modified Atmospheres.* CRC Press, Boca Raton, FL, pp. 187–196.

Dilley, D.R. 2006. Development of controlled atmosphere storage technologies. *Stewart Postharvest Review 2006* 6:5. www.stewartpostharvest.com

Eaves, C.A. 1959. A dry scrubber for CA apple storages. *Transactions of the American Society Agricultural Engineers* 2(1):127–128.

Eaves, C.A. 1960. A Plastic storage for apples with semi-automatic controlled atmospheres. *Canadian Refrigeration Air Conditioning* 26(12):29–33.

Eaves, C.A. 1963. Atmosphere generators for CA apple storages. *Annual Report Nova Scotia Fruit Growers Association 1963* 100:107–109.

El-Goorani, M.A. and N.F. Sommer. 1981. Effects of modified atmospheres on postharvest pathogens of fruits and vegetables. *Horticulture Review* 3:412–461.

Fidler, J.C., B.L. Wilkinson, K.L. Edney, and R.O. Sharples. 1973. *The Biology of Apple and Pear Storage.* Research Review 3, Commonwealth Agriculture Bureau, England, pp. 235.

Frenkel, C., I. Klein, and D.R. Dilley. 1968. Protein synthesis in relation to ripening of pome fruits. *Plant Physiology* 43:1146–1153.

Gemma, H., C. Oogaki, M. Fukushima, T. Yamada, and Y. Nose. 1989. Preservation of some tropical fruits with an apparatus of low pressure storage. *Journal of Japanese Society of Food Science and Technology* 36:508–518.

Hill, G.R. 1913. Respiration of fruits and growing plant tissues in certain gases, with reference to ventilation and fruit storage. *Cornell University of Agricultural Experimental Station Bulletin* 330:377–408.

Hulme, A.C., M.J.C. Rhodes, T. Gaillard and L.S.C. Wooltorton. 1968. Metabolic changes in excised fruit tissue. IV. Changes occurring in discs of apple peel during the development of the respiration climacteric. *Plant Physiology* 43:1154–1161.

Ilangantileke, S. and V. Salocke. 1989. Low-pressure storage of Thai mango. In: *Other Commodities and Storage Reconditions*, Vol. 2. *Proceedings of the Fifth CA Research Conference* Wenatchee, WA, June 14–16, 1989, pp. 103–117.

Kader, A.A. 1986. Biochemical and physiological basis for effects of controlled and modified atmospheres on fruits and vegetables. *Food Technology* 40(5):99–100, 102–104.

Ke, D., E. Yahia, B. Hess, L. Zhou, and A. Kader. 1995. Regulation of fermentative metabolism in avocado fruit under oxygen and carbon dioxide stress. *Journal of the American Society of Horticultural Sciences* 120(3):481–490.

Ke, D., E. Yahia, M. Mateos, and A. Kader. 1994. Ethanolic fermentation of 'Bartlett' pears as influenced by ripening stage and controlled atmosphere storage. *Journal of American Society for Horticultural Sciences* 119(5):976–982.

Kidd, F. and C. West. 1927a. Atmosphere control in fruit storage. Great Britain Department Scientific Industrial Research Food Investigation Board Report, 1927, pp. 32–33.

Kidd, F. and C. West. 1927b. Gas storage of fruit. Great Britain Department Scientific Industrial Research Report 30, pp. 87.

Kidd, F. and C. West. 1930. The gas storage of fruit. II. Optimum temperatures and atmospheres. *Journal of Pomology and Horticultural Science* 8:67–77.

Kidd, F. and C. West. 1937. Action of carbon dioxide on the respiration activity of apples. Effect of ethylene on the respiration activity and climacteric of apples. Individual variation in apples. Great Britain Department Scientific Industrial Research Food Investigation Board Report, 1937, pp. 101–115.

Kidd, F. and C. West. 1950. The refrigerated gas storage of apples. Great Britain Department Scientific Industrial Research Food Investigation Board Leaflet No. 6 (rev.), pp. 16.

Overholser, E.L. 1928. Some limitations of gas storage of fruits. *Ice and Refrigeration* 74:551.

Plata, M.C., L.C. de Medina, M. Martinez-Cayuela, M.J. Faus, and A. Gil. 1987. Changes in texture, protein content and polyphenoloxidase and peroxidase activities in chirimoya induced by ripening in hypobaric atmospheres or in presence of sulfite (in Spausi)., Rev. Ageoquim Tecnol. Aliment. 27: 215–224.

Spalding, D.H. 1977a. Low pressure (hypobaric) storage of avocados, limes and mangos. In: Dewey, D.H. (Ed.), *Controlled Atmospheres for the Storage and Transport of Perishable Agricultural Commodities. Horticulture Report 28*, Michigan State University, East Lansing, MI, pp. 156–164.

Thatcher, R.W. 1915. Enzymes of apples and their relation to the ripening process. *Journal of Agricultural Research* 5(3):103–105.

Woodruff, R.E. 1977. Use of carbon monoxide in modified atmospheres for fruits and vegetables in transit. In: Dewey, D.H. (Ed.), *Controlled Atmospheres for the Storage and Transport of Perishable Agricultural Commodities. Proceedings 2nd National CA Research Conference, Horticulture Report 28*, Department of Horticulture, Michigan State University, East Lansing, MI, pp. 52–54.

Yahia, E.M. 1994. Apple flavor. *Horticultural Reviews* 16:197–234.

Yahia, E.M. 1997a. Modified/controlled atmospheres for avocado (*Persea americana* Mill). In: Kader, A.A. (Ed.), *Fruits Other than Apples and Pears*, Vol. 3. *CA'97 Proceedings*. University of California, Davis, CA, pp. 97–103.

Yahia, E.M. 1997b. MA/CA for mango (*Mangifera indica* L.). In: Kader, A.A. (Ed.), *Fruits Other than Apples and Pears*, Vol. 3. *CA'97 Proceedings*. University of California, Davis, CA, pp. 110–116.

Yahia, E.M. 1997c. MA/CA for papaya (*Carica papaya* L.). In: Kader, A.A. (Ed.), *Fruits Other than Apples and Pears*, Vol. 3. *CA'97 Proceedings*. University of California, Davis, CA, pp. 117–120.

Yahia, E.M. 1997d. MA/CA for banana and plantains (*Musa* spp.). In: Kader, A.A. (Ed.), *Fruits Other than Apples and Pears*, Vol. 3. *CA'97 Proceedings*. University of California, Davis, CA, pp. 104–109.

Yahia, E.M. 1997e. Avocado and guava fruits are sensitive to insecticidal MA and/or heat. In: Thompson, J.F. and Mitcham, E.J. (Eds.), *CA Technology and Disinfestation Studies*, Vol. 1. *CA'97 Proceedings*. University of California, Davis, CA, pp. 132–136.

Yahia, E.M. 1998a. Modified and controlled atmospheres for tropical fruits. *Horticultural Reviews* 22:123–183.

Yahia, E.M. 1998b. El uso de atmósferas modificadas y controladas para el control de insectos en postcosecha. *Phytom (Spain)* 97:18–22.

Yahia, E.M. 2004. *Modified and Controlled Atmospheres*. CIAD, Mexico (in Spanish).

Yahia, E.M. 2006. Modified and controlled atmospheres for tropical fruits. *Stewart Postharvest Review 2006* 5:6. www.stewartpostharvest.com

Yahia, E.M. and A. Carrill-Lopez. 1993. Tolerance and responses of avocado fruit to insecticidal O_2 and CO_2 atmospheres. *Food Science and Technology (Lebensmittel Wissenschaft und Technologie)* 26:312–317.

Yahia, E.M. and G. Gonzalez. 1998. Use of passive and semi-active atmospheres to prolong the postharvest life of avocado fruit. *Lebensmittel Wissenschaft und Technologie* 31(7/8):602–606.

Yahia, E.M. and R. Paull. 1997. The future for modified atmosphere (MA) and controlled atmosphere (CA) uses with tropical fruits. *Chronica Horticulturae* 37(4):18–19.

Yahia, E.M., K.E. Nelson, and A.A. Kader. 1983. Postharvest quality and storage life of grapes as influenced by adding carbon monoxide to air or controlled atmospheres. *Journal of the American Society of Horticultural Science* 108(6):1067–1071.

Yahia, E.M. and M. Rivera-Dominguez. 1992. Modified atmosphere packaging of muskmelon. *Food Science and Technology (Lebensmittel Wissenschaft und Technologie)* 25:38–42.

Yahia, E.M., D. Ortega, P. Santiago, and L. Lagunez. 1997. Responses of mango and mortality of *Anastrepha ludens* and *A. obliqua* to modified atmospheres at high temperatures. In: Thompson, J.F. and Mitcham, E.J. (Eds.), *CA Technology and Disinfestation Studies*, Vol. 1. *CA'97 Proceedings*. University of California, Davis, CA, pp. 105–112.

2

Storage Technology and Applications

Ernst Hoehn, Robert K. Prange, and Clément Vigneault

CONTENTS

2.1   Introduction

Controlled atmosphere (CA) storage using low oxygen (O_2) levels and high carbon dioxide (CO_2) levels in the storage atmosphere combined with refrigeration is probably the most successful technology introduced to the fruit and vegetable industry in the twentieth century. In particular, its implementation concerning apples revolutionized storage possibilities of this commodity. Even very early storage practices may have utilized modified atmosphere (MA) enriched with CO_2 and depleted O_2 levels to extend storage life of fruits, vegetables, cereals, and other commodities (Dilley, 2006). First research concerning CA storage began in the early 1800s in France (Dalrymple, 1999). Jacques Etienne Berard at the University of Montpellier in France observed that fruits did not ripen in an atmosphere depleted of O_2 (Dalrymple, 1999). Other studies covered the effects of low levels of O_2 and of high levels of CO_2 on ripening and first attempts of Benjamin Nyce on CA storage followed in 1865 (Dilley, 2006). However, the basis for commercial application of CA was established by Kidd and West (1927, 1930, 1937, 1950) who investigated the effects of O_2, CO_2, and C_2H_4 (ethylene) on respiration and ripening in pome fruits and berries. Consequently, the first commercial CA storage facility was constructed in Kent, England, in 1929. Soon the success of apple storage employing the "gas storage" or CA technique started its worldwide expansion. Eaves (1934) was responsible for the fruit and vegetable storage program at the Experimental Farm in Kentville, Nova Scotia, Canada, and initiated construction of the first commercial CA storage in Canada. The first commercial CA storage facilities in South Africa were installed in the Elgin Valley near Cape Town in 1935 (Dilley, 2006; Graham, 2008). Research on CA storage prompted the further construction of CA storage facilities in New York state in 1940 (Smock, 1941). However, CA usage only became more common after 1950 as CA technology improved. Further developments during the 1990s and ongoing technical innovations are expanding its use worldwide (Prange et al., 2005b).

Initially commercial CA storage was nearly exclusively applied to apples. This was certainly related to its incomparable advantages to refrigeration only for this specific produce. In addition, CA storage offered a means to alleviate some storage disorders and diseases of apples. Although other disorders may be aggravated or induced by CA storage, in general its benefits prevail (Prange et al., 2006). However, research investigating applications for other commodities was pursued from the beginning and led to commercial use and recommended CA conditions for storage of other fruits, vegetables, fresh cut fruits, and vegetables as well as cut flowers and ornamentals (Brecht, 2006; Erkan and Wang, 2006; Kader, 2003; Kupferman, 2003; Saltveit, 2003a; USDA-ARS, 2004). Finally, specific applications of CA include insect control and disinfection (Mitcham et al., 2003).

2.2   CA Methods

The once termed "gas storage" but later more appropriately named CA storage technology evolved steadily since its beginnings. Progress in the construction of CA rooms resulted in better air or gas tightness of the room. The availability of more effective equipment enabled tighter control of O_2 and CO_2 levels in the storage atmosphere. Accurate control of low levels of O_2 was facilitated by the implementation of automated measuring and control systems (Landry et al., 2008; Markarian et al., 2003). In the 1960s, it became evident that lower O_2 levels were advantageous and thus the main thrust of progress focussed on maintaining as low levels as tolerated by the produce. This was reflected in the

TABLE 2.1

CA Methods and Gas Compositions of Storage Atmosphere Storage Method

	O_2 (kPa)	CO_2 (kPa)	N_2 (kPa)
Refrigerated storage (normal atmosphere)	21	0.03	79
Controlled ventilation	6–18	3–15	79
Conventional CA	2–5	2–5	90–92
LO	1.5–2	1–3	95–97.5
ULO	0.8–1.2	0.5–2	96.8–98.7
DCA or DCS	<0.8	<1.5	>98

terminology used for the different CA storage regimes such as LO (low O_2) storage or ULO (ultralow O_2) storage, and more recently dynamic control of CA (DCA) or dynamic control system (DCS) (Table 2.1) (Chapon and Westercamp, 1996; Prange et al., 2005). Control of O_2 and CO_2 determines indirectly the nitrogen (N_2) concentration, which is by far the most abundant CA component, followed by CO_2, O_2, and trace amounts of noble gases, mainly argon (Ar) with about 0.9%. The noble gas components are inert and have no direct effect on plant tissue metabolisms. However, CA effects must be considered as a supplement to refrigeration. Optimal temperature and relative humidity (RH) must be adhered to in conjunction with proper levels of O_2 and CO_2 to assure successful storage outcomes. Hence use of CA does not replace the benefits of optimal temperature, but in some cases, it may alleviate effects of suboptimal temperatures (Brecht et al., 2003).

2.2.1 Controlled Ventilation

This method represents the most basic type of CA. It was used in the beginning and until recently in Northern and Eastern Europe. It is still used successfully in the CA storage of 'Bramley's Seedling' apples in the United Kingdom (Johnson, 2007). The CA is generated by the accumulation of CO_2 from respiration of the fruit or vegetable stored in an airtight room. Typically CO_2 levels are held at 5–10 kPa (about 5%–10% of the total gas composition, which is ~100 kPa) or even higher and are controlled by ventilation with outside air only. The sum of CO_2 and O_2 concentrations of such atmospheres are normally equal to 20%–21%. Consequently, O_2 levels range from about 10 to 16 kPa and are not low enough to exert a sufficient decreasing effect on respiration (Chapon and Westercamp, 1996). Furthermore, many apple cultivars and pears even more so are sensitive to elevated CO_2 concentrations and suffer from CO_2-related storage disorders. In such cases, CO_2 has to be maintained at levels below 3 kPa and consequently at O_2 levels of 18 kPa, which would result in insignificant or modest effects on storage life compared with refrigerated storage. This is in contrast to fruits such as cherries and berries, which tolerate high levels of CO_2. For these fruits, ventilated CA storage may result in a relevant extension of storage life and quality retention, equaling the effect of optimal CA storage conditions but at better cost efficiency (Gasser and Höhn, 2004). Requirements to conduct controlled ventilation storage include gastight rooms or pallet package systems and a device to monitor CO_2 level complemented with means to ventilate with outside air.

2.2.2 Conventional CA

Maintaining low CO_2 (2–3 kPa) and at the same time LO levels (2–3 kPa) by ventilation with outside air is unachievable. However, these are the ranges of CO_2 and O_2 that are effectively reducing respiration of apple and pear and are adhered to in conventional CA.

Since CO_2 evolves from stored fruit, it can accumulate in the storage room to unacceptably high levels. The means to remove excessive CO_2 are thus required. Scrubbers are such devices that chemically or physically remove CO_2 from the storage atmosphere. In addition, monitoring of CO_2 and O_2 levels in the storage rooms requires appropriate measuring systems. In comparison to a controlled ventilation room, a CA room must be relatively more gastight and equipped with reliable remote temperature control. Control accuracy for temperature should not exceed $\pm0.5°C$. The CO_2 as well as O_2 levels must be controlled within ±0.5 kPa of the set values. Such control specifications can be complied with based on daily hand measurements of storage factors and on hand-operated control of equipment such as scrubbers. Thus automatic measurements and control of storage conditions or computerized monitoring and control are not mandatory even though they greatly facilitate the accuracy of control and decrease the duration and intervention frequency required for maintaining the right gas compositions (Markarian et al., 2003). In the manual control system, the frequency of the operations must consider the rates of stored fruit or vegetable O_2 depletion and CO_2 production since these rates vary considerably amongst fruits and vegetables. For example, the initial respiration rate of 'Cox' apples amounts to 5 g t^{-1} h^{-1} whereas 'Golden Delicious' produces only 2 g t^{-1} h^{-1} (Bishop, 1996). This rate of CO_2 production results in a 0.1 kPa CO_2 h^{-1} increase in the storage room for 'Cox' and only 0.04 kPa h^{-1} in the case of 'Golden Delicious.' Operation control such as scrubbing operation duration should be set accordingly to facilitate the adequate maintenance of CO_2 levels.

2.2.3 Low O_2 Storage and Ultralow O_2 Storage

The lowest recommended O_2 for CA storage was for a long time not <2 kPa. However, research began to show that storability of many horticultural produce and preservation of their quality at lower O_2 levels improves without detrimental effects. Thus, in 1965, it was recommended that 'Cox' apples be stored at <1 kPa CO_2 and at 1.25 kPa O_2 (Jameson, 1993; Peppelenbos, 2003). Since then, LO (1.5–2 kPa) or ULO (0.8–1.2 kPa) conditions for storage of different apple cultivars have been successfully implemented worldwide in the apple industry (Dilley, 2006). Furthermore, ULO has been demonstrated to be beneficial for other commodities such as pear, kiwifruit, grapefruit, nectarine, onion, lettuce, and others (Ekman et al., 2005). Maintaining such low levels of O_2 requires airtight rooms exceeding the tightness of common CA rooms. In addition, control of O_2 and CO_2 levels must be within 0.1 kPa. This makes it mandatory to analyze O_2 and CO_2 concentrations every 4–6 h and adjust the gas concentration continuously or very frequently (every 30 min) to maintain the gas concentration with acceptable precision (Markarian et al., 2003), which is more frequent than in conventional CA. To reach such precision, it is highly recommended to use automatic and computerized monitoring and control systems (Chapon and Westercamp, 1996; Markarian et al., 2003).

2.2.4 Rapid CA

Research in the 1960s confirmed that an accelerated reduction of O_2 in a storage room, compared with solely a reduction through respiration, delayed ripening of apples (Dilley, 2006). The beneficial effects of rapid CA include low O_2 levels, reducing C_2H_4 synthesis and increased CO_2 levels diminishing the sensitivity of plant tissue to C_2H_4 (Saltveit, 2003b). In a climacteric fruit such as the apple, ripening can be delayed by low O_2 levels as long as they are harvested in a preclimacteric stage. It is well accepted that C_2H_4 induces ripening processes and it is known that O_2 is required for its synthesis (Yang and Hoffmann, 1984). Thus purging rooms with N_2 to reduce O_2 levels within 1 or 2 days instead of 1–3 weeks was

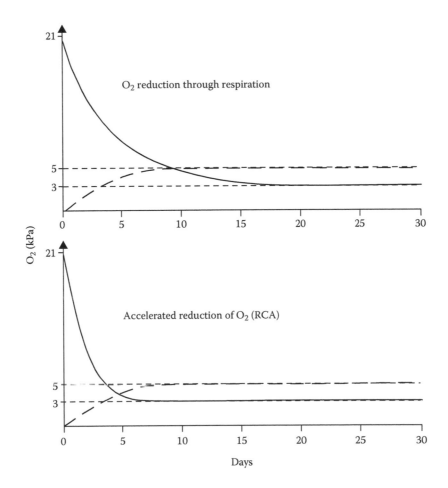

FIGURE 2.1
Evolution of O_2 (—) and CO_2 (---) by respiration and by purging (RCA).

introduced and referred to as rapid CA (RCA) (Figure 2.1). Typically it is recommended to reduce O_2 levels to ≤5 kPa by purging and then achieve further reduction by respiration; however, the cooling process used in postharvest technology for apple is generally room cooling, which is relatively slow compared with forced air or hydrocooling (Vigneault et al., 2008). A commercial application of precooling using cool water and storing the apple within <24 h after harvest generates important benefits in fruit quality, mainly firmness, even within a short period of storage (a few months) (Charles and Vigneault, 2008).

2.2.5 Initial Low O_2 Stress

Initial low O_2 stress (ILOS) is not considered as a storage technique but more as a treatment to generate a positive stress effect on fresh horticultural produce. ILOS involves exposing produce for a limited period of time, up to 2 weeks, at very LO levels of <0.5 kPa. This treatment was specifically introduced as an alternative to treatment of apples with the chemical diphenylamine (DPA) or 1-methylcyclopropene (1-MCP) for the prevention of superficial scald (Matté et al., 2005; Zanella, 2003). Following the time under O_2 stress, the fruits were usually stored under ULO conditions and this apparently completely controls superficial scald of several apple cultivars (Wang and Dilley, 2000b). Other applications of

MA treatments for fresh horticultural produce have been subjected to research, generating promising results and have been discussed recently in the literature (Vigneault and Artés-Hernández, 2007).

2.2.6 Delayed CA

RCA can cause in some apple cultivars ('Braeburn,' 'Fuji') or pears ('Conference') storage disorders such as internal browning (Braeburn browning disorder [BBD]) and cavities. Delaying establishment of CA conditions by 3 weeks prevents or reduces the development of internal browning (Argenta et al., 2000; Curry, 1998; Höhn et al., 1996; Roelofs and de Jager, 1997; Saquet et al., 2003). These findings suggested that fruits can become adapted to lower O_2 levels during the early stages of the storage period when maintained at regular atmosphere. Presumably the higher energy status indicated by higher adenosine triphosphate concentration in the fruit tissue throughout the entire storage period in delayed CA fruit makes them more resistant to browning disorders (Saquet et al., 2003; Verlinden et al., 2002).

2.2.7 Dynamic Control of CA or Dynamic Control System

A multitude of research projects have been conducted to find optimal CA conditions for the storage of fruit and vegetable. Based on such trials, recommendations for CA and MA of fruit and vegetable have been generated. As stated by Saltveit (2003b), such recommendations have been included in the proceedings of the International CA and MA conferences since 1969 and subsequently regularly revised and expanded. The increasing scope of CA recommendations gradually began to account for variation due to factors such as geographical location, cultivar, cultivar strain, orchard factors, harvest date, and storage duration. However, in general, such recommendations remained static in nature and were based on empirical findings derived from storage trials performed in different CA conditions (Gasser et al., 2008; Prange et al., 2005b; Wertheim, 2005). The conditions yielding the best storage results are normally recommended and maintained from the beginning until the end of storage, with a few exceptions such as ILOS or delayed CA. Since stored fresh horticultural commodities are living tissues, it is reasonable to expect their metabolism to be dynamic in nature during storage. Thus, static CA conditions are likely not optimal and even incorrect at some point during the storage period. Saltveit (2003b) highlighted this aspect and pointed out the need for a dynamic CA system. Schouten et al. (1997) reported that dynamic control of ULO conditions in experimental containers (600 L) results in an improvement of quality of 'Elstar' apples. They used ethanol as a marker to control the O_2 levels. First O_2 is pulled down to 4–5 kPa and thereafter further reduced to 1.2 kPa by fruit respiration. After approximately 1 month, O_2 is further lowered at a rate of 0.1 kPa per week until ethanol levels exceeded 500 $\mu L\ L^{-1}$ in the cell air. Then O_2 levels are increased by 0.1 kPa per week until ethanol falls below the critical concentration of 500 $\mu L\ L^{-1}$. In further experiments, the DCS method was optimized and tested in commercial practice as well as an ethanol sensor developed (Veltman et al., 2003). Currently, the DCS method is exclusively used for the storage of 'Elstar' in the Netherlands. Compared with ULO, DCS provides better firmness retention and development of "skin spots" is inhibited (van Schaik, 2008). However, fruit samples must be periodically removed from a DCS storage room and analyzed for internal ethanol content. This course of action is necessary because measurement of ethanol in the storage atmosphere is not sufficiently reliable. This procedure does not meet the characteristics of a dynamic commodity indicator as suggested by Saltveit (2003b). It should be reliable, continuous, nondestructive, and measurable at a distance from the commodity and highly correlated with specific quality indicators.

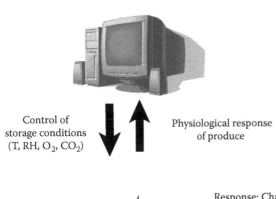

Control of
storage conditions
(T, RH, O_2, CO_2)

Physiological response
of produce

Response: Change of
chlorophyll fluorescence
or ethanol formation

Stress:
Oxygen
reduction

FIGURE 2.2
Concept of dynamic CA. (From Gasser, F.,
Eppler, T., Naunheim, W., Gabioud, S., and
Höhn, E., *Agrarforschung*, 15, 98, 2008. With
permission.)

A promising technique to dynamically control CA appears to be chlorophyll fluorescence measurement (Figure 2.2) (Gasser et al., 2005; Prange et al., 2003, 2005a). Chlorophyll fluorescence is affected by low O_2 and high CO_2 (DeEll et al., 1995, 1998) and research indicates that the technique detects the lowest level of O_2 that is tolerated by stored fruit and vegetable. This O_2 concentration would correspond to the anaerobic compensation point (ACP) (Figure 2.3). The ACP is the O_2 concentration at which CO_2 production is at a minimum (Boersig et al., 1988). According to the most recent static CA recommendations,

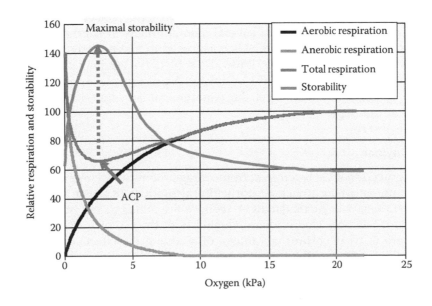

FIGURE 2.3
Effect of O_2 concentration on O_2 consumption and CO_2 production and ACP. (From Prange, R.K., DeLong, J.M., and Harrison, P.A., *J. Am. Soc. Hort. Sci.*, 128, 603, 2003. With permission.)

TABLE 2.2

LO Threshold of Apple Cultivars Determined by Chlorophyll Fluorescence
and Range of O_2 Employed in Storage

Cultivar	LO Threshold (kPa)	O_2 Settings for Storage (kPa)
Braeburn[a]	0.4	0.5–0.6
Cortland[b]	0.5	0.6–0.8
Delicious[b]	0.4	0.5–0.8
Elstar[a]	0.3	0.3–0.6
Golden Delicious[b]	0.5	0.5–0.8
Golden Delicious[b]	0.3	0.3–0.6
Honeycrisp[b]	0.4	0.5–0.8
Idared[c]	0.4	0.5–0.8
Jonagold[b]	0.5	0.5–0.8
Maigold[a]	0.3	0.4–0.6
McIntosh[b]	0.8	0.9–1.0

[a] Gasser et al. (2007a,b)
[b] DeLong et al. (2004)
[c] Gasser et al. (2005)

the lowest recommended O_2 concentrations is about 1 kPa. Chlorophyll fluorescence research indicates that O_2 threshold concentrations (ACP) for tested apples ranges lower, from 0.4 to 0.8 kPa (Table 2.2). It is recommended that O_2 levels during storage are set at 0.1–0.2 kPa above the detected low O_2 threshold values. In addition, O_2 levels are usually stepwise decreased by 0.2 kPa per week beginning at an O_2 level of 1 kPa. A slow rate of reduction appears to foster adaptation of fruit to LO. As a consequence, lower values for ACP are found than with more rapid rates such as 0.2 kPa per day (Gasser et al., 2008).

First applications of DCA in commercial storage rooms were undertaken in South Tyrol, Italy, in 2004–2005 and increasingly in the following storage seasons. A commercial storage company in Washington state began using DCA in 2004 and, by 2006–2007, it was used on 6400 t storage capacity divided amongst four storage companies in Washington state, primarily on organic scald-sensitive 'Granny Smith' and 'Delicious' apples (Graham, 2008). As of 2006–2007, it is used in several storage rooms in southern Germany (Lake Constance region). In most cases, DCA is being used to maintain the lowest possible O_2 concentration but in several instances, it has also served as a warning alarm to the operator that a malfunction of the O_2 control was occurring. A prerequisite for using DCA is that the storage room construction must meet the requirements for ULO, in particular, the high standards for airtightness.

2.2.8 Low Ethylene

It is well recognized that C_2H_4 removal from storage rooms may beneficially affect storage life of climacteric and nonclimacteric horticultural produce (Knee et al., 1985). As stated before, CA, ULO, and in particular RCA regimes are aimed at minimizing C_2H_4 effects. More recently, it has been clearly demonstrated with the use of 1-MCP in the storage of apple that there is an important advantage of studying the effect of C_2H_4 scrubbing to eliminate 1-MCP treatment (Blankenship and Dole, 2003; Prange and DeLong, 2003). Wills et al. (1999) showed that the storage life of many fruits and vegetables increases linearly with the logarithmic decrease in C_2H_4 concentrations from 10 to <0.005 $\mu L\ L^{-1}$. C_2H_4 levels found during storage, transport, and marketing often exceed this range (Wills et al., 2000). Different systems can be used to remove C_2H_4; however, maintaining a C_2H_4 level below 0.1 $\mu L\ L^{-1}$ seems quite a challenge and may be costly.

2.2.9 Low-Pressure Storage or Hypobaric Storage and Others

Hypobaric storage, which was introduced by Burg and Burg (1966), can not only rapidly remove the heat during the well-known vacuum cooling process (Vigneault et al., 2008), but also reduce the O_2 level and expel the harmful gases that may be released during storage (Wang et al., 2001). Hypobaric storage can adjust the inside temperature and composition of the atmosphere reliably and consistently (Wenxiang et al., 2006). It is a storage system in which the horticultural produce is stored while ventilating with air at less than the atmospheric pressure level. The partial pressure of each gas component in air is reduced in direct proportion to the total pressure. For example, at 10.1 kPa total pressure, the O_2 partial pressure is equivalent to 2.1 kPa at atmospheric pressure.

The use of hypobaric treatments has been reported to delay ripening of some climacteric fruits (Romanazzi et al., 2003) and increase shelf life (Apelbaum et al., 1977). The highly susceptible apple cultivars 'Law Rome' and 'Granny Smith' do not develop superficial scald when placed under hypobaric storage within 1 month after harvest whereas it occurs when stored in air or under CA (Wang and Dilley, 2000a). Hypobaric storage inhibits respiration; prevents loss of chlorophyll, vitamin C, and acidity; improves sensory quality; and delays postharvest senescence of green asparagus (Wenxiang et al., 2006). Cucumbers submitted to 70 kPa for 6 h in the dark to simulate air flight transportation, and then stored in cold storage facilities at normal pressure, have significantly more open stomata, compared with the control, after 96 h of subsequent storage (Laurin et al., 2006). Water can be sprayed to resolve the problem of insufficient RH, causing desiccation during hypobaric storage (Laurin et al., 2006). It is likely that desiccation experienced by fresh fruit and vegetable subjected to hypobaric conditions is due to increased transpiration as affected by the properties and action of stomata (Laurin et al., 2006), lenticels, cuticle, and epidermal cells (Ben-Yehoshua and Rodov, 2003). The beneficial impact of hypobaric storage was also observed on loquat fruit (Gao et al., 2006).

Readers interested in further information on the potential for hypobaric storage should consult Burg (2004). Burg (2004) rebuts the historical pessimism associated with this technology and concludes that this technology is more promising now than at any prior time, due to advances in hypobaric technology and a better understanding of the effects that a low pressure has on stomatal opening, gas and vapor mass transport, heat exchange, disease control, and insect mortality.

In spite of the benefits of hypobaric storage and some advantages, compared with CA storage, it has not gained acceptance in the commercial storage of fruit and vegetable. To our knowledge, there is no current commercial application of this technology but new ideas could be explored to develop a portable or palette-size system that could maintain a hypobaric pressure without generating a prohibitive cost or any danger for the operator of such a system. One should never forget that playing with pressure always includes a security aspect and worker risk.

Its greatest potential is likely associated with intermodal containers, rather than in warehouses. Interestingly, one of the important points this makes is that the early hypobaric storage studies were performed at pressures much higher than optimal thus better results are likely expected at or below the 1.33–2.67 kPa pressure range. This equates to an O_2 concentration of 0.15–0.3 kPa, which is remarkably similar to the low O_2 thresholds determined using DCA (Table 2.2).

As reported, it is advantageous to maintain the O_2 levels in CA at the lowest possible. There have been claims that the opposite, namely superatmospheric O_2 level, enhances the storability of fresh prepared horticultural produce (Barry and O'Beirne, 2000). However, this has not been confirmed in other studies concerning intact fresh horticultural produce (Kader and Ben-Yehoshua, 2000).

2.2.10 Modified Atmosphere

For completeness of storage under gas of different compositions from ambient air, MA must be discussed here. In MA, the required atmosphere is created by respiration or by flushing with single or mixed gases into the horticultural produce container. MA is a passive technology because once the atmosphere is established it is not further controlled by measuring gas concentrations and/or by applying active gas modification. The container may consist of film with specific gas permeability, or impermeable film with a diffusion channel or with specially designed semipermeable windows. These different ways of adjusting the gas composition are static and do not involve any measurement of gas composition and external adjustment. The composition of the MA is thus determined only by the equilibrium between the respiration of the horticultural produce and the static gas diffusion through the permeable films or the diffusion channel incorporated in the container. Since respiration increases more rapidly with temperature than permeability of films or channel gas diffusivity, O_2 levels may decrease and CO_2 levels increase, leading to anaerobic conditions under higher temperature conditions than the condition under which the MA system was designed. Thus temperature control is one of the most important factors along with the selection of plastic films to be considered for reaching the desirable gas composition. Since MA is based on stable conditions over time, which is generally difficult to maintain over long periods, it is generally used for short-term storage and transport applications, in particular for small fruits (Hui et al., 2005). The most important application is however in retail level packages in the form of MA packaging, which is covered in chapters 4 and 18 of this book.

2.3   Storage Facilities and Pallet Package Systems "Palistore"

CA comprises essentially refrigerated and controlled gas composition storage of fruit or vegetable. Storage facilities (Figure 2.4) encompass airtight and thermally insulated rooms or enclosures, refrigeration systems, machinery, and equipment for creating and maintaining the desired gas concentrations in a specific environment and systems for measurement and control of storage factors (Allen, 1998; Bishop, 1990, 1996; Coquinot and Chapon, 1992).

2.3.1 Rooms

One consideration when planning CA facilities is the room size. The major factor determining the size is the speed of loading and unloading. Rooms should ideally be filled within 3–5 days and emptied (grading, packing, and marketing) within 5–7 days (Bishop, 1996). These periods assume that there is additional technology to assure fairly high-quality retention throughout the remaining marketing chain up to the consumer. Where there is a need for more rapid loading and unloading, a new technology has been designed (Charles and Vigneault, 2008) involving a hopper system for allowing storing the produce in the CA room within the 24 h following the harvesting operation and/or removing the produce from the room immediately before marketing.

When not considering the possibility of using the hopper technology, the division of the total storage capacity into smaller units is preferable to installing only a few large units, but this is more expensive. The increase of installation and operation costs must be balanced with the degree of maintenance of quality to justify such divisions; however, the increase in consumer demand for produce of high quality generally compensates for these cost increases. Room sizes used for apples in Europe range from 100 to 200 t, although some

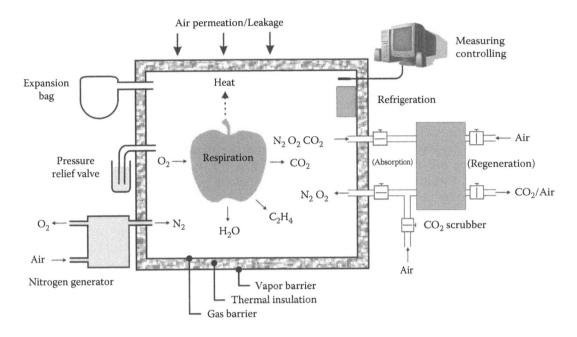

FIGURE 2.4
Scheme of a CA room.

hold up to 500 t. The trend is to smaller sizes of 100–200 t. This trend to smaller rooms applies to the CA industry in Canada and the United States as well (Allen, 1998). The size of CA rooms for apples and pears is declining to ~200–250 t in new rooms. However, in Washington state the room sizes have traditionally been larger compared with other regions and hold up to 570–1000 t. New CA installations being built in nontraditional areas, for example in India, generally comprising rooms of ~150 t each (D. Bishop, 2007, personal communication). In South Africa. the most common size is 400 t. Since 2002, a tendency to smaller rooms of up to 200 t has been observed (D. Graham, 2008, personal communication).

Further considerations during planning of CA storage rooms include taking bin size as well as their spacing into account. Sizes of bins will vary with type of commodities and regions. In addition, proper stacking pattern is important as it determines airflow in the storage room and thus storage room performance.

In terms of thermodynamics, a CA storage room is equal to an airtight refrigerated room. The floor, walls, and ceilings function simultaneously as a thermal insulation, moisture, or vapor barrier from the outside and as an air or gas barrier on the inside. However, airtightness of storage rooms in particular for ULO or DCA must meet very high standards. Hence construction of such rooms needs attention to detail in design and workmanship (Bishop, 1996). This is usually a job best left to experts. Most building materials such as wood, concrete, and steel are still used. The construction method and the selection of materials may vary with local considerations and economic factors (Allen, 1998; Bishop, 1996; Coquinot and Chapon, 1992; Henze and Hansen, 1988; Osterloh, 1996; Raghavan and Gariépy, 1984). According to Bishop (1996), almost all CA storage rooms being built in Europe in the last few years are made from metal-faced insulating panels locked together with proprietary locking systems as seals between panels. Gastightness of the joints is usually increased by taping and coating with a flexible plastic paint. The same process is also carried out on floor–wall and wall–ceiling joints.

In North America, panel-built storage rooms are not so common. The usual practice is to construct storage rooms using timber frames clad with plywood boarding, or with concrete blocks, or tilt-up concrete walls. Rooms are then sealed and insulated with foamed-in-place urethane, which is finished with a fire-retardant coating. A more recent insulation method involves construction of concrete panels complete with an internal insulation layer. Important in terms of airtightness is the gas sealing of entries of cables and piping. Drains of water should be equipped with U-traps to ensure discharge of water without breaking the room seal. Doors merit particular interest because they are common leakage areas (Allen, 1998; Bishop, 1996; Coquinot and Chapon, 1992). Conventional cold storage room doors, even leak tight ones, are not sufficiently airtight. Doors are available, which are specially designed for CA storage rooms and fulfill specifications in terms of airtightness. Special attention has to be taken to the height, width, and position of the door since these items determine the efficiency of the loading and unloading of the storage room. In most of Europe, doors must be fitted with inspection hatches or windows, which can be opened from the inside as well as from the outside. One purpose of this is to secure an escape in case somebody gets trapped inside the room. Thus its size and its position have to be accordingly considered or must meet legal safety requirements. Furthermore, these hatches allow access to fruit for sampling and examination during the storage period. In Europe and elsewhere, it is common to include double- or triple-glazed windows in the wall at a level that allows visual inspection of the fruit in the top bins and to check for any ice buildup on the mechanical refrigeration evaporators.

2.3.2 Pressure Relief Valves

Air or gas pressure fluctuation in a CA room can cause structural damage to rigid and sealed rooms. Causes for changes in relative pressure across room walls include atmospheric pressure changes, wind, room air temperature fluctuations, fan operation, CO_2 scrubber operation, and operation of other storage machineries (Allen, 1998; Bishop, 1996; Coquinot and Chapon, 1992). Atmospheric pressure may vary from a hurricane barometric pressure of 94.3 kPa to a strong high pressure of 103.4 kPa (Ross, 2004). If such a pressure differential is not relieved, it will result in a pressure of 9.1 kPa on the wall and ceiling of the storage room, which is equivalent to the pressure created by 824 mm of water. Even though such an extreme pressure differential is unlikely to happen within a short period of time, moderate pressure differential (0.5–2 kPa) in relatively short time period (1–6 h) is likely to be frequent.

Temperature change entails gas volume change and thus in a CA room pressure changes, since volume of a room does not change with temperature change. The universal gas law relates temperature and volume of gas (air) to its absolute temperature. This relationship means that any increase or decrease in the absolute temperature (K) generates a proportional pressure change, which would add or compensate for any other source of pressure variation depending on the direction of these different pressure modifications. Cooling of an empty room from 25°C (298 K) to 0°C (273 K) will reduce pressure inside an airtight room by 8.5 kPa. Obviously these forces would lead to significant damage of the structure if the pressure differential is not released. Thus it is essential to install a pressure relief valve to limit the pressure on the storage room structure. It is recommended to limit the pressure difference between the inside of the room and the outside to a maximum of ±0.19 kPa (Bishop, 1996), corresponding to 17.2 mm of water. There are basically two types of relief valves in use, either water-based vent systems or mechanical relief valves relying on the weight of a sealed disk (Figures 2.5 and 2.6). Provided that they are properly designed in size and maintained, both systems rapidly equalize pressure difference between the room and outside.

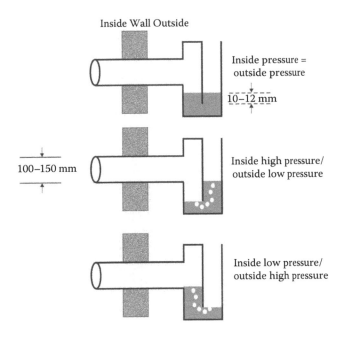

FIGURE 2.5
Scheme of a pressure relief valve: Water vent system.

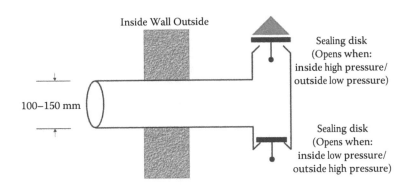

FIGURE 2.6
Scheme of a mechanical pressure relief valve.

2.3.3 Expansion Bags

A temperature decrease of 1°C in a CA room would result in a volume decrease of about 0.35%, which generates a pressure decrease of 370 Pa if there was no gas exchange. As indicated above, a relief valve should open once pressure change exceeds 190 Pa. In this case, the valve opens and outside air would flow into the room. Theoretically, in a well-designed room equipped with a well laid-out refrigeration system, temperature changes of 1°C should not be frequent and the time for it will take hours. In practice, temperature changes of ±0.5°C around the set point are very frequent. Even a well-constructed ULO room is not absolutely airtight and the excess or lack of air volume generated by barometric pressure and/or temperature variations is forced by the pressure differential across the wall and ceiling to be eliminated through air leaks and drains. The CA room is then "breathing" since a significant volume of air or gas is being expelled or drawn in through

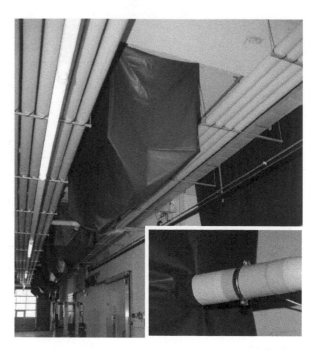

FIGURE 2.7
Expansion bag (breather bag). (From Gasser, F., Crespo, P., Gillard, A., Wernli, S., and Höhn, E., *Schweiz. Z. Obst-Weinbau*, 143, 10, 2007a. With permission.)

pressure relief valves or leaks and drains. This continuous air exchange would destroy the CA gas equilibrium if the system does not adequately compensate for this breathing. Therefore, the fitting of an expansion bag or breather bag would improve the gas operation by maintaining the same air circulating within the breathing system thus easily maintaining low O_2 levels (Figure 2.7). An expansion bag should hold between 0.5% and 1.5% of the empty storage room volume (Bartsch and Blanpied, 1984) and is usually made of polyethylene. The expansion bag is connected to the storage room and will expand due to any positive pressure differential until it is completely filled. When the pressure decreases in the room, the gas in the expansion bag is drawn back into the room. In this way, most of the ingress of O_2 is avoided. Expansion bags also help to reduce leakage into the room since most of the air volume variations inside of the room are compensated by the action of the expansion bag, so the pressure variations that normally drive the air infiltration are eliminated by the controlled breathing effect of the bag. Bags have to be airtight and O_2 permeability must be low (Vigneault et al., 1992). Airtightness and O_2 permeability measuring methods are similar to the methods for determining airtightness of CA rooms (Vigneault et al., 1992). Some storage room operators keep expansion bags partially inflated by giving the storage room a regular injection of N_2 using a gas generator (Bishop, 1996). This constant and slightly positive pressure generally eliminates the air infiltration and generates a constant air leakage from the rooms, maintaining the O_2 level precisely.

2.3.4 Leakage Specifications

The requirements in terms of airtightness as well as permeability to gas of a CA room are formidable. In spite of these stringent requirements, some leakage and gas permeability is expected even into rooms meeting the highest required standards since absolute airtightness in the large volumes encountered in CA storage rooms is impossible. However, it is evident that some fully loaded CA rooms regularly require "air" intake to make up for respiratory O_2 consumption of the stored fresh produce (Bishop, 1996). Coquinot and Chapon (1992) estimate that O_2 permeation into a up-to-standard 240 t CA room equals the respiratory O_2 consumption of 6.2 t of apples and that this is quite acceptable.

Furthermore, there may be leaks in the wall–floor or wall–ceiling seals. Thus a leakage rate should be included in any specification and contract for building a CA room. Specifications should include test procedures and acceptance criteria. A leakage test should also be performed in the yearly inspections of the CA rooms before they are loaded with produce.

Most gastightness tests are performed by monitoring the rate of change of pressure over time in an empty sealed room. To carry out the test, refrigeration should be turned off. The room should be at ambient temperature and empty. All openings should be sealed. It is best to follow a checklist before running the test (Bartsch, 2004b). The room is then pressurized with a vacuum cleaner, a fan blower, or scrubber blower. Great care has to be taken that a pressure of 275 Pa, equivalent to 25 mm of water, is not exceeded, otherwise structural damage may result. In the United States (Bartsch and Blanpied, 1984), the time (minutes) for a 50% pressure drop from 275 Pa is determined by plotting pressure–time data in an appropriate graph (Figure 2.8). Accordingly, three standards for room tightness are indicated, the 12, 20, and 30 min room. For ULO storage, a 30 min room or better is recommended. In the United Kingdom and Europe, the test is slightly different. The time it takes for the pressure to fall from 20 mm water gauge to 13 mm is measured. For a storage room intended for storage at 2.5 kPa O_2, the minimum time is 7 min, which is equivalent to one air change per 30 days (Raghavan et al., 2005). For storage at 2 kPa O_2 and lower, 10 min should be the minimum. For ULO storage rooms, it should be a minimum of 30 min. It is quite common that well-constructed storage rooms exceed 30 min for a pressure drop from 20 to 13 mm water gauge (Bartsch, 2004b; Bishop, 1996). Note that according to Bishop (1996), a 20 min room in North America is equivalent to 12 min in the United Kingdom.

There are a few other methods to determine leakage rates. One possibility represents the diffusion test (Coquinot and Chapon, 1992). It consists of the introduction of CO_2 into the storage room up to 5 kPa. For the retention, an efficiency of 0.97–0.98 should be achieved. This corresponds to a drop of 0.15 kPa per day or after 2 days there should be 4.7 kPa of CO_2 remaining. A further test consists of maintaining a set pressure by an air flow using a variable speed fan. The determined air flow corresponds to the leak rate (Bishop, 1996; Coquinot and Chapon, 1992).

If a room does not meet airtightness standards, remedial action is required. Locating leaks is not an easy task. One way of detecting leaks is placing the room under negative

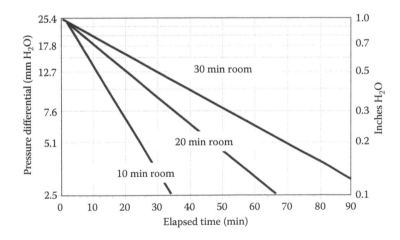

FIGURE 2.8
Pressure–time graph for CA room tightness tests. (From Bartsch, J.A., *Cornell Fruit Handling and Storage Newsletter*, 2004a, pp. 13–15. With permission.)

pressure of 20 mm water gauge. Leaks can then be found by listening in the room for the sound (whistling) of infiltrating air (Bartsch, 2004b; Bishop, 1996).

Since the introduction of on-site N_2 generators and automatic atmosphere control O_2 levels could be maintained by purging with N_2. In spite of this, room airtightness should be up to standard since purging requires energy, resulting in extra costs not required in an airtight room. A further consideration is the fact that N_2 is dry and thus excessive purging means excessive moisture loss from the CA room and water loss from the produce.

2.3.5 MA and CA Pallet Bag Systems (MCA Pallet CA)

For small-scale operations or for MA or CA storage of small fruits, it may not be economical or feasible to store in typical-sized CA rooms. For such applications, a simple bag method named Palistore™ is available, which enables storage of pallets of horticultural produce under MA or CA. These types of storage and handling systems are also named microcontrolled atmosphere (MCA). The system consists of an aluminum pan, which seals the bottom of a pallet (Figure 2.9). Produce crates are stacked on the bottom pan and then an airtight hood attached to an aluminum frame is fitted over the pallet of produce (Gasser et al., 2007a,b). The frame is attached to the bottom pan with clamping levers and forms an airtight seal. The hood is equipped with a pressure relief valve and quick-connect fittings for hookup of external gas tubing. CO_2 and O_2 levels can be monitored and controlled. Control can be accomplished manually or by a complete computer control system that can sample up to 100 pallets. Set points are programmable for each pallet. Pallets are stored in refrigerated rooms at an appropriate temperature. The same type of storage system has been performed by using a water channel airtightness system instead of clamping system, showing the high potential of such systems for both research and commercial applications (Goyette et al., 2002). The components used may vary from each supplier but the performances are quite the same as long as the sealing system and the permeability of the material are high enough to maintain the gas composition at the same level as encountered in a 30 min system.

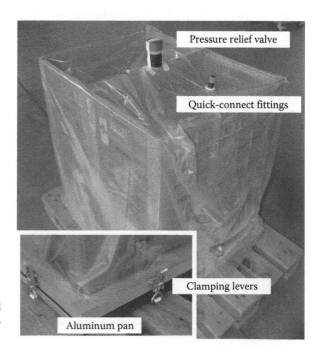

FIGURE 2.9
Pallet CA (Pallistor™). (From Gasser, F. and Höhn, E., *Schweiz. Z. Obst-Weinbau*, 140, 6, 2004. With permission.)

Another system under the trade name "mat tiempo" is available in France. It consists of plastic bins (containers of $100 \times 120 \times 78$ cm), which are fitted with a specially designed lid fitted with "respiration" membranes and fixtures for sampling. According to the supplier, it is recommended for CA storage of apples, kiwis, and cherries (Berger, 2007). A pallet package system using polyethylene film (80–150 μm thick) has been used to wrap pallets and to establish MA for a long time, especially for long distance transport of berries.

2.4 Establishment of Controlled Atmosphere

The following functions need to be carried out for the establishing and maintaining of CA: O_2 removal, excess CO_2 removal, addition of air to replace O_2 consumed by respiration and, in some cases, removal of C_2H_4 as well as addition of CO_2. The need for certain functions as well as the selection of the appropriate devices depends on what horticultural produce is stored and what storage conditions are necessary for each produce. For example, a CA storage system for apples should consist of an N_2 generator, a CO_2 scrubber, and a device to add air. Removal of C_2H_4 may only be indicated for kiwi storage rooms. CO_2 addition may be required for storing berries and cherries (Bishop, 1996).

2.4.1 Oxygen (O_2) Removal

At the beginning of CA, it was common practice to reduce O_2 levels by natural respiration of the horticultural produce. Rapid pull down of O_2 levels in CA rooms is often advantageous and can be achieved by various means (Allen, 1998; Bishop, 1996; Coquinot and Chapon, 1990; Malcolm, 2005; Waelti and Cavalieri, 1990). O_2 levels can be reduced quickly by flushing with N_2. It can be purchased as liquid or as gas; alternatively, it could be produced on site with gas separator systems (N_2 generators).

Use of liquid N_2 may seem expensive. However, capital costs for installations are low unless on-site storage of the N_2 is required.

The amount of N_2 needed can be calculated as

$$V = A \ln O_i/O_f \qquad (2.1)$$

where
  V is the volume of N_2 to be injected
  A is the storage room void space
  O_i is the initial storage room O_2 concentration
  O_f is the final O_2 concentration required

A storage room containing 100 t of apples would have a volume of 350 m³ with a void air volume of about 65%. Thus, it would require 326 m³ of N_2 to reduce the O_2 from air (21 kPa) to 5 kPa O_2 or 432 m³ to reduce it to 3 kPa. It is common practice to reduce O_2 levels to 3–5 kPa and not to the final level by flushing as the respiration can be used for reduction to the final level. The N_2 must be discharged slowly to prevent low-temperature damage to produce. Furthermore, pressure relief valves must be open and care must be taken that pressure does not exceed 190 Pa, otherwise structural damage may result.

Up to about 1990, propane burners were used to convert O_2 to CO_2 by burning with propane ($C_3H_8 + 5O_2 \rightarrow 3CO_2 + 4H_2O$). These external gas generators reduced O_2 levels to 3–5 kPa and were either operated on the "open-flame" or catalytic burner principle.

The catalytic burner was preferable since it could be used in a recycling mode down to 3 kPa O_2. However, there have been safety problems due to incomplete burning of propane. Thus, an explosive gas detector has to be installed. In addition, the large amounts of CO_2 produced have to be removed by the scrubber. Furthermore, some C_2H_4 is also produced, leading to increased levels of C_2H_4 in the storage room. Thus, this equipment is now considered obsolete and is no longer installed on new CA storage rooms. Similarly, an alternative system is no longer in use (Coquinot and Chapon, 1992). It operated based on cracking ammonia (NH_3) at high temperature to N_2 and H_2 ($2NH_3 \rightarrow N_2 + 3H_2$) and then further by consuming O_2 from the storage room to convert H_2 to water ($2N_2 + 6H_2 + 3O_2 \rightarrow 6H_2O + 2N_2$). The advantage of this system compared with the propane burner was that there is no CO_2 and C_2H_4 produced and additionally any C_2H_4 present is also destroyed.

During the 1980s, a new era began with the introduction of air or gas separators, usually named N_2 generators (Bishop, 1996; Coquinot and Chapon, 1992; Dilley, 2006; Malcolm, 2005). These installations separate air into O_2 and N_2. Two types are still commercially available. One is based on membrane technology and known as the hollow fiber membrane (HFM) system (Figure 2.10). The alternative is based on adsorption technology and termed pressure swing adsorber (PSA) (Figure 2.11). Both the HFM and PSA systems require a compressed air feed stream in the range of 800–1300 kPa. In the HFM system, the pressure differential between the inside of the membrane fiber and the outside of the fiber is the driving force for gases to permeate across the fiber wall. Each gas has a characteristic permeation rate. O_2, H_2O, and CO_2 are "fast" gases and diffuse at fast rates whereas N_2 is a "slow" gas with a slow diffusion rate. In this way, N_2 is separated from O_2 and other gases. Permeation occurs continuously and no regeneration of membrane fibers is necessary. Hence HFM systems are continuous operations.

PSA systems operate in a discontinuous mode and use completely different physical properties. They are made of two beds of carbon molecular sieve. At the high pressure of the compressed air, stream gases such as O_2 and CO_2 are adsorbed by the molecular sieve. When the pressure is reduced, these gases are released by the molecular sieves, which are regenerated. Generating the air pressure is the principal operating cost. However, the output of both systems parallels operating pressure. Reducing operating costs by reduction of pressure is feasible but would result in an increase of capital costs because it would require an increase of membrane surface or of molecular sieve to produce the same amount of N_2. Thus, the trade-off is lower operating costs at the expense of capital cost. The utilization of vacuum is commercially viable for PSA systems. Vacuum is used during the regeneration. A PSA, which utilizes "vacuum" during regeneration, is referred to as vacuum pressure swing adsorber (VPSA) or vacuum swing adsorber (Malcolm, 2005).

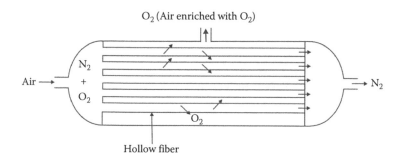

FIGURE 2.10
Scheme of a HFM nitrogen generator.

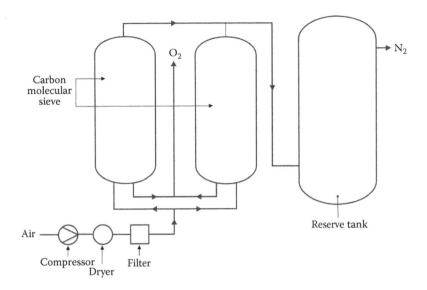

FIGURE 2.11
Scheme of a PSA nitrogen generator.

The adsorption cycle pressure reaches between 100 and 200 kPa and the regeneration cycle pressure is 10 kPa. Suppliers claim that up to 40% savings of energy are attainable with VPSA compared with PSA or HFM generators. However, a comparison of N_2 generator equipment in terms of total costs must include costs for base equipment, ancillary equipment, installations, operating cost, maintenance costs, and furthermore operating flexibility as well as reliability of the systems must be considered (Malcolm, 2005). Both technologies are widely used in the CA industry. Careful evaluation is indicated and requirements in terms of output flow and purity are accounted for. The purity of the N_2 output depends on the concentration of remaining O_2, which varies with the output flow. The higher the output flow the higher is the O_2 concentration. Therefore, the capacity of a N_2 generator is usually indicated for a given O_2 residue in the output. However, based on Equation 2.1, pull-down time depends on the flow rate and the void volume of the storage room and can be calculated as

$$t = \frac{A}{f} \ln\left(\frac{O_i - O_s}{O_f - O_s}\right) \tag{2.2}$$

where
  t is the pull-down time
  A is the storage room void space
  f is the flow rate of gas supply
  O_i is the O_2 initial storage room concentration
  O_s is the O_2 concentration in the N_2 purge gas
  O_f is the final O_2 concentration required

Assuming that the void volume of a CA storage room is 70% and the N_2 purge gas contains 2 kPa O_2, then resulting pull-down times depicted by Figure 2.12 may serve to determine the required output of an N_2 generator in a particular CA operation, based on the time the flushing operation may last.

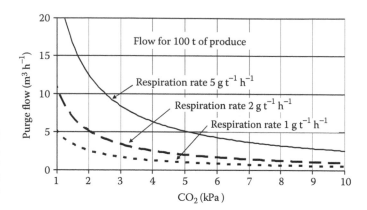

FIGURE 2.12
CO_2 control by N_2 flushing for a 100 t produce capacity (From Bishop, D., *Controlled Atmosphere Storage: A Practical Guide*, David Bishop Design Consultants, Heathfield, East Sussex, England, 1996. With permission.)

2.4.2 Carbon Dioxide (CO_2) Removal

CO_2 control is generally achieved using scrubber systems to remove CO_2 from the storage atmosphere. Caustic soda (NaOH) dissolved in water was one of the first reagents used for commercial CA storage. The solution was circulated in open tubes and absorbed CO_2 produced according to the following chemical reaction: $2NaOH + CO_2 \rightarrow Na_2CO_3 + H_2O$. The residence time was controlled according to the required removal rate (Raghavan et al., 2005). About 25 kg of NaOH was required per tonne of apples and a storage time of 5–6 months. The use of NaOH and K_2CO_3 (potash) (Osterloh, 1996) was discontinued due to their corrosiveness and potential danger in handling and disposal. Eaves and Lightfoot (Eaves, 1959) discovered in 1950 that newly laid cement absorbed CO_2, leading to the use of lime in place of caustic soda. This is the simplest method of CO_2 removal and it is based on utilizing hydrated lime [$Ca(OH)_2$] becoming limestone ($CaCO_3$) in the absorption process [$Ca(OH)_2 + CO_2 \rightarrow CaCO_3 + H_2O$]. Freshly hydrated lime can absorb 0.4–0.5 kg of CO_2 per kg of $Ca(OH)_2$. It is available either in chemical or agricultural grade. The Ca as well as Mg content is usually stated in terms of calcium oxide (CaO) and magnesium oxide (MgO), respectively. Both types are suitable but in view of efficiency, they should contain more than 70%–75% CaO or less than 20%–25% MgO and should be a particle size small enough to get 95% of the mass passing through a 100 mesh sieve (Bartsch, 2004a; Vigneault et al., 1994). These specifications are usually listed on the bags. Lime should be packaged in 23 kg maximum weight paper bags without polyethylene liner. Bags with plastic liners must be punctured before use. Only fresh lime should be used since lime absorbs CO_2 from ambient air. If a 23 kg bag exceeds 25 kg at delivery, it should be rejected since it has already lost about 20% of its capacity. A 23 kg bag will weigh approximately 34 kg when maximum of CO_2 is absorbed. Lime may be used in a scrubber or to supplement other scrubbing methods by placing it directly inside the CA room. A lime scrubber consists of an airtight and insulated box (plywood or other material) externally connected to the CA room. Air flows to the scrubber by natural convection or regulated by a blower. To keep CO_2 below 2 kPa about 12 kg of lime per tonne of apples is recommended for each 3–4 months of storage. Fifty percent of the recommended lime for the anticipated storage period may be placed inside the rooms to supplement other scrubbing methods. At the end of the storage period, the lime can be used as agricultural fertilizer but any other disposal may pose problems. Lime scrubbing is still in use but it is being replaced gradually by activated charcoal scrubbers.

An activated charcoal scrubber consists usually of two cylindrical beds or chambers filled with activated charcoal. Air from the storage room is circulated through one unit where CO_2 is absorbed. The CO_2-depleted air is returned to the storage room. When the charcoal is saturated, it is regenerated by circulating outside air through the activated charcoal and back to the outside. During the regeneration of one unit, the other is operated in the absorption mode. The twin bed design allows, in contrast to a single-bed type, virtually continuous scrubbing. Due to the regeneration with air, the scrubber can introduce some O_2 into the CA rooms. This O_2 introduction may be a drawback for ULO storage; however, there are various ways of overcoming this problem. Some manufacturers recommend the use of air from the CA storage room, held in a breather bag, or N_2 from a gas generator to flush the scrubbers after regeneration (Bishop, 1996). Although activated charcoal can be regenerated, it is recommended to replace it every 5 years.

It is common practice that one scrubber is used for several rooms. There are different scrubber sizes available to match the needs of any size of storage. The capacity required depends on the horticultural produce being stored and storage conditions. The nominal capacity of a scrubber is usually stated based on its adsorption rate under a 3 kPa CO_2 condition. However, the user must be aware that the efficiency of any charcoal scrubber decreases as the CO_2 level decreases. Insufficient scrubbing capacity may entail long scrubber times, leading to O_2 incorporation into the room and making it unfeasible to maintain ultralow levels of O_2. This applies in particular for ULO and DCA since under higher O_2 storage conditions, the stored horticultural produce is generally able to eliminate by respiration the small quantity of O_2 incorporated by the scrubbing operation.

Other molecular sieve scrubbers have also been used to adsorb CO_2. These are porous materials such as sodium or aluminum silicate zeolites. For regeneration of these types of scrubber, a heating process is required. This supplementary energy demand may be one reason for the limited use of these types of scrubber.

CO_2 levels may also be controlled by flushing with N_2 (Bishop, 1996; Cavalieri et al., 1989). If this is done, required O_2 may be brought into the room by operation of a vent and fan, which brings in outside air. This approach allows the use of only a N_2 generator to control both CO_2 and O_2 levels independently. However, energetically, it is less efficient than a CO_2 scrubber. The purging flow rate is determined by the respiration rate of the horticultural produce, the required levels of CO_2 and O_2 in the storage room, the airtightness of the room and N_2 purging gas purity. Examples of purging flow rate as a function of respiration rate and set level of CO_2 reveal that a 100 t room capacity and a respiration rate of 1 and 5 g CO_2 t^{-1} h^{-1} require purging flow rates of 3 and 12 m^3 h^{-1}, respectively, to maintain CO_2 at a set level of 2 kPa (Figure 2.12). Based on this, Bishop (1996) compared energy requirements of CO_2 control by flushing and CO_2 scrubber (Table 2.3). His results showed that the energy requirements for flushing are 10–50 times those of scrubbers.

Other methods for CO_2 removal exist. One system uses water to control CO_2 levels. The drawback of water scrubbers is their high demand of water, for example, about

TABLE 2.3

CO$_2$ Removal by Flushing versus CO$_2$ Scrubber and Energy Consumption

Produce	Respiration Rate (g t^{-1})	Set Points Levels		CO$_2$ Scrubber (kW day^{-1})	N$_2$ Generator (kW day^{-1})
		CO$_2$ (kPa)	O$_2$ (kPa)		
'Cox' apples	5	0.9	1.2	30.0	1550
'Golden delicious' apples	2	3.0	2.0	4.0	148
Onions	1	5.0	3.0	1.5	37
Cabbage	2	5.0	3.0	3.0	74

Source: From Bishop, D., *Controlled Atmosphere Storage: A Practical Guide*, David Bishop Design Consultants, Heathfield, East Sussex, England, 1996. With permission.

$100 \, \text{L} \, \text{h}^{-1} \, \text{t}^{-1}$ of apples. This is due to their low CO_2 adsorption capacity of 20 L of $CO_2 \, \text{m}^{-3}$ of water (Vigneault et al., 1994). In addition, a water scrubber returns O_2 to the room because when the water is aerated outside the room to release CO_2, it adsorbs O_2, which is then released in the room.

Finally CO_2 may be removed from storage atmosphere by diffusion units as proposed by Marcellin and Leteinturier (1967). The Marcellin system consists of gas diffusion panels in an airtight container with two separate air flow paths. The gas diffusion unit is equipped with semipermeable membranes made of silicon rubber. According to Osterloh (1996), 0.6–$1.2 \, \text{m}^2$ diffusion area is required to exchange 50–60 g of CO_2 per day. In spite of simplicity of construction and modest energy use, it has not gained industrial acceptance (Raghavan et al., 1984).

2.4.3 Ethylene (C_2H_4) Removal

The challenge of C_2H_4 removal lies in the fact that even at a very low concentration (<0.1 ppm), it may induce ripening or cause physiological disorders in some horticultural produce. Two types of C_2H_4 scrubbers are commercially available. One approach is based on catalytic oxidation of C_2H_4 to water and CO_2 (Wojciechowski, 1989). The alternative approach makes use of C_2H_4 absorbing beads. C_2H_4 levels in CA storage rooms are also influenced by the type of CO_2 scrubbing system. When N_2 flushing systems are used to remove the CO_2, it also decreases the C_2H_4 concentration. This way may exert sufficient control of C_2H_4 for horticultural produce that produce little C_2H_4. CO_2 scrubbers also absorb some C_2H_4. However, it has not been efficient enough for apple storage (Vigneault et al., 1994).

It may be possible to oxidize C_2H_4 using ultraviolet radiation or ozone gas techniques but so far it has not gained acceptance in commercial CA storage operations based on limited success in refrigerated storage. C_2H_4 levels in apple CA storage rooms may amount up to 200 ppm (Wojciechowski, 1989) but are much lower in storage rooms containing apples treated with 1-MCP. This is in contrast to C_2H_4 levels ranging from 20 up to 1500 ppm, which are found in conventional cold storage rooms, wholesale markets, distribution centers, supermarket storage rooms, and even in domestic refrigerators (Wills et al., 2000).

Swingtherm®, a catalytic converter for removal of C_2H_4 and other volatile products, has been sometimes recommended for use in refrigerated storage rooms and CA storage rooms of apple and other fruits. The suppliers claim that conversion rates attained an efficiency of 97% in the concentration range of 0.1–50 ppm. Furthermore, they also state that it is feasible to maintain C_2H_4 level at <0.02 ppm in kiwifruit storage rooms or at <0.05 ppm in citrus fruit, pear, and vegetable storage rooms. The process consists of cyclically reversing the gas flow direction through a platinum catalyst bed maintained at a temperature of 180°C–250°C. The average temperature difference between inlet and outlet gas ranges from 10°C to 15°C but may reach 30°C. This results in heat input into the storage room, which must be compensated by additional refrigeration.

C_2H_4-absorbing bead scrubbers consist of small spherical particles impregnated with potassium permanganate. Several porous inert matrix materials such as silica gel, zeolite, alumina, and others have been utilized (Jayaraman and Raju, 1992; Wills and Warton, 2004). The efficacy of adsorbents is affected by a range of environmental conditions. For example, efficiency of C_2H_4 adsorption decreases with increasing RH, at 90% RH, the absorption is 50% less than that at 70% RH (Wills and Warton, 2004). The beads are usually loaded into a sealed cartridge through which the air from the storage room is circulated. Regeneration of beads is not possible and therefore scrubbers must be checked frequently and spent beads replaced. Apparently, the relatively high cost of the potassium permanganate may exceed the advantage of controlling the C_2H_4 using this method, explaining the limited use of this technology.

2.5 Control of Storage Atmosphere

Successful CA storage relies on maintaining an appropriate and stable composition of the atmosphere in the storage room. However, this is difficult to achieve due to constantly changing conditions. Control of the storage atmosphere is affected by the physiological response of the stored horticultural produce, CO_2 and O_2 regulating equipment, cyclic operation of the refrigeration system, and changes in the weather. Thus O_2 and CO_2 levels must be measured at regular time intervals and corrective measures taken if the measured concentrations deviate from the set points. For control of conventional CA conditions, it may be sufficient to perform just one measurement per day and thus it is feasible to carry out manual measurements and controls. However, in modern and large CA facilities, automated control systems have been adopted. They have become essential for control of the storage conditions but in addition ensure optimum quality and efficiency (Mittal, 1997; Raghavan et al., 2005; van Doren, 1998). The control systems aim to reduce the differences between set points and measured level of the variable and decide whether or not to operate the particular regulation equipment. Devices used to operate equipment like scrubbers include on/off switches, and proportional (P), proportional-integrated (PI), or proportional-integral-derivative (PID) controllers. PID controllers are most often installed in CA operations. There are other more advanced systems, which are more intelligent and could be described as self-learning or self-tuning because they can adjust controller parameters automatically. Alternatively personal computer (PC)-based systems can provide direct online control of O_2 and CO_2 levels in CA storage rooms. Such systems consist of a computer, control software, a data acquisition system, communication ports, and switching devices. PC systems feature great flexibility; in addition to controlling storage conditions, they can be programmed to compute online parameters to manage interactions between process variables and automatically log information for later retrieval and analysis (Raghavan et al., 2005). Several dedicated systems are commercially available. Most suppliers provide technical support and assist in selecting the most suitable equipment and systems for each particular application. Usually the proposed systems can be interfaced with a PC to allow the operator to enter the required control parameters and to monitor the process, and even in a remote mode. An essential prerequisite for accurate control of gas composition is proper instrumentation for measurement of O_2 and CO_2 levels in the CA room. These instruments should provide reliable and accurate measurements as well as long-term stability. Therefore they must meet required specifications in terms of accuracy and should be regularly calibrated according to the recommendation of the supplier. In addition, checking readings of fixed systems with a second and usually portable system is highly recommended.

2.5.1 Oxygen (O_2)

O_2 measurements are most critical especially in ULO storage. Chemical analyzers such as a Fyrite analyzer or Orsat analyzer are based on the reaction of O_2 with pyrogallol and of CO_2 with alkaline solutions. Volume changes of a gas sample taken from the storage room are utilized to determine concentrations of O_2 and CO_2. The Fyrite system is inexpensive and portable but its accuracy is insufficient for O_2 levels lower than 2 kPa. The Orsat analyzer is usually centrally installed and gas samples are aspirated by pump directly from the rooms via a permanent tubing system. The capital cost is low and the accuracy is acceptable at higher O_2 levels. However, the time needed for a measurement is considerable and requires a skilled operator. In addition, it is unsuitable for automated systems.

O_2 analyzers based on paramagnetic sensors have proven to be accurate, reliable, and exhibit longevity. O_2 is highly paramagnetic and this property is used for measurement.

Since other common gases lack this property, this property is highly specific for O_2. Alternatively there are instruments using electrochemical cells for O_2 measurement. In recent years, there have been many improvements and thus electrochemical cells match accuracy and resolution of parametric sensors. Electrochemical cells have to be replaced regularly at about every 2 years. However, for both types of O_2-measuring instruments, regular calibration is required. It is recommended to use N_2/CO_2 mixtures to set the zero reading of the sensor, and a standard O_2 concentration of 21 kPa in air for full-scale O_2 measurement. Furthermore, to calibrate against a known O_2 concentration, it is preferable to use an $O_2/CO_2/N_2$ calibration gas mixture close to the one encountered in the particular CA operation. Modern computer-based systems are generally programmed to automatically calibrate the sensor at the zero, the full scale, and the operation level using the calibration gas on a regular basis.

2.5.2 Carbon Dioxide (CO_2)

As mentioned above, CO_2 levels can be determined with the chemically based Fyrite or Orsat methods in which CO_2 measurement is based on its absorption by alkaline solutions. The drawbacks of these methods are requirement of the long time and constricted accuracy, and unsuitability for automatic control. Thus, measurement of CO_2 in CA rooms is nearly exclusively done by infrared CO_2 sensor. Infrared absorbance at a wavelength of $4,260 \pm 20$ nm or at 15,000 nm is specific for CO_2 and proportional to its concentration and has been used to measure its concentration in CA rooms. As in the case of O_2 analyzers, CO_2 measuring instruments have to be calibrated regularly. Most suppliers suggest using air for zero and a N_2/CO_2 mixture of a known concentration for full-scale adjustment.

In terms of good storage practice, monitoring O_2 and CO_2 levels must be performed regularly using a fixed measuring system. Portable dual gas analyzers are convenient for detecting any deviation of the fixed measuring system to avoid any surprises at the end of the storage season. These portable systems are normally capable of measuring O_2 ranging from 0 to 25 kPa as well as CO_2 usually in a range between 0 and 10 kPa.

2.5.3 Ethylene (C_2H_4)

C_2H_4 measurement in CA storage rooms is rarely practiced. Normally low concentrations exert its effects on fruits and make measuring difficult. For control purposes at higher levels, there are disposable tubes filled with chemicals that change color, indicating C_2H_4 levels (Dräger). In recent years, C_2H_4 analyzers or C_2H_4 detectors are available on the market. These instruments are equipped with electrochemical sensors and cover concentration ranges of 0–100 ppm with a minimum resolution of 0.2 ppm. The best method for C_2H_4 measurements is the gas chromatograph. However, this is a costly laboratory instrument and must be operated by well-trained personnel.

2.5.4 Automation and Control Systems

The conventional basic principle behind any automated computer-based CA control system is measuring the instantaneous gas composition and adjusting the control system parameters to adequately activate the gas modification system to reach the desired composition. Measurements of gas concentrations are automatically acquired at a preset time interval and sent to a controller for evaluation. In an automated control system, the controller is essentially the portion of the algorithm that makes use of the sensed information to manipulate the system parameters to reach the desired objectives. In the controller algorithm, the process error is initially computed by subtracting the desired gas levels from the acquired gas levels. According to this difference, the controller computes the corrective action required. Thus, if the O_2 level is lower than the set point, the controller computes the amount of air required to reach the desired O_2 level and opens the fresh air valves for the calculated period. Similarly, if the CO_2 level is too high, the controller computes the time required to scrub the atmosphere in the storage room to reach the desired level and opens the scrubbing valves as well. Finally, these decisions are executed in the control process and data is recorded for future reference.

The ability of the control system to effectively compute the required times for O_2 injection, CO_2 scrubbing, or CA room gas flushing depends on the type of control strategy implemented in the controller algorithm. A properly designed control system should be reasonably stable, and provide a fairly fast response and proper damping. The controller attempts to reduce the difference between the set points and the process variables (gas concentration, temperature) to a minimum, normally near zero. Various types of controllers are used in process control. Although the applications of conventional on–off control techniques are frequently unsuitable in agricultural systems (Stone, 1991), these types of controllers are still common. Some of the control systems used in CA room include on–off, PI, or PID controllers. The on–off controller simply turns on or off the devices according to the computed process error. The PI and PID controllers take into consideration the integral of the error over time, which eliminates offset. In addition, the PID controller takes into account the derivative of the error, which acts as a process stabilizer. These conventional controllers are designed for static operations. Since stored horticultural produce are living organisms that go through a series of physiological changes over the storage period, the need for a dynamic storage environment poses a control challenge. For this reason, before implementing a control system, the process must be carefully studied. The control system should be able to respond to potential changes of the stored commodity. When choosing a control strategy for CA control, special considerations should be given to the four following factors: potential disturbances during the storage period, efficiency of the CA room flushing system, stabilization time required reequilibrate the gas composition when facing any stored produce respiration rate change, and compliance to different horticultural produce physiological mechanisms.

Conventional controllers are generally modified to encompass the dynamic physiological activities of the stored horticultural produce. Although the use of more advanced controllers is possible in CA control systems, most commercial CA control systems have not adapted such systems because of their high complexity. Advanced control techniques include adaptive, fuzzy logic, knowledge-based, and artificial intelligent controllers. Some of these advanced techniques have been implemented in greenhouse environmental control (Gauthier, 1991; Hashimoto, 1993). Advanced controllers, such as self-learning and self-tuning controllers, are generally more intelligent and can dynamically optimize the parameters by deduction processes (Seborg et al., 1983).

Most of the control systems available commercially are microprocessor based, with embedded software that can be interfaced to a computer. The advantage of such systems is the reliability the microprocessor offers. If the computer is unexpectedly interrupted, the microprocessor continues the control process. These systems have a high cost; however, with the advancement of the modern computer they can be safely used to provide direct online control for smaller operations. A computer control system used with a data acquisition system can be interfaced with all the necessary devices and instrumentation (Landry et al., 2008). Software allows the communication between each component used to perform the control. In addition to its lower cost, this type of computer control system offers a greater flexibility in facilities used for research because custom-made software can be developed for any specific need (Markarian et al., 2006).

When designing a CA control system, future growth and upgrading should be antici-pated. The control system should permit new control needs with minimal cost. A modular architecture is highly recommended as it would simplify new control needs by allowing a change to a portion of the system only without any disruption to other parts (Mittal, 1997). It is also important to plan for contingencies in cases where emergencies or unexpected equipment failure are encountered. The downtime of the system should be minimized to prevent any spoilage considering tons of horticultural produce are generally involved in such CA rooms. Power failures should be considered and fast solutions should be integrated in the software to reactivate the system as fast as possible; however, the software should allow manual control of all equipment when required. In case of major disturbances, this option will let an experienced operator deduce a solution and override any automatic control. An alarm system should be incorporated to notify of any critical increases or decreases in gas levels in the storage room. The implementation of automatic emergency paging features is recommended. To prevent any data loss, the computer should be regularly backed up, preferably performed automatically through a network connection. When a network connection is available, remote control and monitoring are highly desirable.

2.5.5 Monitoring Produce Response and Adaptive Control

Adaptive control of CA conditions entails monitoring the response of horticultural produce to changes of storage parameters or more precisely to stressful conditions. The onset of stressful conditions concurs with the lowest acceptable respiration rate. Options to deter-mine the lowest rate may include three nondestructive methods. Measuring respiration to find the lowest CO_2 production rate would yield the ACP. This type of measurement is only accomplishable under laboratory conditions but not practical in commercial practice. A second indicator represents the fermentation threshold. It can be monitored and deducted from increases of acetaldehyde or ethanol in the storage atmosphere. Devices to monitor ethanol have been developed but their reliability does not always meet the desired standards. However, it was applied in DCS storage of 'Elstar' apples in the Netherlands (Veltman et al., 2002). Based on the principle of chlorophyll fluorescence, the HarvestWatch^{TM} system has been developed and subsequently patented (Prange et al., 2007). It is known that light entering a plant is absorbed by chlorophyll as well as related pigments and used for carbon fixation. Not all the light can be used and some of it is converted into energy or is emitted at a longer wavelength. This is called fluorescence. This can be observed in chlorophyll-containing solutions, which are illuminated with shorter wavelength light and then emit light of longer wavelength (690 and 710 nm). The fluores-cence signature begins immediately after exposure to light and follows a pattern known as the Kautsky curve. If chlorophyll has not been illuminated for at least 30 min, the first fluorescence after exposure to weak red light source is termed F_o. If the light is increased to saturating level, the fluorescence increases to a maximum, which is termed F_m (Figure 2.13). The new HarvestWatch system was designed to generate an approximation of F_o, the initial fluorescence, if horticultural produce was kept in the dark. Instead of calling it F_o it is labeled as F_α, a value generated through an algorithm that is partially influenced by both F_o and F_m (Prange et al., 2008). The HarvestWatch system is designed to operate primarily in dark postharvest environments where there is no extraneous light and the fruit can be sampled for extended periods without disturbance. Its first successful application is the detection of the lowest acceptable O_2 level that can be used in CA storage rooms (Figure 2.14). Thus it can make continuous nondestructive measurements of chloro-phyll fluorescence, measured as F_α and displays it in real time. It is noteworthy that as O_2 concentration declines, F_α does not increase until a certain O_2 concentration threshold is

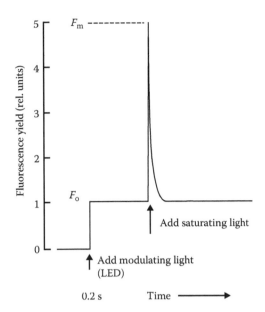

FIGURE 2.13
Typical fluorescence change over time when illuminated in the dark. Under stress either (or both) F_o or F_m can change.

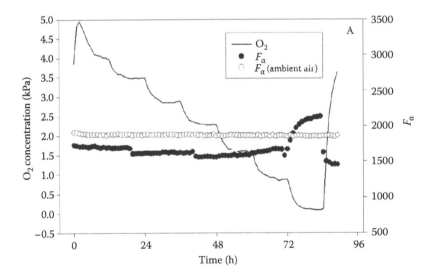

FIGURE 2.14
Example of the F_α fluorescence signal detected in apples held at 20°C in air (open circle) and in a progressively diminished O_2 environment (dark circle). The spike in F_α begins as the chamber O_2 levels fall below 1 kPa at 72 h and continues upward until O_2 concentration is increased at 84 h. (From Prange, R.K., DeLong, J.M., and Harrison P.A., *J. Am. Soc. Hort. Sci.*, 128, 603, 2003. With permission.)

encountered. At this threshold, F_α immediately increases. This increase in F_α is reversible by increasing O_2 concentration above the threshold. The system consists of the fluorescence interactive monitor (FIRM) unit affixed in an upper sampling kennel, apples, or other horticultural produce are placed in the bottom kennel over which the upper kennel

FIGURE 2.15
HarvestWatch fluorescence system showing a central hub (left), the FRIM affixed in an upper sealing kennel (center), and apples in the bottom kennel (right). (From Prange, R.K., Delong, J.M., and Wright, A.H., *Riv. Fruttic.*, 70, 52, 2008.)

housing the FIRM unit is securely fastened (Figure 2.15). In storage, the FIRM units are wired to a hub, which controls the interactions of electronic signals from a central computer to each attached FIRM device. It is advisable to use several units in a storage room. Thus the HarvestWatch system enables dynamic control of O_2 levels. Storing apple just above O_2 threshold concentrations resulted in improved quality retention and it has been reported that this technology maintains apple quality after storage equal to or better than does 1-MCP (Zanella et al., 2005).

2.6 Safety Considerations

Entering operating CA rooms is very dangerous since the O_2 content in such atmospheres is not sufficient to keep a human alive for even a few minutes. Less than 6 kPa O_2 leads in 30–45 s to loss of consciousness, termination of breathing, and is followed by death (Allen, 1998; Bishop, 1996; Vigneault et al., 1994). In addition, CO_2 levels are high enough to be harmful. This is reflected in statutory provisions, which indicate, for example, that upper limits for continuous occupation (8 h) are 0.5 kPa and a 15 min exposure to 1.5 kPa, which are still lower concentrations than the ones encountered in any CA storage. Legal limits of carbon monoxide (CO) exposure for humans may be different in different countries but may be 8 h at 25 ppm, or 30 min per working day at 75 ppm and no admission at levels above 125 ppm. A 1 h exposure to 1500 ppm (0.1 kPa) or higher level causes death. Because CA rooms are gastight, forklift exhausts may lead to accumulation of CO to dangerous levels during loading and unloading. Since propane forklifts produce up to 5 kPa CO of their exhausting gas, they must be avoided in such CA storage rooms and replaced by manual or electric forklifts. A properly tuned catalytic converter on the exhaust may remove 90%–99% of CO, however it is difficult to predict any malfunctioning of the catalytic converter and CO has no odor. CO detectors must be used when using a propane forklift within a closed environment but still any malfunctioning could result in death, which is not an option. Furthermore room ventilation to allow the use of forklift trucks within a CA room would be extremely high, exceeding 150 changes of air per hour to maintain CO at a secure level. To use a propane forklift, a properly tuned catalytic converter must by installed and maintained in optimal operating conditions, a CO detector with a loud signal alarm must be installed and verified very frequently, and sufficient ventilation must be in place and continuously running during the entire period of forklift operation. It may also advisable to monitor CO levels during loading and unloading rooms to evaluate the performance of the ventilation and catalytic converter systems.

All CA storage facilities should be clearly labeled with warning signs and the access doors must be kept locked. A further requirement includes proper instruction of staff involved and responsible for CA operations. For example, each entrance in any CA room must be controlled and supervised. In case of any mechanical failure requiring human intervention during the storage period, self-portable breathing system (similar to the ones used for under sea diving operation) must be used. In such cases, the entering person must be supervised by a trained person already equipped with an independent breathing portable system in such a way that the person could enter the room at any moment to intervene in case of a problem. This precaution may seem to be excessive, but serious accidents resulting in deaths have happened more than once when these precautions were not taken.

2.7 Future Developments

In the last 50–60 years, CA technology has been undergoing continuous development and refinement, resulting in continuing construction of CA facilities worldwide. The major refinements have been associated with improved construction techniques that have improved the airtightness of the rooms, improved sensing technology, improved O_2 and CO_2 scrubbing technology, and computerization, and increased knowledge of produce-specific requirements. Research of the 1980s and 1990s has been implemented now— 10–15 years later—and today's research will likely result in new possibilities in the next 10–15 years.

There are plenty of possibilities. Despite a large body of laboratory-based research literature demonstrating its efficacy in a wide array of fruits and vegetables, most commercial CA storage is still used primarily for extending the storage period of apples. The reason for the slow or nonexistent adoption in the storage of other fruits and vegetables is insufficient or unreliable financial return to justify the cost of CA. Furthermore, since CA is a technology that will not work where there is an unreliable supply of electricity, refrigeration, and/or technical support, its expansion has been constrained by failure to remove these reliability and cost problems, regardless of commodity.

Some fruits, vegetables, and flowers are shipped worldwide immediately after harvest via air without CA. However, a recent editorial by Wainwright (2008) has stated that the current debate on CO_2 emissions and food miles may shift produce transport to sea transport, which emits less CO_2 per kg of produce per mile than airfreight. He cites MA and CA as important technologies that will need to be implemented to help produce survive the longer shipping times associated with sea freight, compared with airfreight. He suggests that target produce for sea shipment might include vegetables and certain flowers such as carnation. Use of CA and MA during sea transport is not new but up to now it has had limited adoption. Perhaps Wainwright's prediction will stimulate an increase in CA and MA research and development appropriate for sea transport and perhaps revive interest in intermodal hypobaric storage.

Empirical research and commercial experience, primarily with apple and pear cultivars, has shown that the timing and degree of CA conditions should be carefully monitored on a cultivar-by-cultivar basis. Thus, there are now recommendations for delaying CA or for rapid CA, depending on the cultivar. These apparently contradictory results indicate the inadequacy of existing knowledge about the physiological basis for responses to CA conditions and the determination of the optimum CA condition during the storage period. This research will be aided by the availability of new technologies such as DCA that are able to dynamically control both the temperature and CA conditions 24 h per day, based entirely on the changing physiology of the produce.

CA equipment refinements can be expected. In an environment of ever-increasing energy cost, a major challenge will be reduction of the electrical energy consumption associated with CA. High-precision sensors for monitoring gas and temperature along with neural network and PID controllers, which are already widely used in other industries, will likely be adapted to CA storage. The development and use of these technologies may represent alternatives to use of chemicals such as 1-MCP or DPA (superficial scald) especially for organic fruits and vegetables.

Other techniques may be incorporated into a CA system, for example, control of storage rots may be assisted in the future by natural antimicrobial compounds delivered as a gas at critical periods during storage or novel gases (NO, N_2O) may be added to increase the benefits of traditional CA.

References

Allen, D. 1998. *Controlled Atmosphere Storage Buildings: Construction and Operational Techniques.* Nova Scotia Fruit Growers' Association, Agricultural Center, Kentville, Canada.

Apelbaum, A., G. Zauberman, and Y. Fuchs. 1977. Subatmospheric pressure storage of mango fruits. *Sci. Hort.* 7:153–160.

Argenta, L., X. Fan, and J. Mattheis. 2000. Delaying establishment in controlled atmosphere or CO_2 exposer reduces Fuji apple injury without excessive fruit quality loss. *Postharvest Biol. Technol.* 20:221–229.

Barry, R.C. and D. O'Beirne. 2000. Novel high oxygen and noble gas modified atmosphere packaging for extending the quality and shelf-life of fresh prepared produce, *Advances in Refrigeration Systems.* Food Technologies and Cold-Chain, Sofia, Bulgaria, September 23–26, 1998, pp. 417–424.

Bartsch, J.A. 2004a. Carbon dioxide control using hydrated lime. *Cornell Fruit Handling and Storage Newsletter*, pp. 13–15. http://www.hort.cornell.edu/department/faculty/watkins/extpubs.html

Bartsch, J.A. 2004b. CA room testing. *Cornell Fruit Handling and Storage Newsletter*, pp. 16–20. http://www.hort.cornell.edu/department/faculty/watkins/extpubs.html

Bartsch, J.A. and G.D. Blanpied. 1984. Refrigeration and controlled atmosphere storage for horticultural crops. The Northeast Regional Agricultural Engineering Services. Cornell University, Ithaca, NY. NRAES J. Paper No. 22, pp. 42.

Ben-Yehoshua, S. and V. Rodov. 2003. Transpiration and water stress. In: J.A. Bartz and J.K. Brecht (Eds.), *Postharvest Physiology and Pathology of Vegetables.* Marcel Dekker, New York, pp. 111–159.

Berger, P. 2007. Une souplesse d'utilisation au verger et à la vente. *L'aboriculture Fruitière* 607 (Avril): 28. http://www.mattiempo.com/fichetech_moduleCap.asp

Bishop, D. 1990. Controlled atmosphere storage. In: C.V.J. Dellino (Ed.), *Cold and Chilled Storage Technology.* Van Nostrand Reinhold, New York, pp. 66–98.

Bishop, D. 1996. *Controlled Atmosphere Storage: A Practical Guide.* David Bishop Design Consultants, Heathfield, East Sussex, England.

Bishop, D. 2007. International Controlled Atmosphere Ltd. Lawrence House, Transfesa Road, Paddock Wood, Kent, UK, TN12 GUT. http://www.ICAstorage.com Personal communication.

Blankenship, S.M. and J.M. Dole. 2003. 1-Methylcyclopropene: A review. *Postharvest Biol. Technol.* 28:1–25.

Boersig, M.R., A.A. Kader, and R.J. Romani. 1988. Aerobic–anerobic respiratory transition in pear fruit and cultured pear fruits cells. *J. Am. Soc. Hort. Sci.* 113:869–873.

Brecht, J.K. 2006. Controlled atmosphere, modified atmosphere and modified atmosphere packaging for vegetables. *Stewart Postharvest Rev.* 5(5):1–6.

Brecht, J.K., K.V. Chau, S.C. Fonseca, F.A.R. Oliveira, F.M. Silva, M.C.N. Nunes, and R.J. Bender. 2003. Maintaining optimal atmosphere conditions for fruits and vegetables throughout the postharvest handling chain. *Postharvest Biol. Technol.* 27:87–101.

Burg, S.P. 2004. *Postharvest Physiology and Hypobaric Storage of Fresh Produce.* CABI Publishing, Walingford, UK.

Burg, S.P. and E.A. Burg. 1966. Fruit storage at subatmospheric pressures. *Science* 153:314–315.

Cavalieri, R.P., W.C. Chiang, and H. Waelti. 1989. Nitrogen scrubbing of CA storages: A simulation study. *Trans. ASAE* 32:1709–1714.

Chapon, J.F. and P. Westercamp. 1996. Entreposage frigorifique des pommes et des poires Tome 2: "Conduite de la conservation." *Ctifl.* Paris, France.

Charles, M.T. and C. Vigneault. 2008. *Qualité et Manutention Post-Récolte de l'asperge.* Centre de référence en agriculture et agroalimentaire du Québec (CRAAQ). Québec, Canada, 52 pp.

Coquinot, J.P. and J.F. Chapon. 1992. Entreposage frigorifique des pommes et des poires Tome 1: "Equipement." *Ctifl.* Paris, France.

Curry, E.A. 1998. Physiology of Braeburn maturity and disorders: A discussion. *Postharvest Inf. Network*, 8. http://postharvest.tfrec.wsu.edu/pgDisplay.php?article=PC98P

Dalrymple, G.D. 1969. The development of an agricultural technology: Controlled atmosphere storage of fruits. *Technol. Cult.* 10(1):35–48.

DeEll, J.R., R.K. Prange, and D.P. Murr. 1995. Chlorophyll fluorescence as a potential indicator of controlled-atmosphere disorders in 'Marshall' McIntosh apples. *HortScience* 30:1084–1085.

DeEll, J.R., R.K. Prange, and D.P. Murr. 1998. Chlorophyll fluorescence techniques to detect atmospheric stress in stored apples. *Acta Hort.* 464:127–131.

DeLong, J.M., R.K. Prange, J.C. Leyte, and P.A. Harrison. 2004. A new chlorophyll fluorescence technology that determines low-O_2 thresholds in ultra low O_2 apple storage. *HortTechnology* 14:262–266.

Dilley, D.R. 2006. Development of controlled atmosphere storage technologies. *Stewart Postharvest Rev.* 6(5):1–8.

Eaves, C.A. 1934. Gas and cold storage as related fruit under Annapolis Valley conditions. *Annu. Rep. Nova Scotia Fruit Growers Assoc.* 71:92–98.

Eaves, C.A. 1959. A dry scrubber for CA apple storages. *Trans. ASAE* 2(1):127–128.

Ekman, J.H., J.B. Golding, and W.B. McGlasson. 2005. Innovation in cold storage technologies. *Stewart Postharvest Rev.* 3(6):1–14.

Erkan, M. and C.Y. Wang. 2006. Modified and controlled atmosphere storage of subtropical crops. *Stewart Postharvest Rev.* 5(4):1–8.

Gao, H.Y., H.J. Chen, W.X. Chen, J.T. Yang, L.L. Song, Y.H. Zheng, and Y.M. Jiang. 2006. Effect of hypobaric storage on physiological and quality attributes of loquat fruit at low temperature. Proc. 4th IC MQUIC. *Acta Hort.* 712:269–274.

Gasser, F. and E. Höhn. 2004. Lagerung von Kirschen in modifizierter Atmosphäre—ein Überblick. *Schweiz. Z. Obst-Weinbau* 140(13):6–10.

Gasser, F., D. Dätwyler, K. Schneider, W. Naunheim, and E. Hoehn. 2005. Effects of decreasing oxygen levels in the storage atmosphere on the respiration and production of volatiles of 'Idared' apples. *Acta Hort.* 682:1585–1592.

Gasser, F., P. Crespo, A. Gillard, S. Wernli, and E. Höhn. 2007a. Kirschen Lagerungsversuche 2005. *Schweiz. Z. Obst-Weinbau* 143(6):10–13.

Gasser, F., T. Eppler, W. Naunheim, S. Gabioud, and E. Hoehn. 2007b. Control of the critical oxygen level during dynamic CA storage of apples by monitoring respiration as well as chlorophyll fluorescence. *COST Action 924. International Conference: Ripening Regulation and Postharvest Fruit Quality.* November 12–13, Weingarten, Germany.

Gasser, F., T. Eppler, W. Naunheim, S. Gabioud, and E. Höhn. 2008. Control of critical oxygen level during dynamic CA storage of apples. *Agrarforschung* 15:98–103.

Gauthier, L. 1991. Greenhouse environment control using a knowledge-based approach. In: *Automated Agriculture for the 21st Century: Proceedings of the 1991 ASAE Symposium, Chicago, Illinois.* St. Joseph, MI, December, pp. 468–477.

Goyette B., C. Vigneault, N.R. Markarian, and J.R. DeEll. 2002. Design and implementation of an automated controlled atmosphere storage facility for research. *Can. Biosyst. Eng.* 44:3.35–3.40

Graham, D. 2008. Gas At Site Ltd., http://GasAtSite.com Personal communication.

Hashimoto, Y. 1993. *Computer Integrated System for the Cultivating Process in Agriculture and Horticulture. The Computerized Greenhouse.* Academic Press, New York, pp. 175–196.

Henze, J. and H. Hansen 1988. Lagerräume für Obst und Gemüse. *KTBL-Schrift* 327:113–130.

Höhn, E., M. Jampen, and D. Dätwyler. 1996. Kavernenbildung in Conférence—Risikoverminderung. *Schweiz. Z. Obst-Weinbau* 132(7):180–181.

Hui, K.P.C., C.F. Forney, J.R. DeEll, and N.R. Markarian. 2005. In: Vigneault, C. (Sc. Ed.), *Postharvest Handling of Small Fruits for Fresh Market*. Ontario Berry Growers' Association, Vineland, Canada. pp. 43.

Jameson, J. 1993. CA storage technology in the 1990s. *Postharvest News Inf.* 4(1):16N–17N.

Jayaraman, K.S. and P.S. Raju. 1992. Development and evaluation of a permanganate-based ethylene scrubber for extending shelf life of fresh fruits and vegetables. *J. Food Sci. Technol.* 29(2):77–83.

Johnson, D. 2007. Factors affecting the efficiency of 1-MCP applied to retard apple ripening. *COST Action 924. International Conference: Ripening Regulation and Postharvest Fruit Quality.* November 12–13, Weingarten, Germany.

Kader, A.A. 2003. A summary of CA requirements and recommendations for fruits other than apples. *Acta Hort.* 600:737–740.

Kader, A.A. and S. Ben-Yehoshua. 2000. Effects of superatmospheric oxygen levels on postharvest physiology and quality of fresh fruit and vegetables. *Postharvest Biol. Technol.* 20:1–13.

Kidd, F. and C. West. 1927. Gas storage of fruit. Great Britain Department of Scientific Industrial Research Food Investigation Board Report 30, pp. 87.

Kidd, F. and C. West. 1930. The gas storage of fruit. II. Optimum temperatures and atmospheres. *J. Pomol. Hort. Sci.* 8:67–77.

Kidd, F. and C. West. 1937. Action of carbon dioxide on the respiration activity of apples. Effect of ethylene on the respiration activity and climacteric of apples. Individual variation in apples. Great Britain Department of Scientific Industrial Research Food Investigation Board Report, pp. 101–115.

Kidd, F. and C. West. 1950. The refrigerated gas storage of apples. Variation in apples. Great Britain Department of Scientific Industrial Research Food Investigation Board Leaflet No. 6 (rev.), pp. 16.

Knee, M., F.J. Proctor, and C.J. Dover. 1985. The technology of ethylene control: Use and control in post-harvest handling of horticultural commodities. *Ann. Appl. Biol.* 107:581–595.

Kupferman, E. 2003. Controlled atmosphere storage of apples and pears. *Acta Hort.* 600:729–735.

Landry, J.A., N. Markarian, and C. Vigneault. 2008. Enthalpy based PID controller for horticultural storage facilities. *Appl. Eng. Agric.* (In press).

Laurin, É., M.C.N. Nunes, J.P. Émond, and J.K. Brecht. 2006. Residual effect of low-pressure stress during simulated air transport on Beit Alpha-type cucumbers: Stomata behaviour. *Postharvest Biol.Technol.* 41:121–127.

Malcolm, G.I. 2005. Advancements in the implementation of CA technology for storage of perishable commodities. *Acta Hort.* 682:1593–1597.

Marcellin, P. and J. Leteinturier. 1967. Premières applications industrielles des membranes en caoutchouc de silicone à l'entreposage des pommes en atmosphere contrôlée. *Inst. Int. Froid. Congr. Intern. Froid.* Madrid, Espagne, pp. 1–9.

Markarian, N.R., C. Vigneault, Y. Gariepy, and T.J. Rennie. 2003. Computerized monitoring and control for a research controlled-atmosphere storage facility. *Comp. Electron. Agric.* 39:23–37.

Markarian, N.R., J.A. Landry, and C. Vigneault. 2006. Model for the simulation of temperature and relative humidity in horticultural storage facilities. *Int. J. Food Agric. Environ.* 4(1): 34–40.

Matté, P., L. Buglia, L. Fadanelli, C. Chistè, F. Zeni, and A. Boschetti. 2005. ILOS + ULO as a practical technology for apple scald prevention. *Acta Hort.* 682:1543–1550.

Mitcham, E.J., T. Lee, A. Martin, S. Zhou, and A.A. Kader. 2003. Summary of CA for Arthropod control on fresh horticultural perishables. *Acta Hort.* 600:741–745.

Mittal, G.S. (Ed). 1997. *Computerized Control Systems in the Food Industry.* Marcel Dekker, New York, pp. 632.

Osterloh, A. 1996. Lagervervahren and Planung, Bau und Ausrüstung von Obstlägern, In: A. Osterloh, G. Ebert, W.-H. Held, H. Schulz, and E. Urban (Eds.), *Lagerung von Obst und Südfrüchten.* Verlag Eugen Ulmer, Stuttgart, Germany, pp. 113–141.

Peppelenbos, H. 2003. How to control the atmosphere? *Postharvest Biol. Technol.* 27:1–2.

Prange, R.K. and J.M. DeLong. 2003. 1-Methylcyclopropene: The "magic bullet" for horticultural products? *Chron. Hort.* 43(1):11–14.

Prange, R.K., J.M. DeLong, and P.A. Harrison. 2003. Oxygen concentration affects chlorophyll fluorescence in chlorophyll-containing fruit and vegetables. *J. Am. Soc. Hort. Sci.* 128:603–607.

Prange, R.K., J.M. DeLong, and P.A. Harrison. 2005a. Quality management through respiration control: Is there a relationship between lowest acceptable respiration, chlorophyll fluorescence and cytoplasmic acidosis? *Acta Hort.* 682:823–830.

Prange, R.K., J.M. DeLong, B.J. Daniels-Lake, and P.A. Harrison. 2005b. Innovation in controlled atmosphere technology. *Stewart Postharvest Rev.* 3(9):1–11

Prange, R.K., J.M. DeLong, B. Daniels-Lake, and P.A. Harrison. 2006. Controlled-atmosphere related disorders of fruits and vegetables. *Stewart Postharvest Rev.* 5(7):1–10.

Prange, R.K., P. DeLong, P. Harrison, J. Leyte, S.D. McLean, J.G.E. Scrutton, and J.J. Cullen. 2007. Method and apparatus for monitoring a condition in chlorophyll containing matter. U.S. Patent # 7,199,376.

Prange, R.K., J.M. DeLong, and A.H. Wright. 2008. La fluorescenza della clorophilla per monitorare frutta ed ortaggi in conservazione. *Riv. Fruttic.* 70(4):52–55.

Raghavan, G.S.V. and Y. Gariépy. 1984. Structure and instrumentation aspects of storage systems. *Acta Hort.* 157:5–40.

Raghavan, G.S.V., Y. Gariépy, R. Thériault, C.T. Phan, and A. Lanson. 1984. System for controlled atmosphere long-term cabbage storage. *Int. J. Refrig.* 7(1):66–71.

Raghavan, G.S.V., C. Vigneault, Y. Gariépy, N.R. Markarian, and Alvo P. 2005. Refrigerated and controlled/modified atmosphere storage. In: D.M. Barett, L. Somogy, and H. Ramaswamy (Eds.), *Processing Fruits.* CRC Press, Boca Raton, FL, pp. 23–52.

Roelofs, F.P.M.M. and A. de Jager. 1997. Reduction of brownheart in Conference pears. *Proc. 7th Int. CA Res. Conf. Davis, California, USA* 2:138–144.

Romanazzi, G., F. Nigro, and A. Ippolito. 2003. Short hypobaric treatments potentiate the effect of chitosan in reducing storage decay of sweet cherries. *Postharvest Biol. Technol.* 29:73–80.

Ross, D.A. 2004. The pressure myth. *Saltwater Fly Fishing Magazine.* http://www.midcurrent. com/articles/science/ross_pressure_myth.aspx

Saltveit, M.E. 2003a. A summary of CA requirements and recommendations for vegetables. *Acta Hort.* 600:723–727.

Saltveit, M.E. 2003b. Is it possible to find an optimal controlled atmosphere? *Postharvest Biol. Technol.* 27:3–13.

Saquet, A.A., J. Streif, and F. Bangerth. 2003. Reducing internal browning disorders in 'Braeburn' apples by delayed controlled atmosphere storage and some related physiological and biochemical changes. *Acta Hort.* 682:453–458.

Schouten, S.P., R.K. Prange, J. Verschoor, T.R. Lammers, and J. Oosterhaven. 1997. Improvement of Elstar apples by dynamic control of ULO conditions. *Proc. 7th Int. CA Res. Conf., Davis, California, USA* 16:71–78.

Seborg, D.E., S.L. Shah, and T.F. Edgar. 1983. Adaptive control strategies for process control: A survey. *AIChE Diamond Jubilee Meeting,* Washington DC, November.

Smock, R.M. and A. Van Doren. 1941. Controlled atmosphere storage of apples. Cornell University Agricultural Experiment Station Bulletin 762, pp. 45.

Stone, M.L. 1991. Control system applications. In: *Automated Agriculture for the 21st century: Proc. 1991 ASAE Symp., Chicago, Illinois.* St. Joseph, MI, December, pp. 163–166.

USDA, ARS. 2004. The commercial storage of fruits, vegetables, and florist and nursery stocks. http://www.ba.ars.usda.gov/hb66/contents.html

van Doren, V.J. 1998. Basics of proportional-integral-derivative control. *Control Eng*. 45(3):135–142.

van Schaik, A.C.R., 2008. Post harvest quality and Technology, AFSG PPO-fruit, Wageningen UR, NL, http://www.afsg.wur.nl Personal communication.

Veltman, R.H., J.A. Verschoor, and J.H. Ruijsch van Dugteren. 2003. Dynamic control system (DCS) for apples (*Malus domestica* Borkh. 'Elstar'): Optimal quality through storage based on product response. *Postharvest Biol. Technol*. 27:79–86.

Verlinden, B.E., A. de Jager, J. Lammertyn, W. Schotsmans, and B. Nicolaï. 2002. Effect of harvest and delaying controlled atmosphere storage conditions on core breakdown incidence in 'Conference' pears. *Biosyst. Eng*. 83:339–347.

Vigneault, C. and F. Artés Hernández. 2007. Gas treatments for increasing phytochemical content of fruits and vegetables. *Stewart Postharvest Rev*. 3(3):8.1–8.9.

Vigneault, C., V. Orsat, B. Panneton, and G.S.V. Raghavan. 1992. Oxygen permeability and air tightness measuring method for breathing bags. *Can. Agric. Eng*. 34(2):183–187.

Vigneault, C., G.S. Raghavan, and R.K. Prange. 1994. Techniques for controlled atmosphere storage of fruits and vegetables. Technical Bulletin 1993–18E. Research Branch. Agriculture and Agri-Food Canada.

Vigneault, C., T.J. Rennie, and V. Toussaint. 2008. Cooling of freshly cut and freshly harvested fruits and vegetables. *Stewart Postharvest Rev*. 4(3):4.1–4.10.

Waelti, H. and R.P. Cavalieri. 1990. Matching nitrogen equipment to your needs. *Wash. State Univ. Tree Fruit Postharvest J*. 1(2):3–13.

Wainwright, H. 2008. Food miles and horticultural trade. *J. Hort. Sci. Biotechnol*. 83:143.

Wang, L., P. Zhang, and S. Wang. 2001. Advances in research on theory and technology for hypobaric storage of fruit and vegetable. *Storage Process* 5:3–6.

Wang, Z. and D.R. Dilley. 2000a. Hypobaric storage removes scald-related volatiles during the low temperature induction of superficial scald. *Postharvest Biol. Technol*. 18:191–199.

Wang, Z. and D.R. Dilley. 2000b. Initial low oxygen stress controls superficial scald in apples. *Postharvest Biol. Technol*. 18:201–213.

Wenxiang, L., M. Zhang, and Y. Han-Qing. 2006. Study on hypobaric storage of green asparagus. *J. Food Eng*. 73:225–230.

Wertheim, S.J. 2005. Fruit storage. In: J. Tromp, A.D. Webster, and S.J. Wertheim (Eds.), *Fundamentals of Temperate Zone Tree Fruit Production*. Backhuys Publishers b.v., Leiden, the Netherlands, pp. 311–324.

Wills, R.B.H. and M.A. Warton. 2004. Efficacy of potassium permanganate impregnated into alumina beads to reduce atmospheric ethylene. *J. Am. Soc. Hort. Sci*. 129:433–438.

Wills, R.B.H., V.V.V. Ku, D. Sholet, and G.H. Kim. 1999. Importance of low ethylene levels to delay senescence of non-climacteric fruit and vegetables. *Austr. J. Exp. Agric*. 39:221–222.

Wills, R.B.H., M.A. Warton, and V.V.V. Ku. 2000. Ethylene levels associated with fruit and vegetables during marketing. *Austr. J. Exp. Agric*. 40:465–470.

Wojciechowski, J. 1989. Ethylene removal from gases by means of catalytic combustion. *Acta Hort*. 258:131–141.

Yang, S.F. and N.E. Hoffmann. 1984. Ethylene biosynthesis and its regulation in higher plants. *Ann. Rev. Plant Physiol*. 35:155–189.

Zanella, A. 2003. Control of apple superficial scald and ripening—A comparison between 1-methylcyclopropene and diphenylamine postharvest treatments, initial oxygen stress and ulta low oxygen storage. *Postharvest Biol. Technol*. 27:69–78.

Zanella, A., P. Cazzanelli, A. Panarese, A. Coser, M. Cecchinel, and O. Rossi. 2005. Fruit fluorescence response to low oxygen stress: Modern storage technologies compared to 1-MCP treatment of apple. *Acta Hort*. 682:1535–1542.

3

Transport Technology and Applications

Patrick E. Brecht, Shawn Dohring, Jeffrey K. Brecht, and Wayne Benson

CONTENTS

3.1 Introduction

Fresh produce has been historically grown in areas where quality and yield can be optimized. Since great distances often separate high population bases from seasonal production areas, transcontinental and transoceanic shipments of produce by air, land, and sea have been necessary.

Advanced cultural and production practices coupled with improved vegetable and fruit cultivars have contributed to superior quality at harvest and increased yields. Accordingly, millions of pounds of fresh fruits and vegetables are produced annually in every corner of the world. Much of this produce can be successfully sold and shipped to foreign markets via land and ocean transportation only if the right blend of climate control technology, services, and all-in landed costs* are available to shippers, buyers, and transporters.

---

* "All-in landed costs" are the total costs associated with importing a product such as the cost of the perishable cargo, marine cargo insurance, import duties, transportation, and other services.

The full impact of these horticultural improvements can only be realized at the wholesale, retail, and consumer levels if product spoilage and deterioration are minimized during distribution. Concomitantly, global demand for fresh fruits and vegetables has continued to grow due to the consumer's appetite for wholesome, fresh produce. Furthermore, the consumer's desire for previously underexploited fruits and vegetables from distant production areas has been augmented by more disposable income and an evolving appreciation for the nutritive value of fresh produce. As a result, hundreds of thousands of loads of fresh produce have been shipped around the world, via land and ocean, in refrigerated intermodal containers called "reefers" (Tanner and Smale, 2005). While the preponderance of these shipments arrives in marketable condition, a considerable number of loads have been plagued by undesirable quality characteristics by the time they reach the consignee.

In this chapter, we explore the use of freshness extending atmosphere management during the transport period as a means of making the delivery time of fresh fruits and vegetables to market less problematic. We also describe the commercial landscape for atmosphere management systems related to fresh air exchange, humidity, modified atmosphere (MA), and controlled atmosphere (CA) technologies. The use and acceptability of atmosphere management systems for land and sea transport of fresh produce to distant markets are discussed.

3.2 Transport Services

One solution to distributing high value and short shelf life fresh produce to world markets is air transport (Pelletier et al., 2005; Thompson et al., 2004). Air transport is quick but it is expensive. By comparison, land and ocean transport, utilizing multimodal refrigerated containers with computer-based controllers and atmosphere management, are far less costly than air shipment, but the transit times to distant markets are much longer than air transport. The extended times in transit require that optimum temperature and atmosphere management be maintained to prolong shelf life and to deliver consistent quality produce to world markets.

Refrigerated transportation for perishable commodities is a mature technology, having been used for over 100 years. The first ice-cooled rail cars departed California for the Eastern United States in 1907 (Daniels, 2000). About 50 years later, mechanical refrigeration was substituted for ice, and since then, technology advances have resulted in evolutionary improvements. In recent years, technological advances in the design of insulated trailers and marine containers and their refrigeration control systems have markedly improved the operation and reliability of the controlled temperature and atmosphere in the transport vehicles (Thompson and Brecht, 2005; Thompson et al., 2000). Many of the improvements have been in information technology and sensors. In the last two decades, temperature control has improved significantly through smarter defrost cycles, improved air distribution, improved reliability, and microprocessor controls. Computer-based controllers now automatically check, diagnose, and record the operating conditions of climate control equipment; and the temperatures, humidities, and atmospheres can be adjusted based on a product database stored in the computer or available from published guidelines (Anon, 1999; Brecht and Brecht, 2001). Wireless communication technology now allows a company's headquarters staff to monitor and control the operating conditions of climate control equipment during transit.

These technological advances in microprocessor-driven refrigerated containers for transoceanic and overland transport have opened up a new era of produce distribution to overseas markets with transit times often greater than those encountered in domestic trade routes. Additionally, longer transit times, higher energy costs, and dramatically increased volumes of produce have triggered new developments and improvements in climate control technology. Shippers and receivers of fresh produce have witnessed improvements related to supplements to temperature control such as atmosphere, humidity, and ethylene (C_2H_4) management and improved fresh air ventilation and air circulation. The ability to deliver fresh and wholesome produce to consumers by managing the climate within reefer containers can potentially result in financial rewards to growers, exporters, transporters, and importers. On the other hand, sellers and buyers face economic, currency exchange, quarantine, and technological challenges to expanding the global market for fresh produce. The most significant challenge for exporters is to deliver consistently fresh quality produce to distant markets at an affordable and competitive landed cost.

3.3 Refrigerated Transport Basics

Temperature management plays the most significant role for extending the market life of the produce. Bringing the product to its desired carrying temperature as quickly as possible is paramount for global distribution. On the other hand, altering the container's atmosphere to desired levels as a supplement to good temperature control can potentially add days or even weeks to the shelf life of a large number of vegetables and fruits, thereby allowing the produce to travel longer distances and bringing greater selection of produce to consumers around the world. Options for regulating the container environment come at a time when produce trade is flourishing, with agricultural regions from around the world competing for consumer appetites by shipping produce over increasingly longer distances.

When shipping fresh produce in reefer containers, special packaging and loading practices are required (Thompson et al., 2000). Proper product packaging, carton designs, unitization, and stowage are needed to ensure that the conditioned air is capable of evenly and effectively removing heat while maintaining noninjurious or beneficial levels of gases such as carbon dioxide (CO_2), oxygen (O_2), and C_2H_4 within the load. Air, like any gas, will follow the path of least resistance. If the reefer container is stowed in such a way as to allow conditioned air to prematurely return back to the refrigeration unit, the produce stowed progressively toward the rear doors of the reefer container will not receive the climate control protection required. Figure 3.1 is an illustrative example of bottom-airflow in a reefer container.

The reefer container is the platform where atmosphere technologies are found. Highway trailers with atmosphere control technologies are not available. A reefer is made up of an International Standards Organization thermal (insulated) container, typically 20 or 40 ft long, with a refrigeration system that becomes the front wall of the container. Reefer container systems, unlike domestic over-the-road trailer refrigeration systems, deliver conditioned air to the cargo via the container's T-floor. Containers offer superior temperature control compared with highway trailers by delivering pressurized, conditioned air from the T-floor vertically up through the chilled produce load rather than over the top as in a highway trailer. Reefer containers are better insulated and considerably more airtight than highway trailers. As a result of these advantages, fresh produce in reefer containers can be transported in both extreme warm and cold environments. During transit, heat is removed from the container through the circulation of air through the cargo and the use of the refrigeration system's evaporator. Cooling capacity is typically controlled via a combination of suction line modulation and cycling the compressor. Container refrigeration systems are designed to optimally maintain product temperature during transit. Therefore, the product should be fully precooled prior to being loaded into the container. Notwithstanding,

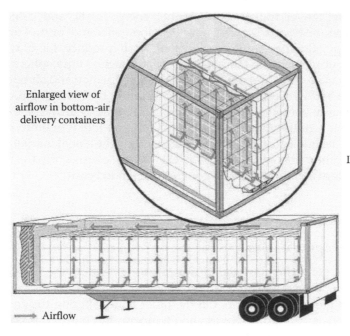

Enlarged view of airflow in bottom-air delivery containers

In bottom-air delivery containers, air is forced through the floor space under the load and up through the cargo. This type of airflow provides the most even temperature management for chilled cargo.

⟶ Airflow

Airflow in bottom-air delivery containers

FIGURE 3.1
Diagram illustrating bottom-airflow within a reefer container. (From Brecht, P.E., *Shipping Special Commodities*, American President Lines, Oakland, CA, 1992. With permission.)

container refrigeration systems are routinely tasked with reducing the cargo temperatures of fresh produce loaded "hot." Produce loaded "hot" will lose quality attributes, shelf life, and weight during the potentially slow temperature pull-down process.

3.4 Atmosphere Management Basics

MA and CA technologies have been in existence since the 1920s as a result of the classic work of Kidd and West (Smock, 1979). The terms CA and MA signify that the atmospheric composition surrounding fresh produce is different from that of normal air in a transit vehicle. MA and CA generally involve the manipulation of CO_2, O_2, and nitrogen (N_2). MA differs from CA only in how precisely the gas partial pressures are controlled. CA is more precise than MA in that CA systems employ feedback control of the gas concentrations (Brecht, 1980).

In the commercial reefer container trade, there are four basic approaches to altering reefer container atmospheres during transit, namely CA, MA, automated fresh air management, and manual fresh air exchange. Although MA and CA systems and services have been in use for decades, the commercial use of these systems during transport has been primarily limited to international movements of selected vegetables and fruits. The use of manually operated fresh air exchange, as a means of altering the atmosphere, has been available to reefer container operators and shippers for many years and is the most common means of altering the atmosphere. Automated fresh air management was introduced less than a decade ago and is growing in popularity. There have been no significant changes to commercially available technologies in the last few years.

As a general rule, atmosphere modification techniques involving reefer containers utilize atmospheric generation and/or commodity-MAs. With atmospheric generation, O_2 partial pressures can be purged with N_2 and/or CO_2 to desired set points. For commodities that derive benefits from elevated CO_2, this gas is metered into the reefer container. Commodity-MAs are established when actively respiring and metabolizing produce reduces the O_2 and increases the CO_2 partial pressures in the ambient air within a sealed reefer container, primarily when barriers and restrictions to gas exchange exist. Ethylene and CO_2 scrubbers are often used when either of these components is considered potentially injurious to the commodity quality.

Regardless of the systems, atmosphere levels within a reefer container can vary greatly according to many variables including the leak rate of the container, variable airspace volume (void volume), loading patterns, load mix, packaging, carton design, time in transit, product maturity or ripeness stages, and the amount, type, and temperature of the product. Partially loaded containers are associated with inferior temperature control, larger void volumes, and reduced accumulation of respiratory gases, which are not conducive to optimizing atmosphere management using today's most common MA or CA systems.

Shippers and importers who procure atmosphere services rely on the postharvest handling experience and expertise of atmosphere service companies and/or postharvest subject matter experts. Service companies like TransFresh Corporation (TransFresh), Salinas, California and Mitsubishi Australia (MAXtend™), Melbourne, Australia do their own research, initiate the atmosphere settings, and take some level of responsibility for the product outturn at the destination (Mitsubishi Australia Ltd., 2008; TransFresh Corporation, 2008). Equipment manufacturers like Carrier Corporation (Carrier), Syracuse, New York and Thermo King Corporation (Thermo King), Minneapolis, Minnesota sell atmosphere systems that are integrated into their climate control container equipment. The manufacturer's systems come with factory guidance and "how-to" guidebooks that have been compiled by third party subject matter experts. In most cases, transportation companies that purchase atmosphere technologies need a higher level of in-house expertise to sell and provide atmosphere services. In an effort to minimize the need for in-house expertise, Thermo King has incorporated expert systems into the reefer unit's microprocessor that automatically establish the atmosphere and temperature settings when a commodity is entered.

3.5 Atmosphere Services: Buyer Psychology

There are buyers of selected commodities and handlers operating in some trade lanes that have fully embraced the utilization of atmosphere management technologies and services. However, the use of these services remains a niche market. The cost benefit of atmosphere management, the level of risk involved, and the knowledge and experience of the participants utilizing these technologies are the biggest obstacles to helping growers, shippers, and traders take full advantage of atmosphere management services. Where shipping produce via land and ocean rather than air may indeed lower shipping costs, it can change the cash flow of the companies buying and selling the product and introduce an element of risk; "Will the currency exchange change during extended transit to the distant market" and/or "Will the product arrive safely and on time with the quality level required for marketing?"

Since atmosphere technologies and services add to the landed cost of produce transported to overseas markets, the prospective purchaser of these services expects a favorable cost benefit. To cost justify atmosphere services, a compelling case needs to be made to exporters and importers that there are tangible cost benefits of the service. Buyers of atmosphere services want evidence of proven results and the wherewithal to justify the cost of the atmosphere service by improved revenue, reduced shrinkage, increased market share, fewer adjustments and claims, enhanced brand recognition, consistent quality, penetration of new and existing markets, pipelining the product into the market, and/or extending the shipping season. Due to the different types and costs of atmosphere technologies and services available to the market, the value of these technologies is rarely fully understood by shippers, importers, and transporters.

3.6   Humidity Management

Although humidity management has been shown to be a beneficial means of supplementing good temperature control, shippers do not frequently utilize special humidity management services. Optimal humidity can be maintained in a reefer container by adding moisture or dehumidifying the conditioned space. Proper sizing of the refrigeration system is the first key to limiting unwanted dehumidification. Misting and ultrasonic fogging systems allow for the addition of moisture into the conditioned space. For dehumidifying the cargo space when shipping produce like onions, garlic, ginger, and winter squash that require reduced humidity levels, all refrigeration systems will do some dehumidification while some are specially equipped to further decrease humidity levels by increasing the heat load on the evaporator or by managing the evaporator fan speed. Desiccants can also be used. Carrier and Thermo King offer humidity and dehumidification control systems. Carrier has incorporated a humidification feature called NatureFresh into their CA reefer units since there is a risk of dehydration of the produce when dry N_2 gas is injected into CA containers.

3.7   Manual Fresh Air Exchange

Reefer containers are routinely fitted with manually adjustable fresh air exchange ports. The fresh air exchange can be set to manage the amount of outside air that is allowed to enter and leave the container through the vent ports. The purpose of manual fresh air exchange is to protect fruits and vegetables from injurious levels of O_2, CO_2, and/or C_2H_4 that may develop in the interior of reefer containers. These fresh air exchange systems are designed to introduce fresh air into the reefer container while replenishing O_2 and removing CO_2, C_2H_4, and trace volatiles from the cargo space.

Fresh air exchange is a simple and inexpensive means of altering the atmosphere and maintaining a safe environment. However, mistakes in the adjustment of the fresh air exchange are common, potentially leading to undesirable consequences. Moreover, the effectiveness of the fresh air exchange and the actual number of air exchanges per unit time within the cargo space are highly dependent on uncontrolled variables such as the void volume within the container, the incoming pulp temperature of the produce, the amount of cargo, and the leak rate of the reefer container. Furthermore, the rates of O_2 consumption and CO_2 and C_2H_4 production by fresh fruits and vegetables changes over time, while the setting of the fresh air exchange is selected and set at the time the container is loaded with product and is not typically changed at any time during shipping.

Furthermore, excessive fresh air exchange settings, particularly in warm and humid tropical environments, can cause a great deal of the refrigeration capacity to be used to cool and condense moisture out of the incoming air, thereby causing elevated carrying temperatures and fluctuating humidity levels within the container. Many reefer containers operating in tropical areas are used to ship uncooled fruit and all the refrigeration capacity is needed to cool the fruit.

The fresh air exchange vent settings are usually approximations based on prior experience and published guidelines. The air exchange occurs when there is differential pressure across the evaporator coil of the refrigeration unit and the fresh air exchange doors are open. The fresh air exchange, with its door opening set manually, does not precisely "control" gas levels within the container. However, it does alter the atmosphere within the cargo space of the reefer container by facilitating air exchange. When the fresh air exchange vents are closed, the respiratory activity of the fresh produce and the production of C_2H_4 can potentially result in the development of injurious levels of CO_2, O_2, and C_2H_4 in the cargo space. Figure 3.2 shows the components and functioning of a manual fresh air exchange system.

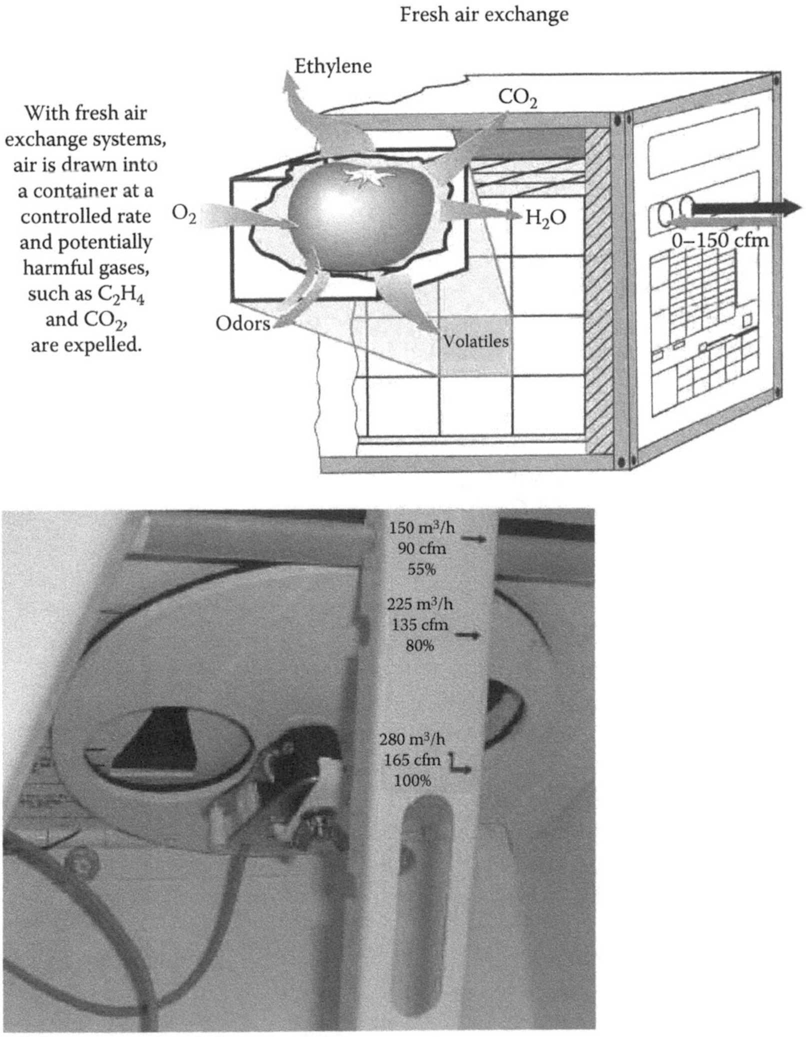

FIGURE 3.2
Diagram illustrating manual fresh air exchange (top) and picture of a manual fresh air exchange system (bottom). (From Brecht, P.E., *Shipping Special Commodities*, American President Lines, Oakland, CA, 1992; and Thermo King Corporation. With permission.)

3.8   Automated Fresh Air Exchange

Less than a decade ago, Thermo King attempted to address the manual fresh air exchange problems with its advanced fresh air management system (AFAM+) by introducing a feedback component that controls when and how much the fresh air exchange door opens (Brecht and Brecht, 2001). The AFAM+ system was primarily developed to minimize the refrigeration load contributed by infiltration of outside air into the conditioned container space, as well as to maintain high humidity levels in the container in order to minimize product shrivel. The AFAM+ system was also developed as an alternative to MA and CA systems, which are expensive to purchase and operate, thereby limiting mass-market appeal. Transportation companies were interested in a low-cost alternative to MA and CA that was user-friendly and enabled them to provide a value-added service to their customers and their customer's customers. Exporters, importers, and retailers, in turn, were looking for a low-cost alternative to MA or CA for shipping a broader range of produce items.

Although many commodities can derive a benefit from CA systems, shippers and receivers often cannot justify the added cost of CA for the preponderance of the fruits and vegetables that they ship in international trade. Additionally, many international transportation companies have been unable to justify the capital and operating costs of acquiring specialized CA containers because of the small volume of fruits and vegetables that derive a cost benefit from using CA. Reportedly, <2% of the world's fleet of reefer containers is fitted with "stand-alone" CA technology. However, thanks to improved box construction, ~25% of today's fleet is capable of MA or CA services (Dohring, 2006).

Carrier and Mitsubishi Australia subsequently released similar systems. The Carrier system is called enhanced automated fresh air on-demand ventilation system (eAutoFresh) and the Mitsubishi Australia system is called MAXtend regulated atmosphere (MAXtend RA). The Thermo King and Carrier systems utilize a CO_2 sensor to control a stepper motor that opens and closes the fresh air makeup vent and maintain selected CO_2 values once the CO_2 set point is reached. In contrast, the Mitsubishi Australia system utilizes an O_2 controller. The Thermo King and Carrier microprocessors monitor and record the O_2 and CO_2 readings whereas the Mitsubishi Australia system only records O_2. Each system functions to maintain the fresh air exchange in the closed position as much as possible, opening the fresh air exchange door only when necessary to avoid exposure of the product being carried to a potentially injurious atmosphere. The three types of automated fresh air exchange systems are compared in Table 3.1.

Like a CA system, automated fresh air exchange systems utilize mechanical ventilation of the container to control accumulation of respiratory CO_2. Once the maximum CO_2 or minimum O_2 set point is sensed, the system opens the mechanized fresh air exchange ports that draw outside ambient air into the container, thereby replenishing O_2 and removing excess CO_2. Since automated fresh air exchange systems depend on ambient atmospheres (normal air), and fruits and vegetables typically respire equal amounts of O_2 and CO_2, the summation of the O_2 and CO_2 concentrations inside the container will be about 21%.

For shippers to decide between the two mechanical systems, automated fresh air exchange and CA, an analysis of the cost benefit of the services and the tolerances of the produce being shipped is required. Furthermore, the physiological response of the product being shipped in an automated fresh air exchange-equipped system is primarily affected by the CO_2 concentration that develops. This is because when O_2 and CO_2 concentrations change equally from ambient levels, in almost all cases, the CO_2 tolerance limit of the product will be reached before the O_2 tolerance limit (i.e., anaerobic compensation point) is

TABLE 3.1

Automated Fresh Air Exchange Systems Used Commercially in Conjunction with Reefer Container Cooling Systems

Commercial Air Exchange Offering	O_2 Concentration Changes			CO_2 Concentration Changes			Notes
	Sensor	Increase	Decrease	Sensor	Increase	Decrease	
Thermo King's AFAM+ and Carrier's eAutoFresh	Yes (see note)	Yes	Yes (see note)	Yes	Yes (passive, see note)	Yes	Atmosphere generation, increases in CO_2 and decreases in O_2 gas concentrations, relies solely on product respiration. CO_2 gas can only be removed by allowing fresh air into the container. However, this in turn causes an increase in the O_2 gas concentration. Thermo King offers an optional secondary O_2 controller as a backup to the CO_2 controller.
Mitsubishi Australia's MAXtend RA (marketed as a regulated atmosphere)	Yes	Yes	Yes (passive, see note)	No	Yes (passive, see note)	Yes (passive, see note)	Atmosphere generation, increases in CO_2 and decreases in O_2 gas concentrations, relies solely on product respiration. A CO_2 scrubber is used to passively remove CO_2.

Source: Adapted from Dohring, S., *Stewart Postharvest Rev.*, 2(5), 1, 2006.

reached. Therefore, a fruit or vegetable that potentially derives more benefit from a reduced O_2 atmosphere than from an elevated CO_2 atmosphere will benefit only from avoidance of injurious CO_2 concentrations in an automated fresh air exchange system. The proper CO_2 set point for a product depends on that product's tolerance to elevated CO_2 for the expected duration of the transport. The complexities involved in successfully applying automated fresh air exchange systems requires that the steamship lines have a level of expertise in their sales, operations, and maintenance departments. Brecht and Brecht (2001) provided guidelines for application of AFAM+ technology to a wide range of fruits and vegetables.

3.8.1   AFAM+ and eAutoFresh

AFAM+ and eAutoFresh derive desirable atmospheric modifications in properly sealed reefer containers by taking advantage of the O_2 consumed and CO_2 produced by the commodities as part of the process of respiration. Figure 3.3 shows the components of an AFAM+ system. All actions and climate control readings are recorded in the data logger on the unit, ensuring that proper settings were made, and that openings actually did happen at the appropriate times. Commodities that derive their primary benefit from high CO_2 with less benefit from low O_2 include asparagus, blackberry, blueberry, raspberry, strawberry, broccoli, cantaloupe, cherry, durian, fig, freesia, lily, mango, mushroom, okra, Anjou pear (short term), chili pepper, grape, nectarine, and peach, and these are the best candidates for deriving quality benefits from automated fresh air exchange systems. Likewise, sensitive commodities that can be damaged by exposure to elevated CO_2 including various cultivars of apples (Boskoop, Braeburn, Cox, Cortland, Elstar, Empire, Fuji, Gala, Gloster, Granny Smith, Idared, Red Delicious, Spartan), Asian pears, Bosc pears, plums, iceberg lettuce, cucumbers, and eggplants are also good candidates because the system ensures avoidance of injurious CO_2 concentrations. Some high-respiring commodities like durian and asparagus that derive benefits from both low O_2 and high CO_2 are also potentiality good candidates if a desirable MA environment can be established in properly sealed refrigerated containers.

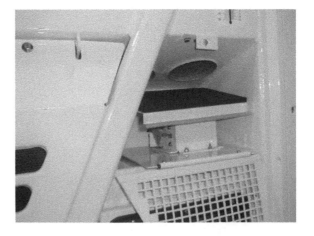

FIGURE 3.3
AFAM+ system components include a controller unit with screen and touchpad on the right side of the picture [left] and a motor-driven fresh air exchange door (right). (From Thermo King Corporation. With permission.)

The fresh air exchange door is controlled by the system's microprocessor controller according to the commodity or set point entered by the operator, which controls the rate of air exchange primarily based on the level of CO_2 in the container. The ventilation rate is restricted when CO_2 is below a set level and increased when the set point concentration is reached, which is based on commodity tolerance to elevated CO_2 levels. This allows the container to exclude outside air completely during the period immediately following loading, until the CO_2 concentration reaches the set point, and minimize air exchange during the remainder of the transit period. The fresh air exchange vent can be fully opened or closed in about 10 s, allowing realtime control. The system can also monitor O_2 concentration. Settings for the optimum gas concentrations and temperatures for each commodity that may be carried are stored in the microprocessor controller. The AFAM+ controller permits the driver or user to scroll down a list of commodities and select the name of the product being carried and the controller will automatically set the correct temperature and atmosphere for that product. This eliminates the need to find the set point temperature or atmosphere in a reference book and also eliminates the human errors caused by misinterpreting the proper temperature and atmosphere settings.

3.8.2 MAXtend RA

The MAXtend RA service requires logistical coordination between parties at the port of origin and destination. Unlike the TransFresh MA service described later, Mitsubishi does not leak check containers prior to loading. Mitsubishi reportedly asserts that 3-year-old or newer reefer containers consistently meet their maximum allowable air leakage requirements. After loading cargo into the container, a passive CO_2 scrubber, consisting of lime enclosed in a membrane of preselected permeability to CO_2 is installed (Figure 3.4). A passive C_2H_4 scrubber (KMnO$_4$ [potassium permanganate]) is placed in the container

FIGURE 3.4
MAXtend passive CO_2 scrubber system is secured to the top of cartons. (From Mitsubishi Australia Ltd., http://www.maxtend.com.au/newsite/ra_system_01.html [accessed June 13, 2008], 2008. With permission.)

according to the need of the commodity being shipped. Then a curtain is installed at the rear doors of the container. Mitsubishi does not flush the container with gases to generate an initial atmosphere; the atmosphere is self-generating, over time, via the product's respiration. The MAXtend RA system uses a controller, an O_2 sensor, and a purge valve to influence the O_2 level in the container. Even though the system does not have a CO_2 sensor, Mitsubishi has a patented CO_2 scrubber design that reportedly allows the system to predict the CO_2 level in a container based on observed changes in the O_2 level detected by the O_2 sensor.

3.9 Modified Atmosphere Systems

MA systems and services are designed for low-respiring fresh produce items that are transported over short distances to overseas markets. For a reefer container to be suitable for MA use, it needs to be equipped with a purge port assembly (Figure 3.5). As with other atmosphere systems, the reefer container must be properly sealed. Desired gas mixtures are injected into the container prior to the outset of the trip. After the gases are introduced, the container is sealed off for shipping.

The MA process lacks precise control of atmospheric gases compared with CA and, as a result, MA systems are intended only to maintain the O_2 and CO_2 concentrations within an acceptable range during transit. Actually, the gas environment in MA-equipped reefer containers typically continues to change during transit due to the changing respiratory activity of the produce, the performance of CO_2 and C_2H_4 scrubbers, if used, and the leak rate of the container. However, MA is a less costly investment and operating alternative than CA.

FIGURE 3.5
TransFresh single purge port located below the condenser coil and between the unit's step-up transformer (left) and compressor (right) on a Carrier reefer unit. This port is used after loading and sealing the container to flush the container with a beneficial mixture of N_2 and CO_2.

3.9.1 TransFresh TECTROL© MA Service

TransFresh has decades of experience providing TECTROL MA service. They have representatives who communicate with both the shippers and receivers of product to ensure that desirable results are achieved. To utilize this service, a logistical coordination between all parties (TransFresh, the packinghouse, and the ocean carrier) at the port of origin is needed. Reefer containers that pass an air leakage test are dispatched for loading. Once the fresh produce is loaded, passive CO_2 (hydrated lime bed) and C_2H_4 (potassium permanganate-infused pellets) scrubber systems may be placed in the container. A plastic curtain is then installed to seal the rear doors, and the container is flushed, via the purging port, with a beneficial mixture of N_2 and CO_2. With this service, the entire container essentially becomes one large MA package. From this point on, any changes to the atmosphere happen passively and without direct control (Table 3.2).

3.10 Controlled Atmosphere Systems

Passive and active transport CA systems are available for containerized transport of produce items. Active CA systems establish and maintain an atmosphere within a reefer container and are typically built into the refrigeration unit. Passive CA systems maintain an atmosphere inside the reefer container after gas purging and are currently transferable from one reefer container to another. CA systems use N_2 from the outside ambient air to establish a desirable balance of O_2 and CO_2 within a reefer container.

Currently available transport CA systems control and maintain O_2 and CO_2 levels over a wide spectrum of potential settings. They utilize N_2 and/or CO_2 to adjust the gas environment within the reefer container and have the ability to precisely manage the atmospheres for a broad range of produce items, especially when optimum O_2 ranges of $<5\%$ with CO_2 ranges of $<10\%$ are desired. Transport CA systems are particularly useful for shippers who ship sensitive produce to distant markets.

There are three main CA systems that are commercially available to exporters and importers (Table 3.3). Carrier's EverFresh system is an active CA system that is integral to the reefer unit and sold as an option at the time of initial reefer unit purchase. The TransFresh and Mitsubishi Australia systems include third party services and are passive systems that are fastened to the reefer unit prior to a shipment. The TransFresh system requires hardware and container provisions prior to use. The Mitsubishi service has a container box age requirement.

3.10.1 Carrier's EverFresh CA System

Carrier's EverFresh CA system is the only active technology on the market that is capable of completely controlling O_2 and CO_2 levels within a container from door-to-door. The EverFresh system's main component is its semipermeable membrane, which is used to generate a N_2-rich air stream by separating O_2 and CO_2 from ambient, outside air (Figure 3.6). Nitrogen-rich air is then injected by the system into the container to lower O_2 and CO_2 levels in the cargo space.

Since temperature control remains the most important variable for maintaining the quality of fresh produce, the EverFresh CA system does not operate until after the loaded container's doors are shut and the air temperature is within range of its set point.

TABLE 3.2

MA System Used Commercially in Conjunction with Reefer Container Cooling Systems

Commercial MA Offering	O_2 Concentration Changes			CO_2 Concentration Changes			Notes
	Sensor	Increase	Decrease	Sensor	Increase	Decrease	
TransFresh's TECTROL MA (service)	No	No	Yes (via initial gas flushing at origin then passive, see note)	No	Yes (via initial gas flushing at origin then passive, see note)	Yes (passive see note)	The O_2 level within the container is decreased and the CO_2 level is increased by flushing the container with a mixture of N_2 and/or CO_2 after loading and sealing. The atmosphere is maintained passively via the product's respiration, air leakage, and a CO_2 scrubber.

Source: Adapted from Dohring, S., *Stewart Postharvest Rev.*, 2(5), 1, 2006.

TABLE 3.3

CA Systems and Services Used Commercially in Conjunction with Reefer Container Cooling Systems

Commercial CA System	O_2 Control Capability			CO_2 Control Capability			Notes
	Sensor	Increase	Decrease	Sensor	Increase	Decrease	
Carrier Transicold's EverFresh CA system	Yes	Yes	Yes	Yes	Yes (see note)	Yes	EverFresh is an active system. When product respiration does not generate enough CO_2, levels can be maintained and controlled with an external bottle of CO_2 connected to the system.
Mitsubishi Australia's MAXtend CA (service)	Yes	Yes	Yes. Gas flushing completed to attain beneficial level for commodity (see note)	No	Yes. Gas flushing required (see note)	Yes. Passive CO_2 scrubber	MAXtend is a passive system that does not generate CO_2 or supply N_2 to reduce O_2 levels during transit. Increases in CO_2 or decreases to O_2 gas levels after initial flushing relies on product respiration. O_2 sensors are used to control the system. O_2 levels are recorded. CO_2 is not recorded.
TransFresh TECTROL CA (service)	Yes	Yes	Yes. Gas flushing completed to attain beneficial level for commodity (see note)	Yes	Yes. Gas flushing required (see note)	Yes. Active CO_2 scrubber	TECTROL is a passive system that does not generate CO_2 or supply N_2 to reduce O_2 levels during transit. Increases in CO_2 or decreases to O_2 gas levels after initial flushing relies on product respiration.

Source: Adapted from Dohring, S., *Stewart Postharvest Rev.*, 2(5), 1, 2006.

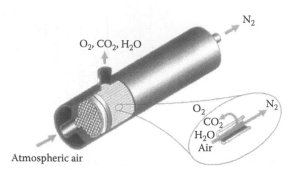

N₂

O₂, CO₂, H₂O

O₂
CO₂
H₂O
Air

N₂

FIGURE 3.6
Semipermeable membrane that separates other gases from N_2 in Carrier's EverFresh CA system. (From Carrier Corp. 2008. With permission.)

Atmospheric air

The protocol for controlling the atmosphere within an EverFresh unit is as follows: Once the reefer unit is in range of its set point temperature, the CA system begins working to control first the O_2 level and then the CO_2 level within the container.

If the O_2 gas concentration is high, a N_2-rich air stream is continuously injected into the container until the O_2 is lowered and in range of set point ($\pm$ a user-selected tolerance). Once the O_2 is in range, the system begins to manage the CO_2 level. If the O_2 level becomes dangerously low, or if the CO_2 level gets dangerously high, the system adjusts to correct the situation immediately.

If the CO_2 level is low and a CO_2 cylinder is being used as the source of CO_2, then the controller would energize a solenoid valve directing CO_2 gas into the container. Because the use of CO_2 cylinders, while in transit, has proven to create logistical and operational issues, commodities which require high CO_2 to be shipped successfully but do not generate enough CO_2 are typically avoided.

A microprocessor controller, independent of the cooling unit, uses O_2 and CO_2 sensors to monitor and record gas levels. Two calibration tests need to take place at the pretrip inspection prior to loading the container. One test uses a special calibration gas (95% N_2 and 5% CO_2) and one uses fresh air (air-cal). The special calibration gas is used for the low end of the O_2 sensor and high end of the CO_2 sensor. The air-cal spans the highest level of O_2 (21%) and lowest level of CO_2 (0%). During transit, the system periodically runs automatic air calibrations.

The advantages of this system over other systems are (1) the positive pressure builds up within the container and (2) the use of N_2 to remove CO_2. Positive pressure within the container keeps O_2-rich air from leaking in while in transit. The use of N_2 purging to lower CO_2 levels eliminates the need for CO_2 scrubbers, which are essentially bags of lime. Lime scrubbers have a cost per use, occupy valuable cargo space, and can become saturated, reducing their effectiveness, and have a disposal issue at unloading.

In order to remove C_2H_4 from the atmosphere within the container, potassium permanganate filters are typically placed in the container. When required, this would be the case for all CA systems presented.

Due to the requirements of the membrane itself, N_2 injected into the container is extremely dry. With this, there is a risk of dehydration. However, a significantly high percentage of EverFresh systems in the world fleet were built with a humidification feature.

With the EverFresh system's O_2 and CO_2 control capabilities, this system is best suited to using atmosphere recommendations directly from research findings. However, the optimum CA for each product varies according to the storage (transit) period, which means that better results may be obtained by using an atmosphere that differs from what research has determined to be the best for the longest possible storage of that product (Brecht et al., 2003). Similar to AFAM+, this system requires that the steamship lines have a level of expertise in their sales, operations, and maintenance departments in order to properly utilize the technology.

3.10.2 TransFresh TECTROL CA Service

The TransFresh TECTROL CA service is similar to their MA service described earlier. However, with TECTROL CA, a controller fitted with O_2 and CO_2 sensors is used to actively maintain a desired atmosphere. In addition to the one gas-purging port and a curtain track at the rear door, this service requires a reefer fitted with hardware to accept the controller and a communications port. Personnel-wise, the service requires similar logistical coordination between parties at the port of origin and destination.

Reefer containers that pass an air leakage test are dispatched for loading fresh produce. Once the container is loaded, both an active CO_2 scrubber and a passive C_2H_4 scrubber may be placed in the container. The active CO_2 scrubber consists of several boxes of lime contained within a larger box (Figure 3.7). The scrubber controls the flow of air through the lime via the controller actively turning on and off a fan that in turn pulls air through the lime bed as scrubbing is required.

A plastic curtain is installed to seal the rear doors and the container is flushed, via the purging port, with a beneficial mixture of N_2 and CO_2 (Figure 3.8).

TransFresh technicians program the unit's microprocessor controller with optimum O_2 and CO_2 set points for the cargo being shipped. If the O_2 level falls below desired level, a valve is triggered, allowing fresh air into the container. Once serviced and in transit, this system is limited in its control of O_2 and CO_2; O_2 can be added, not removed, and CO_2 can be removed, but not added. After gas purging, the system relies on the respiration of the product to maintain O_2 and CO_2 levels. This is why a leak-tight container is essential to the system's success (Figure 3.9).

When the container arrives at its destination, the CO_2 scrubber needs to be disposed of and the controller needs to be retrieved and sent to TransFresh for data downloading and review.

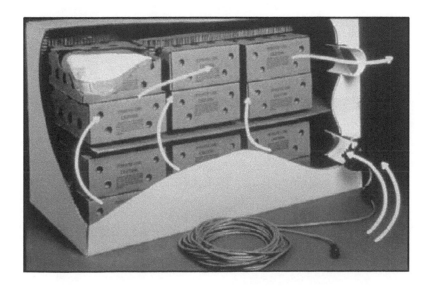

FIGURE 3.7
Active CO_2 scrubber system components and airflow diagram for a TransFresh TECTROL CA system. (From TransFresh Corporation, 2008. With permission.)

FIGURE 3.8
Curtain installation for a TransFresh atmosphere load. With the exception of the Thermo King AFAM+ system, curtains are recommended or used with all atmosphere services (From TransFresh Corporation, 2008. With permission.)

FIGURE 3.9
Container leak check setup.

3.10.3 Mitsubishi Australia: MAXtend CA Service

Since no distinct specifications or preinstalled hardware are required to use MAXtend atmosphere services, MAXtend provides flexibility and lower costs, in terms of both equipment and equipment control, to steamship lines. As with Mitsubishi's MAXtend RA service, the only box requirement at this time is that the reefer is 3 years old or newer.

The MAXtend CA service is similar to their RA service, but with the CA service an initial atmosphere is established at the origin by flushing the container with gas. There is also an added level of active atmosphere control.

Since air leakage is more of a concern with this CA service, prior to dispatching containers they are leak checked via the fresh air vents and must meet set criteria (Figure 3.9). Gas purging is facilitated in a similar fashion, through the fresh air exchange. After loading fresh produce into the container, the passive CO_2 scrubber is installed and when necessary, a C_2H_4 scrubber system is also placed in the container.

A curtain is used to provide an air seal at the rear doors and the controller is then installed. Mitsubishi programs the controller according to the product being shipped. Finally the container flushing takes place to generate an initial atmosphere.

The MAXtend CA system uses a controller (Figure 3.10), an O_2 sensor, and two purge valves together to control the O_2 and CO_2 levels within the container. The controller is battery operated and continues to function even when the reefer unit is off power. This battery operation is an advantage over both the TransFresh CA and EverFresh CA systems.

The maintenance of the atmosphere with the MAXtend CA system after gas flushing is similar to that of the TransFresh CA system, with the exception that the scrubbing of CO_2 remains passive.

Even though the system does not have a CO_2 sensor, Mitsubishi has a patented CO_2 scrubber design, which allows the system to predict the CO_2 concentration in a container based on observed changes to the O_2 concentration as detected by the O_2 sensor.

Oxygen data is recorded by the controller and can be downloaded and analyzed to ensure proper operation and to troubleshoot the operation as a whole. Carbon dioxide data is not monitored and recorded to validate adequacy of the CO_2 scrubber and the prediction of the CO_2 level in a container.

FIGURE 3.10
MAXtend's external CA controller. (From Mitsubishi Australia Ltd., http://www.maxtend.com.au/newsite/ra_system01.html [accessed June 13, 2008], 2008. With permission.)

3.11   The Future of Transport and Climate Control Technology

Internationalization of commerce in fresh fruits, vegetables, and flowers is introducing new products to people in every part of the world. The international transport industry has reached a point now where the next advances will be in how perishable food products can be economically marketed with the best possible retention of quality for consumers no matter where they reside. Future reefer containers will most likely be complete climate control systems equipped with expert feedback systems and traceability. The containers will in all probability manage and record O_2, CO_2, N_2, temperature, humidity, and volatile organic compounds including C_2H_4 as well as mold spores and bacteria.

Understanding that the end goal of atmosphere management is to extend the shelf life of wholesome fresh fruits and vegetables, there is significant opportunity for future developments in climate control management to establish prediction models for managing shelf life of produce. Refrigeration and atmosphere control systems that respond to environmental and physiological changes in the product during transport will work in concert with those shelf life models to maintain the optimum environment for the product at every stage during distribution.

The interface between the refrigeration and atmosphere control systems and the product being carried may be biosensors that detect ethylene (to detect stress or onset of ripening), ethanol (indicative of fermentative metabolism and stress), chlorophyll fluorescence (another stress indicator), microbe-generated volatiles (related to decay and food safety), and others presently unimagined. Expert system software residing in the refrigeration system's microprocessor will control the system response to these biological cues. Relating sensor data to commodity physiology will allow early detection of physiological or pathological disorders, as well as residual shelf life prediction modeling with models that are unique to each product. However, this vision will require research to establish all of the possible limiting quality factors for each product, which can change for different temperatures, varieties, origins, and maturity/ripeness stages.

Microbiological quality of transported food is a nonnegotiable expectation of customers and consumers at every stage of food distribution. In the near future, transport refrigeration and atmosphere control systems will minimize if not eliminate food safety risks by utilizing technologies for sanitizing the container and cargo based on the recorded bioload. To illustrate, sanitizing the air that is circulating within the conditioned space of a reefer container could reduce the presence of food spoilage and food poisoning organisms from the air. One such technology utilizes the localized creation of reactive O_2 species (ROS) to eliminate bacteria, mold, and viruses in the air and on the surface of the produce, product packaging, and refrigeration equipment as well as to remove C_2H_4 by oxidizing it to CO_2. ROS are created by partially reducing molecular O_2 to various intermediaries. These intermediaries then react with themselves, O_2, and any carbon-based elements that pass by the ROS generator. The result is a conditioned space where mold, bacteria, and viruses are reduced and C_2H_4 is scrubbed. This occurs without any negative effect on the commodity or humans as long as the reaction is sized correctly and monitored and controlled. A few of the ROS are longer lasting and could be used as a secondary cleaner for surfaces. In addition to minimizing the growth of mold on the product surface, this secondary cleaning action can minimize the growth of biofilm on the refrigeration coil, thereby improving the refrigeration performance and reducing the risk of distributing pathogens into the air and cargo space.

In the future, the condition of the product and the performance of the refrigeration and atmosphere control systems during transport will be continuously available to stakeholders in real time with transmission by radio frequency identification (RFID) and satellite communication. Future CA systems will be dynamically controlled to change as needed in response to physiological and microbiological cues from the cargo to continuously maintain the environmental conditions that best maintain the product quality. When problems do arise, the availability of real-time information will allow operators to initiate manual interdiction to avert a problem, take corrective actions to adjust the system, or reroute the cargo to nearer destinations. Programs that allow prediction of remaining product shelf life will allow receivers to make accept/reject decisions in real time, without delayed downloads as with traditional temperature data loggers.

References

Anonymous 1999. *Controlled Atmosphere Handbook. A Guide for Shipment of Perishable Cargo in Refrigerated Containers*, 2nd ed. Carrier Transicold Division, Carrier Corporation, Syracuse, NY.

Brecht, P.E. 1980. Use of controlled atmosphere to retard deterioration of produce. *Food Technol.* 34: 45–50.

Brecht, P.E. 1992. *Shipping Special Commodities*. American President Lines, Oakland, CA.

Brecht, P.E. and J.K. Brecht. 2001. *VFD AFAM/AFAM+ Setting Guide*. Thermo King Corporation, Minneapolis, MN.

Brecht, J.K., K.V. Chau, S.C. Fonseca, F.A.R. Oliveira, F.M. Silva, M.C.N. Nunes, and R.J. Bender. 2003. Maintaining optimal atmosphere conditions for fruits and vegetables throughout the postharvest handling chain. *Postharvest Biol. Technol.* 27: 87–101.

Carrier Corporation 2008. http://www.container.carrier.com/details/0,2806,CLI1_DIV9_ETI653,00.html (accessed June 13, 2008).

Daniels, R. 2000. *Across the Continent: North American Railroad History*. Indiana University Press, Bloomington, IN.

Dohring, S. 2006. Modified and controlled atmosphere reefer container transport technologies. *Stewart Postharvest Rev. [online]* 2(5): 1–8.

Mitsubishi Australia Ltd. 2008. http://www.maxtend.com.au/newsite/ra_system_01.html (accessed June 13, 2008).

Pelletier, W., M.C.N. Nunes, and J.P. Émond. 2005. Air transportation of fruits and vegetables: An update. *Stewart Postharvest Rev. [online]* 1: 1–7.

Smock, R.M. 1979. Controlled atmosphere storage of fruits. *Hort. Rev.* 1: 301.

Tanner, D. and N. Smale. 2005. Sea transportation of fruits and vegetables: An update. *Stewart Postharvest Rev. [online]* 1: 1–9.

Thermo King Corporation. 2008. http://www.thermoking.com/tk/index.asp (accessed June 13, 2008).

Thompson, J.F. and P.E. Brecht. 2005. Innovations in transportation. In: S. Ben-Yehoshua (Ed.), *Environmentally Friendly Technologies for Agricultural Produce Quality*. Taylor & Francis, New York, pp. 439–445.

Thompson, J.F., P.E. Brecht, T. Hinsch, and A.A. Kader. 2000. Marine container transport of chilled perishable produce. University of California, Division of Agriculture and Natural Resources Publication 21595.

TransFresh Corporation. 2008. http://www.transfresh.com/index.asp (accessed June 13, 2008).

Thompson, J.F., C.F.H. Bishop, and P.E. Brecht. 2004. Air transport of perishable products. University of California, Division of Agriculture and Natural Resources Publication 21618.

4

Modified and Controlled Atmosphere Packaging Technology and Applications

Jeffrey S. Brandenburg and Devon Zagory

CONTENTS

Modified atmosphere packaging (MAP) is one of the key technologies associated with the shelf life extension of fresh produce. This chapter will discuss the technology, materials, and applications of MAP. Discussions will include the symbiotic relationship between the packaging and the produce and the importance of produce physiology when specifying packaging systems. Specific polymers, films, and structures will be analyzed with respect to their individual properties and subsequent impact on MAP. An analysis of packaging types and formats will also be detailed. Additive technologies such as antimists, slip, and antimicrobials, and their impact on the functionality of MAP will be explored. Common reasons for the failure of MAP will also be reviewed. The chapter will conclude with some of the newest emerging MAP technologies, such as microwave steam technology, and multicomponent trays along with topics where additional research is warranted.

MAP, when combined with proper postharvest handling procedures and temperature control management, can have a positive impact on the quality and shelf life of fresh produce. It is important to begin by stating under what conditions MAP would not work. First and foremost, MAP is only effective if there is consistent temperature management throughout the entire life cycle of the produce. This includes processing as well as the entire distribution channel. Lack of temperature control will result in physiological variations of the produce, which will impact the effectiveness of the packaging system. In addition, MAP will never improve the quality of the incoming raw material product. Under ideal circumstances, the best that can be achieved is to maintain the existing quality level throughout the desired shelf life. In real-world applications, often MAP will maintain quality for the majority of the targeted shelf life, but due to parameter variations during distribution, quality will suffer at the very end of the desired shelf life. Since MAP will never improve incoming product quality, the need for optimal postharvest handling procedures is paramount. Postharvest handling will be discussed in detail in other chapters.

4.1 Relationship between Packaging and Product Physiology

When designing MAP, the convergence of three unique and separate sciences must take place. These are the sciences of produce physiology, polymer engineering, and converting technology (Figure 4.1). Effective MAP design can only take place within the intersection of these three disciplines. Surrounding these scientific disciplines is the impact and requirements of

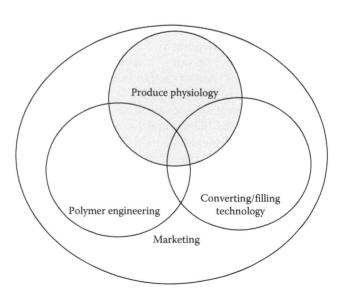

FIGURE 4.1
Overlapping Sciences Impacting Fresh Produce MAP.

marketing and the consumer. What makes fresh produce packaging unique is the impact and requirement of the produce physiology. Unlike almost all other packaging applications, produce MAP involves packaging a living product. Therefore, in order to effectively design a package an understanding of the physiological properties and requirements is a critical parameter.

4.1.1 Gas Technology

Most extended shelf life packaging strategies involve the use of atmospheric gases in proportions different from those found in air. In order to understand the uses of the various gases, it is first necessary to understand the properties of the gases and how they interact with and affect the packaged foods. The physical and chemical properties of the gases will be affected by their concentrations and especially by the temperature. In general, the gases will not react with each other except when exposed to heat or metabolic catalysts. The partial pressures, diffusion properties, and permeation rates of each gas are independent of the other gases present. This simplifies matters enormously since we do not have to be concerned with the gas interactions.

4.1.1.1 Oxygen (O₂)

Oxygen is a reactive gas comprising about 20.9% of the atmosphere. It occurs most commonly in its diatomic form O_2, but can also be present as ozone (O_3). Oxygen is present as a constituent of nearly all organic molecules, especially carbohydrates, and can form compounds with virtually any chemical element. Oxygen is somewhat soluble in water (4.89 cm^3/100 mL at 0°C), but not nearly as soluble as is carbon dioxide. As is the case with all gases, the solubility in water increases as temperature decreases.

Most of the reactions with food constituents involving oxygen are degradation reactions involving the oxidative breakdown of foods into their constitutive parts. Because of this, many packaging strategies seek to exclude oxygen and thus slow these degradation processes. Many spoilage microorganisms require oxygen and will grow and cause off-odors in the presence of sufficient oxygen. Oxygen is necessary for the normal respiratory metabolism of fresh fruits and vegetables and normal atmospheric concentrations of oxygen encourage and facilitate senescence and degradation of quality.

Because O_2 acts as the terminal electron acceptor in many metabolic reactions, the rates of some essential metabolic processes are sensitive to O_2 concentration. Reducing O_2 concentrations below about 10 kPa around many fresh fruits and vegetables slows their respiration rate and indirectly slows the rates at which they ripen, age, and decay. Reducing the O_2 concentration can, in some cases, reduce oxidative browning reactions, which can be of particular concern in precut leafy vegetables. Reduced O_2 can delay compositional changes such as fruit softening, pigment development, toughening of some vegetables (such as asparagus and broccoli), and development of flavor (Kader, 1986). Finally, there is a great deal of interest in the use of low O_2 as a quarantine treatment to disinfest fresh produce of insects and insect larvae. Proper combinations of low O_2, low temperature, and time may be effective against some of the most troublesome insect pests of concern in international commerce (Ke and Kader, 1992).

However, O_2 is required for normal metabolism to proceed. O_2 concentrations below 1–2 kPa can lead to anaerobic (sometimes called fermentative) metabolism and associated production of ethanol and acetaldehyde resulting in off-flavors, off-odors, and loss of quality. Of even greater concern is the potential growth of anaerobic bacteria, some of which are pathogenic to humans under low oxygen conditions. The proper O_2 concentration will depend upon the fruit or vegetable and its tolerance to low O_2, the temperature (which will affect the product's tolerance to low O_2), and the time that the product will be exposed to low O_2.

Oxygen permeates through plastic polymers at various rates depending on the polymer, but it generally permeates through more slowly than carbon dioxide. The permeability rate of oxygen (and all gases) in plastics increases as temperature increases. Similarly, the chemical reactivity of oxygen with food constituents increases as temperature increases.

Ozone is highly reactive and is inhibitory to many microorganisms, particularly bacteria. Ozone has been used to sterilize water and ozone generators have found some use in cold storage for fruits and vegetables. But ozone is very reactive and breaks down to O_2 rapidly and so is not used in MAP of fruits or vegetables.

4.1.1.2 Carbon Dioxide (CO_2)

Carbon dioxide is present in the atmosphere in low levels, typically about 0.03%, but is an important product of combustion and so is easily produced. It is very soluble in water, especially in cold water (179.7 cm^3/100 mL at 0°C), and will thus be absorbed by high-moisture foods. When CO_2 dissolves in water, it produces carbonic acid, which will cause a drop in pH and an acidifying effect. This acidification, as well as direct antimicrobial effects, can suppress the growth of many spoilage microorganisms and for this reason is essential in many extended shelf life packages.

Carbon dioxide also has a minor suppression effect on the respiration of some fresh fruits and vegetables and thus can help extend their shelf life. At concentrations above 1–2 kPa, CO_2 reduces the sensitivity of plant tissues to the ripening hormone ethylene. Ethylene is a colorless, odorless, tasteless gas that has many effects on plant physiology and is active in such small amounts (parts per million) that it is considered a plant hormone. Ethylene has many effects on plant tissues. Ethylene can cause premature ripening, fruit softening, yellowing of leafy vegetables, increased respiration rate, and senescence of many fruits and vegetables. The prevention of these ethylene effects is important in the maintenance of quality attributed to MAP. With some fruits, such as bananas and tomatoes, ethylene is routinely applied as a postharvest treatment to ensure rapid and uniform ripening. With many other fruits, vegetables, and ornamental plants, it is important to prevent exposure to ethylene and its subsequent deleterious effects.

Ethylene is normally produced by many kinds of ripening fruit. In addition, ethylene is produced by any aerobic combustion such as fires, auto exhaust or diesel, and propane forklifts. Such sources should be eliminated from areas where produce is stored or handled. Elevated CO_2 (greater than $\sim$2 kPa) can help reduce the damaging effects of ethylene by rendering plant tissues insensitive to ethylene (Herner, 1987; Kader et al., 1988). This may be one of the primary benefits of modified atmospheres for many commodities.

Elevated CO_2 can, like reduced O_2, slow respiratory processes thereby extending shelf life. Although the effects of elevated CO_2 on respiration are not as dramatic as those of low O_2, high CO_2, and low O_2 together can, in some cases, reduce respiration more than either gas alone (Kader et al., 1988).

CO_2 at relatively high concentration (>10 kPa) has been shown to suppress the growth of a number of decay-causing fungi and bacteria. For example, 15–20 kPa CO_2 is routinely applied around strawberries during shipment primarily to suppress growth of the mold *Botrytis cinerea*, which would otherwise greatly reduce the postharvest life of strawberries. However, these levels of CO_2 do not suppress some human pathogenic bacteria of potential concern on fresh produce. For example, *Clostridium botulinum* and *Listeria monocytogenes* are relatively resistant to the effects of CO_2 (Farber, 1991). There is some concern that elevated CO_2 could suppress spoilage microorganisms that would otherwise signal microbial growth and product spoilage while allowing potentially hazardous pathogens to continue to grow. For this reason, MAP should always work in conjunction with an excellent program of sanitation and quality assurance. In addition, too much CO_2 can be damaging to plant tissues and individual fruits and vegetables differ in their tolerance to CO_2. Carbon dioxide typically permeates most packaging materials more rapidly than other atmospheric gases.

4.1.1.3 Nitrogen (N₂)

Nitrogen is the most abundant component in air ($\sim 79\%$) and can be used in either gaseous or liquid form. It is physiologically inert in its gaseous and liquid forms and is used in packaging primarily as a filler and to exclude other more active gases. In its N_2 form, it does not participate in any physiological reactions within plant tissues, nor does it effect the growth of microorganisms except to the degree that it significantly displaces O_2. It is sparingly soluble in water ($2.33 \text{ cm}^3/100 \text{ mL}$ at $0°C$).

4.1.1.4 Carbon Monoxide (CO)

Carbon monoxide is a colorless, odorless, tasteless, very toxic gas, which has been shown to be very effective as a microbial inhibitor. As low as 1 kPa, CO will inhibit many bacteria, yeasts, and molds. It can also delay oxidative browning of fruits and vegetables when combined with low O_2 (2–5 kPa) and has found limited use commercially for this purpose. However, due to the toxicity of the gas, and its explosive nature at 12.5–74.2 kPa in air, CO must be handled using special precautions and so is used little, if at all, in MAP of fruits or vegetables.

4.1.1.5 Sulfur Dioxide (SO₂)

Sulfur dioxide has been used to control growth of mold and bacteria on a number of soft fruits, particularly grapes and dried fruits. It has also found use in the control of microbial growth in fruit juices, wines, shrimp, pickles, and some sausages. Sulfur dioxide is very chemically reactive in aqueous solution and forms sulfite compounds, which are inhibitory to bacteria in acid conditions (pH < 4). However, a significant minority of the population displays hypersensitivity to sulfite compounds in foods and the use of sulfites has come under public and regulatory scrutiny in recent years. Sulfur dioxide is used during storage and shipment of table grapes in order to retard fungal spoilage.

4.1.2 Fruit and Vegetable Physiology and Deterioration Processes

The plant tissues in fresh fruits and vegetables are still living after harvest and even after fresh-cut processing. To stay alive, their metabolic processes must derive energy, primarily through the process of respiration. Respiration involves the consumption, using atmospheric oxygen (O_2), of carbohydrates and organic acids and the consequent production of metabolic energy, heat, carbon dioxide (CO_2), and moisture vapor. Different fruits and vegetables, and even different varieties of a given fruit or vegetable, will vary in their rates of respiration. Those that have high respiration rates (such as asparagus, mushrooms, strawberries, and broccoli) tend to be most perishable while those with low respiration rates (such as nuts, apples, onions, and potatoes) tend to be least perishable. Respiration rate also strongly depends on temperature and may more than double for every increase of 10°C.

The best way to reduce respiratory metabolism and thus conserve the plants stores of carbohydrate, acids and moisture, is to reduce the temperature. All biological processes proceed more slowly at lower temperatures. Most fruits and vegetables will maintain their best quality at temperatures near 0°C. Exceptions include fruits and vegetables of tropical or subtropical origin (such as tomatoes, bananas, and papayas), which should be kept at 10°C–13°C to avoid chilling injury. In any case, keeping fresh produce at the lowest possible temperature without causing freezing or chilling injury is the surest way to maintain quality and shelf life. As a supplement to good temperature control, MAP can further extend quality and shelf life.

Many of the effects of MAP on produce are based on the often observed slowing of plant respiration in low O_2 environments. Respiration is typically measured as the amount of CO_2 produced (or O_2 consumed). They are approximately equal to milliliters per kilogram of product per hour (mL/kg h). Respiration is based on the oxidative consumption of carbohydrates and organic acids. Carbohydrates and organic acids are storage compounds in plant cells and are important flavor and texture components as well as being essential to the normal metabolism of the plant tissues. When they are rapidly consumed, the freshness of the tissues is rapidly lost. When respiration rate is reduced, fruits and vegetables become less perishable. For example, some apples may be stored up to 10 months in the proper low temperature and controlled atmosphere conditions.

Air is about 21% O_2, 0.03% CO_2, 0.9% argon, and the remainder nitrogen (N_2). If the concentration of O_2 falls below about 10 kPa, plant respiration starts to slow (Figure 4.1). This suppression of respiration continues until O_2 reaches about 1–3 kPa for most fruits or vegetables. If O_2 gets lower than 1–3 kPa (depending on the product and the temperature), anaerobic (fermentative) metabolism replaces normal aerobic metabolism and large amounts of CO_2, off-flavors, off-odors, and undesirable volatile compounds are produced (Figure 4.2). Similarly, as CO_2 increases above the 0.03 kPa found in air, a suppression of respiration results for some commodities (Figure 4.3). If CO_2 reaches higher levels, the production of undesirable volatiles and physiological injury occur. The amount of CO_2 that is injurious varies by commodity. The relationships, for hypothetical commodities, between O_2 concentration and respiration rate and CO_2 concentration and respiration rate are shown in Figures 4.2 and 4.3. Reduced O_2 and elevated CO_2 together can reduce

FIGURE 4.2
Effect of oxygen levels on fruit or vegetable respiration rate. (Adapted from Zagory, D., Physiology and microbiology of fresh produce in modified atmosphere packages. *Society of Manufacturing Engineers Symposium, Fundamentals of Modified Atmosphere Packaging*, December 4–5, Monterey, CA, 1996, 3.)

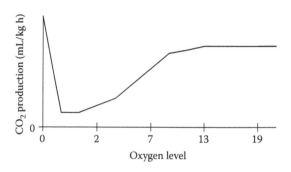

FIGURE 4.3
Effect of carbon dioxide levels on fruit or vegetable respiration rate. (Adapted from Zagory, D., Physiology and microbiology of fresh produce in modified atmosphere packages. *Society of Manufacturing Engineers Symposium, Fundamentals of Modified Atmosphere Packaging*, December 4–5, Monterey, CA, 1996, 3.)

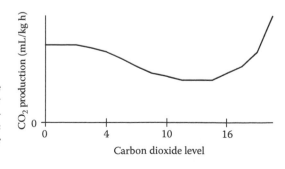

respiration more than each one of them alone (Kader et al., 1988). In addition, elevated CO_2 suppresses plant tissue sensitivity to the effects of the ripening hormone ethylene. Brown (1922) reported that 10 kPa CO_2 or more can retard the sporulation and/or growth of numerous fungal decay organisms, while O_2 typically has little effect on plant pathogen growth or survival at levels appropriate for use in MAP. Under hypobaric (low pressure) conditions, extremely low O_2 partial pressures have been shown to be effective in decay control (Burg, 2004). The effect of superatmospheric O_2 (>21 kPa) on pathogens varies depending on the product, but it is usually less effective than CO_2 (Kader and Ben-Yehoshua, 2000). Additional benefits of MAP include preservation of vitamins, particularly vitamin C (Kader et al., 1988).

Due to the slowing of oxidative metabolism associated with reduced O_2 and elevated CO_2, the ripening processes of many fruits and vegetables are retarded. Ethylene, a colorless, odorless, and tasteless gas is a plant hormone that has multiple physiological effects when present at levels as low as one part per million (ppm). Ethylene can induce rapid and irreversible ripening and softening of climacteric fruits such as avocados, tomatoes, kiwifruit, apples, stone fruit, mangoes, and others. Ethylene may also cause yellowing of many vegetables such as spinach, broccoli, cucumbers, and celery. Ethylene causes russet spotting of iceberg lettuce, consisting of small, oval brown spots that develop on the lettuce midribs. Ethylene may also induce formation of bitter tasting isocoumarins in carrots, sprouting of potatoes, and toughening of asparagus (Reid, 1985). Elevated CO_2 can prevent or delay all of these effects by reducing the sensitivity of plant tissues to ethylene. In addition, low O_2 (below 2–4 kPa) can reduce ethylene production by plant tissues. Synthesis of some anthocyanin pigments is also retarded by reduced O_2, which slows color development of some fruits such as tomatoes (Table 4.1).

The usefulness of MAP for specific fruits or vegetables depends upon the purpose. If fruits are softening during distribution or storage due to exposure to ethylene, then MAP may reduce the sensitivity of the tissues of kiwifruit, banana, apple or other fruits, and thereby retard the softening. Very low oxygen MAP is widely used in the fresh-cut salad industry to retard oxidative browning. MAP can slow the breakdown of green chlorophyll pigments in spinach, broccoli, cucumbers, and other vegetables. Elevated CO_2 can slow the

TABLE 4.1

Physiological Effects of Reduced O_2 and Elevated CO_2 on Fruits and Vegetables

| | General Effects of | |
Cause of Deterioration	Reduced O_2	Elevated CO_2
Respiration rate	(>1%) − (<1%) +	(<15%–20%) − (>15%–20%) +
Ethylene action	−	−
Chlorophyll degradation	−	−
Anthocyanin development	−	−
Carotenoid biosynthesis	−	−
Enzymatic browning	− (Near 0%)	−
Off-flavors	+ (<1%)	+ (>15%–20%)
Vitamin C loss	−	−
Fungal growth	− (<1%)	− (>10–15%)
Bacterial growth	− or 0	− or 0

Source: Kader, A.A., D. Zagory, and E.L. Kerbel. 1988. *CRC Crit. Rev. Food Sci. Nutr.* 28(1):1–30.

Notes: −, Decrease or inhibit; 0, no effects; +, stimulate or increase.

growth of certain spoilage bacteria and molds and is widely used in strawberry shipping for this purpose. Low-oxygen MAP inhibits sprouting in onions. However, MAP cannot enhance the safety of fresh fruits and vegetables. Most of the human pathogens of concern are either not affected by MAP, or their growth may actually be enhanced by MAP (see Chapter 10 on microbial safety of MAP applications). MAP will not improve the quality of poor quality products. At best, MAP can slow the processes of deterioration and so extend quality and shelf life, but MAP cannot reverse those processes.

4.1.3 Resources for Information on Appropriate Target-Modified Atmospheres

There are many research reports on appropriate atmospheres for specific fruits and vegetables. However, much of this literature is dispersed and difficult to access for those without access to a large University library. Fortunately much of this literature has been summarized and is available online on several Web sites, such as of the Postharvest Technology Research and Information Center (http://postharvest.ucdavis.edu/) at the University of California, Davis. Most of the Produce Facts sheets are available in English, Spanish, French, and Arabic and contain much useful information. Many contain recommendations on appropriate O_2 and CO_2 levels for MAP, injurious levels of O_2 and CO_2, and respiration rates at a series of temperatures.

Information specifically regarding MAP of fresh-cut fruits and vegetables is also available online and in print, such as in the Proceedings of the Seventh International Controlled Atmosphere Conference (Gorny, 1997), held at the University of California, Davis. For example, the Transicold container division of the Carrier Corporation published a guide to CA shipment of fruits and vegetables that contains a wealth of information about proper handling of a variety of fruits and vegetables, including recommendations for appropriate atmospheres (Zagory and Kader, 2003).

4.1.4 Respiration Rate Quantification

MAP depends upon the respiratory activity of the enclosed product as a driving force for atmosphere modification and the permeability of the packaging material to maintain atmospheres within desired limits. It is the continued depletion of O_2 and/or the release of CO_2 (and water vapor) by the product that enables the modified atmosphere to become established within a sealed package. Factors that must be controlled or incorporated include film permeability, film area, film thickness, temperature, and the respiratory behavior (responses to O_2 and temperature) of the product.

The respiratory dependence of MAP requires a quantitative understanding of the respiration of specific fresh produce. Produce respiration is determined through testing. The condition, age, and temperature of the raw material product significantly impact the outcome of the respiration test. There are a number of independent resources for product respiration calculations including University of California at Davis (http://postharvest. ucdavis.edu/) and The JSB Group, LLC (www.jsbgroup.com).

The primary method for determining the required MAP oxygen transmission rate (OTR) is seen in the following equation. Parameters, in addition to respiration rate (RR), include package surface area, product weight, structure thickness, and target final modified atmosphere. A thorough understanding of the equation and how each parameter can affect the package OTR is critical for optimal package design. It is the interplay of these parameters that mandates that there is no one package that fits all uses. MAP must be designed on an individual application basis. Once the produce respiration is quantified and the balance of parameters are determined, the target OTR can be calculated. Establishment of the target OTR allows the design process to move from the physiological research into the polymer science research. OTR for a target atmosphere is

$$OTR = RR_{O_2} \times t \times W/A \times (O_{2\,air} - O_{2\,pkg})$$

where

OTR is the film O_2 permeability (oxygen transmission rate) per mil
RR is the respiration rate (O_2 consumption rate in mL/kg h)
t is the film thickness (mil)
W the product weight (kg)
A the film surface area (cm^2)
$O_{2\,pkg}$ is the desired O_2 level in the package (kPa O_2 target atmosphere)

OTR in this example is calculated on a per mil basis. Therefore, when a 2 mil thickness film is used, the OTR on a per mil basis must double to achieve the equivalent package OTR.

In order to achieve the target modified atmosphere, the packaging films must be permeable to gases. Specifically they must have the required gas transmission properties to achieve the targeted OTR. The movement of gases across films depends on several physical factors that are related through Fick's law as follows:

$$J_{gas} = \frac{A \times \Delta C_{gas}}{R}$$

where

J_{gas} is the total flux of gas (cm^3/s)
A is the surface area of the film (cm^2)
ΔC_{gas} is the concentration gradient across the film
R is the resistance of the film to gas diffusion (s/cm)

The gas flow across a film increases with increasing surface area and with increasing concentration gradient across the film. The gas flow across the film decreases with increasing film resistance to gas diffusion. Gases diffuse through polymeric films at different rates. Carbon dioxide diffuses between two to five times faster than oxygen. The ratio, within a polymer, film or structure of CO_2 transmission rate (CO_2TR) to OTR is termed the beta value. Polymeric films therefore have a beta value of 2–5:1, with an average of 3:1. The beta value of a modified atmosphere package will have a direct bearing on the final modified atmosphere achieved within the package. For example, in polymer films, it is possible to achieve a low, 2 kPa, O_2 level in combination with a mid-level 7 kPa, CO_2. The impact of the beta value is dependent on the specific ideal modified atmosphere for each individual product.

4.2 Polymer Engineering

4.2.1 Polymers

There are a variety of polymers used in fresh-cut produce MAP. A portion of these polymers is used in primarily flexible packaging structures, a portion is used in primarily rigid packaging structures, and a portion is found in both applications. Each specific polymer has physical, chemical, and gas transmission rate properties that are unique to

TABLE 4.2

Common Polymers Used in MAP and Some of Their Characteristics

Polymer	Abbreviation	Characteristic	Typical OTR (cc/100 in.^{2}/ mil/atm/day)	Application Most Commonly Used
Low-density polyethylene	LDPE	General-purpose polymer	450–500	Both
Linear low-density polyethylene	LLDPE	Increased stiffness	480–500	Both
Linear medium-density polyethylene	LMDPE	Increased stiffness, lower OTR, decreased clarity	300–350	Both
High-density polyethylene	HDPE	Relatively stiff, opaque	150	Rigid
Ultralow-density polyethylene	ULDPE	High OTR, decreased stiffness	900	Flexible
Plastomer metallocenes	—	Very high OTR, soft	1100	Flexible
Amorphous polyethyleneterapthalate	APET	Clear, rigid	5	Rigid
Polyvinyl chloride	PVC	Clear, rigid	10	Rigid
Polypropylene	PP	Decreased clarity	300	Both
Ethylene vinyl acetate	EVA	Sealability	600–900	Flexible
Polystyrene	PS	Stiffness	350	Rigid

that polymer. The design of a packaging structure entails matching the specific polymer properties to the requirements of the MAP application. For the physiological portion of fresh-cut modified atmosphere applications, a polymers gas transmission rate, specifically, OTR, and CO_2TR are key attributes. Table 4.2 lists many of the common polymers used in produce MAP.

4.2.2 Films and Structures

As with most food packaging, produce MAP contains not only individual polymers but also combinations of polymers in the form of films and structures. These films can consist of blended monolayer films, coextruded films, or combinations of both laminated together. The determination of the gas transmission rate therefore requires knowledge of not only the individual polymers but also how they are combined. A blend of polymers within a single layer of plastic film, referred to as a blend, will yield an OTR that is the weighted average of the OTR of the individual polymer components. Individual distinct layers of pure polymers, or blends, created within a single total structure, are referred to as a coextrusion. Coextrusions can be used as is or combined with additional coextrusions or polymer films to create a lamination. Since gases will move through each independent layer sequentially, the OTR of each independent layer of polymer must be determined and inserted into the equation in Figure 4.4 to calculate the overall structure OTR.

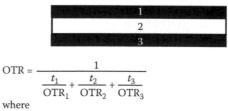

$$OTR = \cfrac{1}{\cfrac{t_1}{OTR_1} + \cfrac{t_2}{OTR_2} + \cfrac{t_3}{OTR_3}}$$

where

FIGURE 4.4
Calculating the OTR.

t is the thickness of the individual layer
OTR is the oxygen transmission rate of the individual layer at 1 mil

Based upon this equation, the total structure OTR can never be higher than the lowest individual layer OTR. This limiting factor establishes a practical ceiling for polymer MAP. Depending on the type of structure and the all of the converting and marketing requirements, this ceiling may be as low as 175 cc/100 in.^{2}/mil/atm/day (2713 cc/m^2/mil/atm/day) or as high as 900–1000 cc/100 in.^{2}/mil/atm/day (13,950–15,500 cc/m^2/mil/atm/day).

The specific combination of polymers, blends, coextrusions, and laminations are governed by numerous parameters in addition to the physiological requirements mandated by the product being packaged. These additional functional requirements often act in an opposing manor. Therefore to effectively design a modified atmosphere package, it is necessary to have a full understanding of all the desired features of the package. The requirements impacting polymer and film choice can include stiffness, sealability, esthetics clarity, graphics, dimensions, economics, sustainability, runnability, packaging format, coefficient of friction (COF), antimicrobial additives, and thickness. Examination of just one of the parameters demonstrates the results of these opposing parameters. As is listed in Table 4.2, polymers that exhibit increased stiffness properties also exhibit lower gas transmission rate properties. Therefore if structure stiffness is of paramount importance, then the overall gas transmission rate of the package may be low. So low in fact that it does not meet the requirements of the packaged produce. We can see that this parameter alone has a significant impact on the overall package design. Rarely though does one have the luxury in examining one parameter in isolation, therefore it is critical to have a thorough understanding of all applicable parameters and their relative importance.

4.2.3 Perforations

The increase in demand for higher respiring fresh produce, outside the traditional leafy greens market, has necessitated a significant increase in MAP gas transmission rates. A method for achieving high OTR packaging structures that is not limited by the upper end of polymer gas transmission rates, and stiffness constraints is microperforation technology. This technology employs the science of placing microholes in the packaging structure. With microperforation technology, the gas transmission rate of the modified atmosphere package is governed by the configuration of holes and their individual geometry and size. The hole size and configuration can vary with the specific perforation method but all microperforations are not visible to the naked eye and range from 40 to 200 μm in diameter. It is essential to have a complete understanding of the package geometry so that holes are not blocked or obstructed in any way. This is critical to the success and control of gas transmissions. Microperforations also have transmission rate limitations, however for microperforation technology the limitations are at the lower end of the range. Typical OTR for microperforated packaging are from 250 cc/100 in^2/mil/atm/day and above. Gas transmission rates of microperforated structures are determined by the gas diffusion properties through the combined effect of the individual microholes, their corresponding placement, and in certain cases the OTR of the perforated structure. As the targeted OTR requirements are lower, the number of holes decreases. Since microperforated structures cannot have less than one hole, the gas transmission rate through a single hole dictates the lower transmission rate level. Modified atmosphere microperforation packaging with only one hole can be problematic. With two or more holes, there is less risk from hole blockage as well as there is more uniform gas flow throughout the package. In order to avoid only one microperforated hole, alterations to the other key OTR control parameters, such as product weight, or package dimension, should be considered.

The diffusion rates of various gases through microperforations are very similar. In effect, the diffusion rates of CO_2 and O_2 are virtually the same, a beta value of 1. Therefore for a targeted 2 kPa O_2, a19 kPa CO_2 level will be achieved. The beta value difference between polymeric and microperforated structures creat significantly different final modified atmospheres. It is not possible with a microperforated film to achieve a modified atmosphere consisting of low O_2 levels and low to moderate CO_2 levels. Conversely, if low O_2 levels in combination with high CO_2 levels are required, then packaging comprising of engineered polymers is not suitable. This parameter needs to be accounted for in the packaging design process.

Macroperforations, which are visible to the naked eye, should not be mistaken for microperforations. The gas movement through the larger visible holes utilized in macroperforation technology is too great to consistently modify and control the gas level within the package. Therefore, attempting to create a low O_2 modified atmosphere is not feasible. This does not mean however there is not a function and need for macroperforated structures. Since the gas transmission is so high, a macroperforated structure will virtually never become anaerobic with O_2 levels falling below 0 kPa, even under temperature abuse situations. If having a fresh-cut produce package not become anaerobic under any circumstances is the highest priority and atmospheric levels of O_2 do not significantly impact shelf life then macroperforation technology may be applicable. Historically, this has been the technology of choice for mushroom packaging. Recent advances in microperforation technology have provided mushroom growers and processors alternatives.

4.2.4 Gas Flushing

Fresh-cut MAP relies on the relationship between produce respiration and package transmission rate to alter the atmosphere within the package. This process generally takes a number of days to reach the target atmosphere and equilibrium. For produce items that are prone to enzymatic browning reactions, "pinking," which can be exacerbated by O_2 levels above 3 kPa, this gradual descent may be too long. Gas flushing of fresh-cut MAP establishes an initial low alternative atmosphere within the package, which can be beneficial in reducing enzymatic browning reactions, or "pinking." Gas flushing however is not a substitute for proper package design or leakers. Gas flushing is only optimized when it is employed in combination with proper package design and a leak-free package. Since the primary goal of gas flushing is to reduce the initial O_2 level within the package, N_2 is the most effective and economic gas.

4.3 Converting Technology

4.3.1 Packaging Format

As previously mentioned, effective modified atmosphere package design understands the balance of the physiological, polymer, converting, and marketing requirements. Issues such as package configuration, package stiffness, graphics, filling method, economics, environmental impact as well as ancillary requirements such as cook-in, antimist, resealability, and compostability all impact the design and makeup of the MAP, many of which are similar to those found in traditional packaging applications. The significant difference with MAP is the impact that these requirements have on the gas transmission rate properties of the package. The majority of these requirements fall under the science of converting technology. The science of converting technology combines the raw material polymers, films, adhesives, inks and additives in the proper sequence to create the desired package. Depending upon the format, packaging can be developed separately or in

TABLE 4.3

Common Packaging Formats

Package Format	Suitable for MAP	Natural Asperation	Sealing Formats Prone to Leakers	Common Products Packaged
Side weld premade bag	Somewhat	Somewhat	Yes	Carrots
Premade pouch	Yes	No	No	
VFFS bag	Yes	No	Structure dependent	Leafy greens
Premade SUP	Yes	No	No	
VFFS SUP	Yes	No	No	
Thermoformed tray with attachable lid	No	Yes	N/A	In-store fresh-cut fruit
Clamshell tray	No	Yes	N/A	Berries
Thermoformaed tray with sealable lid	Yes	No	Lidding dependent	
Tray with overwrap	Package dependent	Yes	Yes	Cut squash
Macroperforations	No	Yes	N/A	Mushrooms

combination with automatic filling equipment. The result of these various converting requirements is the development of a wide variety of packaging formats. There is no right or wrong format. Rather it is important to choose the format with optimum suitability based on all of the outlined requirements. Packaging formats can range from the very basic monolayer preformed side weld bag to the very complex multilayer coextruded reverse print lamination and thermoformed multilayer tray with peelable lidding structure. Table 4.3 lists some of the more common packaging formats. It is important to note that this is not an all-inclusive list as there are many variations to these common formats, which optimize specific desired properties to the specific produce packaged.

4.3.2   Flexible versus Rigid Packaging

As the names imply, the fundamental difference between flexible versus rigid packaging relates to the stiffness of the respective packaging configuration. The relative stiffness of a given package is controlled by the choice of polymers and their respective stiffness properties as well as the thickness of the structure and geometry. In MAP, there is another significant difference between rigid and flexile packaging. This relates to the effective surface area available to gas diffusion. In rigid packaging, the polymers required to maintain the rigid form of the package exhibit low gas transmission rates. These low transmission rates in combination with the thickness required to maintain the package form effectively create a gas barrier structure. This dictates that when determining the effective "breathable" surface area of the package, the rigid portion of the package cannot be included in the calculations. This means that the flexible lidding material sealed to the top of the tray must take on the entire burden of gas transmission. Therefore when comparing a flexible package with a rigid tray type package, the gas transmission level of the effective surface area of the flexible package can have a significantly lower overall transmission rate.

When a rigid tray is used in combination with a rigid lid, both components of the package are effective gas barriers. Therefore if a hermetic seal is created between lid and tray, effective gas transmission will not take place and depending upon the type and quantity of produce anaerobic conditions will rapidly develop.

Often a rigid tray and lid combination is not designed to create a complete seal. In this situation, the package is not a true modified atmosphere package but rather "natural aspiration" package, meaning that depending on how effective the seal the atmosphere inside could range from anaerobic to ambient. When designing packaging for the produce market, it is important to decide if you are designing a modified atmosphere package or a "natural aspiration" package. Both can be effective packaging formats depending upon the initial requirements and desired outcome. Shelf life optimization can only be achieved through MAP technology.

Irrespective of the format, flexible or rigid, chosen, if proper MAP is desired, then the controlled and quantifiable transmission of gases through the package in concert with the physiological characteristics of the produce being packaged is necessary. If there is no control or quantification of gas transmission rates, then optimal modified atmospheres cannot be guaranteed and the packaging system is thereby out of control. There are a number of reasons for an out-of-control packaging system including improperly quantified produce physiological properties, an improperly specified package, and out-of-specification raw materials. However, the most common reason is due to a leaking package. If the package does not have a leak-free seal, then gases will immediately begin to pass through the leak. Depending upon the size of the leak, the impact could range from missing the optimal target modified atmosphere to allowing the package to remain at ambient conditions. In either case, optimal shelf life and quality will not be achievable.

Therefore, one of the most important packaging parameters that must be considered is the selection of the sealant layer polymer and configuration. The choice of the correct sealing polymer and format is dependent upon operating parameters including package machine type, filling speed, package configuration, seal configuration, as well as product type and weight. Potential sealant layer polymers exhibit a wide range of seal characteristics. Common choices of sealant polyolefin sealant polymers include low-density polyethylene, ethyl vinyl acetate, ultralow-density polyethylene, and plastomer metallocenes. Each polymer has its own sealing characteristics including ultimate seal strength, hot tack strength, seal initiation temperature, ability to seal through contamination, and OTR. Careful consideration should be given in order to optimize the specific polymers characteristics to the specific requirements of the package necessary properties.

In certain circumstances, a peelable seal is desired. This can be in both a bag or tray configuration; however, it is much more common in a tray configuration. There are a number of technologies that can be employed to achieve an optimal peelable seal. As Gorny and Brandenburg (2003) point out, peelable lidding stock technologies are generally complex; therefore, a dialogue with a packaging expert is recommended. The three most commonly available peelable technologies are controlled contamination, dissimilar resins, and controlled delamination, with each having its own advantages and disadvantages. Gorny and Brandenburg (2003) describe the basic attributes of the three technologies:

1. *Controlled contamination.* This technology utilizes a small amount of contaminant resin within the sealant layer to achieve a controlled weak seal or "peel seal." Common contaminant resins are various grades of polybutylene. This technology is used predominately for applications where the film is peelable to itself. Through engineered blending technology, the film can be designed to be peelable within a specific temperature range. This results in peelable areas, as well as nonpeelable areas within the same finished packaging material.

2. *Dissimilar resins.* This technology utilizes a dissimilar resin between the peelable lidstock and the bottom (thermoformed) tray stock. This technology is most commonly used in lidstock applications where a peelable lid is sealed to a rigid or flexible thermoformed bottom web or rigid container. The choice of peelable sealant layers is directly dependent upon the sealant layer of the bottom material. Whenever possible, the top sealant layers should be designed together so the optimum match is made and the desired peel seal is achieved.

3. *Controlled delamination.* This technology involves achieving an initial fusion seal between the sealant layers of the lidstock and the bottom web. The "peel seal" is then achieved through the delamination of the coextruded sealant film. The delamination within the coextrusion occurs between the sealant layer and the next inner layer (often the core layer). It is possible to achieve a frosty or "tamper evident" seal as well as reseal structure. This technology is often found in medical packaging. As Gorny and Brandenburg (2003) conclude, it is critical to get the correct match between the peelable lidding material and the bottom web. There are a wide variety of polymers and constructions available, depending on the type of bottom web sealant layer, peel seal requirements, as well as the look and feel of the total structure. Companies familiar with peelable lidding materials should be consulted so that all of the necessary variables are taken into account.

4.3.3 Packaging Equipment

Packaging equipment used for MAP is as varied as the packaging types themselves. Packaging equipment can range from the basic handheld impulse sealer to processor-controlled automatic form fill and seal lines. Within this plethora of options however are two fundamental categories: vertical and horizontal. Figures 4.5 and 4.6 are examples of a vertical and horizontal filing line, respectively. Both styles of packaging equipment can be configured for either manual or automatic operation.

Although there are exceptions, generally vertical packaging machines are designed to run flexible style packaging and horizontal machines are designed to run rigid style packaging. Within the vertical configuration, there are two seal configurations: fin and lap as seen in Figure 4.7 (Brody and Marsh, 1977). Determining which seal type will be

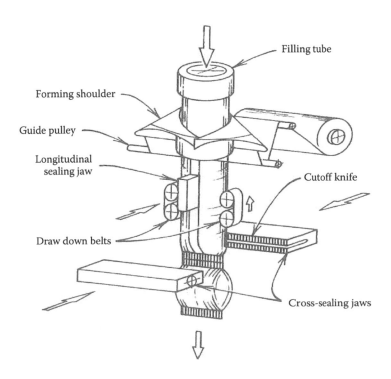

FIGURE 4.5
Examples of a vertical filing line.

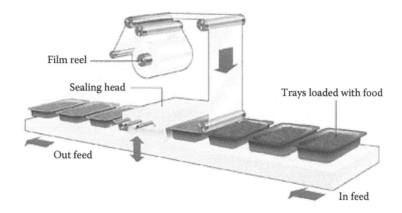

FIGURE 4.6
Examples of a horizontal filing line.

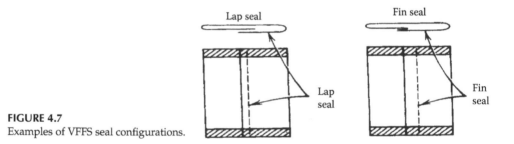

FIGURE 4.7
Examples of VFFS seal configurations.

used is a critical step in the package design since the seal type can impact polymer selection, graphics, and material usage. The most significant difference between the two types is that the fin seal configuration seals only the inside of the package whereas the lap seals the inside to the outside of the package. Therefore when a lap seal configuration is used the polymers on both the inside and outside of the package must be sealable.

4.3.4 Additives

Additive technology can have a very significant impact on MAP design. There are a number of different additives that are frequently incorporated into MAP applications. Each has its own distinctive purpose and function. These can include slip and antiblock, antifog, UV inhibitors, antimicrobials, pigments, as well as absorbents.

4.3.4.1 Antifog

As Gorny and Brandenburg (2003) discuss antifog technology is often incorporated into produce packaging to prevent water condensation on the inside of the package, which would obscure a potential purchaser's view of the product. The functionality of antifog technologies is accomplished by coating the interior surface of the flexible packaging material with compounds that reduce water surface tension or reduce the ability of the water to adhere to the packaging material and thus cause the condensed water to run off the interior surface of the package. There are two categories of antifog technology: applied coatings and sealant layer incorporation with each yielding distinct advantages and disadvantages. Applied coatings are applied to the sealing surface of the finished package and historically have been superior to the incorporation method in that they provide superior optics and antifog properties. Within the category of applied antifog coatings, there are two predominant formats: registered applied and flood coating. In the register applied system, the coating placement is controlled so that the antifog coating does not fall into the sealing area, thereby not impacting the sealing properties of the structure. In order to utilize this system, specialized equipment is necessary to achieve registration of the antifog coating to the package surface, as well as to achieve good adhesion without damaging the sealing surface. Not all converting companies have the equipment to properly apply and register these coatings. Flood coating does not apply the antifog in register but rather coats the entire sealing surface, thereby eliminating the need for the specialized register application equipment.

The main disadvantage with applied coatings is that the coatings are applied to the nontreated surface, and most applied coatings require a surface treatment to guarantee strong adhesion to the packaging structure. Therefore, antifog coating pickoff may occur. Coordination and good communication with both the antifog supplier and the converter are critical when planning to use this type of antifog coating.

Antifog sealant layer blends incorporates or blends an antifog compound into the sealant layer during the film manufacturing process. This system significantly reduces the cost of adding antifog technology to produce packages since no specialized converting equipment is needed, and thus has advantages both for the converter and end user. The major disadvantage to antifog sealant layer blends is that the antifog, by default, is in the sealing area of the bag. Therefore, weaker seals and potentially package leakers are a greater possibility. However, altering sealant layer formulations can often overcome this issue. The optics and performance of this type of antifog technology may be inferior to coated antifogs. In addition, in order to properly function, the incorporated antifog must "bloom" to the service and thus can cause a significant reduction in the gas transmission rates of the package.

4.3.4.2 Slip and Antiblock

Slip and antiblock are key components of most packaging, especially flexible packaging. The addition of slip and antiblock compounds into the polymer film structures reduces the natural tackiness of certain polymers as well as increases the film's resistance to blocking. This is especially important if vertical form fill and seal (VFFS) equipment is used to form and seal the package. The ability of the packaging structure to slide over the metal forming collars and tubes is essential for proper structure formation. Like antifog additives, slip and antiblock can significantly alter the gas transmission rate properties of the packaging structure. Care must be taken to properly quantify and minimize the effect upon gas transmission rates. Slip and improper application of antiblock additives can also impact package forming and sealability, leading to an increased incidence of improperly sealed packages and leakers.

4.3.4.3 Antimicrobial Films

Similar to the aforementioned additive applications, antimicrobial additives also bloom to the surface of the inside layer of the package. Therefore the issues of gas transmission rate reduction and sealing ability are equally present. Their method of operation is by contact of the antimicrobial component to the packaged product. Although commonplace in protein MAP, they are not widely employed in produce MAP. This is due to the relatively high amount of product surface area in relation to package surface area in fresh produce applications. This ratio imbalance prevents the majority of the product to come in contact with the microbial agent, thus making these systems relatively ineffective. New technologies are currently under development with potentially exciting results and effectiveness possible.

In all additive applications, careful consideration must be given when considering the incorporation of any additive into a modified atmosphere package. Although the addition may enhance or add certain desirable features or properties, they will almost surely detract or negatively impact other properties or functions. The key property in fresh produce applications that is most likely to be negatively affected will be the overall gas transmission rate properties of the packaging system.

4.4 Applications

As the fresh produce market has matured beyond the initial category of leafy green packaging, the technology of MAP has responded. New and novel types of vegetables, fruit, unique combinations of different types of fresh produce, ready meals, microwave, and steam-in microwave applications are all examples of new applications and markets. As with all new developments, each new application challenges the science and technology of MAP and adds its own demands and requirements.

4.4.1 Microwave and Steam-In

As Gorny and Brandenburg (2003) point out, one of the fastest growing segments of fresh-cut produce packaging is the use of microwaveable and steam-in microwaveable packages. In addition to the functional requirements generated by cooking product inside the package, there is an additional and often hidden requirement of migratory compliance. Polymers that are direct food contact compliant and act as "functional barriers" at ambient temperatures may or may not remain so under elevated temperatures. In addition, as the internal package temperature rises, fewer and fewer polymers remain both functional and regulatory compliant. It is therefore critical to engage in a detailed dialogue with the packaging supplier and to make sure that all applicable regulations have been identified. The following sections of the U.S. Code of Federal Regulations (CFRs) may be applicable when dealing with microwave and steam-in microwave applications:

   21CFR § 177.1350
   21CFR § 177.1390
   21CFR § 177.1395
   21CFR § 177.1520

It is important to assure that, before developing a microwave product, an in-depth discussion with the packaging supplier is held to ensure that the above-mentioned issues have been addressed and that the product is fully regulatory compliant. Not all packaging materials are suitable for all microwave applications.

4.4.2 Varietal Blends and Novel Produce Combinations

The growth of a variety of blends and novel combinations is a direct result of processors continually searching and investigating ways to vary their product line. This need to differentiate has led to an explosion of vegetable and leafy green combinations, vegetable stew and soup mixes, varieties and combinations of tropical fruits. These new opportunities place additional challenges and requirements on the technology of MAP. Specifically each separate and unique raw material has its own physiological properties combined with its own individual optimal target atmosphere. Therefore not all fresh-cut fruits and vegetables are computable. In other words, what may be an optimal target for one product may be very detrimental to another. Compromises in atmosphere targets will most likely be required, however, it is critical to minimize the compromises. Instead, companies are quantifying the physiological properties of their individual raw materials and then grouping like products in order to create desirable blends and combinations without having to sacrifice quality and shelf life.

4.4.3 Ready Meals

The combination of produce, protein, and carbohydrate into a fresh ready meal format has increased significantly over the past few years. A technology that is still more widespread in Europe than in other world markets is rapidly gaining mainstream popularity. This type packaging requires different types of modified atmosphere technology for each type of food category. Hence these applications generally require multicompartment trays to separate each food product with unique target atmospheres in each compartment. Since within a single package there are different and unique food groups, effective package design must crossover and combine often competing requirements. Coordination between the packaging designer and food scientists from each of the participating food groups is essential.

One requirement that has come out of all of these new applications is the need to incorporate and coordinate the package design process into and with the new product development process. Leaving the package design until the end of the product development process prevents the optimization of both product and package. As historically this has not been the norm, reeducation and enhanced communication between the new product development group within food companies and packaging designers is required.

4.5 Future Research Directions

It is an exciting time within the MAP industry. There are emerging technologies and opportunities that will have far reaching impact on the marketplace. Issues such as sustainability in packaging and the impact that packaging has on current food safety issues are already providing both tremendous challenges and opportunities. The challenge will be how to incorporate all of the desired requirements into MAP without diluting its fundamental purpose. A package that tries to become all things to all applications becomes mediocre at best with respect to any one requirement.

References

Brody, A.L. and K.S. Marsh (Eds.), 1997. *The Wiley Encyclopedia of Packaging Technology*, 2nd ed. Wiley, New York.

Brown, W. 1922. On the germination and growth of fungi at various temperatures and in various concentrations of oxygen and carbon dioxide. *Ann. Bot.* 36:257–283.

Burg, S.P. 2004. *Postharvest Physiology and Hypobaric Storage of Fresh Produce*. CAB International, Wallingford, UK.

Farber, J.M. 1991. Relative effect of CO_2 on the growth of food-borne microorganisms. *J. Food Protec.* 54(1):58–70.

Gorny, J.R. 1997. A summary of CA and MA requirements and recommendations for fresh-cut (minimally processed) fruits and vegetables. *Seventh International Controlled Atmosphere Research Conference*, Davis, CA, 5:30–66.

Gorny, J.R. and J. Brandenburg. 2003. *Packaging Design for Fresh-cut Produce*. International Fresh-cut Produce Association, Alexandria, VA.

Herner, R.C. 1987. High CO_2 effects on plant organs. In: J. Weichmann (Ed.), *Postharvest Physiology of Vegetables*. Marcel Dekker, New York, p. 239.

Kader, A.A. 1986. Biochemical and physiological basis for effects of controlled and modified atmospheres on fruits and vegetables. *Food Technol.* 40(5):99–100, 102–104.

Kader, A.A. and S. Ben-Yehoshua. 2000. Effects of superatmospheric oxygen levels on postharvest physiology and quality of fresh fruits and vegetables. *Postharvest Biol. Technol.* 20:1–13.

Kader, A.A., D. Zagory, and E.L. Kerbel. 1988. Modified atmosphere packaging of fruits and vegetables. *CRC Crit. Rev. Food Sci. Nutr.* 28(1):1–30.

Ke, D. and A.A. Kader. 1992. Potential of controlled atmospheres for postharvest insect disinfestations of fruits and vegetables. *Postharvest News Info.* 3(2):31N–37N.

Reid, M.S. 1985. Ethylene in postharvest technology. In: A.A. Kader, R.F. Kasmire, F.G. Mitchell, M.S. Reid, N.F. Sommer, and J.F. Thompson (Eds.), *Postharvest Technology of Horticultural Crops*. University of California, Cooperative Extension Special Publication 3311, p. 192.

Zagory, D. 1996. Physiology and microbiology of fresh produce in modified atmosphere packages. *Society of Manufacturing Engineers Symposium, Fundamentals of Modified Atmosphere Packaging.* December 4–5. Monterey, CA, p. 3.

Zagory, D. and A.A. Kader. 2003. *Controlled Atmosphere Handbook: A Guide for Shipment of Perishable Cargo in Refrigerated Containers*, 2nd ed. Carrier.

5

Gas Exchange Modeling

Bart M. Nicolaï, Maarten L.A.T.M. Hertog, Q. Tri Ho, Bert E. Verlinden, and Pieter Verboven

CONTENTS

5.1 Introduction

In a controlled atmosphere storage, the O_2 partial pressure is typically reduced while that of CO_2 is often increased. The purpose of this procedure is to lower the respiration rate and, consequently, slow down those associated metabolic pathways that negatively affect the quality of the stored product. The respiration rate is therefore a good indicator of the physiological stage of the fruit and its storage potential, and has been measured in the past for many fruit and vegetables at various storage temperatures and gas compositions. As the respiration rate depends on the species and even the cultivar, season, development stage, climate, and many more factors, it needs to be determined for every new cultivar before it is introduced in the market.

Many efforts have been done to model respiration or gas exchange of fruit and vegetables in general. A first advantage is that such mathematical models can be used to condense many respiration measurements carried out at various temperatures and O_2 and CO_2 partial pressures into a few material properties, which have biochemical or physical meaning. A more advanced use is, however, to use these models for *in silico* experiments for increased understanding of the mechanisms of gas exchange of fruit and vegetables, or even to design controlled atmosphere storage procedures just like an automobile or an aircraft is designed almost entirely using computer models describing the aerodynamics, mechanical deformation, and vibrations. Such models could potentially reduce the large amount of experiments required to optimize controlled atmosphere storage conditions considerably.

In this chapter, we will review models for gas exchange of fresh fruit and vegetables. First, we will review enzyme kinetics-based models to describe the respiratory activity of intact fruit. Such models do not take into account the gas transport resistance of tissue, notably the epidermis. More advanced models based on Fick's first and second laws will therefore be discussed as well. Finally, we will review models to relate quality changes to respiration rate.

5.2 Respiration

Respiration is responsible for energy production and synthesis of many biochemical precursors essential for growth and maintenance of living cells. Respiration is essentially the enzymatic oxidation of a wide variety of compounds such as starch, sugars, and organic acids by means of molecular oxygen to water and carbon dioxide. The most prominent substrate is glucose, and its overall oxidation can be described as

$$\text{Glucose} + O_2 + 38\text{ADP} + 38\text{Pi} \rightarrow 6CO_2 + 6H_2O + 38\text{ATP} \qquad (5.1)$$

The chemical energy released during this process is eventually captured in the form of adenosine triphosphate (ATP) in which form it can be used for many processes requiring energy, including biosynthesis and transportation of molecules across cell membranes.

Increasing the partial pressure of CO_2 or decreasing the partial pressure of O_2 will shift reaction (Equation 5.1) to the left and reduce the respiration rate. However, if the O_2 concentration becomes too low, the respiration metabolism may be inhibited completely and fermentation will take place. Fermentation is an anaerobic metabolic route, which is much less efficient from the energetic point of view and which has ethanol as an end product:

$$\text{Glucose} + 2\text{ADP} + 2\text{Pi} \rightarrow 2CO_2 + 2\text{ Ethanol} + 2\text{ATP} \qquad (5.2)$$

The amount of energy captured as ATP during fermentation (Equation 5.2) is 19 times less as compared to aerobic respiration. This is because a great deal of chemical energy remains captured within the ethanol. As ethanol causes off-flavors in fruit and vegetables, fermentation is to be avoided. Also, many physiological disorders such as core breakdown in pear have been shown to be associated with fermentation.

5.2.1 Factors Affecting Respiration

As already indicated, respiration is inhibited by decreasing availability of O_2 inside the mitochondria where the actual oxidation step takes place. The influence of CO_2 on the respiration rate is not so clear as it depends on the type and developmental stage of the commodity, the CO_2 concentrations, and the time of exposure (Fonseca et al., 2002). CO_2 might have some direct controlling effect on the respiration metabolism or some indirect effects through ethylene production affecting ripening. In the end, the in situ O_2 and CO_2 levels are the combined result of the metabolic activity of the product, the atmospheric levels applied, and the diffusion characteristics of the tissue.

Temperature is generally recognized as the most important external factor influencing respiration. Biological reactions generally increase two- or threefold for every 10°C rise in temperature. The respiration rate is usually maximal at moderate temperatures between 20°C and 30°C, but decreases considerably to almost zero around 0°C, depending on the genus, species, and even cultivar. If the temperature reaches chilling injury levels, an increase in respiration rate can be observed (Fidler and North, 1967). At high temperatures over 30°C, enzymatic denaturation may occur and, hence, reduce respiration rates.

The respiration rate is affected by the development stage of the fruit as well. For climacteric fruit like pears and apples, the ripening process is associated with a rise in respiratory activity, often denoted as the climacteric rise. This rise is triggered by the plant hormone ethylene, which is autocatalytically produced. In nonclimacteric fruit (e.g., citrus, pineapple), the respiration rate does not increase but progressively declines during senescence until microbial or fungal invasion.

5.2.2 Model Approaches to Respiration

Several attempts have been made to model the gas exchange by either empirical models (Jurin and Karel, 1963; Henig and Gilbert, 1975; Hayakawa et al., 1975; Yang and Chinnan, 1988; Cameron et al., 1989; Raghavan and Gariépy, 1989; Talasila et al., 1992) or strongly simplified fundamental or kinetic models using, for instance, a single Arrhenius equation (Mannapperuma and Singh, 1994). Chevillotte (1973) introduced a more fundamental approach using Michaelis–Menten kinetics to describe respiration at the cell level. Lee et al. (1991) introduced and extended this approach in the postharvest field to describe the respiration of whole fruit. After him, several other authors successfully applied this Michaelis–Menten approach to a wide range of products (Fonseca et al., 2002) and extended the original Michaelis–Menten equation to include different types of CO_2 inhibitions (Peppelenbos and Van't Leven, 1996; Hertog et al., 1998).

Although respiration is known to change with time due to processes like ripening, aging, and wounding, till date no systematic model approaches have been developed to cope with these time effects. Genard and Gouble (2005) developed a climacteric ethylene production model for apple, but this was not linked to the corresponding climacteric respiration response. Brash et al. (1995) applied a simple exponential decay to describe the change in respiration rate of asparagus after harvest. Zhu et al. (2001) observed a linear relationship between the respiration rates and the degree of cutting injury for rutabaga, but still ignored the time effect. Overall, respiration is generally treated as time-invariable.

5.2.3 Michaelis–Menten Approach

Modified or extended Michaelis–Menten kinetics are widely used to describe the relationship between O_2 and CO_2 partial pressures on one hand and O_2 consumption and CO_2 production rates on the other hand. The whole respiration pathway is assumed to be determined by one rate-limiting enzymatic reaction (Equation 5.1). In this case, the O_2 consumption rate can be described using Michaelis–Menten kinetics as

$$r_{O_2} = \frac{r_{max,O_2} \cdot P_{O_2}}{K_{m,O_2} + P_{O_2}} \tag{5.3}$$

with r_{max,O_2} the maximum oxidative O_2 consumption rate (mol kg^{-1} s^{-1}) unconstrained by O_2; the Michaelis–Menten constant for O_2 consumption (kPa); P_{O_2} the O_2 partial pressure (kPa). The parameter K_{m,O_2} can be interpreted as the O_2 partial pressure at which the oxygen consumption rate becomes half its maximal value. This simple Michaelis–Menten equation in Equation 5.3 is not suitable to describe the inhibition of respiration by CO_2.

Peppelenbos and Van't Leven (1996) evaluated four types of inhibition for modeling the influence of CO_2 levels on O_2 consumption of fruits and vegetables as compared to no influence of CO_2. They introduced an equation (Equation 5.4) describing the O_2 consumption rate as inhibited by CO_2 both in a competitive and in an uncompetitive way. This combined type of inhibition of O_2 consumption was formulated as

$$r_{O_2} = \frac{r_{max,O_2} \cdot P_{O_2}}{K_{m,O_2} \cdot \left(1 + \frac{P_{CO_2}}{K_{mc,CO_2}}\right) + P_{O_2} \cdot \left(1 + \frac{P_{CO_2}}{K_{mu,CO_2}}\right)} \tag{5.4}$$

where K_{mc,CO_2} and K_{mu,CO_2} (all in kPa) are the Michaelis constants for respiration and the competitive and uncompetitive inhibition of respiration by CO_2, respectively. The combined type of inhibition is a comprehensive and flexible formulation that, depending on the values of K_{mc,CO_2} and K_{mu,CO_2}, can describe all generally distinguished types of inhibitions on the rate of O_2 consumption (competitive, noncompetitive, and uncompetitive inhibition of O_2 consumption by CO_2; Chang, 1981) including the case of no inhibition.

The behavior of the respiration model from Equation 5.4 is illustrated in Figure 5.1 and schematically shows the effect of O_2 and the influence of the rate constants from Equation 5.4 indicating the different types of inhibition by CO_2.

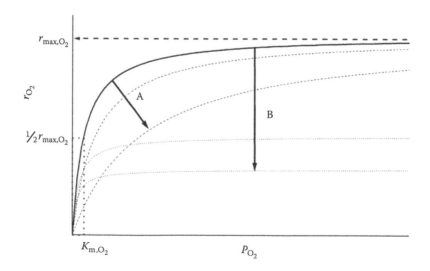

FIGURE 5.1
Model behavior for r_{O_2} (Equation 5.4). Starting from the case of only O_2 limitation (solid line) increasing K_{m,O_2} results in a shift following arrow A. Introducing competitive inhibition by CO_2 results in a comparable shift along arrow A. The higher the level of CO_2 and the lower the value of the competitive inhibition constant the larger this shift. Uncompetitive inhibition by CO_2 results in a shift along arrow B. The higher the level of CO_2 and the lower the value of the uncompetitive inhibition constant the larger this shift.

The specific CO_2 production rate is often assumed to be proportional to the specific O_2 production rate (Equation 5.5), neglecting a possible contribution of fermentative CO_2 production:

$$r_{CO_2} = RQ \cdot r_{O_2} \tag{5.5}$$

However, CO_2 production results from both oxidative and fermentative processes simultaneously and can be described according Peppelenbos et al. (1996):

$$r_{CO_2} = RQ_{ox} \cdot r_{O_2} + \frac{r_{max,CO_2(f)}}{1 + \frac{P_{O_2}}{K_{mc,O_2(f)}}} \tag{5.6}$$

with $r_{max,CO_2(f)}$ the maximum fermentative CO_2 production rate ($mol\,kg^{-1}\,s^{-1}$) unconstrained by O_2; $K_{mc,O_2(f)}$ (in kPa) the Michaelis constants for the inhibition of fermentation by O_2 and RQ_{ox} the respiration quotient for oxidative respiration. Peppelenbos et al. (1998) extended this approach even further including an inhibition of fermentative CO_2 production by CO_2 to describe gas exchange of mungbean sprouts (equation not shown). The model behavior of the fermentative CO_2 production model from Equation 5.6 is depicted in Figure 5.2.

The dependency of respiration and fermentation on temperature is often modeled by means of the Arrhenius equation. Such temperature dependency can theoretically be assigned to both the K_m and the $r_{max,(C)O_2}$ values (Song et al., 1992; Ratti et al., 1996) to account for the observed temperature effects. However, most of the times the temperature effect on the K_m is not statistically significant and it is enough to apply a temperature dependency to the $r_{max,(C)O_2}$ values only (Hertog et al., 1998).

5.2.4 Outlook: Metabolic Flux Models

As described above, Michaelis–Menten kinetics have been successfully applied to describe rates of gas exchange of horticultural produce. From a fundamental point of view, the single enzyme representation used in the Michaelis–Menten kinetics is an oversimplification of the actual energy metabolism and can only be used at a global modeling level. To get better insight into the kinetics of the underlying pathways like the glycolosis and the Krebs cycle and in the regulatory systems involved, a metabole network approach might be more appropriate.

Metabole network models were used to model microbial growth and product synthesis (Stephanopoulos et al., 1998). Such metabole network models are based on the

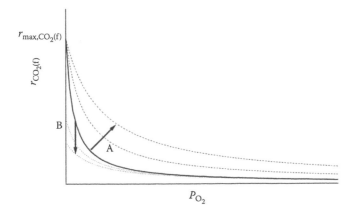

FIGURE 5.2
The model behavior for the fermentative part of CO_2 production (Equation 5.6). Starting from the case of only O_2 inhibition (solid line) an increasing $K_{mc,O_2(f)}$ results in a shift following arrow A. When competitive inhibition by CO_2 is included this results in a shift along arrow B. The higher the level of CO_2 and the lower the competitive inhibition constant the larger this shift.

stoichiometry of all relevant cellular biochemical reactions taking place. Roughly three different approaches can be identified: metabole flux analysis (MFA), kinetic models, and metabole control analysis (MCA).

In the case of MFA, the metabolic network is described by a system of algebraic equations from which the unknown intracellular fluxes can be determined. Starting from the system of algebraic equations, experimental data on some main externally measurable substrate fluxes (consumption of sugars or O_2 or the production of CO_2 or fermentation products like ethanol) can be gathered to calculate the unknown intracellular fluxes. Enough external fluxes need to be determined before one is able to solve the algebraic system. Alternatively, reactions from the pathway can be pooled to simplify the metabole network model at the cost of losing information. A better alternative would be to invest in measuring more in vivo fluxes to reduce the number of unknown fluxes to be solved.

Measuring such fluxes can be done either directly or indirectly—directly, by measuring turnover rates of the intermediates themselves; indirectly, by measuring the activity of the enzymes involved in the turnover reactions of the relevant intermediates. Preferably this is done for enzymes having a regulatory function in the energy metabolism. The in vivo activity of these enzymes will be directly related to the fluxes they catalyze. Therefore a metabolic flux chart can be assembled based on the fluxes derived from enzymatic data. By preparing such flux chart for produce stored under different gas conditions and temperatures, more detailed conclusions can be drawn on their effect on the energy metabolism.

So far MFA was mainly applied to steady-state growth cultures of microorganisms (Marx et al., 1996). The application of MFA to eukaryotes is being hindered by the intracellular compartmentalization. MFA was applied to lower eukaryotes like yeast and fungi (Stephanopoulos et al., 1998) and higher eukaryotes like plant cells (Roscher et al., 2000). Although these models were not yet applied to model respiration and fermentation of fruits and vegetables, they do provide the right tool to get better fundamental insight in the metabolic processes underlying gas exchange.

5.3 Gas Transport

5.3.1 Lumped Gas Transport

Gas transport in fruit and vegetables is a complex process. The gas transport depends on gas concentration gradients inside the tissue and between the tissue and the environment. Consequently, these concentrations are different in different places of the fruit or vegetable and are altered by their biological activity. However, in certain circumstances, which in part can be established by choosing the right experimental conditions, the gas transport equations can be simplified and a simple lumped model can be applied. In the so-called efflux or effusion method, the assumption is made that the internal atmosphere of a fruit is uniform, respiratory rates are at equilibrium, and the fruit is a hollow sphere.

5.3.1.1 *Effusion or Efflux Method*

The effusion or efflux method is a simple non steady-state method, based on kinetic analysis of the efflux of a specific gas (Cameron and Yang, 1982). This method involves measuring changes in the concentration of an inert gas (e.g., neon) in a jar caused by diffusion of the gas out of a preloaded fruit. The change in concentration can be described using Fick's first law of diffusion. The speed of the method was enhanced by Banks (1985) and used by several other authors to determine skin resistance to gas diffusion

(Solomos, 1987; Emond et al., 1991; Knee, 1991; de Wild and Peppelenbos, 2001). Schotsmans et al. (2002) combined the accuracy of the original method with the speed of the enhanced method to measure the diffusion properties of apple. The fruit is placed in an airtight jar and after closing the jar, neon gas is injected and the fruit are stored overnight to allow the gas to diffuse into the fruit. Subsequently, at equilibrium, the concentration in the jar is measured and the fruit is quickly transferred to a new jar of equal volume, filled with air without neon gas. The concentration change of neon in the jar is then monitored over time. This technique is based on Fick's first law of diffusion (Equation 5.7), which states that the flux of a gas, J (mol m^{-2} s^{-1}), diffusing through a barrier, is dependent on the diffusivity of the gas, D (m^2 s^{-1}), and the concentration gradient over this barrier, $\partial C/\partial x$ (mol m^{-3} m^{-1}):

$$J = -D\frac{\partial C}{\partial x} \tag{5.7}$$

It is assumed that the change in concentration in the barrier (the skin of the fruit) is linear; hence, in Equation 5.7 the concentration gradient $\partial C/\partial x$ can be replaced by $\Delta C/\Delta x$. Since the skin resistance to gas diffusion, R_s (s m^{-1}), is the parameter of interest, $\Delta x/D$ is replaced with this more commonly used resistance to gas diffusion:

$$J = -\frac{1}{R_s}\Delta C \tag{5.8}$$

The concentration change in the free external volume with time ($dC_e(t)/dt$) of the jar in which a fruit was placed after loading overnight is given by

$$\frac{dC_e(t)}{dt} = -\frac{A}{R_s V_e}[C_e(t) - C_i(t)] \tag{5.9}$$

where
V_e is the free external volume (m^3)
A is the surface area (m^2)
$C_i(t)$ is the gas concentration inside the fruit (on the inside of the barrier being the skin) (Banks, 1985)

A constant mass during the experiment can be assumed since the amount of gas taken out by sampling is very small and can therefore be neglected. Additionally, the second jar is initially free ($C_e(0)=0$) of the inert gas. The initial concentration of inert gas in the fruit, or more specific in the internal free gas space volume in the fruit (V_i in m^3), is $C_i(0)$ and after complete equilibration, the concentrations of the inert gas in the fruit and in the free space outside the fruit are equal ($C_i(\infty)=C_e(\infty)$). The change in concentration of the inert gas in the jar, $dC_e(t)/dt$ is then given by Equation 5.10, which can be integrated to yield

$$\frac{dC_e(t)}{dt} = \frac{A(V_i + V_e)}{R_s V_i V_e}[C_e(\infty) - C_e(t)] \tag{5.10}$$

$$C_e(t) = C_e(\infty)\left(1 - e^{-A(V_i+V_e)t/R_s V_i V_e}\right) \tag{5.11}$$

The gas concentration typically changes exponentially in time, with an initial, virtually linear phase, and this been confirmed experimentally by Cameron and Yang (1982) and Banks (1985). In the original method, the measurements are log-transformed, resulting in a linear function (Equation 5.12) when plotted against time (Cameron and Yang, 1982):

$$-\ln\left(1 - \frac{C_e(t)}{C_e(\infty)}\right) = \frac{A(V_i + V_e)}{R_s V_i V_e}t \tag{5.12}$$

From the slope (S) of the resulting linear function, the resistance value (R) is calculated as

$$R_s = \frac{A(V_i + V_e)}{V_i V_e S} \tag{5.13}$$

In an attempt to find a faster method and because the log-transformation of the data did not always result in a linear plot, for instance in case of potato, Banks (1985) suggested using the nontransformed efflux curve. When a sufficient amount of samples is taken during the short initial, virtually linear phase of the efflux process, the resistance value (R_s) is obtained from the slope (S_t) of the tangent of the nontransformed efflux curve at $t = 0$ as given by

$$R_s = \frac{C_i(0)A}{V_e S_t} \tag{5.14}$$

However, when the sampling frequency is a restricting factor, it is impossible to limit sampling within the short initial linear part. To overcome this problem, a nonlinear parameter estimation procedure is used where the parameter (R_s) is directly estimated from Equation 5.11 using a nonlinear least squares method (Schotsmans et al., 2002).

5.3.1.2 *Limitations*

The measurement of skin resistance to gas transport (Banks, 1985; Solomos, 1987; Emond et al., 1991; Knee, 1991; Schotsmans et al., 2002) is nondestructive but based on the assumption that gas transport limitations in the fruit flesh are negligible. Although the skin represents the main barrier to gas exchange and gas transport in the fruit flesh is 10–20 times faster, it does not exclude the fruit flesh as a possible barrier (Solomos, 1989; Banks and Nicholson, 2000; Lammertyn et al., 2003a).

Streif (1999) used diffusion of an inert gas through a tissue sample and the diffusivity of oxygen and carbon dioxide was then recalculated using Graham's law stating a constant relation between the diffusivity of two gases; however, this law has since shown not to apply to biological tissue (Lammertyn et al., 2001; Schotsmans et al., 2003). In these methods, respiration effects were not taken into account, although they do affect the measurements as shown by Lammertyn et al. (2001), who measured the respiration of the tissue and the diffusivity of the respiratory gases (O_2, CO_2) separately.

5.3.2 Reaction–Diffusion

The lumped gas transport model assumes that there are no gas concentration gradients inside the fruit. However, such gradients may appear because of consumption of O_2 and production of CO_2. Fick's first law is not capable of describing spatial gas concentration gradients. Several authors (Mannapperuma et al., 1991; Lammertyn et al., 2003a,b) therefore developed reaction–diffusion models to describe the exchange of O_2 and CO_2 inside fruit of different plant species. Diffusion properties of fruit tissue were determined by measuring gas exchange through small tissue samples (Schotsmans et al., 2003, 2004; Ho et al., 2006a,b, 2007). The results showed that the CO_2 diffusivity in apple and pear tissue was much higher than the O_2 diffusivity. However, as this may cause the outflow of CO_2 to be larger than the inflow of O_2, a pressure difference between the inside of the fruit and the external atmosphere may develop. Hence, besides gas diffusion driven by concentration gradients, gas exchange in the fruit may occur by permeation due to pressure gradients in the fruit tissues. A reaction–diffusion–permeation model is therefore suitable.

5.3.2.1 Model

Two phases are assumed to constitute the tissue structure of fruit, namely the intracellular liquid phase of the cells and the air-filled intercellular space. Henry's law applies to find the concentration of the compound in the liquid phase of fruit tissue if we assume local equilibrium at a certain concentration of the gas component i in the gas phase $C_{i,g}$ (mol m^{-3}). For tissue with a porosity ε, the volume-averaged concentration $C_{i,\text{tissue}}$ (mol m^{-3}) of species i is then defined as

$$C_{i,\text{tissue}} = \varepsilon \cdot C_{i,g} + (1 - \varepsilon) \cdot R \cdot T \cdot H_i \cdot C_{i,g} \tag{5.15}$$

where
  H_i (mol m^{-3} kPa^{-1}) is Henry's constant of component i (i is O_2, CO_2 or N_2)
  R (8.314 J mol^{-1} K^{-1}) is the universal gas constant
  T (K) is the temperature

Equation 5.15 leads to the following expression (Equation 5.16) for the gas capacity α_i of the component i of the tissue:

$$\alpha_i = \varepsilon + (1 - \varepsilon)R \cdot T \cdot H_i = \frac{C_{i,\text{tissue}}}{C_{i,g}} \tag{5.16}$$

Henry's constant was reported by Lide (1999). The porosity of 'Conference' pear is less than 10% (Ho et al., 2008).

To describe the diffusion and permeation processes in pear tissue, a permeation–diffusion–reaction model needs to be constructed for the three major atmospheric gases O_2, CO_2, and N_2. Equations for transport of O_2, CO_2, and N_2 were established by Ho et al. (2006b). The transport of gas i in those two phases of the tissue is governed by the following equations:

$$\varepsilon \frac{\partial C_{i,g}}{\partial t} + \nabla \cdot \varepsilon \mathbf{u}_g C_{i,g} = \nabla \cdot \varepsilon D_{i,g} \nabla C_{i,g} \tag{5.17}$$

$$(1 - \varepsilon) \frac{\partial C_{i,l}}{\partial t} = \nabla \cdot (1 - \varepsilon) D_{i,l} \nabla C_{i,l} + (1 - \varepsilon) r_{i,l} \tag{5.18}$$

where
  $C_{i,g}$ (mol m^{-3}) is the concentration of i in the gas phase
  $D_{i,g}$ (m^2 s^{-1}) is the diffusion coefficient in the gas phase
  $\mathbf{u}_g$ (m s^{-1}) is the velocity vector in the gas phase of tissue
  $C_{i,l}$ (mol m^{-3}) is the concentration of i in the liquid phase
  $D_{i,l}$ (m^2 s^{-1}) is the diffusion coefficient in the liquid phase of tissue
  $r_{i,l}$ (mol m^{-3} s^{-1}) is the respiration rate (mol m^{-3} s^{-1})
  t (s) is the time

Since gas transfer in the intercellular spaces was considered to be in equilibrium with the liquid phase of the cells, the mass transport of component i in the liquid phase in Equation 5.18 could be rewritten using Henry's law. Adding the two equations then leads to

$$\alpha_i \frac{\partial C_i}{\partial t} + \nabla \cdot \mathbf{u} C_i = \nabla \cdot D_i \nabla C_i + r_i \tag{5.19}$$

and, at the boundary,

$$C_i = C_{i,\infty} \tag{5.20}$$

where
    the subscript "g" has been dropped for convenience
    ∇ (m^{-1}) is the gradient operator
    t (s) is the time

The index ∞ refers to the gas concentration of the ambient atmosphere. D_i (m^2 s^{-1}) is the apparent diffusion coefficient (Equation 5.21), $\mathbf{u}$ (m s^{-1}) the apparent velocity vector (Equation 5.22), and r_i (mol m^{-3} s^{-1}) the effective production term of the gas component i related to O_2 consumption or CO_2 production (Equation 5.23).

$$D_i = \varepsilon \cdot D_{i,g} + (1 - \varepsilon)D_{i,l} \cdot R \cdot T \cdot H_i \tag{5.21}$$

$$\mathbf{u} = \varepsilon \cdot \mathbf{u}_g \tag{5.22}$$

$$r_i = (1 - \varepsilon)r_{i,l} \tag{5.23}$$

r_i can be modeled by the equations outlined in Section 5.2. The first term in Equation 5.19 models the accumulation of gas i, the second term permeation transport driven by an overall pressure gradient, the third term molecular diffusion due to a partial pressure gradient, and the last term consumption or production of gas i because of respiration or fermentation. The apparent diffusion coefficients are not physical properties as such but rather phenomenological parameters that depend on both the actual gas diffusion properties and fruit microstructure.

    The permeation through the barrier of tissue by the pressure gradient can be described by Darcy's law (Equation 5.24; Geankoplis, 1993):

$$\mathbf{u} = -\frac{K}{\mu}\nabla P = -\frac{K \cdot R \cdot T}{\mu}\nabla\left(\sum c_i\right) \tag{5.24}$$

where
    K (m^2) is the permeation coefficient
    P (Pa) is the pressure
    μ (Pa s) is the viscosity of the gas

The relation between gas concentration and pressure was assumed to follow the ideal gas law ($P = CRT$).

5.3.2.2 Parameter Estimation

Measurements of gas diffusivity in the fruit flesh are needed. Burg and Burg (1965) developed a method based on diffusion of gas through a tissue sample. A tissue sample is placed in between two chambers, one chamber is flushed with air containing CO_2 and the other is flushed with CO_2 free air, both gas streams are then sampled and the composition determined. A similar method to determine oxygen diffusivity was used by Zhang and Bunn (2000) and Lammertyn et al. (2001). The latter used an adaptation of the method of Burg and Burg (1965) in that one chamber was constantly flushed with a gas mixture with constant composition; the concentration change in the other chamber was then followed in time. In the method developed by Schotsmans et al. (2003), neither

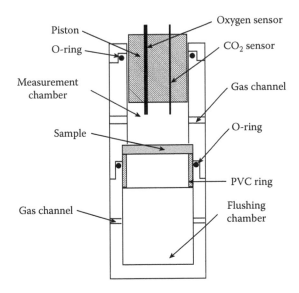

FIGURE 5.3
Schematic view of cross-section of the experimental setup for measurement of gas exchange properties of plant tissues. (Reprinted from Ho, Q.T., Verlinden, B.E., Verboven, P., and Nicolaï, B.M., *Postharvest Biol. Technol.*, 41, 113, 2006a.)

chamber was constantly flushed and the respiration of the tissue and the diffusivity of the respiratory gases (O_2, CO_2) were measured separately.

Ho et al. constructed a setup consisting of two chambers (measurement chamber and flushing chamber) separated by a disk-shaped tissue sample (Figure 5.3; Ho et al., 2006a) to measure gas transport properties of fruit tissue. A difference in gas partial pressure between the two chambers resulted in diffusion through the tissue sample. The method involved measuring gas partial pressure profiles as a function of time in the measurement chamber closed from the environment with the tissue disk. In the other chamber, gas was flushed to maintain a constant partial pressure of the gases. The transport properties were then estimated by fitting a gas transport model to the measurements using a least squares procedure.

To measure diffusion coefficients, a total pressure difference is to be avoided as this would result in permeation gas transport in addition to diffusion transport. Hereto the pressure in the measurement setup needs to be adjusted to get a pressure difference between the measurement and flushing chamber. Finally, the inlet and outlet valves of the measurement chamber were closed, and the decrease in pressure of the measurement chamber was monitored as a function of time. Diffusion and permeation coefficients were eventually estimated by fitting a finite element solution of the gas transport model to measured O_2 and CO_2 concentrations (Ho et al., 2006b).

5.3.2.3 Numerical Solution

The coupled nonlinear gas exchange equations are solved using the finite element method. The O_2 concentration can become negative due to O_2 consumption ($r_{O_2} \leq 0$). The model needs to be limited to avoid this nonphysical behavior. The respiration term in the permeation–diffusion–reaction model can be modified for the O_2 and CO_2 in the solution:

If $C_{O_2,g} < 0$ then $r_{O_2} = 0$ and $r_{CO_2} = r_{max,CO_2(f)}$;

If $C_{CO_2,g} \geq 0$ then r_{O_2} and r_{CO_2} are described by equations in Section 5.1.

Analytically, there is no O_2 consumption when O_2 reaches zero. Therefore, the O_2 concentration should never become negative. It is clear that when the O_2 concentration is positive, the reaction terms are the same as the original model. The solution, therefore, will be physically consistent. Other techniques to guarantee a physical solution are described in Ho et al. (2008).

5.3.2.4 Case Study

Gas exchange was simulated for four 'Conference' pear shapes with different equatorial radii of 2.6, 3.2, 3.4, and 3.7 cm. At the boundary, 20 kPa O_2, 0 kPa CO_2, and 80 kPa N_2 were applied at 5°C (Ho et al., 2008). The respiratory gas partial pressure profiles in the four pears are shown in Figure 5.4. Due to the gas exchange barrier properties of the cortex and epidermis tissue, the partial pressures of O_2 and CO_2 in the center of the smallest pear (2.72 and 4.57 kPa) were significantly higher and lower, respectively, than those of the largest pear (1.4×10^{-2} and 6.3 kPa). The values of the other pears were in between these extremes. This indicates that smaller fruit are less likely to experience the adverse effects of low O_2 and high CO_2 partial pressures, which is in line with the results of Lammertyn et al. (2000) who found that heavy pear fruit are more susceptible to disorders related to anerobiosis.

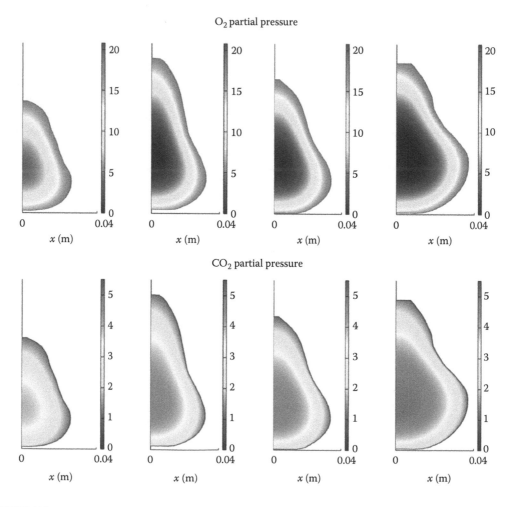

FIGURE 5.4
Respiratory gas partial pressure distribution at different pear geometry. Simulations were carried out at 5°C, 20 kPa O_2, 0 kPa CO_2 at the ambient atmosphere. (Reprinted from Ho, Q.T., Verboven, P., Verlinden, B.E., Lammertyn, J., Vandewalle, S., and Nicolaï, B.M., *PLoS. Comput. Biol.*, 4, e1000023, 2008.)

5.3.3 Outlook: Multiscale Analysis of Gas Exchange

The microstructure of fruit tissue is complex since it consists of randomly distributed cells connected together by the cell wall network and free spaces. The intercellular free space may greatly affect the gas exchange in the fruit tissue as it offers a low resistance route for gas supply. The hypothesis is that O_2 flows through the intercellular space system and subsequently permeates through the cellular membrane to the cytoplasm. Then, the O_2 diffuses within the cytoplasm into the mitochondria. Through the metabolic process of respiration O_2 is reduced to water and CO_2 is produced. CO_2 essentially follows the reversed path to the ambient. The solubility of CO_2 in water is higher than that of O_2 and comparable to its concentration presenting in air; therefore, the exchange mechanisms of the two gases may be significantly different. The relative importance of intra- and intercellular gas exchange rates and metabolic reaction rates has not yet been quantified. While the presented model provides a tool to investigate the global gradients in gas concentrations, the approach has two limitations. One, the parameters in the model are apparent values that need to be measured and are difficult to relate to the tissue structure. Two, the model is not able to predict the intracellular concentrations. Rather, it predicts an average concentration on a coarser mesoscale. However, the local concentrations are important for understanding of physiological disorders in plant tissues.

Microscale gas exchange in leaves has been investigated using theoretical models (Denison, 1992; Aalto and Juurola, 2002). Denison (1992) developed a reaction–diffusion model for oxygen diffusion and respiration in idealized legume root nodules and found large effects of flooding of the intercellular space on the O_2 permeability. Aalto and Juurola (2002) constructed a three-dimensional leaf model consisting of basic geometrical elements such as spheres and cylinders and calculated CO_2 transport using a computational fluid dynamics code. Recently, geometrical models of the microstructures of pome fruit parenchyma tissue were developed using Voronoi tessellation and ellipse tessellation algorithms on microscopic images (Mebatsion et al., 2006a,b). The geometrical models generated by ellipse tessellation algorithms (Mebatsion et al., 2006b) showed that a model microstructure consisting of truncated ellipses fills up the entire space with the same number of cells as that of microscopic images and with similar area, orientation, and aspect ratio distribution. Statistical analysis showed that the geometry generated with this approach yielded spatially equivalent geometries to that of real plant microstructures. Models solved on this geometry will provide the necessary intracellular information for evaluation of the metabolic activity (Figure 5.5). This opens the way for multiscale

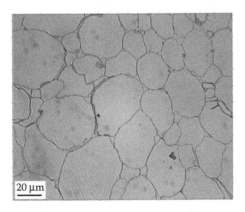 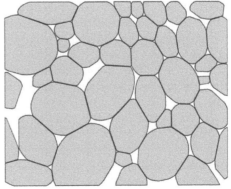

FIGURE 5.5
Microscale geometry of pear cortex tissue.

analysis of gas exchange: using the macroscale apparent models, the effect of fruit shape and external conditions is predicted and then microscale models provide the intracellular concentrations. The reversed analysis is also possible: using the microscale model, the apparent gas exchange properties of tissue can be calculated without the need for experiments.

5.4 Relating Metabolic Rate to Rate of Quality Change

The quality of horticultural products is largely based on subjective consumer evaluation of a complex of quality attributes (like taste, texture, color, appearance), which are based on specific product properties (like sugar content, volatile production, cell wall structure; Sloof et al., 1996; Shewfelt, 1999). These product properties generally change with time, as part of the normal metabolism of the product. Although much research has been done to define optimum atmosphere conditions for a wide range of fresh food products (Gross et al., 2002), the underlying mechanisms for the action of controlled or modified atmospheres are still only superficially understood. The application of controlled or modified atmospheres generally involves reducing O_2 levels and elevating levels of CO_2 to reduce the respiratory metabolism (Kader, 1992). Parallel to the effect on the respiratory metabolism, the energy produced to support other metabolic processes, and consequently these processes themselves, will be affected accordingly (Brash et al., 1995). This still covers only part of the story of how the storage atmosphere composition can affect the metabolism of the packaged produce. The physiological effects of the storage atmosphere can be diverse and complex (Burton, 1978).

Temperature and the levels of O_2 and CO_2 affecting the rate of respiration and fermentation also affect the amount of energy produced. These quality changes that are either directly influenced by O_2 or CO_2 or driven by the energy supplied by respiration or fermentation will thus be affected by the storage conditions applied. Some quality degrading processes are affected more than others due to the way they depend on these storage conditions. In spite of the large volume of research on modified and controlled atmospheres showing the general effects on the respiration rate as such, data on the quantification of the rates of quality loss in relation to the applied conditions is still limited.

In spite of the conceptual difficulties in linking the metabolic rate to the rate of quality change, several successful case studies have been published (Hertog et al., 1999, 2001, 2003, 2004, 2007). These case studies range from linking the rate of spoilage of strawberry to the overall gas exchange rate as expressed by the total CO_2 production (Hertog et al., 1999), to stem growth of Belgian endive (Hertog et al., 2007), showing that the concept of linking rates of gas exchange to rates of quality change also works for such a primary process as growth. The case studies support the hypothesis that quality attributes changes are often driven by the energy status of the tissue as indicated by the rates of gas exchange.

One could argue that the rate of gas exchange is mainly serving to convey the temperature effect. And of course, temperature is the most important factor in postharvest and thus also in controlled and modified atmosphere systems. However, in all examples, a distinction was made between the direct effect of temperature on the rates of gas exchange, the additional temperature effect on the rate of quality decay, and the additional effect of O_2 and/or CO_2 on the rates of gas exchange. With that, the developed modeling approach goes beyond the step of only using gas exchange as an indicator of temperature.

To further improve the understanding of the behavior of certain external quality aspects, a good understanding of the underlying product physiology is needed. The examples outlined clearly illustrate the link between the effects of storage conditions on metabolic rate on one hand and the effects of storage conditions on external quality aspects on the other hand. Subsequent evidence needs to be gathered to identify the exact relationships, whether gas conditions affect quality-degrading processes directly through their involvement as a reactant, or indirectly through their involvement in ATP production.

Mechanistic models are only the first step in interpreting experimental data to determine the likelihood of the possible underlying mechanisms and to direct future research to elucidate these mechanisms at a physiological and biochemical level. Even though exact details are still to be unraveled, the simplified approach of directly linking metabolic rate to the rate of quality breakdown has already proven successful in describing the effects of modified atmosphere on external quality attributes through their known effects on metabolic rate.

5.5 Conclusions

The gas exchange models presented are based on simple enzyme kinetics; it is clear that such models are phenomenological and do not really help to explain the underlying physiology of respiration. Many fruit and vegetables have a finite size, and limited gas transport is an important issue which needs to be taken into account. The gas transport models presented account for physical phenomena such as diffusion and permeation but require extensive computations. While they have provided a better understanding of gas transport in the fruit, they are still too crude to fully understand the consequences of gas transport on respiration and other associated physiological phenomena.

5.6 Future Research Needs

For future research, major advances are expected in the area of multiscale modeling, in which the gas transport is described at multiple spatial scales; the models operating at different scales are then coupled through *in silico* experiments. Such models require advances in 3-D imaging techniques to construct geometric models of tissue at micrometer scales and beyond. They would, for the first time, allow investigating the effects of storage gas composition on (limited) gas transport at the cellular scale which we believe to be associated to the occurrence of some storage disorders. Metabolic modeling is another emerging area, which aims at modeling the biochemistry of respiration at a level of detail unseen previously. In combination with multiscale gas transport models, such metabolic models would allow to predict the effect of storage gas composition on cellular metabolic fluxes and the consequence on quality changes or the development of physiological storage disorders. Clearly, high-throughput data from transcriptomics, proteomics, and metabolomics are required to calibrate and validate such models. Computational postharvest physiology is a novel discipline that aims at combining such advanced models with novel high-throughput experimental techniques. It is expected to boost our understanding of controlled atmosphere and pave the way for innovative and improved controlled atmosphere storage procedures.

Acknowledgments

The K.U. Leuven (project OT 04/31, IRO PhD scholarship of author Q.T. Ho; Industrial Research Fund) and the Institute for the Promotion of Innovation by Science and Technology in Flanders (IWT project 050633) are gratefully acknowledged for financial support.

References

Aalto, T. and E. Juurola. 2002. A three-dimensional model of CO_2 transport in airspaces and mesophyll cells of a silver birch leaf. *Plant Cell Environ.* 25: 1399–1409.

Banks, N.H. 1985. Estimating skin resistance to gas diffusion in apples and potatoes. *J. Exp. Bot.* 36: 1842–1850.

Banks, N.H. and S.E. Nicholson. 2000. Internal atmosphere composition and skin permeance to gases of pepper fruit. *Postharvest Biol. Technol.* 18: 33–41.

Brash, D.W., C.M. Charles, S. Wright, and B.L. Bycroft. 1995. Shelf-life of stored asparagus is strongly related to postharvest respiratory activity. *Postharvest Biol. Technol.* 5: 77–81.

Burg, S.P. and E.A. Burg. 1965. Gas exchange in fruits. *Plant Physiol.* 18: 870–874.

Burton, W.G. 1978. Biochemical and physiological effects of modified atmospheres and their role in quality maintenance. In: Hultin H.O. and Milner M. (Eds.), *Postharvest Biology and Biotechnology.* FNP, Westport, CT.

Cameron, A.C. and S.F. Yang. 1982. A simple method for the determination of resistance to gas diffusion in plant organs. *Plant Physiol.* 70: 21–23.

Cameron, A.C., W. Boylan-Pett, and J. Lee. 1989. Design of modified atmosphere packaging systems: Modeling oxygen concentrations within sealed packages of tomato fruits. *J. Food Sci.,* 54: 1413–1416, 1421.

Chang, R. 1981. *Physical Chemistry with Applications to Biological Systems,* 2nd ed. Macmillan, New York.

Chevillotte, P. 1973. Relation between the reaction cytochrome oxidase-oxygen and oxygen uptake in cells in vivo. The role of diffusion. *J. Theor. Biol.* 39: 277–295.

De Wild, H.P.J. and H.W. Peppelenbos. 2001. Improving the measurement of gas exchange in closed systems. *Postharvest Biol. Technol.* 22: 111–119.

Denison, R.F. 1992. Mathematical modeling of oxygen diffusion and respiration in legume root nodules. *Plant Physiol.* 98: 901–907.

Emond, J.P., F. Castaigne, C.J. Toupin, and D. Desilets. 1991. Mathematical modelling of gas exchange in modified atmosphere packaging. *Trans. ASAE* 34: 239–245.

Fidler, J.C. and C.J. North. 1967. The effect of conditions of storage on the respiration of apples. I. The effects of temperature and concentrations of carbon dioxide and oxygen on the production of carbon dioxide and uptake of oxygen. *J. Hort. Sci.,* 42: 189–206.

Fonseca, S.C., F.A.R. Oliveira, and J.K. Brecht. 2002. Modelling respiration rate of fresh fruits and vegetables for modified atmosphere packages: a review. *J. Food Eng.* 52: 99–119.

Geankoplis, J.C. 1993. *Transport Processes and Unit Operations.* Prentice-Hall, Englewood Cliffs, NJ.

Genard, M. and B. Gouble. 2005. ETHY. A theory of fruit climacteric ethylene development. *Plant Physiol.* 139: 531–545.

Gross, K.C., C.Y. Wang, and M. Saltveit. 2002. *The Commercial Storage of Fruits, Vegetables, and Florist and Nursery Crops. USDA Agriculture Handbook 66.* http://usna.usda.gov/hb66/ Retrieved March 2008.

Hayakawa, K., Y.S. Henig, and S.G. Gilbert. 1975. Formulae for predicting gas exchange of fresh produce in polymeric film package. *J. Food Sci.* 40: 186–191.

Henig, Y.S. and S.G. Gilbert. 1975. Computer analysis of the variables affecting respiration and quality of produce packaged in polymeric films. *J. Food Sci.* 40: 1033–1035.

Hertog, M.L.A.T.M., H.W. Peppelenbos, R.G. Evelo, and L.M.M. Tijskens. 1998. A dynamic and generic model on the gas exchange of respiring produce: The effects of oxygen, carbon dioxide and temperature. *Postharvest Biol. Technol.* 14: 335–349.

Hertog, M.L.A.T.M., H.A.M. Boerrigter, G.J.P.M. Van den Boogaard, L.M.M. Tijskens, and A.C.R. Van Schaik. 1999a. Predicting keeping quality of strawberries (cv 'Elsanta') packed under modified atmospheres: An integrated model approach. *Postharvest Biol. Technol.* 15: 1–12.

Hertog, M.L.A.T.M., S.E. Nicholson, and N.H. Banks. 2001. The effect of modified atmospheres on the rate of firmness change in 'Braeburn' apples. *Postharvest Biol. Technol. 2001.* 23: 175–184.

Hertog, M.L.A.T.M., S.E. Nicholson, and K. Whitmore. 2003. The effect of modified atmospheres on the rate of quality change in 'Hass' avocado. *Postharvest Biol. Technol.* 29: 41–53.

Hertog, M.L.A.T.M., S.E. Nicholson, and P.B. Jeffery, 2004. The effect of modified atmospheres on the rate of firmness change of 'Hayward' kiwifruit. *Postharvest Biol. Technol.,* 31: 251–261.

Hertog, M.L.A.T.M., N. Scheerlinck, J. Lammertyn, and B.M. Nicolaï. 2007. The impact of biological variation on postharvest behaviour of Belgian endive: The case of multiple stochastic variables. *Postharvest Biol. Technol. 2007.* 43: 78–88.

Ho, Q.T., B.E. Verlinden, P. Verboven, and B.M. Nicolaï. 2006a. Gas diffusion properties at different positions in the pear. *Postharvest Biol. Technol.* 41:113–120.

Ho, Q.T., B.E. Verlinden, P. Verboven, S. Vandewalle, and B.M. Nicolaï. 2006b. A permeation–diffusion–reaction model of gas transport in cellular tissue of plant materials. *J. Exp. Bot.* 57: 4215–4224.

Ho, Q.T., B.E. Verlinden, P. Verboven, S. Vandewalle, and B.M. Nicolaï. 2007. Simultaneous measurement of oxygen and carbon dioxide diffusivities in pear fruit tissue using optical sensors. *J. Sci. Food. Agric.* 87: 1858–1867.

Ho, Q.T., P. Verboven, B.E. Verlinden, J. Lammertyn, S. Vandewalle, and B.M. Nicolaï. 2008. A continuum model for gas exchange in pear fruit. *PLoS. Comput. Biol.* 4(3): e1000023. doi:10.1371/journal.pcbi.1000023.

Jurin, V. and M. Karel. 1963. Studies on control of respiration of McIntosh apples by packaging methods. *Food Technol.* 17: 104–108.

Kader, A.A. 1992. Modified atmospheres during transport and storage. In: Kader, A.A. (Ed.), *Post-Harvest Technology of Agricultural Crops.* 2nd ed. Publication 3311 University of California. California, chap. 11.

Knee, M. 1991. Rapid measurement of diffusion of gas through the skin of apple fruits. *HortScience* 26: 885–887.

Lammertyn, J., M. Aerts, B.E. Verlinden, W. Schotsmans, and B.M. Nicolaï. 2000. Logistic regression analysis of factors influencing core breakdown in 'Conference' pears. *Postharvest Biol. Technol.* 20: 25–37.

Lammertyn, J., N. Scheerlinck, B.E. Verlinden, W. Schotsmans, and B.M. Nicolaï. 2001. Simultaneous determination of oxygen diffusivity and respiration in pear skin and tissue. *Postharvest Biol. Technol.* 23: 93–104.

Lammertyn, J., N. Scheerlinck, P. Jancsók, B.E. Verlinden, and B.M. Nicolaï. 2003a. A respiration–diffusion model for 'Conference' pears I: Model development and validation. *Postharvest Biol. Technol.* 30: 29–42.

Lammertyn, J., N. Scheerlinck, P. Jancsók, B.E. Verlinden, and B.M. Nicolaï. 2003b. A respiration–diffusion model for 'Conference' pears II: Simulation and relation to core breakdown. *Postharvest Biol. Technol.* 30: 43–55.

Lee, D.S., P.E. Haggar, J. Lee, and K.L. Yam. 1991. Model for fresh produce respiration in modified atmospheres based on principles of enzyme kinetics. *J. Food Sci.* 56: 1580–1585.

Lide, D.R. 1999. *Handbook of Chemistry and Physics.* CRC Press, New York.

Mannapperuma, J.D. and R.P. Singh. 1994. Modeling of gas exchange in polymeric packages of fresh fruits and vegetables. In: Singh, R.P. and Oliveira, F.A.R. (Eds.), *Process Optimization and Minimal Processing of Foods.* CRC Press, Boca Raton, FL, pp. 437–458.

Mannapperuma, J.D., R.P. Singh, and M.E. Montero. 1991. Simultaneous gas diffusion and chemical reaction in foods stored in modified atmospheres. *J. Food Eng.* 14: 167–183.

Marx, A., A.A. Graaf, W. Wiechert, L. Eggeling, and H. Sahm. 1996. Determination of the fluxes in the central metabolism of *Corynebacterium glutamicum* by nuclear magnetic resonance spectroscopy combined with metabolite balancing. *Biotechnol. Bioeng.* 49: 111–129.

Mebatsion, H.K., P. Verboven, B.E. Verlinden, Q.T. Ho, T.A. Nguyen, and B.M. Nicolaï. 2006a. Microscale modelling of fruit tissue using Voronoi tessellations. *Comput. Electron. Agric.* 52: 36–48.

Mebatsion, H.K., P. Verboven, Q.T. Ho, F. Mendoza, B.E. Verlinden, T.A. Nguyen, and B.M. Nicolaï. 2006b. Modelling fruit microstructure using novel ellipse tessellation algorithm. *CMES* 14: 1–14.

Peppelenbos, H.W., L.M.M. Tijskens, J. Van 't Leven, and E.C. Wilkinson. 1996. Modelling oxidative and fermentative carbon dioxide production of fruits and vegetables. *Postharvest Biol. Technol.* 9: 283–295.

Peppelenbos, H.W. and J. Van't Leven. 1996. Evaluation of four types of inhibition for modelling the influence of carbondioxide on oxygen consumption of fruits and vegetables. *Postharvest Biol. Technol.*, 7:27–40.

Peppelenbos, H.W., L. Brien, and L.G.M. Gorris. 1998. The influence of carbon dioxide on gas exchange of mungbean sprouts at aerobic and anaerobic conditions. *J. Sci. Food Agric.* 76: 443–449.

Raghavan, G.S.V. and Y. Gariépy. 1989. Respiration activity of vegetables under CA. Presentation at the ASAE International Summer Meeting, June 1989, Québec, Canada.

Ratti, C., G.S.V. Raghavan, and Y. Gariépy. 1996. Respiration model and modified atmosphere packaging of fresh cauliflower. *J. Food Eng.* 28: 297–306.

Roscher, A., N.J. Kruger, and R.G. Ratcliffe. 2000. Strategies for metabolic flux analyses in plants using isotope labelling. *J. Biotechnol.* 77: 81–102.

Schotsmans, W., B.E. Verlinden, J. Lammertyn, A. Peirs, P. Jancsók, N. Scheerlinck, and B.M. Nicolaï. 2002. Factors affecting skin resistance measurements in pipfruit. *Postharvest Biol. Technol.* 25: 169–179.

Schotsmans, W., B.E. Verlinden, J. Lammertyn, and B.M. Nicolaï. 2003. Simultaneous measurement of oxygen and carbon dioxide diffusivity in pear tissue. *Postharvest Biol. Technol.* 29: 155–166.

Schotsmans, W., B.E. Verlinden, J. Lammertyn, and B.M. Nicolaï. 2004. The relationship between gas transport properties and the histology of apple. *J. Sci. Food Agric.* 84: 1131–1140.

Shewfelt, R.L. 1999. What is quality? *Postharvest Biol. Technol.* 15: 197–200.

Sloof, M., L.M.M. Tijskens, and E.C. Wilkinson. 1996. Concepts for modelling the quality of perishable products. *Trends Food Sci. Tech.* 7: 165–171.

Solomos, T. 1987. Principles of gas exchange in bulky plant tissues. *HortScience* 22: 766–771.

Solomos, T. 1989. A simple method for determining the diffusivity of ethylene in McIntosh apples. *Sci. Hortic.* 39: 311–318.

Song, Y., H.K. Kim, and K.L. Yam. 1992. Respiration rate of blueberry in modified atmosphere at various temperatures. *J. Am. Soc. Hort. Sci.* 117: 925–929.

Stephanopoulos, G., A. Aristidou, J. Nielsen, and J. Nielson. 1998. *Metabolic Engineering: Principles and Methodologies*. Academic Press, Burlington, VA.

Streif, J. 1999. Gasdiffusionsmessungen an Früchten. Annual meeting of the DGQ, Freising-Weihenstephan, Germany, XXXIV.

Talasila, P.C., K.V. Chau, and J.K. Brecht. 1992. Effects of gas concentrations and temperature on O_2 consumption of strawberries. *Trans. ASAE,* 35: 221–224.

Yang, C.C. and M.S. Chinnan. 1988. Modeling the effect of O_2 and CO_2 on respiration and quality of stored tomatoes. *Trans. ASAE,* 31: 920–925.

Zhang, J. and J.M. Bunn. 2000. Oxygen diffusivities of apple flesh and skin. *Trans. ASAE* 43: 359–363.

Zhu, M., C.L. Chu, S.L. Wang, and R.W. Lencki. 2001. Influence of oxygen, carbon dioxide, and degree of cutting on the respiration rate of rutabaga. *J. Food Sci.* 66: 33–37.

6

Effects on Nutritional Quality

Adel A. Kader

CONTENTS

6.1 Introduction

Fruits, nuts, and vegetables play a very essential role in human nutrition and health, especially as sources of vitamins, minerals, dietary fiber, and phytonutrients (phytochemicals). Phytonutrients, which can lower the risk of heart disease, cancer, and other diseases, include carotenoids and flavonoids (anthocyanins, phenolic acids, polyphenols). The antioxidant capacity of fruits is related to their contents of anthocyanins, phenolic compounds, carotenoids, ascorbic acid (AA), and vitamin E. Although antioxidant capacity varies greatly among fruits and vegetables, it is better to consume a variety of commodities rather than limiting consumption to a few with the highest antioxidant capacity. There is increasing evidence that consumption of whole foods is better than isolated food components (such as dietary supplements).

Examples of the phytochemicals in fruits and vegetables that have established or proposed positive effects on human health and their important sources are shown in Tables 6.1 and 6.2. Some changes in these tables are likely as the results of additional studies on effects of phytochemicals and their bioavailability on human health become available in the next few years. Meanwhile, it is important to evaluate the validity and dependability of the results of every study before reaching conclusions for the benefit of consumers (Kader et al., 2004).

Many pre- and postharvest factors influence the composition and quality of fruits and nuts. These include genetic factors (selection of cultivars, rootstocks used for fruit species), preharvest environmental factors (climatic conditions and cultural practices), maturity at harvest, harvesting method, postharvest handling procedures, and processing and cooking methods.

The selection of the genotype with the highest flavor and nutritional quality for a given commodity is much more important factor than climatic conditions and cultural practices

TABLE 6.1

Nutritive Constituents of Fruits and Vegetables That Have a Positive Impact on Human Health and Their Sources

Constituent	Sources	Established or Proposed Effects on Human Wellness
Vitamin C (AA)	Broccoli, cabbage, cantaloupe, citrus fruits, guava, kiwifruit, leafy greens, pepper, pineapple, potato, strawberry, tomato, watermelon	Prevents scurvy, aids wound healing, healthy immune system, cardiovascular disease
Vitamin A (carotenoids)	Dark-green vegetables (such as collards, spinach, and turnip greens), orange vegetables (such as carrots, pumpkin, and sweet potato), orange-flesh fruits (such as apricot, cantaloupe, mango, nectarine, orange, papaya, peach, persimmon, and pineapple), tomato	Night blindness, prevention chronic fatigue, psoriasis, heart disease, stroke, cataracts
Vitamin K	Nuts, lentils, green onions, crucifers (cabbage, broccoli, Brussels sprouts), leafy greens	Synthesis of procoagulant factors, osteoporosis
Vitamin E (tocopherols)	Nuts (such as almonds, cashew nuts, filberts, macadamias, pecans, pistachios, peanuts, and walnuts), corn, dry beans, lentils and chickpeas, dark-green leafy vegetables	Heart disease, LDL oxidation, immune system, diabetes, cancer
Fiber	Most fresh fruits and vegetables, nuts, cooked dry beans and peas	Diabetes, heart disease
Folate (folicin or folic acid)	Dark-green leafy vegetables (such as spinach, mustard greens, butterhead lettuce, broccoli, Brussels sprouts, and okra), legumes (cooked dry beans, lentils, chickpeas and green peas), asparagus	Birth defects, cancer, heart disease, nervous system
Calcium	Cooked vegetables (such as beans, greens, okra and tomatoes), peas, papaya, raisins, orange, almonds, snap beans, pumpkin, cauliflower, rutabaga	Osteoporosis, muscular/skeletal, teeth, blood pressure
Magnesium	Spinach, lentils, okra, potato, banana, nuts, corn, cashews	Osteoporosis, nervous system, teeth, immune system
Potassium	Baked potato or sweet potato, banana, and plantain, cooked dry beans, cooked greens, dried fruits (such as apricots and prunes), winter (orange) squash, and cantaloupe	Hypertension (blood pressure) stroke, arteriosclerosis

Source: Kader, A.A., Perkins-Veazie, P., and Lester, G.E., Nutritional quality of fruits, nuts and vegetables and their importance in human health. U.S. Dept. Agric., Agric. Handbook 66, 2004. http://www.ba.ars.usda.gov/hb66/index.html

in producing the best quality for consumers of that commodity. Producers should use an integrated crop management system to optimize yield and quality of each commodity. Buyers and consumers should be willing to pay more for higher flavor and nutritional quality products because often the producer sacrifices some yield to produce better quality fruits and vegetables.

Maturity at harvest has a major impact on quality and postharvest life potential of fruits. All fruits, with a few exceptions like avocados, bananas, and pears reach their best eating quality stage when fully ripened on the tree. However, since such ripe fruits cannot survive the postharvest handling system, they are usually picked mature but not ripe. It is better to pick fruits partially ripe than mature but not ripe (mature green) to provide the consumer with better flavor and nutritional quality fruits.

TABLE 6.2

Nonnutritive Plant Constituents That May Be Beneficial to Human Health

Constituent	Compound	Sources	Established or Proposed Effects on Human Wellness
Phenolic compounds			
Proanthocyanins	Tannins	Apple, grape, cranberry, pomegranate	Cancer
Anthocyanidins	Cyanidin, malvidin, delphinidin, pelargonidin, peonidin, petunidin	Red, blue, and purple fruits (such as apple, blackberry, blueberry, cranberry, grape, nectarine, peach, plum and prune, pomegranate, raspberry, and strawberry)	Heart disease, cancer initiation, diabetes, cataracts, blood pressure, allergies
Flavan-3-ols	Epicatechin, epigallocatechin, catechin, gallocatechin	Apples, apricots, blackberries, plums, raspberries, strawberries	Platelet aggregation, cancer
Flavanones	Hesperetin, naringenin, eriodictyol	Citrus (oranges, grapefruit, lemons, limes, tangerine)	Cancer
Flavones	Luteolin, apigenin	Celeriac, celery, peppers, rutabaga, spinach, parsley, artichoke, guava, pepper	Cancer, allergies, heart disease
Flavonols	Quercetin, kaempferol, myricetin, rutin	Onions, snap beans, broccoli, cranberry, kale, peppers, lettuce	Heart disease, cancer initiation, capillary protectant
Phenolic acids	Caffeic acid, chlorogenic acid, coumaric acid, ellagic acid	Blackberry, raspberry, strawberry, apple, peach, plum, cherry	Cancer, cholesterol
Carotenoids			
Lycopene		Tomato, watermelon, papaya, Brazilian guava, Autumn olive, red grapefruit	Cancer, heart disease, male infertility
α-carotene		Sweet potatoes, apricots, pumpkin, cantaloupe, green beans, lima beans, broccoli, Brussels sprouts, cabbage, kale, kiwifruit, lettuce, peas, spinach, prunes, peaches, mango, papaya, squash and carrots	Tumor growth
β-carotene		Cantaloupes, carrots, apricots, broccoli, leafy greens (lettuce, Swiss chard), mango, persimmon, red pepper, spinach, sweet potato	Cancer
Xanthophylls	Lutein, zeaxanthin, β-cryptoxanthin	Sweet corn, spinach, corn, okra, cantaloupe, summer squash, turnip greens	Macular degeneration
Monoterpenes	Limonene	Citrus (grapefruit, tangerine)	Cancer
Sulfur compounds	Glucosinolates, isothiocyanates, indoles, allicin, diallyl disulfide	Broccoli, Brussels sprouts, mustard greens, horseradish, garlic, onions, chives, leeks	Cancer, cholesterol, blood pressure, diabetes

Source: Kader, A.A., Perkins-Veazie, P., and Lester, G.E., *Nutritional Quality of Fruits, Nuts and Vegetables and Their Importance in Human Health. U.S. Dept. Agric., Agric. Handbook 66,* 2004. http://www.ba.ars.usda.gov/hb66/index.html

Keeping vegetables, fruits, and nuts within their optimum ranges of temperature and relative humidity is the most important factor in maintaining their quality and minimizing postharvest losses. Above the freezing point (for nonchilling-sensitive commodities) and the minimum safe temperature (for chilling-sensitive commodities), every 10°C increase in temperature accelerates deterioration and the rate of loss in nutritional quality by two- to threefold. Delays between harvesting and cooling or processing can result in quantitative losses due to water loss and decay and qualitative losses in flavor and nutritional quality (Lee and Kader, 2000). The limited published research on impact of atmospheric modification on compositional changes, including vitamins (mainly AA content) of fruits and vegetables has been reviewed by Weichmann (1986), Wang (1990), and Lee and Kader (2000).

Postharvest opportunities for enhancing the quantity and quality of essential nutrients present in fruits and vegetables include (1) increasing overall consumption of fruits and vegetables, (2) improving bioavailability of nutrients, (3) increasing levels of essential nutrients through fortification methods, and (4) reducing nutrient losses (Beuscher et al., 1999).

6.2 Vitamin C

In general, atmospheric modification within the ranges tolerated by the commodity reduces physiological and chemical changes, including losses of vitamin C of fruits and vegetables during storage. For most commodities tested, 1–4 kPa O_2 generally slows AA degradation through prevention of oxidation. The effects of carbon dioxide on AA may be positive or negative depending upon the commodity, CO_2 concentration and duration of exposure, and temperature (Lee and Kader, 2000; Weichmann, 1986).

Loss of AA can be reduced by storing apples in a reduced oxygen atmosphere (Delaporte, 1971).Vitamin C content of fresh-cut kiwifruit slices kept in 0.5, 2, or 4 kPa O_2 at 0°C decreased by 7%, 12%, or 18%, respectively, after 12 days storage. Vitamin C content in kiwifruit slices kept in air +5, 10, or 20 kPa CO_2 decreased by 14%, 22%, or 34%, respectively, of their initial vitamin C contents (Agar et al., 1999). Generally, high CO_2 concentration in the storage atmosphere caused degradation of vitamin C in fresh-cut kiwifruit slices. Enhanced losses of vitamin C in response to air +10 and 20 kPa CO_2 may be due to their stimulating effects on oxidation of AA and/or inhibition of dehydroascorbic acid (DHA) reduction to AA (Agar et al., 1999). Bangerth (1977) found accelerated AA losses in apples and red currants stored in elevated CO_2 atmospheres. Storage in 2 kPa O_2 + 10 kPa CO_2 resulted in 60% loss in AA content of 'Conference' pears (Veltman et al., 1999). Storage for 6 days in CO_2-enriched atmospheres resulted in a reduction in AA content of sweet pepper kept at 13°C (Wang, 1977) but an increase in its content in broccoli kept at 5°C (Wang, 1979). Vitamin C content was reduced by high CO_2 (10–30 kPa), particularly in strawberries, while losses were moderate in black currants and blackberries, and very low in raspberries and red currants (Agar and Streif, 1996; Agar et al., 1997). Reducing the O_2 concentration in the storage atmosphere in the presence of high CO_2 had little effect on the vitamin C content. AA was more diminished at high CO_2 than DHA (Agar et al., 1997). High CO_2 may stimulate the oxidation of AA, probably by ascorbate peroxidase. Mehlhorn (1990) demonstrated an increase in ascorbate peroxidase activity in response to ethylene. High CO_2 at injurious levels for the commodity may reduce AA by increasing ethylene production and, thus, the activity of ascorbate peroxidase.

Wright and Kader (1997a) studied the fresh-cut products of strawberries and persimmons for 8 days controlled atmosphere (CA) (2 kPa O_2, air + 12 kPa CO_2, or 2 kPa O_2 + 12 kPa CO_2) at 0°C and found that the postcutting life based on visual quality ended before significant losses of vitamin C occurred. Nutritional quality of broccoli florets was evaluated during storage at 4°C in air or CA of 2 kPa O_2 + 6 kPa CO_2. Retention of vitamin C was slightly greater in CA than in air. Returning the samples to ambient conditions for 24 h after storage in either condition resulted in chlorophyll and vitamin C losses (Paradis et al., 1996). Wang (1983) noted that 1 kPa O_2 retarded AA degradation in Chinese cabbage stored for 3 months at 0°C. In contrast, treatment with 10 or 20 kPa CO_2 for 5 or 10 days was without any effect, and 30 or 40 kPa CO_2 increased AA decomposition. The effect of elevated CO_2 on AA content varied among commodities and was dependent on CO_2 level and storage temperature and duration (Weichmann, 1986). Modified atmosphere packaging (MAP) of broccoli resulted in better maintenance of AA, chlorophyll, and moisture retention compared to broccoli stored in air (Barth et al., 1993a,b; Barth and Zhuang, 1996). There was no significant difference in activity of AA oxidase in response to package conditions. Greater humidity inside the packages possibly served to better preserve vitamin C content in packaged broccoli spears. Retention of vitamin C in jalapeno pepper rings after 12 days storage at 4.4°C, and an additional 3 days at 13°C was 83% in MAP and 56% in air. MAP retarded the conversion of AA to DHA that occurred in air-stored peppers. Other quality attributes of peppers were maintained better in MAP than in air (Howard and Hernandez-Brenes, 1998; Howard et al., 1994). Initially, fresh-cut spinach contained AA as a predominant form of vitamin C. However, a decrease in AA and an accumulation of DHA were observed during storage. The increase in DHA was more prominent in MAP and resulted in a higher vitamin C value of spinach in MAP than air. An increase in the pH of spinach stored in MAP was also observed (Gil et al., 1999). In contrast, fresh-cut products of Swiss chard contained only DHA, and vitamin C was better preserved in air than MAP (Gil et al., 1998). Barker and Mapson (1952) reported that AA content in potato tubers kept in 100 kPa O_2 was lower than in those stored in air.

6.3 Vitamin A

Fresh fruits and vegetables are an important dietary source of vitamin A (retinol), which is essential for normal growth, reproduction, and resistance to infection; severe deficiency may lead to irreversible blindness. Plant materials do not contain vitamin A, but provide carotenoids that are converted to vitamin A after ingestion. Provitamin A carotenoids found in significant quantities in fruits and vegetables include β-carotene, β-cryptoxanthin, and α-carotene. These and other carotenoids, such as lycopene, have a role as antioxidants.

CA conditions that delay ripening of fruits result in delayed synthesis of carotenoids, such as lycopene in tomatoes and β-carotene in mangoes, but the synthesis of these pigments resumes upon transfer of the fruits to air at ripening temperatures (15°C–25°C). Weichmann (1986) reported that low oxygen atmospheres enhanced the retention of carotene in carrots, air + 5 kPa CO_2 caused a loss of carotene, while air + 7.5 kPa CO_2 or higher appeared to cause de novo synthesis of carotene. The carotene content of leeks was found to be higher after storage in 1 kPa O_2 + 10 kPa CO_2 than after storage in air (Weichmann, 1986). Barth and Zhuang (1996) found that MAP retained carotenoids in broccoli florets. Howard and Hernandez-Brenes (1998) found that the retention of β-carotene in jalapeno pepper rings after 12 days at 4.4°C plus 3 days at 13°C was 87% in MAP (5 kPa O_2 + 4 kPa CO_2) and 68% in air. Retention of α-carotene was 92% in MAP and 52% in air after 15 days. Wright and Kader (1997b) reported that peach slices kept in air + 12 kPa CO_2 had a lower content of β-carotene and β-cryptoxanthin (retinol equivalent) than slices kept in air, 2 kPa O_2, or 2 kPa O_2 + 12 kPa CO_2 for 8 days at 5°C. Storage of persimmon slices in 2 kPa O_2 or air + 12 kPa CO_2 resulted in slightly lower retinol equivalent after 8 days at 5°C, but the loss was insignificant in slices kept under 2 kPa

$O_2 + 12$ kPa CO_2. For sliced peaches and persimmons, the limit of shelf life based on sensory quality was reached before major losses of carotenoids occurred.

6.4 Phenolic Compounds and Antioxidant Activity

CA conditions that delay ripening result in delayed synthesis of anthocyanins in some fruits, such as nectarines, peaches, and plums, but the synthesis of these pigments resumes upon transfer of the fruits to air at ripening temperatures (15°C–25°C). Lin et al. (1989) reported that very high CO_2 levels (>73 kPa) as a result of MAP destabilized cyaniding derivatives in the skin of 'Starkrimson' apples. Gil et al. (1995) found similar effects of MAP on delphinidin derivatives in pomegranates. Gil et al. (1997) reported that elevated CO_2 (10, 20, or 40 kPa in air) had a minimal effect on the anthocyanin content of external tissues of strawberries, but induced a remarkable decrease in anthocyanin content of internal tissues after storage for 10 days at 5°C. Phenolic compounds increased in strawberries during storage at 5°C for 10 days, but were not affected by the storage atmosphere (Holcroft and Kader, 1999). The increase in pH and decrease of titratable acidity were enhanced in internal tissues of strawberries by the CO_2 treatments, and may in turn have influenced anthocyanin expression (Holcroft and Kader, 1999). Allende et al. (2007) reported that when compared with storage in air, strawberries stored under superatmospheric oxygen and carbon dioxide-enriched concentrations showed lower total phenolic contents after 5 days and a vitamin C reduction after 12 days of storage, accompanied by a more pronounced conversion from reduced to oxidized forms under superatmospheric oxygen. Holcroft et al. (1998) observed that the arils of pomegranates stored for 6 weeks at 10°C in air were deeper red and had higher level of anthocyanins than the initial controls and than those stored in 10 or 20 kPa CO_2-enriched air, possibly due to suppressed anthocyanin biosynthesis.

Curry (1997) reported up to 10-fold increases in antioxidant content of "Delicious" and 'Granny Smith' apples during the first two months of storage at −1°C and significant decreases during the following 4 months of storage.

Vigneault and Artes-Hernandez (2007) reviewed the literature on gas treatments for increasing the phytochemical content of fruits and vegetables and concluded that much more research is needed before guidelines for choosing gas treatments to enhance the phytochemical content of fresh fruits and vegetables can be developed.

6.5 Summary

Fresh fruits and vegetables play a very significant role in human nutrition and health, especially as sources of vitamins (vitamin C, vitamin A, vitamin B_6, thiamine, niacin), minerals, and dietary fiber. Other constituents that may lower risk of cancer and other diseases include flavonoids, carotenoids, polyphenols, and other phytonutrients. Postharvest losses in nutritional quality, particularly vitamin C content, can be substantial and are enhanced by physical damage, extended storage duration, high temperatures, low relative humidity, and chilling injury of chilling-sensitive commodities. Responses to atmospheric modification vary greatly among plant species, organ type and developmental stage, and duration and temperature of exposure. Maintaining the optimal ranges of oxygen, carbon dioxide, and ethylene concentrations around the commodity extends its postharvest life by about 50%–100% relative to air control, including its nutritional quality.

In general, low O_2 (2–4 kPa) atmospheres reduce losses of AA in fresh intact and fresh-cut fruits and vegetables. Elevated CO_2 atmospheres up to 10 kPa also reduce AA losses in commodities that tolerate such levels of carbon dioxide without physiological damage, but higher CO_2 concentrations can accelerate these losses in most commodities. Also, elevated carbon dioxide atmospheres reduce the rate of postharvest synthesis of phenolic compounds in fruits and vegetables, which can be desirable (no increase in browning potential and astringency) or undesirable (no increase in antioxidant activity).

6.6 Future Research Needs

The evaluation of the impact of atmospheric modification on quality of fruits and vegetables and their products should include flavor and nutritional quality. Changes in nutritional quality should cover both the essential nutrients (listed in Table 6.1) and the phytochemicals (listed in Table 6.2) that are present in the commodity. Since consumption is influenced by flavor acceptability, it is important to determine postharvest life based on flavor and include such information in any recommendation for use of modified atmospheres to maintain quality of fresh fruits and vegetables and their fresh-cut products.

References

Agar, I.T., R. Massantini, B. Hess-Pierce, and A.A. Kader. 1999. Postharvest CO_2 and ethylene production and quality maintenance of fresh-cut kiwifruit slices. *J. Food Sci.* 64:433–440.

Agar, I.T. and J. Streif. 1996. Effect of high CO_2 and controlled atmosphere (CA) storage on the fruit quality of raspberry. *Gartenbauwissenschaft* 61:261–267.

Agar, I.T., J. Streif, and F. Bangerth. 1997. Effect of high CO_2 and controlled atmosphere on the ascorbic and dehydroascorbic acid content of some berry fruits. *Postharv. Biol. Technol.* 11:47–55.

Allende, A., A. Marin, B. Buendia, F. Tomas-Barberan, and M.I. Gil. 2007. Impact of combined postharvest treatments (UV-C light, gaseous O_3, superatmospheric O_2 and high CO_2) on health promoting compounds and shelf-life of strawberries. *Postharv. Biol. Technol.* 46:201–211.

Bangerth, F. 1977. The effect of different partial pressures of CO_2, C_2H_4, and O_2 in the storage atmosphere on the ascorbic acid content of fruits and vegetables. *Qual. Plant.* 27:125–133.

Barker, J. and L.W. Mapson. 1952. The ascorbic acid content of potato tubers. III. The influence of storage in nitrogen, air and pure oxygen. *New Phytol.* 51:90–115.

Barth, M.M., E.L. Kerbel, S. Broussard, and S.J. Schmidt. 1993a. Modified atmosphere packaging protects market quality in broccoli spears under ambient temperature storage. *J. Food Sci.* 58:1070–1072.

Barth, M.M., E.L. Kerbel, A.K. Perry, and S.J. Schmidt. 1993b. Modified atmosphere packaging affects ascorbic acid, enzyme activity and market quality of broccoli. *J. Food Sci.* 58:140–143.

Barth, M.M. and H. Zhuang. 1996. Packaging design affects antioxidant vitamin retention and quality of broccoli florets during postharvest storage. *Postharv. Biol. Technol.* 9:141–150.

Beuscher, R., L. Howard, and P. Dexter. 1999. Postharvest enhancement of fruits and vegetables for improved human health. *HortScience* 34:1167–1170.

Curry, E.A. 1997. Effect of postharvest handling and storage on apple nutritional status using antioxidants as a model. *HortTechnology* 7:240–243.

Delaporte, N. 1971. Effect of oxygen content of atmosphere on ascorbic acid content of apple during controlled atmosphere storage. *Lebens. Wissen. Technol.* 4:106–112.

Gil, M.I., F. Ferreres, and F.A. Tomas-Barberan. 1998. Effect of modified atmosphere packaging on the flavonoids and vitamin C content of minimally processed Swiss chard (*Beta vulgaris* Subspecies *cycla*). *J. Agric. Food Chem.* 46:2007–2012.

Gil, M.I., F. Ferreres, and F.A. Tomas-Barberan. 1999. Effect of postharvest storage and processing on the antioxidant constituents (flavonoids and vitamin C) of fresh-cut spinach. *J. Agric. Food Chem.* 47:2213–2217.

Gil, M.I., D.M. Holcroft, and A.A. Kader. 1997. Changes in strawberry anthocyanins and other polyphenols in response to carbon dioxide treatments. *J. Agric. Food Chem.* 45:1662–1667.

Gil, M.I., J.A. Tudela, J.G. Marin, and F. Artes. 1995. Effects of high CO_2 and low O_2 on colour and pigmentation of MAP stored pomegranates (*Punica granatum* L.). In C. Garcia-Viguera et al. (eds.), *Current Trends in Fruit and Vegetable Phytochemistry*. CSIC, Madrid, Spain.

Holcroft, D.M., M.I. Gil, and A.A. Kader. 1998. Effect of carbon dioxide on anthocyanins, phenylalanine ammonia lyase and glucosyltransferase in the arils of stored pomegranates. *J. Amer. Soc. Hort. Sci.* 123:136–140.

Holcroft, D.M. and A.A. Kader. 1999. Carbon dioxide-induced changes in color and anthocyanin synthesis of stored strawberry fruit. *HortScience* 34:1244–1248.

Howard, L.R. and C. Hernandez-Brenes. 1998. Antioxidant content and market quality of jalapeno pepper rings as affected by minimal processing and modified atmosphere packaging. *J. Food Qual.* 21:317–327.

Howard, L.R., R.T. Smith, A.B. Wagner, B. Villalon, and E.E. Burns. 1994. Provitamin A and ascorbic acid content of fresh pepper cultivars (*Capsicum annum*) and processed jalapenos. *J. Food Sci.* 59:362–365.

Kader, A.A., P. Perkins-Veazie, and G.E. Lester. 2004. Nutritional quality of fruits, nuts and vegetables and their importance in human health. U.S. Dept. Agric., Agric. Handbook 66. http://www.ba.ars.usda.gov/hb66/index.html

Lee, S.K. and A.A. Kader. 2000. Preharvest and postharvest factors influencing vitamin C content of horticultural crops. *Postharv. Biol. Technol.* 20:207–220.

Lin, T.Y., P.E. Koehler, and R.L. Shewfelt. 1989. Stability of anthocyanin in the skin of 'Starkrimson' apples stored unpackaged, under heat shrinkable wrap and in-package modified atmosphere. *J. Food Sci.* 54:405–407.

Mehlhorn, H. 1990. Ethylene-promoted ascorbate peroxidase activity protects plants against hydrogen peroxide, ozone, and paraquat. *Plant Cell Environ.* 13:971–976.

Paradis, C., F. Castaigne, T. Desrosiers, J. Fortin, N. Rodrigue, and C. Willemot. 1996. Sensory, nutrient and chlorophyll changes in broccoli florets during controlled atmosphere storage. *J. Food Qual.* 19:303–316.

Veltman, R.H., M.G. Sanders, S.T. Persijn, H.W. Peppelenbos, and J. Oosterhaven. 1999. Decreased ascorbic acid levels and brown core development in pears (*Pyrus communis* L. cv. Conference). *Physiol. Plant.* 107:39–45.

Vigneault, C. and F. Artes-Hernandez. 2007. Gas treatments for increasing the phytochemical content of fruits and vegetables. *Stewart Postharvest Rev.* 3:8. www.stewartpostharvest.com

Wang, C.Y. 1977. Effects of CO_2 treatment on storage and shelf-life of sweet pepper. *J. Amer. Soc. Hort. Sci.* 102:808–812.

Wang, C.Y. 1979. Effect of short-term high CO_2 treatment on the market quality of stored broccoli. *J. Food Sci.* 44:1478–1482.

Wang, C.Y. 1983. Postharvest responses of Chinese cabbage to high CO_2 treatment or low O_2 storage. *J. Amer. Soc. Hort. Sci.* 108:125–129.

Wang, C.Y. 1990. Physiological and biochemical effects of controlled atmosphere on fruits and vegetables. In: Calderon, M. and R. Barkai-Golan (eds.), *Food Preservation by Modified Atmospheres*. CRC Press, Boca Raton, FL, pp. 197–223.

Weichmann, J. 1986. The effect of controlled-atmosphere storage on the sensory and nutritional quality of fruits and vegetables. *Hort. Rev.* 8:101–127.

Wright, K.P. and A.A. Kader. 1997a. Effect of slicing and controlled-atmosphere storage on the ascorbate content and quality of strawberries and persimmons. *Postharv. Biol. Technol.* 10:39–48.

Wright, K.P. and A.A. Kader. 1997b. Effect of controlled-atmosphere storage on the quality and carotenoid content of sliced persimmons and peaches. *Postharv. Biol. Technol.* 10:89–97.

7

Effects on Flavor

Charles F. Forney, James P. Mattheis, and Elizabeth A. Baldwin

CONTENTS

7.1   Introduction

The flavor of fresh fruits and vegetables is an important factor in determining quality and consumer satisfaction. However, there is often dissatisfaction among consumers concerning the flavor of fruits and vegetables. First-time purchases are usually based on appearance and firmness, but repeat buys are dependent on internal quality traits such as texture and flavor (Baldwin, 2002). The loss or lack of flavor following postharvest handling, storage, and marketing of fresh fruits and vegetables often precedes the loss of visual quality, which has been reported for both whole and fresh-cut fruits and vegetables (Gorny, 2005). Ensuring good flavor is critical to encourage increased consumption, which is important for human health and well being as well as strengthening and expanding markets for the horticultural industry. Postharvest flavor change can be affected by many preharvest as well as postharvest factors. However, the impact of postharvest technologies on product flavor is not always appreciated nor are the biochemical and genetic mechanism regulating flavor understood.

Atmosphere modification through traditional controlled atmosphere (CA) storage, modified atmosphere packaging (MAP), or the application of edible coatings can impact the flavor of fresh produce in both positive and negative ways. Altered atmosphere composition can affect metabolic changes in flavor compounds in fresh produce during storage or marketing. In addition, packaging can affect diffusional loss of flavor compounds. In this chapter, we will first discuss what determines flavor of fresh fruits and vegetables and then how CA, MAP, and coating technologies influence postharvest flavor change.

7.2   Flavor of Fruits and Vegetables

The flavor of fruits and vegetables is dependent on human perception of a complex combination of volatile, nonvolatile, and structural components contributing to appearance, aroma, taste, and texture (Drewnowski, 1997). Of these components, volatile compounds are primarily responsible for the unique flavor characteristics that distinguish different fruits and vegetables and determine their desirability to the consumer. Volatile aroma compounds can be detected in parts per billion by olfactory nerve endings in the olfactory epithelium (DeRovira, 1997; Holley, 2006). These receptors send signals to the brain, which processes the information to give an integrated flavor experience or judgment (Baldwin, 2004). Furthermore, smelling an aromatic food through the nose (orthonasal) may produce a different experience than when the aroma is perceived during chewing of food as the aroma goes to the olfactory bulb via the back of the throat (retronasal) (Voirol and Daget, 1987). Nonvolatile compounds contribute to the taste of the product and can influence the perception of volatile compounds (Salle, 2006). Taste of nonvolatile compounds is detected by several types of receptors on the tongue, in parts per hundred, such as polyalcohols (sugars), hydronium ions (acids), sodium ions, glucosides, and alkaloids. These correspond to the perception of sweet, sour, salty, and bitter tastes in food. Temperature, viscosity, and polarity of the food can affect relative vapor pressure and aroma release in the mouth (Land, 1994; Taylor and Linforth, 1994). In addition, changes in product texture can also effect the olfactory perception of volatile compounds in a food product (Lubbers, 2006). Texture can influence flavor in that mealiness affects juiciness, which can affect flavor perception (Baldwin, 2002). Texture is perceived through tactile and auditory senses, which is integrated with appearance, taste, and olfactory response to determine perceived flavor.

7.2.1 Sensory Assessment

Human perception of flavor is complex and difficult to measure sensorily. Chemical compounds responsible for flavor can be analyzed, which generates a lot of data on sugars, acids, sugar/acid ratios, and especially aroma volatiles. However, all this data means little for flavor quality unless it is related to sensory perception. There are a variety of sensory tests designed to answer different questions. Questions concerning sensory input include basic information about flavor, differences between samples or products, magnitude of differences, and consumer preference or the "likeability" of a product. This decision-making process runs from the objective to the subjective (Baldwin and Plotto, 2007).

For fundamental flavor information, a panel is trained on a food, determining the different flavor aspects, and providing descriptors for them. The panel is then trained to objectively rate the intensity of these descriptors, and in this way, acts like an instrument. Generally a trained panel should have at least 10–12 panelists, and for horticultural crops, typically is used to understand what comprises flavor, texture, and aroma using a 15 mm line scale (Meilgaard et al., 1999; O'Mahony, 1995). Overall difference tests, such as the triangle and duo-trio tests, are used to determine whether a general difference is perceived between two samples. The triangle test is widely used, but it is not appropriate for heterogeneous samples such as fruits and vegetables (Aust et al., 1985; Baldwin and Plotto, 2007). The duo-trio test is easier to perform than the triangle test because a reference sample is provided with two of the coded samples, one of which matches the reference sample. The simplest test is the difference-from-control test when it is important to know the size of the difference between samples for a particular attribute (Baldwin and Plotto, 2007). These types of panels should have at least 20 panelists for detection of differences and at least 50 for detection of similarities (Baldwin and Plotto, 2007; Meilgaard et al., 1999). Consumer panels should be quite large and untrained, although the panelists should like to eat the type of food they are sampling. Generally, consumer panels should consist of at least 30 panelists for an "experienced" panel, where the panelists often rate the type of food to be paneled, or 50–100 for inexperienced consumers. These panels can rate "likeability" of an attribute, "preference" of one product over another, or rank samples for a particular attribute, like sweetness. In the latter test, a limit of six samples is recommended (Baldwin and Plotto, 2007), since the panelist's flavor discriminating ability can become overwhelmed with large numbers of samples, and 16–30 panelists are acceptable (Baldwin and Plotto, 2007; Meilgaard et al., 1999). Results from consumer panels can vary depending on socioeconomic, ethnic, and geographical background. This necessitates the segmenting of subpopulations for a particular study (Baldwin, 2002; O'Mahony, 1995).

For the difference-from-control test, it is assumed that panelists are at least familiar with the scale and the attribute of interest. Often a scale of 1–10 is used. On the other hand, ranking tests, which are used to determine the difference between samples for one attribute (e.g., flavor, sweetness, off-flavor, preference), do not quantify the difference. The simple ranking test is well adapted for three to six samples (Baldwin and Plotto, 2007). For consumer panels determining likeability or acceptance, often a nine-point hedonic or "just right" scale is used. Sometimes a simple three-point scale is used, for example, denoting an attribute to be outstanding, acceptable, or unacceptable for a tomato (*Lycopersicon esculentum*, Mill) evaluation (Baldwin et al., 1995a). Even a two-point scale can be employed as in the case of one study where adaptation of logistic regression from medical science proved useful (the mouse lived or died in response to a treatment gave a 1 or 0, respectively). In this case, a 0 or 1 indicated whether the consumer would or would not purchase a mango (*Mangifera indica* L.) based on flavor, which was then related to flavor chemical constituents (Malundo et al., 2001b).

Sensory studies for fruits and vegetables can be used to identify optimal harvest maturity (Tandon et al., 2000), evaluate flavor quality in breeding material (Baldwin et al., 1998), determine optimal storage (Maul et al., 2000) and handling conditions (Bai et al., 2003b; Bett et al., 2000; Hagenmaier, 2002), assess effects of disinfestation or preconditioning techniques on flavor quality (Bai et al., 2004; Plotto et al., 2006), and measure flavor quality over the postharvest life of the product (Baldwin, 2002).

7.2.2 Chemical Flavor Constituents

In addition to sensory analysis, the chemical constituents that impact flavor can be characterized through various analytical methods. Volatile compounds can be analyzed using various gas chromatography methodologies. Using these methods, more than 6900 different volatile compounds have been identified in food products (Misry et al., 1997). These compounds represent a diverse range of chemistries including esters, terpenes, alcohols, aldehydes, ketones, lactones, and sulfur compounds. Each commodity has a unique profile of compounds, which can vary within a species depending on maturity and cultivar (Baldwin, 2004). In addition to these primary naturally occurring volatiles present in fruits and vegetables, additional secondary volatiles may be produced as a result of tissue disruption. Rupturing of cells, caused by bruising, cutting, chewing, etc., results in the mixing of enzymes and substrates and the production of a variety of compounds, many of which contribute to flavor (Beaulieu and Baldwin, 2002). A prime example is onions (*Allium cepa* L.), which produce a variety of reactive organosulfur flavor compounds when S-alk(en)yl cysteine sulfoxides are hydrolyzed by the enzyme alliinase upon cellular disruption (Järvenpää et al., 1998; Randle and Lancaster, 2002). The diverse chemistry and metabolic pathways involved in the synthesis of both primary and secondary flavor compounds create an ever-changing flavor profile that makes the understanding of aroma volatile chemistry complex. Aspects of the metabolic processes involved in volatile bio-synthesis have been reviewed (Baldwin et al., 2000; Beaulieu and Baldwin, 2002; Fellman et al., 2000; Sanz et al., 1997).

In addition to the volatile components, a wide range of nonvolatile compounds also contribute to the flavor of fresh produce. The most prominent of these are sugars and acids. Sugars and sugar alcohols provide the sweet taste in most fruits and some vegetables. Sugars commonly found are glucose, fructose, and sucrose (Beaulieu and Baldwin, 2002). In some fruit and vegetables, sugar alcohols such as sorbitol and mannitol contribute to sweetness. In cherries (*Prunus avium* L.), for example, up to 10% of the total sugars, depending on cultivar, is sorbitol (Usenik et al., 2008). While in celery (*Apium graveolens* L.) stalks, mannitol comprises about a third of the total carbohydrate

(Rupérez and Toledano, 2003). Organic acids contribute sourness and their composition varies among fruits and vegetables. Common organic acids found include citric in citrus (*Citrus* sp.) and tomatoes, malic in apples (*Malus sylvestris* (L.) var. *domestica* (Borkh.) Mansf.), tartaric in grapes (*Vitis vinifera* L.), and quinic in cranberry (*Vaccinium macrocarpon* Ait.) (Baldwin, 2004). Other minor components also can impact flavor such as polyphenols in which (+)-catechin and (−)-epicatechin as well as polymeric flavan-3-ols impart bitterness (Serra Bonvehí and Ventura Coll, 1997).

7.2.3 Linking Sensory and Chemical Analysis

7.2.3.1 Gas Chromatography-Olfactometry

Determining the contribution of individual volatile compounds to the flavor of fresh fruits and vegetables requires sensory evaluation of individual compounds to determine their odor activity, which has been achieved by gas chromatography-olfactometry (GC-O) (van Ruth, 2001). GC-O combines chemical and sensory analyses. This method employs a GC with a split column with some flow going to the chemical detector and some to a sniff port and a "human" sensor. The panelist at the sniff port determines which peaks from the chemical detector have odor activity and can also rank their intensity. Using this method, only 30–40 of the more than 400 volatile compounds produced by apple fruit were found to contribute significantly to flavor (Cunningham et al., 1986; Dirinck et al., 1989). Descriptive terms can be assigned to the respective peaks on the chromatogram that have odor activity (Acree, 1993). However, the drawback to this method is that the interactive effects of the volatile compounds with each other and with sugars and acids, both chemically and in terms of human perception, cannot be determined (Baldwin, 2004). GC-O has evolved into three basic techniques. In the Osme method (McDaniel et al., 1990), samples are evaluated by three or four trained panelists, who rate odor-active compounds on a 15-point scale. This method has been used to quantify changes in odor-active volatiles for fresh produce during storage (Da Silva et al., 1993; Plotto et al., 2000). In CharmAnalysis or aroma extract dilution analysis (AEDA), one or two panelists evaluate samples in successive dilutions until no aroma is perceived to determine compound potency and threshold in air (Ferreira et al., 2002). In the frequency method, six to nine panelists indicate when an odor-active compound is present during a sniffing run, which takes advantage of a larger panel to reduce bias due to panelist differences (Abbott et al., 1993; Acree and Barnard, 1994; Marin et al., 1988). However, this method is less precise because information about compound potency is lost.

7.2.3.2 Statistical Analyses

Statistics can be used to relate sensory attributes, preferences, and intensity to chemical components in foods (Bett, 1993; Martens et al., 1994). Once chemical and sensory analyses have been conducted, the data can be compared using simple correlations (Baldwin et al., 1998; Guadagni et al., 1966), linear or multiple regression (stepwise forward or backward), logistic regression (Baldwin et al., 1998; Malundo et al., 2001a), or multiple regression modeling (Abegaz et al., 2004). Multivariate statistics, such as principal component analysis (PCA), is used to create a perceptual map with each principal component representing a linear combination of covarying attributes that hopefully explain much of the variation in each dimension. Multiple factor and discriminant analyses are also used as well as partial least square (Baldwin and Plotto, 2007; Martens and Martens, 1986). Correlations between instrumental flavor data and a trained panel were used for tomato and mango flavor analysis (Baldwin et al., 1998, 2004; Malundo et al., 2001b), logistic regression was used to relate instrumental to consumer panel data for mango (Malundo et al., 2001a), and PCA was used to relate spiked aroma volatile compounds and their levels to trained panel data in tomato (Baldwin et al., 2004).

7.2.4 Prestorage Factors Affecting Flavor

To understand flavor changes that occur in fresh fruits and vegetables when submitted to CA storage or MAP, one must appreciate the many different factors that can influence product flavor. In addition to the effects of CA storage, MAP, or the application of coatings, primary factors that affect fresh fruit and vegetable flavor include genetic, preharvest environment, harvest maturity, and postharvest treatments. The complex interactions of these factors will influence how the product's flavor responds to atmosphere modification during storage and/or marketing and the ultimate flavor experienced by the consumer. Therefore, the role of each of these factors must be understood and properly managed to optimize product flavor.

7.2.4.1 Genetics

The genetic makeup of the crop will determine its potential flavor when it reaches harvest maturity. As discussed above, flavor is the integrated perception of many chemical components in the plant tissue. The expression of each of these components is under genetic control. Therefore, genetic diversity within a crop can result in a diverse range of flavors among cultivars or wild genomes within a species. For example, when fruit from 28 sweet cherry (*Prunus avium* L.) cultivars and selections were evaluated by a sensory panel, substantial differences in sweetness, sourness, and cherry flavor were identified (Dever et al., 1996). Similarly, sensory panels found large flavor differences among different apple (Stebbins et al., 1991) and strawberry (*Fragaria* × *ananassa* Duch.) (Podoski et al., 1997) cultivars. Cultivar differences in flavor can be correlated to differences in chemical composition (Baldwin et al., 1991). Genetic differences among cultivars can also affect their response to postharvest environments including atmosphere modification. While genetics plays a large role in determining the flavor of freshly harvested fruits and vegetables, genetic expression is developmentally regulated and can be modified by many pre- and postharvest environmental factors.

7.2.4.2 Preharvest Environment

The production environment can have a large effect on the quality and flavor of fruits and vegetables. Many factors such as sunlight, water availability, fertilization, and chemical applications can affect crop growth and development including flavor development. Sulfur fertility of soils as well as other environmental factors can affect the pungency of onions (Randle and Lancaster, 2002). Excessive water prior to harvest can dilute flavor compounds in tomato (Baldwin, 2004). Controlling stress to the plant, by applying miticides, resulted in sweeter and more flavorful strawberries when compared with fruit subjected to mite infestation (Podoski et al., 1997).

7.2.4.3 Maturity

Maturity at harvest of fruit, and to a lesser extent vegetables, determines the postharvest expression and development of flavor. For optimum flavor, fruits and vegetables should be harvested at a point when full flavor has been developed and they are at their best eating quality. However, harvest maturity is often determined to minimize physical damage during shipping and handling in order to optimize shelf life based on appearance and integrity of the product. Therefore, many fruit are normally harvested under-ripe at physiological maturity, which limits their flavor potential in the market.

7.2.4.4 Postharvest Treatments

After harvest, many fruits and vegetables are subjected to various treatments to delay ripening, control decay, or eliminate pests (quarantine requirements) in order to expand market access. Many of these treatments can impact the flavor of the treated product. Treatment of many fruit with 1-methylcyclopropene (1-MCP), an ethylene-action inhibitor, inhibits fruit ripening and the formation of aroma volatiles, particularly esters, which are responsible for much of the fruity flavor (Watkins, 2006). Heat treatments, used to control decay, reduce microbial contamination, or disinfest insects can alter product flavor (Lurie, 1998). Heat treatments can reduce fruit acidity and alter volatile production and texture. Peeling and cutting of fresh fruits and vegetables to produce fresh-cut, ready-to-eat products also can affect flavor (Beaulieu and Baldwin, 2002). Tissue disruption caused by these processes can induce secondary volatile production, invoke stress responses, and enhance flavor loss.

7.3 Dynamics of Postharvest Flavor Change

After harvest, the flavor of fresh fruits and vegetables can change substantially during storage and postharvest handling. The two primary mechanisms of flavor change in harvested fresh produce are metabolic and diffusional (mass transfer) (Voilley and Souchon, 2006). Metabolic changes in flavor can be the result of the synthesis or catabolism of either flavor compounds or compounds responsible for off-flavors. These metabolic processes are dependent on product physiology, which is influenced by maturity and a variety of environmental factors. Diffusional changes in product flavor are a result of diffusion and mass transfer of volatile compounds into and out of the commodity. Diffusional rates are determined by the volatility of the compound, its concentration gradient, and diffusional barriers in the fruit or vegetable or packaging materials. The role of each of these mechanisms depends on the product and the environment in which it is held, which is illustrated in Figure 7.1.

7.3.1 Metabolic Changes

Metabolic processes that affect postharvest flavor change vary among commodities due to the diversity of flavor compounds and physiology among different fruits and vegetables. In a single commodity, many metabolic pathways are involved in producing the complex mixture of compounds responsible for its flavor. Among fruit, climacteric fruit are prime examples of commodities that actively synthesize flavor compounds following harvest, while nonclimacteric fruit have more limited postharvest flavor development. Climacteric fruits, including apples, tomatoes, and bananas (*Musa* sp.), initiate flavor volatile synthesis in association with a burst of respiration and ethylene synthesis, which can occur pre- or postharvest. When harvested under-ripe, these fruit can be stimulated to ripen and develop flavor by exposing the fruit to the ripening hormone ethylene (Lelièvre et al., 1997). However, most under-ripe fruits are not able to obtain full flavor through post-harvest ripening. Tomato fruit harvested at the immature green stage produced ripe fruit with lower aroma volatile levels than mature green-harvested fruit, while fruit harvested

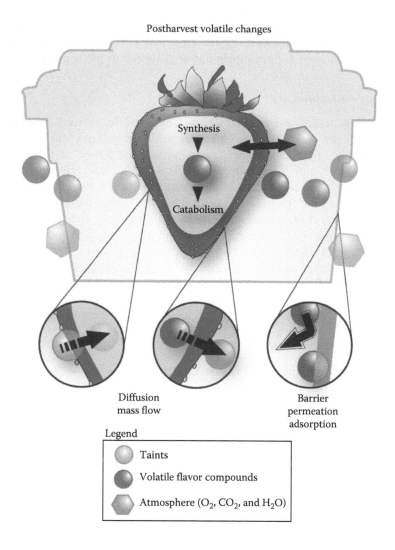

Postharvest volatile changes

Synthesis

Catabolism

Diffusion
mass flow

Barrier
permeation
adsorption

Legend

Taints

Volatile flavor compounds

Atmosphere (O_2, CO_2, and H_2O)

FIGURE 7.1

An illustration of factors affecting the aroma and flavor of whole and fresh-cut fruits and vegetables during storage and marketing in controlled atmosphere storage or modified atmosphere packaging. (Illustration by Natacha Sangalli, AAFC.)

at the table ripe stage had the highest intensity of flavor components (Watada and Aulenbach, 1979). The climacteric stage at which apple fruit are harvested affects ester formation during postharvest ripening (Fellman et al., 2000). 'Golden Delicious' apples harvested during the climacteric rise in respiration had a 2.5-fold greater rate of volatile production than preclimacteric fruit harvested 1 week earlier, but after 18 days at 20°C volatile production of the preclimacteric fruit increased six-fold (Song and Bangerth, 1996). Nonclimacteric fruit, such as blueberries (*Vaccinium corymbosum* L.) or strawberries, demonstrate minimal postharvest flavor development (Forney, 2001; Forney et al., 2000b; Miszczak et al., 1995). Volatile concentrations in juice expressed from 'Valencia' oranges (*Citrus sinensis* L.), a nonclimacteric fruit, do not show a rise in volatiles during fruit storage. Instead, the total concentration of seven volatiles that contribute to orange flavor decreased 12% following 21 days of storage at 21°C (Baldwin et al., 1995b).

Metabolic processes that affect postharvest flavor in fresh produce are influenced by the postharvest environment and duration of storage/marketing. Temperature affects the metabolic activity of the product and directly impacts changes in flavor compound synthesis and catabolism. Therefore, elevated temperatures can speed up both quantitative and qualitative changes in flavor compounds thus affecting flavor. Ester metabolism in stored strawberry fruit was affected by temperature, where ethyl butanoate and ethyl hexanoate increased during storage at $1°C$, while methyl butanoate and methyl hexanoate increased during storage at $15°C$ (Forney et al., 2000b). Postharvest environmental stresses, such as heat, freezing, ozone, radiation, or chemical fumigation, can also induce changes in volatile production or alter metabolism of flavor compounds (Forney and Jordan, 1998; Forney et al., 2000a, 2007; Song et al., 1996, 2001). Physical handling, including cutting of fresh fruits and vegetables to add value and convenience, may also induce metabolic flavor changes. Cutting 'Nairobi' carrot (*Daucus carota* L.) slices with a dull blade enhanced fresh carrot aroma loss, judged by a sensory panel, during the first 5 days of storage at $8°C$ in oriented polypropylene (OPP) bags when compared to razor blade cut slices (Barry-Ryan and O'Beirne, 1998). As mentioned before, cutting of fresh onions induced rapid formation of secondary aroma volatiles (Järvenpää et al., 1998). In addition, atmosphere modification in storage and packaging of fresh product can have significant effects on altering flavor metabolism and will be discussed in the following sections.

7.3.2 Diffusional Changes

7.3.2.1 Flavor Loss

Diffusional loss of flavor volatiles is determined by the volatility of the compound, its partitioning coefficient in the product matrix, diffusional barriers within and surrounding the product, and the volatile concentration surrounding the product, which is determined by packaging and storage conditions (Voilley and Souchon, 2006). While many of these processes have been well documented in processed food products and model food systems (Boelrijk et al., 2006; Bylaite et al., 2003; de Roos, 2006; Juteau-Vigier et al., 2007), little work has been conducted on fresh fruits and vegetables. Unlike processed food products, the properties of fresh products, including permeation rates and volatile concentration, may change due to metabolic and physiological processes, making them more challenging to study.

7.3.2.2 Natural Barriers

Whole fruits and vegetables are surrounded by a cuticle that acts as a natural diffusional barrier. The cuticle is composed of cutin, which is a three-dimensional polymer of C_{16} and C_{18} hydroxy fatty acids cross-linked by ester and other bonds and has permeance values for water similar to many 1 μm synthetic polymer films (Riederer and Schreiber, 2001). For apples, the O_2 diffusivity of the flesh is estimated to be only 10- to 20-fold greater than that of the skin/cuticle, suggesting that concentration gradients could develop within the flesh (Solomos, 1987). Since the diffusivity in this study was determined for O_2, values for flavor volatiles would vary depending on molecular size, polarity, and other chemical properties. Cellular walls and membranes also may pose resistance to volatile movement and need to be considered. For fruits and vegetables that have effective intact natural barriers, the diffusional loss of flavor volatiles should be relatively slow.

Recent development of "fresh-cut" products has resulted in the reduction or elimination of natural barriers to flavor loss, resulting in an increase of flavor volatile diffusion into the package/storage atmosphere. Thinly cut melon (*Cucumis melo* L.) slices stored in Petri plates demonstrated a rapid loss of 50%–80% of their volatile esters after 1 day of storage at either 4°C or 22°C (Beaulieu, 2007; Lamikanra and Richard, 2002). The thin melon slices had a large surface area to volume ratio and large concentration gradient between the product and the air, which enhanced diffusional loss of volatile compounds. Further supporting the diffusional loss of esters was the greater loss of low-molecular-weight esters compared to those of higher molecular weight and the lack of evidence of enzymatic ester breakdown (Beaulieu, 2007). Diffusional loss of volatile compounds from fresh-cut melon with a lower surface-to-volume ratio (i.e., cubes) would be less than the thin slices, however no experiments have been conducted in which volatile loss was quantified from melon cubes in ambient air atmospheres. Aroma of honeydew and cantaloupe pieces was found to decline after 6 days when the cut-fruit was held at 5°C in permeable cheese cloth-covered jars (Portela and Cantwell, 1998, 2001). In other commodities, thin slices of pineapple (*Ananas comosus* (L.) Merr.) held in Petri plates at 4°C also lost 47% of esters and 27% on monoterpenes following 24 h of storage (Lamikanra and Richard, 2004). Sensory tests determined that fresh-cut orange slices stored at 4°C in a closed plastic box that maintained ambient atmospheres lost flavor during 5 days of storage as a result of the loss of flavor components in the fruit (Rocha et al., 1995).

The contribution of diffusional losses of flavor volatiles in fresh-cut fruits and vegetables would vary depending on the commodity, nature of cutting, surface area/volume ratios, and the storage environment. The significance of this flavor loss mechanism is poorly understood, and is difficult to separate from catabolic-driven changes. However, diffusional loss should be considered, especially in the design of MAP for fresh-cut products.

7.3.2.3 *Artificial Barriers*

Diffusional loss can be reduced by the addition of artificial barriers surrounding the product, which can reduce concentration gradients of flavor volatiles between the product and its storage environment. This is normally in the form of packaging or coatings. Most packaging materials provide a good barrier to diffusional loss assuming the package is well sealed and leak free. Total esters in fresh-cut orange-fleshed cantaloupe sealed in polyethylene terephthalate (PET) bowls at 4°C increased as much as 40% after the first 7 days of storage and then decreased. Nonacetate esters increased more than acetate esters, which changed little during the first 7 days (Beaulieu, 2006). These results suggest that the sealed package minimized diffusional volatile loss described in Section 7.3.2.2. A discussion of the properties of packaging materials and how they may interact with flavor volatiles is presented below.

Edible coatings also can act as a barrier to flavor volatile loss and may provide additional options for reducing flavor loss in fresh produce. A discussion of the effects of coating technology is also presented. Use of edible coatings affects flavor by creating a modified atmosphere (MA) similar to MAP. This in turn can affect levels of volatile flavor compounds due to an induced change in volatile synthesis in reaction to the MA, or due to the coating permeability to aroma volatiles.

7.3.2.4 *Taints/Migration*

Volatile compounds from the storage environment or packaging materials can also migrate (diffuse) into the fresh product, resulting in the development of off-odors and flavors. Within a storage room, off-flavors can originate from a variety of sources including chemicals, building materials, packaging, microorganisms (fungi, bacteria), and other produce (Baigrie, 2003). Depending on the chemistry (i.e., the partitioning coefficient in the product matrix) of these volatile compounds in the storage environment, many can migrate into fresh fruits or vegetables. Chloroanisoles, a musty off-flavor that has been found in a variety of foods, can be produced by microbial methylation of chlorophenols, which are commonly found in treated lumber, recycled paper products, and adhesives (Tindale et al., 1989). Recently chloroanisoles were found to cause off-flavors in potatoes (*Solanum tuberosum* L.) (Daniels-Lake et al., 2007). Mixed commodity storage can also lead to cross transfer of odors among incompatible commodities (Hardenburg et al., 1986). For instance, apples and pears (*Pyrus communis* L.) can acquire an earthy taste when stored with potatoes, citrus can develop off-flavors when stored with strong-scented vegetables, and many fruits and vegetables will develop undesirable flavors when stored with onions.

While strong off-flavors are not normally acquired during the proper postharvest handling of fresh produce, fresh products may acquire various taints that contribute to "stored" or "old" flavors. These off-flavors may become more apparent as desirable flavors are lost. The contribution of taints to the loss of desirable flavor during storage has received little attention, but could be an important factor affecting product flavor-life.

7.4 Atmosphere Effects on Flavor

Produce storage in sealed environments where gas composition is actively maintained at nonambient concentrations is known as CA storage. Typically, CA environments have increased CO_2 and reduced O_2 pressures compared to air, and storage in CA can have profound impacts on fruit and vegetable quality including flavor (Isenberg, 1979). As well, CA environments alter humidity and may incorporate other novel gases in the storage atmosphere. Under CA storage conditions that result in decreased rates of ripening and/or senescence, quality loss can be slowed, resulting in an extension of the period of acceptable consumer perception (Bangerth, 1973; Parsons et al., 1970). Although CA can extend the marketable period of many fruits and vegetables compared to storage in air, the patterns of ripening and senescence after removal from CA may be altered. A residual effect of CA on chlorophyll loss, softening, and production of volatile compounds, contributing to flavor has been observed for apple fruit (Patterson et al., 1974). A period of storage in air after removal from CA can partially to completely restore volatile production to amounts typical of fruit stored in air (Kidd and West, 1936; Streif and Bangerth, 1988; Yahia et al., 1990). Storage under CA conditions outside the tolerance range for CO_2 and/or O_2 can accelerate quality loss including the stimulation of off-flavor development (Fidler and North, 1971).

7.4.1 Oxygen

Low-oxygen CA can have an inhibitory impact on production of volatile compounds that contribute to aroma and flavor of many fruit including apple (Griffith and Potter, 1949; Mattheis et al., 1995, 1998; Patterson et al., 1974; Streif and Bangerth, 1988; Yahia, 1991; Yahia et al., 1990, 1991), black currant (*Ribes nigrum* L.) (Harb et al., 2008), kiwifruit (*Actinidia deliciosa* [A. Chev.] C.F. Liang et A.R. Ferguson var. *deliciosa*) (Burdon et al., 2005), pear (Fidler and North, 1969; Lopez et al., 2001; Paillard, 1975; Rizzolo et al., 1991), and strawberry (Almenar et al., 2006; Ke et al., 1994; Perez and Sanz, 2001). The decrease in volatile production following low O_2 CA is enhanced with increased storage duration and decreased O_2 concentration (Lidster et al., 1983; Lopez et al., 2001; Mattheis et al., 1995; Patterson et al., 1974; Rizzolo et al., 1991; Streif and Bangerth, 1988; Willaert et al., 1983; Yahia et al., 1990, 1991). Recovery of volatile production occurs during storage in air after removal from low O_2 (Patterson et al., 1974; Plotto et al., 1999; Streif and Bangerth, 1988; Yahia, 1991; Yahia et al., 1990, 1991), but the amount of recovery decreases with increased storage duration, decreased CA O_2 concentration, and post-CA storage temperature (Hansen et al., 1992; Streif and Bangerth, 1988; Yahia, 1991; Yahia, et al., 1990, 1991). Fruit storage in low O_2 environments delays or reduces production of many volatile compounds including alcohols, aldehydes, esters (Chervin et al., 2000; Echeverria et al., 2004a,b; Fellman et al., 1993; Hatfield and Patterson, 1975; Lidster et al., 1983; Lopez et al., 2001, 2007; Mattheis et al., 1991, 1995; Rizzolo et al., 1991; Yahia, 1991; Yahia et al., 1990, 1991), terpenoids (Aaby et al., 2002; Chervin et al., 2000), ketones (Plotto et al., 2000), and a benzenoid (Mattheis et al., 1998). The impact of low O_2 concentrations can differ across chemical groups as well as within a group. The production of many esters with a straight C-chain alcohol moiety is reduced following low O_2 storage (Hansen et al., 1992; Yahia, 1991; Yahia et al., 1990, 1991) although ethyl- and butyl pentanoate production have been observed to increase following CA (Yahia et al., 1990, 1991).

Ester production during fruit ripening utilizes many different substrates with the major sources being amino acids, carbohydrates, and lipids. The availability of alcohols is a factor limiting fruit ester production. An increase in apple ester emission following exposure to alcohol vapors occurs regardless of the developmental stage at the time of alcohol treatment (Knee and Hatfield, 1981). Both branched and straight C-chain alcohols are utilized for ester production. Branched C-chains are largely the product of amino acid metabolism and straight C-chain alcohols primarily arise from β-oxidation of lipids (Drawert, 1975). Reduced production of esters synthesized using straight C-chain alcohols implies low O_2 storage may impede lipid metabolism (Hansen et al., 1992). Ethanol is a substrate for ester synthesis and can arise from pyruvate via activity of pyruvate decarboxylase (PDC) and alcohol dehydrogenase (ADH). The activity of PDC and ADH increased during low O_2 storage of 'Doyene du Comice' pears (Lara et al., 2003), although enzyme product and activity relationships were not apparent. Lipoxygenase (LOX) activity, which contributes to the production of straight C-chain esters, is reduced following low O_2 storage (Lara et al., 2003, 2007). Alcohol acyltransferase (AAT, EC 2.3.1.84) is the terminal enzyme in the ester synthesis pathway and catalyzes reactions between various alcohols and acyl-CoA to produce esters. AAT activity is present after storage in low O_2, but activity decreases with decreased O_2 concentration (Fellman et al., 1993) and increased storage duration (Lara et al., 2007). AAT activity increases when fruit are held in air after low O_2 storage, but Fellman et al. (1993) found this increase to be transient.

Sensory studies conducted with apple have demonstrated low O_2 CA to have variable impacts on components of fruit flavor and aroma. 'Gala' apples stored 24 weeks in CA with 1.2 to 5 kPa O_2 were more crisp, firm, juicy, and sour, had fewer off-flavors, and greater overall acceptability than fruit stored in air (Cliff et al., 1998). Ratings for "fruity flavor" were highest for apples stored in 1.2 or 2.5 kPa O_2 compared to 5 kPa O_2 or air. Another sensory study evaluating 'Gala' apples found intensity of compounds with fruity and floral aroma descriptors to be lower for fruit stored in CA compared to air over a 20 week period (Plotto et al., 1999). Aroma of cut CA apples retained high vegetative and citrus characters but less anise aroma throughout the 20-week storage period compared to apples stored in air. Sourness and astringency were higher and sweetness lower for CA fruit, although soluble solids content was similar for air and CA fruit after 20 weeks of storage. 'Fuji' apples stored in 1 or 2 kPa O_2 emitted lower amounts of esters compared to fruit stored in air over a 7 month period, but sensory flavor scores were similar when fruit were evaluated 1 day after removal from CA (Echeverria et al., 2004a). Flavor and overall acceptability scores were not consistently different in relation to air, CA O_2 concentration, or days held at 20°C after removal from storage. After 7 months plus 10 days at 20°C, air-stored fruit had the lowest sensory scores for acceptability, firmness, acidity, sweetness, and appearance, but flavor scores for fruit stored in air or 1 kPa O_2 were similar. Apple fruit acceptability is determined by a number of quality attributes and flavor often does not rank as the predominant factor governing consumer preference (Daillant-Spinnler et al., 1996).

Low O_2 conditions that exceed the range of tolerance induce anaerobic metabolism and the accumulation of acetaldehyde and ethanol (Thomas and Fidler, 1933). The availability of these compounds for volatile production results in synthesis of ethyl esters in amounts atypical of normal ripening (Ke et al., 1991, 1994; Larsen and Watkins, 1995; Shaw et al., 1990). Enhanced production of ethyl esters also results in reduced production of other volatile esters (Mattheis et al., 1991). This perturbation of typical ester production can have little or no impact on flavor (Ke et al., 1991) or can generate market limiting off-flavors as documented for apple (Fidler and North, 1971), mango (Bender et al., 2000), strawberry (Ke et al., 1991), and papaya (*Carica papaya* L.) (Yahia et al., 1992). Short-term exposures to nitrogen or low oxygen sufficient to induce anaerobiosis have been shown to enhance flavor of grapefruit (Bruemmer and Roe, 1970), orange (Shaw et al., 1992), apple (Dixon and Hewett, 2001), and feijoa (*Feijoa sellowiana* Berg) (Pesis et al., 1991). As sugar and acid content as well as other compounds contribute to flavor, a simple relationship between ethanol, ethyl acetate, acetaldehyde, and flavor, particularly off-flavor, has not been demonstrated (Ke et al., 1991; Larsen and Watkins, 1995). Loss of acidity during ripening contributes to reduced flavor and taste (Anderson and Penney, 1973; Gorin, 1973), however, low O_2 storage can also reduce acid loss (Kader, 1997; Ulrich, 1966).

High O_2 concentrations have been assessed as a means to regulate ripening and senescence, and to control decay of fruits and vegetables (Day, 1996; Kader and Yehoshua, 2000). Impacts of high O_2 on flavor and aroma can occur via alterations in respiratory metabolism and volatile compound production. Acetaldehyde, ethanol, and ethyl acetate all accumulate in strawberry during storage in 40–100 kPa O_2 to amounts greater than fruit stored in air (Perez and Sanz, 2001; Wszelaki and Mitcham, 2000). Ethanol and ethyl acetate content continued to increase upon return of strawberries to air, perhaps reflecting conversion of acetaldehyde to ethanol and utilization of ethanol for ethyl acetate synthesis (Wszelaki and Mitcham, 2000). The presence of anaerobic products of high O_2 storage also resulted in the diminished production of other esters by strawberries (Perez and Sanz, 2001). While high O_2 exposure reduced strawberry sensory quality (Wszelaki and Mitcham, 2000), aroma and flavor ratings for table grapes held in high O_2 increased relative to fruit stored in air (Deng et al., 2005), possibly reflecting differences in compounds responsible for characteristic flavors for different species. Work with apples exposed to 100 kPa O_2 after storage in 3 kPa O_2 + 3 kPa CO_2 (Yahia, 1991) and blueberries packaged in 40 kPa O_2 (Rosenfeld et al., 1999) did not result in changes to volatile production or sensory quality compared to fruit held in air.

7.4.2 Carbon Dioxide

Elevated CO_2 can also impact volatile production and flavor. During CA storage of 'Golden Delicious' apple, the impact of CO_2 on volatile production is dependent in part on the CA O_2 concentration (Streif and Bangerth, 1988). With 3 kPa or higher O_2, high CO_2 concentrations result in reduced fruit volatile production, and the CO_2 effect increases with increased storage duration. The CO_2 effect on volatile production following CA storage has been suggested to result in part from diminished carboxylic acid metabolism and ADH activity (De Pooter et al., 1987). Production of branched C-chain esters decreases with increased CO_2 content in CA at a constant O_2 concentration (Brackmann et al., 1993), suggesting amino acid metabolism is altered by high CO_2.

A sensory study of a limited (1 or 3 kPa) CO_2 range with 1 kPA O_2 CA storage of 'Fuji' apples indicated fruit acceptability scores for all CA treatments were similar through 7 months storage plus up to 10 days at 20°C (Echeverria et al., 2004a). However, sensory flavor scores were higher for fruit stored in 3 kPa CO_2 after 3, 5, or 7 months plus 10 days in air at 20°C. The emission of volatile compounds detected was highest throughout the study for fruit stored in 3 kPa O_2 although no qualitative results were presented. The results with 'Fuji' and 'Golden Delicious' may indicate cultivar differences in response to CO_2 during CA storage.

High CO_2 storage environments can result in accumulation of fermentative products that can alter flavor. Dark fixation of CO_2 to malic acid can be enhanced in a high CO_2 storage environment (Pesis and Ben-Arie, 1986). Malic acid can be converted to acetaldehyde (Clijsters, 1965; Pesis and Ben-Arie, 1986) and subsequently to ethanol via ADH. Strawberries stored in 20–80 kPa CO_2 in 21 kPa O_2 accumulate more acetaldehyde and ethanol than fruit stored in 100 kPa N_2 (Ke et al., 1991). Strawberries held in 50 or 80 kPa CO_2 developed off-flavors when stored for 8 or 10 days at 0°C or 5°C, but fruit stored in 20 kPa had no off-flavors (Ke et al., 1991). Off-flavors in strawberries held in 20 kPa CO_2 at 0°C dissipated during holding in air (Larsen and Watkins, 1995). The decrease in off-flavor was associated with an increase in ethyl butanoate rather than a decrease in ethyl acetate.

High CO_2 treatments can enhance or alter flavor of a number of other fruit. High CO_2 exposure of 'Hamlin' oranges enhanced content of acetaldehyde and ethanol as well as several ethyl esters known to contribute to orange flavor (Shaw et al., 1992). Cultivar and maturity were demonstrated to be factors influencing table grape response to high CO_2. Off-flavors developed in 'Redglobe' table grapes held at 10 or 15 kPa CO_2 with 3–12 kPa O_2 (Crisosto et al., 2002a), however, no off-flavors developed in 'Thompson Seedless' grapes harvested at a late maturity and stored in 15 kPa CO_2 with 3–12 kPa O_2 (Crisosto et al., 2002b). High CO_2 treatments are the basis of carbonic maceration of red vinifera grapes prior to crushing. The process, whereby grapes are held in 99–100 kPa CO_2, results in increased production of many volatile compounds as well as free amino acids (Dourtoglou et al., 1994). Astringency in persimmon (*Diospyros kaki* L.) can be eliminated by storage in high CO_2 (Gore and Fairchild, 1911). This treatment results in a significant decrease in soluble tannins without deleterious effects on softening or other aspects of fruit quality (Matsuo and Ito, 1977).

7.4.3 Humidity

Storage in environments with high humidity reduces mass loss and can delay ripening, and senescence, as well as maintain fruit turgidity (Paull, 1999). Apple volatile emission is altered differentially by storage humidity (Wills and McGlasson, 1970). Emission of butyl-, hexyl-, and isopentyl acetates increased as humidity decreased, while emission of butanol, hexanol, and isopentanol increased in high humidity. No difference in sweet cherry flavor was observed when fruit were held in 90%–94% RH compared to 95%–99% RH (Sharkey and Peggie, 1984). Flavor ratings for peach (*Prunus persica* (L.) Batsch) fruit decreased when fruit were stored in 95%–99% RH perhaps due in part to increased woolliness occurring in the same environment (Sharkey and Peggie, 1984). 'LeLectier' pears stored in 55% or 75% RH at 0°C prevented ripening and typical flavor development compared to fruit stored in 95% RH (Murayama et al., 1995).

7.4.4 Ethylene

Ethylene, a two-carbon alkene gas, is naturally produced by plant tissues and modulates many physiological processes including ripening, senescence, and responses to environmental stress (Abeles, 1992; Lelièvre et al., 1997). Climacteric fruit ripening requires ethylene production and action, and endogenous ethylene has been suggested to act as a rheostat for ethylene-dependent processes (Theologis, 1992). Processes that limit ethylene production and/or ethylene action slow ripening (Burg and Burg, 1969) and form the basis for commercial management of climacteric fruit (Watkins, 2002). Ethylene also influences ripening and senescence of nonclimacteric fruit and vegetables but typically the effects are deleterious to product quality (Saltveit, 1999). Endogenous as well as exogenous sources of ethylene promote ripening and senescence, and commercial strategies to limit or utilize product ethylene exposure, which includes CA storage, are important tools for postharvest quality management (Saltveit, 1999; Watkins, 2002).

In general, climacteric fruit flavor and taste are enhanced during ripening due in part to starch conversion to sugars, increased acid utilization, and production of volatile compounds that contribute to aroma and flavor (Watada, 1986). The progression of climacteric fruit ripening and flavor development is influenced by fruit maturity at harvest. Fruit harvested physiologically mature but unripe ripen in response to ethylene (Watada et al., 1984) but do not develop flavor and aroma with the intensity of fruit harvested at a later stage of maturity. Tomato fruit harvested over a range of mature but unripe stages then ripened were less sweet, more sour, and had less typical tomato flavor and more off-flavors compared to fruit harvested table ripe (Kader et al., 1977). Volatile production by 'Golden Delicious' apples is closely related to dynamics of respiratory rate and ethylene production (Song and Bangerth, 1996). For apples harvested 4 weeks prior to the climacteric rise, respiration rate, ethylene production, and total synthesis of compounds contributing to aroma were lower during ripening than that for fruit harvested at the climacteric onset. The production of straight- and branched C-chain esters was impacted similarly by harvest maturity with the exception of ethyl-2-methylbutanoate, which increased in the later stage of ripening. The duration between harvest and maximum volatile production also decreased as harvest approached the optimum date.

Stimulation of ripening by ethylene of fruit picked mature but unripe is commercially practiced for a number of fruit including avocado (*Persea americana* Mill.), banana, pear, and tomato (Gross et al., 2004). The postharvest life of these fruit is greatly extended by harvest prior to the tree-ripe stage as unripe fruit withstands the rigors of the commercial harvest, handling, storage, and distribution systems much better than tree-ripe fruit. Exposure to exogenous ethylene prior to distribution enhances quality by stimulating ripening including softening, color change, and production of volatile compounds. Ethylene treatment does not, however, completely overcome the impact of harvest maturity on flavor development. Tomatoes picked mature green and stimulated to ripen with ethylene had higher volatile emission than fruit not exposed to ethylene, but emission was less than that of vine-ripened fruit (Stern et al., 1994). Ethylene production and ripening of 'd'Anjou' pears are induced by exposure to low temperature (Blankenship and Richardson, 1985). Unchilled 'd'Anjou' pears exposed to ethylene soften and develop buttery and juicy texture, however, ripened fruit lack high flavor compared to fruit that have been stored at −1°C for 30 days (Ma et al., 2000).

Ethylene mediates responses to stress and can stimulate accumulation of phytoalexins that enhance resistance to pathogen attack (Niskanen and Dris, 2006). Some of these compounds contribute to flavor deterioration in vegetables by imparting bitterness. Phenolic phytoalexins in carrot and parsnip roots accumulate in response to ethylene (Sarkar and Phan, 1974; Seljasen et al., 2001; Shattuck et al., 1988). In carrot, isocoumarin accumulation correlates with the degree to which respiration increased in response to ethylene, carrot developmental stage, and storage history (Lafuente et al., 1996). Isocoumarin accumulation is inversely related to maturity, with more accumulation occurring in immature carrots in response to ethylene. Isocoumarin accumulation also decreased when carrots were stored (5°C for 30 days) prior to ethylene exposure. Bitterness can develop in cabbage during storage in air or CA (Toivonen et al., 1982). Ethylene removal during storage reduced or eliminated bitterness for cabbage stored in air or CA, respectively.

Ethylene production and action are reduced by low O_2 and/or high CO_2 generated as a CA (Gorny and Kader, 1997) or under hypobaric conditions (Burg and Burg, 1969). The lack of ethylene action induced by CA or hypobaric conditions reduces ripening during storage as well as after removal from storage (Bangerth, 1988). The residual effect on production of volatiles contributing to aroma, particularly those associated with ripening, can contribute to decreased flavor compared to fruit stored in air (Plotto et al., 1999). Post-CA volatile production can be enhanced by storage under conditions conducive to ethylene production. Production of volatile compounds by 'McIntosh' apples previously stored in CA occurs more readily when fruit are stored in air at warm (20°C) compared to cold (3.3°C) temperature (Yahia, 1991). Ethylene treatment of fruit previously stored in CA can also enhance volatile production. Production of esters and alcohols by 'Gala' apples previously stored in CA was enhanced by ethylene treatment after removal from CA (Mattheis et al., 2005). The response to ethylene decreased with storage duration for esters and alcohols but increased for the volatile phenylpropanoid 1-methoxy-4-(2-propenyl)benzene, which imparts a spicy, anise aroma to apple fruit (Williams et al., 1977).

7.4.5 1-Methylcyclopropene

Research characterizing ethylene receptor biology resulted in identification of 1-MCP as a potent inhibitor of ethylene action (Sisler and Blankenship, 1996; Sisler and Serek, 1997). Similar to ethylene, 1-MCP is a low-molecular-weight unsaturated hydrocarbon gas that readily diffuses through plant tissues. Plant tissues exposed to 1-MCP lose the capacity to respond to ethylene possibly due to a direct interaction between 1-MCP and ethylene receptors (Sisler and Serek, 1997, 2003). The nontoxic chemical properties of 1-MCP allowed its commercialization for food crops in 2002 (Blankenship and Dole, 2003).

Perception and modulation of the ethylene signal is required at all stages of climacteric fruit ripening (Hoebrichts et al., 2002), and 1-MCP has proven to be a potent ripening inhibitor (Watkins, 2006). Climacteric fruit exposed to 1-MCP exhibit reduced respiration rate and ethylene production as well as a diminution or inhibition of many ripening processes including acid loss, color change, softening, and volatile production (Blankenship and Dole, 2003). The duration and magnitude of 1-MCP responses are dependent on species, cultivar, stage of development, 1-MCP treatment conditions, and concentration, and post-treatment storage conditions and duration (Watkins, 2006). Exogenous ethylene treatments to stimulate ripening of 1-MCP-treated fruit has not proven to be consistently effective with results ranging from no response to differential stimulation of various processes of ripening (Argenta et al., 2003; Calvo and Sozzi, 2004; Ekman et al., 2004; Golding et al., 1998; Jeong and Huber, 2004).

The production of volatile compounds is reduced by climacteric fruit after 1-MCP treatment. Volatile production by banana was reduced following exposure to 1-MCP (Golding et al., 1998). The reduction was due primarily to reduced ester production as 1-MCP treatment enhanced alcohol production. Generation of ripe aroma by 'Beauty' and 'Shiro' plums was delayed following 1-MCP treatment (Abdi et al., 1998). Apple ester production is inhibited by 1-MCP but effects on other compounds vary with cultivar and storage conditions. Production of esters and alcohols was reduced by 1-MCP (Defilippi et al., 2004; Fan and Mattheis, 1999). Aldehydes, acetic acid, and 1-methoxy-4-(2-propenyl) benzene were also reduced by 1-MCP treatment in intact 'Gala' apples (Mattheis et al., 2005), but Defilippi et al. (2004) found no impact of 1-MCP on aldehyde content in homogenates from 'Greensleeves' apple. Ester production by 'Golden Delicious,' 'Jonagold,' and

'Delicious' apples was reduced by 1-MCP (Ferenczi et al., 2006). The duration necessary for initiation of ester production by 'Delicious' and 'Jonagold' stored at 22°C after removal from 0°C in air decreased with increased storage duration. However, a similar reduction in duration to initiate ester production after CA storage of 'Jonagold' fruit was not observed. In all cases, maximal ester production by 1-MCP-treated fruit after storage did not exceed 50% of that of untreated fruit. Ester and alcohol production by 'd'Anjou' pear fruit was delayed by 1-MCP with straight C-chain compounds impacted more than branched C-chain compounds (Argenta et al., 2003). At an equivalent stage of ripening, based on firmness, of untreated and 1-MCP-treated fruit, no differences in volatile production were observed.

Ester production is clearly linked to climacteric fruit ripening, but a simple relationship between ester concentration and flavor of 1-MCP-treated fruit is not apparent. For the summer apple 'Anna,' 1-MCP treatment reduced production of esters but aldehydes and alcohols were enhanced (Lurie et al., 2002). While sensory evaluation indicated control fruit developed more fruity, ripe, and overall aroma compared to apples treated with 1 μL L^{-1} 1-MCP, 1-MCP-treated apples with less ripe aroma had higher preference ratings. 'Gala' apples treated with 1-MCP and stored in CA had less volatile production and perceived aroma but higher firmness compared to nontreated fruit also stored in CA (Moya-Leon et al., 2007). In spite of less aroma, 1-MCP-treated fruits were preferred over controls. In another sensory study with 'Gala' apples, untrained panelists could differentiate between control fruit stored in CA, new crop fruit from Chile that had not been stored, and 1-MCP-treated fruit stored in air or CA (Marin et al., 2009). The fruit had been matched for firmness with a nondestructive sensor prior to evaluation, and ester production by 1-MCP-treated fruit was less than that of untreated controls. Liking scores for all treatments were similar, however, consumers who identified themselves as "regular 'Gala' eaters" preferred control-CA over 1-MCP-CA fruit.

The ethylene-induced accumulation of 8-methoxy-3,4-dihydro-isocoumarin in carrot, which imparts bitterness, is prevented when carrots are treated with 1-MCP prior to exposure to ethylene (Fan and Mattheis 2000). However, treatment of 'Shamouti' oranges with 1-MCP did not prevent an accumulation of acetaldehyde and ethanol induced by exogenous ethylene (Porat et al., 1999).

7.4.6 Other Gases

7.4.6.1 Argon

Gases that alter atmosphere physical properties or that have biochemical impacts have been evaluated for use in CA or MAP systems. Diffusivity of O_2, CO_2, and C_2H_4 is enhanced in atmospheres with enhanced argon (Ar) (Burg and Burg, 1965). Enhanced diffusivity of these gases was suggested to result in a lower critical O_2 concentration compared to that occurring for low O_2 in N_2. Mature-green tomatoes stored in a CA with 3 kPa O_2 in Ar ripened slower compared to fruit stored in 3kPa O_2 with N_2 (Lougheed and Lee, 1991). Typical tomato flavor developed after fruit were removed from CA conditions regardless of whether Ar or N_2 was present. For other commodities, Ar-supplemented atmospheres have not proven to impact produce quality beyond that obtained with conventional CA or MAP systems (Jamie and Saltveit, 2002; Rocculi et al., 2005; Zhang et al., 2008).

7.4.6.2 Carbon Monoxide

Atmospheres with carbon monoxide (CO) can inhibit discoloration of cut produce surfaces as well as enhance decay control (Kader and Klaustermeyer, 1977; Woodruff, 1977).

CO added to air can also act as an ethylene mimic, causing accelerated ripening (Burg and Burg, 1969), but CO in a low O_2 atmosphere delays ripening (Kader et al., 1977). Responses of tomato fruit to storage in atmospheres containing carbon monoxide (CO) varied with atmosphere composition and fruit maturity at harvest (Kader et al., 1978). Ripening of tomatoes harvested at the mature green stage was enhanced after removal from atmospheres containing 5 or 10 kPa CO in air or 4 kPa O_2. Tomatoes harvested at later stages (breaker, turning, or pink) stored under the same conditions did not exhibit enhanced ripening. CO environments did not alter sugars or acids compared to fruit stored in air. CO in combination with low O_2 increased marketable weight of celery (Reyes and Smith, 1987) and lettuce (*Lactuca sativa* L.) (Stewart and Uota, 1976), but no impacts on flavor were observed.

7.4.6.3 Nitrous Oxide

The presence of nitrous oxide (N_2O) has been demonstrated to interfere with ethylene production by climacteric fruit (Gouble et al., 1995). While exposed to a mixture of 80 kPa N_2O and 20 kPa O_2, ethylene production by preclimacteric tomato and avocado fruit was delayed compared to controls held in air, and ethylene production was lower in tomatoes during ripening. Subsequent studies using climacteric and nonclimacteric fruit have demonstrated efficacy for decay suppression (Qadir and Hashinaga, 2001) and color, texture, pH, and sugar content (Benkeblia and Varoquaux, 2003; Rocculi et al., 2005), although Qadir and Hashinaga (2001) and Palomer et al. (2005) found no effects on quality of onion, strawberry, or banana. None of these studies investigated impacts of N_2O on flavor.

7.4.6.4 Nitric Oxide

Nitric oxide (•NO) was reported to have activity as an endogenous ethylene antagonist (Leshem and Pinchasov, 2000; Leshem et al., 1998). Exogenous •NO applied as a gas or generated from •NO donors reportedly enhances storage life of bok choy (*Brassica rapa* L., Chinensis group), green bean (*Phaseolus vulgaris* L.), (Soequarto and Wills, 2004), lettuce (Wills et al., 2008), broccoli (*Brassica oleracea* L., Italica group), strawberry, kiwifruit (Leshem et al., 1998; Wills et al., 2000), pear (Sozzi et al., 2003), apple slices (Pristijono et al., 2006), and peach (Zhu and Zhou, 2006). The effects include delayed browning on cut surfaces, delayed softening, and changes in color. No evaluation of •NO effects on flavor have been reported.

7.4.7 Precursor Atmospheres

The capacity of apple fruit to utilize volatile compounds present in the storage atmosphere for ester synthesis (Knee and Hatfield, 1981) has been evaluated as means to enhance fruit flavor. Apples previously stored in CA that were exposed to a mixture of volatile alcohols, aldehydes, and carboxylic acids had an enhanced aroma compared to untreated fruit (Kollmannsberger and Berger, 1992). A mixture of aliphatic alcohols in the precursor atmosphere resulted in a better balanced apple aroma than one or two alcohols alone. Sensory panelists could detect a pear-like note after exposure to precursor atmosphere not present in untreated fruit. Precursor atmospheres applied to 'Golden Delicious' apples, using aldehydes and carboxylic acids, selectively enhanced volatile production (De Pooter et al., 1983), however, significant organoleptic improvement of the fruit was not observed. Whole apples exposed to hexanal vapor had increased emission of hexanol, hexyl acetate, hexyl butanoate, and hexyl hexanoate compared to untreated fruit, but enhanced emission of these compounds was limited to 4–7 days (Fan et al., 2006).

7.5 Packaging Effects

The development of MAP for fresh fruits and vegetables has focused on designing packages to establish and maintain beneficial atmospheres to reduce decay and maintain quality with little concern about effects on flavor. In recent years with the increased marketing of fresh-cut produce, there have been renewed efforts to develop beneficial packaging for these new products. The development of MAP for fresh-cut products has been primarily focused on reducing browning of cut surfaces, minimizing dehydration, reducing decay, and preventing contamination of the product (Forney, 2007; Martín-Belloso and Soliva-Fortuny, 2006).

7.5.1 Passive MAP

Properly designed packages can facilitate the development and maintenance of beneficial concentrations of O_2 and CO_2 within the package. Passive MAP relies on the respiration rate of the product and the gas transmission properties of the package to regulate the atmosphere composition within the package. The permeability of the package therefore must be balanced with the respiration rate of the product to ensure a favorable atmosphere. Many different types of polymer films are available for use in MAP with a wide range of CO_2 and O_2 permeabilities (Al-Ati and Hotchkiss, 2002). In addition, many models have been developed to predict equilibrium package atmospheres for passive MAP (Fonseca et al., 2002; Tanner et al., 2002). However, none of these models can fully account for the variability in product respiration rates that may occur due to variation in holding temperature, product age, interactions with atmosphere composition, product biology, and effects of handling (Fonesca et al., 2002).

When packages are unable to maintain a beneficial atmosphere composition as a result of one or more of these factors, injurious concentrations of CO_2 and/or O_2 may develop, resulting in physiological injury to the product. Understanding both the potential benefits of atmosphere modification, as well as the negative effects that can occur, is important to determine viable MAP applications. Appreciating MAP effects on product flavor is also critical to optimize quality. The response of different fruits and vegetables to atmosphere modification varies considerably and therefore the response of each commodity must be determined. In this section, examples of both positive and negative effects of MAP on the flavor of packaged fruits and vegetables are presented.

7.5.1.1 Effects on Flavor

When anaerobic atmospheres are avoided, MAP environments that maintain product quality can have a varied effect on flavor. In whole fruit, many studies have identified MAP technologies that maintain appearance and marketability and in many cases flavor. 'Flame Seedless' table grapes sealed in packages of OPP that developed atmospheres of 5 kPa O_2 and 5 kPa CO_2 maintained marketable quality and developed no off-odors during 53 days of storage at 1°C, while unpackaged control grapes and those in perforated bags developed off-flavors (Martínez-Romero et al., 2003). Litchi (*Litchi chinensis* Sonn.) stored in microperforated packages that maintained atmospheres of 17 kPa O_2 and 6 kPa CO_2 maintained flavor and eating quality following 34 days at 2°C and 2 days at 14°C, while other packages that created greater atmosphere modification developed off-odors (Sivakumar and Korsten, 2006). Shrink-wrap plastic over whole melons also maintained better flavor and overall preference than unwrapped melons after 28 days storage at 4°C (Lester, 2006). However, the flavor of sweet cherries held at 0°C in sealed polyethylene (PE) bags with atmospheres of 4–6 kPa O_2 and 4–10 kPa CO_2 gradually lost flavor and became unacceptable after 4 weeks even though appearance of the fruit remained acceptable (Meheriuk et al., 1997).

Holding fresh-cut fruit in MAP also has been successful in maintaining quality and flavor. Fresh-cut mango packaged in MAP and held at 10°C developed atmospheres ranging from 5 to 15 kPa O_2 and 4 to 10 kPa CO_2 after 15 days, which maintained sensory characteristics and sugar and acid content (González-Aguilar et al., 2000). Pineapple chunks that were held in MAP for 14 days at 0°C or 5°C and developed atmospheres of 1.5 kPa O_2 and 11 kPa CO_2 or <1 kPa O_2 and 15 kPa CO_2, respectively, maintained good appearance and developed no off-flavors (Marrero and Kader, 2006). However, a slight off-odor, associated with the anaerobic atmosphere in the 5°C stored packages, was detected when packages were first opened. 'Gala' apple slices stored for 14 days in barrier bags maintained acceptable flavor and developed no off-flavors or fermented flavors as judged by a sensory panel, although there were significant changes in esters (Bett et al., 2000).

When changes in flavor volatile content of fresh-cut fruit were measured during MAP storage there tended to be a significant loss, but this compositional change was not detected by sensory analysis. Minimally processed durian (*Durio zibethinus* L.) fruit pulp, stored on trays overwrapped with a low-density polyethylene (LDPE) cling film at 4°C, lost 53% of total volatiles, including 77% of esters, after 7 days of storage (Voon et al., 2007). However, sensory evaluation did not detect a significant decrease in fruity, sulfury, or nutty aromas until 28 days, at which time only 1% of the volatile esters remained. When 'Gala' apple fruit slices were stored in perforated PE bags at 5.5°C for 7 days, ethyl ester concentration increased 11-fold, while acetate ester concentration declined about 50% (Bai et al., 2004). No significant changes in apple-like flavor were associated with these changes in ester composition. Sensory analysis of fresh-cut 'Sol Real' cantaloupe cubes stored at 4°C showed minimal changes in flavor during 14 days storage with the exception of mustiness, which increased after 12 days (Beaulieu et al., 2004). However, nonacetate esters increased 87% during storage while acetate esters decrease 66%, which changed the ester balance. Unlike these previous studies where there seemed to be little relationship between volatile composition and perceived flavor, in fresh-cut mangos, volatile terpene concentration was related to product aroma. Total terpene concentration of mango held in MAP at 4°C decreased to 21% of initial values after 11 days at which time they were considered unmarketable due to lack of aroma (Beaulieu and Lea, 2003).

MAP has also extended the storage life of a variety of vegetables. Snow peas (*Pisum sativum* L. var. *saccharatum*) held in MAP at 5°C for 28 days with steady-state atmospheres of about 5 kPa O_2 and 5 kPa CO_2 maintained quality and developed no off-flavors unlike peas held in ambient atmospheres that became unmarketable (Pariasca et al., 2000). Carrot disks stored in PA-60 bags that developed atmospheres of 5–7 kPa O_2 and 12–15 kPa CO_2 also maintained good aroma and flavor (Cliffe-Byrnes and O'Beirne, 2007). MAP of trimmed green onions reduced the loss of soluble solids and thiosulfinates responsible for pungency after 14 days of storage at 5°C with package atmospheres of 0.2 kPa O_2 and 7.5 kPa CO_2 being most effective (Hong et al., 2000). Aroma and taste of fresh-cut tomato slices stored in MAs at temperatures of 0°C–5°C for up to 10 days were better than that of air-stored slices (Gil et al., 2002; Hakim et al., 2004).

7.5.1.2 Off-Flavor Development

When MAP atmospheres become anaerobic, the primary response of fresh fruits and vegetables is to produce elevated concentrations of ethanol and to a lesser extent acetaldehyde. The impact of elevated levels of these compounds on product flavor is variable depending on the commodity. The odor threshold for ethanol is high (100 ppm), so elevated concentrations of ethanol alone may not be detrimental to product flavor. More often, the effect of ethanol on the metabolism of other aroma volatiles is what affects product flavor. Song et al. (1997) found volatile production by apple fruit held in MAP was dominated by ethanol, acetaldehyde, and ethyl acetate when O_2 was <0.5 kPa, while at higher O_2 concentration esters responsible for normal flavor predominated. This alteration of flavor chemistry under anaerobic conditions often leads to off-flavor development. Anaerobic atmospheres that developed in microperforated packages caused strawberries to accumulate ethanol, acetaldehyde, and ethyl acetate and develop off-flavor (Sanz et al., 1999). Fresh-cut 'Nairobi' carrot slices sealed in OPP bags lost fresh carrot aroma during the first 5 days of storage at 8°C, which was followed by the development of off-odors associated with tissue decay and atmospheres of 1–2 kPa O_2 and 30 kPa CO_2 (Barry-Ryan and O'Beirne, 1998). Anaerobic atmospheres also induced off-flavor formation in fresh-cut salad savoy (*Brassica oleracea* L., Capitata group) (Kim et al., 2004) and snow peas (Pariasca et al., 2000), rendering them inedible.

The accumulation of ethanol associated with anaerobic atmospheres does not always render products unmarketable. In a survey of fresh-cut mixed salads in MAP, most packages contained <5 kPa O_2 and >2 Pa ethanol, an atmosphere that prevents browning but causes a slight fermented odor (Cameron et al., 1995). However, the odor was not objectionable and the salads remained edible after adding salad dressing. MAP of persimmons induced ethanol production reaching up to 600 μg L^{-1} in the headspace during 84 days of storage at 1°C followed by 5 days at 24°C, however, no off-odors or off-flavors were detected and the fruit were judged to be of good quality (Cia et al., 2006). The aroma of fresh-cut mango fruit held in MAP was not affected by anaerobic atmospheres that increased production of ethanol, ethyl acetate, and ethyl butanoate, and overall sensory ratings were similar to fruit held in aerobic packaging (Beaulieu and Lea, 2003).

Strong sulfur-based odors can be produced by some *Brassica* vegetables when subjected to anaerobic atmospheres. Broccoli is the worst case, producing significant quantities of methanethiol, hydrogen sulfide, dimethyl disulfide, and dimethyl trisulfide, resulting in severe off-odors and off-flavors (Forney et al., 1991; Gillies et al., 1997). However, when 12 different *Brassica* vegetables were subjected to anaerobic atmospheres by N_2 flushing, many, including cauliflower (*Brassica oleracea* L., Botrytis group), Chinese cabbage (*Brassica rapa* L., Pekinensis group), and kohlrabi (*Brassica oleracea* L., Gongylodes group), produced low levels of these compounds making them a low risk to develop off-flavors under anaerobic atmospheres that may develop in passive MAP (Forney and Jordan, 1999).

7.5.2 Active MAP

The limitations of passive MAP to maintain beneficial atmospheres throughout marketing have prompted the development of active packaging approaches. Active packaging includes features that respond to environmental changes, flushing packages with beneficial gas mixtures, and adding sachets that adsorb or release gases or volatile compounds (Forney, 2007).

7.5.2.1 Interactive Features

Ideally the permeation properties of a package should be able to respond to changes in the environment, such as temperature or humidity, changes in atmosphere composition in the package, and/or physiological changes in the product. Such changes may be indicated by the evolution of volatile compounds such as ethanol or ethylene. Active modification of package properties could be beneficial to optimize flavor of products throughout marketing. However, practical and economical technologies to meet these goals have been limited. Some development has occurred to create films that are more responsive to temperature so that permeation properties can parallel temperature-dependent respiration rates of a specific commodity (Butler, 2001). Other proposed features include pores that open to increase gas transmission in response to temperature, changes in O_2 or CO_2, and ethanol concentrations inside the package (Cameron et al., 1995). Incorporation of package features that respond to physiological changes occurring in the product may help to regulate package gas exchange in the future and could provide new tools for flavor management.

7.5.2.2 Gas Flushing

Rapid establishment of a MA can be achieved by flushing packages with a desirable gas mixture. Flushing displaces air in the package replacing it with the desired atmosphere composition for more rapid atmosphere establishment than can be achieved using passive MAP. Packages can be flushed with optimum concentrations of O_2 and CO_2, elevated concentrations of O_2, or N_2 to rapidly exclude O_2.

Flushing packages containing 'Athena' melon cubes with 4 kPa O_2 and 10 kPa CO_2 for rapid atmosphere establishment, maintained a mild melon aroma after 12 days at 5°C, while melon in perforated packages and passive MAP developed a faint off-odor that were associated with fungal growth (Bai et al., 2001). Similarly, fresh-cut pineapple flushed with 4 kPa O_2 and 10 kPa CO_2 maintained sensory quality for 5–7 days at 4°C versus only 3–5 days for pineapple held in passive MAP (Liu et al., 2007). Flushing packages of fresh-cut 'Conference' pears with N_2 reduced microbial growth and did not induce any off-odors during 3 weeks of storage at 4°C, whereas packages flushed with 7 kPa CO_2 developed undesirable odors (Soliva-Fortuny and Martín-Belloso, 2003). Evacuating packages has also been used to exclude O_2. When a vacuum was drawn for 24–48 h on bananas placed in a PE bag at 20°C, prior to venting with microperforations, ripening was delayed, good yellow color maintained, and banana flavor enhanced (Pesis et al., 2005). Flushing with high concentrations of O_2 has also shown potential to maintain product quality and flavor. Flushing packages of grated celeriac (*Apium graveolens* L., Rapaceum group) and sliced mushrooms (*Agaricus bisporus* (Lge.) Sing.) with 95 kPa O_2 maintained fresh taste and smell after 7 days at 7°C, when compared to product held in passive MAP containing 3 kPa O_2 and 5 kPa CO_2 (Jacxsens et al., 2001).

7.5.2.3 Sachets

Sachets also can be incorporated into packages. Sachets, containing a variety of substances that can absorb or release gases, provide another mechanism for regulating atmosphere composition and product quality (Ozdemir and Floros, 2004). Sachets containing sorbitol and $KMnO_4$ can adsorb volatiles, such as ethanol and acetaldehyde, from the package atmosphere, reducing the development of off-odors and off-flavors in stored broccoli (DeEll et al., 2006). In a different study, broccoli stored in LDPE film containing an ethylene scrubber sachet had flavor and appearance closest to fresh broccoli when compared to broccoli packaged in other films without sachets (Jacobsson et al., 2004). Sachets of granulated activated carbon alone or with impregnated palladium placed in MAP with sliced tomatoes prevented the development of off-flavors and after 14 days gave the highest scores for sweetness, juiciness, odor, and flavor when compared with packages without the sachet (Bailén et al., 2006). The sachets adsorbed volatiles, including ethylene, and slowed tomato ripening. In another study, sachets containing an ethylene and a CO_2 scrubber were placed in MAP bags containing bananas that developed an atmosphere of

12 kPa O_2 and 4 kPa CO_2 during storage at 10°C (Nguyen et al., 2004). After 18 days, bananas in the MAP bags with the sachets had less chilling injury and better ratings for pulp softness, sweetness, and flavor than fruit held in corrugated boxes.

7.5.3 Interaction with Packaging Films

As introduced above, loss of flavor during storage and marketing of fresh fruits and vegetables can be the result of diffusional losses of volatile flavor compounds from the product. Diffusion of each volatile compound is driven by its concentration gradient between the product and the surrounding atmosphere. If the volatile concentration in the package atmosphere is in equilibrium with the product, there will be no net diffusional loss from the product. Therefore, it is important to understand the chemical and physical properties of the packaging material and how they interact with volatile compounds released from the product that contributes to flavor.

Packaging can serve as a barrier to the loss of volatile compounds from the atmosphere surrounding the product. The volatile concentration in the package is determined by the interaction of the packaging material with the specific volatiles released by the product. Polymer films used in packaging of processed food products have been shown to interact with product flavor through scalping and permeation (Risch, 1998). Understanding the properties of packaging materials in relation to flavor volatiles could identify additional methods to preserve product flavor and aroma.

7.5.3.1 *Scalping and Permeation*

The interactions of flavor volatiles with different packaging films differ substantially and may inhibit or enhance flavor loss. In a recent review, Dury-Brun et al. (2007) discuss the principles underlying the mass transfer of volatile flavor compounds into and through food packaging. The permeation of flavor volatiles through a packaging film is a three-step process that is dependent on the chemistry of both the volatile and the film. The initial interaction of the volatile with the film results in its sorption into the film, it then diffuses through the film driven by chemical potential differences, and finally is desorbed from the film by evaporation. While the permeability of packaging films to O_2 and water vapor have been extensively studied, limited information is available concerning their permeability to other molecules.

As part of the permeation process, volatile compounds can be removed from the package headspace through sorption (scalping) and thus enhance flavor loss from the product (Brody, 2002; Nielsen and Jägerstad, 1994). The process of sorption involves the adsorption of the volatile compound onto the polymer film and is dependent on the solubility (partitioning) of the compound in the packaging material. PEs and polypropylenes (PPs), which are widely used in the packaging of fresh fruits and vegetables, are highly nonpolar and therefore have a high affinity for nonpolar volatile compounds, whereas polyesters including the biobased polymer polylactic acid are more polar (Sajilata et al., 2007). Since most volatile flavor compounds are nonpolar, PE and PP have a high potential to scalp these compounds. In studies conducted on the scalping of D-limonene from orange juice, strips of LDPE reduced D-limonene concentrations from 40% to 60% in 6 h (Mannheim et al., 1987). The rates of sorption of flavor compounds tend to decrease with increasing polarity. Affinity of volatile compounds to LDPE was greatest with hydrocarbons followed by ketones, esters, aldehydes, and alcohols (Fayoux et al., 1997). The sorption also increases with molecular size (carbon chain length) and branching. When a solution of aliphatic esters were placed in a PE-lined bag, the distribution ratio increased from 0.02 to 5.77 as carbon number increased from 6 to 10

(Shimoda et al., 1988). The distribution ratio represented the ratio of the amount sorbed to the film to the amount remaining in solution. Linssen et al. (1991) observed that branched molecules were sorbed to a greater extent than linear molecules by high-density polyethylene (HDPE). Therefore, the differential effects of sorption could alter volatile profiles and thus affect flavor.

Environmental factors can also affect the interaction of flavor volatiles with packaging materials. In MAP, the typical low-temperature and high-humidity environment may enhance flavor loss (Dury-Brun et al., 2007). Unlike liquids, the rate of sorption of flavor volatiles from the gas phase increases with reduced temperature (Paik and Writer, 1995). As well, high RH may enhance flavor sorption (Fayoux et al., 1997).

Similar to sorption, the permeation of volatile compounds though the film is dependent on many of the same properties described above. When a variety of polymer films were compared, the permeability of D-limonene varied substantially from $<10^{-20}$ $kg\ m^{-1}\ s^{-1}\ Pa^{-1}$ for ethylene vinyl alcohol (EVOH) and PET to $10^{-13}\ kg\ m^{-1}\ s^{-1}\ Pa^{-1}$ for LDPE and ethylene vinyl acetate (EVA) (Dury-Brun et al., 2007). The movement of a compound through a film is linearly related to its concentration gradient across the film. However, the significance of either flavor scalping or permeation to flavor change of fresh packaged produce is yet to be determined.

7.5.3.2 Perforations

A larger factor that contributes to the loss of volatiles from a package is the presence of perforations or leaks. Both macro- and microperforations are commonly used in the packaging of whole and fresh-cut produce to provide adequate exchange of O_2 and CO_2 to avoid anaerobic conditions. Diffusion of O_2 and CO_2 through perforations has been modeled and follows a modified Fick's equation (González et al., 2008). Perforations can greatly enhance the movement of these gasses through the package. Similarly, perforations can greatly enhance loss of flavor volatiles from the package atmosphere and thus increase diffusional loss of flavor volatiles from the product. The magnitude of volatile loss through microperforations was demonstrated by Del-Valle et al. (2004), who showed that three 100 μm pores increased the rate of ethanol permeation through a 60 μm thick PP film 186-fold. They suggest that the rapid permeation of volatiles through perforations could lead to organoleptic deterioration of foods. Similarly, small leaks in package seals or micropores present in some materials can greatly enhance volatile loss. Dury-Brun et al. (2007) discuss the role of film porosity in the movement of flavor volatiles.

7.6 Interactions of Coatings

Use of packaging and edible coatings can create a MA with reduced O_2 and elevated CO_2 levels, similar to that of CA (Baldwin, 1994; Smith et al., 1987). As discussed already, lowering O_2 and raising CO_2 levels can delay ripening of climacteric fruits or vegetables, undesirable color changes in nonclimacteric fresh produce, and maintain the quality of many fruits and vegetables for extended periods. However, exposure of fresh produce to O_2 levels below their tolerance level can increase anaerobic respiration and lead to the development of off-flavor (Kader, 1986) in part due to ethanol production. Volatiles like ethanol and other alcohols have greater water solubility and tend to stay in the flesh of fruits, while esters tend to evaporate having a lower Henry's law coefficient than alcohols (Bai et al., 2002). However, the presence of a barrier coating may retain less water-soluble volatiles can affect fruit flavor. Coating-induced changes have been noted in apple

(Bai et al., 2002; Saftner, 1999; Saftner et al., 1999), citrus (Baldwin et al., 1995a; Cohen et al., 1990; Hagenmaier, 2002), and mango fruits (Baldwin et al., 1999a,b). A coating barrier can induce anaerobic respiration and the subsequent synthesis of ethanol and acetaldehyde, and then entrap volatiles, including ethanol and acetaldehyde (Baldwin et al., 1995a,b), leading to a fermented off-flavor.

For citrus, coatings that have higher amounts of shellac or wood rosin, which are the least permeable of the film-formers used on fruits and vegetables, end up causing the development a more MA. The more severe MA created by resin coatings results in more off-flavor compared to fruit coated with the more permeable carnauba or PE waxes due to an increase in ethanol and a change in the volatile profile (Baldwin et al., 1995a; Hagenmaier, 2002; Hagenmaier and Shaw, 2002; Nisperos-Carriedo et al., 1990). Shellac and wood rosin are moderately effective at reducing water loss; however, it is the gloss or shine that they impart to the fruit that is most desired by the fresh fruit industry. The waxes are very effective at preventing water loss and are more permeable to gases compared to the resins, and therefore, induce less off-flavor in fruits, while imparting moderate shine (Baldwin, 1994). Flavor of oranges and grapefruit (*Citrus paradisi* M.) are affected by coatings (Hagenmaier and Baker, 1994; Nisperos-Carriedo et al., 1990), but tangerines (*Citrus reticulata* Blanco) are especially susceptible to anaerobiosis due to coating treatments (Hagenmaier and Shaw, 2002; Mannheim and Soffer, 1996). Some new coating materials have been tested on citrus including hydroxypropyl methyl cellulose (HPMC) and beeswax in different ratios. It was found that the higher the solids content in these coatings, the higher the internal CO_2, ethanol, and off-flavor. Coatings developed from corn zein (Bai et al., 2003a) and polyvinyl acetate (PVA) (Hagenmaier and Grohmann, 2000) also make shiny coatings that are more permeable than the resins, and in the case of PVA, have no problems with discoloration often associated with shellac (Hagenmaier and Grohmann, 1999, 2000). Composite coatings of candelilla, PE, or carnauba waxes and shellac, which are more permeable than coatings made of shellac or wood rosin alone, add gloss with less flavor alteration (Alleyne and Hagenmaier, 2000; Hagenmaier, 2002; Mannheim and Soffer, 1996).

Apples are another fruit often coated for high gloss or shine, which helps to boost sales. Shellac and carnauba wax, or a mixture of the two, are often used. The less permeable wood rosin and the more permeable PE wax are not allowed for use on apples, so choices for apple coatings are restricted compared to citrus. Shellac-coated 'Golden Delicious' apples accumulated ethanol and ethyl acetate when they were held at 20°C (Saftner, 1999). Storage temperature affects the modification of a fruit's internal atmosphere by a coating in that the warmer the storage or holding temperature (such as is encountered during the marketing period), the greater the fruit respiration, requirement for O_2, and production of CO_2, as is seen with MAP. Other coatings, made from starch, corn zein, and PVA were tested on apples in comparison with carnauba wax and shellac (Bai et al., 2002, 2003a,b) and were found to be similar to shellac for gloss. Carnauba wax, zein, and PVA coatings, however, resulted in higher internal fruit O_2 concentrations than did shellac or starch coatings, especially during a simulated marketing period at 20°C. Apples coated with starch, shellac, or a carnauba-polysaccharide composite coating accumulated ethanol and ethyl esters, but these volatiles decreased after transfer of fruit to 21°C (Bai et al., 2002). The reaction of apples to coating permeability was also found to be dependent on cultivar (Bai et al., 2002, 2003b), in part perhaps due to their anatomical structure (number of stomates and lenticels or pores) (Bai et al., 2003b). Shellac coatings caused excessive modification of internal gases and induced an abrupt rise of the respiratory quotient accompanied by accumulation of ethanol in 'Braeburn' and 'Granny Smith' apples, as well as flesh browning at the blossom end of 'Braeburn' fruit stored at 20°C. 'Fuji' apples were less affected by the shellac coating and 'Delicious' fruit were the least affected. 'Delicious' apples happen to have more pores, compared to an apple like 'Granny Smith,' which may explain why they can tolerate coatings like shellac with low permeability to gases, while 'Granny Smith' cannot (Bai et al., 2003b).

Mangoes are another good example of a fruit whose flavor can be affected by a coating. Like tangerines, mango fruit easily go anaerobic. Low-permeability coatings cause a rise in internal ethanol, which results in a fermented off-flavor as with citrus and apple. Mangoes, however, cannot tolerate low-temperature storage (below 12°C–16°C), thus coatings have more effect on respiration and subsequently on flavor, although they are useful to retard ripening/softening and to extend shelf life of this climacteric fruit (Hoa et al., 2002). Low-solid shellac, cellulose, and carnauba wax coatings delayed ripening and extended shelf life of mangoes, and although the shellac and cellulose coatings caused elevated levels of ethanol, they did not result in significant flavor differences from uncoated fruit in sensory tests (Hoa et al., 2002). Cellulose-coated mangoes also exhibited increased concentrations of flavor volatiles compared to uncoated controls (Baldwin et al., 1999a).

Finally, edible coatings have been shown to affect flavor of fresh-cut fruit. Edible coatings can prevent water loss from fresh-cut fruit, be carriers of antibrowning or antimicrobial agents, and prevent surface drying and perhaps volatile off-gassing. However, cut apple slices dipped in soybean oil emulsion or carboxy methylcellulose coatings with aqueous antioxidants actually lost some important aroma compounds compared to slices dipped in aqueous antioxidant solutions alone (Bai and Baldwin, 2002), although they exhibited reduced browning and water loss. Mango pieces, coated with carboxy methylcellulose or maltodextrin, maintained good visual quality, had less drying of the cut surface, and were not different in flavor compared to uncoated controls. Both the aqueous antioxidant and the antioxidants + coatings treatments showed some volatile retention compared to uncoated pieces. Mango pieces coated with chitosan, starch, whey protein, or soybean oil emulsion, however, resulted in poor flavor (Plotto et al., 2004).

7.7   Future Research Needs

Many opportunities exist to further our understanding of postharvest flavor change and how it is affected by atmosphere modification through storage, packaging, and coatings. Areas of research that could lead to improved CA and MAP technologies to optimize produce flavor and ultimately consumer satisfaction include greater integration of sensory and compositional studies, impact of atmosphere composition on flavor biochemistry, and interactions of packaging and coating materials with volatile flavor compounds.

Sensory evaluation of flavor changes occurring as a result of atmosphere modification need to be integrated with compositional analysis to gain a better linkage of the impact of flavor compounds on product flavor. Establishment of optimum or critical concentrations of compounds responsible for flavor would aid in interpreting the effects of atmosphere modification on product flavor. In addition, a better understanding of the interaction of flavor compounds, such as sugars, acids, and volatiles, with each other and with other parameters including texture and appearance, would aid in relating quality and compositional changes to their impact on flavor. Sensory studies are also needed to identify and characterize yet to be discovered compounds responsible for fruit and vegetable flavor.

Metabolic pathways responsible for the synthesis and catabolism of flavor compounds are poorly defined. Research is needed to define the biochemistry responsible for flavor in many fruits and vegetables and the effects atmosphere composition has on this process. The effects and interactions of O_2, CO_2, H_2O, and other volatile compounds that accumulate in the storage atmosphere on flavor biochemistry are yet to be determined for many fruits and vegetables.

In addition, molecular mechanisms responsible for flavor metabolism need to be identified to provide new tools to maintain and optimize flavor throughout storage and marketing.

The use of packaging or edible coatings on whole and fresh-cut fruits and vegetables can modify internal concentrations of O_2 and CO_2, prevent loss of water, and alter volatile flavor components. Wide arrays of both packaging and coating materials are available, but their effects on product flavor are poorly understood. In addition to the effects O_2 and CO_2 modification has on flavor metabolism, the chemical properties of polymer films and coating materials may directly interact with volatile flavor components, reducing, altering, or enhancing their loss. Properties of packaging and coating materials, including rates of permeation and scalping of flavor compounds, need to be determined. The significance of these processes on flavor loss needs to be assessed in order to predict the effect packaging or coatings will have on flavor quality.

7.8 Conclusions

The flavor of fresh fruits and vegetables is a complex trait that is affected by the genetics of the product as well as many pre- and postharvest factors. To understand the influence of atmosphere modification and packaging on flavor, many chemical and physical components of the product must be considered. It is the complex interaction of these components that determine the human perception of flavor and ultimately the desirability and market value of the product. Manipulating the postharvest environment to maintain or enhance flavor to optimize consumer satisfaction is a challenge.

The use of CA to preserve fresh produce can have numerous effects on flavor. The development of CA technologies has primarily focused on the preservation of product appearance and texture with little consideration of flavor. Altering concentrations of O_2, CO_2, or other gasses can alter metabolism, including biosynthesis of flavor compounds. This can result in a loss or altering of flavor during storage, which may not be totally reversible when the product is returned to air. Atmosphere modification can also induce anaerobic conditions that induce fermentation of the product. Fermentation results in the production of ethanol, acetaldehyde, and a variety of other volatile compounds that can result in off-odors and off-flavors.

While MAP can affect the flavor of fresh produce by altering O_2 and CO_2 concentrations in a manner similar to CA storage, packaging materials can have additional effects on flavor. The diversity of polymer films that can be used in packaging have a wide range of chemical and physical properties that may affect product flavor. Packaging films can act as a barrier to the diffusional loss of volatile flavor compounds from the product. This can be particularly important with fresh-cut products where natural diffusional barriers have been removed. The chemical interactions of the packaging film with flavor volatiles, resulting in flavor scalping can affect their rate of loss. Perforations in the packaging film can also contribute significantly to diffusional losses. Understanding these processes may lead to improved packaging that will ensure acceptable product flavor.

The use of edible coatings on fresh fruits and vegetables can have effects similar to MAP. Coatings can result in modification of internal concentrations of O_2 and CO_2 as well as prevent diffusional loss of volatile flavor components. Wide arrays of coating materials are available with many formulations including waxes, shellacs, carbohydrates, and proteins. The effects of coating on product flavor are dependent on the chemical properties of the coating material and their interaction with respiratory gasses and volatile flavor components.

The characterization of the impacts atmosphere modification and packaging have on the flavor of fresh fruits and vegetables is an area that requires additional research. Increasing our understanding of atmosphere effects on flavor metabolism and the interaction of packaging materials and coatings on flavor loss could lead to new technologies to optimize flavor quality of fresh produce and thus improve consumer satisfaction.

References

Aaby, K., K. Haffner, and G. Skrede. 2002. Aroma quality of Gravenstein apples influenced by regular and controlled atmosphere storage. *Food Sci. Technol.* 35:254–259.

Abbott, N.P., P. Etiévant, S. Issanchou, and D. Langlois. 1993. Critical evaluation of two commonly used techniques for the treatment of data from extract dilution sniffing analysis. *J. Agr. Food Chem.* 41:1698–1703.

Abdi, N., W.B. McGlasson, P. Holford, M. Williams, and Y. Mizrahi. 1998. Responses of climacteric and suppressed-climacteric plums to treatment with propylene and 1-methylcyclopropene. *Postharvest Biol. Technol.* 14:29–39.

Abegaz, E.G., K.S. Tandon, J.W. Scott, E.A. Baldwin, and R.L. Shewfelt. 2004. Partitioning of taste and aromatic flavor notes from fresh tomato (*Lycopersicon esculentum*, Mill) to develop predictive models as a function of volatile and nonvolatile components. *Postharvest Biol. Technol.* 34:227–236.

Abeles, F.B. 1992. Ethylene in plant biology. San Diego: Academic Press.

Acree, T.E. 1993. Bioassays for flavor. In: T.E. Acree and R. Teranishi (Eds.), *Flavor Science: Sensible Principles and Techniques*, ACS Books, Washington DC, pp. 1–20.

Acree, T.E. and J. Barnard. 1994. Gas chromatography-olfactometry and CharmAnalysis™. In: H. Maarse and D.G. Van Der Heij (Eds.), *Trends in Flavour Research*, Elsevier, Amsterdam, the Netherlands, pp. 211–220.

Al-Ati, T. and J.H. Hotchkiss. 2002. Application of packaging and modified atmosphere to fresh-cut fruits and vegetables. In: O. Lamikanra (Ed.), *Fresh-Cut Fruits and Vegetables Science, Technology and Market*, CRC Press, Boca Raton, FL, pp. 305–338.

Alleyne, V. and R.D. Hagenmaier. 2000. Candeililla-shellac: An alternative formulation of coating apples. *HortScience* 35:691–963.

Almenar, E., P. Hernandez-Munoz, J.M. Lagaron, R. Catala, and R. Gavara. 2006. Controlled atmosphere storage of wild strawberry fruit (*Fragaria vesca* L.). *J. Agr. Food Chem.* 54:86–91.

Anderson, R.E. and R.W. Penney. 1973. Quality of "Stayman Winesap" apples stored in air, controlled atmospheres, or controlled atmospheres followed by storage in air. *HortScience.* 8:507–508.

Argenta, L.C., X. Fan, and J.P. Mattheis. 2003. Influence of 1-methylcyclopropene on ripening, storage life, and volatile production by d'Anjou cv. pear fruit. *J. Agr. Food Chem.* 51:3858–3864.

Aust, L.B., M.C. Gacula Jr., S.A. Beard, and R.W. Washam II. 1985. Degree of difference test method in sensory evaluation of heterogeneous product types. *J. Food Sci.* 50:511–513.

Bai, J., V. Alleyne, R.D. Hagenmaier, J.P. Mattheis, and E.A. Baldwin. 2003a. Formulation of zein coatings for apples (*Malus domestica* Borkh). *Postharvest Biol. Technol.* 28:259–268.

Bai, J. and E.A. Baldwin. 2002. Postprocessing dip maintains quality and extends the shelf life of fresh-cut apple. *Proc. Fla. State Hort. Soc.* 115:297–300.

Bai, J., E.A. Baldwin, and R.H. Hagenmaier. 2002. Alternative to shellac coatings provide comparable gloss, internal gas modification and quality for 'Delicious' apple fruit. *HortScience* 37:599–563.

Bai, J., E.A. Baldwin, R.C. Soliva-Fortuny, J.P. Mattheis, R. Stanley, C. Perera, and J.K. Brecht. 2004. Effect of pretreatment of intact 'Gala' apple with ethanol vapor, heat or 1-methylcyclopropene on quality and shelf life of fresh-cut slices. *J. Amer. Soc. Hort. Sci.* 129:583–593.

Bai, J., R.D. Hagenmaier, and E.A. Baldwin. 2003b. Coating selection of 'Delicious' and other apples. *Postharvest Biol. Technol.* 28:381–390.

Bai, J.-H., R.A. Saftner, A.E. Watada, and Y.S. Lee. 2001. Modified atmosphere maintains quality of fresh-cut cantaloupe (*Cucumis melo* L.). *J. Food Sci.* 66(8):1207–1211.

Baigrie, B. 2003. *Taints and Off-Flavours in Food*. Woodhead Publishing Ltd., Cambridge, England, pp. 203.

Bailén, G., F. Guillén, S. Castillo, M. Serrano, D. Valero, and D. Martínez-Romero. 2006. Use of activated carbon inside modified atmosphere packages to maintain tomato fruit quality during cold storage. *J. Agr. Food Chem.* 54:2229–2235.

Baldwin, E.A. 1994. Edible coatings for fresh fruits and vegetables: Past present and future. In: J. Krochta, E.A. Baldwin, and M.O. Nisperos-Carriedo (Eds.). *Edible Coatings and Films to Improve Food Quality*, Technomic Publishing Co., Lancaster, PA, pp. 25–64.

Baldwin, E.A. 2002. Fruit flavor, volatile metabolism and consumer perceptions. In: M. Knee (Ed.), *Fruit Quality and its Biological Basis*, CRC Press, Boca Raton, FL, pp. 89–106.

Baldwin, E.A. 2004. Flovor. In: K.C. Gross, C.Y. Wang, and M. Saltveit (Eds.), The commercial storage of fruits, vegetables, and florist and nursery stocks. Agricultural Handbook Number 66. http://www.ba.ars.usda.gov/hb66/023flavor.pdf

Baldwin, E.A., J.K. Burns, W. Kazokas, J.K. Brecht, R.D. Hagenmaier, R.J. Bender, and E. Pesis. 1999a. Effect of two edible coatings with different permeability characteristics on mango (*Mangifera indica* L.) ripening during storage. *Postharvest Biol. Technol.* 17:215–226.

Baldwin, E.A., K. Goodner, A. Plotto, K. Pritchett, and M. Einstein. 2004. Effect of volatiles and their concentration on perception of tomato descriptors. *J. Food Sci.* 69:S310–318.

Baldwin, E.A., T.M.M. Malundo, R. Bender, and J.K. Brecht. 1999b. Interactive effects of harvest maturity, controlled atmosphere and surface coatings on mango (*Mangifera indica* L.) flavor quality. *HortScience*: 34:514.

Baldwin, E.A., M.O. Nisperos-Carriedo, R. Baker, and J.W. Scott. 1991. Quantitative analysis of flavor parameters in six Florida tomato cultivars (*Lycopersicon esculentum* Mill). *J. Agr. Food Chem.* 39:1135–1140.

Baldwin, E.A., M.O. Nisperos-Carriedo, P.E. Shaw, and J.K. Burns. 1995a. Effect of coatings and prolonged storage conditions on fresh orange flavor volatiles, degrees Brix, and ascorbic acid levels. *J. Agr. Food Chem.*: 43:1321–1331.

Baldwin, E.A. and A. Plotto. 2007. Shelf-life versus flavour-life for fruits and vegetables: How to evaluate this complex trait. www.stewartpostharvest.com Stewart Postharvest Solutions (UK) Ltd.

Baldwin, E.A., J.W. Scott, M.A. Einstein, T.M.M. Malundo, B.T. Carr, R.L. Shewfelt, and K.S. Tandon. 1998. Relationship between sensory and instrumental analysis for tomato flavor. *J. Am. Soc. Hort. Sci.* 12:906–915.

Baldwin, E.A., J.W. Scott, and R.L. Shewfelt. 1995b. Quality of ripened mutant and transgenic tomato cultigens. *Tomato Quality Workshop Proc.* December 11–14, 1995, Davis, CA.

Baldwin, E.A., J.W. Scott, C.K. Shewmaker, and W. Schuch. 2000. Flavor trivia and tomato aroma: Biochemistry and possible mechanisms for control of important aroma components. *HortScience* 35(6):1013–1022.

Bangerth, F. 1984. Changes in the sensitivity for ethylene during storage of apple and banana fruits under hypobaric conditions. *Sci. Hort.* 24:151–163.

Bangerth, F. 1988. The effect of ethylene on the physiology and ripening of apple fruits at hypobaric conditions. *Facteurs et Regulation de la Matuaratio des Fruits. Coll. Int. CNRS, Paris*, 238:183–187.

Barry-Ryan, C. and D. O'Beirne. 1998. Quality and shelf-life of fresh cut carrot slices as affected by slicing method. *J. Food Sci.* 63(5):1–6.

Beaulieu, J.C. 2006. Volatile changes in cantaloupe during growth, maturation, and in stored fresh-cuts prepared from fruit harvested at various maturities. *J. Amer. Soc. Hort. Sci.* 131 (1):127–139.

Beaulieu, J.C. 2007. Effects of UV Irradiation on cut cantaloupe: Terpenoids and esters. *J. Food Sci.* 72(4):S272-S281.

Beaulieu, J.C. and E.A. Baldwin. 2002. Flavor and aroma of fresh-cut fruits and vegetables. In: O. Lamikanra (Ed.), *Fresh-Cut Fruits and Vegetables Science, Technology and Market*, CRC Press, Boca Raton, FL, pp. 391–425.

Beaulieu, J.C., D.A. Ingram, J.M. Lea, and K.L. Bett-Garber. 2004. Effect of harvest maturity on the sensory characteristics of fresh-cut cantaloupe. *J. Food Sci.* 69(7):S250-S258.

Beaulieu, J.C. and J.M. Lea. 2003. Volatile and quality changes in fresh-cut mangos prepared from firm-ripe and soft-ripe fruits, stored in clamshell containers and passive MAP. *Postharvest Biol. Technol.* 30:15–28.

Bender, R.J., J.K. Brecht, S.A. Sargent, and D.J. Huber. 2000. Mango tolerence to reduced oxygen levels in controlled atmosphere storage. *J. Amer. Soc. Hort. Sci.* 125:707–713.

Benkeblia, N. and P. Varoquaux. 2003. Effect of nitrous oxide (N_2O) on respiration rate, soluble sugars and quality attributes of onion bulbs Allium cepa cv. Rouge Amposta during storage. *Postharvest Biol. Technol.* 30:161–168.

Bett, K. 1993. Measuring sensory properties of meat in the laboratory. *Food Technol.* 47(11):121–125.

Bett, K.L., D.A. Ingram, C.C. Grimm, S.W. Lloyd, A.M. Spanier, J.M. Miller, K.C. Gross, E.A. Baldwin, and B.T. Vinyard. 2000. Flavor of fresh-cut Gala apples in barrier film packaging as affected by storage time. *J. Food Qual.* 24:141–146.

Blankenship, S.M. and J.M. Dole. 2003. 1-Methylcyclopropene: A review. *Postharvest Biol. Technol.* 28:1–25.

Blankenship, S.M. and D.G. Richardson. 1985. Development of ethylene biosynthesis and ethylene induced ripening in 'd'Anjou' pears during the cold requirement for ripening. *J. Am. Soc. Hort. Sci.* 110:520–523.

Boelrijk, A.E.M., G. Smit, and K.G.C. Weel. 2006. Flavour release from liquid food products, p. 260–286. In: A. Voilley and P. Etiévant (Eds.), *Flavour in Food*, Woodhead Publications Ltd., Cambridge, England.

Brackmann, A., J. Streif, and F. Bangerth. 1993. Relationship between a reduced aroma production and lipid metabolism of apples after long-term controlled-atmosphere storage. *J. Amer. Soc. Hort. Sci.* 118:243–247.

Brody, A.L. 2002. Flavor scalping: Quality loss due to packaging. *FoodTechnology* 56(6):124–125.

Bruemmer, J.H. and B. Roe. 1970. Biochemical changes in grapefruit during anaerobic metabolism. *Proc. Fla. State Hort. Soc.* 83:290–294.

Burdon, J., N. Lallu, D. Billing, D. Burmeister, C. Yearsley, M. Wang, A. Gunson, and H. Young. 2005. Carbon dioxide scrubbing systems alter the ripe fruit volatile profiles in controlled-atmosphere stored 'Hayward' kiwifruit. *Postharvest Biol. Technol.* 35:133–141.

Burg, S.P. and E.A. Burg. 1965. Gas exchange in fruits. *Physiol. Plant.* 18:870–884.

Burg, S.P. and E.A. Burg. 1969. Interaction of ethylene, oxygen and carbon dioxide in the control of fruit ripening. *Qualitas Plant Mater. Veg.* 19:185–200.

Butler, P. 2001. Smart packaging—intelligent packaging for food, beverages, pharmaceutical and household products. *Mater. World* 9(3):11–13.

Bylaite, E., A.S. Meyer, and J. Adler-Nissen. 2003. Changes in macroscopic viscosity do not affect the release of aroma aldehydes from a pectinaceous food model system of low sucrose content. *J. Agr. Food Chem.* 51:8020–8026.

Calvo, G. and G.O. Sozzi. 2004. Improvement of postharvest storage quality of 'Red Clapp's' pears by treatment with 1-methylcyclopropene at low temperature. *J Hort. Sci. Biotechnol.* 79:930–934.

Cameron, A.C., P.C. Talasila, and D.W. Joles. 1995. Predicting film permeability needs for modified-atmosphere packaging of lightly processed fruits and vegetables. *HortScience* 30(1):25–34.

Chervin, C., J. Speirs, B. Loveys, and B.D. Patterson. 2000. Influence of low oxygen storage on aroma compounds of whole pears and crushed pear flesh. *Postharvest Biol. Technol.* 19:279–285.

Cia, P., E.A. Benato, J.M.M. Sigrist, C. Sarantopóulos, L.M. Oliveira, and M. Padula. 2006. Modified atmosphere packaging for extending the storage life of 'Fuyu' persimmon. *Postharvest Biol. Technol.* 42:228–234.

Cliff, M.A., O.L. Lau, and M.C. King. 1998. Sensory characteristics of controlled atmosphere- and air-stored 'Gala' apples. *J. Food Qual.* 21:239–249.

Cliffe-Byrnes, V. and D. O'Beirne. 2007. The effects of modified atmospheres, edible coating and storage temperatures on the sensory quality of carrot discs. *Intl. J. Food Sci.* 42:1338–1349.

Clijsters, H. 1965. Malic acid metabolism and initiation of the internal breakdown in Jonathan apples. *Physiol. Plant.* 18:85–93.

Cohen, E., Y. Shalom, and I. Rosenberger. 1990. Postharvest ethanol buildup and off-flavor in Murcott tangerine fruits. *J. Amer. Soc. Hort. Sci.* 115:775–778.

Crisosto, C.H., D. Garner, and G. Crisosto. 2002a. Carbon-dioxide-enriched atmospheres during cold storage limit losses from Botrytis but accelerate rachis browning of 'Red globe' table grapes. *Postharvest Biol. Technol.* 26:181–189.

Crisosto, C.H., D. Garner, and G. Crisosto. 2002b. High carbon dioxide atmospheres affect stored 'Thompson Seedless' table grapes. *HortScience* 37:1074–1078.

Cunningham, D.G., T.E. Acree, J. Barnard, R.M. Butts, and P.A. Braell. 1986. Charm analysis of apple volatiles. *Food Chem.* 19:137–147.

Daillant-Spinnler, B., H.J.H. MacFie, P.K. Beyts, and D. Hedderley. 1996. Relationships between perceived sensory properties and major preference directions of 12 varieties of apples from the Southern hemispheres. *Food Qual. Pref.* 7:113–126.

Daniels-Lake, B., R.K. Prange, S.O. Gaul, K.B. McRae, R. de Antueno, and D. McLachlan. 2007. A musty "off" flavor in Nova Scotia potatoes is associated with 2,4,6-trichloroanisole released from pesticide-treated soils and high soil temperature. *J. Am. Soc. Hort. Sci.* 132:112–119.

Da Silva, M.A.A.P., V. Elder, C.L. Lederer, D.S. Lundahl, and M.R. McDaniel. 1993. Flavor properties and stability of a corn-based snack: Relating sensory, gas-chromatography, and mass-spectrometry data. In: G. Charalambous (Ed.), *Shelf Life Studies of Foods and Beverages*, Elsevier, Amsterdam, pp. 707–738.

Day, B.P.F. 1996. High oxygen modified atmosphere packaging for fresh prepared produce. *Postharvest News Info.* 7:31N–34N.

DeEll, J.R., P.M.A. Toivonen, F. Cornut, C. Roger, and C. Vigneault. 2006. Addition of sorbitol with $KMnO_4$ improves broccoli quality retention in modified atmosphere packages. *J. Food Qual.* 29:65–75.

Defilippi, B.G., A.M. Dandekar, and A.A. Kader, 2004. Impact of suppression of ethylene action or biosynthesis on flavor metabolites in apple (*Malus domestica Borkh*) fruits. *J. Agr. Food Chem.* 52:5694–5701.

Del-Valle, V.E. Almenar, P. Hernández-Muñoz, J. Lagarón, R. Catala, and R. Gavara. 2004. Volatile organic compound permeation through porous polymeric films for modified atmosphere packaging of foods. *J. Sci. Food Agr.* 84:937–942.

Deng, Y., Y. Wu, and Y. Li. 2005. Effects of high O_2 levels on post-harvest quality and shelf life of table grapes during long-term storage. *Eur. Food Res. Technol.* 221:392–397.

DePooter, H.L., J.P. Montens, G.A. Willaert, P.J. Dirinck, and N.M.O. Schamp. 1983. Treatment of Golden Delicious apples with aldehydes and carboxylic acids: effect on the headspace composition flavor. *J. Agr. Food Chem.* 31:813–818.

DePooter, H.L., M.R. Van Acker, and N.M. Schamp. 1987. Aldehyde metabolism and the aroma quality of stored Golden Delicious apples. *Phytochemistry* 26:89–92.

de Roos, K.B. 2006. Modelling aroma interactions in food matrices. In: A. Voilley and P. Etiévant (Eds.), *Flavour in Food*. Woodhead Publications Ltd., Cambridge, England, pp. 229–259.

DeRovira, D. 1997. *Flavor Nomenclature Workshop: An Odor Description and Sensory Evaluation Workshop*, Flavor Dynamics Manual, Inc. Summerset, NJ.

Dever, M.C., R.A. MacDonald, M.A. Cliff, and W.D. Lane. 1996. Sensory evaluation of sweet cherry cultivars. *HortScience* 31(1):150–153.

Dirinck, P., H. De Pooter, and N. Schamp. 1989. Aroma development in ripening fruits. In: *ACS Symp. Ser. Amer. Chem. Soc.* 388:23–34.

Dirinck, P., H. DePooter, and N. Schamp. 1989. Aroma development in ripening fruits. In: Flavor Chemistry: Trends and Developments. ACS Symp. Ser. 388:23–34. → Dirinck, P., H. De Pooter, and N. Schamp. 1989. Aroma development in ripening fruits. *ACS Symp. Ser. Amer. Chem. Soc.* 388:23–34.

Dixon, J. and E.W. Hewett. 2001. Exposure to hypoxia conditions alters volatile concentrations of apple cultivars. *J. Sci. Food Agr.* 81:22–29.

Dourtoglou, V.G., N.G. Yannovits, V.G. Tychopoulos, and M.M. Vamvakias. 1994. Effect of storage under CO_2 atmosphere on the volatile, amino acid, and pigment constituents in red grape (*Vitis vinifera* L. var. *Agiorgitiko*). *J. Agr. Food Chem.* 42:338–344.

Drawert, F. 1975. Biochemical formation of aroma components. In: M.H. Maarse, P.J. Growenen, (Eds.), Aroma Research, Proceedings of the International Symposium on Aroma Research, Pudoc, Wageningen, the Netherlands.

Drewnowski, A. 1997. Taste preferences and food intake. *Annu. Rev. Nutr.* 17:237–253.

Dury-Brun, C., P. Chalier, S. Desobry, and A. Voilley. 2007. Multiple mass transfers of small volatile molecules through flexible food packaging. *Food Rev. Int.* 23:199–255.

Echeverría, G., I. Lara, T. Fuentes, M.L. Lopez, J. Graell, and J. Puy. 2004a. Assessment of relationships between sensory and instrumental quality of controlled-atmosphere-stored 'Fuji' apples by multivariate analysis. *J. Food Sci.* 69:S368–S375.

Echeverría, G., T. Fuentes, J. Graell, I. Lara, and M.L. Lopez. 2004b. Aroma volatile compounds of 'Fuji' apples in relation to harvest date and cold storage technology: A comparison of two seasons. *Postharvest Biol. Technol.* 32:29–44.

Ekman, J.H., M. Clayton, W.V. Biasi, and E.J. Mitcham. 2004. Interactions between 1-MCP concentration, treatment interval and storage time for 'Bartlett' pears. *Postharvest Biol. Technol.* 31:127–136.

Fan, X. and J.P. Mattheis. 1999. Impact of 1-methylcyclopropene and methyl jasmonate on apple volatile production. *J. Agr. Food Chem.* 47:2847–2853.

Fan, X. and J.P. Mattheis. 2000. Reduction of ethylene-induced physiological disorders of carrots and iceberg lettuce by 1-methylcyclopropene. *HortScience* 35:1312–1314.

Fan, L., J. Song, R.M. Beaudry, and P.D. Hildebrand. 2006. Effect of hexanal vapor on spore viability of *Penicillium expansum*, lesion development on whole apples and fruit volatile biosynthesis. *J. Food Sci.* 71:M105-M109.

Fayoux, S.C., A.M. Seuvre, and A.J. Voilley. 1997. Aroma transfers in and through plastic packagings: Orange juice and D-limonene. A review. Part II: Overall sorption mechanisms and parameters. *Packaging Technol. Sci.* 10:145–160.

Fellman, J.K., D.S. Mattinson, B.C. Bostick, J.P. Mattheis, and M.E. Patterson. 1993. Ester biosynthesis in 'Rome' apples subjected to low-oxygen atmospheres. *Postharvest Biol. Technol.* 3:201–214.

Fellman, J.K., T.W. Miller, D.S. Mattinson, and J.P. Mattheis. 2000. Factors that influence biosynthesis of volatile flavor compounds in apple fruits. *HortScience* 35(6):1026–1033.

Ferenczi, A., J. Song, M. Tian, K. Vlachonasios, D. Dilley, and R. Beaudry. 2006. Volatile ester suppression and recovery following 1-methycyclopropene application to apple fruit. *J. Am. Soc. Hort. Sci.* 131:691–701.

Ferreira, V., J. Pet'Ka, and M. Aznar. 2002. Aroma extract dilution analysis: Precision and optimal experimental design. *J. Agr. Food Chem.* 50:1508–1514.

Fidler, J.C. and C.J. North. 1969. Production of volatile organic compounds by pears. *J. Sci. Food Agr.* 20:518–520.

Fidler, J.C. and C.J. North. 1971. The effect of periods of anaerobiosis on the storage of apples. *J. Hort. Sci.* 46:213–221.

Fidler, J.C., B.G. Wilkinson, K.L. Edney, and R.O. Sharples. 1973. The biology of apple and pear storage. Farnham Royal Slough, England: Commonwealth Agricultural Bureaux.

Fonseca, S.C., F.A.R. Oliveira, and J.K. Brecht. 2002. Modelling respiration rate of fresh fruits and vegetables for modified atmosphere packages: A review. *J. Food Eng.* 52(2):99–119.

Forney, C.F. 2001. Horticultural and other factors affecting aroma volatile composition of small fruit. *HortTechnology* 11:529–538.

Forney, C.F. 2007. New innovations in the packaging of fresh-cut produce. *Acta Hort.* 746:53–60.

Forney, C.F. and M.A. Jordan. 1998. Induction of volatile compounds in broccoli by postharvest hot-water dips. *J. Agr. Food Chem.* 46(12):5295–5301.

Forney, C.F. and M.A. Jordan. 1999. Aerobic production of methanethiol and other compounds by Brassica vegetables. *HortScience* 34(4):696–699.

Forney, C.F., M.A. Jordan, K.U.K.G. Nicholas, and J.R. DeEll. 2000a. Volatile emissions and chlorophyll fluorescence as indicators of freezing injury in apple fruit. *HortScience* 35:1283–1287.

Forney, C.F., W. Kalt, and M.A. Jordan. 2000b. The composition of strawberry aroma is influenced by cultivar, maturity, and storage. *HortScience* 35:1022–1026.

Forney, C.F., J.P. Mattheis, and R.K. Austin. 1991. Volatile compounds produced by broccoli under anaerobic conditions. *J. Agr. Food Chem.* 39(12):2257–2259.

Forney, C.F., J. Song, P.D. Hildebrand, L. Fan, and K.B. McRae. 2007. Interactive effects of ozone and 1-methylcyclopropene on decay resistance and quality of stored carrots. *Postharvest Biol. Technol.* 45:341–348.

Gil, M.I., M.A. Conesa, and F. Artés. 2002. Quality changes in fresh cut tomatoes as affected by modified atmosphere packaging. *Postharvest Biol. Technol.* 25:199–207.

Gillies, S.I., M.A. Cliff, P.M.A. Toivonen, and M.C. King. 1997. Effect of atmosphere on broccoli sensory attributes in commercial MAP and microperforated packages. *J. Food Qual.* 20:105–115.

Golding, J.B., D. Shearer, S.G. Wyllie, and W.B. McGlasson. 1998. Application of 1-MCP and propylene to identify ethylene-dependent ripening processes in mature banana fruit. *Postharvest Biol. Technol.* 14:87–98.

González-Aguilar, G.A., C.Y. Wang, and J.G. Buta. 2000. Maintaining quality of fresh-cut mangoes using antibrowning agents and modified atmosphere packaging. *J. Agr. Food Chem.* 48:4204–4208.

González, J., A. Ferrer, R. Oria, and M.L. Salvador. 2008. Determination of O_2 and CO_2 transmission rates through microperforated films for modified atmosphere packaging of fresh fruits and vegetables. *J. Food Eng.* 86:194–201.

Gore, H.C. and D. Fairchild. 1911. Experiments on the processing of persimmons to render them nonastringent. U.S. Department of Agriculture, Bureau of Chemistry, Bull. 141.

Gorin, N. 1973. Several compounds in Golden Delicious apples as possible parameters of acceptability. *J. Agr. Food Chem.* 21:670–673.

Gorny, J.R. 2005. Leveraging innovative fresh-cut technologies for competitive advantage. *Acta Hort.* 687:141–147.

Gorny, J.R. and A.A. Kader. 1997. Low oxygen and elevated carbon dioxide atmospheres inhibit ethylene biosynthesis in preclimacteric and climacteric apple fruit. *J. Am. Soc. Hort. Sci.* 122:542–546.

Gouble, B., D. Fath, and P. Soudain. 1995. Nitrous oxide inhibition of ethylene production in ripening and senescing climacteric fruits. *Postharvest Biol. Technol.* 5:311–321.

Griffith, D.G. and N.A. Potter. 1949. Effects of the accumulation of substances produced by apples in gas storage. *J. Hort. Sci.* 25:10–18.

Gross, K.C., C.Y. Wang, and M. Saltveit. 2004. *The Commercial Storage of Fruits, Vegetables and Florist and Nursery Stocks. USDA Agricultural Handbook 66.* http://www.ba.ars.usda.gov/hb66/contents.html

Guadagni, D.G., S. Okano, R.G. Buttery, and H.K. Burr. 1966. Correlation of sensory and gas–liquid chromatographic measurements of apple volatiles. *Food Technol.* 20(4):166–169.

Hagenmaier, R.D. 2002. The flavor of mandarin hybrids with different coatings. *Postharvest Biol. Technol.* 24:79–87.

Hagenmaier, R.D. and R.A. Baker. 1994. Internal gases, ethanol content and gloss of citrus fruit coated with polyethylene wax, carnauba wax, shellac or resin at different application levels. *Proc. Fla. State Hort. Soc.* 107:261–265.

Hagenmaier, R.D. and K. Grohmann. 1999. Polyvinyl acetate as a high-gloss edible coating. *J. Food Sci.* 64:1064–1067.

Hagenmaier, R.D. and K. Grohman. 2000. Edible food coatings containing polyvinyl acetate. U.S. Patent Number 6,162,475.

Hagenmaier, R.D. and P.E. Shaw. 2002. Changes in volatile components of stored tangerines and other specialty citrus fruits with different coatings. *J. Food Sci.* 67:1742–1745.

Hakim, A., M.E. Austin, D. Batal, S. Gullo, and M. Khatoon. 2004. Quality of fresh-cut tomatoes. *J. Food Qual.* 27:195–206.

Hansen, K., L. Poll, C.E. Olsen, and M.J. Lewis. 1992. The influence of oxygen concentration in storage atmospheres on the post-storage volatile ester production of 'Jonagold' apples. *Lebensm. Wiss. u-Technol.* 25:457–461.

Harb, J., R. Bisharat, and J. Streif. 2008. Changes in volatile constituents of blackcurrants (*Ribes nigrum* L. cv. '*Titania*') following controlled atmosphere storage. *Postharvest Biol. Technol.* 47:271–279.

Hardenburg, R.E., A.E. Watada, and C.Y. Wang. 1986. *The Commercial Storage of Fruits, Vegetables, and Florist and Nursery Stocks. USDA Agricultural Handbook 66.* 136 pp.

Hatfield, S.G.S. and B.D. Patterson. 1975. Abnormal volatile production by apples during ripening after controlled atmosphere storage. *Colloques Intl. C.N.R.S.* 238:57–62.

Hoa, T.T., M.N. Ducamp, M. Lebrun, and E.A. Baldwin. 2002. Effect of different coating treatments on the quality of mango fruit. *J. Food Qual.* 25:471–486.

Hoebrichts, F.A., L.H.W. Van Der Plas, and E.J. Woltering. 2002. Ethylene perception is required for the expression of tomato ripening-related genes and associated physiological changes even at advanced stages of ripening. *Postharvest Biol. Technol.* 26:125–133.

Holley, A. 2006. Processing information about flavour. In: A. Voilley and P. Etiévant (Eds.), *Flavour in Food*, CRC Press, Boca Raton, FL, pp. 36–61.

Hong, G., G. Peiser, and M.I. Cantwell. 2000. Use of controlled atmospheres and heat treatment to maintain quality of intact and minimally processed green onions. *Postharvest Biol. Technol.* 20:53–61.

Isenberg, F.M.R. 1979. Controlled atmosphere storage of vegetables. *Hort. Rev.* 1:337–394.

Jacobsson, A., T. Nielsen, I. Sjöholm, and K. Wendin. 2004. Influence of packaging material and storage condition on the sensory quality of broccoli. *Food Qual. Pref.* 15:301–310.

Jacxsens, L., F. Devlieghere, C. Van der Steen, and J. Debevere. 2001. Effect of high oxygen modified atmosphere packaging on microbial growth and sensorial qualities of fresh-cut produce. *Intl. J. Food Microbiol.* 71:197–210.

Jamie, P. and M.E. Saltveit. 2002. Postharvest changes in broccoli and lettuce during storage in argon, helium, and nitrogen atmospheres containing 2% oxygen. *Postharvest Biol. Technol.* 26:113–116.

Järvenpää, E.P., Z. Zhang, R. Huopalahti, and J.W. King. 1998. Determination of fresh onion (*Allium cepa* L.) volatiles by solid phase microextraction combined with gas chromatography-mass spectroscopy. *Z. Lebensm. Uters Forsch. A.* 207:39–43.

Jeong, J. and D.J. Huber. 2004. Suppression of avocado (*Persea americana* Mill.) fruit softening and changes in cell wall matrix polysaccharides and enzyme activities: Differential responses to 1-MCP and delayed ethylene application. *J. Am. Soc Hort. Sci.* 129:752–759.

Juteau-Vigier, A., S. Atlan, I. Deleris, E. Guichard, I. Souchon, and I.C. Trelea. 2007. Ethyl hexanoate transfer modeling in carrageenan matrices for determination of diffusion and partition properties. *J. Agr. Food Chem.* 55:3577–3584.

Kader, A.A. 1986. Biochemical and physiological basis for effects of controlled and modified atmospheres on fruits and vegetables. *Food Technol.* 40(5):99–104.

Kader, A.A. 1997. Biological basis of O_2 and CO_2 effects on postharvest life of horticultural perishables. *Proceedings of the Seventh International Controlled Atmosphere Research Conference, Vol. 4: Vegetables and Ornamentals*, 13–18 July, 1997, Davis, CA, pp. 160–163.

Kader, A.A. and S. Ben-Yehoshua. 2000. Effects of superatmospheric oxygen levels on postharvest physiology and quality of fresh fruits and vegetables. *Postharvest Biol. Technol.* 20:1–13.

Kader, A.A., G.A. Chastagner, L.L. Morris, and J.M. Ogawa. 1978. Effects of carbon monoxide on decay, physiological response, ripening, and composition of tomato fruits. *J. Am. Soc. Hort. Sci.* 103:665–670.

Kader, A.A. and J.A. Klaustermeyer. 1977. Physiological responses of some vegetables to carbon monoxide. In: D.H. Dewey (Ed.), Controlled *Atmospheres for the Storage and Transport of Perishable Agricultural Commodities*. Hort. Rep. 28, Michigan State University, East Lansing, MI, pp. 197–202.

Kader, A.A., M.A. Stevens, M. Albright-Holton, L.L. Morris, and M. Algazi. 1977. Effect of fruit ripeness when picked on flavor and composition in fresh market tomatoes. *J. Am. Soc. Hort. Sci.* 102:724–731.

Ke, D., L. Goldstein, M. O'Mahony, and A.A. Kader. 1991. Effects of short-term exposure to low O_2 and high CO_2 atmospheres on quality attributes of strawberries. *J. Food Sci.* 56:50–54.

Ke, D., L. Zhou, and A.A. Kader. 1994. Mode of oxygen and carbon dioxide action on strawberry ester biosynthesis. *J. Am. Soc. Hort. Sci.* 119:971–975.

Kidd, P. and C. West. 1936. Gas storage of fruit. IV. Cox's Orange Pippin apples. *J. Pomol.* 14:276–294.

Kim, J.G., Y. Lua, and K.C. Gross. 2004. Effect of package film on the quality of fresh-cut salad savoy. *Postharvest Biol. Technol.* 32:99–107.

Knee, M. and S.G.S. Hatfield. 1981. The metabolism of alcohols by apple fruit tissue. *J. Sci. Food Agr.* 32:593–600.

Kollmannsberger, H. and R.G. Berger. 1992. Precursor atmosphere storage induced flavour changes in apples cv. *Red Delicious. Chem. Mikrobiol. Technol. Lebensm.* 14:81–86.

Lafuente, M.T., G. Lopez-Galzez, M. Cantwell, and S.F. Yang. 1996. Factors influencing ethylene-induced isocoumarin formation and increased respiration in carrots. *J. Am. Soc. Hort. Sci.* 121:537–542.

Lamikanra, O. and O.A. Richard. 2002. Effect of storage on some volatile aroma compounds in fresh-cut cantaloupe melon. *J. Agr. Food Chem.* 50:4043–4047.

Lamikanra, O. and O.A. Richard. 2004. Storage and ultraviolet-induced tissue stress effects on fresh-cut pineapple. *J. Sci. Food Agr.* 84:1812–1816.

Land, D.G. 1994. Savory flavors—An overview. In: J.R. Piggott and A. Paterson (Eds.), *Understanding Natural Flavors*. Blackie Academic & Professional, Chapman & Hall, New York, pp. 298–306.

Lara, I., G. Echeverria, J. Graell, and M.L. Lopez. 2007. Volatile emission after controlled atmosphere storage of Mondial Gala apples (*Malus domestica*): Relationship to some involved enzyme activities. *J. Agr. Food Chem.* 55:6087–6095.

Lara, I., R.M. Miro, T. Fuentes, G. Sayez, J. Graell, and M.L. Lopez. 2003. Biosynthesis of volatile aroma compounds in pear fruit stored under long-term controlled-atmosphere conditions. *Postharvest Biol. Technol.* 29:29–39.

Larsen, M. and C.B. Watkins. 1995. Firmness and aroma composition following short-term high carbon dioxide treatments. *HortScience* 30:303–305.

Lelièvre, J.M., A. Latché, B. Jones, M. Bouzayen, and J.C. Pech. 1997. Ethylene and fruit ripening. *Physiol. Plant.* 101:727–739.

Leshem, Y.Y. and Y. Pinchasov. 2000. Non-invasive photoacoustic spectroscopic determination of relative endogenous nitric oxide and ethylene content stoichiometry during the ripening of strawberries *Fragaria anannasa* (Duch.) and avocados *Persea americana* (Mill.). *J. Exp. Bot.* 51:1471–1473.

Leshem, Y.Y., R.B.H. Wills, and V.V.V. Ku. 1998. Evidence for the function of the free radical gas—nitric oxide(NO)—as an endogenous maturation and senescence regulating factor in higher plants. *Plant Physiol. Biochem.* 36:825–833.

Lester, G. 2006. Consumer preference quality attributes of melon fruits. *Acta Hort.* 712:175–181.

Lidster, P.D., H.J. Lightfoot, and K.B. McRae. 1983. Production and regeneration of principal volatiles in apples stored in modified atmospheres and air. *J. Food Sci.* 48:400–402.

Linssen, J.P.H., A. Verheul, J.P. Roozen, and M.A. Posthumus. 1991. Absorption of flavor compounds by packaging material: Drink yogurts in polyethylene bottles. *Intl. Dairy J.* 1:33–40.

Liu, C.L., C.K. Hsu, and M.M. Hsu. 2007. Improving the quality of fresh-cut pineapples with ascorbic acid/sucrose pretreatment and modified atmosphere packaging. *Packaging Technol. Sci.* 20:337–343.

Lopez, M.L., R. Miro, and J. Graell. 2001. Quality and aroma production of Doyenne du Comice pears in relation to harvest date and storage atmosphere. *Food Sci. Technol. Intl.* 7:293–500.

Lopez, M.L., C. Villatoro, T. Fuentes, J. Graell, I. Lara, and G. Echeverria. 2007. Volatile compounds, quality parameters and consumer acceptance of 'Pink Lady®' apples stored in different conditions. *Postharvest Biol. Technol.* 43:55–66.

Lougheed, E.C. and R. Lee. 1991. Ripening, CO_2 and C_2H_4 production, and quality of tomato fruits held in atmospheres containing nitrogen and argon. *Proceedings of the Fifth International Controlled Atmosphere Research Conference, 1989*, Wenatchee, Washington, vol. 2, pp. 141–150.

Lubbers, S. 2006. Texture-aroma interactions, In: A. Voilley and P. Etiévant (Eds.), *Flavour in Food*. Woodhead Publishing Ltd., Cambridge, England, pp. 327–344.

Lurie, S. 1998. Postharvest heat treatments. *Postharvest Biol. Technol.* 14:257–269.

Lurie, S, C. Pre-Aymard, U. Ravid, O. Larkov, and E. Fallik. 2002. Effect of 1-methylcyclopropene on volatile emission and aroma in cv. Anna apples. *J. Agr. Food Chem.* 50:4251–4256.

Ma, S.S., P.M. Chen, D.M. Varga, and S.R. Drake. 2000. Ethylene capsule promotes early ripening of 'd'Anjou' pears packed in modified atmosphere bags. *J. Food Qual.* 23:245–259.

Malundo, T.M.M., R.L. Shewfelt, G.O. Ware, and E.A. Baldwin. 2001a. Sugars and acids influence flavor properties of mango (*Mangifera indica*). *J. Amer. Soc. Hort. Sci* 126:115–121.

Malundo, T.M.M., R.L. Shewfelt, G.O. Ware, and E.A. Baldwin. 2001b. Alternative methods for consumer testing and data analysis used to identify critical flavor properties of mango (*Mangifera indica* L.). *J. Sens. Stud.* 16:119–214.

Mannheim, C.H. and T. Soffer. 1996. Permeability of different wax coatings and their effect on citrus fruit quality. *J. Agr. Food Chem.* 44:919–923.

Mannheim, C.H., J. Milt, and A. Letzter. 1987. Interaction between polyethylene laminated cartons and aseptically packed citrus juices. *J. Food Sci.* 52(3):737–740.

Marin, A.B., T.E. Acree, and J. Barnard. 1988. Variation in odor detection thresholds determined by charm analysis. *Chem. Senses* 13:435–444.

Marin, A.B., A.E. Colonna, K. Kudo, E.M. Kupferman, and J.P. Mattheis. 2009. Measuring consumer response to 'Gala' apples treated with 1-methylcyclopropene (1-MCP). *Postharvest Biol. Technol.* 51:73–79.

Marrero, A. and A.A. Kader. 2006. Optimal temperature and modified atmosphere for keeping quality of fresh-cut pineapples. *Postharvest Biol. Technol.* 39:163–168.

Martens, M. and H. Martens. 1986. Partial least square regression. In: J.R. Piggott (Ed.), *Statistical Procedures in Food Research*. Elsevier Applied Science, Oxford, pp. 293–359.

Martens, M., E. Risvik, and H. Martens. 1994. Matching sensory and instrumental analyses. In: J.R. Piggott and A. Paterson (eds.), *Understanding Natural Flavors*, Blackie Academic & Professional, Chapman & Hall, New York, pp. 60–76.

Martín-Belloso, O. and R. Soliva-Fortuny. 2006. Effect of modified atmosphere packaging on the quality of fresh-cut fruits. *Stewart Postharvest Rev.* 1:3 www.stewartpostharvest.com

Martínez-Romero, D., F. Guillén, S. Castillo, D. Valero, and M. Serrano. 2003. Modified atmosphere packaging maintains quality of table grapes. *J. Food Sci.* 68(5):1838–1843.

Matsuo, T. and S. Ito. 1977. On mechanism of removing astringency in persimmon fruits by carbon dioxide treatment. I. Some properties of the two processes in the deastringency. *Plant Cell Physiol.* 18:17–25.

Mattheis, J.P., D.A. Buchanan, and J.K. Fellman. 1991. Change in apple fruit volatiles after storage in atmospheres inducing anaerobic metabolism. *J. Agr. Food Chem.* 39:1602–1605.

Mattheis, J.P., D.A. Buchanan, and J.K. Fellman. 1995. Volatile compound production by Bisbee Delicious apples after sequential atmosphere storage. *J. Agr. Food Chem.* 43:194–199.

Mattheis, J.P., D.A. Buchanan, and J.K. Fellman. 1998. Volatile compounds emitted by 'Gala' apples following dynamic atmosphere storage. *J. Amer. Soc. Hort. Sci.* 123:426–432.

Mattheis, J.P., X. Fan, and L.C. Argenta. 2005. Interactive responses of Gala apple fruit volatile production to controlled atmosphere storage and chemical inhibition of ethylene action. *J. Agr. Food Chem.* 53:4510–4516.

Maul, F., S.A. Sargent, C.A. Sims, E.A. Baldwin, M.O. Balaban, and D.J. Huber. 2000. Storage temperatures affect tomato flavor and aroma quality. *J. Food Sci.* 65:1228–1237.

McDaniel, M.R., R. Miranda-Lopez, B.T. Watson, N.J. Micheals, and L.M. Libbey. 1990. Pinot noir aroma: A sensory/gas chromatographic approach. In: G. Charalambous (Ed.), *Flavors and Off-Flavors*, Elsevier, Amsterdam, pp. 23–36.

Meheriuk, M., D.L. McKenzie, B. Girard, A.L. Moyls, S. Weintraub, R. Hocking, and T. Kopp. 1997. Storage of 'Sweetheart' cherries in sealed plastic film. *J. Food Qual.* 20:189–198.

Meilgaard, M., G.V. Civille, and B.T. Carr. 1999. *Sensory Evaluation Techniques*, 3rd ed. CRC Press, Boca Raton, FL, pp. 387.

Misry, B.S., T. Reineccius, and L.K. Olson. 1997. Gas chromatography-olfactometry for the determination of key odorants in foods. In: R. Marsili (Ed.), *Techniques for Analyzing Food Aroma*, Marcel Dekker, New York, pp. 265–292.

Miszczak, A., C.F. Forney, and R.K. Prange. 1995. Development of aroma volatiles and color during postharvest ripening of 'Kent' strawberries. *J. Am. Soc. Hort. Sci.* 120(4):650–655.

Moya-Leon, M.A., M. Vergara, C. Bravo, M. Pereira, and C. Moggia. 2007. Development of aroma compounds and sensory quality of 'Royal Gala' apples during storage. *J. Hort. Sci. Biotech.* 82:403–413.

Murayama, H., D. Satoh, Y. Ohta, and T. Fukushima. 1995. Effect of relative humidity on ripening of 'LeLectier' pear fruit. *Acta Hort.* 398:187–194.

Nguyen, T.B.T., S. Ketsa, and W.G. van Doorn. 2004. Effect of modified atmosphere packaging on chilling-induced peel browning in banana. *Postharvest Biol. Technol.* 31:313–317.

Nielsen, T. and M. Jägerstad. 1994. Flavour scalping by food packaging. *Trends Food Sci. Technol.* 5:353–356.

Niskanen, R. and R. Dris. 2006. Stress responses of fruits and vegetables during storage. *J. Food Agr. Environ.* 4:202–208.

Nisperos-Carriedo, M.O., P.E. Shaw, and E.A. Baldwin. 1990. Changes in volatile flavor components of pineapple orange juice as influenced by the application of lipid and composite films. *J. Agr. Food Chem.* 38:1382–1387.

O'Mahony, M. 1995. Sensory measurement in food science: Fitting methods to goals. *Food Technol.* 49 (4):72–82.

Ozdemir, M. and J.D. Floros. 2004. Active food packaging technologies. *Critical Rev. Food Sci. Nutr.* 44:185–193.

Paik, J.S. and M.S. Writer. 1995. Prediction of flavor sorption using the Flory–Huggins equation. *J. Agr. Food Chem.* 43:175–178.

Paillard, N. 1975. Les aromes des fruits. *Proc. Symp. Intern. 'Les Aromes Alimentaires' Paris*, APRIA, Paris, pp. 37–46.

Palomer, X., I. Roig-Villanova, D. Grima-Calvo, and M. Vendrell. 2005. Effects of nitrous oxide (N_2O) treatment on the postharvest ripening of banana fruit. *Postharvest Biol. Technol.* 36:167–175.

Pariasca, J.A.T., T. Miyazaki, H. Hisaka, H. Nakagawa, and T. Sato. 2000. Effect of modified atmosphere packaging (MAP) and controlled atmosphere (CA) storage on the quality of snow pea pods (*Pisum sativum* L. var. *saccharatum*). *Postharvest Biol. Technol.* 21:213–223.

Parsons, C.S., R.E. Anderson, and R.W. Penny. 1970. Storage of mature-green tomatoes in controlled atmospheres. *J. Amer. Soc. Hort. Sci* 95:791–794.

Patterson, B.D., S.G.S. Hatfield, and M. Knee. 1974. Residual effects of controlled atmosphere storage on the production of volatile compounds by two varieties of apples. *J. Sci. Food Agr.* 25:843–849.

Paull, R.E. 1999. Effect of temperature and relative humidity on fresh commodity quality. *Postharvest Biol. Tech.* 15:263–277.

Perez, A.G. and C. Sanz. 2001. Effect of high-oxygen and high-carbon dioxide atmospheres on strawberry flavor and other quality traits. *J. Agr. Food Chem.* 49:2370–2375.

Pesis, E. and R. Ben-Arie. 1986. Carbon dioxide assimilation during postharvest removal of astringency from persimmon fruits. *Physiol. Plant.* 67:644–648.

Pesis, E., R. Ben Arie, O. Feygenberg, and F. Villamizar. 2005. Ripening of ethylene-pretreated bananas is retarded using modified atmosphere and vacuum packaging. *HortScience* 40(3):726–731.

Pesis, E., G. Zauberman, and I. Avissar. 1991. Induction of certain aroma volatiles in feijoa by postharvest application of acetaldehyde or anaerobic conditions. *J. Sci. Food Agr.* 54:329–337.

Plotto, A., M.R. McDaniel, and J.P. Mattheis. 1999. Characterization of 'Gala' apple aroma and flavor: Differences between controlled atmosphere and air storage. *J. Amer. Soc. Hort. Sci.* 124:416–423.

Plotto, A., J. Bai, J.A. Narciso, J.K. Brecht, and E.A. Baldwin. 2006. Ethanol vapor prior to processing extends fresh-cut mango storage by decreasing spoilage, but does not always delay ripening. *Postharvest Biol. Technol.* 39:134–145.

Plotto, A., K.L. Goodner, E.A. Baldwin, J. Bai, and N. Rattanapanone. 2004. Effect of polysaccharide coatings on quality of fresh cut mangoes (*Mangifera indica*). *Proc. Fla. State Hort. Soc.* 117:382–388.

Plotto, A., J.P. Mattheis, and M.R. McDaniel. 2000. Characterization of changes in 'Gala' apple aroma during storage using Osme analysis, a gas chromatography-olfactometry technique. *J. Amer. Soc. Hort. Sci.* 125:714–722.

Podoski, B.W., C.A. Sims, S.A. Sargent, J.F. Price, C.K. Chandler, and S.F. Okeefe. 1997. Effects of cultivar, modified atmosphere, and pre-harvest conditions on strawberry quality. *Proc. Fla. State Hort. Soc.* 110:246–252.

Porat, R., B. Weiss, L. Cohen, A. Daus, R. Goren, and S. Droby. 1999. Effects of ethylene and 1-methylcyclopropene on the postharvest qualities of 'Shamouti' oranges. *Postharvest Biol. Tech.* 15:155–163.

Portela, S.I. and M.I. Cantwell. 1998. Quality changes of minimally processed honeydew melons stored in air or controlled atmosphere. *Postharvest Biol. Technol.* 14:351–357.

Portela, S.I. and M.I. Cantwell. 2001. Cutting blade sharpness affects appearance and other quality attributes of fresh-cut cantaloupe melon. *J. Food Sci.* 66(9):1265–1270.

Pristijono, P., R.B.H. Wills, and J.B. Golding. 2006. Inhibition of browning on the surface of apple slices by short term exposure to nitric oxide (NO) gas. *Postharvest Biol. Technol.* 42:256–259.

Qadir, A. and F. Hashinaga. 2001. Inhibition of postharvest decay of fruits by nitrous oxide. *Postharvest Biol. Technol.* 22:279–283.

Randle, W.M. and J.E. Lancaster. 2002. Sulphur compounds in Alliums in relation to flavour quality. In: H.D. Rabinowitch and L. Currah (Eds.), *Allium Crop Science: Recent Advances*. CAB International, Wallingford, UK, pp. 329–356.

Reyes, A.A. and R.B. Smith. 1987. Effect of oxygen, carbon dioxide, and carbon monoxide on celery in storage. *HortScience* 22:270–271.

Riederer, M. and L. Schreiber. 2001. Protecting against water loss: Analysis of the barrier properties of plant cuticles. *J. Expt. Bot.* 52(363):2023–2032.

Risch, S.J. 1998. Flavor and aroma interactions between volatile organic compounds and packaging materials. *TAPPI Proc. 1998 Polymer, Laminations and Coating Conf.*, pp. 1157–1159.

Rizzolo, A., C. Sodi, and A. Polesello. 1991. Influence of ethylene removal on the volatile development in Passa Crassana pears stored in a controlled atmosphere. *Food Chem.* 42:275–285.

Rocculi, P., S. Romani, and M.D. Rosa. 2005. Effect of MAP with argon and nitrous oxide on quality maintenance of minimally processed kiwifruit. *Postharvest Biol. Technol.* 35:319–328.

Rocha, A.M.C.N., C.M. Brochado, R. Kirby, and A.M.M.B. Morais. 1995. Shelf-life of chilled cut orange determined by sensory quality. *Food Control* 6(6):317–322.

Rosenfeld, H.J., K. Roed Meberg, K. Haffner, and H.A. Sundell. 1999. MAP of highbush blueberries: Sensory quality in relation to storage temperature, film type and initial high oxygen atmosphere. *Postharvest Biol. Technol.* 16:27–36.

Rupérez, P. and G. Toledano. 2003. Celery by-products as a source of mannitol. *Eur. Food Res. Technol.* 216(3):224–226.

Saftner, R.A. 1999. The potential of fruit coating and film treatments for improving the storage and shelf-life qualities of 'Gala' and 'Golden Delicious' apples. *J. Amer. Soc. Hort. Sci.* 124:682–689.

Saftner, R.A., W.S. Conway, and C.E. Sams. 1999. Postharvest calcium infiltration alone and combined with surface coating treatments influence volatile levels, respiration, ethylene production, and internal atmospheres of 'Golden Delicious' apples. *J. Amer. Soc. Hort. Sci.* 124:553–558.

Sajilata, M.G., K. Savitha, R.S. Singhal, and V.R. Kanetkar. 2007. Scalping of flavors in packaged foods. *Compre. Rev. Food Sci. Food Safety* 6:17–35.

Salle, C. 2006. Odour-taste interactions in flavour perception. In: A. Voilley and P. Etiévant (Eds.), *Flavour in Food*. Woodhead Publishing Ltd., Cambridge, England, pp. 345–368.

Saltveit, M.E. 1999. Effect of ethylene on quality of fresh fruits and vegetables. *Postharvest Biol. Technol.* 15:279–292.

Sanz, L.C., J.M. Olías, and A.G. Pérez. 1997. Aroma biochemistry of fruits and vegetables. In: F.A. Tomás-Barberán and R.J. Robins (Eds.), *Phytochemistry of Fruits and Vegetables*. Clarendon Press. Oxford, pp. 125–155.

Sanz, L.C., A.G. Pérez, R. Olías, and J.M. Olías. 1999. Quality of strawberries packed with perforated polypropylene. *J. Food Sci.* 64(4):748–752.

Sarkar, S.K. and C.T. Phan. 1974. Effect of ethylene on the qualitative and quantitative composition of the phenol content of carrot roots. *Physiol. Plant.* 30:72–76.

Seljasen, R., H. Hoftun, and G.B. Bengtsson. 2001. Sensory quality of ethylene-exposed carrots (*Daucus carota* L., cv. 'Yukon') related to the contents of 6-methoxymellein, terpenes and sugars. *J. Sci. Food Agr.* 81:54–61.

Serra Bonvehí, J. and F. Ventura Coll. 1997. Evaluation of bitterness and astringency of polyphenolic compounds in cocoa powder. *Food Chem.* 60(3):365–370.

Shattuck, V.I., R. Yada, and E.C. Lougheed. 1988. Ethylene-induced bitterness in stored parsnips. *HortScience* 23:912.

Sharkey, P.J. and I.D. Peggie. 1984. Effects of high-humidity storage on quality, decay and storage life of cherry, lemon and peach fruits. *Sci. Hort.* 23:181–190.

Shaw, P.E., R.D. Carter, M.G. Moshonas, and G. Sadler. 1990. Controlled atmosphere storage of oranges to enhance aqueous essence and essence oil. *J. Food Sci.* 55:1617–1619.

Shaw, P.E., M.G. Moshonas, M.O. Nisperos-Carriedo, and R.D. Carter. 1992. Controlled-atmosphere treatment of freshly harvested oranges at elevated temperature to increase volatile flavor components. *J. Agr. Food Chem.* 40:1041–1045.

Shimoda, M., T. Ikegami, and Y. Osajima. 1988. Sorption of flavour compounds in aqueous solution into polyethylene film. *J. Sci. Food Agr.* 42:157–163.

Sisler, E.C. and S.M. Blankenship. 1996. Methods of counteracting an ethylene response in plants. U.S. Patent Number 5,518,988.

Sisler, E.C. and M. Serek. 1997. Inhibitors of ethylene responses in plants at the receptor level: Recent developments. *Physiol. Plant.* 100:577–582.

Sisler, E.C. and M. Serek. 2003. Compounds interacting with the ethylene receptor in plants. *Plant Biol.* 5:473–480.

Sivakumar, D. and L. Korsten. 2006. Influence of modified atmosphere packaging and postharvest treatments on quality retention of litchi cv. *Mauritius. Postharvest Biol. Technol.* 41:135–142.

Smith, S., J. Geeson, and J. Stow. 1987. Production of modified atmospheres in deciduous fruits by the use of films and coatings. *HortScience* 22:772–776.

Soequarto, L. and R.B.H. Wills. 2004. Short term fumigation with nitric oxide gas in air to extend the postharvest life of broccoli, green bean, and bok choy. *HortTechnology* 14:538–540

Soliva-Fortuny, R.C. and O. Martín-Belloso. 2003. Microbiological and biochemical changes in minimally processed fresh-cut Conference pears. *Eur. Food Res. Technol.* 217:4–9.

Solomos, T. 1987. Principles of gas exchange in bulky plant tissues. *HortScience* 22(5):766–771.

Song, J. and F. Bangerth. 1996. The effect of harvest date on aroma compound production from 'Golden Delicious' apple fruit and relationship to respiration and ethylene production. *Postharvest Biol. Technol.* 8:259–269.

Song, J., W. Deng, L. Fan, J. Verschoor, and R. Beaudry. 1997. Aroma volatiles and quality changes in modified atmosphere packaging. In: J. Gorny (Ed.). *Proceedings of the seventh International Controlled Atmosphere Research Conference, vol. 5. Fresh-cut fruits and vegetables and MAP.* University of California Postharvest Horticultural Series 19, pp. 89–95.

Song, J., L. Fan, C.F. Forney, and M.A. Jordan. 2001. Using volatile emissions and chlorophyll fluorescence as indicators of heat injury in apple fruit. *J. Amer. Soc. Hort. Sci.* 126:771–777.

Song, J., R. Leepipattanawit, W. Deng, and R.M. Beaudry. 1996. Hexanal vapor is a natural, metabolizable fungicide: Inhibition of fungal activity and enhancement of aroma biosynthesis in apple slices. *J. Amer. Soc. Hort. Sci.* 121:937–942.

Sozzi, G.O., G.D. Trinchero, and A.A. Fraschina. 2003. Delayed ripening of 'Bartlett' pears treated with nitric oxide. *J. Hort. Sci. Biotechnol.* 78:899–903.

Stebbins, R.L., A.A. Duncan, O.C. Compton, and D. Duncan. 1991. Taste ratings of new apple cultivars. *Fruit Varieties J.* 45(1):37–44.

Stern, D.J., R.G. Buttery, R. Teranishi, L. Ling, K. Scott, and M. Cantwell. 1994. Effect of storage and ripening on fresh tomato quality. Part I. *Food Chemistry*, vol. 49, Elsevier Applied Science, Essex, pp. 225–231.

Stewart, J.K. and M. Uota. 1976. Postharvest effect of modified levels of carbon monoxide, carbon dioxide, and oxygen on disorders and appearance of head lettuce. *J. Am. Soc. Hort. Sci.* 101:382–384.

Streif, J. and F. Bangerth, 1988. Production of volatile aroma substances by 'Golden Delicious' apple fruits after storage for various times in different CO_2 and O_2 concentrations. *J. Hort. Sci.* 63:193–199.

Tandon, K.S., E. Abegaz, R.L. Shewfelt, E.A. Baldwin, and J.W. Scott. 2000. Interrelationship of sensory descriptors and chemical composition as affected by harvest maturity and season on fresh tomato flavor. *Proc. Fla. State Hort. Soc.* 113:289–294.

Tanner, D.J., A.C. Cleland, L.U. Oparab, and T.R. Robertson. 2002. A generalised mathematical modelling methodology for design of horticultural food packages exposed to refrigerated conditions: Part 1, formulation. *Intl. J. Refrig.* 25(1):33–42.

Taylor, A.J. and R.S.T. Linforth. 1994. Methodology for measuring volatile profiles in the mouth and nose during eating. In: J. Maarse and D.G. van der Heij (Eds.), *Trends in Flavour Research*, Elsevier Science, New York, pp. 3–14.

Theologis, A. 1992. One rotten apple spoils the whole bushel: The role of ethylene in fruit ripening. *Cell* 70:181–184.

Thomas, M. and J.C. Fidler. 1933. Studies to zymasis: IV. Zymasis by apples in relation to oxygen concentration. *Biochem J.* 27:1629–1642.

Tindale, C.R., F.B. Whitfield, S.D. Levingston, and T.H.L. Nguyen. 1989. Fungi isolated from packaging materials: Their role in the production of 2,4,6-trichloroanisole. *J. Sci. Food Agr.* 49:437–447.

Toivonen, P., J. Walsh, E.C. Lougheed, and D.P. Murr. 1982. Ethylene relationships in storage of some vegetables. In: D.G. Richardson and M. Meheriuk, (Eds.). *Controlled Atmospheres for Storage and Transport of Perishable Agricultural Commodities*. Symposium Series No. 1, Oregon State University School of Agriculture. Timber Press, Beaverton, OR, pp. 299–307.

Ulrich, R. 1966. Qualite des fruits conserves en atmosphere controlee. *Qual. Plant. Materiae Veg.* 13:246–266.

Usenik, V.J. Fabcic, and F. Štamper. 2008. Sugars, organic acids, phenolic composition and antioxidant activity of sweet cherry (*Prunus avium* L.). *Food Chem.* 107:185–192.

van Ruth, S.M. 2001. Methods for gas chromatography-olfactometry: A review. *Biomol. Eng.* 17:121–128.

Voilley, A. and I. Souchon. 2006. Flavour retention and release from the food matrix: An overview. In: A. Voilley and P. Etiévant (Eds.), *Flavour in Food*, Woodhead Publishing Ltd., Cambridge, England, pp. 117–132.

Voirol, E. and N. Daget. 1987. Nasal and retronasal olfactory perception of a meat aroma. In: M. Martens, G.A. Dalen, and H. Russwurm (Eds.), *Flavor Science and Technology*, Wiley, New York, pp. 309–316.

Voon, Y.Y., N.S.A. Hamid, G. Rusul, A. Osman, and S.Y. Quek. 2007. Volatile flavour compounds and sensory properties of minimally processed durian (*Durio zibethinus* cv. D24) fruit during storage at 4°C. *Postharvest Biol. Technol.* 46:76–85.

Watada, A.E. 1986. Effects of ethylene on the quality of fruits and vegetables. *Food Technol.* 40:82–85.

Watada, A.E. and B.B. Aulenbach. 1979. Chemical and sensory qualities of fresh market tomatoes. *J. Food Sci.* 44:1013–1016.

Watada, A.E., R.C. Herner, A.A. Kader, R.J. Romani, and G.L. Staby. 1984. Terminology for the description of developmental stages of horticultural crops. *HortScience* 19:20–21.

Watkins, C.B. 2002. Ethylene synthesis, mode of action, consequences and control. In: M. Knee (Ed.), *Fruit Quality and Its Biological Basis*, Academic Press, Boca Raton, FL, pp. 180–224.

Watkins, C.B. 2006. The use of 1-methylcyclopropene (1-MCP) on fruits and vegetables. *Biotechnol. Adv.* 24:389–409.

Willaert, G.A., P.J. Dirinck, H.L. DePooter, and N.N. Schamp. 1983. Objective measurement of aroma quality of Golden Delicious apples as a function of controlled atmosphere storage time. *J. Agr. Food Chem.* 31:809–813.

Williams, A.A., O.G. Tucknott, and M.J. Lewis. 1977. 4-methoxyallylbenzene: An important aroma component of apples. *J. Sci. Food Agr.* 28:185–190.

Wills, R.B.H., V.V.V. Ku, and Y.Y. Leshem. 2000. Fumigation with nitric oxide to extend the post-harvest life of strawberries. *Postharvest Biol. Technol.* 18:75–79.

Wills, R.B.H. and W.B. McGlasson. 1970. Loss of volatiles by apples in cool storage: A differential response to increased water loss. *J. Hort. Sci.* 46:115–120.

Wills, R.B.H., P. Pristijono, and J.B. Golding. 2008. Browning on the surface of cut lettuce slices inhibited by short term exposure to nitric oxide (NO). *Food Chem.* 107:1387–1392.

Woodruff, R.E. 1977. Use of carbon monoxide in modified atmospheres for fruits and vegetables, in transit. In: D.H. Dewey (Ed.), *Controlled Atmospheres for the Storage and Transport of Perishable Agricultural Commodities*. Hort. Rep. 28, Michigan State University, East Lansing, MI, pp. 52–54.

Wszelaki, A.L. and E.J. Mitcham. 2000. Effects of superatmospheric oxygen on strawberry fruit quality and decay. *Postharvest Biol. Technol.* 20:125–133.

Yahia, E.M. 1991. Production of some odor-active volatiles by 'McIntosh' apples following low-ethylene controlled-atmosphere storage. *HortScience* 26:1183–1185.

Yahia, E.M., F.W. Liu, and T.E. Acree. 1990. Changes of some odor-active volatiles in controlled atmosphere-stored apples. *J. Food Qual.* 13:185–202.

Yahia, E.M., F.W. Liu, and T.E. Acree. 1991. Changes of some odor-active volatiles in low-ethylene controlled atmosphere stored apples. *Lebensm. Wiss. u-Technol.* 24;145–151.

Yahia, E.M., M. Rivera, and O. Hernandez. 1992. Response of papaya to short-term insecticidal oxygen atmosphere. *J. Amer. Soc. Hort. Sci.* 117:96–99.

Zhang, M., Z.G. Zhan, S.J. Wang, and J.M. Tang. 2008. Extending the shelf-life of asparagus spears with a compressed mix of argon and xenon gases. *Lebensm. Wiss. u-Technol.* 41:686–691.

Zhu, S. and J. Zhou. 2006. Effects of nitric oxide on fatty acid composition in peach fruits during storage. *J. Agr. Food Chem.* 54:9447–9452.

8

Effects on Physiological Disorders

Wendy C. Schotsmans, John M. DeLong, Christian Larrigaudière, and Robert K. Prange

CONTENTS

8.1 Introduction

Controlled and modified atmosphere storage of fruits and vegetables is employed to maintain the highest level of fruit and vegetable quality for relatively long periods of time by slowing down the respiration-related processes of senescence, and by providing environments unfavorable for pathogen or insect survival (low O_2 and/or high CO_2 at low, but above-freezing temperatures) (Prange and DeLong, 2006). Generally speaking, the capacity for controlled atmosphere (CA) or modified atmosphere (MA) storage gives unprecedented advantages to food-producing nations in securing long-term food supply for regional and national populations, supplier-controlled export options, and a level of food quality that is historically unparalleled. In light of these major advantages, CA and MA technologies are industry standards in developed countries that grow crops suitably stored in these environments. With MA in particular, crops can be sealed in the MA packaging immediately following harvest, shipped to ports of destination, and be ready for sale without significant loss of quality. Hence, MA (and in some cases, CA) technology is not necessarily confined to a singular site for the duration of the established storage atmosphere.

While the benefits of crop storage in CA and MA environments are many, a major concern does exist for the potential loss of crop quality and related economic revenue through the development of storage disorders, which are defined as areas of tissue degradation not directly caused by insects, diseases, or mechanical damage. At least five categories of physiological disorders exist, and are related to (1) nutrition, (2) temperature (usually low), (3) respiration, (4) senescence, and (5) miscellaneous (Wills et al., 2007). Each of these disorder types can be alleviated, exacerbated, or induced by CA or MA environments. Thus, it is imperative for storage physiologists generally, and storage operators specifically, to be familiar with the potential for disorder development in those crops being managed over the duration of CA or MA storage. An additional problem is that symptom description can be very similar for different physiological disorders and it is often very difficult to distinguish amongst them. It is one thing to identify the symptoms but agreement on what the disorder is and identifying the underlying mechanism based on the symptoms can be difficult, especially when working with a disorder that may vary amongst different cultivars or different species.

The sensitivity to many disorders is determined by cultivar and/or preharvest conditions, which include genetic factors, cultural practices, and climatic factors (Watkins and Rao, 2003). Cultural practices and their influence on disorder development have been studied extensively and include rootstock, soil management, nutrition, training and pruning practices, crop load, application of growth regulators, and final product size (Emongor et al., 1994; Ferguson et al., 1999; Kruger and Truter, 2003). Climatic factors such as temperature, solar irradiance, and soil and water characteristics play an important role in the regional propensities for disorder development and in season-to-season variation within a particular region (Ferguson et al., 1999; Lau, 1998; Park et al., 1997; Paull and Reyes, 1996; Thomai et al., 1998). For some species (specifically apples, pears, peaches, and nectarines) and some of their disorders, cultivar sensitivity may be well known, as well as the cultural and climatic circumstances that increase this sensitivity. For other species and disorders, this is still speculative.

The identification of disorder symptoms is often difficult because some disorders occur simultaneous and thus show a multitude of symptoms (e.g., superficial scald and internal breakdown, watercore, and internal breakdown). Also, the terminology used to describe symptom expression is not standardized across all growing regions. Along with the problem of correct symptom diagnosis is the need to know the preharvest and storage conditions (e.g., temperature, atmosphere, duration) that exacerbate disorder expression. This knowledge is helpful for correct disorder identification and for possible amelioration. For specific symptom description as well as the storage temperatures and duration associated with specific disorders for individual horticultural crops, several good references are the USDA Agriculture Handbook Number 66 (Gross et al., 2004), the postharvest produce fact sheets (UC Davis, 2008), the apple disorder guide (Jones et al., 2007), and AAFC publication 1737/E (Meheriuk et al., 1994).

Generally speaking, there are at least three levels of investigation when attempting to understand the development of postharvest physiological disorders: (1) identification of the preharvest conditions that predispose a fruit or vegetable towards disorder development; (2) identification of symptoms and the physiological processes that alleviate or exacerbate symptoms; and (3) research to determine the underlying mechanisms, the initiating process, and the first triggering events. Much of the emphasis in this chapter will be on these last two levels, emphasizing disorders that are (1) alleviated or exacerbated and (2) induced by CA and MA environments.

8.2 Unifying Physiological Concept of Disorder Development

Many research groups now agree (and this is evident from the publications on the topic) that a specific stress (cold, atmosphere, senescence) induces either an ethylene-related response or an oxidative response (Figure 8.1) that will result in damage to cell membranes, cellular decompartmentation, and all the physiological processes (e.g., browning, texture changes) responsible for the symptoms that develop as evidenced in the following discussion.

8.2.1 Ethylene-Related Response

Since the first mention of the importance of ethylene in plant physiology at the beginning of the twentieth century (Neljubow, 1901), considerable research has been done on this plant hormone. Ethylene is a key hormone involved in many aspects of plant physiology (Yang and Hoffman, 1984) including growth, development, maturation, ripening, and senescence. In stress physiology, for example, ethylene is a determinant of plant response to wounding (Abeles et al., 1992) and healing, but is also important in the establishment of plant defense mechanisms against pathogen attack (Boller, 1991). Ethylene production greatly depends on environmental conditions and especially on the temperature and O_2 and CO_2 concentrations surrounding the stored crop (Figure 8.1). Following the Arrhenius model (Bisswanger, 2002), low temperature slows down the activity of enzymes and in the case of ethylene, the activity of 1-aminocyclopropane-1-carboxylic-acid (ACC) synthase (ACS) and ACC oxidase (ACO). As an oxidase, ACO is greatly inhibited by low O_2 levels (Poneleit and Dilley, 1993) but the case of ACS is more complex. Although inhibited by low temperature, ACS residual activity remains sufficient to cause an increase in the amount of ACC (Larrigaudière and Vendrell, 1993) and a significant accumulation of ACS transcripts

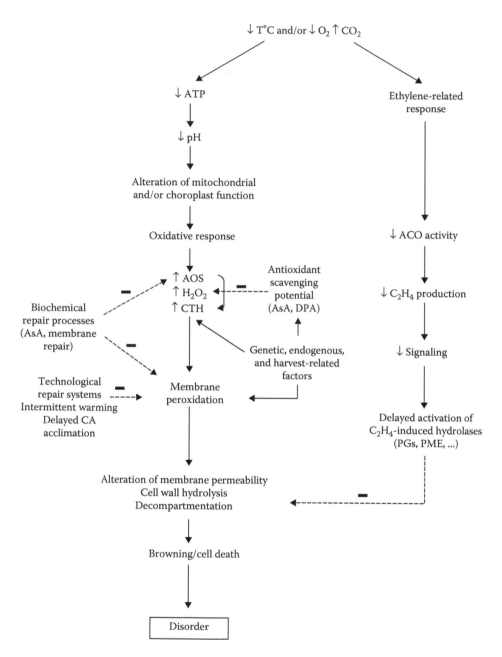

FIGURE 8.1
Stress response pathways for CA/MA-related postharvest storage disorders (▬ indicates inhibition).

(Lelièvre et al., 1997). Both increases explain the burst of ethylene production observed in different fruit species upon removal from storage (Larrigaudière and Vendrell, 1993).

In climacteric fruits, the residual ethylene levels produced during CA and MA storage are not sufficient to trigger the climacteric process. Some ethylene receptors remain in their active state and prevent the activation of transcription factors and genes responsible for the ethylene response (Binder and Bleecker, 2003; Serek et al., 2006). As a consequence, the ethylene-promoted activity of hydrolase enzymes (e.g., polygalacturonase [PG] and pectin methyl esterase [PME]) is delayed. The delay in cell wall hydrolysis will affect the later steps in the process (Figure 8.1) and can eventually delay or prevent disorder occurrence. In non-climacteric fruits and in vegetables, an opposite behavior may be observed. The ethylene levels are generally low in these products and additional reduction can be harmful. In 'Shamouti' oranges for example, higher chilling incidence occurs when ethylene levels are reduced (Porat et al., 1999).

8.2.2 Oxidative Response

Whereas the ethylene response depends on the ethylene production behavior of the fruit or vegetable, the oxidative response appears to be much more ubiquitous. The current hypothesis is that the first response is located in the chloroplasts and mitochondria. Cytoplasmic acidosis is a common response of plants and animals to stress, for example, anoxia and it determines cell death in plant cells (Perata and Alpi, 1993). Cytoplasmic acidosis is the result of lactate production (Drew, 1997), vacuole leakage of H^+ (De Sousa and Sodek, 2002), or hydrolysis of nucleoside triphosphates such as adenosyl triphosphate (ATP) (Gout et al., 2001). Roberts et al. (1984) studied cytoplasmic acidosis in relation to hypoxia in root meristems of higher plants and suggested the following order of events. When oxidative phosphorylation ceases due to hypoxia, lactic acid production is initiated but does not continue because cytoplasmic acidification will stimulate ethanol production, and inhibit lactic acid production. This will slow down the establishment of cytoplasmic acidosis for several hours. Eventually, cytoplasmic acidosis will occur because of leakage of acid from the acidic vacuoles. This acidification was later confirmed to also happen in pear fruit under very low (0.25 kPa) O_2 conditions (Nanos and Kader, 1993), probably as a result of lactate production or H^+ pump dysfunction. Later, Ke et al. (1994) proposed a model describing how low O_2, high CO_2, and acidification of the cytoplasm stimulate the ethanol fermentation pathway in pears. This could result in the accumulation of toxic levels of ethanol or acetaldehyde (Pintó et al., 2001). Cytoplasmic acidosis also results in chlorophyll fluorescence (CF) shifts in chloroplasts (DeLong et al., 2004b; Prange et al., 2002), a pattern that has been recognized as perhaps one of the first symptoms of plant senescence. During fluorescence studies with fruit (Prange et al., 2002) and vegetables (Prange et al., 2003), they found a rapid increase in F_o and decline in F_v/F_m when the O_2 concentration in the storage atmosphere was decreased to anoxic levels. This O_2 concentration was closely related to the minimum acceptable low O_2 reported in the CA literature. They associate the increase in F_o and decline in F_v/F_m with an inability of the light-harvesting complex in the thylakoid membrane to transfer the exciton energy to the reaction center of Photosystem II; this "extra" energy is subsequently fluoresced (Prange et al., 1997, 2002). The energy in the form of electrons may be channeled into the production of active oxygen species (AOS) as illustrated by the higher levels of hydrogen peroxide (H_2O_2) found immediately after storage (in air but also and especially in CA storage) in pear fruit (Larrigaudière et al., 2001). The production of AOS is a normal, regulated cellular phenomenon with production and removal levels in balance. It is only when AOS generation becomes enhanced that damaging cascade reactions are initiated (Lurie, 2003); at this point, oxidative stress occurs (Hodges, 2003). These cascade reactions result in the production of hydroxyl radicals ($OH^\bullet$) (Lester, 2003) that will denaturate proteins, mutate DNA and peroxidise lipids (Hodges, 2003). Membrane peroxidation, when not balanced by regeneration and repair will result in a loss of membrane integrity, which has far-reaching consequences. These range from diminished regulation of cellular energy generation to loss of membrane semipermeability, and of metabolite and ion compartmentation (Parkin et al., 1989) indicated by increased electrolyte leakage (Pesis et al., 1994). Eventually, this will result in widespread cell wall hydrolysis and decompartmentation and eventually browning and cell death.

These last few steps in the process have been widely discussed in the literature as will be evident in the following paragraphs.

8.2.3 Defense against Oxidative Response

As the model in Figure 8.1 suggests, plants do have several defense mechanisms in place to reduce or prevent oxidative stress. These defense mechanisms prevent the accumulation of AOS and H_2O_2, prevent membrane peroxidation, or ensure continued membrane repair.

Plants have also developed several enzymatic and nonenzymatic antioxidant defense mechanisms to prevent oxidative damage. An antioxidant can be defined as any compound that can quench reactive oxygen species without being converted to a destructive radical (Masia, 2003). Some examples of nonenzymatic antioxidants involved in the destruction of free radicals are α-tocopherol, carotenoids, ascorbate, and glutathione. The best studied of these are glutathione and ascorbate; they basically scavenge AOS in a similar fashion. Their reduced forms, GSH and AsA, respectively, react with AOS producing their oxidized forms, GSSH, monodehydroascorbate, and dehydroascorbate, which are stable (Lurie, 2003). Glutathione and ascorbate are also involved in the ascorbate–glutathione cycle found in chloroplasts, mitochondria, and the cytoplasm where they undergo continuous oxidation and reduction (Foyer et al., 1994). The relation between a decrease in ascorbate and the development of browning disorders has been widely discussed in the literature (Franck et al., 2003b, 2007; Larrigaudière et al., 2001; Lentheric et al., 1999; Veltman et al., 2003; Zerbini et al., 2002). Plants that have higher levels of endogenous antioxidants, especially higher ascorbate (being the most abundant and most reactive) at harvest (Lentheric et al., 1999) and during CA storage (Veltman et al., 1999), are more likely to resist the negative effects of oxidative stress (Lentheric et al., 1999; Veltman et al., 2003).

Effective destruction of AOS also requires the action of several antioxidant enzymes in addition to nonenzymatic antioxidant compounds. The enzymatic antioxidant system involves a wide range of enzymes including superoxide dismutase (SOD), catalase (CAT), peroxidase (POX), and ascorbate peroxidase (APX). SOD is involved in the conversion of the superoxide radical ($O_2^{\bullet-}$) to H_2O_2, which may in turn be toxic for the cells were it not for the action of CAT, POX, and APX, which convert H_2O_2 to water. Plants normally increase the production of one or more of the components of the antioxidant system in response to stress and rely on a sufficient supply of antioxidants to be effective (Noctor and Foyer, 1998) as illustrated by the reduced capacity to prevent oxidative damage with a decrease in ascorbate or in antioxidant enzymatic potential (Larrigaudière et al., 2004c; Noctor and Foyer, 1998).

Continued membrane repair is further guaranteed by ensuring the energy status (ATP) and ATP production rate are high enough to sustain the basic metabolic requirements for maintenance of membrane function (Saquet et al., 2001).

8.2.4 Enzymatic Browning

Enzymatic browning is one of the most important color reactions that affect fruits and vegetables. It is catalyzed by the enzyme polyphenol oxidase (PPO), which oxidizes polyphenol compounds in fruits and vegetables to quinones. These are then polymerized to dark, insoluble melanin pigments (Friedman, 1996; Mathew and Parpia, 1971; Mayer, 1987). Ascorbate, a natural antioxidant, can prevent this by converting *o*-quinones back to diphenols (Franck et al., 2007). In healthy intact fruits and vegetables, enzymes and substrates are separated by membranes and thus compartmentalized in different organelles (Larrigaudière et al., 1998). The vast majority of phenolic compounds in higher plant cells are located in the vacuole. PPO is normally located solely in the plastids but does exist free in the cytoplasm of senescing or degenerating tissue, for example, ripening fruit, harvested vegetables due to altered membrane permeability, and membrane damage (Vaughn and Duke, 1984). Thus, PPO and phenolics can only mix if the membranes separating them are damaged. Therefore, the ultimate cause of enzymatic browning must be sought in processes affecting membrane integrity (Franck et al., 2007).

In anthocyanin-containing tissue, there seems to be an additional browning process (Jiang et al., 2004). In litchi, pericarp browning is due to the oxidation of phenolics as well as degradation of red pigments by PPO (Tian et al., 2005) and POX (Gong and Tian, 2002). PPO cannot directly oxidize anthocyanins, but it can oxidize products of anthocyanin degradation. Zhang et al. (2001) propose anthocyanase hydrolyses anthocyanins to form anthocyanidin. Under high pH conditions, anthocyanidin can transform into an *o*-phenol similar in structure to catechol, which is oxidized by PPO (Tian et al., 2005) and/or POX (Gong and Tian, 2002) and the oxidative products then accelerate anthocyanidin degradation (Jiang et al., 2004).

8.3 Disorders Occurring in Air Storage that Are Alleviated or Aggravated by CA/MA

8.3.1 Low-Temperature or Chilling Injury

Chilling injury (CI) is a physiological disorder due to exposure to low but not freezing temperatures for a suitable duration during which tissue is damaged (Lurie and Crisosto, 2005). CI is most common in tropical and subtropical plants but temperate zone plants are also affected. A biochemical basis to explain the CI mechanism has not been fully established mainly because of the variety of symptoms and responses seen in different fruits and vegetables upon chilling (Parkin et al., 1989). It is now clear that CI is genetically determined and preharvest factors govern the plant's sensitivity to chilling stress (Lurie and Crisosto, 2005).

8.3.1.1 Symptoms of Chilling Injury

There are basically three types of CI symptoms: external symptoms (most obvious), flesh browning (often a signal of a more advanced stage of the disorder), or textural change. The latter has been described in detail for plum, peach, and nectarines (Lurie and Crisosto, 2005).

The external skin or peel-related symptoms are described as pitting, sunken spots, water-soaked lesions, scalding, brown staining, surface bronzing, skin browning, discoloration, rind breakdown, and irregular (blotchy) color development. CI generally starts with some pitting and skin discoloration (scalding), but these often deteriorate further into water-soaked sunken lesions on the skin. These areas are important entry points for secondary infections, which often result in other symptoms typical of increased decay sensitivity. Examples of fruits with mainly external symptom are zucchini squash (Wang and Buta, 1999), grapefruit (Ezz et al., 2004; Vakis et al., 1970), pomegranate (Elyatem and Kader, 1984), chili pepper (Kader et al., 1975), avocado (Pesis et al., 2002), melon (Flores et al., 2004), feijoa (Gaddam et al., 2005), okra (Ilker and Morris, 1975), cucumber (Eaks, 1956), eggplant (Rodriguez et al., 2001), carambola (Ali et al., 2004), tomato (Kader and Morris, 1975), kiwifruit (Lallu et al., 2005), and citrus (Porat et al., 2004). Apples and pears also develop external symptoms referred to as superficial scald but they will be discussed in the following section since they present a special case where the browning reaction can be linked to a specific chemical process involving α-farnesene metabolism.

Flesh browning (associated with CI) is described as brown discoloration of the white segments separating the arils and pale color (loss of red color) of the arils in pomegranate (Elyatem and Kader, 1984), flesh, or pit cavity browning in peach and nectarine (Lurie and Crisosto, 2005), internal flesh browning (gray pulp, pulp spot, vascular browning) in avocado (Ali et al., 2004), browning of seeds in tomato (Kader and Morris, 1975), internal browning that begins in the flesh around the pit and radiates outward toward the skin as time progresses in olives (Nanos et al., 2002), core flush, flesh browning, breakdown, soft scald, ribbon scald, deep scald, or diffuse browning of the outer cortex in apples (Meheriuk et al., 1994), and browning of seeds and pulp tissue in eggplant (Rodriguez et al., 2001). Other flesh-related symptoms connected with low-temperature storage include development of a ring or zone of granular, water-soaked tissue in the outer pericarp at the stylar end of kiwifruit (Lallu, 1997) and pulp spot and grey pulp in avocado, which are curtailed by CA (Kruger et al., 1999).

The best documented symptoms of CI are mealiness/woolliness and leatheriness. Either cohesiveness of the flesh digresses during CI, resulting in mealy apples (Arana et al., 2007) or woolly peaches and nectarines (Arana et al., 2007; Lurie and Crisosto, 2005), or the fruit lacks juice and has a hard, dry texture (leatheriness) (Lurie and Crisosto, 2005; Luza et al., 1992). These changes in texture often coincide with, or are followed by, flesh or pit cavity browning and in the more advanced stages, flesh tissue separation, and cavity formation occur (Lurie and Crisosto, 2005).

8.3.1.2   Effect of CA/MA

For some of these crops, CA storage can reduce the degree of CI (Prange and DeLong, 2006). Lowering O_2 ameliorates CI in zucchini squash (Mencarelli et al., 1983), Chinese cabbage (Hermansen and Hoftun, 2005), and some olive cultivars (Nanos et al., 2002). A combination of lowered O_2 (0.5–15 kPa) and increased CO_2 (1.5–1 kPa) reduces CI for MA-packaged (Meir et al., 1998) or CA-stored avocado (Kruger et al., 1999) and 'Charentais' melon (Flores et al., 2004), Chinese cabbage (Hermansen and Hoftun, 2005), apple (DeLong et al., 2006), plum (Sive and Resnizky, 1979), and peach (Tian et al., 2004; Wang et al., 2005).

Often, increasing the CO_2 concentration to 5–20 kPa is more effective in alleviating CI compared with lowering the O_2 concentration. An increased CO_2 concentration during storage reduces CI in zucchini squash (Mencarelli, 1987), chili (hot) pepper (Kader et al., 1975), peach and nectarine (Anderson, 1982; Anderson et al., 1969; Burmeister and Harman, 1998; Wade, 1981; Wang et al., 2005), and okra (Ilker and Morris, 1975). High (>5 kPa) CO_2 treatments before low-temperature storage also reduce pitting and other symptoms of CI in grapefruit (Ezz et al., 2004; Hatton and Cubbedge, 1982; Hatton et al., 1975; Vakis et al., 1970), oranges (Porat et al., 2004) and other citrus (Purvis and Cubbedge, 1990), avocado (Vakis et al., 1970), and okra (Ilker and Morris, 1975) as does intermittent treatment of avocado with 20 kPa CO_2 (Marcellin and Chaves, 1983). Although in most cases CA storage reduces CI, there are cases where increasing the CO_2 concentration accentuates the problem. Examples are cucumber (Eaks, 1956), tomato (Kader and Morris, 1975), orange, and grapefruit at chilling temperatures (Purvis and Cubbedge, 1990), honeydew melon (Lipton, 1990c), some green Greek olive cultivars (Nanos et al., 2002), and some apple cultivars (Meheriuk et al., 1994; Sharples and Johnson, 1987). High CO_2 concentrations during CA storage can also cause external CI symptoms on orange and grapefruit (Purvis and Cubbedge, 1990) and honeydew melon (Lipton, 1990c).

8.3.1.3 *Physiological Background*

Most of the articles about CI deal with treatments to alleviate CI without investigating the causal cellular mechanism responsible for the injury. Nonetheless, it is generally accepted that CI development is linked with membrane damage or dysfunction (Ali et al., 2004; Lurie and Crisosto, 2005; Parkin et al., 1989; Pesis, 2005; Pesis et al., 1994; Wolfe, 1987). This likely happens through an induction of the oxidative response (Figure 8.1) as confirmed by research on cherimoya where chilling temperatures result in cytoplasmic acidosis (Munoz et al., 2001), which would trigger such a response. The loss of subcellular compart-mentation as a result of membrane damage or dysfunction can cause enzymatic browning reactions as described before, leading to skin and flesh discoloration reactions. However, other pigments might also be involved. In pomegranate for instance, changes in carote-noid, acyl lipid, and phenylpropanoid metabolism are associated with scald development and these changes are decreased or prevented by CA storage (Defilippi et al., 2006).

Membrane dysfunction also provides an explanation for mealiness or woolliness, which has been consistently linked with breakdown of the middle lamella causing cell separation rather than breakage or sheering. This is in contrast with leatheriness that is character-ized by a high degree of cell wall thickening (Luza et al., 1992). Numerous studies have dealt with the biochemical process behind the development of mealiness but the results often are contradictory (Lurie and Crisosto, 2005). Mealiness is a texture problem; hence, the likely suspects involved in its development can be found in the enzymes responsible for softening of fruits, i.e., endo-β-1,4-glucanase (EGase), endo-β-1,4-mannase, pectate lyase, PME, PG, and β-galactosidase (Huber et al., 2003). We will not elaborate on these cell wall changes in this discussion and refer to Brummell et al. (2004) and Lurie and Crisosto (2005) for an in-depth discussion of this topic.

Generally, cold storage affects the activity of these cell wall-modifying enzymes, altering the properties of the primary cell wall and middle lamella. The result is that under a strain, the cell walls will separate rather than rupture, which reduces the amount and availability of free juice (Brummell et al., 2004). Ethylene regulates the synthesis of many cell wall-modifying enzymes, including endo-PG and EGase, although others such as PME may be controlled by different factors (Brummell et al., 2004). However, Uthairatanakij et al. (2005) did not find a disruption of ethylene metabolism in peach at chilling temperatures with ACC levels unaffected by low temperatures with or without CA storage. This does not mean ethylene cannot be involved in CI since CA does affect the activity of the enzymes involved in ethylene synthesis in apples. In 'Granny Smith' apples, ACO has been shown to accumulate during chilling and ACS activity is stimulated (Lelièvre et al., 1995) although in this case, it does not lead to CI. It is reasonable to assume similar systems exist in other fruits. CA storage could then interfere with ACO since it requires CO_2 and O_2 for its activity (Poneleit and Dilley, 1993). Therefore, ethylene is inhibited during storage in CA or stimulation of ethylene production would be prevented, but will recover following removal to air.

Pesis (2005) suggests that low levels of acetaldehyde and ethanol (precursors of natural aroma compounds) can reduce CI symptoms by altering membrane function and through a reduction of ethylene production or a decrease in the activity of cell wall enzymes. This could also explain how CA storage alleviates CI. During hypoxic conditions, higher concentrations of acetaldehyde and ethanol are induced and high CO_2 can result in dark CO_2 fixation resulting in increased acetaldehyde production (Pesis, 2005). This is one mechanism through which CA storage could alleviate CI.

CI has also been linked to an increase in putrescine (a diamine linked to various plant stress conditions). Serrano et al. (1996) propose that the effect of polyamines in reducing CI may be a result of the polyamines linking with anionic compounds in the cell membrane and capturing free radicals. This will lower the membrane phase transition temperature fluidity and slow down lipid peroxidation, and thus stabilize the lipid bilayer, prevent membrane deterioration, and will result in increased cell viability (Serrano et al., 1996). The CI-induced increase in putrescine is delayed or prevented by CA/MA storage (McDonald and Kushad, 1986; Rodriguez et al., 2001). CA/MA storage also decreases spermidine levels where they normally remain constant (McDonald and Kushad, 1986; Rodriguez et al., 2001). This can be considered an indication that under CA/MA condition, the low temperature no longer initiates a stress response.

For avocado, the occurrence of foul taste and odor (a symptom of CI) may be caused by the alteration of the internal metabolism, which leads to an increase of the levels of anaerobic respiration and, as a consequence, of abnormal metabolites (Morris, 1982). The reduction of CI-related flesh browning in peach in CA (5 kPa O_2 + 5 kPa CO_2) and high-O_2 CA (70 kPa O_2 + 0 kPa CO_2) has been linked to a delay in the reduction of SOD and CAT during the first 30 days of storage (Wang et al., 2005).

Attempts have been made to show an association between chilling resistance and proline (a main component of extensin). Purvis (1981) found that peel proline levels are higher in grapefruit that are subsequently more resistant to CI (in the form of flavedo pitting) during storage. Ezz et al. (2004) observed that using elevated CO_2 during subsequent storage to reduce flavedo pitting results in lower proline content in grapefruit flavedo, which they speculate may be due to enhanced proline incorporation during storage into proteins, resulting in enhanced stability at lower temperatures.

Research on CA delay of physiological pitting in 'Hayward' kiwifruit has indicated some interesting links with volatiles as three CO_2 removal systems (N_2 flushing, activated C scrubbing, and hydrated lime scrubbing) alter the volatile profiles of the store atmospheres and also affect the incidence of physiological pitting (Lallu et al., 2005).

8.3.1.4 *Superficial or Storage Scald*

In this section, we will focus on superficial scald and not soft scald, which was discussed in the previous section. Superficial scald is characterized by light uneven browning of the skin or more precisely the hypodermis (DeLong and Prange, 2003) and renders the crop unmarketable as fresh fruit (Meheriuk et al., 1994). It starts in the outer hypodermal cell layer and gradually the cell content of all hypodermal layers turns brown (DeLong and Prange, 2003). In severe cases, the cells collapse and sunken areas become apparent. Superficial scald is a typical expression of CI (Watkins et al., 1995) in that it develops during or after extended periods of cold storage.

Since the hypodermal tissue is metabolically active and contains chloroplasts and mitochondria, damage probably starts, as previously described in the general model (Figure 8.1), by a dysfunction of the membrane of these organelles (DeLong and Prange, 2003). Membrane dysfunction will promote the release of AOS, which may promote peroxidation and more specifically the peroxidation of α-farnesene to conjugated trienes or trienols and further to 6-methyl-5-hepten-2-one (MHO) (Ma and Chen, 2003; Mir et al., 1999; Whitaker and Saftner, 2000) that have been recognized as the determining causes of superficial scald. However, other authors make mention of an unknown volatile X as the final causal factor of superficial scald (Wang and Dilley, 2000). The oxidative stress theory is consistent with the results obtained in 'Empire' apples (Burmeister and Dilley, 1995) and in 'Granny Smith' (Abdallah et al., 1997). It is also in accordance with the protective effect obtained with the use of the antioxidants, diphenylamine (DPA) or ethanol (Scott et al., 1995).

Although the oxidative stress theory is plausible, the relationship with the ethylene pathway has been more widely described in the literature and more specifically in an interesting paper of Du and Bramladge (1994). More recently, and coinciding with the use of the ethylene inhibitor 1-methylcyclopropene (1-MCP) on a commercial scale, this evidence has been reinforced and substantiated. 1-MCP reduces the accumulation of MHO (Fan et al., 1999; Rupasinghe et al., 2000a; Watkins et al., 2000), as well as α-farnesene (Apollo Arquiza et al., 2005; Fan et al., 1999; Isidoro and Almeida, 2006; Lurie et al., 2005; Schotsmans et al., 2009; Watkins et al., 2000) through inhibition of upregulation of α-farnesene synthase (Gapper et al., 2006; Lurie et al., 2005), the rate-limiting enzyme in the synthesis of α-farnesene (Rupasinghe et al., 2000b). Inhibited synthesis of α-farnesene may explain why 1-MCP-treated fruits are more resistant to scald, but not why immature fruits that normally have lower levels of α-farnesene are more sensitive. In this last case, other factors have to be taken into account including the higher capability of immature fruit to induce their climacteric during cold storage (Larrigaudière and Vendrell, 1993) and/or possible differences in preventing the oxidation of α-farnesene.

CA storage can decrease superficial scald occurrence (Colgan et al., 1999; Wang and Dilley, 1999) as can initial low oxygen stress (ILOS) (Wang and Dilley, 2000) and dynamic CA (DeLong et al., 2007; Zanella et al., 2005). These storage techniques will reduce respiration rate, ethylene production, and metabolic activity. The first two techniques also result in slightly elevated ethanol production and its subsequent metabolism may suppress the activity of the biochemical mechanism required for scald expression (DeLong and Prange, 2003; Wang and Dilley, 2000).

8.3.2 Ethylene Injury

Russet spotting is a physiological disorder in lettuce caused by ethylene (Gross et al., 2004; Leshuk and Saltveit, 1990) with concentrations as low as 0.1 ppm increasing the incidence and severity of russet spotting (Klaustermeyer et al., 1974). It appears as small brown spots on both sides of the midrib that may spread over the leaf blade in severe cases (Ke and Saltveit, 1988). The main changes that are observed are cell wall thickening and cell discoloration. Through increased phenylalanine ammonia-lyase (PAL) activity and POX activity, ethylene induces the phenylpropanoid pathway (Tomás-Barberán et al., 1997). This pathway is responsible for the synthesis of a wide variety of secondary metabolites, including lignin, cinnamic acid, flavonoids, and chlorogenic acid derivatives (Ke and Saltveit, 1988). The flavonoids and chlorogenic acid derivatives are oxidized by PPO to *o*-quinones, which form brown-colored polymers (Mathew and Parpia, 1971; Mayer, 1987), causing cell discoloration. Furthermore, other derivatives for example, toxic quinones can accumulate and may contribute to cell death, resulting in pit-like depressions (Ke and Saltveit, 1988). Russet spotting is inhibited by lowering O_2 below 8 kPa or increasing CO_2 to 5 kPa or more (Klaustermeyer et al., 1974; Lougheed, 1987; Singh et al., 1972), most likely through decreased ethylene production. This is probably accomplished since ACO requires CO_2 and O_2 for its activity (Poneleit and Dilley, 1993). In practice, increasing the CO_2 concentration is not done for lettuce because it induces brown stain and heart leaf injury, which are known as CO_2-induced disorders.

8.3.3 Senescence-Related Disorders

Senescence is defined as the last phase in the life of a plant, when degradation processes dominate synthesis metabolism. When a vegetable is harvested, the senescence/degradation process starts immediately because biologically the plant part no longer has any useful function. Senescence involves lipid peroxidation, which is the onset of the loss of membrane integrity and leads to membrane leakage and ultimately cell death (Lester, 2003). This is different for fruit, since they will continue the ripening process after harvest, a developmental stage not clearly separated from senescence. Contrary to senescence during which lipid peroxidation occurs, ripening consists of certain developmental changes that occur without loss of membrane integrity (Lester, 2003). Because ripening and senescence overlap, it is often difficult to distinguish between these two developmental phases.

Lester (2003) describes the process by which membranes deteriorate through the enzymatic senescence pathway or autocatalytic lipid degradation as outlined by Paliyath and Drouillard (1992). This results in leakiness of cellular membranes (Dörnenburg and Davies, 1999) and the production of fatty acid hydroperoxides. During normal fruit growth, development, and ripening, these catabolic products of membrane degradation are recycled and the membranes are regenerated. During senescence, membrane regeneration stops (how this shift occurs is unclear) and the fatty acid hydroperoxides undergo several AOS-driven reactions, resulting membrane peroxidation, membrane dysfunction, and organelle decompartmentation (Figure 8.1). Since molecular oxygen is required for the production of AOS, MA and CA environments will reduce oxygen-driven AOS production, including formation of the destructive hydroxyl radical (Lester, 2003).

Concurrently, cold storage and CA/MA storage will delay the ethylene-related response (Figure 8.1) and as a consequence, the ethylene-promoted senescence processes and especially the activity of hydrolase enzymes are delayed or inhibited.

8.3.3.1 Internal Browning in Apple and Pear

There is generally some confusion regarding the terminology used to refer to several disorders in apples and pears involving browning of the cortex and/or core and subsequent or concurrent tissue breakdown. Because the symptoms are not always clearly defined or different, it is often hard to distinguish among them (Franck et al., 2007). In this section we will focus on senescent browning (SB, also referred to in the literature as internal browning, internal breakdown, core browning, and core breakdown), which is the consequence of senescence and fermentation (Larrigaudière et al., 2004b), and can be described as softening and browning of poorly defined areas of cortex tissue with darkening or browning of the skin near the affected areas (Meheriuk et al., 1994). Brown heart, where the cortex is drier and where cavities are present, will be discussed later. It is difficult to differentiate between these two disorders, especially in their earliest stage. Thus, they might be two distinct disorders or a different expression of the same one (Larrigaudière et al., 2004b). They both seem to start with an induction of the fermentative pathway (decarboxylation of pyruvate to acetaldehyde and further to ethanol) under adverse atmospheric conditions but have different physiological bases.

In SB-damaged fruit, local unfavorable atmospheric conditions (low O_2, high CO_2) can cause anoxia and a change in the normal cellular metabolism from aerobic respiration to fermentation (Franck et al., 2007), resulting in ethanol accumulation and cell death. Although a significant negative relationship exists between SB and CAT activity (Larrigaudière et al., 2004b), it is not likely that oxidative processes are important in SB development. SB is likely due to senescence and acetaldehyde accumulation from fermentation. The former is evidenced by the fact that SB can also develop during long-term air storage. SB is also negatively correlated with alcohol dehydrogenase activity, which converts acetaldehyde to ethanol, which is less toxic in the cell. CA storage merely accelerates symptom expression because it can result in local unfavorable atmospheric conditions (Larrigaudière et al., 2004b) when respiration and gas transport are not at equilibrium or the externally applied concentrations are too limiting (low O_2, high CO_2). These unfavorable atmospheric conditions can also occur in prolonged air storage because senescence-related membrane breakdown will occur and the cytoplasm of dying cells leak into the intercellular spaces that would normally be filled with air, thus compromising the gas transport pathway that follows air-filled intercellular spaces (Franck et al., 2007).

Another indication that internal browning and breakdown in pears is a senescence-related disorder is the importance of ethylene (the most important promoter of fruit senescence) in its development, as confirmed in experiments where 1-MCP treatments of pears reduce its incidence (Valentines et al., 2002). Although introduction of low temperatures will delay ethylene-promoted senescence (Figure 8.1) and these disorders (Larrigaudière et al., 2004a), the slowing of respiration by CA conditions is more important. Upon removal from cold storage and exposure to higher temperatures, the disorder can spread very rapidly, indicating a sudden increase in ethylene-related senescence.

8.3.3.2 Flesh Bleeding in Peach

Flesh bleeding or reddening in peach occurs when the anthocyanin normally located in distinctive cells near the pit leaks out of these cells, resulting in red rays extending out from the pit. During long-term cold storage, the color can diffuse outward and the whole area around the pit appears red; in severe cases, the entire flesh is red. Although only a cosmetic problem, which does not affect fruit taste or texture, it can still result in serious economic losses, especially in the peach canning industry (Lurie and Crisosto, 2005). It is often linked with CI but has more recently been associated with fruit senescence, since it is related to a decrease in organic acids in the tissue (Lurie and Crisosto, 2005). Senescence is marked by the development of leakiness in cellular membranes (Dörnenburg and Davies, 1999; Lester, 2003), which could account for the leaking of anthocyanins out of cells and into adjacent cells. Flesh bleeding is prevented by CA storage (Lurie, 1993; Retamales et al., 1992) because CA storage slows down metabolism in general and retards oxidative stress and senescence (Lester, 2003). Flesh reddening has also been linked with impaired ethylene production since 1-MCP treatment aggravates the disorder (Dong et al., 2001) but in this research, flesh reddening occurred together with CI-induced woolliness, which was also aggravated by the 1-MCP treatment. Therefore, it is possible the increased flesh reddening was due to the increased woolliness and the effect of 1-MCP cannot be separated for the two symptoms. Nevertheless, the negative effect of 1-MCP is likely due to blocking the ethylene response pathway, whereas CA/MA storage merely reduces ethylene production, delaying the ethylene response rather than blocking it.

8.3.3.3 Pericarp Browning

Litchi pericarp browning is due to senescence-induced enzymatic browning. Litchi fruit pericarp cells rapidly senesce after harvest at ambient temperatures resulting in enhanced lipid peroxidation, reduced membrane fluidity, and increased membrane permeability (Jiang et al., 2004). As mentioned before, anthocyanins are hydrolyzed to anthocyanidin, which is transformed and oxidized by PPO (Tian et al., 2005) and/or POX (Gong and Tian, 2002). Low-temperature storage decreases the occurrence of this disorder, but CA conditions are more effective in preventing pericarp browning (Duan et al., 2004; Mahajan and Goswami, 2004; Tian et al., 2005) and decreasing fruit decay. CA storage delays senescence, limits PPO and POX activities, reduces total phenol content, and maintains high anthocyanidin levels (Tian et al., 2005) as well as higher acidity and ascorbate content compared with regular air (Mahajan and Goswami, 2004).

Senescence-induced enzymatic browning is also responsible for peel or pericarp browning in longan where the browning index increases with PPO activity during storage (Tian et al., 2002). CA conditions especially those with high CO_2 atmospheres alleviate this disorder significantly by inhibiting PPO activity (Tian et al., 2002). Nonetheless, the initial trigger for the disorder is a loss of compartmentation of enzymes, which finally leads to enzymatic browning (Tian et al., 2005).

Husk scald of pomegranate is also categorized as a browning disorder. Husk scald appears as a superficial browning (limited to the husk) and is similar in appearance to superficial scald of apple, but with a different mechanism as evidenced by the absence of conjugated trienol oxidation products of α-farnesene in the affected fruit. CA storage decreases scald occurrence as well as the production of carotenoids and acyl lipids as well as phenylpropanoid metabolism (Defilippi et al., 2006) likely through a reduction in senescence-induced ethylene production.

8.3.4 Pitting Disorders in Apple and Pear

Bitter pit (Martin et al., 1975), Jonathan spot (Dewey et al., 1982; Dris et al., 1998), cork spot (Green and Smith, 1988), and lenticel blotch pit (Perring, 1984) in apple, and cork spot in pear (Raese and Staiff, 1990) are calcium deficiency-related physiological disorders and have similar symptoms that appear mostly on the fruit surface and in the outer cortex just below the epidermis. Bitter pit in apple is described as a small, brown, somewhat dry, slightly bitter tasting lesion (3–5 mm in cross section) in the fruit flesh (Meheriuk et al., 1994). Although the initiation is clearly during production, symptoms are not always visually detectable at harvest but develop progressively during storage. Fruit with bitter pit display lower CF than healthy fruit (Lötze et al., 2006). Jonathan spot originates as a brownish or nearly black circular spot (2–4 mm in diameter), located mainly on the colored regions of the apple (Meheriuk et al., 1994). Lenticel blotch pit is easily confused with lenticel breakdown due to the similarity in appearance but the lesions are more irregular and on the more sun-exposed side. Additionally, flesh browning may extend deeper, similar to bitter pit. Lenticel breakdown appears as dark brown round pits in the fruit skin around the lenticels mainly on the less sun-exposed side and is limited to the skin (Jones et al., 2007) and not due to calcium deficiency (Kupferman, 2007). The sensitivity is determined by preharvest factors, evidenced by different sensitivities in different orchards (Kupferman, 2007), but the direct cause seems to be chemical injury, most likely CA toxicity (Betts and Bramlage, 1977; Bramlage and Weis, 1994; Moggia et al., 2006).

Pitting disorders (except for lenticel breakdown) are mainly the result of mineral imbalance, with low levels of calcium combined with relatively high concentrations of potassium and magnesium typical in sensitive fruit (Ferguson et al., 1999). Nevertheless, these disorders are also characterized by localized cell breakdown and browning. Thus, the involvement of the ethylene- and oxidative-related processes described in the general model (Figure 8.1) may be important in the development of these disorders. The only difference concerns the spatial localization of the disorder, which appears to be determined by calcium deficiency in the damaged areas. CA storage delays the appearance and severity of bitter pit (Sharples and Johnson, 1987) and Jonathan spot (Meheriuk et al., 1994) probably through a delay of senescence.

8.3.5 Other Disorders

There are many other disorders that are alleviated or aggravated by CA/MA storage but some of these seldom occur or have not been rigorously studied in a way that provides physiological answers. CA storage prevents a type of breakdown in 'Doyenne du Comice' pears described as softening and soggy browning of the cortex and core tissues, with an offensive flavor (Ma and Chen, 2003). CA storage inhibits browning of the cut end and of the external leaves in fennel (Artés et al., 2002), necrotic spot disorder in leek (Lipton, 1987b), "white mealy breakdown" in the cortex of winter squash (Prange and Harrison,

1993), browning in anthocyanin-containing purple carrots (Alasalvar et al., 2005), and red discoloration in the basal parts of the medial leaves of chicory heads (Vanstreels et al., 2002).

CA and MA storage can also aggravate the appearance of some disorders, for example, internal browning in loquat (Ding et al., 1999, 2002), core browning, and late storage corking in apple (Sharples and Johnson, 1987), black midrib and necrotic spot in cabbage (Bérard, 1994), internal and external browning of mushroom (Lopez-Briones et al., 1992), and peel collapse or pitting in citrus (Ke and Kader, 1990; Nelson, 1933; Petracek et al., 1998; Purvis and Cubbedge, 1990; Shellie, 2002).

8.4 Disorders Not Known to Occur or Rarely Occur in Air Storage and Induced by CA/MA

CA-induced physiological disorders have been investigated as long as CA storage has been in use as evidenced by the numerous review articles on the topic, both with (Howard et al., 1994; Lidster et al., 1990; Meheriuk et al., 1994) or without photographs (Gorny, 2003; Herner, 1987; Kader, 2003; Kupferman, 2003; Lougheed, 1987; Saltveit, 2003; Weichmann, 1987).

8.4.1 Blocking Ripening-Related Quality Changes

Although not strictly speaking a disorder, lack of desired color change has important consequences as the color of fruits and vegetables is the first criterion used by consumers to assess quality. Changes in color are due to changes in one or more of four pigment classes: chlorophylls, carotenoids, flavonoids, and betalains.

In some products, loss of chlorophyll is essential to unmask other colored pigments to guarantee consumer acceptability. The desired loss in chlorophyll can sometimes be stopped by low O_2 and/or high CO_2, for example, banana, tomato, cabbage (Menniti et al., 1997; Prange and Lidster, 1991; Wang, 1990). Chlorophyll metabolism is largely dependent on ethylene as evidenced through 1-MCP research (Watkins, 2006). It is likely that CA prevents chlorophyll degradation through a decrease in ethylene production and delay of the ethylene-related response (Figure 8.1).

In CA conditions, fruit epidermal anthocyanin color can change from red to blue or purple either induced by low O_2 (apple, sweet cherries) or high CO_2 (strawberries, sweet cherries) (Lidster et al., 1990). In the case of strawberries, the epidermal darkening is associated with a disappearance of red color in the internal tissue (Holcroft and Kader, 1999a; Lidster et al., 1990) through breakdown of anthocyanin. High CO_2 atmospheres will increase pH by dissolution of CO_2 and/or changes in the metabolism of organic acids (Holcroft and Kader, 1999b). The increase in pH will destabilize anthocyanin to form colorless, yellow, or blue metabolites (Brouillard et al., 1997). Additional to the increased breakdown, blocking the ethylene-related response can also prevent the induction of the phenylpropanoid pathway by ethylene (Tomás-Barberán et al., 1997) and result in a decrease in anthocyanin production.

8.4.2 Cortical Browning in Apple and Pear

As mentioned earlier, there is generally some confusion regarding the terminology used to refer to browning in pome fruit. For example, brown heart can refer to the senescence-related internal browning of the fruit, while at other times, it refers to the CA storage-induced disorder. However, they do not have the same physiological bases. In this section, we will discuss brown heart, flesh browning, and core breakdown, which are caused or aggravated by CA storage. The disorder directly caused by CO_2 injury will be discussed in the next section.

In senescence-related browning and breakdown (referred to here as internal break-down), the cortex decomposes; in CA storage-related browning and breakdown (referred to here as cortical browning), the cortical region is relatively dry with cavities often present. Cortical browning in pears (caused by suboptimal CA conditions) appears to be mainly determined by the oxidative behavior of the fruit (Larrigaudière et al., 2004b), and particularly by the levels of ascorbate both at harvest (Lentheric et al., 1999) and during storage (Veltman et al., 1999). Contrary to senescence-related browning and breakdown, ethylene plays a secondary role in this CA-induced disorder as evidenced by the lack of control by 1-MCP in 'Pink Lady' apple (de Castro et al., 2007).

This CA-induced disorder appears in several apple cultivars, for example, 'Braeburn,' 'Fuji,' 'Bramley's Seedling,' and 'Pacific Rose,' and seems to be CO_2-induced with low O_2 aggravating CO_2 susceptibility (Chung et al., 2005; Colgan et al., 1999; Elgar et al., 1998; Johnson et al., 1998; Lau, 1998; Maguire and MacKay, 2003; Volz et al., 1998). Several pear cultivars, for example, 'Conference,' 'Blanquilla,' 'Rocha,' and 'Doyenne du Comice' develop internal flesh browning, core browning, and cavities under low O_2 conditions, especially in the presence of increased CO_2 (Galvis-Sánchez et al., 2006; Larrigaudière et al., 2001; Ma and Chen, 2003; Pintó et al., 2001; Veltman et al., 2003; Xuan et al., 2001, 2005).

Various authors suggest this disorder is a consequence of membrane damage caused by a combination of maturity, CA conditions, tissue/skin permeability, oxygen-free radical action, loss of antioxidants such as ascorbate, and a lack of maintenance energy (low ATP:ADP ratio) (Larrigaudière et al., 2001; Maguire and MacKay, 2003; Pintó et al., 2001; Saquet et al., 2000; Streif et al., 2003; Veltman et al., 2003). The most comprehensive model for CA-induced browning in pear has been proposed by Pintó et al. (2001), who were first in suggesting cytoplasmic acidification as the first step in this disorder.

In 'Conference' and 'Blanquilla' pears, the cortical browning process has been thoroughly studied, which provided an explanation for disorder initiation and progression (Franck et al., 2007; Larrigaudière et al., 2004b). It appears that initiation of the disorder is determined by ineffective gas exchange (Lammertyn et al., 2000), together with fruit maturity and size. Gas exchange refers to the combination of the respiratory metabolism and gas transport in the fruit. Oxygen entering the fruit passes through the skin, travels through the intercellular spaces, across the cell wall and through the cytoplasm to the mitochondria, whereas carbon dioxide travels in the opposite direction. The diffusivity of these gases in the fruit is not homogenous and decreases through the storage life of the fruit (Schotsmans et al., 2004). When O_2 and CO_2 concentrations in the storage atmosphere are optimal and respiration and gas transport are in equilibrium, the gas partial pressure gradients in the fruit do not cause problems (Franck et al., 2007). However, if any of these factors deviates from the optimal, local anoxic conditions may arise and initiate a shift from the aerobic respiration to the anaerobic (ethanolic) fermentation pathway (Ke et al., 1994). The latter is less efficient and produces an insufficient amount of energy (Ke et al., 1994), resulting in generation of AOS and the oxidative response pathway (Figure 8.1). At this point, the decrease in ascorbate synthesis and/or regeneration becomes very important (Larrigaudière et al., 2004b) because if the antioxidants cannot scavenge the AOS, membrane damage occurs, cellular compartmentalisation is lost, and enzymatic browning results. This is confirmed by the decrease in ascorbate concentration in fruit held in unfavorable conditions (Franck et al., 2003a) and the occurrence of sound spots in the brown tissue zone corresponding to higher ascorbate concentrations. Also, in an ascorbate map, brown tissue is located within the contour line of 0.4 mg ascorbate 100 g^{-1} fresh weight, which confirms the importance of vitamin C (Franck et al., 2003a). Continuing unfavorable conditions result in cell wall breakdown and leaking of cytoplasm into the intercellular space, further reducing the diffusion of metabolic gasses. Moisture then diffuses toward the boundary of the fruit, and cavities are formed (Franck et al., 2007). In addition to these events, the toxic end products of fermentative metabolism (ethanol and acetaldehyde) accumulate and further damage the fruit (Ben-Arie and Sonego, 1985). Although not studied in such detail, the flesh browning found in Asian pears at low (<3 kPa) O_2 concentrations and bronze epidermal discoloration (<1 kPa) found by Kader (1990) are likely similar to what was described here for European pears.

Proteomics has provided some insight, indicating that different genes are expressed in healthy and disordered tissue for energy metabolism, antioxidant systems, and ethylene biosynthesis. In brown tissue, genes involved in energy metabolism and defense mechanisms are upregulated (Pedreschi et al., 2007).

Although much research has been done on CA-induced cortical browning and the general chain of events seems to be well established, there are still outstanding issues. For example, Veltman et al. (2003) stored 'Conference' pears at 0 kPa O_2 with or without CO_2 without any disorder development, which is difficult to explain since hypoxia seems to be the cause of the initiation of the oxidative response.

8.4.3 Carbon Dioxide Injury

Whereas the disorder described in the previous section is caused by hypoxia and increased CO_2 levels may aggravate it, the disorder described next is directly caused by CO_2. A number of synonyms for it exist in literature; for example, in pears, brown core, pithy brown core, and brown heart are all forms of carbon dioxide injury (Blanpied, 1990; Meheriuk et al., 1994). Carbon dioxide injury on apples can present itself as external injury with symptoms similar to superficial scald but affected areas are rough, very well-defined, and somewhat depressed. External CO_2 injury begins at the hypodermis–cortex boundary and spreads from there, leaving the cuticle and epidermis appearing unaffected and unbroken. In time, cell walls collapse and the skin surface sinks (Watkins et al., 1997). Internal symptoms in apples include brown, moist, well-defined areas in cortex and core tissue sometimes associated with the main vascular elements (Meheriuk et al., 1994). In pears, the symptoms first appear as browning of the interior walls of carpels and sometimes the core tissue; cortex tissue may then become injured and turn light brown when more extensive damage occurs (Meheriuk et al., 1994). In grapes, the external symptoms start with rachis browning, which spreads to the berries. This is accompanied by the production of off-flavors (Crisosto et al., 2002).

The harmful effect of high concentrations of CO_2 was already demonstrated in 1804 (de Saussure, 1804). CO_2 has an uncoupling effect on oxidative phosphorylation (Pintó et al., 2001) and also directly interferes with the Krebs cycle, through inhibition of succinic dehydrogenase activity, preventing the conversion of succinic acid (Fernández-Trujillo et al., 2001). Krebs cycle function ceases and the end product of glycolysis (pyruvate) accumulates. Decarboxylation of pyruvate in the fermentative pathway will produce acetaldehyde, which is subsequently reduced to ethanol as evidenced by accumulation of acetaldehyde, ethanol (Deng et al., 2006; Fernández-Trujillo et al., 2001), and ethyl acetate (Park et al., 1999), which are responsible for off-flavors (Crisosto et al., 2002). The subsequent events are very likely similar to what was described for flesh browning, with the actual browning resulting from enzymatic browning reactions confirmed by an increase in PPO and PAL activity (Park, 1999).

8.4.4 Skin Disorders

Burmeister and Dilley (1995) describe a "scald-like" CA disorder in 'Empire' apples, which does not occur in air-stored fruit. It only occurs in fruit held in CA containing high CO_2 and is more prevalent at 1°C compared with 3°C. The symptoms are identical to superficial scald and their hypothesis points toward the oxidative response involving free radical-catalyzed oxidation. This disorder might physiologically be a form of superficial scald but where in other cultivars, the stress of low temperature is sufficient for symptom development, in 'Empire' apples, it is not. The added stress of high CO_2 appears to be required for symptom development. Further research into this disorder would help clarify its relation to superficial scald.

8.4.5 Other Disorders

Several other CA-induced disorders have been mentioned in the literature, but the knowledge of the developmental mechanisms of these disorders is limited. In peaches and nectarines, low O_2 (0 kPa) causes skin browning or skin pitting as well as internal flesh discoloration (Wang et al., 1990); these symptoms are reduced by high CO_2 (5 kPa) (Wang et al., 1990). Other disorders include skin browning in banana (Klieber et al., 2002; Wills et al., 1990), epidermal breakdown in honeydew melon (Lipton, 1990c), hard core in kiwifruit (Arpaia, 1990), discoloration and off-flavors in cauliflower (Lipton, 1990b), blackening of bract and receptacle tissue in artichoke (Saltveit, 1990), water soaking (Kader and Lipton, 1990), and brown stain (Kader and Lipton, 1990; Kim et al., 2005; Lipton, 1987a; Varoquaux et al., 1996; Wang, 1990) in lettuce. Meristem browning or redheart can appear in cabbage (Bérard, 1994; Masters and Hicks, 1990; Menniti et al., 1997), oxheart cabbage (Schouten et al., 1997a), or Brussels sprouts (Lipton, 1990a; Lipton and Mackey, 1987).

8.5 Methods to Avoid CA/MA-Related Disorders

The effect of low O_2 and/or high CO_2 on physiological disorder development is complex and is influenced by a number of preharvest factors that include cultivar, production factors, and harvest maturity (Ferguson et al., 1999). Following harvest, other factors affect the development of physiological disorders, such as duration of exposure to CA, the temperature and relative humidity of the storage area, and the accumulation of biologically active volatiles during the storage period (Prange and DeLong, 2006).

8.5.1 Preharvest Factors and Maturity

Cultivar and production factors have not been discussed so far in this chapter and we would like to refer to a review on the influence of cultural and climatic factors by Ferguson et al. (1999). The preharvest factors they identify as most important are position of the fruit on the tree, characteristics of the fruiting site, crop load, mineral and carbohydrate nutrition of the developing fruit, water relations, and response to temperatures (Ferguson et al., 1999).

CA-induced cortex browning in 'Conference' pears can be reduced by preharvest sprays of boron and calcium, depending on year and site (Xuan et al., 2005).

Harvesting at optimal maturity for long-term storage is not only important to ensure optimal retention of quality attributes but also to prevent physiological disorders.

Immature apples are more likely to develop superficial scald (Emongor et al., 1994). Late harvests are inductive for internal browning in 'Fuji' apples especially if fruit also have watercore (Argenta et al., 2002a,b; Fan et al., 1997; Volz et al., 1998), CA-induced cortex browning in 'Conference' pears (Verlinden et al., 2002) and browning in 'Braeburn' apples (Elgar et al., 1999).

8.5.2 Chlorophyll

The role of chlorophyll has not been widely discussed in this chapter but as was pointed out by Prange and DeLong (2006), more research is needed to elucidate its role in CA-induced disorders in chlorophyll-containing products. In many vegetables, CA-induced damage occurs on tissues without chlorophyll, such as leaf midribs (Lipton, 1977). Within the same species (*Brassica oleracea*), those products without chlorophyll (cauliflower) are less tolerant to CA than those products with chlorophyll (broccoli and Brussels sprouts). The reason for this is unclear and was attributed to greater production of respiratory substrate (Lipton, 1977) but this was refuted because high reducing sugar content at harvest does not decrease the sensitivity of crisphead lettuce to high CO_2 (Forney and Austin, 1988). Prange and DeLong (2006) suggested that exposure of chlorophyll to light after harvest may be important since it reduces CO_2 injury in lettuce (Siriphanich and Kader, 1986) and physiological disorders in cabbage (Prange and Lidster, 1991). Light may help to maintain an active electron transfer through the photosynthetic electron transport chain and prevent chloroplast damage. This is in accordance with our general model (Figure 8.1) and confirms that the chloroplast is likely one of the first structures damaged during stress conditions.

8.5.3 Acclimation

In some cases, postharvest acclimation of the product can reduce the appearance of CA-induced disorders. An acclimation period of at least 2 weeks (preferably 1 month) at the beginning of storage with little or no CO_2 seems to be sufficient to avoid some CO_2-induced disorders in apple (Chung et al., 2005; Colgan et al., 1999; Elgar et al., 1998; Johnson et al., 1998; Prange et al., 1998; Wang et al., 1997; Watkins, 1999). Similarly, apples display acclimation to low O_2 (Blanpied and Jozwiak, 1993; Lidster et al., 1985; Prange et al., 1997).

Delaying CA establishment decreases the CA-induced browning disorder in 'Conference' pears (Höhn et al., 1996; Saquet et al., 2003a; Verlinden et al., 2002), CA-induced browning in 'Braeburn' apples (Saquet et al., 2003b), and CO_2 injury in 'Fuji' apple (Argenta et al., 2000). Delayed CA seems to increase ethylene production, respiration, and ATP concentrations in the fruit tissue of 'Conference' pears and 'Braeburn' apples. A higher energy status was maintained in delayed CA fruit as well as higher concentrations of total and free fatty acids and polar lipids, indicating the energy status of fruit and integrity of membranes play an important role in this adaptation process (Saquet et al., 2001, 2003a,b).

Delayed cooling (7 days) at room temperature (20°C), decreases storage disorders, for example, soft scald and low-temperature breakdown in CA and air storage (DeLong et al., 2004a, 2006; Watkins et al., 2004).

8.5.4 General Adaptation Syndrome

The general adaptation theory states that tolerance to a certain stress can be induced by exposing the plant to another stress at non-damaging levels (Leshem and Kuiper, 1996;

Toivonen, 2003). The fundamental hypothesis is that plant responses to stress generally have similar mechanisms to prevent or lessen oxidative stress. ILOS followed by CA storage can control bitter pit (Pesis et al., 2007) and superficial scald in apple (Pesis et al., 2007; Wang and Dilley, 2000) and inhibit the production of α-farnesene and MHO (Wang and Dilley, 2000). Wang and Dilley (2000) hypothesize that the ILOS treatment could induce a short-term increase in ethanol in the fruit. Since ethanol vapor treatment can control scald (Scott et al., 1995), a short burst in endogenous ethanol concentrations might have the same effect. The precise mechanism behind this is not known. ILOS also reduces CI symptoms in 'Fuerte' avocado (Pesis et al., 1994).

Initial high CO_2 stress (10–15 kPa) can prevent browning in 'Conference' pears likely through a positive effect on antioxidant defence mechanisms evidenced by increases in ascorbate and the activitiy of CAT and POX (Lentheric et al., 2003). However, in 'd'Anjou' pear, high CO_2 (12 kPa) stress increases CO_2 "injury" (Wang and Mellenthin, 1975) and in 'Fuji' apple prestorage high CO_2 (20 kPa) increases flesh browning (Argenta et al., 2002b; Volz et al., 1998). Comparing these treatment is difficult since exposure to high CO_2 was done for 3 days at 1°C (Lentheric et al., 2003), 1–12 days at 20°C (Argenta et al., 2002b; Volz et al., 1998), or 14 or 28 days at −1.1°C (Wang and Mellenthin, 1975).

8.5.5 Antioxidant Application

The importance of tissue antioxidant level has been discussed earlier in this chapter; therefore, it is logical to assume that application of an antioxidant before CA storage may reduce the severity of CA-induced disorders. The antioxidant DPA has been used extensively before CA storage to reduce CO_2-induced external injury in 'Empire' apples (Burmeister and Dilley, 1995) and internal flesh breakdown in 'Fuji' apples (Argenta et al., 2002b), flesh browning in 'Pink Lady' (de Castro et al., 2007), external and internal CO_2 injury in 'Cortland' and 'Law Rome' apples (Fernández-Trujillo et al., 2001). Although the exact way through which DPA reduces these disorders has not been established, it has been shown that DPA treatment enhances SOD and CAT activities and prevents the storage-related increase in POX activity (Abbasi and Kushad, 2006; Zanella, 2003) when used to prevent superficial scald development (Golding, 2004).

8.5.6 1-MCP Treatment

Since the introduction of the ethylene inhibitor 1-MCP on the market, it is being tested for the prevention of many physiological disorders. Depending on the process responsible for the disorder, 1-MCP can have a substantial positive or negative influence as summarized in Watkins and Miller (2005).

1-MCP treatment is more likely to be beneficial in senescence-related disorders through its action on senescence-induced ethylene action. Therefore, it has been mainly investigated as a possible alternative to CA storage for the prevention of senescence-related breakdown (Calvo, 2003; DeLong et al., 2004b; Larrigaudière et al., 2004b) and superficial scald (Apollo Arquiza et al., 2005; Argenta et al., 2003; Fan et al., 1999; Haines et al., 2005; Isidoro and Almeida, 2006; Lurie et al., 2005; Rizzolo et al., 2005; Rupasinghe et al., 2000a; Watkins et al., 2000) in pome fruit as well as CI in pineapple (Selvarajah et al., 2001), avocado fruit (Hershkovitz et al., 2005), and tangerine and grapefruit (Dou et al., 2005).

However, 1-MCP treatment results in a higher activity of antioxidant enzymes (CAT, SOD, and POX) in 'Yali' pears (Fu et al., 2007), and might be useful to increase antioxidant activity during CA storage.

8.5.7 Dynamic Control Atmosphere

It is difficult to predict when a CA environment will induce physiological disorders in a plant organ as the variety of factors that predispose fruits and vegetables to disorders are many. Therefore, the development of nondestructive biosensing techniques that can detect physiological stress and/or the onset of disorder ontogeny while the crop is in storage, is ideal (NRAES, 1997; Prange et al., 1997; Schouten et al., 1997b).

The observation that manual, laboratory-based CF measurements can detect low O_2 and high CO_2 stress in apples (DeEll et al., 1995; Prange et al., 1997), as well as sensitivity to superficial scald development at harvest (DeEll et al., 1997), has led to the development of a nondestructive CF-based biosensing technique called HarvestWatch (Prange et al., 2003). This technology can identify the low-oxygen threshold for any fruit and vegetable that contains chlorophyll (DeLong et al., 2004b; Prange et al., 2003).

In the HarvestWatch system, the fruit are put in CA storage in a kennel that houses the sample (bottom unit), with the fluorescence sensor unit mounted at a fixed distance above the crop (upper unit). The CF of the darkened sample is then measured periodically (typically once every hour), which produces a steady-state baseline fluorescence response. As the O_2 concentration is lowered, an upward spike in the fluorescence signal from the baseline will occur at the low-O_2 threshold. The O_2 concentration is then adjusted to 0.1–0.2 kPa above this threshold without the need to break the room's atmosphere (DeLong et al., 2004b). This technology makes it possible to correct the atmospheric conditions when low-O_2 stress is detected, which mitigates any potential anaerobic damage to the stored crop. It may also help to elucidate the role and timing of cytoplasmic acidosis, the putative first step in the development of many physiological disorders (Prange et al., 2005).

Using the HarvestWatch system, superficial scald in apples can be successfully controlled to the same level as with DPA or 1-MCP treatments (Zanella et al., 2005). This is especially important where the use of chemical treatments for superficial scald control is presently disallowed or where there is a market demand for a reduction in chemical usage. Thus, the HarvestWatch system is a viable option for organic apple producers as a nonchemical means to maintain higher fruit quality during long-term storage (DeLong et al., 2007), and is now used commercially on various CA-stored commodities around the world (Prange and DeLong, 2006; Zanella et al., 2005).

8.6 Conclusions and Future Research Needs

The aim of CA and MA storage is to retain the quality of the fruit or vegetable for as long as possible with the main emphasis on retention of firmness, color, flavor, and elimination of decay. The positive CA/MA effects on physiological disorders control will be increasingly attractive commercially if the present emphasis on organic and chemical-free food production increases. Much of the work describing particular disorder symptoms and control measures has been done in the early to mid-1900s; however, recent decades have seen an increase in research groups attempting to unravel the underlying physiological mechanisms of disorder development.

Although CA/MA environments may alleviate some disorders, imposing these atmospheres on a fruit or vegetable crop induces physiological stress, which will affect the quality of the plant product either positively or negatively. In this chapter, we proposed a general model for the initiation of postharvest physiological disorders. In this model, stress (e.g., senescence, low temperature, hypoxia, high CO_2) triggers an oxidative or ethylene-related response (or a combination of both), which leads to an alteration of membrane permeability, cell wall hydrolysis, and organelle decompartmentation, eventually resulting in browning and/or cellular breakdown. It is this last step that is best understood since it directly results in the appearance of disorder symptoms. Although general pathways have been established for some disorders, many cellular processes and reactions still need elucidation. In addition, the use of 1-MCP (which blocks the ethylene response) will help determine the interaction between the oxidative and ethylene response pathways.

Multidisciplinary research networks will hasten the fundamental understanding of causal processes and subsequent disorder development by studying several aspects (e.g., membrane damage, antioxidants, or gas exchange), simultaneously (rather than one), and at different levels (e.g., metabolites, enzymes, gene regulation and expression). This should produce a broader understanding of disorder development, from the genetic influences and orchard factors, to the specific chemical and cellular events that induce or prevent damage. And with the new techniques presently available for stress detection (e.g., CF), it is now possible to investigate earlier stages of disorder development, which permits timely intervention strategies before actual symptom development occurs.

In this chapter, we have endeavored to demonstrate that disorder development in plant organs can be fundamentally linked to either ethylene chemistry or oxidative stress (or both). Fruits and vegetables are very different and often have specific adaptations/requirements, which are reflected in the CA/MA regimes that result in optimal postharvest quality. Nonetheless, we have attempted to demonstrate from the scientific literature that regardless of how varied the symptoms of physiological disorders are, controlling both the ethylene response, and the development of unchecked oxidation significantly reduces or eliminates disorder expression. Thus, we recommend that future research dealing with the control of disorders in CA and MA environments focus on treatments that directly affect ethylene and/or oxidative chemistry at the cell level.

References

Abbasi, N.A. and M.M. Kushad. 2006. The activities of SOD, POD, and CAT in 'Red Spur Delicious' apple fruit are affected by DPA but not calcium in postharvest drench solutions. *J. Am. Pomol. Soc.* 60: 84–89.

Abdallah, A.Y., M.I. Gil, W.V. Biasi, and E.J. Mitcham. 1997. Inhibition of superficial scald in apples by wounding: Changes in lipids and phenolics. *Postharvest Biol. Technol.* 12: 203–212.

Abeles, F.B., P.W. Morgan, and M.E. Saltveit. 1992. *Ethylene in Plant Biology*, Vol. 15. Academic Press, San Diego, CA.

Alasalvar, C., M. Al-Farsi, P.C. Quantick, F. Shahidi, and R. Wiktorowicz. 2005. Effect of chill storage and modified atmosphere packaging (MAP) on antioxidant activity, anthocyanins, carotenoids, phenolics and sensory quality of ready-to-eat shredded orange and purple carrots. *Food Chem.* 89: 69–76.

Ali, Z.M., L.H. Chin, M. Marimuthu, and H. Lazan. 2004. Low temperature storage and modified atmosphere packaging of carambola fruit and their effects on ripening related texture changes, wall modification and chilling injury symptoms. *Postharvest Biol. Technol.* 33: 181–192.

Anderson, R.E. 1982. Long-term storage of peaches and nectarines intermittently warmed during controlled-atmosphere storage. *J. Am. Soc. Hort. Sci.* 107: 214–216.

Anderson, R.E., C.S. Parsons, and W.L.J. Smith. 1969. Controlled atmosphere storage of eastern grown peaches and nectarines. USDA Market Research Report 836. U.S. Department of Agriculture, Washington DC.

Apollo Arquiza, J.M.R., A.G. Hay, J.F. Nock, and C.B. Watkins. 2005. 1-Methylcyclopropene interactions with diphenylamine on diphenylamine degradation, alpha-farnesene and conjugated trienol concentrations, and polyphenol oxidase and peroxidase activities in apple fruit. *J. Agric. Food Chem.* 53: 7565–7570.

Arana, I., C. Jaren, and S. Arazuri. 2007. Sensory and mechanical characterization of mealy apples and woolly peaches and nectarines. *J. Food Agric. Environ.* 5: 101–106.

Argenta, L.C., X.T. Fan, and J.P. Mattheis. 2000. Delaying establishment of controlled atmosphere or CO_2 exposure reduces 'Fuji' apple CO_2 injury without excessive fruit quality loss. *Postharvest Biol. Technol.* 20: 221–229.

Argenta, L., X. Fan, and J.P. Mattheis. 2002a. Impact of watercore on gas permeance and incidence of internal disorders in 'Fuji' apples. *Postharvest Biol. Technol.* 24: 113–122.

Argenta, L.C., X.T. Fan, and J.P. Mattheis. 2002b. Responses of 'Fuji' apples to short and long duration exposure to elevated CO_2 concentration. *Postharvest Biol. Technol.* 24: 13–24.

Argenta, L.C., X. Fan, and J.P. Mattheis. 2003. Influence of 1-methylcyclopropene on ripening, storage life, and volatile production by d'Anjou cv. pear fruit. *J. Agric. Food Chem.* 51: 3858–3864.

Arpaia, M.L. 1990. Kiwi fruit. In: P.D. Lidster, G.D. Blanpied, and R.K. Prange (Eds.), *Controlled Atmosphere Disorders of Commercial Fruits and Vegetables.* Publication 1847/E. Agriculture and Agri-Food Canada, Kentville, Canada, pp. 41–44.

Artés, F., V.H. Escalona, and F. Artés-Hernández. 2002. Quality and physiological changes of fennel under controlled atmosphere storage. *Eur. Food Res. Technol.* 214: 216–220.

Ben-Arie, R. and L. Sonego. 1985. Modified-atmosphere storage of kiwifruit (*Actinidia chinensis* Planch) with ethylene removal. *Sci. Hort.* 27: 263–273.

Bérard, L.S. 1994. Storage disorders of cabbage. In: R.J. Howard, J.A. Garland, and W.L. Seaman (Eds.)., *Diseases and Pests of Vegetable Crops in Canada.* Canadian Phytopathology Society and Entomology Society of Canada, Ottawa, Canada, pp. 110–112.

Betts, H.A. and W.J. Bramlage. 1977. Uptake of calcium by apples from postharvest dips in calcium chlorine solution. *J. Am. Soc. Hort. Sci.* 102: 785–788.

Binder, B.M. and A.B. Bleecker. 2003. A model for ethylene receptor function and 1-methylcyclopropene action. *Acta Hort.* 628: 177–187.

Bisswanger, H. 2002. *Enzyme Kinetics: Principles and Methods.* Wiley-VCH, Weinheim, Germany.

Blanpied, G.D. 1990. Controlled atmosphere storage of apples and pears. In: M. Calderon and R. Barkai-Golan (Eds.), *Food Preservation by Modified Atmospheres.* CRC Press, Boca Raton, FL, pp. 265–299.

Blanpied, G.D. and Z. Jozwiak. 1993. A study of some orchard and storage factors that influence the oxygen threshold for ethanol accumulation in stored apples. *Proceedings of the 6th International Controlled Atmosphere Research Conference,* Ithaca, New York, pp. 78–86.

Boller, T. 1991. Ethylene in pathogenesis and disease resistance. In: A.K. Mattoo and J.C. Suttle (Eds.), *The Plant Hormone Ethylene.* CRC Press, Boca Raton, FL, pp. 293–314.

Bramlage, W.J. and S. Weis. 1994. Postharvest use of calcium. In: A.B. Peterson and R.G. Stevens (Eds.), *Tree Fruit Nutrition.* Good Fruit Grower, Washington DC, pp. 125–134.

Brouillard, R., P. Figueiredo, M. Elhabiri, and O. Dangles. 1997. Molecular interactions of phenolic compounds in relation to the colour of fruit and vegetables. In: F.A. Tomas-Barberan and R.J. Robins (Eds.), *Phytochemistry of Fruit and Vegetables Proceedings of the Phytochemical Society of Europe.* Clarendon Press, Oxford, U.K., pp. 125–134.

Brummell, D.A., V. Dal Cin, S. Lurie, C.H. Crisosto, and J.M. Labavitch. 2004. Cell wall metabolism during the development of chilling injury in cold-stored peach fruit: Association of mealiness with arrested disassembly of cell wall pectins. *J. Exp. Bot.* 55: 2041–2052.

Burmeister, D.M. and D.R. Dilley. 1995. A scald-like controlled-atmosphere storage disorder of Empire apples—A chilling injury-induced by CO_2. *Postharvest Biol. Technol.* 6: 1–7.

Burmeister, D.M. and J.E. Harman. 1998. Effect of fruit maturity on the success of controlled atmosphere storage of 'Fantasia' nectarines. *Acta Hort.* 464: 363–368.

Calvo, G. 2003. Effect of 1-methylcyclopropene (1-MCP) on pear maturity and quality. *Acta Hort.* 628: 203–211.

Chung, H.S., K.D. Moon, S.K. Chung, and J.U. Choi. 2005. Control of internal browning and quality improvement of 'Fuji' apples by stepwise increase of CO_2 level during controlled atmosphere storage. *J. Sci. Food Agric.* 85: 883–888.

Colgan, R.J., C.J. Dover, D.S. Johnson, and K. Pearson. 1999. Delayed CA and oxygen at 1 kPa or less control superficial scald without CO_2 injury on Bramley's Seedling apples. *Postharvest Biol. Technol.* 16: 223–231.

Crisosto, C.H., D. Garner, and G.M. Crisosto. 2002. High carbon dioxide atmospheres affect stored 'Thompson Seedless' table grapes. *HortScience* 37: 1074–1078.

de Castro, E., W.V. Biasi, E.J. Mitcham, S. Tustin, D.J. Tanner, and J. Jobling. 2007. Carbon dioxide-induced flesh browning in Pink Lady apples. *J. Am. Soc. Hort. Sci.* 132: 713–719.

de Saussure, N.T. 1804. Recherches chimiques sur la végétation. Chez La Ve Nyon, Paris.

De Sousa, C. and L. Sodek. 2002. The metabolic response of plants to oxygen deficiency. *Braz. J. Plant Physiol.* 14: 83–94.

DeEll, J.R., R.K. Prange, and D.P. Murr. 1995. Chlorophyll fluorescence as a potential indicator of controlled atmosphere disorders in 'Marshall McIntosh' apples. *HortScience* 30: 1084–1085.

DeEll, J.R., R.K. Prange, and D.P. Murr. 1997. Chlorophyll fluorescence of Delicious apples at harvest as a potential predictor of superficial scald development during storage. *Postharvest Biol. Technol.* 9: 1–6.

Defilippi, B.G., B.D. Whitaker, B.M. Hess-Pierce, and A.A. Kader. 2006. Development and control of scald on wonderful pomegranates during long-term storage. *Postharvest Biol. Technol.* 41: 234–243.

DeLong, J.M. and R.K. Prange. 2003. Superficial scald—A postharvest oxidative stress disorder. In: D.M. Hodges (Ed.), *Postharvest Oxidative Stress in Horticultural Crops*. The Haworth Press, Inc., New York, pp. 91–112.

DeLong, J.M., R.K. Prange, and P.A. Harrison. 2004a. The influence of pre-storage delayed cooling on quality and disorder incidence in 'Honeycrisp' apple fruit. *Postharvest Biol. Technol.* 33: 175–180.

DeLong, J.M., R.K. Prange, and P.A. Harrison. 2007. Chlorophyll fluorescence-based low-O_2 CA storage of organic 'Cortland' and 'Delicious' apples. *Acta Hort.* 737: 31–37.

DeLong, J.M., R.K. Prange, P.A. Harrison, C.G. Embree, D.S. Nichols, and A.H. Wright. 2006. The influence of crop-load, delayed cooling and storage atmosphere on post-storage quality of 'Honeycrisp'™ apples. *J. Hort. Sci. Biotechnol.* 81: 391–396.

DeLong, J.M., R.K. Prange, J.C. Leyte, and P.A. Harrison. 2004b. A new technology that determines low-oxygen thresholds in controlled-atmosphere-stored apples. *HortTechnology* 14: 262–266.

Deng, Y., Y. Wu, and Y. Li. 2006. Physiological responses and quality attributes of 'Kyoho' grapes to controlled atmosphere storage. *LWT-Food Sci. Technol.* 39: 584–590.

Dewey, D.H., S.A. Sargent, and P. Sass. 1982. Controlling internal breakdown in Jonathan apples by postharvest application of calcium chloride, Research Report, Agricultural Experiment Station, Michigan State University, East Lansing, MI, p. 12.

Ding, C.K., K. Chachin, Y. Ueda, Y. Imahori, and H. Kurooka. 1999. Effects of high CO_2 concentration on browning injury and phenolic metabolism in loquat fruits. *J. Jpn. Soc. Hort. Sci.* 68: 275–282.

Ding, C.K., K. Chachin, Y. Ueda, Y. Imahori, and C.Y. Wang. 2002. Modified atmosphere packaging maintains postharvest quality of loquat fruit. *Postharvest Biol. Technol.* 24: 341–348.

Dong, L., H.W. Zhou, L. Sonego, A. Lers, and S. Lurie. 2001. Ethylene involvement in the cold storage disorder of 'Flavortop' nectarine. *Postharvest Biol. Technol.* 23: 105–115.

Dörnenburg, H. and C. Davies. 1999. The relationship between lipid oxidation and antioxidant content in postharvest vegetables. *Food Rev. Intl.* 15: 435–453.

Dou, H., S. Jones, and M.A. Ritenour. 2005. Influence of 1-MCP application and concentration on post-harvest peel disorders and incidence of decay in citrus fruit. *J. Hort. Sci. Biotechnol.* 80: 786–792.

Drew, M.C. 1997. Oxygen deficiency and root metabolism: Injury and acclimation under hypoxia and anoxia. *Annu. Rev. Plant Physiol. Plant Mol. Biol.* 48: 223–250.

Dris, R., R. Niskanen, and E. Fallahi. 1998. Nitrogen and calcium nutrition and fruit quality of commercial apple cultivars grown in Finland. *J. Plant Nutr.* 21: 2389–2402.

Du, Z.Y. and W.J. Bramladge. 1994. Roles of ethylene in the development of superficial scald in Cortland apples. *J. Am. Soc. Hort. Sci.* 119: 516–523.

Duan, X.W., Y.M. Jiang, X.G. Su, and Z.Q. Zhang. 2004. Effects of a pure oxygen atmosphere on enzymatic browning of harvested litchi fruit. *J. Hort. Sci. Biotechnol.* 79: 859–862.

Eaks, I.L. 1956. Effect of modified atmosphere on cucumbers at chilling and non-chilling temperatures. *Proc. Am. Soc. Hort. Sci.* 67: 473–478.

Elgar, H.J., D.M. Burmeister, and C.B. Watkins. 1998. Storage and handling effects on a CO_2-related internal browning disorder of 'Braeburn' apples. *HortScience* 33: 719–722.

Elgar, H.J., C.B. Watkins, and N. Lallu. 1999. Harvest date and crop load effects on a carbon dioxide-related storage injury of 'Braeburn' apple. *HortScience* 34: 305–309.

Elyatem, S.M. and A.A. Kader. 1984. Post-harvest physiology and storage behavior of pomegranate fruits. *Sci. Hort.* 24: 287–298.

Emongor, V.E., D.P. Murr, and E.C. Lougheed. 1994. Preharvest factors that predispose apples to superficial scald. *Postharvest Biol. Technol.* 4: 289–300.

Ezz, T.M., M.A. Ritenour, and J.K. Brecht. 2004. Hot water and elevated CO_2 effects on proline and other compositional changes in relation to postharvest chilling injury of 'Marsh' grapefruit. *J. Am. Soc. Hort. Sci.* 129: 576–582.

Fan, X., J.P. Mattheis, and S.M. Blankenship. 1999. Development of apple superficial scald, soft scald, core flush, and greasiness is reduced by MCP. *J. Agric. Food Chem.* 47: 3063–3068.

Fan, X.T., J.P. Mattheis, M.E. Patterson, and J.K. Fellman. 1997. Optimum harvest date and controlled atmosphere storage potential of 'Fuji' apple. *Proceedings of the 7th International Controlled Atmosphere Research Conference*, Davis, CA, pp. 42–49.

Ferguson, I., R.K. Volz, and A.B. Woolf. 1999. Preharvest factors affecting physiological disorders of fruit. *Postharvest Biol. Technol.* 15: 255–262.

Fernández-Trujillo, J.P., J.F. Nock, and C.B. Watkins. 2001. Superficial scald, carbon dioxide injury, and changes of fermentation products and organic acids in 'Cortland' and 'Law Rome' apples after high carbon dioxide stress treatment. *J. Am. Soc. Hort. Sci.* 126: 235–241.

Flores, F.B., M.C. Martínez-Madrid, M. Ben Amor, J.C. Pech, A. Latché, and F. Romojaro. 2004. Modified atmosphere packaging confers additional chilling tolerance on ethylene-inhibited cantaloupe Charentais melon fruit. *Eur. Food Res. Technol.* 219: 614–619.

Forney, C.F. and R.K. Austin. 1988. Time of day at harvest influences carbohydrate concentration in crisphead lettuce and its sensitivity to high CO_2 levels after treatment. *J. Am. Soc. Hort. Sci.* 113: 581–583.

Foyer, C.H., P. Descourvières, and K.J. Kunert. 1994. Protection against oxygen radicals: An important defence mechanism studied in transgenic plants. *Plant Cell Environ.* 17: 507–523.

Franck, C., M. Baetens, J. Lammertyn, N. Scheerlinck, M.W. Davey, and B.M. Nicolaï. 2003a. Ascorbic acid mapping to study core breakdown development in 'Conference' pears. *Postharvest Biol. Technol.* 30: 133–142.

Franck, C., M. Baetens, J. Lammertyn, P. Verboven, M.W. Davey, and B.M. Nicolaï. 2003b. Ascorbic acid concentration in cv. Conference pears during fruit development and postharvest storage. *J. Agric. Food Chem.* 51: 4757–4763.

Franck, C., J. Lammertyn, Q.T. Ho, P. Verboven, B.E. Verlinden, and B.M. Nicolaï. 2007. Browning disorders in pear fruit. *Postharvest Biol. Technol.* 43: 1–13.

Friedman, M. 1996. Food browning and its prevention: An overview. *J. Agric. Food Chem.* 44: 631–653.

Fu, L., J. Cao, Q. Li, L. Lin, and W. Jiang. 2007. Effect of 1-methylcyclopropene on fruit quality and physiological disorders in Yali pear (*Pyrus bretschneideri* Rehd.) during storage. *Food Sci. Technol. Intl.* 13: 49–54.

Gaddam, U.S., E.W. Hewett, W.C. Schotsmans, and A.J. Mawson. 2005. *Changes in Physicochemical Attributes of Feijoas During Storage.* Australasian Postharvest Horticulture Conference, Rotorua, New Zealand, p. 39.

Galvis-Sánchez, A.C., S.C. Fonseca, Á. Gil-Izquierdo, M.I. Gil, and F.X. Malcata. 2006. Effect of different levels of CO_2 on the antioxidant content and the polyphenol oxidase activity of 'Rocha' pears during cold storage. *J. Sci. Food Agric.* 86: 509–517.

Gapper, N.E., J.H. Bai, and B.D. Whitaker. 2006. Inhibition of ethylene-induced α-farnesene synthase gene *PcAFS1* expression in 'd'Anjou' pears with 1-MCP reduces synthesis and oxidation of α-farnesene and delays development of superficial scald. *Postharvest Biol. Technol.* 41: 225–233.

Golding, J. 2004. Superficial scald in apples—History, control and alternatives. *Tree Fruit.* 3: 16–18.

Gong, Q.Q. and S.P. Tian. 2002. Partial characterization of soluble peroxidase in pericarp of litchi fruit. *Prog. Biochem. Biophys.* 29: 613–620.

Gorny, J.R. 2003. A summary of CA and MA requirements and recommendations for fresh cut (minimally processed) fruits and vegetables. *Acta Hort.* 600: 609–614.

Gout, E., A.-M. Boisson, S. Aubert, R. Douce, and R. Bligny. 2001. Origin of the cytoplasmic pH changes during anaerobic stress in higher plant cells. Carbon-13 and phosphorous-31 nuclear magnetic resonance studies. *Plant Physiol.* 125: 912–925.

Green, G.M. and C.B. Smith. 1988. Physiological disorders and calcium nutrition in apples. *Compact Fruit Tree* 21: 121–130.

Gross, K.C., C.Y. Wang, and M.E. Saltveit. 2004. *The Commercial Storage of Fruits, Vegetables, and Florist and Nursery Stocks.* Agricultural Handbook Number 66, USDA.

Haines, M.M., D.S. Mattinson, and J.K. Fellman. 2005. Farnesylation of components of the ethylene signaling pathway may be necessary to the development of superficial scald in apples. *Acta Hort.* 682: 313–319.

Hatton, T.T. and R.H. Cubbedge. 1982. Conditioning Florida grapefruit to reduce chilling injury during low temperature storage. *J. Am. Soc. Hort. Sci.* 107: 57–60.

Hatton, T.T., R.H. Cubbedge, and W. Grierson. 1975. Effects of prestorage carbon dioxide treatments and delayed storage on chilling injury of 'Marsh' grapefruit. *Proc. Fla. State Hort. Soc.* 88: 335–338.

Hermansen, A. and H. Hoftun. 2005. Effect of storage in controlled atmosphere on post-harvest infections of *Phytophthora brassicae* and chilling injury in Chinese cabbage (*Brassica rapa* L *pekinensis* (Lour) Hanelt). *J. Sci. Food Agric.* 85: 1365–1370.

Herner, R.C. 1987. High CO_2 effects on plant organs. In: J. Weichmann (Ed.), *Postharvest Physiology of Vegetables.* Marcel Dekker, Inc. New York, pp. 239–253.

Hershkovitz, V., S.I. Saguy, and E. Pesis. 2005. Postharvest application of 1-MCP to improve the quality of various avocado cultivars. *Postharvest Biol. Technol.* 37: 252–264.

Hodges, D.M. 2003. Overview: Oxidative stress and postharvest produce. In: D.M. Hodges (Ed.), *Postharvest Oxidative Stress in Horticultural Crops.* The Haworth Press, Inc., New York, pp. 1–12.

Höhn, E., M. Jampen, and D. Dätwyler. 1996. Kavernenbildung in Conférence—Risikoverminderung. *Schweiz. Z. Obst Weinbau* 7: 180–181.

Holcroft, D.M. and A.A. Kader. 1999a. Carbon dioxide-induced changes in color and anthocyanin synthesis of stored strawberry fruit. *HortScience* 34: 1244–1248.

Holcroft, D.M. and A.A. Kader. 1999b. Controlled atmosphere-induced changes in pH and organic acid metabolism may affect color of stored strawberry fruit. *Postharvest Biol. Technol.* 17: 19–32.

Howard, R.J., J.A. Garland, and W.L. Seaman. 1994. *Diseases and Pests of Vegetable Crops in Canada.* Canadian Phytopathology Society and Entomological Society of Canada, Ottawa, Canada.

Huber, D.J., J. Jeong, and L.C. Mao. 2003. Softening during ripening of ethylene-treated fruits in response to 1-methylcyclopropene application. *Acta Hort.* 628: 193–202.

Ilker, Y. and L.L. Morris. 1975. Alleviation of chilling injury of okra. *HortScience* 10: 324 (abstract).

Isidoro, N. and D.P.F. Almeida. 2006. Alpha-farnesene, conjugated trienols, and superficial scald in 'Rocha' pear as affected by 1-methylcyclopropene and diphenylarnine. *Postharvest Biol. Technol.* 42: 49–56.

Jiang, Y., X. Duan, D. Joyce, Z. Zhang, and J. Li. 2004. Advances in understanding of enzymatic browning in harvested litchi fruit. *Food Chem.* 88: 443–446.

Johnson, D.S., C.J. Dover, and R.J. Colgan. 1998. Effect of rate of establishment of CA conditions on the development of CO_2-induced injury on Bramley's Seedling apples. *Acta Hort.* 464: 351–356.

Jones, W., J. Brunner, E. Kupferman, and C.-L. Xiao. 2007. *Quick Identification Guide to Apple Postharvest Defects & Disorders.* Good Fruit Grower, Wenatchee, WA.

Kader, A.A. 1990. Asian pears. In: P.D. Lidster, G.D. Blanpied, and R.K. Prange (Eds.), *Controlled Atmosphere Disorders of Commercial Fruits and Vegetables.* Publication 1847/E. Agriculture and Agri-Food Canada, Kentville, Canada, pp. 25–26.

Kader, A.A. 2003. A summary of CA requirements and recommendations for fruits other than apples and pears. *Acta Hort.* 600: 737–740.

Kader, A.A. and W.J. Lipton. 1990. Lettuce. In: P.D. Lidster, G.D. Blanpied, and R.K. Prange (Eds.), *Controlled Atmosphere Disorders of Commercial Fruits and Vegetables.* Publication 1847/E. Agriculture and Agri-Food Canada, Kentville, Canada, pp. 45–46, 48.

Kader, A.A. and L.L. Morris. 1975. Amelioration of chilling injury symptoms on tomato fruits. *HortScience* 10: 324 (abstract).

Kader, A.A., J.A. Klaustermeyer, L.L. Morris, and R.F. Kasmire. 1975. Extending storage life of mature green chili peppers by modified atmospheres. *HortScience* 10: 335 (abstract).

Ke, D. and A.A. Kader. 1990. Tolerance of 'Valencia' oranges to controlled atmospheres, as determined by physiological responses and quality attributes. *J. Am. Soc. Hort. Sci.* 115: 779–783.

Ke, D.Y. and M.E. Saltveit. 1988. Plant hormone interaction and phenolic metabolism in the regulation of russet spotting in iceberg lettuce. *Plant Physiol.* 88: 1136–1140.

Ke, D.Y., E. Yahia, M. Mateos, and A.A. Kader. 1994. Ethanolic fermentation of Bartlett pears as influenced by ripening stage and atmosphereic composition. *J. Am. Soc. Hort. Sci.* 119: 976–982.

Kim, J.G., Y. Luo, R.A. Saftner, and K.C. Gross. 2005. Delayed modified atmosphere packaging of fresh-cut Romaine lettuce: Effects on quality maintenance and shelf-life. *J. Am. Soc. Hort. Sci.* 130: 116–123.

Klaustermeyer, J.A., L.L. Morris, and A.A. Kader. 1974. Some factors affecting the occurrence and severity of russet spotting in harvested lettuce. *HortScience* 9: 274 (abstract).

Klieber, A., N. Bagnato, R. Barrett, and M. Sedgley. 2002. Effect of post-ripening nitrogen atmosphere storage on banana shelf life, visual appearance and aroma. *Postharvest Biol. Technol.* 25: 15–24.

Kruger, F.J. and A.B. Truter. 2003. Relationship between preharvest quality determining factors and controlled atmosphere storage in South African export avocados. *Acta Hort.* 600: 109–113.

Kruger, F.J., L. Tait, M. Kritzinger, M. Bezuidenhout, and V. Claassens. 1999. Postharvest browning in South African subtropical export fruits. *Acta Hort.* 485: 225–229.

Kupferman, E. 2007. Plain talk about apple lenticel breakdown. *Good Fruit Grower Magazine*, p. 58.

Kupferman, E.M. 2003. Controlled atmosphere storage of apples and pears. *Acta Hort.* 600: 729–735.

Lallu, N. 1997. Low temperature breakdown of kiwifruit. *Acta Hort.* 444: 579–586.

Lallu, N., J. Burdon, D.P. Billing, D.M. Burmeister, C.W. Yearsley, S. Osman, M. Wang, A. Gunson, and H. Young. 2005. Effect of carbon dioxide removal systems on volatile profiles and quality of 'Hayward' kiwifruit stored in controlled atmosphere rooms. *HortTechnol.* 15: 253–260.

Lammertyn, J., M. Aerts, B.E. Verlinden, W.C. Schotsmans, and B.M. Nicolaï. 2000. Logistic regression analysis of factors influencing core breakdown in 'Conference' pears. *Postharvest Biol. Technol.* 20: 25–37.

Larrigaudière, C. and M. Vendrell. 1993. Short-term activation of the conversion of 1-aminocyclopropane-1-carboxylic acid to ethylene in rewarmed Granny Smith apples. *Plant Physiol. Biochem.* 31: 585–591.

Larrigaudière, C., J. Graell, Y. Soria, and I. Recasens. 2004a. El corazón pardo en peras. Conocimientos actuales y sistemas de prevención. *Fruttic. Prof.* 14: 23–34.

Larrigaudière, C., I. Lentheric, E. Pintó, and M. Vendrell. 2001. Short-term effects of air and controlled atmosphere storage on antioxidant metabolism in Conference pears. *J. Plant Physiol.* 158: 1015–1022.

Larrigaudière, C., I. Lentheric, J. Puy, and E. Pintó. 2004b. Biochemical characterisation of core browning and brown heart disorders in pear by multivariate analysis. *Postharvest Biol. Technol.* 31: 29–39.

Larrigaudière, C., I. Lentheric, and M. Vendrell. 1998. Relationship between enzymatic browning and internal disorders in controlled-atmosphere stored pears. *J. Sci. Food Agric.* 78: 232–236.

Larrigaudière, C., R. Vilaplana, Y. Soria, and I. Recasens. 2004c. Oxidative behaviour of Blanquilla pears treated with 1-methylcyclopropene during cold storage. *J. Sci. Food Agric.* 84: 1871–1877.

Lau, O.L. 1998. Effect of growing season, harvest maturity, waxing, low O_2 and elevated CO_2 on flesh browning disorders in 'Braeburn' apples. *Postharvest Biol. Technol.* 14: 131–141.

Lelièvre, J.M., L. Tichit, P. Dao, L. Fillion, Y.W. Nam, J.C. Pech, and A. Latché. 1997. Effects of chilling on expression of ethylene biosynthetic genes in Passe-Crassane pear (*Pyrus communis* L.) fruit. *Plant Mol. Biol.* 33: 847–855.

Lelièvre, J.M., L. Tichit, L. Fillion, C. Larrigaudière, M. Vendrell, and J.C. Pech. 1995. Cold-induced accumulation of 1-aminocyclopropane 1-carboxylate oxidase protein in Granny Smith apples. *Postharvest Biol. Technol.* 5: 11–17.

Lentheric, I., E. Pintó, J. Graell, and C. Larrigaudière. 2003. Effects of CO_2 pretreatment on oxidative metabolism and core-browning incidence in controlled atmosphere stored pears. *J. Hort. Sci. Biotechnol.* 78: 177–181.

Lentheric, I., E. Pintó, M. Vendrell, and C. Larrigaudière. 1999. Harvest date affects the antioxidative systems in pear fruits. *J. Hort. Sci. Biotechnol.* 74: 791–795.

Leshem, Y. and P.J.C. Kuiper. 1996. Is there a GAS (general adaptation syndrom) response to various types of environmental stress? *Biol. Plant. (Prague)* 38: 1–18.

Leshuk, J.A. and M.E. Saltveit. 1990. Controlled atmosphere storage requirements and recommendations for vegetables. In: M. Calderon and R. Barkai-Golan (Eds.), *Food Preservation by Modified Atmospheres*. CRC Press, Boca Raton, FL, pp. 315–352.

Lester, G.E. 2003. Oxidative stress affecting fruit senescence. In: D.M. Hodges (Ed.), *Postharvest Oxidative Stress in Horticultural Crops*. The Haworth Press, Inc., New York, pp. 113–130.

Lidster, P.D., G.D. Blanpied, and E.C. Lougheed. 1985. Factors affecting progressive development of low oxygen injury in apples. *Proceedings of the 4th National Controlled Atmosphere Research Conference*, Raleigh, NC, pp. 57–69.

Lidster, P.D., G.D. Blanpied, and R.K. Prange. 1990. *Controlled Atmosphere Disorders of Commercial Fruits and Vegetables*. Publication 1847/E. Agriculture and Agri-Food Canada, Kentville, Canada.

Lipton, W.J. 1977. Toward an explanation of disorders of vegetables induced by high CO_2 or low O_2. *Proceedings of the 2nd National Controlled Atmosphere Research Conference*, Michigan State University, East Lansing, pp. 137–141.

Lipton, W.J. 1987a. Carbon dioxide-induced injury of romaine lettuce stored in controlled atmospheres. *HortScience* 22: 461–463.

Lipton, W.J. 1987b. Senescence of leafy vegetables. *HortScience* 22: 854–859.

Lipton, W.J. 1990a. Brussels sprouts. In: P.D. Lidster, G.D. Blanpied, and R.K. Prange (Eds.), *Controlled Atmosphere Disorders of Commercial Fruits and Vegetables*. Publication 1847/E. Agriculture and Agri-Food Canada, Kentville, Canada, pp. 27–29.

Lipton, W.J. 1990b. Cauliflower. In: P.D. Lidster, G.D. Blanpied, and R.K. Prange (Eds.), *Controlled Atmosphere Disorders of Commercial Fruits and Vegetables*. Publication 1847/E. Agriculture and Agri-Food Canada, Kentville, Canada, pp. 31–32.

Lipton, W.J. 1990c. Honeydew melon. In: P.D. Lidster, G.D. Blanpied, and R.K. Prange (Eds.), *Controlled Atmosphere Disorders of Commercial Fruits and Vegetables*. Publication 1847/E. Agriculture and Agri-Food Canada, Kentville, Canada, pp. 39–40.

Lipton, W.J. and B.E. Mackey. 1987. Physiological and quality responses of Brussels sprouts to storage in controlled atmospheres. *J. Am. Soc. Hort. Sci.* 112: 491–496.

Lopez-Briones, G., P. Varoquaux, Y. Chambroy, J. Bouquant, G. Bureau, and B. Pascat. 1992. Storage of common mushroom under controlled atmospheres. *Intl. J. Food Sci. Technol.* 27: 493–505.

Lötze, E., C. Huybrechts, A. Sadie, K.I. Theron, and R.M. Valcke. 2006. Fluorescence imaging as a non-destructive method for pre-harvest detection of bitter pit in apple fruit (*Malus domestica* Borkh.). *Postharvest Biol. Technol.* 40: 287–294.

Lougheed, E.C. 1987. Interactions of oxygen, carbon dioxide, temperature, and ethylene that may induce injuries in vegetables. *HortScience* 22: 791–794.

Lurie, S. 1993. Modified atmosphere storage of peaches and nectarines to reduce storage disorders. *J. Food Qual.* 16: 57–65.

Lurie, S. 2003. Antioxidants. In: D.M. Hodges (Ed.), *Postharvest Oxidative Stress in Horticultural Crops*. The Haworth Press, Inc., New York, pp. 131–150.

Lurie, S. and C.H. Crisosto. 2005. Chilling injury in peach and nectarine. *Postharvest Biol. Technol.* 37: 195–208.

Lurie, S., A. Lers, Z. Shacham, L. Sonego, S. Burd, and B.D. Whitaker. 2005. Expression of alpha-farnesene synthase AFS1 and 3-hydroxy-3-methylglutaryl -coenzyme a reductase HMG2 and HMG3 in relation to alpha-farnesene and conjugated trienols in 'Granny Smith' apples heat or 1-MCP treated to prevent superficial scald. *J. Am. Soc. Hort. Sci.* 130: 232–236.

Luza, J.G., R. Vangorsel, V.S. Polito, and A.A. Kader. 1992. Chilling injury in peaches: A cytochemical and ultrastructural cell wall study. *J. Am. Soc. Hort. Sci.* 117: 114–118.

Ma, S.S. and P.M. Chen. 2003. Storage disorder and ripening behavior of 'Doyenne du Comice' pears in relation to storage conditions. *Postharvest Biol. Technol.* 28: 281–294.

Maguire, K.M. and B.R. MacKay. 2003. A controlled atmosphere induced internal browning disorder of 'Pacific Rose'™ apples. *Acta Hort.* 600: 281–284.

Mahajan, P.V. and T.K. Goswami. 2004. Extended storage life of litchi fruit using controlled atmosphere and low temperature. *J. Food Proc. Pres.* 28: 388–403.

Marcellin, P. and A.R. Chaves. 1983. Effects of intermittent high CO_2 treatment on storage life of avocado fruits in relation to respiration and ethylene production. *Acta Hort.* 138: 155–163.

Martin, D., T.L. Lewis, J. Cerny, and D.A. Ratkowski. 1975. The predominant role of calcium as an indicator in storage disorders in Cleopatra apples. *J. Hort. Sci.* 50: 447–455.

Masia, A. 2003. Physiological effects of oxidative stress in relation to ethylene postharvest produce. In: D.M. Hodges (Ed.), *Postharvest Oxidative Stress in Horticultural Crops.* The Haworth Press, Inc., New York, pp. 165–198.

Masters, J.F. and J.R. Hicks. 1990. Cabbage. In: P.D. Lidster, G.D. Blanpied, and R.K. Prange (Eds.), *Controlled Atmosphere Disorders of Commercial Fruits and Vegetables.* Publication 1847/E. Agriculture and Agri-Food Canada, Kentville, Canada, pp. 29–30.

Mathew, A.G. and H.A.B. Parpia. 1971. Food browning as a polyphenol reaction. *Adv. Food Res.* 19: 75–145.

Mayer, A.M. 1987. Polyphenoloxidase in plants—Recent progress. *Phytochemistry* 26: 11–20.

McDonald, R.E. and M.M. Kushad. 1986. Accumulation of Putrescine during chilling injury of fruits. *Plant Physiol. (Rockv.)* 82: 324–326.

Meheriuk, M., R.K. Prange, P.D. Lidster, and S.W. Porritt. 1994. *Postharvest Disorders of Apples and Pears.* Publication 1737/E. Agriculture and Agri-Food Canada, Ottawa, Canada.

Meir, S., D. Naiman, J.Y. Hyman, A. Akerman, G. Zauberman, and Y. Fuchs. 1998. Modified atmosphere packaging enables prolonged storage of 'Fuerte' avocado fruit. *Acta Hort.* 464: 397–402.

Mencarelli, F. 1987. Effect of high CO_2 atmospheres on stored zucchini squash. *J. Am. Soc. Hort. Sci.* 112: 985–988.

Mencarelli, F., W.J. Lipton, and S.J. Peterson. 1983. Responses of zucchini squash to storage in low-O_2 atmospheres at chilling and nonchilling temperatures. *J. Am. Soc. Hort. Sci.* 108: 884–890.

Menniti, A.M., M. Maccaferri, and A. Folchi. 1997. Physio-pathological responses of cabbage stored under controlled atmospheres. *Postharvest Biol. Technol.* 10: 207–212.

Mir, N.A., R. Perez, and R.M. Beaudry. 1999. A poststorage burst of 6-methyl-5-hepten-2-one (MHO) may be related to superficial scald development in 'Cortland' apples. *J. Am. Soc. Hort. Sci.* 124: 173–176.

Moggia, C.E., J.A. Yuri, and M. Pereira. 2006. Mineral content of different apple cultivars in relation to fruit quality during storage. *Acta Hort.* 265–272.

Morris, L.L. 1982. Chilling injury of horticultural crops. *HortScience* 17: 161–162.

Munoz, T., J. Ruiz-Cabello, A.D. Molina-Garcia, M.I. Escribano, and C. Merodio. 2001. Chilling temperature storage changes the inorganic phosphate pool distribution in cherimoya (*Annona cherimola*) fruit. *J. Am. Soc. Hort. Sci.* 126: 122–127.

Nanos, G.D. and A.A. Kader. 1993. Low O_2-induced changes in pH and energy charge in pear fruit tissue. *Postharvest Biol. Technol.* 3: 285–291.

Nanos, G.D., A.K. Kiritsakis, and E.M. Sfakiotakis. 2002. Preprocessing storage conditions for green 'Conservolea' and 'Chondrolia' table olives. *Postharvest Biol. Technol.* 25: 109–115.

Neljubow, D.N. 1901. Über die horizontale Nutation der Stengel von *Pisum sativum* und einiger anderen Pflanzen. *Beih. Bot. Zentralbl.* 10: 128–139.

Nelson, R. 1933. Some storage and transportational diseases of citrus fruits apparently due to suboxidation. *J. Agric. Res.* 46: 695–713.

Noctor, G. and C.H. Foyer. 1998. Ascorbate and glutathione: Keeping active oxygen under control. *Annu. Rev. Plant Physiol. Plant Mol. Biol.* 49: 249–279.

NRAES. 1997. Sensors for nondestructive testing, measuring the quality of fresh fruits and vegetables. *Proceedings of the Sensors for Nondestructive Testing International Conference,* Orlando, FL.

Paliyath, G. and M.J. Drouillard. 1992. The mechanism of membrane degradation and disassembly during senescence. *Plant Physiol. Biochem. (Paris)* 30: 789–812.

Park, Y.S. 1999. Carbon dioxide-induced flesh browning development as related to phenolic metabolism in 'Niitaka' pear during storage. *J. Korean Soc. Hort. Sci.* 40: 567–570.

Park, Y., H.J. Kweon, H.Y. Kim, and O.H. Ryu. 1997. Preharvest factors affecting the incidence of physiological disorders during CA storage of 'Fuji' apples. *J. Korean Soc. Hort. Sci.* 38: 725–729.

Park, Y.S., C. Pelay, and T. Agar. 1999. Effects of storage temperatures and CA conditions on physiological disorders and volatile production of Asian pears during storage. *J. Korean Soc. Hort. Sci.* 40: 563–566.

Parkin, K.L., A. Marangoni, R.L. Jackman, R.Y. Yada, and D.W. Stanley. 1989. Chilling injury. A review of possible mechanisms. *J. Food Biochem.* 13: 127–153.

Paull, R.E. and M.E.Q. Reyes. 1996. Preharvest weather conditions and pineapple fruit translucency. *Sci. Hort.* 66: 59–67.

Pedreschi, R., E. Vanstreels, S. Carpentier, M.L.A.T.M. Hertog, J. Lammertyn, J. Robben, J.P. Noben, R. Swennen, J. Vanderleyden, and B.M. Nicolaï. 2007. Proteomic analysis of core breakdown disorder in Conference pears (*Pyrus communis* L.). *Proteomics* 7: 2083–2099.

Perata, P. and A. Alpi. 1993. Plant responses to anaerobiosis. *Plant Sci.* 93: 1–17.

Perring, M.A. 1984. Lenticel blotch pit, watercore, splitting and cracking in relation to calcium concentration in the apple fruit. *J. Sci. Food Agric.* 35: 1165–1173.

Pesis, E. 2005. The role of the anaerobic metabolites, acetaldehyde and ethanol, in fruit ripening, enhancement of fruit quality and fruit deterioration. *Postharvest Biol. Technol.* 37: 1–19.

Pesis, E., M. Ackerman, R. Ben-Arie, O. Feygenberg, X. Feng, A. Apelbaum, R. Goren, and D. Prusky. 2002. Ethylene involvement in chilling injury symptoms of avocado during cold storage. *Postharvest Biol. Technol.* 24: 171–181.

Pesis, E., S. Ebeler, and E. Mitcham. 2007. Postharvest low oxygen pretreatment prevented superficial scald and bitter pit symptoms in 'Granny Smith' apples. *HortScience* 42: 882–883.

Pesis, E., R. Marinansky, G. Zauberman, and Y. Fuchs. 1994. Prestorage low-oxygen atmosphere treatment reduces chilling injury symptoms in 'Fuerte' avocado fruit. *HortScience* 29: 1042–1046.

Petracek, P.D., H.T. Dou, and S. Pao. 1998. The influence of applied waxes on postharvest physiological behavior and pitting of grapefruit. *Postharvest Biol. Technol.* 14: 99–106.

Pintó, E., I. Lentheric, M. Vendrell, and C. Larrigaudière. 2001. Role of fermentative and antioxidant metabolisms in the induction of core browning in controlled-atmosphere stored pears. *J. Sci. Food Agric.* 81: 364–370.

Poneleit, L.S. and D.R. Dilley. 1993. Carbon dioxide activation of 1-aminocyclopropane-1-carboxylate (ACC) oxidase in ethylene biosynthesis. *Postharvest Biol. Technol.* 3: 191–199.

Porat, R., B. Weiss, L. Cohen, A. Daus, and N. Aharoni. 2004. Reduction of postharvest rind disorders in citrus fruit by modified atmosphere packaging. *Postharvest Biol. Technol.* 33: 35–43.

Porat, R., B. Weiss, L. Cohen, A. Daus, R. Goren, and S. Droby. 1999. Effects of ethylene and 1-methylcyclopropene on the postharvest qualities of 'Shamouti' oranges. *Postharvest Biol. Technol.* 15: 155–163.

Prange, R.K. and J.M. DeLong. 2006. Controlled-atmosphere related disorders of fruits and vegetables. *Stewart Postharvest Rev.* 5: 1–10.

Prange, R.K. and P.A. Harrison. 1993. Effect of controlled atmosphere and humidity on postharvest physiology of buttercup winter squash, *Cucurbita maxima* Duch. hybrid 'Sweet Mama'. *Proceedings of the 6th International Controlled Atmosphere Research Conference*, Ithaca, NY, pp. 759–766.

Prange, R.K. and P.D. Lidster. 1991. Controlled atmosphere and lighting effects on storage of winter cabbage. *Can. J. Plant Sci.* 71: 263–268.

Prange, R.K., J.M. DeLong, and P.A. Harrison. 1998. Recommended storage conditions for Nova Scotia apples. *Storage Notes for the Apple Industry*, Vol. 5. Nova Scotia Fruit Growers' Association, Kentville, Canada, p. 3.

Prange, R.K., J.M. DeLong, and P.A. Harrison. 2005. Quality management through respiration control: Is there a relationship between lowest acceptable respiration, chlorophyll fluorescence and cytoplasmic acidosis. *Acta Hort.* 682: 823–830.

Prange, R.K., J.M. DeLong, P.A. Harrison, J.C. Leyte, and S.D. McLean. 2003. Oxygen concentration affects chlorophyll fluorescence in chlorophyll-containing fruit and vegetables. *J. Am. Soc. Hort. Sci.* 128: 603–607.

Prange, R.K., J.M. DeLong, J.C. Leyte, and P.A. Harrison. 2002. Oxygen concentration affects chlorophyll fluorescence in chlorophyll-containing fruit. *Postharvest Biol. Technol.* 24: 201–205.

Prange, R.K., S.P. Schouten, and O. Van Kooten. 1997. Chlorophyll fluorescence detects low oxygen stress in 'Elstar' apples. *Proceedings of the 7th International Controlled Atmosphere Research Conference*, Davis, CA, pp. 57–64.

Purvis, A.C. 1981. Free proline in peel of grapefruit and resistance to chilling injury during cold storage. *HortScience* 16: 160–161.

Purvis, A.C. and R.H. Cubbedge. 1990. Citrus fruit. In: P.D. Lidster, G.D. Blanpied, and R.K. Prange (Eds.), *Controlled Atmosphere Disorders of Commercial Fruits and Vegetables*. Publication 1847/E. Agriculture and Agri-Food Canada, Kentville, Canada, pp. 36–38.

Raese, J.T. and D.C. Staiff. 1990. Fruit calcium, quality and disorders of apples (*Malus domestica*) and pears (*Pyrus communis*) influenced by fertilizers. Plant nutrition—Physiology and applications. *Proceedings of the 11th International Plant Nutrition Colloquium*, Dordrecht, the Netherlands, pp. 619–623.

Retamales, J.B., T. Cooper, J. Streif, and J.C. Kania. 1992. Preventing cold storage disorders in nectarines. *J. Hort. Sci.* 67: 619–626.

Rizzolo, A., P. Cambiaghi, M. Grassi, and P.E. Zerbini. 2005. Influence of 1-methylcyclopropene and storage atmosphere on changes in volatile compounds and fruit quality of Conference pears. *J. Agric. Food Chem.* 53: 9781–9789.

Roberts, J.K.M., J. Callis, O. Jardetzky, V. Walbot, and M. Freeling. 1984. Cytoplasmic acidosis as a determinant of flooding intolerance in plants. *Proc. Natl. Acad. Sci. U. S. A.* 81: 6029–6033.

Rodriguez, S.D.C., B. López, and A.R. Chaves. 2001. Effect of different treatments on the evolution of polyamines during refrigerated storage of eggplants. *J. Agric. Food Chem.* 49: 4700–4705.

Rupasinghe, H.P.V., D.P. Murr, G. Paliyath, and L. Skog. 2000a. Inhibitory effect of 1-MCP on ripening and superficial scald development in 'McIntosh' and 'Delicious' apples. *J. Hort. Sci. Biotechnol.* 75: 271–276.

Rupasinghe, H.P.V., G. Paliyath, and D.P. Murr. 2000b. Sesquiterpene α-farnesene synthase: Partial purification, characterization, and activity in relation to superficial scald development in apples. *J. Am. Soc. Hort. Sci.* 125: 111–119.

Saltveit, M.E. 1990. Artichokes. In: P.D. Lidster, G.D. Blanpied, and R.K. Prange (Eds.), *Controlled Atmosphere Disorders of Commercial Fruits and Vegetables*. Publication 1847/E. Agriculture and Agri-Food Canada, Kentville, Canada, pp. 23–24.

Saltveit, M.E. 2003. A summary of CA requirements and recommendations for vegetables. *Acta Hort.* 600: 723–727.

Saquet, A.A., J. Streif, and F. Bangerth. 2000. Changes in ATP, ADP and pyridine nucleotide levels related to the incidence of physiological disorders in 'Conference' pears and 'Jonagold' apples during controlled atmosphere storage. *J. Hort. Sci. Biotechnol.* 75: 243–249.

Saquet, A.A., J. Streif, and F. Bangerth. 2001. On the involvement of adenine nucleotides in the development of brown heart in 'Conference' pears during delayed controlled atmosphere storage. *Gartenbauwissenschaft* 66: 140–144.

Saquet, A.A., J. Streif, and F. Bangerth. 2003a. Energy metabolism and membrane lipid alterations in relation to brown heart development in 'Conference' pears during delayed controlled atmosphere storage. *Postharvest Biol. Technol.* 30: 123–132.

Saquet, A.A., J. Streif, and F. Bangerth. 2003b. Reducing internal browning disorders in 'Braeburn' apples by delayed controlled atmosphere storage and some related physiological and biochemical changes. *Acta Hort.* 628: 453–458.

Schotsmans, W.C., R.K. Prange, and B.M. Binder. 2009. 1-Methylcyclopropene (1-MCP): Mode of action and applications in postharvest horticulture. *Hort. Rev.* 35: 263–313.

Schotsmans, W.C., B.E. Verlinden, J. Lammertyn, and B.M. Nicolaï. 2004. The relationship between gas transport and histological properties of apple. *J. Sci. Food Agric.* 84: 1131–1140.

Schouten, S.P., R.K. Prange, and T.R. Lammers. 1997a. Quality aspects of apples and cabbages during anoxia. *Proceedings of the 7th International Controlled Atmosphere Research Conference*, Davis, CA, pp. 189–192.

Schouten, S.P., R.K. Prange, J. Verschoor, T.R. Lammers, and J. Oosterhaven. 1997b. Improvement of quality of Elstar apples by dynamic control of ULO conditions. *Proceedings of the 7th International Controlled Atmosphere Research Conference*, Davis, CA, pp. 71–78.

Scott, K.J., C.M.C. Yuen, and F. Ghahramani. 1995. Ethanol vapor—A new anti-scald treatment for apples. *Postharvest Biol. Technol.* 6: 201–208.

Selvarajah, S., A.D. Bauchot, and P. John. 2001. Internal browning in cold-storage pineapples is suppressed by a postharvest application of 1-methylcyclopropene. *Postharvest Biol. Technol.* 23: 167–170.

Serek, M., E.J. Woltering, E.C. Sisler, S. Frello, and S. Sriskandarajah. 2006. Controlling ethylene responses in flowers at the receptor level. *Biotechnol. Adv.* 24: 368–381.

Serrano, M., M.C. Martínez-Madrid, G. Martínez, F. Riquelme, M.T. Pretel, and F. Romojaro. 1996. Review: Role of polyamines in chilling injury of fruit and vegetables. *Food Sci. Technol. Intl.* 2: 195–199.

Sharples, R.S. and D.S. Johnson. 1987. Influence of agronomic and climatic factors on the response of apple fruit to controlled atmosphere storage. *HortScience* 22: 763–766.

Shellie, K.C. 2002. Ultra-low oxygen refrigerated storage of 'rio red' grapefruit: Fungistatic activity and fruit quality. *Postharvest Biol. Technol.* 25: 73–85.

Singh, B., C.C. Yang, and D.K. Salunkhe. 1972. Controlled atmosphere storage of lettuce. I. Effects on quality and the respiration rate of lettuce heads. *J. Food Sci.* 37: 48–51.

Siriphanich, J. and A.A. Kader. 1986. Changes in cytoplasmic and vacuolar pH in harvested lettuce tissue as influenced by CO_2. *J. Am. Soc. Hort. Sci.* 111: 73–77.

Sive, A. and D. Resnizky. 1979. Extension of the storage life of 'Red Rosa' plums by controlled atmosphere stora. *Proceedings of the 15th International Congress of Refrigeration*, Venice, Italy, p. 1148 (Abstract No. C1142–1131).

Streif, J., A.A. Saquet, and H. Xuan. 2003. CA-related disorders of apples and pears. *Acta Hort.* 600: 223–230.

Thomai, T., E. Sfakiotakis, G. Diamantidis, and M. Vasilakakis. 1998. Effects of low preharvest temperature on scald susceptibility and biochemical changes in 'Granny Smith' apple peel. *Sci. Hort.* 76: 1–15.

Tian, S.P., A.L. Jiang, Y. Xu, and Y.S. Wang. 2004. Responses of physiology and quality of sweet cherry fruit to different atmospheres in storage. *Food Chem.* 87: 43–49.

Tian, S.P., B.Q. Li, and Y. Xu. 2005. Effects of O_2 and CO_2 concentrations on physiology and quality of litchi fruit in storage. *Food Chem.* 91: 659–663.

Tian, S., Y. Xu, A. Jiang, and Q. Gong. 2002. Physiological and quality responses of longan fruit to high O_2 or high CO_2 atmospheres in storage. *Postharvest Biol. Technol.* 24: 335–340.

Toivonen, P.M.A. 2003. Postharvest treatments to control oxidative stress in fruits and vegetables. In: D.M. Hodges (Ed.), *Postharvest Oxidative Stress in Horticultural Crops*. The Haworth Press, Inc., New York, pp. 225–246.

Tomás-Barberán, F.A., J. Loaiza-Velarde, A. Bonfanti, and M.E. Saltveit. 1997. Early wound- and ethylene-induced changes in phenylpropanoid metabolism in harvested lettuce. *J. Am. Soc. Hort. Sci.* 122: 399–404.

UC Davis. 2008. *Postharvest Produce Facts*, UC Postharvest Technology Research and Information Center, Davis, CA.

Uthairatanakij, A., P. Penchaiya, W.B. McGlasson, and P. Holford. 2005. Changes in ACC and conjugated ACC levels following controlled atmosphere storage of nectarine. *Aust. J. Exp. Agric.* 45: 1635–1641.

Vakis, N., W. Grierson, and J. Soule. 1970. Chilling injury in tropical and subtropical fruit. III. The role of CO_2 in suppressing chilling injury of grapefruit and avocados. *Proc. Trop. Reg. Am. Soc. Hort. Sci.* 14: 89–100.

Valentines, M.C., Y. Soria, I. Recasens, and C. Larrigaudière. 2002. Efecto de un tratamiento con 1-meticiclopropeno sobre las fisiopatias en pera: relación con las capacidades antioxidantes del fruto. *POST 2002: Maduración y Postrecolección 2002. VI Simposio Nacional y III Iberico*, Madrid, Spain, p. 145.

Vanstreels, E., J. Lammertyn, B.E. Verlinden, N. Gillis, A. Schenk, and B.M. Nicolai. 2002. Red discoloration of chicory under controlled atmosphere conditions. *Postharvest Biol. Technol.* 26: 313–322.

Varoquaux, P., J. Mazollier, and G. Albagnac. 1996. The influence of raw material characteristics on the storage life of fresh-cut butterhead lettuce. *Postharvest Biol. Technol.* 9: 127–139.

Vaughn, K.C. and S.O. Duke. 1984. Function of polyphenol oxidase in higher plants. *Physiol. Plant* 60: 106–112.

Veltman, R.H., I. Lentheric, L.H.W. Van der Plas, and H.W. Peppelenbos. 2003. Internal browning in pear fruit (*Pyrus communis* L. cv Conference) may be a result of a limited availability of energy and antioxidants. *Postharvest Biol. Technol.* 28: 295–302.

Veltman, R.H., M. Sanders, S. Persijn, H.W. Peppelenbos, and J. Oosterhaven. 1999. Decreased ascorbic acid levels and brown core development in pears. *Physiol. Plant* 107: 39–45.

Verlinden, B.E., A. de Jager, J. Lammertyn, W.C. Schotsmans, and B.M. Nicolaï. 2002. Effect of harvest and delaying controlled atmosphere storage conditions on core breakdown incidence in 'Conference' pears. *Biosyst. Eng.* 83: 339–347.

Volz, R.K., W.V. Biasi, J.A. Grant, and E.J. Mitcham. 1998. Prediction of controlled atmosphere-induced flesh browning in 'Fuji' apple. *Postharvest Biol. Technol.* 13: 97–107.

Wade, N.L. 1981. Effects of storage atmosphere, temperature and calcium on low temperature injury of peach fruit. *Sci. Hort.* 15: 145–154.

Wang, C.Y. 1990. Physiological and biochemical effects of controlled atmosphere on fruits and vegetables. In: M. Calderon and R. Barkai-Golan (Eds.), *Food Preservation by Modified Atmospheres*. CRC Press, Boca Raton, FL, pp. 197–223.

Wang, C.Y. and J.G. Buta. 1999. Methyl jasmonate improves quality of stored zucchini squash. *J. Food Qual.* 22: 663–670.

Wang, Z.Y. and D.R. Dilley. 2000. Initial low oxygen stress controls superficial scald of apples. *Postharvest Biol. Technol.* 18: 201–213.

Wang, C.Y. and W.M. Mellenthin. 1975. Effect of short-term high CO_2 treatment on storage of 'd'Anjou' pears. *J. Am. Soc. Hort. Sci.* 100: 492–495.

Wang, Z., M. Kossitrakun, and D.R. Dilley. 1997. The effect of acclimatization of fruits on the control of CO2 linked disorder of 'Empire' apples. *Proceedings of the 7th International Controlled Atmosphere Research Conference*, Davis, CA, pp. 193–197.

Wang, C.Y., A.E. Watada, and R.E. Anderson. 1990. Peaches and nectarines. In: P.D. Lidster, G.D. Blanpied, and R.K. Prange (Eds.). *Controlled Atmosphere Disorders of Commercial Fruits and Vegetables*. Publication 1847/E. Agriculture and Agri-Food Canada, Kentville, Canada, pp. 47–49.

Wang, Y.S., S.P. Tian, and Y. Xu. 2005. Effects of high oxygen concentration on pro- and anti-oxidant enzymes in peach fruits during postharvest periods. *Food Chem.* 91: 99–104.

Wang, Z. and D.R. Dilley. 1999. Control of superficial scald of apples by low-oxygen atmospheres. *HortScience* 34: 1145–1151.

Watkins, C.B. 1999. Storage physiology 101: Fundamentals of product response to CA. *Proceedings of the CA Storage: Meeting the Market Requirements Workshop*, Ithaca, NY.

Watkins, C.B. 2006. The use of 1-methylcyclopropene (1-MCP) on fruits and vegetables. *Biotechnol. Adv.* 24: 389–409.

Watkins, C.B. and W.B. Miller. 2005. A summary of physiological processes or disorders in fruits, vegetables and ornamental products that are delayed or decreased, increased, or unaffected by application of 1-methylcyclopropene (1-MCP). Ithaca, NY. http://www.hort.cornell.edu/mcp/

Watkins, C.B. and M.V. Rao. 2003. Genetic variation and prospects for genetic engineering of horticultural crops for resistance to oxidative stress induced by postharvest conditions. In: D.M. Hodges (Ed.), *Postharvest Oxidative Stress in Horticultural Crops*. The Haworth Press, Inc., New York.

Watkins, C.B., W.J. Bramlage, and B.A. Cregoe. 1995. Superficial scald of Granny Smith apples is expressed as a typical chilling injury. *J. Am. Soc. Hort. Sci.* 120: 88–94.

Watkins, C.B., J.F. Nock, S.A. Weis, S. Jayanty, and R.M. Beaudry. 2004. Storage temperature, diphenylamine, and pre-storage delay effects on soft scald, soggy breakdown and bitter pit of 'Honeycrisp' apples. *Postharvest Biol. Technol.* 32: 213–221.

Watkins, C.B., J.F. Nock, and B.D. Whitaker. 2000. Responses of early, mid and late season apple cultivars to postharvest application of 1-methylcyclopropene (1-MCP) under air and controlled atmosphere storage conditions. *Postharvest Biol. Technol.* 19: 17–32.

Watkins, C.B., K.J. Silsby, and M.C. Goffinet. 1997. Controlled atmosphere and antioxidant effects on external CO_2 injury of 'Empire' apples. *HortScience* 32: 1242–1246.

Weichmann, J. 1987. Low oxygen effects. In: J. Weichmann (Ed.), *Postharvest Physiology of Vegetables*. Marcel Dekker, Inc., New York, pp. 231–237.

Whitaker, B.D. and R.A. Saftner. 2000. Temperature-dependent autoxidation of conjugated trienols from apple peel yields 6-methyl-5-hepten-2-one, a volatile implicated in induction of scald. *J. Agric. Food Chem.* 48: 2040–2043.

Wills, R.B.H., A. Klieber, R. David, and M. Siridhata. 1990. Effect of brief pre-marketing holding of bananas in nitrogen on time to ripen. *Aust. J. Exp. Agric.* 30: 579–581.

Wills, R.B.H., W.B. McGlasson, D. Graham, and D.C. Joyce. 2007. *Postharvest: An Introduction to the Physiology and Handling of Fruit, Vegetables and Ornamentals*. CAB International, Wallingford, U.K.

Wolfe, J. 1987. Chilling injury in plants—The role of membrane liquid fluidity. *Plant Cell Environ.* 1: 241–247.

Xuan, H., J. Streif, H. Pfeffer, F. Dannel, V. Romheld, and F. Bangerth. 2001. Effect of pre-harvest boron application on the incidence of CA-storage related disorders in 'Conference' pears. *J. Hort. Sci. Biotechnol.* 76: 133–137.

Xuan, H., J. Streif, A.A. Saquet, V. Romheld, and F. Bangerth. 2005. Application of boron with calcium affects respiration and ATP/ADP ratio in 'Conference' pears during controlled atmosphere storage. *J. Hort. Sci. Biotechnol.* 80: 633–637.

Yang, S.F. and N.E. Hoffman. 1984. Ethylene biosynthesis and its regulation in higher plants. *Annu. Rev. Plant Physiol.* 35: 155–189.

Zanella, A. 2003. Control of apple scald—A comparison between 1-MCP and DPA postharvest treatments, ILOS and ULO storage. *Acta Hort.* 600: 271–275.

Zanella, A., P. Cazzanelli, A. Panarese, M. Coser, M. Cecchinel, and O. Rossi. 2005. Fruit fluorescence response to low oxygen stress: Modern storage technologies compared to 1-MCP treatment of apple. *Acta Hort.* 682: 1535–1542.

Zerbini, P.E., A. Rizzolo, A. Brambilla, P. Cambiaghi, and M. Grassi. 2002. Loss of ascorbic acid during storage of Conference pears in relation to the appearance of brown heart. *J. Sci. Food Agric.* 82: 1007–1013.

Zhang, Z., X. Pang, Z. Ji, and Y. Jiang. 2001. Role of anthocyanin degradation in litchi pericarp browning. *Food Chem.* 75: 217–221.

9

Effects on Decay

Elhadi M. Yahia and Peter L. Sholberg

CONTENTS

9.1 Introduction

Controlled atmosphere (CA) and modified atmosphere (MA) technologies are widely used commercially for storage, transport, and packaging of several horticultural commodities (Ben-Yehoshua et al., 2005). These technologies have many advantages including delay of ripening and senescence, control of some physiological and plant pathological diseases, and control of some insects (Yahia, 1998, 2007, 2006). CA and MA are known to inhibit a wide range of plant pathogens either directly or indirectly (El-Goorani and Sommer, 1981). Atmospheres containing about 2–5 kPa O_2, commonly used in CA storage, suppress decay indirectly by acting upon host resistance, rather than directly on the pathogen since most fungi grow under these conditions (Barkai-Golan, 2001). CA and MA maintain fruit firmness and thus delay or prevent penetration by microorganisms. Atmospheric levels that act directly on microorganisms by inhibiting them and thus reducing decay are generally <1 kPa O_2 and >10 kPa CO_2 (Kader and Ben-Yehoshua, 2000). For this reason, most CA and MA commercial applications effect decay indirectly (Ben-Yehoshua et al., 2005). In recent years, safety of minimally processed foods has become an important issue especially in the interaction of MA with the growth of bacterial pathogens that cause illness or death in humans (Farber, 1991; Phillips, 1996; Werner and Hotchkiss, 2006); this subject is dealt with in detail in Chapter 10. In this chapter, we review the effects of CA and MA in relation to spoilage of horticultural commodities by decay organisms, as impacted by concentration of CO_2 and O_2 and some other gases and their effect on produce quality and microbial growth.

9.2 General Effects of MA and CA on Decay of Horticultural Commodities

CA and MA have been reported to control both fungal and bacterial decay in horticultural commodities. MA is a useful tool in extending the shelf life of berry crops such as strawberries and blueberries (Figure 9.1). Strawberries are characterized by having a very limited postharvest life even when picked at optimum maturity. The effects of low O_2 and high CO_2 atmospheres on biochemical changes and development of fungal pathogens on strawberries were studied by Chambroy et al. (1993). At 20°C, control of fungal development was not possible, regardless of the composition of the MA, and at 10°C, high CO_2 (>10 kPa) markedly reduced development of fungi. MA strongly reduced water loss without incidence of decay. Strawberries in perforated polypropylene bags placed under simulated transport and shelf life conditions developed an internal atmosphere that controlled decay and deterioration with a gas composition close to that recommended for this fruit. On the other hand, the use of MA reduced fruit color and led to off-flavor development (Sanz et al., 1999). Low-density polyethylene (LDPE) films containing 1% antimicrobial agents (extracts of *Rheum palmatum* and *Coptis chinensis*, and Ag-substituted inorganic zirconium matrix) were used for MA packaging (MAP) of fresh strawberries (Sung et al., 1998). Plain LDPE wraps were compared to perforated pinhole packaging impregnated with *R. palmatum*. Packages were stored at 5°C for 13 days and were monitored for packaging atmosphere, microbiological quality, and physiochemical properties of the berries. Antimicrobial films retarded development of both aerobic and lactic acid bacteria, and yeasts, thereby delaying decay. This effect was amplified by hermetically sealing the bags with antimicrobial films to generate atmospheres with low O_2 (4 kPa) and high CO_2 (6.3–9.0 kPa) levels. Blueberries were stored at 4°C in sealed pouches made with intermediate barrier (IB) and high barrier (HB) plastic films to study microbial counts (Day et al., 1990). The IB pouches developed an aerobic (6 kPa O_2) atmosphere with a relatively low CO_2 (4 kPa) content that did not suppress mold spoilage of berries. In the HB pouches,

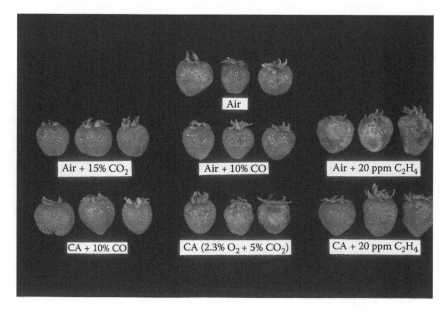

FIGURE 9.1
Effects of CA including CO and ethylene on Botrytis decay in strawberries. (Courtesy of Prof. Adel Kader.)

FIGURE 9.2
Effects of CA on decay control in pomegranates. (Courtesy of Prof. Adel Kader.)

an anaerobic atmosphere developed within 2 weeks, and anaerobic respiration raised the CO_2 level to 70 kPa, preventing microbial spoilage of blueberries for at least 12 weeks.

Other fruit crops such as pomegranates, grapes, papaya, longan fruit, melons, and mangoes have benefited from CA or MA treatments especially if combined with another effective treatment (Yahia, 1998) (Figures 9.2 and 9.3). 'Mollar de Elche' sweet pomegranates stored at 2°C or 5°C for 12 weeks in unperforated polypropylene film of 25 μm thickness reduced the incidence of decay mainly due to *Penicillium* spp. (Artes et al., 2000). Higher decay levels were found in fruit treated at 5°C than those at 2°C. Investigations of CA on growth and development of molds indicate that temperature, especially when just above the minimum growth temperature of the mold reduces growth more than

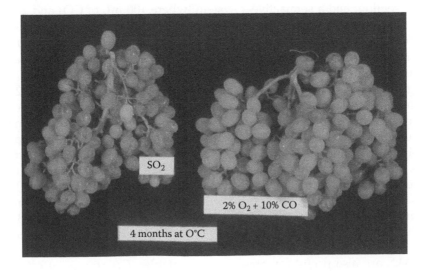

FIGURE 9.3
Effects of CA including CO on preventing decay in grapes, as compared to SO_2 application. (From E.M. Yahia.)

in air (Yackel et al., 1971). Packaging that combines food-packaging materials with antimicrobial substances to control microbial contamination of foods could be useful in controlling foodborne pathogens such as *Listeria monocytogenes* as well as decay-causing organisms (Vermeiren et al., 2002). The addition of 0.5 mL of eugenol, thymol, or menthol improved the beneficial effect of MAP (1.4–2.0 kPa CO_2 and 10.0–14.5 kPa O_2) on 'Crimson Seedless' grapes stored at 1°C for 35 days, delaying rachis deterioration and berry decay, reducing the total viable counts for both yeast and molds (Valverde et al., 2005). Postharvest quality of papaya was significantly enhanced by combining methyl jasmonate and MA treatments reducing postharvest decay caused by fungi such as *Collectotrichum gloeosporiodes* and *Alternaria alternata* (Gonzalez-Aguilar et al., 2003). 'Shixia' Longan fruit treated with CA (4–6 kPa CO_2 + 6–8 kPa O_2) at 1°C for 30 days had better quality and less deterioration than control fruit (Yueming, 1999). The quality of 'Galia' melons stored in CA (10 kPa CO_2 plus 10 kPa O_2) with ethylene absorbent for 14 days at 6°C and an additional 6 days at 20°C was significantly better than control or fruit stored in CA only (Aharoni et al., 1993). This atmosphere also slowed fruit softening and reduced decay incidence. Wickham and Mohammed (1999) investigated the effects of MAP or waxing before storage on quality of immature green mangoes during storage at 10°C, 20°C, or 30°C in sealed LDPE or paper bags. Sixty percent of the fruit stored at 10°C in LDPE bags showed no evidence of pathological infection although severe chilling injury developed in the waxed and control fruit after 14 days.

Tomatoes, peppers, and avocados are high-value crops that can benefit from CA and MA. Bell peppers stored for 15 days at 8°C at four relative humidity (RH) levels (85%, 90%, 95%, and 100%) had an increased incidence of fruit decay with increasing levels of RH, however CA storage at 3 kPa CO_2/3 kPa O_2 reduced the incidence of fruit decay during storage (Polderdijk et al., 1993). Dipping peppers in thiabendazol fungicide solution at 50°C and a successive CA treatment was also reported to inhibit fungal decay in peppers (Yong and Kyung, 1998). Effects of passive and semiactive atmosphere modification on storage life and quality of avocado fruit were evaluated (Yahia and Gonzales-Aguilar, 1998). Avocados (cv. Hass) were sealed in high-density polyethylene (HDPE) or LDPE bags (700 mL in volume) and stored at 5°C and 85% RH for up to 4 weeks. Three MA systems were used: a passive system with no gases introduced other than the initial air; a semiactive-1 system, where 100 mL of CO_2 were injected into each package immediately after sealing; and a semiactive-2 system, where 100 mL of CO_2 and 200 mL of N_2 were injected into the packages immediately after sealing. The results showed that MAP decreased fruit decay and chilling injury, expressed as a general gray discoloration of the mesocarp. MAP will also prolong postharvest life, reduce decay, and preserve quality of muskmelon individually wrapped in different types of LDPE films, with thicknesses ranging from 26 to 75 mm, water vapor permeability from 0.27 to 2.95 g mil^{-1} m^{-2} day atm, and O_2 permeability from 1308 to 9024 mL mil^{-1} m^{-2} day atm (Yahia and Rivera-Dominguez, 1992). Tomatoes have also benefited from MAP. For example breaker tomatoes sealed in polymeric film stored at 15°C for 23 days, where a steady state of about 3.5–4.0 kPa O_2 and CO_2 was established resulted in substantial reduction in fruit weight loss and spoilage compared to fruit without film packaging (Nakhasi et al., 1991). Greenhouse-grown pink tomatoes (cv. Buffalo) were stored in air or CA at 4 kPa O_2 plus 2 kPa CO_2 to study the effect of CA at chilling and nonchilling temperatures on fruit quality characteristics (Nunes et al., 1996). When stored at 12°C in CA for 3 weeks, the tomatoes did not develop any major changes in fruit appearance and had less decay than fruit at 6°C in both air and CA storage.

CA and MA are also useful for preserving vegetables. CA storage at 4°C for 21 days significantly reduced the growth of aerobic microorganisms on broccoli (11 kPa O_2, 10 kPa CO_2, 79 kPa N_2) but had no significant effect on asparagus (15 kPa O_2, 6 kPa CO_2, 79 kPa N_2)

or cauliflower (18 kPa O_2, 3 kPa CO_2, 79 kPa N_2) (Berrang et al., 1990). Sweet corn film-wrapped and stored at 2°C developed a MA of 5–10 kPa CO_2 that inhibited mold growth and prevented fermentation and off-flavor development (Rodov et al., 2000). The produce kept for 2 weeks at 2°C and for 4 additional days at 20°C was found to be preserved with relatively low microbial spoilage and acceptable organoleptic quality.

9.3   Effects of MA and CA on Specific Plant Pathogens

Germination and growth of *Alternaria tenuis, Fusarium roseum, Botrytis cinerea, Cladosporium herbarium,* and *Rhizopus stolonifer* was studied *in vitro* under low O_2 and high CO_2 atmospheres (Wells and Uota, 1970). Mycelial growth decreased with decreasing O_2 levels below 4 kPa. Growth of all the fungi except *F. roseum* decreased linearly with increasing CO_2 atmosphere, and was inhibited about 50% in an atmosphere of 20 kPa CO_2. Low O_2 atmospheres inhibited germination of all fungi tested but CO_2 was variable. El-Goorani and Sommer (1981) studied the effects of MA on postharvest pathogens of fruits and vegetables *in vitro* on solid and liquid media. They tested fungal and one bacterial pathogen on O_2 (23 fungal species) or CO_2 (33 fungal species) using a flow through system. Oxygen atmospheres of about 1 kPa or less were required to obtain appreciable reduction of growth, spore formation, and germination in many of the postharvest fungi. With increasing levels, CO_2 changes from stimulatory to inhibitory. In general growth was not affected by 5 kPa CO_2 but atmospheres over 10 kPa inhibited growth. Carbon dioxide at 13 kPa prevented spores of *Penicillium expansum* from germinating after exposure to the atmosphere for 21 days (Cossentine et al., 2004). However, the treatment had little or no effect on *B. cinerea*. The effect of CA on basidomycetious fungi has not been studied as much as it has been for ascomycetes. A basidiomycete fungus that causes snow mold rot of apples was exposed to CA (1.7 kPa O_2, 1.5 kPa CO_2) conditions for 28 days in a commercial storage at 1°C (Sholberg et al., 1995). CA storage reduced the average colony diameter (26%) and weight of the fungus (66%). Combining low O_2 with high CO_2 does not necessarily result in additive effects (Barkai-Golan, 2001). High CO_2 inhibited growth and sclerotial production of *Sclerotinia minor* at levels of CO_2 >8 kPa with O_2 kept at 21 kPa (Imolehin and Grogan, 1980). Mixing O_2 and CO_2 gases did not add to reducing sclerotial germination but was similar to CO_2 alone. On the other hand, El-Goorani and Sommer (1981) found that *A. alternata* and *P. exspansum* grew more slowly in combined low O_2 and high CO_2 atmospheres than in air or either gas alone.

Apples are generally stored in CA if the goal is to sell them during the winter season after they have been stored for 3 or more months. Apple slices are now commonly sold in MAP at many outlets and have become a fast food item. Decay is an important problem in both CA-stored apples and in MAP-sliced apples. A flow through vapor exposure system was developed to evaluate the effects of 2-nonanone vapor on decay of apple slices inoculated with *P. expansum* and *B. cinerea* (Leepipattanawit et al., 1997). Fungal growth was completely suppressed by 300 $\mu L \cdot L^{-1}$ 2-nonanone vapor; however the fungi regrew when transferred to air. Exposure of *P. expansum* on apple slices to 50 $\mu L \cdot L^{-1}$ 2-nonanone vapor suppressed the fungus and 100 $\mu L \cdot L^{-1}$ apparently killed it at 5°C. However, physiological damage occurred on the peel of treated apples. Another natural fruit volatile, hexanal, was studied for inhibition of fungal activity on apple slices in LDPE film packages (Song et al., 1996). Hexanal concentrations of 18.6 $\mu mol \cdot L^{-1}$ (450 ppm) prevented growth of both *B. cinerea* and *P. expansum* on apple slices. However, permeability data from this study showed that LDPE is a relatively poor barrier to hexanal, which requires approximately 24 h in order to control decay. If this compound were to be used for decay control, a different type of package would be necessary. The authors speculate on the use of a gaseous release mechanism described by Vaughn et al. (1993) that could be used to release a predictable level of vapor into the package to compensate for hexanal lost from the package. Hexanal has been used successfully in stored apples (Sholberg and Randall, 2007) and winter pears (Spotts et al., 2007) to control decay. Biological control of postharvest decay in apples using competitive yeast species must compliment CA storage or it would not be of any practical use for decay control. Three yeast biocontrol agents, *Cryptococcus laurentii* (strain HRA5), *Cryptococcus infirmo-miniatus* (CIM) (strain YY6), and *Rhodotorula glutinis* (strain HRB6), were tested on postharvest diseases of apple and pear in semicommercial and commercial trials. The yeasts were not adversely affected when treated fruits were stored in a CA consisting of 1 kPa O_2 and 99 kPa N_2 (Chand-Goyal and Spotts, 1997). Prestorage heat, CA storage, and pre- and poststorage treatments with the ethylene inhibitor 1-methylcyclopropene (MCP) were tested for their efficacy at inhibiting fungal decay and maintaining quality in 'Golden Delicious' apples stored for up to 5 months at 0°C and 7 days at 20°C (Saftner et al., 2003). Before storage in air at 0°C, pre-climacteric fruits were treated with either MCP at 1 $\mu L \cdot L$ for 17 h at 20°C, 38°C air for 4 days, MCP plus heat, or left untreated. Some sets of untreated fruits were stored in a CA of 1.5 kPa O_2 and 2.5 kPa CO_2 at 0°C, while other sets were removed from cold storage in air after 2.5 or 5 months, warmed to 20°C, and treated with 1 $\mu L \cdot L$ MCP for 17 h. Prestorage MCP, heat, MCP + heat treatments, and CA storage decreased decay severity caused by wound-inoculated *P. expansum*, *B. cinerea*, and *Colletotrichum acutatum*. Poststorage MCP treatment had no effect on decay severity. This study indicates that MCP may provide an effective alternative to CA for reducing decay severity and maintaining quality during postharvest storage of 'Golden Delicious' apples, but prestorage heat to control decay and maintain quality of apples requires further study, especially if used in combination with MCP. *P. expansum* is known to produce the mycotoxin, patulin, in apples and apple juice under certain conditions (McCallum et al., 2002). Moodley et al. (2002) studied the effect of MAP in production of patulin by apples infected with *P. expansum*. They found that patulin production was almost completely prevented in polyethylene-packaged apples under MA conditions of high CO_2.

Brown rot of stone fruit caused by *Monilinia fructicola* is a major disease of all commercially grown *Prunus* spp. in most regions of the world (Ogawa et al., 1995). Integrated control of fruit brown rot of sweet cherry involving preharvest fungicide application, biological control, and MAP was studied by Spotts et al. (2002, 1998). Specifically preharvest application of iprodione was combined with postharvest application of CIM and MAP (50:50 CO_2:N_2) storage at 0.5°C for 42 days in MAP or 2.8°C for 20 days in air (Spotts et al., 1998). Brown rot incidence averaged 61% in untreated fruit stored in air while fruit treated with fungicide plus yeast and MAP only had 1% decay. Brown rot was reduced in sweet cherry by MAP alone and further reduced as a result of a CIM–MAP synergism. The CO_2 level in the MAP was 11.5 kPa after 42 days and may have reduced growth of *M. fructicola* on the one hand and stimulated growth of the yeast on the other hand. An expanded study that used a different preharvest fungicide, propiconazole, and a wettable granular formulation of CIM at two time–temperature regimes for MAP (−0.5°C for 42 days and 2.8°C for 20 days) was conducted on 'Lambert' or 'Lapins' (Spotts et al., 2002). Synergism between the propiconazole fungicide treatment and the yeast biocontrol was observed and was similar to that found for iprodione in the previous study. MA significantly reduced brown rot compared to air-stored fruit. The authors concluded that the integrated decay control system should prove highly effective under commercial conditions with stemmed sweet cherry fruit. In another sweet cherry study using LDPE packaging, control of brown rot improved as the level of CO_2 increased until fungal growth was completely inhibited at 50 kPa for a 7 day period

(De Vries-Paterson et al., 1991). Postharvest gray mold of sweet cherry can be traced to the incidence of blossom infection (Ogawa and English, 1991). Sholberg (1998) showed that acetic and propionic acid fumigation would control decay of sweet cherries caused by *M. fructicola*, *P. expansum*, and *Rhizopus stolonifer* without damage to the fruit. MAP sweet cherries inoculated with spores of *B. cinerea* were fumigated with thymol and acetic acid vapor (Chu et al., 1999). After 10 weeks of cold storage, thymol or acetic acid reduced gray mold rot from 36% to 0.5% or 6%, respectively. However, thymol imparted a medicinal odor to cherries while reducing total soluble solids and increasing titratable acidity. Acetic acid controlled postharvest decay almost as well as thymol but induced stem browning. Acetic acid did not have any significant effect on the MAP atmosphere (15.1 kPa O_2, 3.3 kPa CO_2, 0.020 kPa C_2H_4) when compared to the control, whereas thymol reduced CO_2 to 1.5 kPa. Another approach to using acetic acid vapor with MAP bags was tested at the Pacific Agri-Food Research Centre (PARC) in Summerland, British Columbia, Canada. The objective was to determine if acetic acid vapor would penetrate MAP bags and kill mold spores present on fruit within the bags. The trials were conducted on 'Lambert' and 'Bing' cherries contaminated with *P. expansum* spores. Initial tests with various MAP bags designed to allow 16 kPa O_2 showed that "View Fresh" bags having only 4 pin holes was impermeable to acetic acid while "Ralston" and CryovacPD941 bags were permeable to acetic acid vapor over their entire surface. In these trials, it was found that fumigation through the bags was much more effective at 20°C than at 1°C in killing spores. A starting concentration of 10 mg $\cdot$ L^{-1} yielded about 1 mg $\cdot$ L^{-1} acetic acid vapor in the MAP bag. This rate would effectively kill most *P. expansum* spores on the cherry surface as shown by washing fruit after treating with acetic acid at rates from 0 to 2.6 mg $\cdot$ L^{-1}. In general, the cherry trials were too variable to recommend for commercial use because of problems with condensation, absorption of acetic acid vapor by cardboard, and damage to stems, however the process was promising enough that it should be further studied at temperatures between 20°C and 25°C with bags even more permeable to acetic acid vapor.

CA or MA is not generally used for table grapes but the following studies show that these technologies could be useful for delaying senescence, decreasing stem and berry respiration, reducing stem browning, maintaining berry firmness, and retarding decay development (Yahia et al., 1983). A fumigation technique using brief exposure of fruit to a low concentration of acetic acid vapor was combined with MAP to reduce gray mold decay (*B. cinerea*) and increased shelf life of grapes and strawberries by two or three times normal values (Moyls et al., 1996). It was shown that inoculated table grapes fumigated with 8.0 mg/L acetic acid followed by MAP for 74 days at 0°C combined to reduce the percentage of rotted grapes from 94% to 2%. Optimum CO_2 and O_2 levels designed to control gray mold (*B. cinerea*) on 'Red globe' table grapes without affecting quality were evaluated by Crisosto et al. (2002). Fruits were commercially harvested and field packaged in cluster bags, fumigated with SO_2 during precooling, and stored at 0°C for up to 12 weeks in aluminum tanks under a continuous flow of either air or the desired mixture of CO_2 and O_2 atmospheres. Carbon dioxide at 10 kPa or greater levels combined with 3–12 kPa O_2 limited decay development. Levels >10 kPa CO_2 accelerated stem browning and off-flavor development in early harvested grapes, leading to the recommendation not to use levels >10 kPa CO_2 no <6 kPa O_2 for treatment of grapes. The repackaging losses of grapes kept in CA were similar to grapes kept under commercial conditions even though SO_2 fumigation was used weekly in commercial storage.

Persimmon cv. Triumph is commonly stored commercially at −1°C for up to 3 months. Trials to enhance fruit quality during prolonged storage showed that MAP in LDPE bags delayed decay and softening (Ben-Arie et al., 1991). MAP of persimmon fruit resulted in the accumulation of acetaldehyde to a level of 80 µg/mL, ethanol to a level of 900 µg/mL, and CO_2 up to 30 kPa (Prusky et al., 1997). When fruits were stored at −1°C for 4 months in

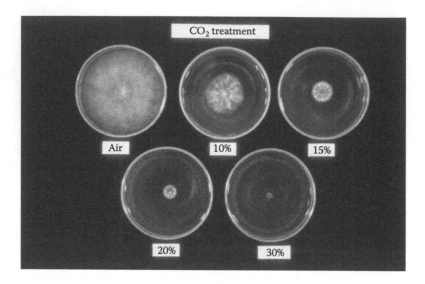

FIGURE 9.4
Carbon dioxide effects on *in vitro* fungal growth. (Courtesy of Prof. Adel Kader.)

such atmospheres, the incidence of black spot disease, caused by *Alternaria alternata*, was reduced. The effects of each of these gases were examined to determine their individual involvement on the inhibition of *A. alternata* development during storage. Inoculated persimmon fruit exposed to different levels of each volatile for 24 h showed that acetaldehyde was the most fungistatic but only at concentrations higher than those that accumulated under MAP; CO_2 was moderately inhibitory at concentrations from 10 to 60 kPa, whereas ethanol had no inhibitory effect. Based on these results and studies with *in vitro* cultures, the authors suggest that the increasing level of CO_2 during storage is the principal factor in the inhibition of black spot disease in MAP-stored persimmon. Figure 9.4 shows the *in vitro* inhibition of fungi by carbon dioxide.

CA storage of vegetables is a desirable technique for maintaining quality or reducing decay but may adversely affect the presence of foodborne pathogens or the occurrence of mycotoxins. CA storage was used to suppress watery soft rot of celery caused by *Sclerotinia sclerotiorum* (Reyes, 1988). Disease suppression at 8°C was greatest in atmospheres of 7.5–30 kPa. Although there is controversy concerning *Aeromonas hydrophila* as a food pathogen, it has been linked to human illness (Gracey et al., 1982). The effects of CA storage on the survival and growth of *A. hydrophila* on fresh asparagus, broccoli, and cauliflower were examined by Berrang et al. (1989). Two lots of each vegetable were inoculated with two strains of *A. hydrophila*, while a third lot served as the uninoculated control. Following inoculation, vegetables were stored at 4°C or 15°C in a CA system previously shown to extend the shelf life of each commodity or in ambient air. Populations of *A. hydrophila* were enumerated at various intervals for 10 days at 15°C or 21 days at 4°C. Without exception, the CA system lengthened the time vegetables were subjectively considered acceptable for consumption. However, CA did not significantly affect populations of *A. hydrophila*, which survived or grew on inoculated vegetables. *Fusarium sporotrichioides* capable of producing T-2 toxin (T-2) was grown on irradiated corn kernels remoistened to 22% under MA conditions (Paster and Menasherov, 1988). The production of T-2 was totally inhibited at 60 kPa CO_2 to 20 kPa O_2, whereas only trace amounts were detected when the gas combination was 40 kPa CO_2–5 kPa O_2 in stored corn. Under all other combinations tested, the amount of T-2 produced was reduced by 25%–50% as compared with the control. Fungal growth was not inhibited by any of the gas mixtures examined, and the growth rate was almost identical to that on kernels in air. Apparently T-2 formation on corn can be inhibited by CO_2 atmospheres less than that required to inhibit fungal growth. The prevalent bacteria on mungbean sprouts and cut chicory endive were determined during storage under CA conditions at 8°C (Bennik et al., 1998). Enumeration of the total mesophilic counts, *Enterobacteriaceae*, *Pseudomonas* species, and lactic acid bacteria indicated that *Enterobacteriaceae* and *Pseudomonas* species constituted the major populations found on these vegetables before and after CA storage. Identification of the predominant species within these populations revealed that on fresh and CA-stored mungbean sprouts, *Enterobacter cloacae*, *Pantoea agglomerans*, *Pseudomonas fluorescens*, *P. viridilivida*, and *P. corrugata* were the prevalent species. On chicory endive, *Rahnella aquatilis* and several *Pseudomonas* species were found on the fresh product, while after CA storage *Escherichia vulneris* and *P. fluorescens* were the main species. In general, these CA conditions did not strongly influence maximum population densities and lag times were not detected. For each of the strains, however, maximum specific growth rates were reduced at increased CO_2 atmospheres, independent of the O_2 level applied. Storage under MA can cause a selective suppression of the outgrowth of different epiphytic populations. Competition between epiphytes and pathogens may retard the growth of pathogens on minimally processed vegetables but the use of MA may unbalance this safety feature. This could lead to the buildup of unforeseen foodborne pathogens so further systematic studies are advised to ensure the safety of minimally processed vegetables under CA or MA. The survival of the virus that causes hepatitis A (HAV) has been studied on lettuce (Bidawid et al., 2001). MAP did not influence HAV survival when present on the surface of produce incubated at 4°C.

9.4 Specific Effects of Low O_2 and High CO_2

The specific effect of low levels of O_2 and high levels of CO_2 on decay-causing microorganisms depends on the gas concentration and the particular microorganism. *Penicillium digitatum*, a major pathogen of citrus fruit, causes green mold in these crops. Green mold was suppressed in grapefruit by reducing the O_2 level in the storage atmosphere from 21 to 0.10–0.05 kPa during short-term (approximately 3 weeks) storage of 'Rio Red' grapefruit (Shellie, 2002). The treatment caused a reduction in quality of early season grapefruit but was considered acceptable for the market. Short-term low O_2 storage of grapefruit may be useful for developing a nonchemical quarantine treatment for postharvest insect control during shipment of citrus fruit. Cherimoya fruit (*Annona cherimola*), native to subtropical South America, was subjected to a CA level of 20 kPa CO_2. The result was to overcome chilling temperature of cherimoya by activation of specific responses such as increased respiration, γ-aminobutyric acid, polyamines, and coinduction of chitinase, and 1,3-β-glucanase by high CO_2 levels (Merodio et al., 1998). The effect of CO_2 on growth of cherimoya microflora was not reported in this study but CO_2 activated the plant defense system and cytoplasmic pH, which likely would inhibit microbial growth. The effect of low O_2 and high CO_2 was studied by Spalding and Reeder (1975) on avocados for the control of anthracnose and chilling injury. The combination of low O_2 and high CO_2 were more effective than either alone for retarding the development of anthracnose decay and chilling injury in 'Fuchs' and 'Waldin' avocados stored at 7.2°C for 3 and 4 weeks, respectively. Spalding and Reeder (1977) also found that less decay developed in mangoes stored at low pressures of 76 or 152 mmHg rather than at normal atmospheric pressure. Pressures below 50 mmHg were required to completely prevent growth in cultures of *P. digitatum*,

Alternaria alternata, Botrytis cinerea, and *Diplodia natalensis* due to low O_2 tension (Apelbaum and Barkai-Golan, 1977). However, normal growth resumed in these fungi when transferred to normal atmospheric pressure, suggesting that no irreversible damage had occurred and that the treatment was fungistatic rather than fungicidal.

The effect of high CO_2 on decay in cranberry and biological control in strawberry have been studied. In cranberry, elevated levels of CO_2 (30 kPa) at 3°C decreased decay in 'Pilgrim' and 'Stevens' cultivars of cranberries (*Vaccinium macrocarpon* Aiton), thereby reducing fruit losses (Gunes et al., 2002). The use of CA storage also reduced the respiration rate of the cranberries that could extend the availability of fresh cranberries with premium quality. Strawberry postharvest life is often extended by carbon dioxide-enriched atmospheres that reduce the incidence and severity of gray mold decay caused by *Botrytis cinerea*. Effects of CA storage, hot water treatment, and biological control were investigated on decay in strawberries inoculated with *B. cinerea* (Wszelaki and Mitcham, 2003). After 1 week, the extent of decay was significantly lower in strawberries treated with a combination of hot water, biocontrol yeast, and CA (15 kPa CO_2) than in the other treatment combinations. After 2 weeks, the above treatment continued to reduce decay but was no better than the other treatment combinations. Overall, the combination treatments did not provide better control than the current commercial CA (15 kPa CO_2) treatment.

High CO_2 and low O_2 levels have been evaluated for the control of fruit brown rot caused by *Monilinia fructicola* in stone fruit crops. Growth of *M. fructicola* significantly declined with increased CO_2 atmospheres, both *in vitro* and *in vivo* in sweet cherries (Tian et al., 2001). Carbon dioxide at 15–25 kPa, provided a significant reduction in cherry lesion size, and 30 kPa CO_2 completely prevented lesion formation at 25°C. *Monilinia fructicola* was more sensitive to high CO_2 levels at 0°C than at 25°C. Brown rot decay was not found on inoculated sweet cherries after 30 days in CA (10–30 kPa CO_2) at 0°C. Injury or off-flavors due to high CO_2 were not detected at any of the CO_2 levels that were tested. The effect of high levels of CO_2 to nectarines infected with *M. fructicola* was studied by Ahmadi et al. (1999). Inoculated fruit were incubated at 20°C for up to 72 h and then transferred to CA (15 kPa CO_2) storage at 5°C for up to 16 days when they were examined for brown rot decay. Nectarine brown rot was influenced by incubation period, storage duration, and storage atmosphere. 'Summer Red' nectarines tolerated a 15 kPa CO_2 atmosphere for 16 days at 5°C. Development of brown rot in fruit inoculated 24 h before 5 or 16 days storage in CA (15 kPa CO_2) at 5°C was arrested. After 3 days ripening in air at 20°C, the progression of brown rot disease was rapid in all inoculated nectarines, demonstrating the fungistatic effect of CA (15 kPa CO_2). It was found in this study that measurement of fungal glucosamine provided a more accurate measurement of brown rot infection than visual measurement of the decayed area on the fruit. The effect of low oxygen on peach decay was studied on 'Fairtime' peaches kept in 0.25 or 0.02 kPa O_2 (Ke et al., 1991b). The low O_2 treatments delayed incidence and reduced severity of chilling injury and decay of peaches stored at 5°C.

Vegetable crops may also benefit from the use of high CO_2 or low O_2 atmospheres. The effect of high levels of CO_2 at 20°C or 40°C for 10 or 16 h on *Botrytis cinerea* storage rot in beans (*Phaseolus vulgaris* L.) was assessed (Cheah and Irving, 1997). Air at 40°C for 16 h gave complete control of *Botrytis* spp. storage rot, but an atmosphere of 60 kPa CO_2 was not effective and adversely affected respiration rate, firmness, and color of the beans. The effect of low levels of O_2 (0.25, 0.5, and 1.0 kPa) was studied on zucchini squash slices treated with calcium chloride (Izumi et al., 1996). Respiration rate, ethylene production, and the development of browning/decay were reduced by low O_2. Low O_2 also improved microbial count, pH, and color at the end of storage of the zucchini slices with or without the use of calcium chloride.

9.5 Combined Effects of Low O_2 and Elevated CO_2 Atmospheres

Hypobaric or low-pressure storage is designed to delay ripening and senescence processes and extend shelf life of fruits and vegetables (Barkai-Golan, 2001). The combined effect of low O_2 and high CO_2 atmospheres is generally better for decay control than either gas alone (Figures 9.5 and 9.6). Banana (*Musa*, AAA) bunch-sections consisting of one hand attached to a short section of the main stalk were inoculated with cultures of fungi originally isolated from diseased bananas. The inoculated bananas were stored in sealed

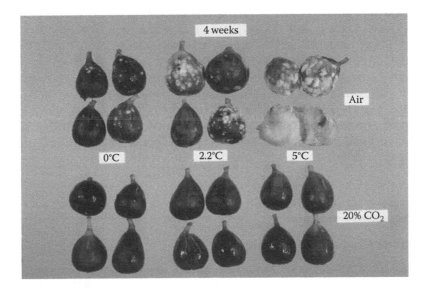

FIGURE 9.5
Effects of CA on decay control in figs. (Courtesy of Prof. Adel Kader.)

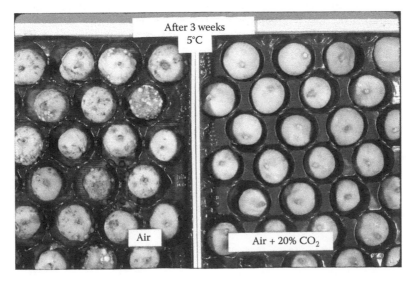

FIGURE 9.6
CA reduces decay on Calimyrna figs. (Courtesy of Prof. Adel Kader.)

polyethylene bags containing 3–7 kPa O_2, 10–13 kPa CO_2, and <0.1 $\mu L \cdot L^{-1}$ ethylene for 40 days at 20°C, then ripened with ethylene in air for 9 days. *Fusarium moniliforme* var. *subglutinans* was a major cause of disease during MA storage, as were *Colletotrichum musae* and *Nattrassia mangiferae* in fingers, and *F. culmorum* in crowns. The pathogenicity of some isolates was greatly affected by MA storage. No disease was caused by *C. acutatum* and *C. gloeosporioides* during or after MA storage, and *F. culmorum* caused only a trace of disease in fingers ripened after MA storage. Two other isolates that caused disease in air but not in or after MA storage were *E. nigrum* and *Nigrospora sphaerica*. The authors hypothesize that wound-healing may have occurred at the inoculation sites during MA storage, so that isolates in which growth and germination were inhibited by MA would have been denied infection sites when returned to air. Strawberries (*Fragaria ananassa* Duch., cv. Selva) stored at 0°C or 5°C for 10 days in low O_2 (0.25, 0.5, 1 kPa) and high CO_2 (20 kPa) atmospheres had reduced decay without detrimental effects on quality (Ke et al., 1991a). If these treatments were followed by transferring to air at 0°C for several days to reduce ethanol and acetaldehyde levels, they could extend the postharvest life of strawberries for several days. Freshly harvested maize grain placed in MA consisting of CO_2 (20, 40, and 60 kPa), in combination with N_2 (30, 50, and 70 kPa) and low O_2 (10 kPa) were stored at 25°C for 45 days to reduce mold growth and prevent loss of food reserves (Janardhana et al., 1998). The MA of 20 kPa CO_2 did not completely eliminate storage molds but by increasing it to 60 kPa, mold growth was further inhibited and loss of food reserves and dry matter was prevented.

9.6 Effects of Carbon Monoxide

Carbon monoxide (CO) has been shown to be very effective for decay control, especially when combined with low O_2 atmospheres (≤ 4 kPa O_2) (Figures 9.7 and 9.8) (El-Goorani and Sommer, 1981; Kader, 1983; Yahia et al., 1983). Carbon monoxide has been found to effectively suppress fungal growth, both *in vitro* and *in vivo* and was very effective in suppressing the growth of *B. cinerea, M. fructicola,* and *P. expansum* if the atmosphere contained <6 kPa O_2 (Barkai-Golan, 1990; Sommer et al., 1982). In air, the most common postharvest fungi are either not affected or only slightly inhibited by

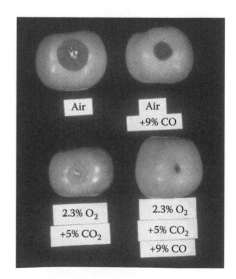

FIGURE 9.7
Effects of CA including CO on decay control in apples. (Courtesy of Prof. Adel Kader.)

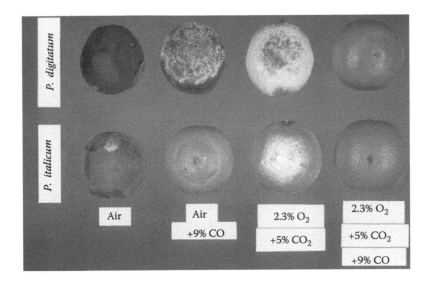

FIGURE 9.8
Effects of CA containing CO on decay control on oranges. (Courtesy of Prof. Adel Kader.)

9 kPa CO (El-Goorani and Sommer, 1979). The minimum level required to achieve decay control is >5 kPa CO in <5 kPa O_2 (Sommer et al., 1982). Carbon monoxide at 5–10 kPa added to reduced O_2 (2–4 kPa) with or without elevated CO_2 (2–5 kPa) completely inhibited growth of brown rot (*M. fructicola*) on inoculated peaches (Kader et al., 1982) and cherries (Sommer et al., 1982). However, Kader et al. (1982) noted some problems with the CO treatment as follows: the normal rate of rot development resumed once inoculated plates and fruits were transferred to air at 20°C and CO mimicked C_2H_4 effects stimulating CO_2 production by fruit stored in air. Stimulation of C_2H_4 was not considered a serious problem because the effect was minimal when fruit were stored at 4 kPa O_2, the optimum condition fungistatic treatment. This tendency of CO to mimic ethylene has also been reported in tomatoes, strawberries, and probably kiwifruit (Barkai-Golan, 1990). The spread of *B. cinerea* in stored kiwifruit can be characterized by stem-end rot in which the fungus penetrates the picking wound at, or soon after harvest and by spreading from fruit to fruit by contact in storage (Droby and Lichter, 2004). Carbon monoxide prevented the formation of aerial mycelium of *B. cinerea* and thus completely eliminated fruit-to-fruit spread in kiwifruit although the fruit was excessively softened by the CO unless counteracted by CO_2. The fungistatic effects of CO on *B. cinerea* have also been demonstrated in tomato fruit (Kader et al., 1978; Morris et al., 1981), grapes (Yahia et al., 1983), and strawberries (El-Kazzaz et al., 1983). Morris et al. (1982) found that 5 kPa CO + 4 kPa O_2 resulted in a significant reduction in decay of tomatoes not only after holding at 12.8°C for up to 7 weeks, but also after an additional 6 days at 20°C in air. Carbon monoxide also suppressed *Sclerotinia sclerotiorum*, the causal agent of watery soft rot in celery (Reyes, 1988). As an effective fungistat, CO was found to be a possible substitute for the conventional SO_2 fumigation treatments for decay control in table grapes (Yahia et al., 1983). The addition of 10 kPa CO to 2 kPa O_2 was very effective in preventing decay and prolonging storage life of 'Thomson Seedless' grapes held at 0°C for up to 4 months (Yahia et al., 1983). Sommer et al. (1982) proposed that CO suppresses fungal growth and lesion development by its inhibitory action on fungal toxins and enzymes. In addition to decay control, CO can provide insect control, and reduce tissue browning and discoloration. The potential use of CO requires the development of safe methods of application so there is no possibility of

accidental inhalation by humans. Its effect on decay as well as on the physiological responses of certain commodities show that this fungistatic gas can be a useful supplement to low O_2 (≤ 4 kPa) atmospheres or to CA, which combines low O_2 and high CO_2 to reduce decay, without causing undesirable side effects on quality (Barkai-Golan, 1990).

9.7   Effects of Superatmospheric Oxygen (High O_2) Atmospheres

Very few studies have dealt with the effect of high O_2 atmospheres on decay in horticultural commodities although there may be potential benefits with this treatment. Superatmospheric O_2 or elevated oxygen has been shown to prevent anaerobic fermentation, off-flavor development, and growth of microbes in some fruit and vegetable crops (Amanatidou et al., 1999, 2000; Tian et al., 2002). In general, exposure to high O_2 alone will not inhibit microbial growth, while CO_2 alone will only reduce growth slightly, but consistent strong inhibition was observed when the two gases were combined (Amanatidou et al., 1999). Thus superatmospheric O_2 concentrations appear to reduce the negative effects of high CO_2 and allow CO_2-sensitive commodities to tolerate fungistatic CO_2 levels (>15 kPa) with minimal CO_2 injury (Kader and Ben-Yehoshua, 2000; Kader and Watkins, 2000).

Some examples of high O_2 MA and its beneficial effects when used in various commodities are given below. Minimally processed carrots with an MA of 50 kPa O_2 and 30 kPa CO_2 were similar or better than those stored at 1 kPa O_2 and 10 kPa CO_2 after 12 days at 8°C (Amanatidou et al., 2000). Adding superatmospheric O_2 to barrier film, MAP alleviated off-odor and tissue injury in addition to reducing aerobic mesophilic bacterial growth in minimally processed baby spinach leaves (Allende et al., 2004). Storage of cucumbers under high O_2 atmosphere increased superoxide dismutase activity and reduced pitting, which was induced by chilling injury at 5°C (Srilaong and Tatsumi, 2003). The effect of different O_2 and CO_2 atmospheres on quality and decay during storage of two longan fruit (*Dimocarpus longan*) cultivars ('Chuliang' and 'Shixia') was studied (Tian et al., 2002). CA with 70 kPa O_2 significantly decreased ethanol production, and maintained a green peel color with lower pH, however, the high O_2 atmosphere was not as effective as the CO_2 atmospheres in reducing decay. Blueberry shelf life is limited by fruit decay, which is often caused by *B. cinerea* (Bristow and Milholland, 1995). Highbush blueberries stored at 5°C that were exposed to 60 and 100 kPa O_2 for 5 weeks had significantly less decay, higher oxygen radical absorbance, and higher total phenolics and anthocyanin levels (Zheng et al., 2003). It was not clear from this study whether the decay was only due to *B. cinerea* or other additional fungi because the authors did not identify the fungi causing decay. Wszelaki and Mitcham (2000) studied the effects of elevated O_2 alone or in combination with elevated CO_2 atmospheres for postharvest decay control on strawberry fruit. Although the high O_2 treatment of 100 kPa reduced decay by *B. cinerea*, its benefits are outweighed by the detrimental effects on flavor quality. High O_2 may be effective against *Penicillium* decay, an important postharvest disease in several crops. In a preliminary experiment, Kader and Ben-Yehoshua (2000) found that 80 kPa O_2 reduced *Penicillium* decay of grape fruit.

High O_2 probably reduces decay by its direct toxic effect on microbial cells that lack mechanisms to protect themselves from damage. In a study of a limited number of prokaryotic and eukaryotic microorganisms, exposure to high O_2 alone had an inhibitory effect on growth in only a few cases and may have a stimulatory effect (Amanatidou et al., 1999).

It is likely that a high O_2 atmosphere generates reactive oxygen species that affect vital cell components and reduce cell viability (Fridovich, 1986). Therefore, microorganisms have developed strategies such as induction of O_2-decomposing enzymes or radical scavengers to avoid lethal damage by oxygen (Amanatidou et al., 1999). Stronger and more consistent inhibition of microbial growth can be obtained when O_2 and CO_2 are used in combination.

9.8 Conclusions

CA and MA are increasingly being used during storage, packaging, and transport of several horticultural commodities to supplement refrigerated storage. A beneficial combination of reduced oxygen and increased carbon dioxide in the atmosphere as a compliment to refrigeration has several positive effects including decay control. Low O_2 and high CO_2 atmospheres cause a reduction in the growth of certain fungi, especially when the atmosphere includes <1 kPa O_2 and/or >10 kPa CO_2. The addition of some other gases to the atmosphere, especially CO at lower levels (<4 kPa) of O_2, can increase the control of decay. Some of the commercial applications of MA/CA for decays control include those used in strawberry and some other berry fruits.

9.9 Future Research

The importance of controlling plant pathogens on fruit and vegetable commodities placed under MAP conditions is often overlooked. The ideal approach is to only include products in MAP free of fungal or bacterial infection; although that is not always practical and achievable, but the technology is now available that would allow testing of the produce within a matter of hours to ensure it is free of potential decay-causing pathogens. Real-time PCR could be used to alert the packer about infection of the produce, the identity of the pathogen, and the risk the pathogen would present for safe storage if found extensively. Pilot studies would be needed to determine exactly how much DNA of a pathogen in a particular crop will lead to decay.

Another problem in MAP storage is contamination of produce after harvest with storage pathogens. In the case of fruit, these are spores and research is needed on how to eliminate them from the surface of the produce either before placing in the MAP bag or when the bag is closed. Research on this contamination problem has been touched upon in this chapter but a more focused approach is needed to develop treatments that are reliable and do not damage the produce or change its organoleptic properties. Fumigants is a good choice because they are able to attack spores in hard-reach places. There are many types of naturally occurring materials that could be considered for this use. Research is needed on how to apply them, such as a component of the MAP bag, as a gas that penetrates the bag leaving after it has killed the spores, or as a pretreatment that could be chemical or physical (such as heat) treatment that kills the pathogen. Research on eliminating disease problems in MAP produce will also make the products safer for human consumption by preventing the production of mycotoxins.

Carbon monoxide application, in combination with low O_2 and/or high CO_2, has been shown to be very effective in controlling decay organisms; however, research is still needed to develop safe measures for its application.

References

Aharoni, Y., A. Copel, and E. Fallik. 1993. Storing 'Galia' melons in a controlled atmosphere with ethylene absorbent. *Hort. Science* 28: 725–726.

Ahmadi, H., W.V. Biasi, and E.J. Mitcham. 1999. Control of brown rot decay of nectarines with 15% carbon dioxide atmospheres. *J. Am. Soc. Hort. Sci.* 124: 708–712.

Allende, A., Y. Luo, J.L. McEvoy, F. Artes, and C.Y. Wang. 2004. Microbial and quality changes in minimally processed baby spinach leaves stored under super atmospheric oxygen and modified atmosphere conditions. *Postharverst Biol. Technol.* 33: 51–59.

Amanatidou, A., R.A. Slump, L.G.M. Gorris, and E.J. Smid. 2000. High oxygen and high carbon dioxide modified atmospheres for shelf-life extension of minimally processed carrots. *J. Food Sci.* 65: 61–65.

Amanatidou, A., E.J. Smid, and L.G.M. Gorris. 1999. Effect of elevated oxygen and carbon dioxide on the surface growth of vegetable-associated micro-organisms. *J. Appl. Microbiol.* 86: 429–438.

Apelbaum, A. and R. Barkai-Golan. 1977. Spore germination and mycelial growth of postharvest pathogens under hypobaric pressure. *Phytopathology* 67: 400–403.

Artes, F., R. Villaescusa, and J.A. Tudela. 2000. Modified atmosphere packaging of pomegranate. *J. Food Sci.* 65: 1112–1116.

Barkai-Golan, R. 1990. Postharvest disease suppression by atmospheric modifications. In: M. Calderon and R. Barkai-Golan (Eds.), *Food Preservation by Modified Atmospheres*. CRC Press, Boca Raton, FL, pp. 237–264.

Barkai-Golan, R. 2001. *Postharvest Diseases of Fruits and Vegetables: Development and Control*. Elsevier Science B.V., Amsterdam, the Netherlands.

Ben-Arie, R., Y. Zutkhi, L. Sonego, and J. Klein. 1991. Modified atmosphere packaging for long-term storage of astringent persimmons. *Postharverst Biol. Technol.* 1: 169–179.

Ben-Yehoshua, S., R.M. Beaudry, S. Fishman, S. Jayanty, and N. Mir. 2005. Modified atmosphere packaging and controlled atmosphere storage. In: S. Ben-Yehoshua (Ed.), *Environmentally Friendly Technologies for Agricultural Produce Quality*. CRC Press, Taylor & Francis Group, Boca Raton, FL, pp. 61–112.

Bennik, M.H.J., W. Vorstman, E.J. Smid, and L.G.M. Gorris. 1998. The influence of oxygen and carbon dioxide on the growth of prevalent Enterobacteriaceae and *Pseudomonas* species isolated from fresh and controlled-atmosphere-stored vegetables. *Food Microbiol.* 15: 459–469.

Berrang, M.E., R.E. Brackett, and L.R. Beuchat. 1989. Growth of *Aeromonas hydrophila* on fresh vegetables stored under a controlled atmosphere. *Appl. Environ. Micobiol.* 55: 2167–2171.

Berrang, M.E., R.E. Brackett, and L.R. Beuchat. 1990. Microbial, color and textural qualities of fresh asparagus, broccoli, and cauliflower stored under controlled atmosphere. *J. Food Prot.* 53: 391–395.

Bidawid, S., J.M. Farber, and S.A. Sattar. 2001. Survival of hepatitis A virus on modified atmosphere-packaged (MAP) lettuce. *Food Microbiol.* 18: 95–102.

Bristow, P.R. and R.D. Milholland. 1995. Botrytis blight. In: F.L. Caruso and D.C. Ramsdell (Eds.), *Compendium of Blueberry and Cranberry Diseases*. APS Press, St. Paul, MN, pp. 8–9.

Chambroy, Y., M.H. Guinebretiere, G. Jacquemin, M. Reich, L. Breuils, and M. Souty. 1993. Effects of carbon dioxide on shelf-life and postharvest decay of strawberry fruit. *Sci. Aliments* 13: 409–423.

Chand-Goyal, T. and R.A. Spotts. 1997. Biological control of postharvest diseases of apple and pear under semi-commercial and commercial conditions using three saprophytic yeasts. *Biol. Control* 10: 199–206.

Cheah, L.H. and D.E. Irving. 1997. Influence of CO_2 and temperature on the incidence of botrytis storage rot in beans. *N. Z. J. Crop Hort. Sci.* 25: 85–88.

Chu, C.L., W.T. Liu, T. Zhou, and R. Tsao. 1999. Control of postharvest gray mold rot of modified atmosphere packaged sweet cherries by fumigation with thymol and acetic acid. *Can. J. Plant Sci.* 79: 685–689.

Cossentine, J.E., P.L. Sholberg, L.B.J. Jensen, K.E. Bedford, and T.C. Shephard. 2004. Fumigation of empty fruit bins with carbon dioxide to control diapausing codling moth larvae and *Penicillium expansum* Link. ex Thom spores. *Hort. Science* 39: 429–432.

Crisosto, C.H., D. Garner, and G. Crisosto. 2002. Carbon dioxide-enriched atmospheres during cold storage limit losses from Botrytis but accelerate rachis browning of 'Red globe' table grapes. *Postharvest Biol. Technol.* 26: 181–189.

Day, N.B., B.J. Skura, and W.D. Powrie. 1990. Modified atmosphere packaging of blueberries: Microbiological changes. *Can. Inst. Food Sci. Technol. J.* 23: 59–65.

De Vries-Paterson, R.M., A.L. Jones, and A.C. Cameron. 1991. Fungistatic effects of carbon dioxide in a package environment on the decay of Michigan sweet cherries by *Monilinia fructicola*. *Plant Dis.* 75: 943–946.

Droby, S. and A. Lichter. 2004. Post-harvest *Botrytis* infection: Etiology, development and management. In: Y. Elad, B. Williamson, P. Tudzynski, and N. Delen (Eds.), *Botrytis: Biology, Pathology and Control*. Kluwer Academic, Dordrecht, the Netherlands, pp. 349–367.

El-Goorani, M.A. and N.F. Sommer. 1979. Suppression of postharvest plant pathogenic fungi by carbon monoxide. *Phytopathology* 69: 834–838.

El-Goorani, M.A. and N.F. Sommer. 1981. Effects of modified atmospheres on postharvest pathogens of fruits and vegetables. *Hort. Rev.* 3: 412–461.

El-Kazzaz, M.K., N.F. Sommer, and R.J. Fortlage. 1983. Effect of different atmospheres on postharvest decay and quality of fresh strawberries. *Phytopathology* 73: 282–285.

Farber, J.M. 1991. Microbiological aspects of modified-atmosphere packaging technology—A review. *J. Food Prot.* 54: 58–70.

Fridovich, I. 1986. Biological effects of the superoxide radical. *Arch. Biochem. Biophys.* 247: 1–11.

Gonzalez-Aguilar, G.A., J.G. Buta, and C.Y. Wang. 2003. Methyl jasmonate and modified atmosphere packaging (MAP) reduce decay and maintain postharvest quality of papaya 'Sunrise.' *Postharverst Biol. Technol.* 28: 361–370.

Gracey, M., V. Burke, and J. Robinson. 1982. Aeromonas-associated gastroenteritis. *The Lancet* 320: 1304–1306.

Gunes, G., R.H. Liu, and C.B. Watkins. 2002. Controlled-atmosphere effects on postharvest quality and antioxidant activity of cranberry fruits. *J. Agric. Food Chem.* 50: 5932–5938.

Imolehin, E.D. and R.G. Grogan. 1980. Effects of oxygen, carbon dioxide, and ethylene on growth, sclerotial production, germination, and infection by *Sclerotinia minor*. *Phytopathology* 70: 1158–1161.

Izumi, H., A.E. Watada, and W. Douglas. 1996. Low O_2 atmospheres affect storage quality of zucchini squash slices treated with calcium. *J. Food Sci.* 61: 317–321.

Janardhana, G.R., K.A. Raveesha, and H.S. Shetty. 1998. Modified atmosphere storage to prevent mould-induced nutritional loss in maize. *J. Sci. Food. Agric.* 76: 573–578.

Kader, A.A. 1983. Physiological and biochemical effects of carbon monoxide added to controlled atmospheres on fruits. *Acta Hort.* 138: 221–226.

Kader, A.A. and S. Ben-Yehoshua. 2000. Effects of superatmospheric oxygen levels on postharvest physiology and quality of fresh fruits and vegetables. *Postharverst Biol. Technol.* 20: 1–13.

Kader, A.A., G.A. Chastagner, L.L. Morris, and J.M. Ogawa. 1978. Effects of carbon monoxide on decay, physiological responses, ripening, and composition of tomato fruits. *J. Am. Soc. Hort. Sci.* 103: 665–670.

Kader, A.A., M.A. El-Goorani, and N.F. Sommer. 1982. Postharvest decay, respiration, ethylene production, and quality of peaches held in controlled atmospheres with added carbon monoxide. *J. Am. Soc. Hort. Sci.* 107: 856–859.

Kader, A.A. and C.B. Watkins. 2000. Modified atmosphere packaging—toward 2000 and beyond. *HortTechnology* 10: 483–486.

Ke, D., L. Goldstein, M. O'mahony, and A.A. Kader. 1991a. Effects of short-term exposure to low O_2 and high CO_2 atmospheres on quality attributes of strawberries. *J. Food Sci.* 56: 50–54.

Ke, D., L. Rodriguez-Sinobas, and A.A. Kader. 1991b. Physiological responses and quality attributes of peaches kept in low oxygen atmospheres. *Sci. Hort.* 47: 295–303.

Leepipattanawit, R., R. Beaudry, and S. Hernandez. 1997. Control of decay in modified-atmosphere packages of sliced apples using 2-nonanone vapor. *J. Food Sci.* 62: 1043–1047.

McCallum, J.I., R. Tsao, and T. Zhou. 2002. Factors affecting patulin production by *Penicillium expansum*. *J. Food Prot.* 65: 1937–1942.

Merodio, C., M.T. Munoz, B. Del Cura, D. Buitrago, and M.I. Escribano. 1998. Effect of high CO_2 level on the titres of γ-aminobutyric acid, total polyamines and some pathogenesis-related proteins in cherimoya fruit stored at low temperature. *J. Exp. Bot.* 49: 1339–1347.

Moodley, R.S., R. Govinden, and B. Odhav. 2002. The effect of modified atmospheres and packaging on patulin production in apples. *J. Food Prot.* 65: 867–871.

Morris, L.L., D. Mansfield, and L. Strand. 1981. The role of CA, including CO, in prolonging the storage life of tomatoes. *National Controlled Atmosphere Research Conference on Controlled Atmospheres for Storage and Transport of Perishable Agricultural Commodities*, Beaverton, OR, pp. 285–287.

Morris, L.L., D. Mansfield, and L. Strand. 1982. The role of CA, including CO, in prolonging the storage life of tomatoes. In: D.G. Richardson and M. Meheriuk (Eds.), *Controlled Atmospheres for Storage and Transport of Perishable Agricultural Commodities*. Timber Press, Beaverton, OR, pp. 285–287.

Moyls, A.L., P.L. Sholberg, and A.P. Gaunce. 1996. Modified-atmosphere packaging of grapes and strawberries fumigated with acetic acid. *HortScience* 31: 414–416.

Nakhasi, S., D. Schlimme, and T. Solomos. 1991. Storage potential of tomatoes harvested at the breaker stage using modified atmosphere packaging. *J. Food Sci.* 56: 55–59.

Nunes, M.C.N., A.M.M.B. Morais, J.K. Brecht, and S.A. Sargent. 1996. Quality of pink tomatoes (cv. Buffalo) after storage under controlled atmosphere at chilling and nonchilling temperatures. *J. Food Qual.* 19: 363–374.

Ogawa, J.M. and H. English. 1991. *Diseases of Temperate Zone Tree Fruit and Nut Crops*. University of California, Division of Agriculture and Natural Resources, Oakland, CA.

Ogawa, J.M., E.I. Zehr, and B.A.R. 1995. Brown rot. In: J.M. Ogawa, E.I. Zehr, G.W. Bird, D.F. Ritchie, K. Uriu, and J.K. Uyemoto (Eds.), *Compendium of Stone Fruit Diseases*. APS Press, St. Paul, MN, pp. 7–10.

Paster, N. and M. Menasherov. 1988. Inhibition of T-2 toxin production on high-moisture corn kernals by modified atmospheres. *Appl. Environ. Micobiol.* 54: 540–543.

Phillips, C.A. 1996. Review: Modified atmosphere packaging and its effects on the microbiological quality and safety of produce. *Intl. J. Food Sci. Technol.* 31: 463–479.

Polderdijk, J.J., H.A.M. Boerrigter, E.C. Wilkinson, J.G. Meijer, and M.F.M. Janssens. 1993. The effects of controlled atmosphere storage at varying levels of relative humidity on weight loss, softening and decay of red bell peppers. *Sci. Hort.* 55: 315–321.

Prusky, D., A. Perez, Y. Zutkhi, and R. Ben-Arie. 1997. Effect of modified atmosphere for control of black spot, caused by *Alternaria alternata*, on stored persimmon fruits. *Phytopathology* 87: 203–208.

Reyes, A.A. 1988. Suppression of *Sclerotinia sclerotiorum* and watery soft rot of celery by controlled atmosphere storage. *Plant Dis.* 72: 790–792.

Rodov, V., A. Copel, N. Aharoni, Y. Aharoni, A. Wiseblum, B. Horev, and Y. Vinokur. 2000. Nested modified-atmosphere packages maintain quality of trimmed sweet corn during cold storage and the shelf life period. *Postharverst Biol. Technol.* 18: 259–266.

Saftner, R.A., J.A. Abbott, W.S. Conway, and C.L. Barden. 2003. Effects of 1-methylcyclopropene and heat treatments on ripening and postharvest decay in 'Golden Delicious' apples. *J. Am. Soc. Hort. Sci.* 128: 120–127.

Sanz, C., A.G. Perez, R. Olias, and J.M. Olias. 1999. Quality of strawberries packed with perforated polypropylene. *J. Food Sci.* 64: 748–752.

Shellie, K.C. 2002. Ultra-low oxygen refrigerated storage of 'Rio Red' grapefruit: Fungistatic activity and fruit quality. *Postharverst Biol. Technol.* 25: 73–85.

Sholberg, P.L. 1998. Fumigation of fruit with short-chain organic acids to reduce the potential of postharvest decay. *Plant Dis.* 82: 689–693.

Sholberg, P.L., P. Haag, and D.A. Gaudet. 1995. Effect of fungicides and controlled-atmosphere storage on Ltb rot of apples. *Can. J. Plant Sci.* 75: 515–520.

Sholberg, P.L. and P. Randall. 2007. Fumigation of stored pome fruit with hexanal reduces blue and gray mold decay. *Hort. Science* 42: 611–616.

Sommer, N.F., M.K. El-Kazzaz, J.R. Buchanan, R.J. Fortlage, and M.A. El-Gooranhi. 1982. Carbon-monoxide suppression of postharvest diseases of fruits and vegetables. In: D.G. Richardson and M. Meheriuk (Eds.), *Controlled Atmospheres for Storage and Transport of Perishable Agricultural Commodities*. Timber Press, Beaverton, OR, pp. 289–297.

Song, J., R. Leepipattanawit, W. Deng, and R.M. Beaudry. 1996. Hexanal vapor is a natural, metabolizable fungicide: Inhibition of fungal activity and enhancement of aroma biosynthesis in apple slices. *J. Am. Soc. Hort. Sci.* 121: 937–942.

Spalding, D.H. and W.F. Reeder. 1975. Low-oxygen high-carbon dioxide controlled atmosphere storage for control of anthracnose and chilling injury of avocados. *Phytopathology* 65: 458–460.

Spalding, D.H. and W.F. Reeder. 1977. Low pressure (hypobaric) storage of mangos. *J. Am. Soc. Hort. Sci.* 102: 367–369.

Spotts, R.A., L.A. Cervantes, and T.J. Facteau. 2002. Integrated control of brown rot of sweet cherry fruit with a preharvest fungicide, a postharvest yeast, modified atmosphere packaging, and cold storage temperature. *Postharverst Biol. Technol.* 24: 251–257.

Spotts, R.A., L.A. Cervantes, T.J. Facteau, and T. Chand-Goyal. 1998. Control of brown rot and blue mold of sweet cherry with preharvest iprodione, postharvest *Cryptococcus infirmo-miniatus*, and modified atmosphere packaging. *Plant Dis.* 82: 1158–1160.

Spotts, R.A., P.L. Sholberg, P. Randall, M. Serdani, and P.M. Chen. 2007. Effects of 1-MCP and hexanal on decay of d'Anjou pear fruit in long-term storage. *Postharvest Biol. Technol.* 44: 101–106.

Srilaong, V. and Y. Tatsumi. 2003. Changes in respiratory and antioxidative parameters in cucumber fruit (*Cucumis sativus* L.) stored under high and low oxygen concentrations. *J. Jpn. Soc. Hort. Sci.* 72: 525–532.

Sung, K.C., H.C. Sung, and S.I. Dong. 1998. Modified atmosphere packaging of fresh strawberries by antimicrobial plastic films. *Korean J. Food Sci. Technol.* 30: 1140–1145.

Tian, S., Q. Fan, Y. Xu, Y. Wang, and A. Jiang. 2001. Evaluation of the use of high CO_2 concentrations and cold storage to control *Monilinia fructicola* on sweet cherries. *Postharverst Biol. Technol.* 22: 53–60.

Tian, S., Y. Xu, A. Jiang, and Q. Gong. 2002. Physiological and quality responses of longan fruit to high O_2 or high CO_2 atmospheres in storage. *Postharverst Biol. Technol.* 24: 335–340.

Valverde, J.M., F. Guillen, D. Martinez-Romero, S. Castillo, M. Serrano, and D. Valero. 2005. Improvement of table grapes quality and safety by the combination of modified atmosphere packaging (MAP) and eugenol, menthol, or thymol. *J. Agric. Food Chem.* 53: 7458–7464.

Vaughn, S.F., G.F. Spencer, and B.S. Shasha. 1993. Volatile compounds from raspberry and strawberry fruit inhibit postharvest decay fungi. *J. Food Sci.* 58: 793–796.

Vermeiren, L., F. Devlieghere, and J. Debevere. 2002. Effectiveness of some recent antimicrobial packaging concepts. *Food Addit. Contam.* 19: 163–171.

Wells, J.M. and M. Uota. 1970. Germination and growth of five fungi in low-oxygen and high-carbon dioxide atmospheres. *Phytopathology* 60: 50–53.

Werner, B.G. and J.H. Hotchkiss. 2006. Modified atmosphere packaging. In: J.R. Gorny, A.E. Yousef, G.M. Sapers (Eds.), *Microbiology of Fruits and Vegetables*. CRC Press, Boca Raton, FL, pp. 437–459.

Wickham, L.D. and M. Mohammed. 1999. Storage of immature green mango (*Mangifera indica*, L.) fruit for processing. *J. Food Qual.* 22: 31–40.

Wszelaki, A.L. and E.J. Mitcham. 2000. Effects of superatmospheric oxygen on strawberry fruit quality and decay. *Postharverst Biol. Technol.* 20: 125–133.

Wszelaki, A.L. and E.J. Mitcham. 2003. Effect of combinations of hot water dips, biological control and controlled atmospheres for control of gray mold on harvested strawberries. *Postharvest Biol. Technol.* 27: 255–264.

Yackel, W.C., A.I. Nelson, L.S. Wei, and M.P. Steinberg. 1971. Effect of controlled atmosphere on growth of mold on synthetic media and fruit. *Appl. Microbiol.* 22: 513–516.

Yahia, E.M. 1998. Modified and controlled atmospheres for tropical fruits. *Hort. Rev.* 22: 123–183.

Yahia, E.M. 2007. Modified and controlled atmospheres. In: R. Troncoso-Rojas, M.E. Tiznado-Hernández, and A. Gonzalez-Leon. (Eds.), *Recent Advances in Postharvest Technologies to Control Fungal Diseases in Fruits and Vegetables*. Transworld Research Network, Kerala, India, pp. 103–125.

Yahia, E.M. 2006. Effect of quarantine treatments on quality of fruits and vegetables. In: G. Gonzalez (Ed.), Storage. *Stewart Rev.* 1(6): 1–18.

Yahia, E.M. and G. Gonzalez-Aguilar. 1998. Use of passive and semi-active atmospheres to prolong the postharvest life of avocado fruit. *Lebens. Wiss. Technol.* 31: 602–606.

Yahia, E.M., K.E. Nelson, and A.A. Kader. 1983. Postharvest quality and storage life of grapes as influenced by adding carbon monoxide to air or controlled atmosphere. *J. Am. Soc. Hort. Sci.* 108: 1067–1071.

Yahia, E.M. and M. Rivera-Dominguez. 1992. Modified atmosphere packaging of muskmelon. *Food Sci. Technol. (Lebensm. Wiss. Technol.)* 25: 38–42.

Yong, J.W. and A. Kyung. 1998. Thiabendazole and CA effects on reduction of chilling injury during cold storage in pepper fruit. *J. Korean Soc. Hort. Sci.* 39: 680–683.

Yueming, J. 1999. Low temperature and controlled atmosphere storage of fruit of longan (*Dimocarpus longan* Lour.). *Trop. Sci.* 39: 98–101.

Zheng, Y., C.Y. Wang, S.Y. Wang, and W. Zheng. 2003. Effect of high-oxygen on blueberry phenolics, anthocyanins, and antioxidant capacity. *J. Agric. Food Chem.* 51: 7162–7169.

10

Microbial Safety of Modified Atmosphere Packaged Fresh-Cut Produce

David O'Beirne and Devon Zagory

CONTENTS

10.1 Microbiology of Fresh-Cut Produce

10.1.1 Introduction

Fresh-cut products provide substrates and environmental conditions, which support the survival and growth of microorganisms. Minimal processing treatments such as peeling and slicing provide a potentially richer source of nutrients than intact produce

(Barry-Ryan and O'Beirne, 1998, 2000; Brackett, 1994). This, combined with high A_w and close to neutral (vegetables) or low acid (many fruits) tissue pH, facilitate microbial growth (Beuchat, 1996).

These products harbor large and diverse populations of microorganisms, and counts of 10^5–10^7 CFU/g are frequently present. Most bacteria present are Gram-negative rods, predominantly *Pseudomonas*, *Enterobacter*, or *Erwinia* species (Brocklehurst et al., 1987; Garg et al., 1990; Magnuson et al., 1990; Manvell and Ackland, 1986; Marchetti et al., 1992; Nguyen-the and Prunier, 1989). Product type, production history, and storage conditions are major factors determining what organisms are present and their popula- tions. Lactic acid bacteria have been detected in mixed salads and grated carrots, and may predominate in salads when held at severe abuse (30°C) temperatures (Manvell and Ack- land, 1986). Yeasts commonly isolated include *Cryptococcus*, *Rhodotorula*, and *Candida* (Brackett, 1994). Webb and Mundt (1978) surveyed 14 different vegetables for molds. The most commonly isolated genera were *Aureobasidium*, *Fusarium*, *Mucor*, *Phoma*, *Rhizopus*, and *Penicillium*.

10.1.2 Food Poisoning Linked to Fresh-Cut Products

A number of important human pathogens have been found in fresh and fresh-cut produce, notably *Listeria monocytogenes*, *Escherichia coli* O157:H7, *Salmonella sp.*, nonproteolytic *Clos- tridium botulinum*, *Yersinia entercolitica*, and *Aeromonas hydrophila*. There are also important emerging threats from viral and protozoan pathogens (Brackett, 1999; Francis et al., 1999). Increasing consumption of fresh produce in the United States has been paralleled by an increase in produce-linked food poisoning outbreaks (NACMCF, 1999). The number of fresh produce associated outbreaks between 1973 and 1997 was 190 (Sivapalasingam et al., 2004); between 1998 and 2002, there were 249 outbreaks. The microbial safety of controlled and modified atmosphere packaging (MAP) of fresh and fresh-cut produce has also been comprehensively reviewed by Farber et al. (2003). The presence of human pathogens on fresh-cut produce is a consequence of contamination during agricultural production, subsequent handling, minimal processing, or cross-contamination by end users after pack opening. Minimal processing, including antimicrobial treatments, cannot be relied upon to eliminate pathogens.

In order to constitute a problem, pathogens must have the opportunity to contaminate the product, and generally must be able to survive and, in the case of bacterial pathogens, grow. Most pathogens can survive and grow on fresh-cut produce stored under proper conditions (Francis and O'Beirne, 2001). The extent of survival or growth can be signifi- cantly affected by processing factors (surface damage, washing treatments, etc.), and by the gas atmospheres, temperatures, and storage times experienced after processing (Francis and O'Beirne, 1997, 1998; Gleeson and O'Beirne, 2005). The use of MAP extends shelf life, enabling, in a worst case scenario, pathogens to grow to significant numbers on otherwise acceptable fresh foods (Berrang et al., 1989b).

In the early years of the development of the fresh-cut industry, there were relatively few links between these products and food poisoning. Those that have been identified include an outbreak of botulism ultimately linked to a modified atmosphere (MA) packaged dry coleslaw product (Solomon et al., 1990) and a *Salmonella* Newport outbreak linked to ready-to-eat salad vegetables (PHLS, 2001). There was also an outbreak of shigellosis linked to shredded lettuce (Davis et al., 1988) though exactly how this product was packaged is unclear. However, in the United States, there have been a series of recent outbreaks of illness due to *E. coli* O157:H7 contamination of bagged spinach and fresh-cut lettuce (Beuchat, 2006); in all, there have been 18 fresh-cut associated outbreaks in the United States between 1998 and 2006.

10.1.3 Human Pathogens of Concern

This section briefly outlines why certain pathogens are of particular concern in relation to fresh-cut products.

10.1.3.1 *Listeria monocytogenes*

L. monocytogenes is a Gram-positive rod that causes several diseases in humans including meningitis, septicemia, stillbirths, and abortions (ICMSF, 1996). It is considered ubiquitous in the environment, being isolated from soil, feces, sewage, silage, manure, water, mud, hay, animal feeds, dust, birds, animals, and humans (Al-Ghazali and Al-Azawi, 1990; Gray and Killinger, 1966; Gunasena et al., 1995; Nguyen-the and Carlin, 1994; Welshimer, 1968). Contamination of vegetables by *L. monocytogenes* may occur through agricultural practices, such as irrigation with polluted water or use of contaminated manure (Geldreich and Bordner, 1971; Nguyen-the and Carlin, 1994). It may also occur during processing. In France (Nguyen-the and Carlin, 1994) and Germany (Lund, 1993) levels of $>10^2$ CFU/g are unacceptable, while in the United Kingdom and United States, the organism must be absent from 25 g samples.

Of particular concern is the organism's ability to grow at refrigeration temperatures; the minimum temperature for growth is reported to be $-0.4°C$ (Walker and Stringer, 1987). It is also facultatively anaerobic, capable of survival/growth under the low O_2 concentrations within MA packages of prepared vegetables. While counts generally remain constant at 4°C (Farber et al., 1998), they can increase to high numbers at mild abuse temperatures (8°C), particularly after antimicrobial dipping treatments or within nitrogen-flushed packages (Francis and O'Beirne, 1997).

However, evidence is emerging that levels of virulence may vary greatly among *L. monocytogenes* strains, and that some serotypes found in MAP produce may be different (and less virulent) than those isolated in food poisoning outbreaks (Beuchat and Ryu, 1997).

10.1.3.2 *Escherichia coli O157:H7*

E. coli, type species of the Enterobacteriaceae genus, *Escherichia*, is a common inhabitant of the gastrointestinal tract of mammals. Despite the commensal status of the majority of strains, pathogenic strains, particularly enterohemorrhagic *E. coli* O157:H7, have emerged as highly significant foodborne pathogens. Gastroenteritis and hemorrhagic colitis are classical symptoms, while complications including thrombocytopenic purpura and hemolytic uremic syndrome have been documented (Martin et al., 1986), the latter potentially leading to renal failure and death in 3%–5% of juvenile cases (Griffin and Tauxe, 1991; Karmali et al., 1983).

The principal reservoir of *E. coli* O157:H7 is believed to be the bovine gastrointestinal tract (Doyle et al., 1997; Wells et al., 1991), though it has been isolated from other animals including deer, goats, pigs, birds, and slugs. Hence, contamination with feces is a significant risk factor. Contamination and survival of the organism in natural water sources make these also potential sources in the distribution of infection, particularly if untreated water is used to wash produce. The potential for cross-contamination during distribution and domestic storage is also of concern. Information regarding contamination rates of MAP prepared vegetables is limited, but there have been several major outbreaks in the United States in recent years (Beuchat, 2006).

10.1.3.3 *Clostridium botulinum*

Cl. botulinum is a member of the genus *Clostridium*, characterized as Gram-positive, rod-shaped, endospore-forming, obligate anaerobes (Varnum and Evans, 1991). *Cl. botulinum* is divided into numerous subdivisions, based on the serological specificity of the neurotoxin produced, and physiological differences between strains. Human botulism is normally attributed to subspecies antigenic types A, B, E, and occasionally type F. Endospores of *Cl. botulinum* are ubiquitous, being distributed in soils, aquatic sediments, and the digestive tract of animals and birds.

Vegetables are potentially contaminated during growth, harvesting, and processing (Rhodehamel, 1992). Despite their ubiquity, a study by Lilly et al. (1996) found that only 0.36% of fresh-cut vegetables were contaminated with *Cl. botulinum* spores. In the case of mushrooms, a much lower incidence of *Cl. botulinum* was reported (Notermans et al., 1989) than had been reported previously (Hauschild et al., 1978), a change attributed to hygienic improvements in growing techniques.

The possibility of growth and toxin production by *Cl. botulinum* before obvious spoilage has long been of concern in overwrapped mushrooms (Sugiyama and Yang, 1975) and in vacuum-packaged prepared potatoes (O'Beirne and Ballantyne, 1987). In addition, sufficiently anoxic conditions are frequently observed in MA packages where the respiration rate of the product is not matched by the permeability of the packaging used. Anoxic conditions may also develop within MAP produce where edible coatings are used (Guilbert et al., 1996). Highly permeable or perforated overwrapping films have been used for fresh mushrooms and low storage temperatures and short shelf lives have been requirements in prepared potato products (IFST, 1990). In the case of other items of vacuum-packaged/fresh-cut produce, the data suggest that spoilage is likely to precede toxin production (Larson et al., 1997; Petran et al., 1995), with a probability of 1 in 10^5 for toxin production to occur prior to obvious spoilage (Larson et al., 1997). However, there is a report linking a botulism outbreak with coleslaw prepared from an MAP dry coleslaw mix (Solomon et al., 1990). The short shelf lives of retail packs and the good control of temperature/modest storage lives of catering packs are likely to minimize such risks, but there is need for vigilance.

10.1.3.4 Salmonella

Salmonella species, a genus of the family Enterobacteriaceae, are characterized as Gram-negative, rod-shaped bacteria. Pathogenic species include *S. typhimurium*, *S. enteritidis*, *S. heidelberg*, *S. saint-paul*, and *S. montevideo*. Salmonellae are mesophiles, with optimum temperatures for growth of 35°C–43°C. The growth rate is substantially reduced at <15°C, while the growth of most salmonellae is prevented at <7°C. Salmonellae are facultatively anaerobic, capable of survival in low O_2 atmospheres.

These organisms are abundant in fecal material, sewage, and sewage-polluted water; consequently they may contaminate soil and crops with which they come into contact. Sewage sludge may contain high numbers of salmonellae and, if used for agricultural purposes, will disseminate the bacterium. Once introduced into the environment, salmonellae remain viable for months (ICMSF, 1996). Potential contamination from workers who handle produce in the field or in processing plants is of great concern. Salmonellae have not generally been found in fresh-cut produce, though they have been isolated from bean-sprouts (20%) in Malaysia (Arumugaswamy et al., 1994) and have been identified as causal agents in several foodborne illness outbreaks associated with tomatoes in the United States (Jablasone et al., 2004).

10.1.3.5 Viruses and Protozoa

The significance of viruses with respect to foodborne disease is clear with the inclusion of Norwalk virus, hepatitis A virus, and "other viruses" within the top 10 causes of foodborne disease outbreaks in the United States (1983–1987; Cliver, 1997). Outbreaks caused by hepatitis A virus, calicivirus, and Norwalk-like viruses have been associated with the consumption of frozen raspberries and strawberries, melons, lettuce, watercress, and diced tomatoes (Beuchat, 1996; Hedberg and Osterholm, 1993; Hutin et al., 1999; Lund and Snowdon, 2000; Rosenblum et al., 1990). Viruses can be transmitted by infected food handlers, through the fecal–oral route, and have been isolated from sewage and untreated water used for crop irrigation. Despite their significance, data regarding the effects of food preparation and storage conditions on the survival and infectivity of viruses is extremely limited, partly through the complexity of viral detection assays. Nonetheless, the potential of several viruses to survive on vegetables for periods exceeding their normal shelf life has been identified (Badawy et al., 1985; Konowalchuk and Speirs, 1975; Sattar et al., 1994). Survival appears to be dependent upon temperature and moisture content (Bidawid et al., 2001; Konowalchuk and Speirs, 1975); however, little information is available on the effects of MAP on virus survival.

The protozoan parasites *Giardia lamblia, Cyclospora cayetanensis*, and *Cryptosporidium parvum* have been the cause of serious foodborne outbreaks involving berries (Herwaldt, 2000; Herwaldt and Ackers, 1997), lettuce and onions (CDC, 1989), and raw sliced vegetables (Mintz et al., 1993). These organisms normally gain access to produce before harvest, usually as a result of contaminated manure or irrigation water and poor hygiene practices by food handlers (Beuchat, 1996). The lack of sensitive methods for determining the survival or inactivation of oocysts has hampered incidence studies and studies focused on the effects of minimal processing and packaging. However, the increase in produce-linked outbreaks due to these organisms indicates that research is needed to examine the behavior of foodborne protozoan parasites on MAP produce.

10.2   Implications of Field Production

10.2.1   Sources of Contamination

Due to the diversity of foodborne pathogens, the diversity of produce types and the diversity of cropping systems, there are many possible points of contamination. However, with a few exceptions (*L. monocytogenes, Cl. botulinum*, and *Staphylococcus aureus*), most foodborne pathogens are associated with human and animal feces. Thus, the primary task of preventing contamination with pathogens is to prevent contamination with feces. Prevention of contamination is essential because, except thorough cooking, there is no process currently available that can reliably remove pathogens once fruits or vegetables have become contaminated. While various washing techniques can remove 90%–99% of pathogen cells, none have been shown to remove all of the cells. And bacterial pathogens can reproduce under proper conditions of temperature and pH and so can replace the cells removed by washing.

Fecal pathogens can contaminate fruits and vegetables through a number of routes (Figure 10.1). Water and improperly composted manure are obvious vectors of pathogens. Wild and feral animals have been implicated in episodes of contamination, though the impact of wild animal feces on safety of fresh produce is unclear. Domestic animals, such as cattle and sheep, are known reservoirs of human pathogens and are suspected vectors of contamination. Insects and wind have been little studied but may spread pathogens to growing fruits and vegetables. Cross contamination by other foods in commercial and domestic kitchens may be responsible for the majority of foodborne illness due to consuming fresh produce, but such events are rarely reported and so are difficult to estimate.

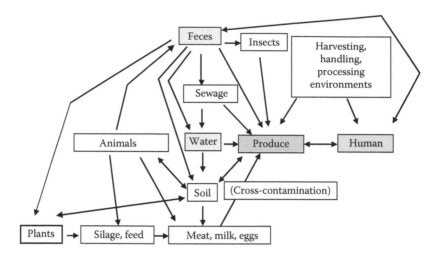

FIGURE 10.1
Multiple paths of contamination of produce.

Finally, workers can be a source of contamination should they be carrying a pathogen. Outbreaks due to hepatitis A and *Shigella* are likely due to infected workers handling fresh produce, as they are almost invariably associated with human, and not animal, reservoirs (FDA, Foodborne Microorganisms and Natural Toxins, 1992).

10.2.2 Implications of Production Systems

Production practices may influence the risk of contamination and exposure to the consumer by human pathogens. Key areas of concern are prior land use, adjacent land use, field slope and drainage, soil properties, crop inputs and soil fertility management, water quality and use practices, equipment and container sanitation, worker hygiene and sanitary facilities, harvest implement and surface sanitation, pest and vermin control, effects of domesticated animal and wildlife on the crop itself or packing area, postharvest water quality and use practices, postharvest handling, transportation and distribution, and documentation and recordkeeping. The role of water quality and manure management practices is particularly critical (Suslow et al., 2003).

The diversity of cropping systems, scale of operation, use and design of equipment, regional and local practices, environmental influences, specifics of on-farm soil-related factors, proximity of animals, and many other production factors complicate attempts to develop uniform approaches to reduction of microbial risk. This was recognized in the evolution of the *Guide to Minimize Microbial Food Safety Hazards for Fresh Fruits and Vegetables* (FDA, 1998), which describes the key areas of presumptive risk for fruits and vegetables in a production environment. Although the available scientific literature is adequate to identify sources of contamination and estimate microbial persistence on plants, the specific influence and interactions among the production environments and crop management practices are not sufficiently understood to provide detailed guidance to growers and shippers (Suslow et al., 2003).

Nevertheless, some production systems may be associated with greater risks than others. Overhead irrigation would be expected to have a greater risk of contamination than drip (Bastos and Mara, 1995; Sadovski et al., 1978) or furrow irrigation. Use of organic amendments containing animal feces may introduce risks that synthetic fertilizers do not. Field coring or other in-field processing may introduce additional risks associated with human handling. Though direct evidence of foodborne illness due to contamination of edible horticultural commodities during commercial production is scant, compelling epidemiological evidence involving these crops has implicated specific production practices, including use of animal waste or manure, fecal-contaminated agricultural water for irrigation or pesticide/crop management applications, and farm labor personal hygiene, as leading to direct contamination (Brackett, 1999). These potential risks have not been quantified and can be reduced through appropriate programs and processes. Water can be tested and, if necessary, alternative sources can be used or water can be sanitized. Organic amendments can be composted to kill vegetative pathogens. The USDA National Organic Program requires specific composting techniques and establishes microbial guidelines for some compost treatments (USDA-AMS, National Organic Program).

10.2.3 Irrigation and Flooding

According to the FDA guidelines (FDA, 1998), "Wherever water comes into contact with fresh produce, its quality may directly determine the potential for pathogen contamination and its persistence." There have been several known cases where fruits or vegetables have become contaminated through contact with water. Mangoes internalized *Salmonella* Newport when they were rinsed in contaminated river water. Parsley and cilantro were thought to have been contaminated with *Shigella* through sprays of canal water. Contaminated water was strongly implicated in contamination of cantaloupes with *Salmonella* Poona in 2000–2002. Suslow et al. (2003) published a comprehensive review of agricultural water quality and associated risks.

The FDA is concerned about the potential for flooded fields to result in contamination of leafy greens. In a letter to California firms that grow and handle lettuce (FDA, November 4, 2005), the FDA stated

> Although it is unlikely that contamination in all 19 outbreaks was caused by flooding from agricultural water sources, we would like to take this opportunity to clarify that FDA considers ready to eat crops (such as lettuce) that have been in contact with flood waters to be adulterated due to potential exposure to sewage, animal waste, heavy metals, pathogenic microorganisms, or other contaminants. FDA is not aware of any method of reconditioning these crops that will provide a reasonable assurance of safety for human food use or otherwise bring them into compliance with the law. Therefore, FDA recommends that such crops be excluded from the human food supply and disposed of in a manner that ensures they do not contaminate unaffected crops during harvesting, storage, or distribution.

In 2003, two outbreaks were suspected to have been caused by lettuce grown on a ranch that had been flooded by Santa Rita Creek in the Salinas Valley of California. Sediments in Santa Rita Creek were shown to harbor *E. coli* O157:H7 (FDA, November 4, 2005).

10.2.4 Training of Field Workers

Humans can be a significant reservoir of human pathogens. Those exhibiting symptoms of foodborne illness should be excluded from handling fruits and vegetables. Such pathogens as *Shigella*, hepatitis A, *S. aureus*, and Norovirus are typically spread by infected humans. Agricultural field workers are often form the lower economic and least educated strata of society. Their access to health services is often limited, particularly those who may be in their host country illegally. Such workers may be expected to harbor higher incidences of human pathogens than those with access to better medical care. Access to health services for all agricultural workers, legal or illegal, is an important public health issue and should be included in efforts to reduce foodborne illness. In addition worker hygiene, and especially proper hand washing, is central to any food safety program. Training workers in personal hygiene and hand washing is very important to food safety assurance. Many, and probably most, produce companies in developed countries and elsewhere include some form of worker training as part of their food safety efforts. Yet the effects of such training are uncertain and very difficult to measure.

10.2.5 Implications of Globalization

As international trade in all commodities increases, movement of fresh fruits and vegetables continues to grow. Contraseasonal produce from the southern hemisphere moves to the north in increasing quantities. The United States and Europe have both become major net importers of fruits and vegetables. With the global movement of these products come concerns of movement of human pathogens. There is particular concern about the introduction of pathogens that are new to the consumers in various countries. There have been outbreaks of foodborne illness linked to imported fruits and vegetables. As examples, Guatemalan raspberries were responsible for a series of outbreaks caused by the parasite *Cyclospora*. Cantaloupes from Mexico contaminated with *Salmonella* caused illness in the United States over several years. Yet most foodborne illness outbreaks due to fruits and vegetables in the United States are caused by domestically produced products (Figure 10.2).

This is not to imply that domestically produced fruits and vegetables in the United States are less safe than imported ones. Rather, most fruits and vegetables are domestically produced and so there have been more incidents associated with them. In fact, a survey of domestic and imported fruits and vegetables performed by the U.S. FDA found higher incidence of contamination with human pathogens on imported products, though the samples were not large enough for the data to be statistically significant. The point is that both domestic and imported fruits and vegetables can be contaminated and that all sources of supply must adhere to food safety standards and employ prevention programs such as good agricultural practices (GAPs), good manufacturing practices (GMPs), global food technical standards such as those of the British Retail Consortium (BRC), Hazard Analysis and Critical Control Points (HACCP) and the like.

10.2.6 Field Production as a Critical Control Point for Contamination

Many of the pathogens of greatest concern are fecal in origin and often associated with animals. For example, *E. coli* O157:H7 has a significant reservoir in cattle, though it has also been isolated from many other animals. The series of foodborne illness outbreaks due to this pathogen on leafy greens and other products almost certainly originated in the field. While poor sanitation practices in packing and processing facilities could allow cross-contamination, the source of contamination is the field. This is probably true for *Salmonella* as well. Many of the strains of *Salmonella* found on fruits and vegetables are typically

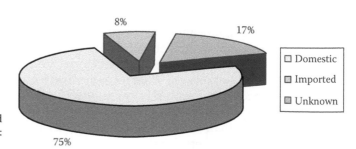

FIGURE 10.2
Percentage of outbreaks in the United States due to fruits and vegetables: 1990–1998, by origin of product.

associated with birds, reptiles, amphibians, or other animals. So while humans can be a reservoir of *Salmonella*, most outbreaks probably had their origin on the farm. Other pathogens, as noted earlier (*Shigella*, hepatitis A, *S. aureus*, etc.) typically are associated with human contact and so do not normally originate on the farm. This points out the importance of performing a hazard analysis for each and every situation to identify and address the food safety risks that are associated with that operation.

10.3 Implications of Minimal Processing

Minimizing damage throughout harvesting and processing reduces the chances of pathogen contamination, penetration, and growth (Liao and Cooke, 2001). Injuries to the wax layer, cuticle, and underlying tissues increased bacterial adhesion and growth (Han et al., 2000, 2001; Seo and Frank, 1999; Takeuchi and Frank, 2001; Takeuchi et al., 2000).

Fresh-cut products are subjected to a series of minimal processing steps. Depending on the commodity, the following may apply: removal of unwanted parts such as outer leaves, preliminary washing to remove soil and debris, peeling and/or slicing as appropriate, antimicrobial dipping (usually in an aqueous chlorine solution), rewashing in freshwater to remove the chlorine residue, and finally packaging within flexible and/or rigid materials with suitable gas barrier properties. When necessary, the final package may be flushed with 100% nitrogen, low oxygen, or other special atmospheres. For many products, however, the respiration rate of the product combined with the gas permeability properties of the packaging are sufficient to modify the package atmosphere.

10.3.1 Cutting and Peeling

Cutting, peeling, and slicing cause the destruction of surface cells, affect product respiration rate and pH, and release nutrients and possibly antimicrobial substances from plant cells (Brackett, 1994). Cutting treatments can allow microorganisms to penetrate below the surface of fresh-cut products, and the severity of these treatments affects the extent and depth of this penetration. We have shown that large populations of *E. coli* O157:H7 can colonize the surface of fresh-cut carrot disks, and penetrate into the tissue to depths of up to 1000 μm (Auty et al., 2005). Slicing with a blunt blade enhanced the penetration by *E. coli* and its subsequent survival during storage, and resulted in consistently higher *E. coli* and *Listeria innocua* counts during storage (Gleeson and O'Beirne, 2005). Cracks and fissures observed in the carrot disks probably enhanced penetration. These data have implications for the quality of cutting surfaces used during processing, but also at harvesting and trimming.

Pathogens can become attached to processing equipment (slicers, shredders) and, once attached (as biofilms) are very difficult to remove by chemical antimicrobial treatments (Bremer et al., 2001; Frank and Koffi, 1990; Garg et al., 1990; Jöckel and Otto, 1990; Nguyen-the and Carlin, 1994). *L. monocytogenes* has been recovered from the environment of processing operations used to prepare fresh-cut vegetables (Zhang and Farber, 1996), highlighting the importance of sanitizing treatments and GMPs generally.

10.3.2 Antimicrobial Treatments

Washing in tap water removes soil and other debris, some of the surface microflora, and cell contents and nutrients released during slicing that help support growth of microorganisms (Bolin et al., 1977). However, water washing had minimal effects on microorganisms on fresh produce (Adams et al., 1989, Beuchat, 1992; Brackett, 1987; Izumi, 1999; Nguyen-the and Carlin, 1994) and, due to the reuse of wash water in industry, may result in cross-contamination of food products and food-preparation surfaces (Beuchat, 1996; Beuchat and Ryu, 1997; Brackett, 1992; Garg et al., 1990).

While a wide range of alternatives have been proposed, the most cost-effective anti-microbial treatment typically includes washing in 100 ppm active chlorine for about 3 min, followed by washing in freshwater. Generally, no more than 2- to 3-$\log_{10}$ reductions have been reported (Adams et al., 1989; Beuchat, 1992, 1999). Environmental and toxicological concerns regarding the use of chlorine have led to the consideration of a range of altern-atives including peroxyacetic acid, acid-electrolyzed water, chlorine dioxide, hydrogen peroxide, various organic acids, ozone, mild thermal treatments, irradiation, etc. (O'Beirne and Francis, 2003). Few have gained significant acceptance by industry to date, except perhaps some use of peroxyacetic acid. None of the sanitizer treatments tested is likely to be totally effective against all pathogens, and the behavior of surviving pathogens during subsequent storage remains unpredictable (Beuchat and Ryu, 1997; Escudero et al., 1999; Park and Beuchat, 1999; Zhang and Farber, 1996).

None of the antimicrobial treatments currently applied can be relied upon to eliminate food pathogens. They are generally effective in reducing microbial counts in the wash water, avoiding the buildup of bacterial numbers, and in reducing cross-contamination (Wilcox et al., 1994). While chlorine can kill bacterial pathogens, these pathogens are not all accessible to an aqueous solution because of the presence of hydrophobic regions, fissures, and possibly other niches, which may harbor pathogens. In addition, the effects of chlorine are neutralized by organic matter and this is also a factor in incomplete decontamination (Beuchat, 1996; Delaquis et al., 2004). Viruses and protozoan cysts are generally more resistant than bacteria to these antimicrobial treatments (Beuchat, 1998).

Use of antimicrobial treatments can introduce increased risks in pathogen-contaminated produce, if the product is subjected to temperature abuse during subsequent storage. Even mild temperature abuse (e.g., 8°C) is a significant hazard. Dipping coleslaw and lettuce in a chlorine solution (100 ppm) reduced initial *L. innocua* and *E. coli* populations, but resulted in enhanced survival during extended storage at 8°C (Francis and O'Beirne, 2002). Dipping in antimicrobial agents reduces microbial counts, but also reduces competition for patho-gens from other organisms. *L. innocua* inoculated onto chlorine-treated lettuce survived poorly at 3°C, but at 8°C grew significantly better than on untreated samples (Francis and O'Beirne, 1997).

These and other data concerning storage temperature for fresh-cut systems point to the importance of low temperature storage (<4°C) and modest shelf life as key components of a HACCP approach to this food system (see below).

10.4 Implications of Modified Atmospheres

When a fresh-cut product is packaged, it continues respiring, modifying the gas atmos-phere inside the package. Ideally, O_2 levels will fall from the 21 kPa found in air to 2–5 kPa, and CO_2 levels will increase to the 3–10 kPa range. The use of such MAs is fundamental to the extension of shelf life in fresh-cut products, mainly through inhibition of enzymatic browning at cut surfaces, and slowing of physiological ageing. A key safety issue is whether MAs increase the opportunities for pathogen survival and growth compared with storage in air.

First MAs might do this by simply increasing the time available for pathogens to grow to significant numbers, and overextension of shelf life particularly at abuse storage temperatures, is a significant risk factor for growth of bacterial pathogens. Second, these atmospheres and low temperatures may inhibit aerobic spoilage organisms, and may thus facilitate the growth of pathogens without obvious signs of spoilage (Daniels et al., 1985; Farber, 1991). A third potential problem could arise where unintended anoxic atmospheres or atmospheres high in CO_2 (>20 kPa) develop due to poor product-package compatibility or temperature abuse. In an extreme case, anoxic conditions could enable growth and toxin production by *Cl. botulinum*, while high CO_2 atmospheres may promote the growth of pathogens, while limiting the growth of spoilage organisms (Carlin et al., 1996). CO_2 had little or no inhibitory effect on growth of *E. coli* O157:H7 on shredded lettuce stored at 13°C or 22°C, and growth potential was increased in an atmosphere of $O_2/CO_2/N_2$: 5/30/65, compared with growth in air (Abdul-Raouf et al., 1993; Diaz and Hotchkiss, 1996). We have recently shown that the glutamate decarboxylase acid resistance mechanism of *L. monocytogenes* assists in its survival and growth in fresh-cut products, notably those with high CO_2 atmospheres (Francis et al., 2007).

When packages were flushed with inert atmospheres to minimize enzymatic browning, contaminating *L. monocytogenes* survived much better than in unflushed packages, when the product experienced mild temperature abuse (Francis and O'Beirne, 1997). As in the case of the antimicrobial treatments discussed earlier, this may be due to the effects of such atmospheres on organisms which compete with *L. monocytogenes*. Various researchers have reported antagonism by the native microflora of vegetables against *Listeria* (Francis and O'Beirne, 1998a,b; Liao and Sapers, 1999). However, in shredded lettuce, these inhibitory effects were dependent on gas atmosphere; in 3% O_2 (balance N_2) the growth of the mixed population was inhibited and *L. monocytogenes* proliferated (Francis and O'Beirne, 1998a).

Effects of gas atmospheres on *L. monocytogenes* and competing microflora were studied using a solid-surface model system (Francis and O'Beirne, 1998b). At 8°C, growth and inhibitory activities of *Enterobacter cloacae* and *Enterobacter agglomerans* were inversely related to the concentration of CO_2. By contrast, the growth and antilisterial activities of *Leuconostoc citreum* increased with elevated CO_2 concentrations. The data suggest that MAs in general seem safe at chill storage temperatures. However, at abuse temperatures, complex interactions between elements of the natural background microflora may have significant effects on survival and growth of pathogens.

10.5 Storage Temperature

Prevention of contamination is the single most important measure for the elimination of pathogens from fresh-cut produce. However if contamination has taken place, storage temperature is a key factor affecting bacterial pathogen survival and growth. Storage of produce at adequate refrigeration temperatures, will limit pathogen growth to those that are psychrotrophic; *L. monocytogenes*, *Y. enterocolitica*, nonproteolytic *Cl. botulinum*, and *A. hydrophila* being amongst the most notable.

Although psychrotrophic organisms, such as *L. monocytogenes*, are capable of growth at low temperatures, reducing the storage temperature (≤4°C) will significantly reduce the rate of growth (Beuchat and Brackett, 1990; Carlin et al., 1995). *L. monocytogenes* populations remained constant or decreased on packaged vegetables stored at 4°C, while at 8°C, growth of *L. monocytogenes* was supported on all vegetables, with the exception of coleslaw mix (Francis and O'Beirne, 2001). Thus even mild temperature abuse during storage permits more rapid growth of psychrotrophic pathogens (Berrang et al., 1989a; Carlin and Peck, 1996; Conway et al., 2000; Farber et al., 1998; García-Gimeno et al., 1996; Rodriguez et al., 2000).

Mesophilic pathogens, such as *Salmonella* and *E. coli* O157:H7, are unable to grow where temperature control is good (i.e., $\leq 4°C$). However, if temperature abuse occurs, they may then grow. Survival of *Salmonella* in produce stored for extended periods in chilled conditions may be of concern (Piagentini et al., 1997; Zhuang et al., 1995); *Salmonella* survived on a range of vegetables for more than 28 days at 2°C–4°C (ICMSF, 1996). *E. coli* O157:H7 populations survived on produce stored at 4°C and proliferated rapidly when stored at 15°C (Richert et al., 2000). Reducing the storage temperature from 8°C to 4°C significantly reduced growth of *E. coli* O157:H7 on MAP vegetables; however, viable populations remained at the end of the storage period at 4°C (Francis and O'Beirne, 2001). For some pathogens, such as *E. coli* O157:H7 or *L. monocytogenes*, the infectious dose can be very low for susceptible populations. In these cases, any contamination can lead to illness. Furthermore, viruses and parasites do not grow and reproduce at all on fresh fruits and vegetables. Therefore, temperature control cannot be the primary strategy for assuring safety. The prevention of contamination is the key step in safety assurance. Once fruits or vegetables become contaminated with pathogens, there is no decontamination step, except cooking, that will assure their safety. While proper temperature control can reduce the growth of bacterial pathogens, and is essential to the maintenance of quality, it cannot, by itself, assure the safety of these products.

10.6 HACCP and Safety Assurance

Over the last decades HACCP has become the *de facto* world standard for safety assurance in the food industry. Nevertheless, HACCP has some limitations, especially when applied to fresh-cut produce. When the U.S. FDA published their "Guide to Minimize Microbial Food Safety Hazards in Fresh Fruits and Vegetables" in 1998 the term "HACCP" was never mentioned. This is because the FDA recognized the lack of critical control points (CCPs) in a farm environment and developed a concept of GAPs as an alternative. In the more recent "Guide to Minimize Microbial Food Safety Hazards of Fresh-Cut Fruits and Vegetables" (FDA, 2008), FDA describes HACCP as proactive approach to ensuring food safety and encourages fresh-cut processors to take a proactive approach. However, the guidance does not specifically recommend nor require HACCP for fresh-cut processors.

In the United States, HACCP became mandatory for the seafood industry in December 1997. In January 1998, the U.S. Department of Agriculture instituted mandatory HACCP for meat and poultry processors. In January 2001, the FDA issued a final rule requiring juice processors to adopt HACCP systems.

HACCP is a system for the prevention of physical, chemical, or microbial contamination of food. The prime function of HACCP is to prevent identified hazards in food preparation through control of the process. HACCP functions as the final stage of an integrated food safety program that includes GAPs, GMPs, and sanitation standard operating procedures. In fact, HACCP can only be effective if these other programs are in place and functioning properly.

Nearly all fresh-cut processors do, in fact, employ HACCP systems. In the United States, because there is no regulatory requirement for HACCP, nor any official arbiter of HACCP, the practice of HACCP can be quite variable. Some operations have many CCPs, while others have none. Some fresh-cut processors combine quality and food safety in HACCP while others use HACCP only for safety, as it is intended. The majority of fresh-cut processors identify two CCPs in their operations; wash water sanitizer concentration or activity, and metal detection. Both of these CCPs fit the concept of a CCP and can be maintained as CCPs. They are points in a process that, when controlled, can significantly reduce or eliminate a hazard, can be measured and can be monitored to ensure that they are within critical limits. There are few, if any, other such points in most fresh-cut processing operations. There are typically no such points in a farming operation.

One of the primary virtues of HACCP in the produce industry is its requirement to develop a systematic hazard analysis. The diversity and variability of operations among fruit and vegetable handlers and processors precludes detailed standards that are broadly applicable. Instead each operator should carefully evaluate their operation to identify and address the significant hazards. This approach is applicable on farms as well, even though there may be no CCPs. A recently adopted Leafy Greens Marketing Agreement in California requires farmers to perform preplant and preharvest hazard evaluations and to address identified hazards. The hazard analysis element of HACCP is broadly applicable and should be adopted throughout the produce industry.

10.7 Challenges and Research Needs

Assuring the microbial safety of fresh-cut produce is challenging in many ways. Produce can become contaminated at any point in the production, handling, processing, and distribution chain. Once contaminated, only cooking will completely assure its safety. This is common to all produce, not only fresh-cut produce. What is unique about fresh-cut produce is that, through size reduction, wounds and entry ports are created that may make such products more felicitous for the attachment, penetration, and growth of food-borne pathogens. Furthermore, the mixing of batches of raw materials in the preparation of bagged salads and other fresh-cut products may lead to larger lots of contaminated products than would typically be the case with whole commodities. Finally, the extension of shelf life through sophisticated MAP and other technologies may provide additional time for proliferation of bacterial pathogens inside packages.

Nevertheless, fruits and vegetables that are free of pathogens are microbiologically safe, whether fresh-cut or not. And fruits and vegetables that are contaminated with human pathogens are potentially unsafe, whether fresh-cut or not. The clear challenge then is to provide fruits and vegetables that are free of human pathogens. This goal may be more worthy of the application of resources than efforts to detect, investigate, or decontaminate fresh-cut products. The role of cutting and packaging in enhancing foodborne illness hazards is, after all, still speculation. There are few data demonstrating that fresh-cut products are inherently less safe than analogous whole commodities. The most prominent recent outbreaks have been associated with fresh-cut products, but in many (or most) of those cases, it appears that the original contamination occurred in the field and so, had those raw materials not been processed, food borne illness may still have resulted.

Research should continue to elucidate more effective means of preventing contamination. There is a need for more detailed knowledge about the sources, movement, survival, and ecology of human pathogens in agricultural environments. Specific areas of endeavor may include

- Relative contributions of wild and domestic animals to on-farm pathogen populations.
- Survival and movement of pathogens in water and wetlands.
- Potential of water quality enhancements to sequester and/or remove pathogens from agricultural environments.

- Genetic profiling of produce microbial ecologies using massively parallel DNA sequencing. Such profiling can, and should, replace the use of single indicator organisms as surrogates for safety.
- Use of animal vaccines to reduce pathogen loads in domestic animals.

There should, of course, also be research into interventions to enhance the safety of fresh-cut produce. These may include such areas as

- Irradiation and other technologies to reduce pathogen populations.
- Better sampling strategies for detecting pathogens.
- Novel sanitizers that can achieve greater log reductions in pathogen populations.
- Validation of existing guidelines, standards, and "metrics" pertaining to safety during production, harvest, handling, processing, and distribution.

10.8   Conclusions

Fresh-cut produce has an extraordinary record of safety, when one considers the number of portions sold and consumed worldwide. Nevertheless, there have been many outbreaks associated with bagged, fresh-cut products, especially salads and other leafy greens. It is unclear if there is a higher incidence of foodborne illness associated with bagged cut produce than with whole commodity produce, or if the outbreaks are merely larger, more often detected and more prominent. If fresh-cut is more prone to safety problems, the reasons may be multiple and have been the subject of some debate. When outbreaks do occur, suspicion and regulatory attention typically focuses first, and often exclusively, on the processor.

Evidence from outbreak source investigations has often pointed to the field as the source of original contamination with human pathogens. Once produce is contaminated, fresh-cut processors cannot completely decontaminate those raw products with currently available processing technologies. For these reasons, food safety assurance must be based on comprehensive prevention programs starting with site analysis of agricultural land and continuing through the entire production and handling chain. Arguing whether the farmer or the processor is responsible for food safety is not relevant. They are both responsible and must cooperate to achieve mutually satisfactory levels of safety using available knowledge and technologies. More inspections, either regulatory or industry sponsored, will, of themselves, not improve produce safety or substantially reduce the incidence of foodborne illness outbreaks. Clear expectations, measurements, and cooperation are the better path to safety.

Initiatives in America, such as the California Leafy Greens Marketing Agreement and the new Center for Produce Safety at the University of California, Davis, California are useful and encouraging and indicate a will to develop and implement better data, better practices and better cooperation throughout the produce industry.

References

Abdul-Raouf, U.M., L.R. Beuchat, and M.S. Ammar. 1993. Survival and growth of *E. coli* O157:H7 on salad vegetables. *Appl. Environ. Microbiol.* 59: 1999–2006.

Adams, M.R., A.D. Hartley, and L.J. Cox. 1989. Factors affecting the efficacy of washing procedures used in the production of prepared salads. *Food Microbiol.* 6: 69–77.

Al-Ghazali, M.R. and S.K. Al-Azawi. 1990. *Listeria monocytogenes* contamination of crops grown on soil treated with sewage sludge cake. *J. Appl. Bacteriol.* 69: 642–647.

Arumugaswamy, R.K., G. Rusul Rahamat Ali, and S. Nadzriah Bte Abd. Hamid. 1994. Prevalence of *Listeria monocytogenes* in foods in Malaysia. *Int. J. Food Microbiol.* 23: 117–121.

Auty, M., G. Duffy, D. O'Beirne, A. McGovern, E. Gleeson, and K. Jordan. 2005. In situ localisation of *Escherichia coli* 0157:H7 in food by confocal scanning laser microscopy. *J. Food Prot.* 68: 482–486.

Badawy, A.S., C.P. Gerba, and L.M. Kelley. 1985. Survival of rotavirus SA-11 on vegetables. *Food Microbiol.* 2: 199–205.

Barry-Ryan, C. and D. O'Beirne. 1998. Effects of slicing method on the quality and storage-life of modified atmosphere packaged carrot discs. *J. Food Sci.* 63: 851–856.

Barry-Ryan, C. and D. O'Beirne. 2000. Effects of peeling method on the quality of ready-to-use carrots. *Int. J. Food Sci. Technol.* 35: 243–254.

Bastos, R.K.X. and D.D. Mara. 1995. The bacterial quality of salad crops drip and furrow irrigated with waste stabilization pond effluent: An evaluation of the WHO guidelines. *Wat. Sci. Technol.* 31(12): 425–430.

Berrang, M.E., R.E. Brackett, and L.R. Beuchat. 1989a. Growth of *Aeromonas hydrophila* on fresh vegetables stored under a controlled atmosphere. *Appl. Environ. Microbiol.* 55: 2167–2171.

Berrang, M.E., R.E. Brackett, and L.R. Beuchat. 1989b. Growth of *Listeria monocytogenes* on fresh vegetables stored under controlled atmosphere. *J. Food Prot.* 52: 702–705.

Beuchat, L.R. 1992. Surface disinfection of raw produce. *Dairy Food Environ. Sanit.* 12: 6–9.

Beuchat, L.R. 1996. Pathogenic microorganisms associated with fresh produce. *J. Food Prot.* 59: 204–216.

Beuchat, L.R. 1998. Surface decontamination of fruits and vegetables eaten raw: A review. World Health Organization, Food Safety Unit, WHO/FSF/FOS/98.2, www.who.int/fsf/fos982~1

Beuchat, L.R. 1999. Survival of enterohemorrhagic *Escherichia coli* O157:H7 in bovine feces applied to lettuce and the effectiveness of chlorinated water as a disinfectant. *J. Food Prot.* 62: 845–849.

Beuchat, L.R. 2006. Report from IAFP's Rapid Respose Symposium: Fresh Leafy Greens–Are they safe enough? *Food Prot. Trends* 26: 942–944.

Beuchat, L.R. and R.E. Brackett. 1990. Survival and growth of *Listeria monocytogenes* on lettuce as influenced by shredding, chlorine treatment, modified atmosphere packaging and temperature. *J. Food Sci.* 55: 755–758, 870.

Beuchat, L.R. and J.-H. Ryu. 1997. Produce handling and processing practices. *Emerg. Infect. Dis.* 3: 459–465.

Bidawid, S., J.M. Farber, and S.A. Sattar. 2001. Survival of hepatitis A virus on modified atmosphere packaged (MAP) lettuce. *Food Microbiol.* 18: 95–102.

Bolin, H.R., A.E. Stafford, A.D. King, Jr., and C.C. Huxsoll. 1977. Factors affecting the storage stability of shredded lettuce. *J. Food Sci.* 42: 1319–1321.

Brackett, R.E. 1987. Antimicrobial effect of chlorine on *Listeria monocytogenes. J. Food Prot.* 50: 999–1003.

Brackett, R.E. 1992. Shelf stability and safety of fresh produce as influenced by sanitation and disinfection. *J. Food Prot.* 55: 804–814.

Brackett, R.E. 1994. Microbiological spoilage and pathogens in minimally processed refrigerated fruits and vegetables. In: Wiley, R.C. (Ed.), *Minimally Processed Refrigerated Fruits and Vegetables*, New York: Chapman & Hall, pp. 269–312.

Brackett, R.E. 1999. Incidence, contributing factors, and control of bacterial pathogens in produce. *Postharvest Biol. Technol.* 15: 305–311.

Bremer, P.J., I. Monk, and C.M. Osborne. 2001. Survival of *Listeria monocytogenes* attached to stainless steel surfaces in the presence or absence of *Flavobacterium* spp. *J. Food Prot.* 64: 1369–1376.

Brocklehurst, T.F., C.M. Zaman-Wong, and B.M. Lund. 1987. A note on the microbiology of retail packs of prepared salad vegetables. *J. Appl. Bacteriol.* 63: 409–415.

Carlin, F. and M.W. Peck. 1996. Growth of and toxin production by non-proteolytic *Cl. botulinum* in cooked pureed vegetables at refrigeration temperatures. *Appl. Environ. Microbiol.* 62: 3069–3072.

Carlin, F., C. Nguyen-the, and A. Abreu da Silva. 1995. Factors affecting the growth of *Listeria monocytogenes* on minimally processed fresh endive. *J. Appl. Bacteriol.* 78: 636–646.

Carlin, F., C. Nguyen-the, A. Abreu da Silva, and C. Cochet. 1996. Effects of carbon dioxide on the fate of *Listeria monocytogenes*, of aerobic bacteria and on the development of spoilage in minimally processed fresh endive. *Int. J. Food Microbiol.* 32: 159–172.

CDC. 1989. Epidemiological notes and reports common source outbreak of giardiasis—New Mexico. *Morb. Mortal Wkly. Rep.* 38: 405–407.

Cliver, D.O. 1997. Foodborne viruses. In: Doyle M.P., Beuchat, L.R., and Montville, T.J. (Eds.), *Food Microbiology—Fundamentals and Frontiers*, Washington DC: ASM Press, pp. 437–446.

Conway, W.S., B. Leverentz, R.A. Saftner, W.J. Janisiewicz, C.E. Sams, and E. Leblanc. 2000. Survival and growth of *Listeria monocytogenes* on fresh-cut apple slices and its interaction with *Glomerella cingulata* and *Penicillium expansum*. *Plant Dis.* 84: 177–181.

Daniels, J.A., R. Krishnamurthi, and S.S.H. Rizvi. 1985. A review of the effect of carbon dioxide on microbial growth and food quality. *J. Food Prot.* 48: 532–537.

Davis, H., J.P. Taylor, J.N. Perdue, J.G.N. Stelma, J.J.M. Humphreys, R. Rowntree, and K.D. Greene. 1988. A shigellosis outbreak traced to commercially distributed shredded lettuce. *Am. J. Epidemiol.* 128: 1312–1321.

Delaquis P., P. Toivonen, K. Walsh, K. Rivest, and K. Stanich. 2004. Chlorine depletion in sanitizing solutions used for apple slice disinfection. *Food Prot. Trends* 24(7): 323–327.

Diaz, C. and J.H. Hotchkiss. 1996. Comparative growth of *E. coli* O157:H7, spoilage organisms and shelf life of shredded iceberg lettuce stored under modified atmospheres. *J. Sci. Food Agri.* 70: 433–438.

Doyle, M.P., T. Zhao, J. Meng, and S. Zhao. 1997. *E. coli* 0157:H7. In: Doyle, M.P., Beuchat, L.R., and Montville, T.J., (Eds.), *Food Microbiology—Fundamentals and Frontiers*, Washington DC: American Society of Microbiology Press, pp. 171–191.

Escudero, M.E., L. Velazquez, M.S. Di Genaro, and A.M.S. de Guzman. 1999. Effectiveness of various disinfectants in the elimination of *Yersinia enterocolitica* on fresh lettuce. *J. Food Prot.* 62: 665–669.

Farber, J.M. 1991. Microbiological aspects of modified atmosphere packaging technology: A review. *J. Food Prot.* 54: 58–70.

Farber, J.M., S.L. Wang, Y. Cai, and S. Zhang. 1998. Changes in populations of *Listeria monocytogenes* inoculated on packaged fresh-cut vegetables. *J. Food Prot.* 61: 192–195.

Farber, J.N., L.J. Harris, M.E. Parish, L.R. Beuchat, T.V. Suslow, J.R. Gorney, E.H. Garrett, and F.F. Busta. 2003. Microbiological safety of controlled and modified atmosphere packaging of fresh and fresh-cut produce. *Compr. Rev. Food Sci. Food Saf.* 2(Suppl.): 142–160.

Food and Drug Administration (FDA), Center for food Safety and Applied Nutrition. 1993. *Foodborne Pathogenic Microorganisms and Natural Toxins*. http://www.cfsan.fda.gov/~mow/badbug.zip

Food and Drug Administration (FDA), Center for Food Safety and Applied Nutrition. 1998. *Guide to Minimize Microbial Food Safety Hazards for Fresh Fruits and Vegetables* [Guidance for Industry]. http://www.foodsafety.gov/~dms/prodguid.html

Food and Drug Administration (FDA), Center for Food Safety and Applied Nutrition. 2005. *Letter to California Firms That Grow, Pack, Process, or Ship Fresh and Fresh-Cut Lettuce*. http://www.cfsan.fda.gov/~dms/prodttr2.html

Food and Drug Administration (FDA), Center for Food Safety and Applied Nutrition. 2008. *Guide to Minimize Microbial Food Safety Hazards of Fresh-Cut Fruits and Vegetables*. http://www.cfsan.fda.gov/guidance.html

Francis, G.A. and D. O'Beirne. 1997. Effects of gas atmosphere, antimicrobial dip and temperature on the fate of *Listeria innocua* and *Listeria monocytogenes* on minimally processed lettuce. *Int. J. Food Sci. Technol.* 32: 141–151.

Francis, G.A. and D. O'Beirne. 1998a. Effects of storage atmosphere on *Listeria monocytogenes* and competing microflora using a surface model system. *Int. J. Food Sci. Technol.* 33: 465–476.

Francis, G.A. and D. O'Beirne. 1998b. Effects of the indigenous microflora of minimally processed lettuce on the survival and growth of *L. monocytogenes*. *Int. J. Food Sci. Technol.* 33: 477–488.

Francis, G.A. and D. O'Beirne. 2001. Effects of vegetable type, package atmosphere and storage temperature on growth and survival of *Escherichia coli* O157:H7 and *Listeria monocytogenes*. *J. Ind. Microbiol. Biotechnol.* 27: 111–116.

Francis, G.A. and D. O'Beirne. 2002. Effects of vegetable type and antimicrobial dipping on survival and growth of *Listeria innocua* and *E. coli*. *Int. J. Food Sci. Technol.* 37: 711–718.

Francis, G.A., C. Thomas, and D. O'Beirne. 1999. Review paper: The microbiological safety of minimally processed vegetables. *Int. J. Food Sci. Technol.* 34: 1–22.

Francis, G.A., J. Scollard, A. Meally, D.J. Bolton, C.G.M. Gahan, P.D. Cotter, C. Hill, C., and D. O'Beirne. 2007. The glutamate decarboxylase acid resistance mechanism affects survival of *Listeria monocytogenes LO28* in modified-atmosphere packaged foods. *J. Appl. Microbiol.* 103: 2316–2324.

Frank, J.F. and R.A. Koffi. 1990. Surface-adherent growth of *Listeria monocytogenes* is associated with increased resistance to surfactant sanitizers and heat. *J. Food Prot.* 53: 550–554.

García-Gimeno, R.M., M.D. Sanchez-Pozo, M.A. Amaro-López, and G. Zurera-Cosano. 1996. Behaviour of *Aeromonas hydrophila* in vegetable salads stored under modified atmosphere at 4°C and 15°C. *Food Microbiol.* 13: 369–374.

Garg, N., J.J. Churey, and D.F. Splittstoesser. 1990. Effect of processing conditions on the microflora of fresh-cut vegetables. *J. Food Prot.* 53: 701–703.

Geldreich, E.E. and R.H. Bordner. 1971. Fecal contamination of fruits and vegetables during cultivation and processing for market: A review. *J. Food Technol.* 34: 184–195.

Gleeson, E. and O'Beirne, D. 2005. Effects of process severity on survival and growth of *Escherichia coli* and *Listeria innocua* on minimally processed vegetables. *Food Control* 16: 677–685.

Gray, M.L. and A.H. Killinger. 1966. *Listeria monocytogenes* and listeric infections. *Bacteriol. Rev.* 30: 309–382.

Griffin, P.M. and R.V. Tauxe. 1991. The epidemiology of infections caused by *E. coli* O157:H7, other enterohaemorrhagic *E. coli*, and the associated haemolytic syndrome. *Epidemiol. Rev.* 13: 60–98.

Guilbert, S., N. Gontard, and L. Gorris. 1996. Prolongation of shelf-life of perishable food products using biodegradable films and coatings. *Lebensm. Wiss. Technol.* 29: 10–17.

Gunasena, D.K., C.P. Kodikara, K. Ganepola, and S. Widanapathirana. 1995. Occurrence of *Listeria monocytogenes* in food in Sri Lanka. *J. Nat. Sci. Counc. Sri Lanka* 23: 107–114.

Han, Y., R.H. Linton, S.S. Nielsen, and P.F. Nelson. 2000. Inactivation of *Escherichia coli* O157:H7 on surface-uninjured and -injured green pepper (*Capsicum annuum* L.) by chlorine dioxide gas as demonstrated by confocal laser scanning microscopy. *Food Microbiol.* 17: 643–655.

Han, Y., R.H. Linton, S.S. Nielsen, and P.E. Nelson. 2001. Reduction of *Listeria monocytogenes* on green peppers (*Capsicum annuum* L.) by gaseous and aqueous chlorine dioxide and water washing and its growth at 7°C. *J. Food Prot.* 64: 1730–1738.

Hauschild, A.H.W., B. Aris, and R. Hilsheimer. 1978. *Cl. botulinum* in marinated products. *Can. Instit. Food Sci. Technology J.* 8: 84–87.

Hedberg, C.W. and M.T. Osterholm. 1993. Outbreaks of foodborne and waterborne viral gastroenteritis. *Clin. Microbiol. Rev.* 6: 199–210.

Herwaldt, B.L. 2000. Cyclospora cayetanensis: a review focusing on the outbreaks of cyclosporiasis in the 1990s. *Clin. Infect Dis.* 31: 1040–1057.

Herwaldt, B.L. and M.L. Ackers. 1997. An outbreak in 1996 of cyclosporiasis associated with imported raspberries. *New Engl. J. Med.* 336: 1548–1556.

Hutin, Y.J.F., V. Pool, E.H. Cramer, O.V. Nainan, J. Weth, I.T. Williams, S.T. Goldstein, K.F. Gensheimer, B.P. Bell, C.N. Shapiro, et al. 1999. A multistate foodborne outbreak of hepatitis A. *N. Engl. J. Med.* 340: 595–602.

International Commission on Microbiological Specifications for Foods (ICMSF). 1996. *Microorganisms in Foods. 5. Microbiological Specifications of Food Pathogens*, London: Blackie Academic & Professional.

Institute of Food Science and Technology (IFST). 1990. *Guidelines for the Handling of Chilled Foods*, 2nd ed., London.

Izumi, H. 1999. Electrolyzed water as a disinfectant for fresh-cut vegetables. *J. Food Sci.* 64: 536–539.

Jablasone, J., L.Y. Brovko, and M.W. Griffiths. 2004. A research note: The potential for transfer of *Salmonella* from irrigation water to tomatoes. *J. Sci. Food Agric.* 84: 287–289.

Jöckel, Von J. and W. Otto. 1990. Technological and hygienic aspects of production and distribution of pre-cut vegetable salads. *Arch. Lebensmittelhyg* 41: 149–152.

Karmali, M.A., B.T. Steele, M. Petric, and C. Lim. 1983. Sporadic cases of haemolytic uremic syndrome associated with faecal cytotoxin and cytotoxin-producing *E. coli. Lancet* i: 619–620.

Konowalchuk, J. and J.L. Speirs. 1975. Survival of enteric viruses on fresh vegetables. *J. Milk Food Technol.* 37: 132–134.

Larson, A.E., E.A. Johnson, C.R. Barmore, and M.D. Hughes. 1997. Evaluation of the botulism hazard from vegetables in modified atmosphere packaging. *J. Food Prot.* 60: 1208–1214.

Liao, C.-H. and P.H. Cooke. 2001. Response to trisodium phosphate treatment of *Salmonella* Chester attached to fresh-cut green pepper slices. *Can. J. Microbiol.* 47: 25–32.

Liao, C.-H. and G.M. Sapers. 1999. Influence of soft rot bacteria on growth of *Listeria monocytogenes* on potato tuber slices. *J. Food Prot.* 62: 343–348.

Lilly, T., H.M. Solomon, and E.J. Rhodehamel. 1996. Incidence of *Clostridium botulinum* in vegetables packaged under vacuum or modified atmosphere. *J. Food Prot.* 59: 59–61.

Lund, B.M. 1993. The microbiological safety of prepared salad vegetables. *Food Technol. Int. Eur.* 196–200.

Lund, B.M. and A.L. Snowdon. 2000. Fresh and processed fruits. In: Lund, B.M., Baird-Parker, T.C., and Gould, G.W. (Eds.), *The Microbiological Safety and Quality of Food*, Vol. 1, Gaithersburg, MD: Aspen, Chapter 27, pp. 738–758.

Magnuson, J.A., A.D. King, Jr., and T. Török. 1990. Microflora of partially processed lettuce. *Appl. Environ. Microbiol.* 56: 3851–3854.

Manvell, P.M. and M.R. Ackland. 1986. Rapid detection of microbial growth in vegetable salads at chill and abuse temperatures. *Food Microbiol.* 3: 59–65.

Marchetti, R., M.A. Casadei, and M.E. Guerzoni. 1992. Microbial population dynamics in ready-to-use vegetable salads. *Ital. J. Food Sci.* 4: 97–108.

Martin, D.L., T.L. Gustafson, J.W. Pelosi, L. Suarez, and G.V. Pierce. 1986. Contaminated produce—A common source for two outbreaks of *Shigella* gastroenteritis. *Am. J. Epidemiol.* 124: 229–305.

Mintz, E.D., M. Hudson-Wragg, P. Mshar, M.L. Cartter, and J.L. Hadler. 1993. Foodborne giardiasis in a corporate office setting. *J. Infect. Dis.* 167: 250–253.

National Advisory Committee on Microbiological Criteria for Foods (NACMCF). 1999. Microbiological safety evaluations and recommendations on fresh produce. *Food Control* 10: 117–143.

Nguyen-the, C. and F. Carlin. 1994. The microbiology of minimally processed fresh fruits and vegetables. *Crit. Rev. Food Sci.* 34: 371–401.

Nguyen-the, C. and J.P. Prunier. 1989. Involvement of pseudomonads in the deterioration of "ready-to-use" salads. *Int. J. Food Sci. Technol.* 24: 47–58.

Notermans, S., J. Dufrenne, and J.P.G. Gerrits. 1989. Natural occurrence of *Cl. botulinum* on fresh mushrooms (*Agaricus bisporus*). *J. Food Prot.* 52: 733–736.

O'Beirne, D. and A. Ballantyne. 1987. Some effects of modified atmosphere packaging and vacuum packaging in combination with antioxidants on quality and storage life of chilled potato strips. *Int. J. Food Sci. Technol.* 22: 515–523.

O'Beirne, D. and G.A. Francis. 2003, Analysing pathogen survival in modified atmosphere packaged (MAP) produce. In: Ahvenainen, R. (Ed.), *Novel Food Packaging Techniques*, London: Woodhead Publishing, Chapter 17.

Park, C.-M. and L.R. Beuchat. 1999. Evaluation of sanitizers for killing *Escherichia coli* O157:H7, *Salmonella* and naturally occurring microorganisms on cantaloupes, honeydew melons and asparagus. *Dairy Food Environ. Sanit.* 19: 842–847.

Petran, R.L., W.H. Sperber, and A.B. Davis. 1995. *Cl. botulinum* toxin formation in romaine lettuce and shredded cabbage: Effects of storage and packaging conditions. *J. Food Prot.* 58: 624–627.

PHLS (Public Health Laboratory System). 2001. *Salmonella* Newport infection in England associated with the consumption of ready to eat salad. *Eurosurveillance Weekly*, June 26.

Piagentini, A.M., M.E. Pirovani, D.R. Güemes, J.H. Di Pentima, and M.A. Tessi. 1997. Survival and growth of *Salmonella hadar* on minimally processed cabbage as influenced by storage abuse conditions. *J. Food Sci.* 62: 616–618, 631.

Rhodehamel, E.J. 1992. FDA's concerns with sous vide processing. *Food Technol.* 46: 73–76.

Richert, K.J., J.A. Albrecht, L.B. Bullerman, and S.S. Sumner. 2000. Survival and growth of *Escherichia coli* O157:H7 on broccoli, cucumber and green pepper. *Dairy Food Environ. Sanit.* 20: 24–28.

Rodriguez, A.M.C., E.B. Alcala, R.M.G. Gimeno, and G.Z. Cosano. 2000. Growth modelling of *Listeria monocytogenes* in packaged fresh green asparagus. *Food Microbiol.* 17: 421–427.

Rosenblum, L.S., I.R. Mirkin, D.T. Allen, S. Safford, and S.C. Hadler. 1990. A multifocal outbreak of hepatitis A traced to commercially distributed lettuce. *Am. J. Public Health* 80: 1075–1079.

Sadovski, A., Y.B. Fattal, and D. Goldberg. 1978. Microbial contamination of vegetables irrigated with sewage effluent by the drip method. *J. Food Prot.* 41: 336–340.

Sattar, S.A., V.S. Springthorpe, and S.A. Ansari. 1994. Rotavirus. In: Hui, Y.H., Gorham, J.R., and Cliver, D.O. (Eds.), *Foodborne Disease Handbook*, Vol. 2, New York: Marcel Dekker, pp. 81–111.

Seo, K.H. and J.F. Frank. 1999. Attachment of *Escherichia coli* O157:H7 to lettuce leaf surface and bacterial viability in response to chlorine treatment as demonstrated by using confocal scanning laser microscopy. *J. Food Prot.* 62: 3–9.

Sivapalasingam, S., C.R. Friedman, L. Cohen, and R. Tauxe. 2004. Fresh produce: A growing cause of outbreaks of foodborne illness in the United States, 1973 through 1997. *J. Food Prot.* 67: 2342–2353.

Solomon, H.M., D.A. Kautter, T. Lilly, and E.J. Rhodehamel. 1990. Outgrowth of *Cl. botulinum* in shredded cabbage at room temperature under modified atmosphere. *J. Food Prot.* 53: 831–833.

Sugiyama, H. and K.H. Yang. 1975. Growth potential of *Cl. botulinum* in fresh mushrooms packaged in semi-permeable plastic film. *Appl. Microbiol.* 30: 964–969.

Suslow, T.V., M.P. Oria, L.M. Beuchat, E.H. Garrett, M.E. Parish, L.J. Harris, J.N. Farber, and F.F. Busta. 2003. Production practices as risk factors in microbial food safety of fresh and fresh-cut produce. *Compr. Rev. Food Sci. Food Saf.*, 2: 38–77 (Suppl.).

Takeuchi, K. and J.F. Frank. 2001. Quantitative determination of the role of lettuce leaf structures in protecting *Escherichia coli* O157:H7 from chlorine disinfection. *J. Food Prot.* 64: 147–151.

Takeuchi, K., C.M. Matute, A.N. Hassan, and J.F. Frank, J.F. 2000. Comparison of the attachment of *Escherichia coli* O157:H7, *Listeria monocytogenes*, *Salmonella typhimurium* and *Pseudomonas fluorescens* to lettuce leaves. *J. Food Prot.* 63: 1433–1437.

Varnum, A.H. and M.G. Evans. 1991. *Foodborne Pathogens—An Illustrated Text*. London: Wolfe Publishing.

USDA-AMS, National Organic Program. http://www.ams.usda.gov/NOP

Walker, S.J. and M.F. Stringer. 1987. Growth of *Listeria monocytogenes* and *Aeromonas hydrophila* at chill temperatures. Campden Food Preservation Research Association Technical Memorandum No. 462, Chipping Campden, U.K: CFPRA.

Webb, T.A. and J.O. Mundt. 1978. Molds on vegetables at the time of harvest. *Appl. Environ. Microbiol.* 35: 655–658.

Wells, J.G., L.D. Shipman, K.D. Greene, E.G. Sowers, J.H. Green, D.N. Cameron, F.P. Downes, M.L. Martin, P.M. Griffin, S.M. Ostroff, M.E. Potter, R.V. Tauxe, and I.K. Wachsmuth. 1991. Isolation of *E. coli* O157:H7 and other Shiga-like toxin producing *E. coli* from dairy cattle. *J. Clin. Microbiol.* 29: 985–989.

Welshimer, H.J. 1968. Isolation of *Listeria monocytogenes* from vegetation. *J. Bacteriol.* 95: 300–303.

Wilcox, F., P. Tobback, and M. Hendrickx. 1994. Microbial safety assurance of minimally processed vegetables by implementation of the hazard analysis critical control point system. *Acta. Aliment.* 23: 221–238.

Zhang, S. and J.M. Farber. 1996. The effects of various disinfectants against *Listeria monocytogenes* on fresh-cut vegetables. *Food Microbiol.* 13: 311–321.

Zhuang, R.-Y., L.R. Beuchat, and F.J. Angulo. 1995. Fate of *Salmonella montevideo* on and in raw tomatoes as affected by temperature and treatment with chlorine. *Appl. Environ. Microbiol.* 61: 2127–2131.

11

Effects on Insects

Lisa G. Neven, Elhadi M. Yahia, and Guy J. Hallman

CONTENTS

11.1 Introduction

International trade of horticultural commodities has become increasingly important, but phytosanitary restrictions continue to limit its growth. Many insect pests are of quarantine importance in different regions in the world because they are either absent from an importing region in the same country or a different country, or the importing region or country has a "zero tolerance" for all live insects whether or not they are economically important. Some important quarantine insect pests are listed in Table 11.1. A systems approach is commonly needed to achieve an effective insect control, including different control measures in the field and surrounding areas, control of the commodity's maturity stages at harvest, inspection during arrival to packing station, the legal quarantine treatment or system, and certification prior to shipment.

TABLE 11.1

Some Examples of Insects Considered as Quarantine Pests in Some Regions or Countries

Common Name	Scientific Name	Common Hosts	Distribution
Codling moth	*Cydia pomonella* (L.)	Apples, pears, quince, walnut, *Prunus* spp.	Worldwide, except Japan
Sweet potato weevil	*Cylas formicarius* (Fab.)	Sweet potato	Asia, Africa, Australia, Hawaii
Mango seed weevil	*Stemochaetus mangiferae* (Fab.)	Mango	Asia, Africa, Australia, Hawaii
Red-legged earth mite	*Halotydeus destructor* (Tucker)	Leafy vegetables	Africa, Australia, New Zealand
Pineapple mealy bug	*Dysmicoccus brevipes* (Ckll.)	Pineapple	Asia, Africa, Australia, South America, Pacific Islands
Fruit flies			
Apple maggot fly	*Rhagoletis pomonella* (Walsh)	Apple	USA, Canada
Caribbean fruit fly	*Anastrepha suspensa* (Loew)	Tropical and subtropical fruits	Caribbean, southern Florida
Mexican fruit fly	*Anastrepha ludens* (Loew)	Citrus, mango, some other tropical and subtropical fruits	Mexico, Central America
European cherry fruit fly	*Rhagoletis cerasi* (L.)	Cherry, *Lonicera* spp.	Europe
Mediterranean fruit fly	*Ceratitis capitata* (Wied.)	Deciduous, subtropical, and tropical fruits	Southern Europe, Africa, Central America, South America, Western Australia, Hawaii
Melon fruit fly	*Batrocera cucurbitae* Coq.	Cucurbits, tomato, several other fleshy fruits	Hawaii, Asia, Papua New Guinea, Africa
Oriental fruit fly	*Batrocera dorsalis* (Hendel)	Most fleshy fruits and vegetables	Asia, Hawaii
Queensland fruit fly	*Batrocera tryoni* (Froggatt)	Deciduous, subtropical, and tropical fruits	Australia, Pacific islands

Quarantine treatments are often required to disinfest host commodities of economically important pests before they are moved through market channels to areas where the pest is not present. The goal of the quarantine treatments or systems is to kill, remove, or sterilize infesting (real or in certain instances, potential) insects to meet quarantine requirements, and the effects of these can impact commodity quality. A common system used to evaluate the effectiveness of quarantine treatments and systems is the "probit 9" concept (Baker, 1939), which refers to complete control of a test population of about 93,600 insects. Some countries use a variation of this concept that requires no survivors from a treated population of 30,000 target insects. Several treatments have been developed, but traditionally, the most common postharvest quarantine treatments for several years have involved the use of chemical fumigants such as ethylene dibromide (EDB) and methyl bromide (MeBr) and cold storage. In 1984, EDB was banned by the U.S. Environmental Protection Agency (EPA) due to its identification as a carcinogen, which caused major restrictions and difficulties in the trade of many commodities from several countries. Since then MeBr became the treatment of choice for achieving quarantine security for several fresh horticultural crops. However, in 1992, MeBr was determined to be an ozone-depleting substance (Anonymous, 1992) and thus restrictions on its future use were established. In response to this determination, the U.S. EPA included MeBr as a chemical to be removed from use by 2001 (Clean Air Act, 1993, Title VI, Section 602). In October 1998, the U.S. Congress, under the 1999 Appropriations Bill, altered the U.S. Clean Air Act (Section 764.a) to reflect the restrictions of the Montreal Protocol. This meant that postharvest phytosanitary uses of MeBr in the United States would be exempt from restrictions.

The increase in the demand and export of fresh horticultural crops, along with concerns about the safety of the food supply and about the impact of agricultural chemicals on the environment, including increased restrictions in the use of chemical fumigants as quarantine treatments, have increased interest of the industry and research activities for the development of alternative physical (nonchemical) quarantine treatments. As a result of this, several alternative quarantine treatments and systems have been developed using low and high temperatures, modified atmospheres (MA) and controlled atmospheres (CA is a subset of MA (Calderon, 1990)), ionizing radiation, microwaves, ultraviolet radiation, infrared radiation, electricity, radio frequency, and combinations of some of these. The type of physical treatment or system investigated or selected has been largely determined by commodity tolerance. Several quarantine physical treatments or systems are now available as alternatives to fumigants, although commercial application is still limited. For example, current quarantine treatments for oriental fruit moth include either MeBr fumigation or cold storage for several weeks (Hallman, 2004).

Quarantine treatments and systems should control the insect pests without negatively affecting the quality of the crop. The challenge for research is to develop nonchemical quarantine treatments and systems that do not harm the consumer or the environment, are relatively low cost, and can be applied either in permanent installations or aboard container ships. Quarantine security can be achieved with individual treatments, but most effectively should be developed as combination of a systems approach and direct treatment in order to facilitate the use of a less severe treatment, which might cost less or be less likely to damage the commodity.

MA treatments have been employed to control stored product pests for centuries. A very early example is the storage of grains in ancient Egypt, where the cribs were sealed tightly to obtain a low-oxygen environment, which would prevent the propagation and growth of insects. Historically, MA treatments were designed to preserve commodity quality during long-term storage. The secondary effect of providing some level of insect control was serendipitous. MA treatments help store a commodity at low temperatures for prolonged periods of time, and the low temperatures have an effect on insect mortality; however, in some cases MA can have a direct effect on insects. Reported effects of MA on insect physiology include reduction of NADPH levels, reduction in energy charge as a result of slower production of ATP, reduced glutathione, and the inhibition of the regeneration of choline to acetylcholine (Friedlander, 1983). There is also an observed reduction of high-temperature tolerance in insects exposed to anoxic environments (Mitcham et al., 1999; Neven, 2003, 2004, 2005, 2008; Neven and Mitcham, 1996; Neven and Rehfield-Ray, 2006a,b; Neven et al., 2006).

Insect control by MA has commercially been used for grains and dried fruits (Banks, 1984), but not yet fully developed for fresh horticultural commodities. Atmospheres needed for this application should contain very low partial pressures of oxygen (<1 kPa) or/and very high levels of CO_2 (up to 50–80 kPa). These insecticidal atmospheres can eliminate insects within a period of 2 to 4 days at room temperature (Ke et al., 1995; Yahia, 1998a, 2006a; Yahia and Carrillo, 1993; Yahia and Tiznado-Hernandez, 1993; Yahia and Vazquez-Moreno, 1993), and within a period of only 2–4 h at higher temperatures (Neven, 2004, 2005, 2008; Neven and Mitcham, 1996; Neven and Rehfield-Ray, 2006a,b; Neven et al., 2006; Ortega-Zaleta and Yahia, 2000a; Yahia and Ortega-Zaleta, 2000a,b). Potential problems that hinder the possibility for developing insecticidal MA (IMA) for fresh horticultural commodities include fermentation due to the use of anaerobic gas mixtures, and heat injury when MA is used in combination with heat. The advantages of using this system at room temperature include the low energy input and the avoidance of heat injury. However, the advantage of using it at high temperatures is to accomplish the mortality of insects in a very short period, and thus to increase throughput and reduce the accumulation of respiration products before treatments. Another advantage of IMA is that it can be used either in permanent stores or in marine transport containers.

MA can be achieved in different forms, such as in storage rooms and in marine transport containers through the introduction of nitrogen, carbon dioxide, and other atmospheric gases., and through the use of semipermeable membranes, called MA packaging (MAP), which can be applied individually to fruits, vegetables, or portions of these. MAP restricts the movement of oxygen and carbon dioxide (Dentener et al., 1992; Hallman, 1997; Jang, 1990) and can also be generated using film wraps, film coverings, and coatings. MAP is generally performed at temperatures between 0°C and 20°C, at 1–18 kPa O_2, or/and 0 to 10 kPa CO_2, for few days to several weeks. The commodity consumes O_2 and increases CO_2 production during normal respiration processes. Normally, the reduction of O_2 and elevation of CO_2 are not as severe in an aerobic atmosphere, but if it is changed to an anaerobic atmosphere the level of O_2 will no longer support aerobic commodity respiration and/or the level of elevated CO_2 can become inhibitory to commodity respiration. Therefore, MAP with aerobic atmospheres often takes a longer time to kill the target pest (Carpenter and Potter, 1994; Hallman, 1994; Neven, 2008). Low pressures (hypobaric atmospheres) are predominantly a low O_2 treatment, which can also affect insects, with the mode of action depending primarily on the lack of oxygen and, less on the physical effect of low pressure or desiccation (Burg, 2004; Mbata and Phillips, 2001).

Hallman (1994) identified a number of MA treatments that showed promise against several tortricids, two mites, two thrips, green peach aphis (*Myzus persicae*), San Jose scale (*Quadraspidiotus perniciosus*), and sweetpotato weevil (*Cylas formicarius elegantulus*), on apple, asparagus, strawberry, sweet potato, and walnut. Considerable additional research has been done since then. However, only one commercial MA treatments has been used for fresh horticultural commodities: a single shipment of 225 kg of asparagus from New Zealand to Japan that was disinfested of New Zealand flower thrips (*Thrips obscuratus*) and green peach aphid (Carpenter and Potter, 1994). The treatment consisted of holding asparagus in air with 60 kPa CO_2 (which resulted in about 8 kPa O_2)

for 4.5 days at 0°C–1°C. Although no live quarantine pests were found and asparagus quality was not significantly affected, Japan did not permit additional shipments with this treatment.

Efficacy of MA treatments at temperatures near 0°C may be due more to the low temperature than to low oxygen or high CO_2 (Hallman, 1994; Neven, 2003). For example, synergism between MA and low temperature applied at the same time does not seem to occur. However, a 20 h exposure of Mediterranean fruit fly (*Ceratitis capitata*)-infested mandarins to 95 kPa CO_2 at 25°C before cold treatment at 1.5°C in ambient atmosphere reduced time to 100% mortality to less than half of that required without the MA pretreatment (Alonso et al., 2005). The 20 h pretreatment alone and the control, an approximation of the full treatment regime, had about 30% mortality.

One reason why MA treatments may not have been used commercially for insect control in fresh horticultural commodities is that they are difficult to research and implement on a commercial scale, and gas and temperature stress required to control insects in short period can also be injurious to some fresh horticultural commodities. Of the commonly studied phytosanitary treatments, MA may be the most difficult to study. For example, although Hallman (1994) identified several possible treatments that might work on a commercial scale, he noted that further research was needed for all of them before they could be confidently used. One of the complicating factors for MA and research is the response of insects to elevated CO_2 (Fleurat-Lessard, 1990). Hallman (1994) cited several examples among a tephritid fruit fly, stored product weevils, and tortricids where increased reaction to increasing CO_2 levels did not follow a typical dose–response curve. Green peach aphid response to increasing CO_2 concentrations was steeper at lower than at higher concentrations (Epenhuijsen et al., 2002). The authors compared their results with others for grain weevils, crickets, thrips, and other aphids and hypothesized that there is an acute response to CO_2 in the 10–20 kPa range and a gradual response in the 20–60 kPa range of CO_2 atmosphere. Response of pests to O_2 levels may be no less complicated than response to CO_2 and results of combinations of modified levels of both gases can be exceedingly difficult to examine, model, and predict (Fleurat-Lessard, 1990). However, because the modes of actions of hypercarbia and hypoxia are different, the combination should be more effective than either alone. Another complicating factor in MA research is temperature, with increasing temperature usually related to increasing pest mortality as well as increasing damage to fresh commodities (Ortega and Yahia, 2000a,b; Yahia, 1998a; Yahia and Ortega, 2000a,b).

Challenges to the commercial application of MA treatments include product tolerance and overall logistical suitability. One advantage for MA disinfestation is the absence of residues, which characterize chemical fumigants. However, potential safety hazards exist for operational staff of MA rooms and transport containers because MA typically contains little if any oxygen.

IMA at high temperatures can eliminate many insects within a period of 2–4 h at >40°C (Neven, 2004, 2008; Neven and Mitcham, 1996; Neven and Rehfield-Ray, 2006a,b; Yahia and Ortega, 2000a,b). Potential problems that hinder the possibility for developing this system for fresh horticultural commodities are possible fermentation (due to the use of anaerobic gas mixtures), and heat injury. However, the advantage of using this system at high temperatures is to achieve complete mortality of insects in a very short period. The potential development of quarantine insect control systems using MA at room or higher temperatures in several crops has been tested for several horticultural commodities (Delate and Brecht, 1989; Lay-Yee and Whiting, 1996; Mitchell et al., 1984; Navarro et al., 1998; Neven, 2004, 2008; Neven and Mitcham, 1996; Neven and Rehfield-Ray, 2006a, b; Pesis and Ben-Arie, 1986; Shellie et al., 1997b; Yahia and Ortega-Zaleta, 1999; Yahia et al., 1997, 1989). Different fruits and vegetables respond very differently to this stress. Mangoes were found to be sufficiently resistant to both gas and heat stress, and therefore this technique can be developed commercially for this fruit (Ortega-Zaleta and Yahia, 2000a,b; Yahia, 1998a; Yahia and Tiznado-Hernandez, 1993; Yahia and Vazquez-Moreno, 1993), however, several fruits such as papaya, avocado, guava, and pears were found to be sensitive to both types of stress (Ke et al., 1994a, 1995; Yahia, 1998a; Yahia and Carrillo-López, 1993; Yahia et al., 1992).

Combination of high temperature and MA treatments have been developed for a number of internal feeding pests of apples, pears, peaches, nectarines, and sweet cherries (APHIS, 2008; Mitcham et al., 1999; Neven, 2004, 2005, 2008; Neven and Mitcham, 1996; Neven and Rehfield-Ray, 2006a, b; Neven et al., 2006; Shellie et al., 2001). These treatments were performed in an atmosphere containing 1 kPa O_2, 15 kPa CO_2 with the balance being N_2. All but the cherry treatments were performed under linear heating rates from 12°C to 100°C/h. These linear rates of heating were used to preserve commodity quality of these temperate fruits. Cherries respond to heat quite differently in that their internal heating rates on the tree often approach over 50°C/h (Brown, 2006). Therefore, treatments for sweet cherries follow a flow-through method where they are placed into the final treatment conditions immediately, not following the ramp normally employed for other types of fruit.

To date, there are only four MA treatments in the USDA, APHIS treatment manual (APHIS, 2008) (see Table 11.2). These treatments employ a combination of MA and hot forced air called controlled atmosphere temperature treatment system (CATTS). Other MA treatments have been developed (see Table 11.3), but are not yet in the APHIS treatment manual.

Adoption of MA quarantine treatments by industry has been slow because of the wide availability of chemical fumigants like MeBr. MeBr was identified as an ozone-depleting substance under the Montreal Protocol, and most uses in developed countries are being phased out. However, its use for quarantine purposes is still unrestricted (Anon, 1993). The cost of MeBr in the United States has rapidly increased from $5 for a 45 kg tank in 1995 to nearly $1000 in 2007. As fumigation costs rise, alternative quarantine treatments should gain industry support and become more common.

TABLE 11.2

USDA- APHIS Approved Controlled Atmosphere Treatments for Fresh Fruits as of 2008

Commodity	Treatment	Pest	Levels (kPa) O_2	CO_2	Temperature (°C)	Heating Rate (°C/h)	Total Time
Apple	CATTS[a]	Codling moth and oriental fruit moth	1	15	46	12	3 h
Cherry	CATTS	Codling moth and Western cherry fruit fly	1	15	47	>200	25 min
Cherry	CATTS	Codling moth and Western cherry fruit fly	1	15	45	>200	45 min
Nectarines and peaches	CATTS	Codling moth and oriental fruit moth	1	15	46	24	2.5 h
Nectarines and peaches	CATTS	Codling moth and oriental fruit moth	1	15	46	12	3.0 h

[a] CATTS (Controlled Atmosphere Temperature Treatment System).

TABLE 11.3

Treatment Conditions Used to Control Various Quarantine Pests

Group	Stage	Commodity	O_2/CO_2 (kPa)	Temperature (°C)	Duration	Notes
Fruit Flies						
Anastrepha suspensa	5 d larvae	Citrus	0/0	22–23	60 h	Benschoter et al. (1981)
Anastrepha suspensa	Eggs and larvae	In vitro	10–50/20–80	10 and 15.6	7 days	Benschoter (1987)
Rhagoletis mendax	Larvae	Blueberry	Var/50–100	21	48 h	Prange and Lidster (1992)
Rhagoletis pomonella	Larvae	Apple	Var/15–19	10	14 days	Agnello et al. (2002)
Rhagoletis indifferens	Eggs and larvae	Cherry	1/15	45	45 min	Neven and Rehfield-Ray (2006a)
Rhagoletis indifferens	Eggs and larvae	Cherry	1/15	47	25 min	Neven and Rehfield-Ray (2006a)
Lepidoptera						
Cydia pomonella	Larvae	Apple	2.2–3/1.6	0	<13 weeks	Toba and Moffitt (1991)
Cydia pomonella	larvae	Walnut	Var/98	39–45	33–48 h	Gaunce et al. (1982)
Cydia pomonella	Eggs and larvae	Apple	1/15	46	3 h	Neven and Rehfield-Ray (2006b)
Cydia pomonella	Eggs and larvae	Nectarine	1/15	46	2.5–3 h	Neven et al. (2006)
Cydia pomonella	Eggs and larvae	Cherry	1/15	45 and 47	45 and 25 min	Neven (2005)
Graphlitha molesta	Eggs and larvae	Apple	1/15	46	3 h	Neven and Rehfield-Ray (2006b)
Graphlitha molesta	Eggs and larvae	Nectarine	1/15	46	2.5–3 h	Neven et al. (2006)
Epiphyas postvittana	Eggs and larvae	Apple	1/1	40	>16 h	Whiting and Hoy (1998)
Platynota stultana	All	Grape	Var/45	0	4.1 days	Mitcham et al. (1997b)
Coleoptera						
Cylas formicarius elegantulus	Eggs and larvae	Sweet potato	8/40–60	30	4–8 days	Delate et al. (1990)
Cylas formicarius elegantulus	Eggs and larvae	Sweet potato	2–4/40–60	25	2–8 days	Delate et al. (1990)
Hemiptera						
Quadraspidiotus perniciosu	All	Apple	<1/>90	12	2 days	Carpenter and Potter (1994)
Quadraspidiotus perniciosu	All	Apple	2.6–3/1.0–1.1	1 and 3	31–34 weeks	Chu (1992)
Pseudococcus longispinosus	All	Apple	Var/18	0	2 weeks	Potter et al. (1990)

(continued)

TABLE 11.3 (continued)

Treatment Conditions Used to Control Various Quarantine Pests

Group	Stage	Commodity	O$_2$/CO$_2$ (kPa)	Temperature (°C)	Duration	Notes
Pseudococcus affinis	All	Apple	1/NA	45	14 h	Whiting and Hoy (1997)
Heteroptera						
Nysius huttoni	All		Var/9 and 18	0 and 20	2 weeks	Potter et al. (1990)
Thysanoptera						
Frankliniella occidentalis	All	Strawberry	2/90	2.5	2 days	Aharoni et al. (1981)
Frankliniella occidentalis	All	Grapes	Var/45	0	5.8	Mitcham et al. (1997)
Thrips obscuratus	Adults	Flowers	2/18	15	4 days	Potter et al. (1994)
Thrips obscuratus	Adults	Flowers	2/18	10–20	6 days	Potter et al. (1994)
Thrips obscuratus	Adults	Flowers	2/9	0 or 20	8 days	Potter et al. (1994)
Psocidae						
Liposcelis bostrychophila	All		1/35			Wei et al. (2002)
Aphidae						
Myzus persicae	All	Asparagus	Var	0–2	4 days	Carpenter (1995)
Myzus persicae	3 and 4 instar	*Hydrangea* sp.	Var/40–95	12	≤12 days	Epenhuijsen et al. (2002)
Mites						
Panonychus ulmi	Overwintering adult	Apple	0/60 0/100 0/0	21–24	2 days	Gaunce et al. (1982)
Tetranychus mcdanieli	Overwintering adult	Apple	0/100	21–24	>7 days	Gaunce et al. (1982)
Tetranychus urticae	Diapausing adult	Apple	1.3–0.4/5–20	20–40	112–15 h	Whiting and van den Heuvel (1995)
Tetranychus pacificus	All	Grapes	Var/45	0	8.1 days	Mitcham et al. (1997)

11.2 Effects of Low O_2 and High CO_2 Levels on Arthropods

Atmospheres with very low O_2 content and/or very high CO_2 levels have insecticidal effects (Neven, 2003; Stewart et al., 2005; Yahia, 1998b, 2006a; Yahia and Ortega-Zaleta, 2000a). Insect control with MA depends on the O_2 and CO_2 concentration, temperature, RH, insect species and development stage, and duration of the treatment. In general, the lower the O_2 concentration, the higher the CO_2, the higher the temperature, and the lower the RH, the shorter will be the time necessary for insect control. MA has several advantages in comparison with other means for insect control. These are physical treatments that do not leave toxic residues on the fruit, and are competitive in costs with chemical fumigants (Aegerter and Folwell, 2001). MA does not accelerate fruit ripening and senescence as compared to the use of high temperatures, and have better consumer acceptance as compared to the use of irradiation and fumigation in that they are accepted by organic growers and consumers (Stewart et al., 2005; Yahia, 2006a). Disadvantages of MA treatments are their complexity, long treatment times at low temperatures, and damage to fresh commodities including development of off-flavors.

11.2.1 Low Oxygen

The most common types of controlled atmosphere treatments are those that employ low oxygen environments. In the case of apples, oxygen levels used vary from 1 to 5 kPa, and carbon dioxide levels used vary from 0.3 to 3 kPa (Neven, 2003), with the balance being nitrogen. Other treatments use ultra low levels of oxygen, 0.0025 kPa (Liu, 2003, 2005, 2007), as is the case with broccoli and lettuce. These treatments are performed at temperatures well below 5°C.

Low-oxygen treatments can be effective in killing insects provided that the temperature is high enough to put a stress on the metabolic system of the insect. Reduced O_2 consumption leads to a decreased rate of ATP production (Hoback and Stanley, 2001). As a result of energy insufficiency, the membrane ion pumps fail, leading to K^{2+} efflux, Na^+ influx, and membrane depolarization (Fleurat-Lessard, 1990; Friedlander, 1983). The voltage-dependent Ca^{2+} gates are then opened, causing Ca^{2+} influx. The high Ca^{2+} concentration in the cytosol activates phospholipases and leads to increased membrane phospholipid hydrolysis (Fleurat-Lessard, 1990; Friedlander, 1983). The cell and mitochondrial membranes become further permeable, causing cell damage or death (Zhou et al., 2000, 2001).

Omnivorous leafroller pupae use metabolic arrest as a major response to hypoxia (Zhou et al., 2000, 2001). Pupal O_2 consumption rate and metabolic heat rate decrease slightly with decreasing O_2 concentration until a critical concentration is attained, below which the metabolic arrest sequence is initiated and the decrease becomes rapid (Zhou et al., 2000, 2001). The critical concentration points are 10, 8, and 6 kPa at 30°C, 20°C, and 10°C, respectively. Although pupal metabolism decreases quickly below the critical concentration points, the pupae do not initiate anaerobic metabolism until the O_2 concentration is below 2 kPa at 20°C (Zhou et al., 2000, 2001). Concentrations of O_2 below the anaerobic compensation point appear to be in the insecticidal range.

11.2.2 High Carbon Dioxide

Some treatments use carbon dioxide levels of 60 kPa or higher, with the oxygen level not regulated. These treatments have been used at temperatures within the normal growing range of commodities being treated (10°C–40°C). High carbon dioxide treatments have been shown to be very effective in controlling mites and diapausing insects. However, when elevated carbon dioxide is used in combination with low oxygen levels, the result on insect mortality has been variable. A combination of high temperature and controlled atmosphere treatments used effectively against lepidopteran pests do not work as effectively against fruit flies. This may be due to the differences in the respiratory systems and regulatory mechanisms of a terrestrial (Lepidoptera) and semiaquatic (fruit fly larvae) insects. At temperatures near 0°C, mortality of the moth *Platynota stultana* is greater with 45 kPa O_2 + 11.5 kPa O_2 (air) compared with 45 kPa CO_2 + 0.5 kPa O_2. In some cases, when only a small amount of CO_2 is present in an O_2-deficient atmosphere, it can enhance mortality by up to 10-fold (Zhou et al., 2000, 2001). Elevated CO_2 in the hemolymph (hypercapnia) can reduce the rate of insect respiration. High levels of CO_2 can reduce oxidative phosphorylation by inhibiting respiratory enzymes such as succinate dehydrogenase and malic enzyme (Friedlander, 1983). Reduced oxidative phosphorylation leads to reduced ATP generation, which in turn, leads to a failure of membrane ion pumps, membrane depolarization, and eventual cell death, as described for hypoxia (Friedlander, 1983).

Elevated CO_2 levels can decrease pH through the formation of carbonic acid (Fleurat-Lessard, 1990). Reduced pH can increase intercellular Ca^{2+} concentration, which causes the cell and mitochondrial membranes to become more permeable, suggesting that high CO_2 can increase membrane permeability (Fleurat-Lessard, 1990). High CO_2 levels can alter the ratio of pyruvate to lactate by 25% of normal, changing the redox potential and causing a lesion in the electron transport chain, presumably by a modification in the permeability of mitochondrial membranes (Friedlander, 1983).

11.3 Factors Affecting Efficacy of MA and CA Treatments for Arthropod Control

Temperature is the major factor affecting the efficacy of modified atmosphere phytosanitary treatments. In general, increasing the temperature increases efficacy. At low temperatures, the lethal effect may be due more to the cold temperature than the modified atmosphere. For example, Mitcham et al. (1997a) achieved higher mortality of western flower thrips (*Frankliniella occidentalis*) and the omnivorous leafroller (*Platynota stultana*) under controlled atmospheres at 0°C compared with 5°C. The Pacific spider mite (*Tetranychus pacificus*) suffered greater mortality in both modified atmosphere and air treatments at 5°C compared with 0°C. This may be the case where metabolism of the mite was elevated enough at 5°C for the MA to be more lethal as compared to the metabolism of the other two pest species. This might also be a case where the effects of temperature were more severe than the MA effects.

The relationship between modified atmospheres and temperatures near 0°C is not clear; the relationship is most likely additive with cold providing significant mortality. At cool temperatures that are not generally lethal to insects but low enough to inhibit development, the cool temperature seems to be antagonistic to the efficacy of modified atmospheres (Neven, 2003; Soderstrom et al., 1991). This seems logical in that insects metabolizing at a very low rate may be less susceptible to deficiencies or excesses in atmospheric components. Knowing where the threshold for antagonism ends above the cool temperature range and where mortality increases again below it (whether due to simply the cold temperature or some interaction of cold temperature and modified atmosphere) is essentials to optimizing modified atmosphere phytosanitary treatments.

Although the relationship between MA and temperatures near 0°C has not been adequately elicited, at the other end of the phytosanitary temperature treatment range, heat, it is better understood in terms of treatment efficacy, but not perhaps on physiology, and has been taken advantage of extensively in research. Treatments using MA at elevated temperatures have been developed that are ready for commercial implementation (APHIS, 2008).

In in vitro studies Whiting et al. (1992a,b) developed a potential treatment consisting of 0.4% O_2 and 5% CO_2 at 40°C and requiring an estimated 4.2 h to achieve 100% mortality of four tortricids. At 30°C and 20°C, respectively, the LT_{99} for fifth instar light brown apple moth (*Epiphyas postvittana*) was about 8 and 22 times what it was at 40°C (Whiting et al., 1991). Large-scale confirmatory testing is needed to set the minimum treatment time duration before this treatment could be recommended for commercial use. Within an insect species, tolerance to MA treatments commonly varies among developmental stages. In general, tolerance increases as development progresses, but there are exceptions that may depend upon differences in metabolic rate in relation to physiological state (i.e., diapause). Whiting et al. (1992a, 1995) found that the most tolerant stage was the fifth instar and the most tolerant of the four species was the codling moth. Nondiapausing fifth instar codling moth was slightly more tolerant than diapausing ones.

When Whiting et al. (1995) exposed eggs and first, third, and fifth instars of six leafrollers (Tortricidae) to three different atmospheres (1.2% O_2 and 5% CO_2; 4.2% O_2 and 5% CO_2, and air) at 40°C, fifth instars were not always the most tolerant. However, the most tolerant of the six species was *E. postvittana*, and the fifth instar was more tolerant than any stage tested of any of the six species. Thus, a treatment for fifth instar *E. postvittana* should control all stages of concern of the other species as well, although one could raise a question about second and fourth instars, which were not tested.

In other studies, (Neven and Rehfield-Ray, 2006a,b; Neven et al., 2006) found that the addition of a controlled atmosphere (1% O_2, 15% CO_2) to a heat treatment of fruit using a linear heating rate of 12°C and 24°C/h to core temperatures of 44.5°C effectively blocked the thermal acclimation process that normally is observed in heat treatments under regular atmospheres. Subsequent studies showed that low oxygen environments block the synthesis of many heat shock proteins (Follett and Neven, 2006; Neven, unpublished).

Use of high temperatures for control of diapausing codling moth (*Cydia pomonella* (L.)) larvae in walnuts was investigated alone or in conjunction with low O_2 or enriched CO_2 atmospheres (Soderstrom et al., 1996a,b). The temperatures used were 39°C, 41°C, 43°C and 45°C, and were selected as being appropriate for the drying of walnuts. Atmospheres investigated were 98 kPa CO_2 in air, 0.5 kPa O_2 in N_2, and 0.5 kPa O_2 and 10 kPa CO_2 in N_2, with normal air as a control. High temperature alone and in conjunction with modified atmospheres controlled codling moth larvae effectively. Effects of temperature were modeled using an asymptotic model based on LT_{95}s estimations for each temperature–atmosphere combination. Increasing temperature gave a more rapid kill of larvae. The most effective atmosphere for control of codling moth larvae in walnuts was 98 kPa CO_2, followed by the two 0.5 kPa O_2 atmospheres, which were more effective than the control atmosphere.

In the three previously cited in vitro studies (Whiting et al., 1992a,b, 1995) there appeared to be an interaction between tolerance to O_2 levels and the developmental stage for some tortricid leafrollers. When the O_2 level was 0.4% the fifth instar of *Ctenopseustis obliquana* and *Planotortrix octo* was most tolerant. When the O_2 level was 1.2 kPa, 4.2 kPa or ambient, other stages, especially the third instars, were often more tolerant.

Heating rate may affect the efficacy of heat/modified atmosphere treatments. As the time to heat the chamber to operating temperatures (= ramp time) increased in a treatment consisting of 1 kPa O_2 and 1 kPa CO_2 at 40°C the estimated time required at the final temperature to achieve 99% mortality of light brown apple moth decreased until it passed 7.5 h of ramping time, after which it increased again (Whiting and Hoy, 1998).

This indicated that ramping times of 7.5 h or greater resulted in acclimation of the insect to the heat treatment. Under this treatment scenario, the preferred treatment should be the one that achieved quarantine security in the shortest amount of total time (heat-up plus treatment time).

Coatings of fresh commodities designed to prolong shelf life may modify the atmosphere inside and have also been shown to synergize heat treatments. Hallman et al. (1994) reduced by half the heated air treatment time required to kill Caribbean fruit fly in grapefruits by coating the fruit before heating.

11.3.1   Hypobaric Storage as MA/CA Treatment

The shelf life of fresh commodities can be increased by using hypobaric storage, created by using pumps to reduce atmospheric pressure to around 15 to 35 mmHg. With this technique, MA conditions can be achieved without necessarily introducing any gases (Burg, 2004). In a number of cases, hypobaric storage treatments designed for increasing shelf life of commodities also have a detrimental effect on arthropod pests infesting the commodities. The mode of action is considered to be due to reductions in O_2 levels with insignificant physical effects of low pressure per se or dehydration (Mbata and Phillips, 2001; Mbata et al., 2005; Navaro and Calderon, 1979). Advantages of hypobaric treatments over other forms of modified atmospheres are that gases need not be created, introduced, or monitored. The key variable is maintenance of the desired low pressure.

Hypobaric storage may be an ideal disinfestation technique for some difficult to treat fresh commodities, such as lettuce. Partial vacuum is already used to cool lettuce after harvest. Complete control of two aphids was achieved in four days at 5°C using a vacuum to initially remove air followed by insertion of 6 kPa CO_2, although the treatment achieved only 95% control of larvae of the leafminer (*Liriomyza langei*) (Liu, 2003). Previous research with lettuce has usually found the commodity to be intolerant of even small amounts of CO_2. The technique should also be investigated for cut flowers where few viable disinfestation alternatives to fumigation have been found.

All Caribbean fruit fly (*Anastrepha suspense*) eggs and larvae in agar diet were killed in about 9 days upon exposure to a vacuum of 15 mmHg at 13°C (Davenport et al., 2006). In small-scale tests, mangoes, carambolas, and guavas survived the treatment well.

11.4   Physiological Effects of MA on Insects

11.4.1   Target Pest Groups

Pest groups may respond differently to MA (Table 11.3). More progress has been made and more conclusions can be drawn about pest response for stored product pests (Fleurat-Lessard, 1990; Mbata et al., 2004). However, responses of stored product pests may differ significantly from pests infesting fresh commodities because the latter are less tolerant to low RH. Also, the durable commodities infested by stored product pests tolerate more extreme treatments than fresh commodities.

Reducing the availability of O_2 during a heating stress hinders the insects' ability to support elevated metabolic demands due to the heat load. There is also evidence that a heat treatment under anoxic conditions reduces the production of heat shock proteins in insects (Thomas and Shellie, 2000, Neven, unpublished), and elevated CO_2 atmospheres may interfere with the insects' ability to produce ATP (Friedlander, 1983; Zhou et al., 2000, 2001). Heat treatments combined with an anoxic environment can provide quarantine security more rapidly than a heat treatment or a MA treatment alone (Lay-Yee and Whiting, 1996; Moss and Jang, 1991; Neven, 2005; Neven and Mitcham, 1996; Neven and Rehfield-Ray, 2006a,b; Neven et al., 2006; Soderstrom et al., 1992; Whiting and Hoy, 1997, 1998; Whiting and Van Den Heuvel, 1995; Whiting et al., 1991, 1992a,b, 1995, 1996; Yocum and Denlinger, 1994).

11.4.2 Diptera

Tephritid fruit flies are the most important group of quarantine pests for fresh fruit traded across quarantine barriers. However, this pest group has not received as much attention from modified atmosphere researchers as their importance would indicate.

Benschoter et al. (1981) seem to be the first to have explored the effects of modified atmospheres on a tephritid, the Caribbean fruit fly (*Anastrepha suspense*). Five-day-old larvae died after 60 h exposure at 22°C–23°C in 100 kPa nitrogen in vitro. Benschoter (1987) assessed the response of Caribbean fruit fly eggs and larvae in vitro to modified atmospheres of 20, 50, or 80 kPa CO_2 and 10, 20 or 50 kPa O_2 (balance N_2) at 10°C and 15.6°C. Increased mortality generally coincided with the highest carbon dioxide concentration regardless of O_2 concentration. At the lowest CO_2 concentration, however, increased mortality coincided with lower O_2 level. Mortality was somewhat increased at the higher of the two temperatures. Complete mortality of 150 insects tested occurred in 7 days with some of the treatment combinations.

Prange and Lidster (1992) achieved up to 90% mortality of blueberry maggot (*Rhagoletis mendax*), after 48 h in various levels of CO_2 at 21°C. Mortality reached a peak at about 70 kPa CO_2 and then declined until less mortality was achieved at 100 kPa CO_2 than at 50 kPa CO_2.

Complete kill of apple maggot, *Rhagoletis pomonella*, larvae in apples was accomplished in 14 days at 10°C in atmospheres with 15 or 19 kPa CO_2 and the balance nitrogen (Agnello et al., 2002).

Coatings applied to fruit to enhance appearance and prolong shelf life may kill tephritid immatures in fruit and the mode of action is most likely modification of the atmosphere inside the fruit (Hallman, 1997; Hallman et al., 1994). The technique, not totally efficacious as a single treatment, might be incorporated as part of a phytosanitary system to reduce risk of infestation. The added benefit to a phytosanitary system is that no additional action may be required where coatings are already used except to document and quantify any risk reduction imparted by the coating.

Modified atmospheric packaging was studied as a phytosanitary treatment against fruit flies over 10 years ago but was susceptible to interruption of the seal, leading to variation in results (Hallman, 1994). Nonetheless, modified atmospheric packaging might form part of a phytosanitary system in a like manner as coatings.

Research on temperate fruit flies indicates a slightly different response to MA treatments under elevated temperatures (Neven, 2004; Neven and Rehfield-Ray, 2006a). It was found that the combination of a low oxygen, high carbon dioxide environment (1 kPa O_2, 15 kPa CO_2) effectively controlled apple maggot in a 12°C/h heat treatment of apples to a chamber temperature of 44°C or 46°C. These MA levels also controlled western cherry fruit fly (*Rhagoletis indifferens*) in sweet cherries using a flow-through treatment to 45°C and 47°C using heating rates of 264°C/h and 146°C/h, respectively. Treatments under regular atmospheres were not effective in controlling these pests.

11.4.3 Lepidoptera

The largest group of pests studied with modified atmosphere phytosanitary treatments has been Lepidoptera, especially larvae of the fruit-boring family Tortricidae.

Fruits of the family Rosaceae (e.g., apples) customarily tolerate storage under modified atmosphere for many months. Modified atmospheres may have the greatest potential as a disinfestation treatment against lepidopterous pests infesting these types of fruits. Codling moth (*Cydia pomonella*) is the most important quarantine pest of apples internationally. Toba and Moffitt (1991) found that mortality of all nondiapausing larvae of codling moth occurred in <13 weeks at 0.8–1.6 kPa carbon dioxide and 2.2–3 kPa oxygen at 0°C. These are the commercial modified atmosphere storage conditions for conserving apple quality for several months. Nineteen weeks were required for complete mortality of diapausing larvae (Moffitt and Albano, 1972). Low temperature and not so much the modified atmosphere may have been the chief cause of mortality in these cases. Additional work by Hansen et al. (2006) indicated that it is the low temperature that is the active component in causing codling moth mortality during storage rather than the effects of the controlled atmospheres.

Gaunce et al. (1982) found that complete mortality of diapausing and nondiapausing codling moth larvae might be achieved after 33–48 h in 95 kPa carbon dioxide at 27°C, but injury to some apples was observed. Soderstrom et al. (1996a,b) tested atmospheres of 0.5 kPa O_2 in N_2, 0.5 kPa O_2 plus 10 kPa CO_2 in air or 98 kPa CO_2 in air against stages of codling moth infesting stored walnuts at 39°C–45°C with the 98 kPa CO_2 atmosphere being the most efficacious.

Other tortricid pests include the light brown apple moth (*Epiphyas postvittana*) and a number of other leafrollers native to New Zealand. These leafrollers cause surface damage to fruit as well as attacking foliage. Whiting et al. (1991, 1995) found that at temperatures $\geq$20°C, atmospheres low in O_2 (1.2–4.2 kPa) and with moderate CO_2 (5 kPa) levels would achieve disinfestation of fruit.

Mortality responses of 3-day-old eggs and first, third, and fifth instars of three leafroller (Lepidoptera: Tortricidae) pests, infesting kiwifruit (*Actinidia deliciosa*), after exposure to a high-temperature MA (2 kPa O_2 with 5 kPa CO_2 at 40°C) were determined by Hoy and Whiting (1998). All four life stages of *Cnephasia jactatana* (Walker) had similar mortality responses to MA treatment at the 99% mortality level. Fifth instars of *Ctenopseustis obliquana* (Walker) were more tolerant than the other three life stages tested. *Epiphyas postvittana* (Walker) showed a trend of increasing exposure time to achieve 99% mortality as development progressed. *E. postvittana* was more tolerant of MA treatment than *C. obliquana* and *C. jactatana*. An exposure time of 6 h resulted in control (greater than 99% mortality) of *C. jactatana* and *C. obliquana* however, did not control all life stages of *E. postvittana*. These authors suggested a potential to combine a short high-temperature MA treatment with a longer period of low-temperature MA storage to control all three species.

Research on codling moth and oriental fruit moth (Neven, 2008; Neven and Rehfield-Ray, 2006b; Neven et al., 2006) found that for deciduous tree fruits the addition of a controlled atmosphere (1 kPa O_2 15 kPa CO_2) to the heat treatment to internal temperatures of 43.5°C–45.5°C, greatly increased treatment efficacy while blocking the ability of the insects to acclimate to the heat load (Figure 11.1).

11.4.4 Coleoptera

Adult sweetpotato weevils (*Cylas formicarius elegantulus* (Summers)) and sweet potato roots (*Ipomoea batatas* L. (Lam.)), infested with immature stages of the sweetpotato weevil were exposed to controlled atmospheres containing low O_2 and increased concentrations of CO_2 with a balance of N_2 for up to 10 days at 25°C and 30°C (Delate et al., 1990). Adult sweetpotato weevils were killed within 4–8 days when exposed to 8 kPa O_2 plus 40–60 kPa CO_2 at 30°C. At 25°C, exposure to 2 or 4 kPa O_2 plus 40 or 60 kPa CO_2 at 25°C killed all the adult insects within 2–8 days. Exposure of sweet potato roots infested with

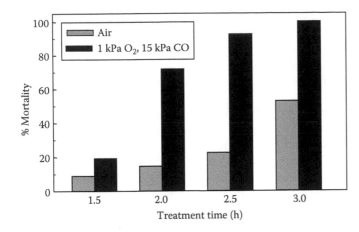

FIGURE 11.1
Percent corrected mortality of fourth instar codling moth in apples using a heating rate of 12°C/h to a final chamber temperature of 46°C under air (▨) and CA (■) conditions (1 kPa O_2, 15 kPa CO_2). (From Neven, L.G. and Rehfield-Ray, L.M., *J. Econ. Entomol.*, 99, 658, 2006a. With permission.)

weevils to 8 kPa O_2 plus 30–60 kPa CO_2 for 1 week at 30°C failed to kill all the sweetpotato weevils. However, no adult sweetpotato weevils emerged from infested roots treated with 4 kPa O_2 plus 60 kPa CO_2 or 2 kPa O_2 plus 40 or 60 kPa CO_2 for 1 week at 25°C.

11.4.5 Hemiptera

Mealybugs and scale insects are the main groups of hemipterans requiring phytosanitary disinfestation measures because they are frequently found on citrus and pome fruits and are not completely removed or killed during the washing process. Other groups in this insect order that are often quarantine pests are aphids (Aphididae), whiteflies (Aleyrodidae), seed bugs (Lygaeidae), and leafhoppers (Cicadellidae).

Hard scales (Diaspididae) are considered to be difficult to kill with modified atmospheres relative to other insect groups (Carpenter and Potter, 1994). Gaunce et al. (1982) recorded complete mortality of San Jose scale (*Quadraspidiotus perniciosus*), a phytosanitary pest of apples and other pome fruits, in a modified atmosphere of <1 kPa O_2 and >90 kPa CO_2 at 12°C after 2 days but only 50% mortality at the commercial storage temperature of 1°C after 5 days.

'Red Delicious' apples (*Malus sylvestris domestica* (Mill.)) infested with San Jose scale (SJS) were subjected to cobalt-60 gamma radiation at 0–300 Gy and then stored in either regular cold storage (−0.5°C), controlled-atmosphere storage 1.5 kPa O_2, 1.5 kPa CO_2 (−0.5°C), or at room temperature, and scale survival measured at various intervals after treatment declined with time and the magnitude of the radiation dose received, but was not influenced by method of storage (Angerilli and Fitzgibbon, 1990).

Storage conditions of regular MA (2.6–3.0 kPa O_2 and 2.4–2.5 kPa CO_2) and low-oxygen MA (1.5–1.7 kPa O_2 and 1.0–1.1 kPa CO_2) at 1°C or 3°C and air at 1°C were used to control postharvest infestation of SJS on 'McIntosh' and 'Delicious' apples (Chu, 1992). Apples were sampled periodically from storage rooms to evaluate the mortality of SJS. When examined after 6–18 weeks of storage for scale on 'McIntosh' apples and 8–21 weeks for scale on 'Delicious' apples, storage temperature, and atmosphere had noticeable effects on the reduction of live scales. The differences in mortality rate caused by various storage conditions diminished as the storage period was extended to 24 weeks. Complete elimination of SJS was achieved by storing infested apples in regular MA or low-oxygen MA at 1°C or 3°C for 31–34 weeks plus an additional week at 20°C and 50%–60% RH.

Contrary to most other results on the relationship between modified atmosphere and temperature, mortality of the longtailed mealybug (*Pseudococcus longispinosus*) decreased with an increase in temperature from 0°C to 20°C in atmospheres containing 0, 9, or 18 kPa

CO_2 plus 2 kPa O_2 with the balance nitrogen (Potter et al., 1990). Complete mortality after 2 weeks was achieved only at 0°C, 18 kPa CO_2 while 18 kPa CO_2 at 20°C gave 52% mortality. This result may be due to cold overriding the effect of the modified atmosphere. Under the same atmospheres and temperatures the New Zealand wheat bug (*Nysius hutton*) was easier to kill; complete mortality was achieved with 0°C, 18 kPa CO_2 and 20°C, 9 and 18 kPa CO_2 after 2 weeks.

Complete mortality of the obscure mealybug (*Pseudococcus affinis*) from apples could be achieved in a number of hours combining low oxygen with heat (Whiting and Hoy, 1997). As the treatment temperature approached 45°C, differences in O_2 concentration had less effect on time needed. For example, at 40°C the time to achieve 99% kill of adult females at 5 kPa O_2 (38 h) was almost three times that required at 1 kPa O_2 (14 h) while at 45°C there was essentially no difference in time required (6 h). In this case, it is possible that the lethal effect of the heat at 45°C was the chief cause of mortality and that whether the oxygen level was 1 or 5 kPa was not the critical factor.

11.4.6 Thysanoptera

Western flower thrips (*Frankliniella occidentalis*) is one of the most important quarantine thrips for trade in cut flowers and fruits. Controlled atmospheres of low O_2 and high CO_2 have been used with some success against this species on strawberry fruits although some off-flavors developed (Aharoni et al., 1981). An atmosphere that might work on a commercial scale to disinfest strawberries of the thrips is 2 days with 2 kPa O_2 and 90 kPa CO_2 at 2.5°C.

Potter et al. (1994) identified several combinations of CO_2, temperature, and time (O_2 was fixed at 2kPa) that provided complete mortality of a total of 300 thrips per treatment combination of *Thrips obscuratus* adults. Among all combinations of five temperatures (0°C–20°C) and three CO_2 levels (0–18 kPa) used the following achieved compete mortality: 15°C, 18 kPa CO_2 in 4 days; 10–20°C, 18 kPa CO_2 in 6 days; 0°C or 20°C, 9 kPa CO_2 in 6 days; all temperatures at 18 kPa CO_2 in 8 days, and 0–5 and 20°C, 9 kPa CO_2 in 8 days. Large-scale confirmatory testing would be required before any of these combinations could be used commercially.

Page et al. (2002) achieved 100% control of *Thrips tabaci* in onions at CO_2 levels ≥ 30 kPa (balance air) after 24 h at 20°C. However, the control thrips suffered high mortality, reaching >70% in 12 h, casting doubt onto the efficacy of the modified atmosphere system in achieving high levels of control by itself.

11.4.7 Psocidae

The psocid *Liposcelis bostrychophila* is a common insect pest, which infests a variety of processed and unprocessed dry foods in households, granaries, and warehouses (Wei et al., 2002). Psocids are highly resistant to most forms of pest control, and their populations in stored foods have increased alarmingly in some regions of Asia. Alternating MA and insecticide (DDVP; a dichlorvos formulation) applications were used by Wei et al. (2002) to reduce population growth and development of resistance of psocids. A stock psocid colony was established and maintained at 28°C, 80% RH; after 11 weeks, the population of this colony increased 48.1-fold. Subsequent exposure of divided colonies to either MA (35 kPa CO_2, 1 kPa O_2) or DDVP (0.3 mg/mL) failed to control population growth in the colony. However, alternating MA and DDVP treatments resulted in a significant increase in psocid mortality compared with these treatments alone. After six exposures, resistance to MA and DDVP treatments increased 1.8- and 2-fold, respectively; probit analysis of data suggested that potential for further increased resistance existed with

MA or DDVP alone. Results of this study have indicated that alternating MA with insecticide applications could be a more effective management measure for control of psocids in stored foods than use of these treatments alone.

11.4.8 Aphidae

Effects of CO_2 treatment on mortality of green peach aphids (*Myzus persicae*) were studied by Epenhuijsen et al. (2002). Experiments were designed to determine the effects of variation in CO_2 concentration from 0–95 kPa on green peach aphid mortality, with fixed time and temperature. Experimental conditions were selected to give little or no mortality at 0 kPa CO_2 and 100% mortality at 95 kPa CO_2; this design was selected to facilitate development of a mathematical model that could be used to design the CO_2 component of a controlled atmosphere treatment without extensive empirical testing. With CO_2 concentration in the range 20–60 kPa, 30% of treated aphids were moribund 2 days after treatment; all moribund aphids died within the following 10 days. Models were fitted to curves for mortality due to CO_2 concentration. Mortality curves deviated from a standard dose–response after 2 days; this was attributed in part to the occurrence of moribund aphids. At CO_2 concentration greater than 40 kPa, mortalities of greater than 90% were recorded 12 days after treatment, indicating the presence of a physiological threshold at that point. It is considered that the occurrence of high numbers of moribund aphids resulting from CO_2 treatment poses challenges to standard quarantine inspections, even though all aphids are likely to die within 12 days as a result of the treatment. It is proposed that optimization of temperature and exposure time with respect to pest mortality and stored commodity quality is required.

Carpenter (1995) investigated several modified atmosphere/temperature combinations against green peach aphid (*Myzus persicae*) on asparagus and found that cold treatment alone (0°C–2°C in air for 4 days) was best for efficacy against the aphid and asparagus quality.

11.4.9 Acari (Mites and Their Kin)

Gaunce et al. (1982) found increased mortality of European red mite (*Panonychus ulmi*) and McDaniel spider mite (*Tetranychus mcdanieli*) on apples stored in modified atmosphere versus air cold storage over periods of 2–5 months. However, mortality was not complete after 5 months.

Whiting and van den Heuvel (1995) studied modified atmospheres against diapausing two-spotted spider mites (*Tetranychus urticae*). Reducing the O_2 concentration from 1.3 to 0.4 kPa reduced the LT_{99} by 50%. Increasing the CO_2 concentration from 5 to 20 kPa had a similar effect. Temperature had a great effect on mortality reducing the LT_{99} from 112 to 15 h as the temperature increased from 20°C to 40°C.

Mites are among the most resistant type of arthropod to MA at low (<5°C) temperatures. Mitcham et al. (1997b) found that the LT_{99} for the most resistant life stages of Pacific spider mite (*Tetranychus pacificus*) to an exposure to 45 kPa CO_2 at 0°C was 8.1 days. This was considerably higher than those recorded for western flower thrips and omnivorous leaf-roller (*Platynota stultana*), of 5.8 and 4.1 days, respectively (Mitcham et al., 1997b).

11.5 Insect Quarantine Systems Using MA

Low temperatures (0°C–2.2°C for 10–16 days) can be used for control of the Mediterranean fruit fly. However, these temperatures cannot be used for many tropical fruits because of their chilling sensitivity. Hot water treatments have been used in several countries to control fruit flies in mangoes (46.1°C for 65–90 min) and papayas (two-stage heating process with temperatures of 42°C for 30 min and 49°C for 20 min). Vapor heat treatments have been developed and used. Injury has been reported in both mango and papaya fruits treated with heat (Stewart et al., 2005; Yahia, 2006a,b). Irradiation has proved potentially applicable for insect control in some tropical fruits such as mango and papaya, however, commercial application is still very limited due to several problems including possible injury to the fruit, high costs, and consumer concerns. MA ($\geq$1.0 kPa O_2 and/or $\leq$50 kPa CO_2) have insecticidal and fungicidal effects, and have been developed as quarantine treatments (Neven, 2005; Neven and Rehfield-Ray, 2006a,b; Neven et al., 2006; Stewart et al., 2005; Yahia, 1998a, 2006a,b).

The control of insects by MA has commercially been tried for grains and for dried fruits and vegetables (Banks, 1984), but not yet fully developed for fresh fruits and vegetables. Atmospheres needed for this application should contain very low levels of oxygen (less than 1 kPa) or/and very high levels of CO_2 (up to 50–80 kPa). These insecticidal atmospheres can eliminate insects within a period of 2–4 days at room temperature (Ke et al., 1995; Yahia, 1998b, 2006a; Yahia and Carrillo-López, 1993; Yahia and Tiznado-Hernandez, 1993; Yahia and Vazquez-Moreno, 1993), and within a period of only 2–4 h at higher temperatures (Neven, 2004, 2005, 2008; Neven and Mitcham, 1996; Neven and Rehfield-Ray, 2006a,b; Neven et al., 2006; Ortega-Zaleta and Yahia, 2000a; Yahia and Ortega-Zaleta, 2000a,b). Although these treatments are not in commercial use now, commercial units have been developed and are currently being tested (Figures 11.2 and 11.3) (Neven, 2008; Neven et al., 2006). Potential problems that hinder the possibility for developing IMA for fresh fruits and vegetables include possible fermentation due to the use of anaerobic gas mixtures, and heat injury. The advantages of using this system at room temperature include the low energy input and the avoidance of heat injury. However, the advantage of using it at high temperatures is to accomplish the mortality of insects in a very short period, and thus eliminate the accumulation of products before treatments. In addition, IMA can be used either in permanent stores or in transport marine containers.

FIGURE 11.2
Commercial CATTS chamber at the USDA-ARS laboratory in Parlier, CA. Chamber is designed to treat two pallets (2 tons) of commercially packed peaches and nectarines. The chamber controls heating rate, reversible air flow, humidity, dew point, O_2 and CO_2 levels.

FIGURE 11.3
Commercial CATTS chamber at a packing house in George, WA. This chamber is designed to treat 4 commercial bins of fruit (2 tons). The chamber controls heating rate, reversible air flow, humidity, dewpoint, O_2 and CO_2 levels.

11.6 Effects of Insecticidal MA/CA on Horticultural Commodities

MA atmospheres that are insecticidal commonly consist of very low O_2 and/or very high CO_2 levels. Some horticultural commodities may tolerate these extreme atmospheres for short periods of time, especially at low temperatures, but many commodities do not tolerate them for long duration, especially at high temperatures. Extreme (IMA) atmospheres can inhibit aerobic respiration and promote the initiation of anaerobic respiration, and thus can lead to tissue injury and fermentation. Table 11.4 summarizes the tolerance of different horticultural commodities to different atmospheres.

'Hass' avocado was found to be very sensitive to IMA, tolerating no more than 1 day in a MA consisting of 0.1 kPa O_2 and up to 88 kPa CO_2 at 20°C, after which fermentation and decay were obvious upon fruit ripening (Yahia, 1994a,b, 1997a,b; Yahia and Carrillo-López, 1993; Carrillo-López and Yahia, 1990). Guava was also evaluated and shown to be sensitive to IMA (Yahia, 1997b; Yahia and Balderas, 2000; Yahia and Paull, 1997; Yahia et al., 2000a). Storage of 'Hass' avocado fruit in IMA consisting of 0.25 kPa O_2 (balance N_2) and 0.25 kPa $O_2 + 80$ kPa CO_2 (balance N_2) at 20°C for 3 days causes the accumulation of acetaldehyde and ethanol and increased NADH concentration, but decreased NAD level (Ke et al., 1995). Lactate concentration increased in the low O_2 atmosphere but not in the combination of low O_2 and high CO_2 atmosphere. Activities of pyruvate decarboxylase (PDC) and lactate dehydrogenase (LDH) were slightly enhanced and a new isozyme of alcohol dehydrogenase (ADH) was induced by 0.25 kPa O_2, 20 kPa $O_2 + 80$ kPa CO_2, or 0.25 kPa $O_2 + 80$ kPa CO_2.

Storage of 'Bartlett' pears in IMA consisting of 0.25 kPa O_2, 20 kPa $O_2 + 80$ kPa CO_2 or 0.25 kPa $O_2 + 80$ kPa CO_2 from 1, 2, or 3 days at 20°C increased the accumulation of acetaldehyde, ethanol, and ethyl alcohol (Ke et al., 1994a). Postclimacteric pears had higher

TABLE 11.4

Tolerance of Some Horticultural Commodities to Different Atmospheres

Commodity	Temperature (°C)	Atmosphere (kPa) O₂	Atmosphere (kPa) CO₂	Tolerance (days)	Reference
Apples	0	0.02		12	Chen et al. (1985)
	0	0.25		23	Chu (1992)
	0	0.5		100	Couey and Olsen (1977)
	5	0.02		9	Little et al. (1982)
	5	0.25		11	Morgan and Gaunce (1975)
Apricot	0	0.3		12	Folchi et al. (1995)
	6	0.3		5	Botondi et al. (2000)
Avocado	20	0.25	80	3	Ke et al. (1995)
	20	0.1–0.4	50–75	1	Yahia and Carrillo-López (1993)
Blueberry	5	0		14–21	Prange and Lister (1992)
	21			2	Smittle and Miller (1988)
Broccoli	0	0.5	54	21	Zheng et al. (1993)
Carnation	5		30–50	6	Zheng et al. (1993)
	0		60	3	Irving and Honnor (1994)
	5		60	1	Shelton et al. (1997)
Cherry	0	0.02		25	Chen et al. (1981)
	0	0.25		35–44	Folchi et al. 1995
	5	0.02		21	Patterson (1982)
	5	0.25		38	Patterson and Melsted (1977)
	6	0.3		18	Patterson and Melsted (1977)
	0		60	7	Patterson and Melsted (1977)
	0		80–100	4–7	Patterson and Melsted (1977)
	47	1		25 min	Shellie et al. (2000)
Chrysanthemum	0.5	0.5		4	Shelton et al. (1997)
Dates	22–28	4–7	60–85	145	Navarro et al. (1998)
Grapes	5	0.5	35–45	6	Ahumada et al. (1996)
Grapefruit	10	0.05	0	21	Shellie et al. (1997a)
	46	1	20	0.13	Shellie et al. (1997b)
Kiwifruit	40	0.4	20	0.3	Lay-Yee and Whiting (1996)
Lettuce	0	0.5		21	Zheng et al. (1993)
	5	0.02		11	Zheng et al. (1993)
	0		30–50	5	Zheng et al. (1993)
	5		30–50	3–4	Zheng et al. (1993)
Mango	20	0.2–0.3	50	5	Yahia (1993, 1994b)
	20	2	70–80	5	Yahia et al. (1989)
	20	0.5		4	Yahia and Tiznado-Hernandez (1993)
	43	0	50	160 min	Yahia and Ortega-Zaleta (1999)
	48	0.8	67	160 min	Yahia et al. (1997)
	12	1	30–50	3	Leon et al. (1997)
Nectarine	0	0.02		14	Folchi et al. (1995)
	5	0.5		3–6	Ke et al. (1994b)
	6	0.3		30	Mitchell et al. (1984)
	10	0.02		10	Smilanick and Fouse (1989)
	10	0.25		13	Soderstrom et al. (1987)
	0	4.2	80	10	Soderstrom et al. (1987)
	0	1	99	6	Soderstrom et al. (1987)
Orange	5	0.02		16	Ke and Kader (1989, 1990)
	5	0.25		23	Ke and Kader (1989, 1990)
	10	0.02		15	Ke and Kader (1989, 1990)
	5	8.4	60	5	Ke and Kader (1989, 1990)
Orchid	0.5	0.1		2	Shelton et al. (1997)

TABLE 11.4 (continued)

Tolerance of Some Horticultural Commodities to Different Atmospheres

Commodity	Temperature (°C)	Atmosphere (kPa) O_2	CO_2	Tolerance (days)	Reference
Papaya	20	0.2–0.4		2–3	Yahia (1995a) Yahia et al. (1989, 1992)
Peach	0	0.02		40	Carpenter et al. (1992)
	0	0.25		>40	Ke et al. (1991c)
	5	0.02		14	Ke et al. (1994b)
	5	0.25	60	14	Mitchell et al. (1984)
	0	21		19	Mitchell et al. (1984)
			99		
	0	0.21		5	Mitchell et al. (1984)
	20	0.25	99	3–5	Mitchell et al. (1984)
	20	0.21		4	Mitchell et al. (1984)
Pear	0	0.02		>10	Chen and Mellenthin (1992)
Pear, Asian	0	0.02		14	Ke et al. (1991b)
	0	0.25		14	Zagory et al. (1989)
	5	0.02		11	Zagory et al. (1989)
	5	0.25		13	Zagory et al. (1989)
Persimmon	20	0–4	80–100	2–5	Mitcham et al. (1997a)
	20	0.25	40	7	Pesis and Ben-Arie (1986)
Plum	0	8	60	7	Carpenter et al. (1992)
	5	0.02		32	Folchi et al. (1995)
	5	0.25		41	Ke et al. (1991b)
	6	0.3		15	Mitchell et al. (1984)
	10	0.02		9	Mitchell et al. (1984)
	10	0.25		14	Mitchell et al. (1984)
			99		
	0	1		5	Mitchell et al. (1984)
	5	4.2	80	6	Mitchell et al. (1984)
Rose	0	0.5		>21	Zheng et al. (1993)
	5	0.02		7–14	Zheng et al. (1993)
	5	0.5		>21	Zheng et al. (1993)
	0		30–50	2–4	Zheng et al. (1993)
	5		30–50	1–2	Zheng et al. (1993)
	0.5		60	1	Shelton et al. (1997)
Strawberry	0	0.02		6	Ke et al. (1991a)
	0	0.25		10	Ke et al. (1991a)
	5	0.25		8	Ke et al. (1991a)
	0–5	4.2–10	50–80	8	Ke et al. (1991a)
Sweet potato	25	2	40	7–10	Delate and Brecht (1989)

Source: Adapted from information compiled by T. L. A. Martin, S. Zhou, and A. A. Kader, University of California, Davis, CA.

activities of PDC and higher concentrations of fermentative volatiles than those of the preclimacteric fruit. Low O_2 (0.25 kPa) atmosphere dramatically increased ADH activity, which was largely due to the enhancement of one ADH isozyme. The authors have suggested that ethanolic fermentation in 'Bartlett' pears could be induced by low O_2 and/or high CO_2, via: (1) increased amount of PDC and ADH, (2) PDC and ADH activation caused by decreased cytoplasmic pH, or (3) ADH activation or more rapid fermentation due to increased concentrations of their substrates (pyruvate, acetaldehyde, or NADH).

Storage of several California-grown nectarine cultivars in 0.5 kPa O_2 (balance N_2) at 15°C for 3 days (for their utility in insect control) caused abnormal ripening, suppressed ethylene production, elevated ethanol and acetaldehyde contents, and increased internal browning (Smilanick and Fouse, 1989). Treatment in 0.5 kPa O_2 for 6 days at 5°C caused similar injuries, but these were less severe than those at 15°C. Presoftening in air or posttreatment storage for 2 weeks at 2.5°C did not reduce the injuries induced by the 0.5 kPa O_2 treatment. Since these fruit were damaged after only 3 days' treatment at 15°C, the atmospheres used to control insects probably cannot be tolerated by nectarines. Increased alcohol content and reduced ethylene production were the most sensitive indicators of MA-induced changes.

Sweet potatoes (*Ipomoea batatas* L.) were exposed to low O_2 and high CO_2 for 1 week during curing or subsequent storage to evaluate the use of MA as insecticidal treatments for control of sweetpotato weevil (*Cylas formicarius elegantulus*) (Delate and Brecht, 1989). Sweet potato roots tolerated 8 kPa O_2 during curing, but, when exposed to 2 kPa or 4 kPa O_2 or to 60 kPa CO_2 plus 21 kPa or 8 kPa O_2, they were unsaleable within 1 week after curing, mainly due to decay. Exposure of cured sweet potatoes to 2 kPa or 4 kPa O_2 plus 40 kPa CO_2 or 4 kPa O_2 plus 60 kPa CO_2 for 1 week at 25°C had little effect on postharvest quality. However, exposure to 2 kPa O_2 plus 60 kPa CO_2 resulted in increased decay, less sweet potato flavor, and more off-flavor. These results indicate that exposure of sweet potatoes to O_2 and CO_2 levels required for insect control is not feasible during curing, but that cured sweet potatoes are capable of tolerating MA treatments that have potential as quarantine procedures.

'Sunrise' papaya fruit exposed to a continuous flow of an atmosphere containing <0.4kPa O_2 (balance N_2) for 0 to 5 days at 20°C had decay and some fruit had developed off-flavors after 3 days in low O_2 plus 3 days in air at 20°C (Yahia et al., 1992). The intolerance of papaya fruit to low O_2 correlates with an increase in the activity of PDC and LDH, but not with the activity of ADH. Therefore, the authors suggested that insecticidal low O_2 (<0.4 kPa) atmospheres can be used as a quarantine insect control treatment in 'Sunrise' papaya for periods <3 days at 20°C without the risk of significant fruit injury.

The tolerance of 'Fuyu' persimmons to IMA was investigated by Mitcham et al. (1997a). Freshly harvested 'Fuyu' persimmon fruit were exposed to air, 0.25 kPa O_2, 40 kPa CO_2 in air, and 0.25 kPa $O_2 + 40$ kPa CO_2 at 20°C for 3, 5, and 7 days. Fruit treated with IMA exhibited slight external injury and fruit treated with 40 kPa CO_2, regardless of the O_2 concentration, were softer than fruit treated with air or 0.25 kPa O_2 only 5 days after removal from the treatment. Ethanol and acetaldehyde concentrations were higher in treated fruit compared with air controls upon removal from IMA, with the highest levels in persimmons exposed to 0.25 kPa $O_2 + 40$ kPa CO_2. Treatment with 40 kPa CO_2 resulted in higher levels of ethanol and acetaldehyde in the fruit than treatment with 0.25 kPa O_2 (with 0.03 kPa CO_2 and the rest being N_2). Ethanol and acetaldehyde concentrations decreased 30% and 20%, respectively, after 5 days storage in air at 10°C following IMA. Informal sensory evaluations indicated no loss in quality as a result of the IMA treatments. Therefore, these authors suggested the potential of IMA for developing quarantine treatments for insect quarantine in 'Fuyo' persimmons.

Non-SO_2-fumigated 'Thompson Seedless' table grapes were stored at 5°C or 20°C for 6 and 4.5 days, respectively, in air or one of four IMA: 0.5 kPa $O_2 + 35$ kPa CO_2; 0.5 kPa $O_2 + 45$ kPa CO_2; 0.5 kPa $O_2 + 55$ kPa CO_2; or 100 kPa CO_2, and fruits were evaluated for weight loss, berry firmness, soluble solids concentration (SSC), titratable acidity, berry shattering, rachis browning, berry browning, and anaerobic volatiles (Ahumada et al., 1996). Fruit quality was not affected at 5°C with the exception of greater rachis browning in fruit treated with 0.5 kPa $O_2 + 45$ kPa CO_2. At 20°C, IMA treatments maintained greener rachis compared to the air control; however, SSC was reduced in the fruit treated with 55 kPa and 100 kPa CO_2. At both temperatures, IMA induced the production of high levels of acetaldehyde and ethanol. Ethanol concentrations were two-thirds lower at 5°C than at

20°C. Consumer preference was negatively affected by some IMA treatments for grapes kept at 20°C, but not by any of the treatments at 5°C.

'Keitt' Mango was found to be very tolerant to IMA containing as low as <0.2 kPa and as high as 80 kPa CO_2 for up to 5 days at 20°C without any type of injury upon ripening, although fermentative odors could be noted while the fruit is under stress (Yahia, 1993, 1994a, 1995a, 1997d, 2004; Yahia and Ortega-Zaleta, 2000c). Some other mango cultivars were later evaluated and shown to be also tolerant (Yahia, 1998a).

One of the major concerns related to developing the commercial application of IMA is the duration of the treatment. Previously developed and used chemical quarantine systems, and currently used hot water treatments are commonly done for a few (2–4) h, and therefore developing the IMA system for several days (at 20°C or less) would not be accepted nor practical for commercial application. One efficient method to significantly reduce the duration of the treatment is to increase the temperature, although it would add another (heat) type of stress.

Several fruits and vegetables have been tested for their tolerance/sensitivity to heat stress, and mango (being tolerant to gas stress) was found to be also tolerant to heat, while other fruits that were found to be sensitive to gas stress (such as guava, avocado, and pears) have also been found to be sensitive to heat stress (Yahia, 2004, 1997b; Yahia and Ariza, 2001; Yahia et al., 2000a). IMA (≤ 0.5 kPa $O_2 + \geq 50$ kPa CO_2) at different temperatures (from 20°C to 55°C) and RH was tested on the sensitivity/tolerance of several fruits including different cultivars of mango, avocado, guava and papaya. Only mango fruit were found to be sufficiently tolerant to both types of stress, and therefore this technique was thought to have potential to be developed commercially for this fruit. IMA at 43°C and 50% RH for 160 min achieved insect mortality in mango (probit 9) (Yahia, 1998a; Yahia and Ortega-Zaleta, 2000a,b; Yahia et al., 2000b), without causing negative effects on the fruit (Ortega-Zaleta and Yahia, 2000a; Yahia, 1995b; Yahia et al., 2004).

MA containing 1 kPa O_2 and 10–15 kPa CO_2 at 25°C, 30°C and 35°C for 24 to 72 h caused 100% mortality of *Cadra cautella* in dried figs, without causing negative effects on the fruit (Damarh et al., 1998).

Apple fruit quality was evaluated after a potential quarantine treatment and storage in MA (Neven and Drake, 2000; Warner, 1998). The fruit was heat-treated after harvest at 44°C or 46°C, placed in regular MA storage for 90 days, and then assessed for quality. 'Red Delicious' apples treated at certain temperature were redder than untreated fruit, while 'Granny Smith' and 'Golden Delicious' apples maintained their color. The balance of acids and soluble solids was better, indicating that treated fruit should have a better flavor. There was a large reduction in storage scald and in decay organisms and combining the treatment with MA of 1 kPa O_2 and 15 kPa CO_2 could considerably reduce treatment time. Good results were obtained with 'Gala' apples and some problems appeared on 'Fuji' apples.

'Royal Gala' apples (*Malus sylvestris domestica* Bork.) subjected to a high-temperature MA (HTCA) consisting of 2 kPa O_2 and 5 kPa CO_2 at 40°C or hot air (HA) at 40°C for 6 or 8 h, were either cooled by immersion in water at ambient temperature, or not cooled in water, and then exposed to cold storage (CS) at 0°C for 7 weeks, before being transferred to 20°C for 7 days and assessed for quality (Whiting et al., 1999). No significant damage was observed in control, HTCA- or HA-treated apples not cooled in water, though a very low incidence of cavitation and breakdown was observed in water-cooled apples.

Neven and Drake (1998) described quarantine treatments for sweet cherries as an alternative to methyl bromide treatments to meet export quarantine restrictions, including irradiation, cold treatments, high temperature, heat plus MA, and microwaves. Cherries were reported to withstand an atmosphere of 0.5 kPa O_2 and a CO_2 level as high as 15 kPa, at a temperature of up to 47°C, for long enough to kill the larvae codling moth, but with some problems of stem browning and a small amount of pitting. Two MA treatments (45 kPa CO_2 + 0.5 kPa O_2 and 45 kPa CO_2 + 11.5 kPa O_2) at 0°C for 10 or 15 days, which significantly increased mortality of omnivorous leafroller (*Platynota stultana* Walsingham); Pacific spider mites (*Tetranychus pacificus* McGregor); and western flower thrips (*Frankliniella occidentalis* Pergande) had no effects on table grapes quality (Mitcham et al., 1997b).

Yellow- and white-fleshed peach and nectarine (*Prunus* spp.) of mid- and late-season maturity classes were subjected to combined MA-temperature treatments using heating rates of either 12°C/h (slow rate) or 24°C/h (fast rate) with a final chamber temperature of 46°C, while maintaining a MA of 1 kPa O_2 and 15 kPa CO_2 (Obenland et al., 2005). Fruit seed surface temperatures generally reached 45°C within 160 min and 135 min for the slow and fast heating rate, respectively. The total duration of the slow heating rate treatment was 3 h, while 2.5 h was required for the fast heating rate treatment. Following treatment the fruit were stored at 1°C for 1, 2, or 3 weeks followed by a ripening period of 2–4 days at 23°C and subsequent evaluation of fruit quality. Fruit quality was similar for both heating rate treatments. Compared with the untreated fruit treated ones displayed higher surface injury, although increased injury was only an important factor to marketability in cultivars that had high amounts of surface injury before treatment. The percentage of free juice in the flesh was slightly less in treated fruit early in storage but was often greater in treated fruit toward the end of the storage period. Slower rates of softening during fruit ripening were apparent in treated fruit. Soluble solids, acidity, weight loss and color were either not affected or changed to a very small degree as a result of the treatment. A sensory panel preferred the taste of untreated fruit over treated fruit but the ratings of treated and nontreated fruit were generally similar and it is unclear whether an average consumer could detect the difference.

The response of 'Hayward' kiwifruit to high-temperature MA treatments for control of two-spotted spider mite (*Tetranychus urticae* Koch) was investigated by Lay-Yee and Whiting (1996). 'Hayward' kiwifruit were subjected to 40°C for 7 or 10 h in 20 kPa CO_2 (treatments identified as giving 100% mortality for nondiapausing and diapausing two-spotted spider mites, respectively) or in air, and following treatment, fruit were cooled in ambient water or ambient air, stored at 0°C in air for 8 weeks, then held at 20°C overnight and assessed for quality. Relative to nontreated controls, no significant damage was observed with fruit subjected to 40°C air treatments. No significant damage was observed with fruit treated for 7 h with 20 kPa CO_2 followed by hydrocooling. However, the treatment without hydrocooling and 10 h treatments with hydrocooling showed only slight damage, while the 10 h MA without hydrocooling had moderate fruit damage. Following storage, flesh firmness of fruit treated at 40°C in air for 10 h or in 20 kPa CO_2 for 7 and 10 h, with and without hydrocooling, was lower than that of nontreated controls.

Quarantine treatments, using MA at high temperatures, were also tried to control codling moth in sweet cherries (Neven, 2005), using treatments at 45°C for 45 min and 47°C for 25 min at 1 kPa O_2, 15 kPa CO_2, −2°C dew point environment. These treatments have been shown to provide control of all life stages of codling moth while preserving commodity market quality. 'Bing' sweet cherries exposed to 45°C or 47°C in 1 kPa O_2 with 15 kPa CO_2 (balance nitrogen) were heated to a maximum center temperature of 44°C or 46°C in 41 or 27 min, respectively, had similar incidence of pitting and decay, and similar preference ratings after 14 days of storage at 1°C as nonheated or MeBr fumigated fruit (Shellie et al., 2001). Heated cherries and MeBr fumigated cherries were less firm after 14 days of cold storage than nonheated, control fruit. The stems of MeBr fumigated cherries were less green than heated or nonheated cherries. Cherries exposed to 45°C had lower titratable acidity than nonheated cherries, fumigated cherries, or cherries exposed to 47°C.

Cherry quality after 14 days of cold storage was not affected by hydrocooling before heating (5 min in water at 1°C) or by method of cooling after heating (hydrocooling, forced air cooling, or static air cooling). Cherries stored for 14 days at 1°C in 6 kPa O_2 with 17 kPa CO_2 (balance nitrogen) had similar market quality as cherries stored in air at 1°C. This work has suggested that 'Bing' sweet cherry can tolerate heating in an atmosphere of low O_2 containing elevated CO_2 at doses that may provide quarantine security against codling moth (*Cydia pomonella*) and western cherry fruit fly (*Rhagoletis indifferens*).

Traditionally, heat treatments have been conducted to reach the target temperature as quickly as possible (Armstrong, 1994; Hallman and Armstrong, 1994; Mangan and Ingle, 1994; Mangan et al., 1998; Sharp, 1994; Shellie and Mangan, 1994; Shellie et al., 1993, 1996). This strategy may work effectively for some tropical and subtropical fruits, but not for temperate tree fruits (Neven et al., 1996). Studies on apples and pears indicated the rate of heating directly impacted fruit quality (Neven and Drake, 2000). By controlling the rate of heating, the fruit can compensate for the heat load. Some studies have indicated that the slower the rate of heating, and the lower the final treatment temperature, the longer the total treatment needs to control the insect pest (Neven, 2000; Neven and Rehfield, 1995; Neven et al., 1996).

It is possible that the sensitivity/tolerance of fruits and vegetables to gas and heat stress is controlled genetically, and therefore the genetic manipulation of sensitive commodities to increase their resistance/tolerance to these types of stress can probably help in the establishment of the commercial application of this system for several fruits and vegetables. A possible mechanism for the injury of sensitive fruits by gas stress has been suggested (Ke et al., 1994a, 1995; Swan and Watson, 1998; Yahia, 1998a; Yahia and Balderas, 2000; Yahia and Carrillo-López, 1993; Yahia et al., 2000a). Exposure of sensitive fruits to insecticidal gas stress reduces cellular pH, which inhibits aerobic enzymes such as PDH and activates (and possibly synthesizes) anaerobic enzymes such as PDC, ADH, and LDH. The production of fermentative metabolites (such as lactate, acetaldehyde, and ethanol) is increased significantly, and energy production is decreased significantly. The accumulation of fermentative metabolites also increases the activity of some of these fermentative (allosteric) enzymes, which further increases the accumulation of toxic fermentative metabolites. It is possible that the sensitivity of some fruits to heat stress is due to a lack or reduced capacity in the synthesis of heat-shock proteins (Yahia et al., unpublished). On the other hand, it is possible that the tolerance of mango fruit to heat stress is due to its capacity for the synthesis of such proteins, some of which can provide protection against this stress. An accelerated protein synthesis in mango (tolerant to heat stress) exposed to heat treatment, compared to guava (sensitive to heat stress) has been demonstrated (Yahia et al., unpublished data). Some sugars, such as trehalose, are associated with the resistance of some microorganisms to some environmental stresses (Swan and Watson, 1998), but some reports have been skeptical about the presence of trehalose in plant tissues (Mueller et al., 1995). It is possible that resistance to heat stress is related to the presence of trehalose, and sensitivity to this stress is due to its absence or reduced levels. Trehalose is metabolized by trehalase. The presence of trehalose and trehalase in fruits and vegetables has been detected and correlated with the tolerance/sensitivity of fruits and vegetables to heat stress (Yahia and Balderas, 2000; Yahia et al., 2000a). In general, fruits and vegetables contain very low concentrations of trehalose and very low activity of trehalase, and in many cases it was difficult to detect them. Some of the fruits and vegetables that are sensitive to heat stress usually contain lower concentration or no trehalose, and relatively higher activity of trehalase compared to tolerant fruits and vegetables. It is possible that some of the proteins synthesized by heat-tolerant fruits as a response to heat stress are responsible for the production of protecting agent(s), which might be in the form of sugars such as trehalose. It is also possible that fruit and vegetables with relatively higher activity of trehalase would not accumulate sufficient amounts of trehalose and therefore that would render them sensitive to stress. The understanding of the mechanism behind the sensitivity/tolerance of different fruits and vegetables to gas and to heat stress can help in the manipulation of sensitive fruits, and thus increasing their tolerance and allowing for the establishment of these techniques on a commercial basis.

11.7 Conclusions

Quarantine treatments are needed to facilitate trade of horticultural commodities. The importance of the development of new treatments has increased significantly in recent years because of the increase in the trade of fresh horticultural crops and the restrictions in the use of chemical fumigants. Major interest and significant research activities have been exerted toward developing physical (nonchemical) treatments and systems, because of interest in reducing health and environmental problems. Modified atmospheres may kill insects and have been extensively researched as phytosanitary treatments against quarantine pests. Atmospheres that can control insects efficiently are commonly composed of higher CO_2 (>10 kPa) and lower O_2 ($\leq$1 kPa) levels than those commonly used for fruit storage and transport, hence, they may present a risk of damage to these commodities. The time required for 100% mortality varies with arthropod species and its developmental stage, temperature, O_2 and CO_2 concentration, and relative humidity. Several treatments and systems have been developed and proved effective, without causing negative effects on horticultural commodities. However, the responses of the different commodities to the different quarantine treatments and systems are variable and depend on several factors such as the type and condition of the commodity and type of and severity of the treatment, among others. Research trends continue in developing adequate treatments and systems for different commodities, especially with heat, cold, MA, irradiation, and combinations of these. MA may have the potential to be efficacious phytosanitary treatments against quarantine pests of fruits, vegetables and cut flowers. A disadvantage for using MA on fresh commodities is the long time required at low temperatures for complete pest mortality, which is similar to cold treatments. Indeed, at temperatures near 0°C, it may be the low temperature more than the MA that is causing mortality. The complicated nature of researching the different factors and their effect on efficacy is an obstacle to development of MA treatments. Treatment times can be alleviated by applying MA at higher temperatures, but that may have negative consequences for commodity quality.

11.8 Future Research Needs

The tolerance of the different horticultural commodities to IMA at low and at high temperatures is required before they can be suggested for commercial application for any commodity. The differential tolerance between horticultural commodities to extreme gas atmospheres and high temperatures and possible mechanisms for this tolerance need to be investigated so that possible means to alleviate injury can be developed.

Although many MA treatments have been developed for a number of insect species, it is still unclear what are the optimal atmospheric gas concentrations, temperatures and durations for optimal insect control. The development of model systems may help elucidate these parameters, but will probably never totally circumvent treatment efficacy and confirmatory trials of infested commodities.

It is relatively easy to develop treatments in the laboratory, where the scale is small and conditions easily controlled. However, commercialization requires a significant increase in commodity load, and not all treatment parameters can be as easily controlled. Adequate engineering is crucial for commercialization of MA treatments to make them efficacious and economic. The successful commercial use of MA treatments that have been developed sufficiently for phytosanitary purposes should stimulate increased interest in developing and using more treatments.

References

Aegerter, A.F. and R.J. Folwell. 2001. Selected alternatives to methyl bromide in the postharvest and quarantine treatment of almonds and walnuts: An economic perspective. *Food Process. Preserv.* 25(6): 389–410.

Agnello, A.M., S.M. Spangler, E.S. Minson, T. Harris and D.P. Kain. 2002. Effect of high-carbon dioxide atmospheres on infestations of apple maggot (Diptera: Tephritidae) in apples. *J. Econ. Entomol.* 95: 520–526.

Aharoni, Y., J.K. Stewart, and D.G. Guadagni. 1981. Modified atmospheres to control western flower thrips on harvested strawberries. *J. Econ. Entomol.* 74: 338–340.

Ahumada, M.H., E.J. Mitcham and D.G. Moore. 1996. Postharvest quality of 'Thompson Seedless' grapes after insecticidal controlled-atmosphere treatments. *HortScience* 31(5): 833–836.

Alonso, M., M.A. Del Río, and J.-A. Jacas. 2005. Carbon dioxide diminishes cold tolerance of third instar larvae of *Ceratitis capitata* Wiedemann (Diptera: Tephritidae) in 'Fortune' mandarins: Implications for citrus quarantine treatments. *Postharv. Biol. Technol.* 36: 103–111.

Angerilli, N.P.D. and F. Fitzgibbon. 1990. Effects of cobalt gamma radiation on San Jose scale (Homoptera: Diaspididae) survival on apples in cold and controlled-atmosphere storage. *J. Econ. Entomol.* 83: 892–895.

Anonymous. United Nations Environmental Programme. 1992. *Methyl Bromide Atmospheric Science, Technology and Economics.* U.N. Headquarters, Ozone Secretariat. Nairobi, Kenya.

Anonymous. U.S. Clean Air Act. 1993. Federal Register 58(239): 65554.

APHIS (United States Department of Agriculture Animal and Plant Health Inspection Service) 2008. Treatment Manual.

Armstrong J.W. 1994. Heat and cold treatments. In: R.E. Paull and J.W. Armstrong (eds). *Insect Pests and Fresh Horticultural Products: Treatments and Responses,* Wallingford, U.K.: CAB International, pp. 103–119.

Baker A.C. 1939. The basis for treatment of products where fruit flies are involved as a condition for entry into the United States. U.S. Department of Agriculture Circular 551. 8pp.

Banks, H.J. 1984. Current methods and potential systems for production of controlled atmospheres for grain storage. *Dev. Agric. Eng.* 5: 523–542.

Benschoter, C.A. 1987. Effects of modified atmospheres and refrigeration temperatures on survival of eggs and larvae of the Caribbean fruit fly (Diptera: Tephritidae) in laboratory diet. *J. Econ. Entomol.* 80: 1223–1225.

Benschoter, C.A., D.H. Spalding, and W.F. Reeder. 1981. Toxicity of atmospheric gases to immature stages of *Anastrepha suspensa. Florida Entomol.* 64: 543–544.

Botondi, R., A. Crisa, R. Massantini, and F. Mencarelli. 2000. Effect of low oxygen short-term exposure at 15°C on postharvest physiology and quality of apricots harvested at two ripening stages. *J. Hort. Sci. Biotechnol.* 75(2): 202–208.

Brown, G. 2006. Maintaining apple fruit quality with CATTS disinfestation treatment. In: Proceedings of the 2006 Annual International Research Conference on Methyl Bromide Alternatives and Emissions Reductions. Orlando, FL.

Burg, S.P. 2004. *Postharvest Physiology and Hypobaric Storage of Fresh Produce.* Wallingford, U.K.: CABI Publishing, 654pp.

Calderon, M. 1990 Introduction. In: M. Calderon and R. Barkai-Golan (eds). *Food Preservation by Modified Atmospheres*, Boca Raton, FL: CRC Press, pp. 3–8.

Carpenter, A. 1995. Implementation of controlled atmosphere disinfestation of export asparagus. In: *Proceedings of the 48th New Zealand Plant Protection Conference*, pp. 318–321.

Carpenter, A. and M. Potter. 1994. Controlled atmospheres. In: J.L. Sharp and G.J. Hallman (eds). *Quarantine Treatments for Pests of Food Plant*. Boulder, CO: Westview Press, pp. 171–198.

Carpenter, A., A. Stocker, and G. Van der Mespel. 1992. Impact of CO_2 levels on New Zealand flower thrips and on stone fruit quality. *N Z J. Crop Hort. Sci.* 20: 256.

Carrillo-Lopez, A., and E.M. Yahia. 1990. Tolerance of avocado var 'Hass' to insecticidal O_2 and CO_2 and their effect on the anaerobic respiration (In Spanish). *Tecnología de Alimentos (México)* 1990: 25(6): 13–18.

Chen, P.M. and W.M. Mellenthin. 1992. Storage behavior of d'Anjou pears in low oxygen and air. In: D.G. Richardson and M. Meheriuk (eds). *Controlled Atmospheres for Storage and Transport of Perishable Agricultural Commodities*, Beaverton, OR: Timber Press, pp. 139–148.

Chen, P.M., K.I. Olsen, and M. Meheriuk. 1985. Effects of low oxygen atmosphere on storage scale and quality preservation of "Delicious" apples. *J. Am. Soc. Hort. Sci.* 110: 16–20.

Chen, P.M., W.M. Mellenthin, S.B. Kelly, and T.J. Facteau. 1981. Effects of low oxygen and temperature on quality retention of "Bing" cherries during prolonged storage. *J. Am. Soc. Hort. Sci.* 106: 533–535.

Chu, C.L. 1992. Postharvest control of San Jose scale on apples by controlled atmosphere storage. *Postharv. Biol. Technol.* 1: 361–369.

Couey, M. and K. Olsen. 1977. Commercial use of prestorage carbon dioxide treatment to retain quality in Golden Delicious apples. In: Dewey, D.H. (ed). *Controlled Atmospheres for Storage and Transport of Perishable Agricultural Commodities*, Horticultural Report No. 28. East Lansing, MI: Michigan State University, pp. 165–169.

Damarh, E., G. Gun, G. Ozay, S. Bulbul, and P. Oechsle. 1998. An alternative method instead of methyl bromide for insect disinfestation of dried figs: Controlled atmosphere. *Acta Hort.* 480: 209–214.

Davenport, T.L., S.P. Burg, and T.L. White. 2006. Optimal low-pressure conditions for long-term storage of fresh commodities kill Caribbean fruit fly eggs and larvae. *HortTechnology* 16: 98–104.

Delate, K.M. and J.K. Brecht. 1989. Quality of tropical sweet potatoes exposed to controlled-atmosphere treatments for postharvest insect control. *J. Am. Soc. Hort. Sci.* 114: 963–968.

Delate, K.M., J.K. Brecht, and J.A. Coffelt. 1990. Controlled atmosphere treatments for control of sweetpotato weevil (Coleoptera: Curculionidae) in stored tropical sweet potatoes. *J. Econ. Entomol.* 83: 461–465.

Dentener, P.R., S.M. Peetz, and D.B. Birtles. 1992. Modified atmospheres for the postharvest disinfestation of New Zealand persimmons. *N Z J. Crop Hort. Sci.* 20: 203–208.

Epenhuijsen, C.W. van, A. Carpenter, and R. Butler. 2002. Controlled atmospheres for the postharvest control of *Myzus persicae* (Sulzer) (Homoptera: Aphididae): Effects of carbon dioxide concentration. *J. Stored Prod. Res.* 38: 281–291.

Fleurat-Lessard, F. 1990. Effect of modified atmospheres on insects and mites infesting stored products. In: M. Calderon and R. Barkai-Golan (eds). *Food Preservation by Modified Atmospheres*, Boca Raton FL: CRC Press, pp. 21–38.

Folchi, A., G.C. Pratella, S.P. Tian, and P. Bertolini. 1995. Effect of low oxygen stress in apricot at different temperatures. *Ital. J. Food Sci.* 7: 245–254.

Follett, P. and L.G. Neven. 2006. Current Trends in Quarantine Entomology. *Ann. Rev. Entomol.* 51: 359–385.

Friedlander A. 1983. Biochemical reflections on a non-chemical control method: The effect of controlled atmosphere on the biochemical processes in stored products insects. In: *Proceedings of the 3rd International Working Conference on Stored Products Entomology*. Manhattan, KS: Kansas State University, 1983. pp. 471–486.

Gaunce, A.P., C.V.G. Morgan, and M. Meherhuik. 1982. Control of tree fruit insects with modified atmospheres. In: D.G. Richardson and M. Meherhuik (eds). *Controlled Atmospheres for Storage and Transport of Perishable Agricultural Commodities*. Symposium Series 1. Beaverton, OR: Timber Press, Oregon State University School of Agriculture, pp. 383–390.

Hallman, G.J. 1994. Controlled atmospheres. In: R.E. Paull and J.W. Armstrong (eds). *Insect Pests and Fresh Horticultural Products: Treatments and Responses*. Wallingford, U.K.: CAB International, pp. 121–136.

Hallman, G.J. 1997. Mortality of Mexican fruit fly (Diptera: Tephritidae) immatures in coated grapefruits. *Florida Entomol.* 80: 324–328.

Hallman, G.J. 2004. Ionizing irradiation quarantine treatment against oriental fruit moth (Lepidoptera: Tortricidae) in ambient and hypoxic atmospheres. *J. Econ. Entomol.* 97: 824–827.

Hallman, G.J. and J.W. Armstrong. 1994. Heated air treatments. In: J.L. Sharp and G.J. Hallman (eds). *Quarantine Treatments for Pest of Food Plants*. Boulder CO: Westview Press, pp. 149–164.

Hallman, G.J., M.O. Niperos-Carriedo, E.A. Baldwin, and C.A. Campbell. 1994. Mortality of Caribbean fruit fly (Diptera: Tephritidae) immatures in coated fruits. *J. Econ. Entomol.* 87: 752–757.

Hansen, J.D., M.A. Watkins, M.L. Heidt, and P.A. Anderson. 2006. Cold storage to control codling moth larvae in fresh apples. *HortTechnology* 17: 195–198.

Hoback, W.W. and D.W. Stanley. 2001. Insects in hypoxia. *J. Insect Physiol.* 47: 533–542.

Hoy, L.E. and D.C. Whiting. 1998. Mortality responses of three leafroller (Lepidoptera: Tortricidae) species on kiwifruit to a high-temperature controlled atmosphere treatment. *NZ J. Crop Hor. Sci.* 26: 11–15.

Irving, D.E. and L. Honnor. 1994. Carnations: Effects of high concentrations of carbon dioxide on flower physiology and longevity. *Postharv. Biol. Technol.* 4: 281–287.

Jang, E.B. 1990. Fruit fly disinfestation of tropical fruits using semipermeable shrinkwrap film. *Acta Hort.* 269: 453–458.

Ke, D. and A. Kader. 1989. Tolerance and responses of fresh fruits to oxygen levels at or below 1%. In: *Proceedings of the 5th International Controlled Atmospheres Research Conference*, Wenatchee, WA. June 14–16, 1989, pp. 209–216.

Ke, D. and A. Kader. 1990. Tolerance of 'Valencia' oranges to controlled atmospheres as determined by physiological responses and quality attributes. *J. Am. Soc. Hort. Sci.* 115: 779–783.

Ke, D., L. Goldstein, M. O'Mahony, and A.A. Kader. 1991a. Effects of short term exposure to low O_2 and high CO_2 atmospheres on quality attributes of strawberries. *J. Food Sci.* 56: 50–54.

Ke, D., L. Rodriguew-Sinobas, and A.A. Kader. 1991b. Physiology and prediction of fruit tolerance to low oxygen atmospheres. *J. Am. Soc. Hort. Sci.* 116: 253–260.

Ke, D., L. Rodriguew-Sinobas, and A.A. Kader. 1991c. Physiological responses and quality attributes of peaches kept in low oxygen atmospheres. *Scientia Hort.* 47: 295–303.

Ke, D., E. Yahia, M. Mateos, and A. Kader. 1994a. Ethanolic fermentation of 'Bartlett' pears as influenced by ripening stage and controlled atmosphere storage. *J. Am. Soc. Hortic. Sci.* 119: 976–982.

Ke, D., F. El-Wazir, B. Cole, M. Mateos, and A.A. Kader. 1994b. Tolerance of peach and nectarine fruits to insecticidal controlled atmospheres as influenced by cultivar, maturity and size. *Postharv Biol. Technol.* 4: 135–146.

Ke, D., E. Yahia, B. Hess, L. Zhou, and A. Kader. 1995. Regulation of fermentative metabolism in avocado fruit under oxygen and carbon dioxide stress. *J. Am. Soc. Hort. Sci.* 120: 481–490.

Lay-Yee, M. and D.C. Whiting. 1996. Response of 'Hayward' kiwifruit to high-temperature controlled atmosphere treatments for control of two-spotted spider mite (*Tetranychus urticae*). *Postharv. Biol. Technol.* 7: 73–81.

Leon, D.M., J. de la Cruz, K.L. Parkin, and H.S. Garcia. 1997. Effect of controlled atmospheres containing low O_2 and high CO_2 on chilling susceptibility of Manila mangoes. *Acta Hort.* 455: 635–642.

Little, C.R., J.D. Fragher, and H.J. Taylor. 1982. Effects of initial oxygen stress treatment in low oxygen modified atmosphere storage of "Granny Smith" apples. *J. Am. Soc. Hort. Sci.* 107: 320–323.

Liu, Y.-B. 2003. Effects of vacuum and controlled atmosphere treatments on insect mortality and lettuce quality. *J. Econ. Entomol.* 96: 1100–1107.

Liu, Y.-B. 2005. Ultralow oxygen treatment for postharvest control of *Nasonovia ribisnigri* (Homoptera: Aphididae) on iceberg lettuce. *J. Econ. Entomol.* 98: 1899–1904.

Liu, Y.-B. 2007. Ultralow oxygen treatment for postharvest control of western flower thrips on broccoli. *J. Econ. Entomol.* 100: 717–722.

Mangan, R.L. and S.J. Ingle. 1994. Forced hot-air quarantine treatment for grapefruit infested with Mexican fruit-fly (Diptera:Tephritidae). *J. Econ. Entomol.* 87: 1574–1579.

Mangan, R.L., K.C. Shellie, S.J. Ingle, and M.J. Firko. 1998. High temperature forced-air treatments with fixed time and temperature for 'Dancy' tangerines, 'Valencia' oranges, and 'Rio Star' grapefruit. *J. Econ. Entomol.* 91: 933–939.

Mbata, G.N. and T.W. Phillips. 2001. Effects of temperature and exposure time on mortality of stored-product insects exposed to low pressure. *J. Econ. Entomol.* 94: 1302–1307.

Mbata, G.N., T.W. Phillips, and M. Payton. 2004. Mortality of eggs of stored-product insects held under vacuum: Effects of pressure, temperature, and exposure time. *J. Econ. Entomol.* 97: 695–702.

Mbata, G.N., M. Johnson, T.W. Phillips, and M. Payton. 2005. Mortality of life stages of cowpea weevil (Coleoptera: Bruchidaw) exposed to low pressure at different temperatures. *J. Econ. Entomol.* 98: 1070–1075.

Mitcham, E.J., M.M. Attia, and W. Biasi. 1997a. Tolerance of 'Fuyu' persimmons to low oxygen and high carbon dioxide atmospheres for insect disinfestations. *Postharv. Biol. Technol.* 10: 155–160.

Mitcham, E.J., S. Zhou, and V. Bikoba. 1997b. Controlled atmospheres for quarantine control of three pests of table grape. *J. Econ. Entomol.* 90: 1360–1370.

Mitcham, E.J., L. Neven, and B. Biasi. 1999. Effect of high-temperature controlled-atmosphere treatments for insect control in 'Bartlett' pear fruit. *HortScience* 34: 527.

Mitchell, F.G., A.A. Kader, G. Cristoso, and G. Mayer. 1984. The tolerance of stone fruits to elevated CO_2 and low O_2 levels. Report to the California Tree Fruit Agreement. Sacramento, CA, 11 pp.

Moffitt, H.R. and D.J. Albano. 1972. Effects of commercial fruit storage on stages of the codling moth. *J. Econ. Entomol.* 65: 770–773.

Morgan, C.V.G. and A.P. Gaunce. 1975. Carbon dioxide as a fumigant against the San Jose scale (Homotera: Diaspididae) on harvested apples. *Can. Entomol.* 107: 935–936.

Moss, J.I. and E.B. Jang. 1991. Effects of age and metabolic stress on heat tolerance of Mediterranean fruit fly (Diptera: Tephritidae) eggs. *J. Econ. Entomol.* 84: 537–541.

Mueller, J., T. Boller, and A. Wiemken. 1995. Trehalose and trehalase in plants: Recent developments. *Plant Sci.* 112: 1–9.

Navarro, S.E. and M. Calderon. 1979. Mode of action fo low atmospheric pressures on *Ephestia cautella* (Wlk.) pupae. *Experimentia* 35: 620–621.

Navarro, S., E. Donahaye, M. Rindner, and A. Azieli. 1998. Storage of dried fruits under controlled atmosphere for preservation and control of nitidulid beetles. *Acta Hort.* 480: 221–226.

Neven, L.G. 2000. Insect physiological responses to heat. *Postharv. Biol. Technol.* 21: 103–111.

Neven, L. 2003. Effects of physical treatments on insects. *HortTechnology* 13: 272–275.

Neven, L.G. 2004. Hot forced air with controlled atmospheres for disinfestation of fresh commodities. In: R. Dris and S.M. Janin (eds). *Production Practices and Quality Assessment of Food Crops*, Vol. 4: *Post Harvest Treatments*. Dordrecht: Kluwer Academic, pp. 297–315.

Neven, L.G. 2005. Combined heat and controlled atmosphere quarantine treatments for control of codling moth in sweet cherries. *J. Econ. Entomol.* 98: 709–715.

Neven, L.G. 2008. Organic quarantine treatments for tree fruits. *HortScience* 43: 22–26.

Neven, L.G. and S.R. Drake. 1998. Quarantine treatment for sweet cherries. *Good Fruit Grower* 49: 43–44.

Neven, L.G. and S.R. Drake. 2000. Effects of the rate of heating on apple and pear fruit quality. *J. Food Qual.* 23: 317–325.

Neven, L. and E. Mitcham. 1996. CATTS (Controlled Atmosphere/Temperature Treatment System): A novel tool for the development of quarantine treatments. *Am. Entomol.* 42: 56–59.

Neven, L. and L. Rehfield. 1995. Comparison of pre-storage heat treatments on fifth instar codling moth (Lepidoptera: Tortricidae) mortality. *J. Econ. Entomol.* 88: 1371–1375.

Neven, L.G. and L.M. Rehfield-Ray. 2006a. Combined heat and controlled atmosphere quarantine treatment for control of western cherry fruit fly in sweet cherries. *J. Econ. Entomol.* 99: 658–663.

Neven, L.G. and L. Rehfield-Ray. 2006b. Confirmation and efficacy tests against codling moth, *Cydia pomonella* and oriental fruit moth, *Grapholitha molesta*, in apples using combination heat and controlled atmosphere treatments. *J. Econ. Entomol.* 99: 1620–1627.

Neven, L.G., L.M. Rehfield, and K.C. Shellie. 1996. Moist and vapor forced air treatments of apples and pears: Effects on the mortality of fifth instar codling moth (Lepidoptera: Tortricidae). *J. Econ. Entomol.* 89: 700–704.

Neven, L.G., L. Rehfield-Ray, and D. Obenland. 2006. Confirmation and efficacy tests against codling moth and oriental fruit moth in peaches and nectarines using combination heat and controlled atmosphere treatments. *J. Econ. Entomol.* 99: 1610–1619.

Obenland, D., P. Nepp, B. Mackey, and L. Neven. 2005. Peach and nectarine quality following treatment with high-temperature forced air combined with controlled atmosphere. *HortScience* 40: 1425–1430.

Ortega-Zaleta, D. and E.M. Yahia. 2000a. Tolerance and quality of mango fruit exposed to controlled atmospheres at high temperatures. *Postharv Biol. Technol.* 20: 195–201.

Ortega-Zaleta, D. and E.M. Yahia. 2000b. Mortality of eggs and larvae of *Anastrepha obliqua* (Macquart) and *A.* ludens (Lowe) (Diptera: Tephritidae) with controlled atmospheres at high temperature in mango (Mangifera indica) cv Manila. *Folia Entomologica Mexicana (in Spanish).* 109: 43–53.

Page, B.B.C., M.J. Bendall, A. Carpenter, and C.W. van Epenhuijsen. 2002. Carbon dioxide fumigation of *Thrips tabaci* in export onions. *N Z Plant Protect.* 55: 303–307.

Patterson, M.E. 1982. CA storage of cherries. In: Richardson, D.G. and M. Meheriuk (eds). *Controlled Atmosphere for Storage and Transport of Perishable Agricultural Commodities.* Beaverton, OR: Timber Press, pp. 149–154.

Patterson, M.E. and J.L. Melsted. 1977. Sweet cherry handling and storage alternatives. In: Dewey, D.H. (ed). *Controlled Atmospheres for Storage and Transport of Perishable Agricultural Commodities.* Horticultural Report No. 28. East Lansing, MI: Michigan State University, pp. 55–59.

Pesis, E. and R. Ben-Arie. 1986. Carbon dioxide assimilation during postharvest removal of astringency from persimmon fruit. *Physiol. Plant.* 67: 644–648.

Potter, M.A., A. Carpenter, A. Stocker, and S. Wright. 1994. Controlled atmospheres for the postharvest disinfestations of *Thrips obsucartus* (Thysanoptera:Thripidae). *J. Econ. Entomol.* 87: 1251–1255.

Potter, M.A., A. Carpenter, and A. Stocker. 1990. Response surfaces for controlled atmosphere and temperature by species: Implications for disinfestation of fresh produce for export. In: Beattie, B.B. (ed). *Managing Postharvest Horticulture in Australasia.* Occasional Publication No 46. Melbourne: Australian Institute of Agricultural Science, pp. 183–190.

Prange, R.K. and P.D. Lidster. 1992. Controlled-atmosphere effects on blueberry maggot and lowbrush blueberry fruit. *HortScience* 27: 1094–1096.

Sharp, J.L. 1994. Hot water immersion. In: Sharp J.L. and G.J. Hallman (eds). *Quarantine Treatments for Pest of Food Plants.* Boulder, CO: Westview Press, pp. 133–148.

Shellie, K.C. and R.L. Mangan. 1994. Postharvest quality of Valencia orange after exposure to hot, moist, forced-air for fruit-fly disinfestations. *HortScience* 29: 1524–1527.

Shellie, K.C., M.J. Kirko, and R.L. Mangan. 1993. Phytotoxic response of Dancy tangerine to high-temperature, moist, forced-air treatment for fruit fly disinfestations. *J. Am. Soc. Hortic. Sci.* 118: 481–485.

Shellie, K.C., R.L. Mangan, and S.J. Ingle. 1996. Tolerance of grapefruit and Mexican Fruit Fly larvae to high-temperature controlled atmospheres. *HortScience* 31: 637–637.

Shellie, K.C., J. Cumaragunta, and R.L. Mangan. 1997a. Hypoxic and hypercarbic cold storage for disinfecting grapefruit of Mexican fruit fly. In: *CA'97 Proceedings.* Vol. 1 Postharvest Horticulture Series No. 15. University of California Postharvest Outreach Program, Davis, CA, pp. 98–104.

Shellie, K.C., R.L. Mangan, and S.J. Ingle. 1997b. Tolerance of grapefruit and Mexican Fruit Fly larvae to heated controlled atmospheres. *Postharv Biol. Technol.* 10: 179–186.

Shellie, K.C., L.G. Neven, and S.R. Drake. 2000. Controlled atmosphere temperature treatments for disinfestations of 'Bing' sweet cherry. *HortScience* 35(3): 412.

Shellie, K.C., L.G. Neven, and S.R. Drake. 2001. Assessing 'Bing' sweet cherry tolerance to a heated controlled atmosphere for insect pest control. *HortTechnology* 11: 308–311.

Shelton, M., A. Carpenter, and C.W. Van Epenhuijsen. 1997. Sequential CA: Effects on aphids, spider mites and cut flower vase life. In: *CA'97 Proceedings.* Vol. 1 Postharvest Horticulture Series No. 15. University of California Postharvest Outreach Program, CA, pp. 127–131.

Smilanick, J.L. and D.C. Fouse. 1989. Quality of nectarines stored in insecticidal low-O2 atmospheres at 5 and 15°C. *J. Am. Soc. Hort. Sci.* 114: 431–436.

Smittle, D.A. and W.R. Miller. 1988. Rabbiteye blueberry storage life and fruit quality in controlled atmospheres and air storage. *J. Am. Soc. Hort. Sci.* 113: 723–728.

Soderstrom, E.L., D.G. Brandl, and J.L. Smilanick. 1987. Controlled atmospheres for postharvest control of codling moth on fresh tree fruits. Report to California Tree Fruit Agreement, Sacramento, CA.

Soderstrom, E.L., D.G. Brandl, and B. Mackey. 1991. Responses of *Cydia pomonella* (L.) (Lepidoptera: Tortricidae) adults and eggs to oxygen deficient or carbon dioxide enriched atmospheres. *J. Stored Prod. Res.* 27: 95–101.

Soderstrom, E.L., D.G. Brandl, and B. Mackey. 1992. High temperature combined with carbon dioxide enriched or reduced oxygen atmospheres for control of *Tribolium castaneum* (Herbst) (Coleoptera: Tenebrionidae). *J. Stored Prod. Res.* 28: 235–238.

Soderstrom, E.L., D.G. Brandl, and B.E. Mackey. 1996a. High temperature alone and combined with controlled atmospheres for control of diapausing codling moth (Lepidoptera: Tortricidae) in walnuts. *J. Econ. Entomol.* 89(1): 144–147.

Soderstrom, E.L., D.G. Brandl, and B.E. Mackey. 1996b. High temperature and controlled atmosphere treatment of codling moth (Lepidoptera: Torticidae) infested walnuts using a gas-tight treatment chamber. *J. Econ. Entomol.* 89: 712–714.

Stewart, O.J., G.S.V. Raghavan, and Y. Gariépy. 2005. MA storage of Cavendish bananas using silicone membrane and diffusion channel systems. *Postharv. Biol. Technol.* 35: 309–317.

Swan, T.M. and K. Watson. 1998. Stress tolerance in a yeast sterol auxotroph: Role of ergosterol, heat shock proteins and trehalose. *FEMS Microbiol. Lett.* 169: 191–197.

Thomas, D.B. and K.C. Shellie. 2000. Heating rate and induces thermotolerance in Mexican fruit fly (Diptera: Tephritidae) larvae, a quarantine pest of citrus and mangoes. *J. Econ. Entomol.* 93: 3173–1379.

Toba, H.H. and H.R. Moffitt. 1991. Controlled-atmosphere cold storage as a quarantine treatment for nondiapausing codling moth (Lepidoptera: Tortricidae) larvae in apples. *J. Econ. Entomol.* 84: 1316–1319.

Warner, G. 1998. Quarantine treatment enhances fruit quality. *Good Fruit Grower* 49: 32.

Wei, D., J.J. Wang, Z.M. Zhao, and J.H. Tsai. 2002. Effects of controlled atmosphere and DDVP on population growth and resistance development by the psocid, *Liposcelis bostrychophila* Badonnel (Psocoptera: Liposcelididae). *J. Stored Prod. Res.* 38: 229–237.

Whiting, D.C. and L.E. Hoy. 1997. High-temperature controlled atmosphere and air treatments to control obscure mealybug (Hemiptera: Pseudococcidae) on apples. *J. Econ. Entomol.* 90: 546–550.

Whiting, D.C. and L.E. Hoy. 1998. Effect of temperature establishment time on the mortality of *Epiphyas postvittana* (Lepidoptera: Tortricidae) larvae exposed to a high-temperature controlled atmosphere. *J. Econ. Entomol.* 91: 287–292.

Whiting, D.C. and J. Van Den Heuvel. 1995. Oxygen, carbon dioxide, and temperature effects on mortality responses of Diapausing *Tetranychus urticae* (Acari: Tetranychidae). *J. Econ. Entomol.* 88: 331–336.

Whiting, D.C., S.P. Foster, and J.H. Maindonald. 1991. Effects of oxygen, carbon dioxide and temperature on the mortality responses of *Epiphyas postvittana* (Lepidoptera: Tortricidae). *J. Econ. Entomol.* 84: 1544–1549.

Whiting, D.C., S.P. Foster, J. van den Heuvel, and J.H. Maindonald. 1992a. Comparative mortality responses of four tortricid (Lepidoptera) species to a low oxygen-controlled atmosphere. *J. Econ. Entomol.* 85: 2305–2309.

Whiting, D.C., J. van den Heuvel, and S.P. Foster. 1992b. Potential of low oxygen/moderate carbon dioxide atmospheres for postharvest disinfestation of New Zealand apples. *N Z J. Crop Hort. Sci.* 20: 217–222.

Whiting, D.C., G.M. O'Connor, J. Van Den Heuvel, and J.H. Maindonald. 1995. Comparative mortalities of six tortricid (Lepidoptera) species to two high-temperature controlled atmospheres and air. *J. Econ. Entomol.* 88: 1365–1370.

Whiting, D.C., G.M. O'Connor, and J.H. Maindonald. 1996. First instar mortalities of three New Zealand leafroller species (Lepidoptera: Tortricidae) exposed to controlled atmosphere treatments. *Postharv. Biol. Technol.* 8: 229–236.

Whiting, D.C., L.E. Jamieson, K.J. Spooner, and M. Lay-Yee. 1999. Combination high-temperature controlled atmosphere and cold storage as a quarantine treatment against *Ctenopseustis obliquana* and *Epiphyase postvittana* on 'Royal Gala' apples. *Postharv. Biol. Technol.* 16: 119–126.

Yahia, E.M. 1993. Responses of some tropical fruits to insecticidal atmospheres. *Acta Hort.* 343: 371–376.

Yahia, E.M. 1994a. The potential use of insecticidal atmospheres for mango, avocado, and papaya fruit. In: B.R. Champ, E. Highley, and G.I. Johnson (eds) *Postharvest Handling of Tropical Fruits.* Australian Centre for International Agricultural Research (ACIAR) Proceedings No. 50., pp. 373–374.

Yahia, E.M. 1994b. The use of modified and controlled atmospheres for tropical fruits. In: *Proceedings of the 1st Regional Symposium on Postharvest Handling of Tropical Products*, University of Costa Rica, San José, Costa Rica, July 25–29 (in Spanish), pp. 1–11.

Yahia, E.M. 1995a. Insecticidal atmospheres for tropical fruits. In: Kushwaha, R. Serwtoski, and R. Brook (eds). *Harvest and Postharvest Technologies for Fresh Fruits and Vegetables.* ASAE, Guanajuato, Gto, Mexico, February 20–24, pp. 282–286.

Yahia, E.M. 1995b. Application of differential scanning calorimetry in the study of avocado and mango fruit responses to hypoxia. In: A. Ait-Oubahou and M. El-Otmani (eds). *Postharvest Physiology, Pathology and Technologies for Horticultural Crops: Recent Advances.* Agadir, Morocco: Institut Agronomique and Vétérinaire Hassan II, pp. 206–209.

Yahia, E.M. 1997a. Modified/controlled atmospheres for avocado. In: A.A. Kader (ed). *CA'97 Proceedings*, Vol. 3: Fruits Other Than Apples and Pears. University of California, Davis, CA, pp. 97–103.

Yahia, E.M. 1997b. Guava and avocado fruits are sensitive to insecticidal MA and/heat. In: J.F. Thompson and E.J. Mitcham (eds). *CA'97 Proceedings*, Vol. 1 CA Technology and Disinfestation Studies. University of California, Davis, CA, pp. 132–136.

Yahia, E.M. 1997d. Modified/controlled atmospheres for mango. In: A.A. Kader (ed). *CA'97 Proceedings*, Vol. 3 Fruits Other Than Apples and Pears. University of California, Davis, CA, pp. 110–116.

Yahia, E.M. 1998a. Modified and controlled atmospheres for tropical fruits. *Hort. Rev.* 22: 123–183.

Yahia, E.M. 1998b. The use of modified and controlled atmospheres for insect control (In Spanish). *Phytoma (Spain)* 97: 18–22.

Yahia, E.M. 2004. The development of quarantine insect control system for tropical fruits using controlled atmospheres at high temperatures. In: *Proceedings of the Frigair 2004, International Refrigeration and Air Conditioning Conference*, Vol. 1 Cape Town, South Africa, June 2–4, 2004, pp. 7–16.

Yahia, E.M. 2006a. Effects of insect quarantine treatments on the quality of horticultural crops. *Stewart Postharv. Rev.* 1: 1–18.

Yahia E.M. 2006b. Handling tropical fruits. In: *Scientists Speak Proceedings of World Food Logistics Organization Advisory Council*, Walt Disney World Swan Hotel, Orlando, FL, pp. 5–9.

Yahia, E.M. and R. Ariza. 2001. Physical treatments for the postharvest handling of horticultural products (In Spanish). *Horticultura Internacional* (Spain). Extra 80–88: 153.

Yahia, E.M. and M. Balderas. 2000. Response differences to hot air treatments between tolerant (mango) and sensitive (guava) fruits and the role of sugar content and trehalase activity. In: F. Artes, M.I. Gil, and M.A. Conesa (eds), *Improving Postharvest Technologies of Fruits, Vegetables, and Ornamentals*, Vol. 2, International Institute of Refrigeration. pp. 746–752.

Yahia, E.M. and A. Carrillo-López. 1993. Tolerance and responses of avocado fruit to insecticidal O_2 and CO_2 atmospheres. *Food Sci. Technol. (Lebens. Wiss. u-Technol.)* 26: 312–317.

Yahia, E.M. and D. Ortega-Zaleta. 1999. Effects of insecticidal controlled atmospheres at high temperature on the quality of mango fruit. *HortScience* 34(3): 512.

Yahia, E.M. and D. Ortega. 2000a. Mortality of eggs and third instar larvae of *Anastrepha ludens* and *A. obliqua* with insecticidal controlled atmospheres at high temperatures. *Postharv. Biol. Technol.* 20: 295–302.

Yahia, E.M. and D. Ortega. 2000b. The use of controlled atmospheres at high temperature to control fruit flies (*Anastrepha ludens and A. oblique*) and their effect on mango quality. In: W.J. Florkowski, S.E. Prussia, and R.L. Shewfelt (eds). *Proceedings International Multidisciplinary*

Conference: Integrated View of Fruit and Vegetable Quality. Lancaster, Basel: Technomic Publishing Co, Inc. pp. 143–153.

Yahia, E.M. and D. Ortega-Zaleta. 2000c. Responses and quality of mango fruit treated with insecticidal controlled atmospheres at high temperatures. *Acta Hort.* 509: 479–486.

Yahia, E.M. and R. Paull. 1997. The future for modified atmosphere (MA) and controlled atmosphere (CA) uses with tropical fruit. *Chronica Hort.* 37: 18–19.

Yahia, E.M. and M. Tiznado-Hernandez. 1993. Tolerance and responses of harvested mango to insecticidal oxygen atmospheres. *HortScience* 28: 1031–1033.

Yahia, E.M. and L. Vazquez-Moreno. 1993. Tolerance and responses of mango to insecticidal oxygen and carbon dioxide atmospheres. *Food Sci. Tech. (Lebens. Wiss. u-Technol.)* 26: 42–48.

Yahia, E.M., F. Medina, and M. Rivera. 1989. The tolerance of mango and papaya to atmospheres containing very high levels of CO_2 and/or very low levels of O_2 as a possible insect control treatment. In: Fellman, J.K. (ed). In: *Proceedings of the 5th International Controlled Atmosphere Research Conference*, Vol. 2 Wenatchee, WA, June 14–16, 1989, pp. 77–89.

Yahia, E.M., M. Rivera, and O. Hernandez. 1992. Responses of papaya to short-term insecticidal oxygen atmosphere. *J. Am. Soc. Hort. Sci.* 117: 96–99.

Yahia, E.M., D. Ortega-Zaleta, P. Santiago, and L. Lagunez. 1997. Responses of mango and mortality of *Anastrepha ludens* and *A. obliqua* to modified atmospheres at high temperatures. In: J.F. Thompson and E.J. Mitcham (eds). *CA'97. Proceedings*, Vol. 1 CA Technology and Disinfestation Studies. University of California, Davis, CA, 1997, pp. 105–112.

Yahia, E.M., G. Villagomez, and A. Juarez. 2000a. Tolerance and responses of different fruits and vegetables to hot air treatments at 43°C or 48°C and 50% RH for 160 minutes. In: Artes, M.I. Gil and M.A. Conesa (eds). *Improving Postharvest Technologies of Fruits, Vegetables, and Ornamentals*, Vol. 2F. International Institute of Refrigeration, pp. 714–718.

Yahia, E.M., D. Ortega-Zaleta, A. Martinez, and P. Moreno. 2000b. The mortality of artificially infested third instar larvae of *Anastrepha ludens* and *A. obliqua* in mango fruit with insecticidal controlled atmospheres at high temperatures. *Acta Hort.* 509: 833–839.

Yahia, E.M., C. Barry-Ryan, and R. Dris. 2004. Treatments and techniques to minimize the postharvest losses of perishable food crops. In: R. Dris and S.M. Jain (eds). *Production Practices and Quality Assessment of Food Crops*, Vol. 4 *Postharvest Treatment and Technology*. Finland: Kluwer Academic, pp. 95–133.

Yocum, G.D. and D.L. Denlinger. 1994. Anoxia blocks thermotolerance and the induction of rapid cold hardening in the flesh fly, *Sarcophaga crassipalpis*. *Physiol. Entomol.* 19: 152–158.

Zagory, D., D. Ke, and A. Kader. 1989. Long-term storage of 'Early Gold' and 'Shinko' Asian pears in low oxygen atmospheres. In: Fellman, J.K. (ed). *Proceedings of the 5th International Controlled Atmosphere Research Conference*, Vol. 1, Wenatchee, WA, June 14–16, 1989, pp. 353–358.

Zheng, J., M.S. Reid, D. Ke, and M. Cantwell. 1993. Atmosphere modification for postharvest control of thirps and aphids on flowers and green leafy vegetables. In: *Proceedings of the 6th International Controlled Atmosphere Research Conference*, Ithaca, NY, June 1993, pp. 394–401.

Zhou, S., R.S. Crittle, and E.J. Mitcham. 2000. Metabolic response of *Platynota stultana* pupae to controlled atmospheres and its relation to insect mortality response. *J. Insect Physiol.* 46: 1375–1385.

Zhou, S., R.S. Crittle, and E.J. Mitcham. 2001. Metabolic response of *Platynota stultana* pupae during and after extended exposure to elevated CO_2 and reduced O_2 atmospheres. *J. Insect Physiol.* 47: 401–409.

12

Pome Fruits

Jinhe Bai, Robert K. Prange, and Peter M.A. Toivonen

CONTENTS

Controlled atmosphere (CA) and modified atmosphere (MA) are most widely used for the commercial storage and transportation of apples and pears. After it was first described in the U.K. in the 1920s and actively researched in several countries in the 1930s and 1940s, the commercial use of CA storage for apples rapidly expanded in the 1950s and has subsequently been adopted worldwide. Since its initial commercialization in the 1950s, several improvements in CA technology have been made. These include rapid CA systems, which allow atmosphere set-points to be reached within 1 or 2 days; use of low oxygen concentrations (0.7–1.5 kPa) that can be accurately monitored and controlled; ethylene-free CA, programmed (or stepwise) CA, and dynamic CA where levels of O_2 and CO_2 are modified as needed, based on monitoring specific attributes of produce quality, such as ethanol concentration and chlorophyll fluorescence. MA packaging of pome fruits includes pallet covers, consumer bagging, and most commonly for packaging fresh-cut apples and pears. A generally used MA approach for intact pome fruits is the application of edible coatings (waxing) in the United States and many other countries.

1-Methylcyclopropene (1-MCP), an ethylene inhibitor, was approved by U.S. EPA in 2002 and produced by AgroFresh Inc. (Springhouse, PA) with the commercial trademark of SmartFresh[SM]. The use of 1-MCP in combination with CA can further improve storability of fruits and this has led to changes in CA storage management, particularly for apples.

12.1 Apple

12.1.1 Maturity

In general, an apple fruit harvested for CA storage must be less mature than fruit destined for earlier marketing after harvest. Overmature fruit have poor storability, and short storage life; fruits lose firmness and acidity quickly, and certain physiological disorders become more obvious, e.g., senescent breakdown and watercore. However, fruit harvested at an immature stage have poor eating quality upon ripening with little or no typical apple/fruit flavor and taste, and lack of juiciness and crispness. Most cultivars harvested at an immature stage can be more susceptible to physiological disorders such as bitter pit and superficial scald. The most widely used maturity indicators for apples include flesh firmness, starch content, sugar content (soluble solids content (SSC); expressed as % or °Brix), fruit color, and internal ethylene concentration. The following attributes are also used as supplemental maturity indicators: titratable acidity content, days from full bloom, and temperature accumulation. Fruit harvested at optimum maturity and handled properly have good storability and good eating quality. Although the benefit from CA storage is clear for fruit quality maintenance, in commercial practice, only fruits, which will be stored for 3 months or longer, are stored in CA because of cost. 'Gala' apples sometimes are stored in CA for less than 3 months due to the poor storability of that cultivar and its high relative financial return. A proper harvest prediction program ensures that representative fruit are tested for maturity within each orchard and each block by experienced personnel. The testing should begin in the orchard on a regular basis at least 2 to 3 weeks prior to the anticipated harvest date.

A "harvest window" for many cultivars is only a few days and an even narrower period is acceptable for long term CA storage. It has been suggested that ethylene production or internal ethylene concentration (IEC) should be a major determinant of harvest decisions (Lau, 1985). However, because the relationship between ethylene production and optimum harvest dates may not always be high (Watkins et al., 2004) and because ethylene measurements require expensive equipment, flesh firmness, starch iodine tests and surface color are most commonly used as maturity indices in commercial practice.

Production areas, yearly climate differences, tree age and vigor, and field practices influence maturity. Therefore, maturity estimation is usually based on a combination of regional maturity indicators, advice of local extension personnel, and experience of packinghouse technical personnel. Field production practices can complicate the maturity determination. For example, the background color of many apple cultivars is strongly influenced by nitrogen fertilization levels. Another example is that 'Fuji' apples growing on light crop trees mature earlier than well-cropped trees by as much as 10 days (Kupferman, 1997). Heavily cropped 'Honeycrisp' trees produce poorly colored, low-quality fruit (Embree et al., 2007). Thus fruit color is not a good maturity indicator.

The Streif index, a maturity determination method named after its developer, Dr. Josef Streif from Germany, appears to be less sensitive to year and location variations, giving it good potential for regional utilization (Streif, 1996). It is calculated with the formula: firmness/(soluble solids × starch index).

There are three features which make this approach appealing: its simplicity; all three measurements are accepted already as important individual measurements; these measurements are rapid, inexpensive and easily done by any orchard manager. The application of the Streif index has been studied on various popular European cultivars grown in different regions of Europe and other countries (Prange and DeLong, 1998).

ReTain^{TM} (Valent BioScience, Libertyville, Illinois) is the commercial trademark for the plant growth regulator aminoethoxyvinylglycine (AVG), which blocks the production of ethylene (Byers, 1997; Greene, 2005). When ReTain is applied to apples, several ripening processes are slowed, including preharvest fruit drop, fruit flesh softening, starch disappearance, and red color formation (Byers, 1997; Greene, 2005). Fruit treated with ReTain can be picked during the normal harvest period for enhanced retention of firmness in regular cold or CA storage, or harvest may be delayed without significantly sacrificing fruit firmness, allowing the fruit to continue to grow and develop red color for an extended time. But one of the greatest benefits of using ReTain is reduction in preharvest fruit drop, often by as much as 30% for 'McIntosh' (Greene, 2005).

Fruits, which are to be treated with 1-MCP after harvest, can be harvested when they are more mature than fruits that will not be treated. However, fruit must be at the preclimacteric stage. The expected result is that these later-harvested fruit have a better eating quality without loss of storability. By harvesting more mature fruit, it may have time to grow larger in size and improve the pack-out returns. See Section 12.1.3 for detailed information about the interaction of 1-MCP and CA.

12.1.2 CA Regimes

12.1.2.1 Temperature Control in CA Room

CA rooms should be precooled to the desired set-point before product loading begins. Most, but not all, cultivars will retain fruit quality better if rapidly cooled to 0°C–1°C within 3 days after harvest (Table 12.1). Delayed cooling is associated with shortened storage life, flesh softening, increased physiological disorders, and storage disease. However, there are some cultivars that do not respond well to rapid cooling. A new cultivar, 'Honeycrisp,' develops several disorders unless cooling is delayed for at least 7 days at 20°C (Delong et al., 2004b, 2006). Some cultivars are sensitive to low temperature, which causes chilling injury. Depending on the cultivar and the growing region, recommended storage temperatures can be as high as 4°C (Kupferman, 2003). 'Braeburn' and 'Fuji' require a stepwise cooling: fruit are loaded at 2°C–3°C and gradually cooled to 1°C with 2–3 weeks. 'Pink Lady' requires cooling down slowly to 0.5°C–1°C over a 5–7-day period. Delayed cooling can reduce internal browning of 'Delicious' fruits that have severe watercore at harvest. Watercore is a physiological disorder of apple fruit characterized by water-soaked tissue around the vascular bundles or core area due to the spaces between cells becoming filled with sorbitol-rich fluid instead of air (Marlow and Loescher, 1984). Recovery from watercore can occur if the fruits are handled gently, i.e., with gradual or delayed cooling.

Other factors can influence response of fruit to low temperature. When decreasing temperature from 1°C to 0°C, maintenance of the RH >90% becomes much more difficult, and therefore this last degree of cooling leads to an increased risk of fruit dehydration. For 'McIntosh' and other low-temperature sensitive cultivars, a temperature below 3°C may cause injury during long-term storage, but the risk is low for short-term storage of 2 or 3 months.

TABLE 12.1

Storage Characteristics of Several Apple Varieties in Air and CA

Cultivar	Temperature Control		Air Storage Life (months)	CA				
	Temperature (°C)	Cooling Rate[a]		CO$_2$ (kPa)	O$_2$ (kPa)	Storage Life (months)	Rapid CA Availability[a]	CO$_2$ Sensitivity
Braeburn	1	Stepwise	3–4	0.5	1.5–2	8–10	Slow	Sensitive
Delicious	0	Rapid	3	2	0.7–2	12	Rapid to moderate[b]	
Empire	2	Slow	2–3	2–3	2	5–10	Slow	Sensitive
Fuji	0–1	Stepwise	4	0.5	1.5–2	12	Slow	
Gala	0–1	Rapid	2–3	2–3	1–2	5–6	Rapid	
Golden Delicious	0–1	Rapid	3–4	2–3	1–2	8–10	Rapid	
Granny Smith	1	Rapid	3–4	0.5	1.5–2	10–11	Slow	Sensitive
Jonagold	0	Rapid	2	2–3	1–1.5	5–7	Rapid	
Pink Lady	1	Slow	3–4	1	2	9	Slow	

Sources: Modified from Kupferman, E., *Acta Hort.*, 600, 729, 2003; Swindeman, A.M. Fruit packing and storage loss prevention guidelines, http://postharvest.tfrec.wso. edu/REP2002D.pdf, 2002; Watkins, C.B., Kupferman, E., and Rosenberger, D.A., Apple, Gross, K.C., Wang, C.Y., and Saltveit, M. (Eds.), *The Commercial Storage of Fruits, Vegetables, and Florist and Nursery Crops.* USDA, ARS, *Agriculture Handbook 66,* http://www.ba.ars.usda.gov/hb66/027apple.pdf. Prange, R.K., DeLong, J.M., and Harrison, P.A., *J. Am. Soc. Hort. Sci.*, 128, 603, 607, 2003.

[a] Cooling rate and rapid CA availability (O$_2$ pulldown rates): Rapid = within 3 days; Slow = 5–7 days; Stepwise = 2°C–3°C during loading, 2°C at sealing, and 1°C after 2–3 weeks of CA establishment.

[b] Fruit for long-term CA are recommended to use rapid CA, but water-cored fruit should be stored at high oxygen (2–2.5 kPa) to prevent internal breakdown.

12.1.2.2 Control of Gas Components

After loading fruit, the temperature is reduced to below 10°C, hydrated lime may be placed in the room to remove extra CO_2 and then CA rooms are sealed. After the room temperature is subsequently reduced to below 5°C, nitrogen is injected to pull down the oxygen levels. Until the mid-1970s, 8–10 days were often required to load a CA room and a further 15–20 days were needed for fruit respiration to lower O_2 to 2.5–3 kPa. New technologies, which were developed in the 1980s, such as pressure swing adsorption (PSA) and selective gas-permeable membranes for separating nitrogen, oxygen, carbon dioxide, make it possible to pull down O_2 to 1 kPa in 1 or 2 days. Adoption of rapid and low O_2 CA has significantly improved the storage and marketing quality of fruit.

The recommended gas conditions vary among cultivars, growing regions, and years. General recommendations include 1–2 kPa O_2 and 0.5–2 kPa CO_2 (Kupferman, 2003; Watkins et al., 2004).

The recommendations have been changed over time due to research progress in plant physiology and improvements in gas control technology. Suggested CA storage oxygen levels for 'Delicious' apples in the U.S. Pacific Northwest was 2–3 kPa O_2 until the early 1980s. Lau (1983) and Chen et al. (1985) reported that it looked as though 1 kPa is adequate. British Columbia-produced 'Delicious' can be stored safely at 0.7 kPa O_2 (Lau, 1997), but the same cultivar from other growing regions may show injury at such a low oxygen (Lau et al., 1998). Recently, a chlorophyll fluorescence-based low-O_2 CA system, so-called "dynamic CA" has been developed, which can indicate the lowest acceptable oxygen. Under this system, low oxygen problems can be potentially avoided (Prange et al., 2003).

Table 12.1 shows current recommendations for the major U.S. cultivars. Kupferman (2003) has produced a much longer list including 33 cultivars from over 11 growing regions worldwide.

Low ethylene CA storage (<1 ppm) was evaluated as a method to enhance storage performance by slowing softening and reducing superficial scald (Blanpied, 1990). However, Lau (1999) concluded ethylene scrubbing offers no firmness and scald benefits to 'Golden Delicious,' 'Delicious,' and 'Spartan' apples in low-oxygen CA storage. Therefore, there has been a limited commercial use of this approach.

In addition to slowing ethylene production, respiration, ripening, and senescence of apple fruits and decreasing decay incidence, CA also plays an important role in controlling superficial scald, a severe postharvest physiological disorder on some apple cultivars. In British Columbia, Canada, 0.7 kPa O_2 storage is used as a substitute for DPA treatment, the commonly used antioxidant to control scald (Lau, 1997; DeLong et al., 2004a, 2007). A stress level of low O_2 at 0.25–0.5 kPa for up to 2 weeks, before regular CA storage, controls superficial scald of 'Granny Smith,' 'Delicious,' and 'Law Rome' (Little et al., 1982; Wang and Dilley, 2000; Zanella, 2003). Although it has been subjected to commercial trials, the extent of commercial adoption of this practice is not known.

12.1.2.3 Dynamic CA

Apples in storage are living, and respiration and other metabolic processes are dynamic. However, most recommendations for certain cultivars are static from the beginning to the end of storage. Therefore, it is expected that the recommended CA conditions are not the best fit to the fruit at all points during the storage period. Dynamic CA has been developed to maintain CA conditions in storage over time to best fit the needs of living fruit at any point in time during the storage. A dynamic control of CA through monitoring ethanol production has been proposed (Veltman et al., 2003) but not yet commercialized. Currently commercialized dynamic CA technology is based on a continual measurement

TABLE 12.2

Low-O_2 Thresholds Determined for Each Cultivar by HarvestWatch and the Subsequent Range of O_2 Levels Employed in Dynamic CA Storage

Cultivar	Low-O_2 Threshold (kPa)	O_2 Setting for Each Cultivar (kPa)
Cortland	0.5	0.6–0.8[a]
Delicious	0.4	0.5–0.8
Golden Delicious	0.5	0.5–0.8
Honeycrisp	0.4	0.5–0.8
Jonagold	0.5	0.5–0.8
McIntosh	0.8	0.9–1.0

Sources: Delong, J.M., Prange, R.K., Leyte, J.C., and Harrison, P.A., *HortTechnology*, 14, 262, 2004a.

[a] O_2 levels reflect the ideal setting (0.1–0.2 kPa above the detected low-O_2 threshold value) and the system variation encountered during the storage period.

of chlorophyll fluorescence; i.e., the HarvestWatch™ system designed by Satlantic (Nova Scotia, Canada).

DeEll et al. (1995) first reported that chlorophyll fluorescence was affected by low oxygen and high CO_2 in CA storage of 'McIntosh' apples. DeEll et al. (1998) and Prange et al. (2002, 2003) confirmed the phenomenon. DeLong et al. (2004a) reported that with this chlorophyll fluorescence technology, the lowest acceptable O_2 concentration in CA is much lower than the recommendations in Table 12.1. The range is 0.4–0.8 kPa (Table 12.2) and closer to the theoretical lower limit for aerobic respiration. Using HarvestWatch to dynamically control O_2 concentrations just above the O_2 threshold concentrations in CA-stored apples results in improved quality retention and eliminates the occurrence of superficial scald in susceptible apple cultivars (DeLong et al., 2004a; Zanella et al., 2005).

The principle underlying chlorophyll fluorescence analysis is relatively straightforward. Light energy absorbed by chlorophyll molecules in an apple can undergo one of three fates: it can be used to drive photosynthesis (photochemistry), excess energy can be dissipated as heat, or it can be re-emitted as light-chlorophyll fluorescence. These three processes occur in competition, such that any increase in the efficiency of one will result in a decrease in the yield of the other two. Hence, by measuring the yield of chlorophyll fluorescence, information about changes in the efficiency of photochemistry, and heat dissipation can be gained (Maxwell and Johnson, 2000). There is a consistent emission spike of the chlorophyll fluorescence parameter F_α, which occurs when fruit is exposed to oxygen levels at the anaerobic threshold in CA-stored fruits and vegetables (Figure 12.1, Prange et al., 2003, 2005a). The reason for the change in chlorophyll fluorescence due to low O_2 and high CO_2 has been proposed as an occurrence of cytoplasmic acidosis (Prange et al., 2005a,b). Gout et al. (2001) have shown that as anoxia is imposed, cytoplasmic pH drops (acidosis), and is correlated with the hydrolysis of adenosine triphosphate (ATP) and other nucleoside triphosphate (NTPs), which generate phosphoric acid, e.g., $ATP + H_2O \rightarrow ADP + H_3PO_4$. After returning the cells to normal air, the cytoplasmic pH increases; however, both the vacuolar and extracellular pH decreases. This suggests that cytoplasmic acidosis is reduced by transport of H^+ out of the cytoplasm into the vacuole and cell wall (Gout et al., 2001).

The needs of "organic" apples are increasing, and it has been a challenge to store apples for an extended time without 1-MCP and other chemicals. Dynamic CA increases technology options for the apple industry, especially as a promising tool for the "organic" apple industry (DeLong et al., 2007).

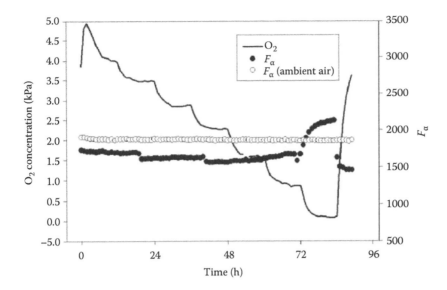

FIGURE 12.1

An example of the F_α fluorescence signal detected in Summerland McIntosh apples held at 20°C in air (open circle) and in a progressively diminished oxygen environment (dark circle). The spike in F_α begins as the chamber oxygen levels fall below 1 kPa at 72 h and continues upward until the oxygen concentration is increased at 84 h. (Reprinted from Prange, R.K., DeLong, J.M., and Harrison, P.A., *J. Am. Soc. Hort. Sci.* 128, 603, 2003.)

12.1.3 1-MCP and CA

In recent years, a very effective agent, 1-MCP, for blocking the ethylene receptor has been widely used in the apple industry for extending storage life (Watkins, 2006; Mattheis, 2008). 1-MCP is an analogue of ethylene, which occupies ethylene receptors such that ethylene cannot bind and elicit action. The affinity of 1-MCP for the receptor is approximately 10 times greater than the affinity of ethylene (Sisler and Serek, 1997). Compared with ethylene, 1-MCP is active at much lower concentrations. 1-MCP also influences ethylene biosynthesis in some species through feedback inhibition. Although Harvesta[SM] (AgroFresh, Springhouse, Pennsylvania), a preharvest sprayable application of 1-MCP is being researched and may be added to the label in the United States, we will only discuss SmartFresh, which is applied after harvest, in this chapter.

Generally, apples treated with SmartFresh are stored in CA for long-term storage. Research shows that the combination of 1-MCP and CA is more effective in maintaining fruit quality than when either 1-MCP-treated apples are stored in air or non-1-MCP-treated fruit are stored in CA. Due to the interaction between CA and 1-MCP, SmartFresh has brought an additional dimension to storage management decisions, which must respect 1-MCP interaction with cultivar, harvest maturity, storage temperature, and atmosphere, and the entire storage strategy.

SmartFresh application has increased the need for accurate maturity information, as optimizing results with SmartFresh technology depends on firmness levels of incoming fruit. Current SmartFresh apple use recommendations are mean flesh firmness 76 N for 'Cripps Pink,' 71 N for 'Delicious,' 'Fuji,' 'Gala,' and 'Granny Smith,' and 67 N for 'Golden Delicious' and 'Jonagold' (AgroFresh Inc., 2007).

Some storage operators have altered CA conditions based on the retarded rates of fruit ripening after SmartFresh treatment. The use of higher O_2 and higher temperature has increased due to the lower risk of low O_2 and/or low temperature injuries, and also to help reduce operational expenses and energy use (Mattheis, 2008). Another change in CA operations following the availability and use of SmartFresh is opening and resealing CA rooms to remove individual orchard lots. Because 1-MCP-treated fruit are less sensitive to short interruptions of CA and/or low-temperature storage conditions, some warehouses have utilized this practice successfully to pack and market individual lots to optimize quality and/or fulfill market demands.

Without SmartFresh treatment, apples ripen faster after being removed from CA and/or low temperature storage. SmartFresh slows the ripening of fruit, thus allowing a longer marketing time (Fan et al., 1999; Bai et al., 2005; Watkins, 2006; Mattheis, 2008).

Some special attention should be paid when applying SmartFresh. For example, the presence of watercore may limit the use of SmartFresh as 1-MCP can delay the dissipation of watercore (Mattheis, 2008).

12.1.4 Modified Atmosphere Packaging

The use of polymeric films for packaging apples using pallet or bin covers, box polyliners, and consumer bagging or wrapping continues to increase. The marketability of fresh-cut apples has expanded rapidly, in part due to improvements in modified atmosphere packaging (MAP) technology. When fruit are sealed in a polymeric film, the gas composition inside the film will be modified by respiration of fruit to low O_2 and/or high CO_2. O_2, and CO_2 concentrations depend on respiration rate, product mass, gas permeability of film, film surface area, and film thickness.

Fruit retains a stable and low respiration rate in cold storage. However, at nonrefrigerated retail display or room temperature conditions, respiration rates generally are high and the onset of climacteric rise can dramatically increase the respiration rates over 10 times. As temperature increases, permeability of film also increases gradually. For instance, O_2 permeability through low-density polyethylene (LDPE) can increase by twofold from 0°C to 15°C (Moyls et al., 1998).

LDPE film is one of the most popular films in produce packaging. The oxygen transmission rate (OTR) of LDPE film is about 7000–7500 mL m^{-2} 24 h^{-1} and CO_2 transmission rate is usually two to six times greater than the OTR (Brody, 2005). New technologies are now allowing the manufacture of very high OTR ($>$15,000 mL m^{-2} 24 h^{-1}) films for application to very high respiration rate commodities. Microperforated and microporous films are alternative approaches to providing high OTR. However, for perforated films, the diffusion to CO_2 is equal to that of O_2 (Brody, 2005). As a result, it is impossible to achieve low O_2 (1%–5%) without accumulating high CO_2 (15%–20%). Thus, these films are applicable only for those products that tolerate high CO_2. Regular perforated films, such as those with 2–10 mm diameter holes, usually cause a very limited atmosphere modification.

The range of steady-state O_2 and CO_2 levels in the packaging varies from 1 to 15 kPa and from 1 to 30 kPa, respectively (Ueda et al., 1993; Soliva-Fortuny et al., 2005; Rojas-Graü et al., 2007). Fresh-cut apples generally are handled under cold chain, allowing a low O_2 concentration to prevent fruit from browning and to reduce microbial growth if there is assurance that the product will not have a risk of exposure to warm temperature during distribution.

Recently, a side-chain polymer technology was developed that allows the film OTR to increase rapidly as temperature increases, thereby avoiding anaerobic conditions subsequent to loss of temperature control. These polymers also provide an adjustable CO_2/O_2 permeability ratio and a range of moisture vapor transmission rates (Zagory, 1997).

Accumulations of organic volatiles, such as alcohols, aldehydes, and jasmonates may have physiological and/or quality effects on fruits. Certain volatiles, such as hexanal, are of overall benefit, while others can lead to decline in quality and storability (e.g., ethanol and acetaldehyde) (Toivonen, 1997).

12.1.5 Edible Coatings

Coatings may be applied to apples just before marketing, to improve appearance to protect against water loss and shrinkage and modify the internal atmospheres that benefit quality by decreasing the rate of metabolism and senescence (Baldwin, 1994). Coatings can increase internal CO_2 and decrease O_2 partial pressures because they are gas barriers in a manner similar to MAP (Banks et al., 1993). There is a wide range of gas permeabilities for edible coatings; O_2 permeabilities ranged from 470 to 21,600, and CO_2 from 1700 to 175,000 mL mil m^{-2} day^{-1} atm^{-1} at 50% RH for 19 commercial fruit coatings (Table 12.3, Hagenmaier and Shaw, 1992). The five coatings available for apples have O_2 permeabilities ranging from 470 to 4400, and CO_2 from 1700 to 16,000 mL mil m^{-2} day^{-1} atm^{-1}, respectively. Coatings applied to the surfaces of fruit and vegetables are commonly called waxes, whether or not any component in the product is actually a wax. The application of coatings to apples, prior to marketing, is standard practice in many countries, e.g., the United States. Freshly harvested apples have their own waxy coating, however, they are washed at the fruit packing sheds to remove dust and chemical residues and this washing removes about half of the original apple wax.

Major apple coatings are carnauba-based microemulsions, shellac-based solutions, or the mixture of the two coatings. Both carnauba and shellac have been used on a

TABLE 12.3

Gas Transmission Rates (mL mil m^{-2} day^{-1} atm^{-1}) of 19 Commercial Fruit Coatings, a Shrink-Wrap Film, and Four Experimental Coatings at 50% or 92% RH

Major Ingredients	O_2 50% RH	CO_2 50% RH	C_2H_4 50% RH	Water Vapor 92% RH
Waxes, natural and synthetic, and fatty acids	21,600	175,000	88,000	813,369
Polyethylene and shellec	11,200	37,000	17,000	4,256,000
Carnauba and fatty acids	10,300	49,000	14,000	13,240,889
Polyethylene-vinyl asetate copolymer (shrink-wrap film D955)[a]	9,100	27,000	12,000	208,071
Carnauba and shellac[b]	4,400	16,000	7,400	756,622
Shellac, carnauba and fatty acids[b]	4,300	14,000	5,000	2,931,911
Carnauba, fatty acids and shellac[b]	2,700	9,800	3,100	2,175,289
Coumarone-indene resin	2,600	7,600	940	1,418,667
Rosin and carnauba	2,300	7,200	930	2,837,333
Hydrocarbon resins and fatty acids	2,200	4,600	670	10,403,556
Rosin, oleic acid and shellac	2,000	3,600	450	4,256,000
Carnauba, morpholine and oleic acid[a]	2,000	7,800	3,200	567,467
Shellac and rosin	1,700	3,800	610	3,404,800
Shellac and fatty acids	1,000	3,500	360	8,228,267
Shellac and rosin	1,000	2,600	270	13,524,622
Carnauba, shellac and rosin	1,000	2,800	400	9,457,778
Sucrose esters and carboxymethyl cellulose[b]	800	4,500	1,980	8,228,267
Rosin, oleic acid and ammonia[a]	780	2,200	170	28,373,333
Shellac and rosin	730	1,800	180	10,403,556
Shellac	640	1,900	170	5,485,511
Shellac, rosin and morpholine	550	1,700	140	11,349,333
Shellac[b]	470	1,700	180	9,079,467
Shellac and morpholine[a]	370	1,100	90	17,969,778
Modified maleic resin and morpholine[a]	250	910	310	3,404,800

Sources: Modified from Hagenmaier, R.D. and shaw, P.E., *J. Am. Soc. Hort. Sci.*, 117, 105, 1992.
[a] Noncommercial coatings or shrink-wrap film.
[b] Coatings available for apples.

variety of foods for decades. Sucrose fatty acid ester coatings are approved for fruits, however, only a few apple packinghouses use them due to unsatisfactory gloss appeal that results.

'Delicious' apples have been a key apple cultivar in the development of apple coating formulations and technology, and because this cultivar is relatively tolerant to high gas barriers, the coatings developed have tended to emphasize improvement of visual gloss with little need for other effects on the fruit that might result from a high barrier to gas exchange (Baldwin, 1994). A shellac coating seems an excellent fit for dark red 'Delicious' apples, because it imparts high gloss, hides bruises and forms a MA condition that tends to preserve firmness and prolong shelf life in this cultivar (Bai et al., 2002a,b, 2003). Shellac has a problem with flaking (Hagenmaier and Shaw, 1992; Baldwin, 1994), which occurs when fruits are removed from cold storage to ambient air. Humidity in the air condenses on the fruit surface and melts part of the coating. After the condensation dries out, white spots remain on the surface. This whitening or chalky appearance limits marketability.

Carnauba use on apple coating has been increasing in the last decade. Carnauba wax does not discolor, but has less gloss than shellac. Bai et al. (2002a) reported that application of carnauba coatings results in less modification of the internal atmosphere in coated 'Delicious' apples compared with shellac, and is more effective in preventing weight loss.

It is well known that when fruit is separated by a barrier, such as a coating or packaging, from exchange of gases with the atmosphere there is the possibility for the respiration to become anaerobic, leading to the associated development of off-flavor. Coatings and packages developed for one type of fruit may not be suitable for another. Bai et al. (2002b, 2003) applied several experimental coatings, with a very wide range of gas permeabilities, to four apple cultivars (Figure 12.2). Shellac coating results in maximum fruit gloss, lowest internal O_2, highest CO_2, and least loss of flesh firmness for all of the cultivars. 'Granny Smith' with shellac has very low internal O_2 (<2 kPa), and freshly harvested 'Braeburn' has very high internal CO_2 (25 kPa). These excessive modifications of internal gas induces an abrupt rise of the respiratory quotient, prodigious accumulation of ethanol in both 'Braeburn' and 'Granny Smith', and flesh browning at the blossom end of 'Braeburn'. In addition, the shellac coating results in an unusually high accumulation of ethanol in freshly harvested and 5-month-stored 'Fuji.' Candelilla and carnauba–shellac coatings maintain better internal O_2 and CO_2 and better quality for 'Fuji,' 'Braeburn,' and 'Granny Smith,' although even these coatings may present too much of a gas barrier for 'Granny Smith.' In general, the gas permeabilities of the coatings are useful as indicators of differences in coating barrier properties, but do not account for differences in pore blockage, which is expected to play a more important role in the gas exchange for the whole fruit (Bai et al., 2002b, 2003.). Candelilla wax coating, which has lower permeability than the carnauba–shellac coating, results in higher values in internal O_2 and lower CO_2. We can speculate that the carnauba–shellac coating has a greater tendency to block pores in the fruit skin, but have insufficient evidence to conclude that is the case (Bai et al., 2003).

'Braeburn' and 'Granny Smith' apples are sensitive to high CO_2 and also differ from 'Delicious' in the porosity of the peel and the structure of blossom- and stem-ends, and thus the same coating may result in a different modified internal atmospheres, and physiological reactions to a given internal gas composition may also differ. These considerations suggest it is appropriate to once again determine if coatings developed for 'Delicious' are optimum for other cultivars. There also seems to be a possibility that the trend in consumer preference for more 'natural' products might lead to less preference for high glossy coatings. In such cases, candelilla wax may be acceptable, which offers a moderate gas barrier and a natural gloss (Bai et al., 2003).

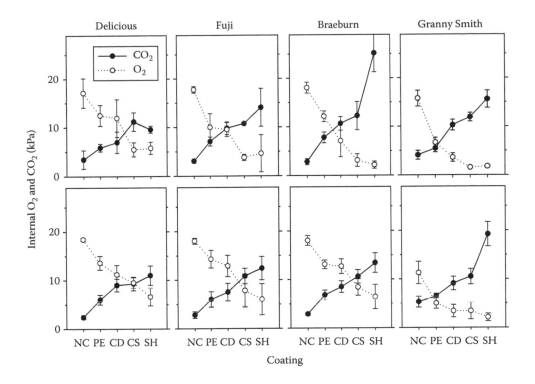

FIGURE 12.2
Internal CO_2 and O_2 partial pressures of apples with the different coating treatments at 20°C. Freshly harvested (upper) or 20-week commercially stored (bottom) apples of four varieties were coated with five coating treatments and held at 20°C for 2 weeks. Abbreviations for coating treatments: NC, noncoated; PE, polyethylene (not approved for apples); CD, candelilla; CS, carnauba–shellac; and SH, shellac. (Data from Bai, J., Hagenmaier, R.H., and Baldwin. E.A., *J. Agric. Food Chem.* 50, 7660, 2002.)

12.2 Pear

Pears are similar to apples regarding responses to environment temperature, humidity, and atmosphere. Thus many CA/MA storage technologies, which were developed for apples first, have been adapted for pear storage later. For instance, rapid cooling, rapid CA, and low O_2 CA have been adopted by the pear industry successfully, and dynamic CA is a potential tool for pears. On the other hand, pears are different from apples in many respects. The most significant is that pears generally are consumed after they are fully ripened with a juicy and buttery texture and distinct pear aroma and taste. Thus any storage and handling technologies, which may consequently disrupt the ripening ability, should be avoided. A good example is 1-MCP technology; postharvest application of 1-MCP has been shown in apples to be very effective at reducing superficial scald and extending storage life, however, in pear fruit, 1-MCP prevents the fruit from properly ripening.

12.2.1 Maturity

Pear fruits are capable of developing good dessert quality upon ripening only when they are harvested at proper maturity. Pear fruits harvested with improper maturity are more susceptible to physiological disorders and have a shorter storage life. Immature pear fruits are more susceptible to superficial scald, shriveling, and friction-generated discoloration. However,

TABLE 12.4

Storage Characteristics of Several Pear Varieties in Air and CA at −1.4°C to 0°C (Modified from Chen, 2004; Kader, 2007; Kupferman, 2003; Sugar, 2007.)

Cultivar	Harvest Flesh Firmness (N)	Air Storage Life (Months)	CA		
			CO$_2$ (kPa)	O$_2$ (kPa)	Storage Life (Months)
Abate Fetel	58–67	3–4	0–1	1–2	4–6
Anjou	58–67	6–7	0–0.5	0.7–2	8–10
Bartlett	67–84	2–3	0–0.5	1–2	3–6
Bosc	62–71	4	0.5–1.5	1–2.5	5–7
Comice	49–58	4	0.5–4	1.5–4	5–6
Concorde	49–58	4–5	0–1	0.5–2	6–7
Conference	49–58	5–6	0.5–1.5	1–2.5	7–8
Forelle	58–67	4–5	0–1.5	1–2	6–7
Packham's Triumph	62–71	6	1.5–2.5	1.5–2	7–8
Winter Nelis	58–67	7–8	0–1	1–3	8–10

Source: Modified from Chen, P.M., Pear, Gross, K.C., Wang, C.Y., and Saltveit, M. (Eds.), *The Commercial Storage of Fruits, Vegetables, and Florist and Nursery Crops, USDA, ARS, Agriculture Handbook 66*, http://www.bu.ars.usda.gov/hb66/107pear.pdf, 2004; Kader, A.A., Controlled atmosphere, Mitcham, E.J. and Elkins, R.B. (Eds.), *Pear Production and Handling Manual*, University of California Agriculture and Natural Resources Publication 3483, 2007, 175–177; Kupferman, E., *Acta Hort.*, 600, 729, 2003; Sugar, D., Postharvest handling of winter pears, Mitcham, E.I., Elkins, R.B. (Eds.), *Pear Production and Handling Manual*, University of California Agriculture and Natural Resources Publication 3483, 2007, 171–174.

overmature fruits tend to have higher incidence of core breakdown, CO$_2$ injury, storage decay, and mealy texture upon ripening (Chen, 2004). Flesh firmness is a reliable maturity indicator (Chen, 2004), but sometimes a starch test is necessary, because in certain seasons, firmness will remain high for a time while the starch content changes (Kupferman, 2002).

The measurement of pear flesh firmness uses the same penetrometer as used for apples: however, an 8 mm plunger is used for pears rather than the 11 mm plunger used for apples. Recommended ranges of firmness for harvesting different pear cultivars are as follows: 'd'Anjou,' 58 to 67 N; 'Bartlett,' 67 to 84 N, 'Bosc,' 62 to 71 N, and 'Comice,' 49 to 58 N (Table 12.4; Sugar, 2007).

12.2.2 CA Regime

12.2.2.1 Temperature

Pears are very sensitive to temperature. The storage life of 'd'Anjou' and 'Bartlett' pears is 35%–40% longer at −1°C than at 0°C (Porritt, 1964). Most pears in the Pacific Northwest are stored at −1°C with RH of 90%–94% (Hansen and Mellenthin, 1979). As long as the total soluble solids content in the core area is at least 11%, pears can be successfully stored at −1°C without freezing. Pears must be at final storage temperature before the oxygen is lowered. Many packing houses use thermocouples in the air and in fruit to determine temperatures at selected areas in the storage. Precise temperature control is needed to prevent freezing when pears are stored at these low temperatures. Rapid removal of field heat and prompt cooling of harvested pears are essential for long-term storage. During the pull-down period, room temperatures of −3.5°C to −2.0°C can be used but should be raised to −1°C as the fruit temperatures approach the target temperature (Hansen and Mellenthin, 1979). Delay in cooling shortens storage life. It has been recommended to rapidly cool down core temperature within 3 days of harvest for 'd'Anjou' pears (Swindeman, 2002).

Pears lose moisture rapidly; hence, it is advisable to hold the RH >90%. Polyethylene bin liners are effective in controlling moisture loss.

12.2.2.2 CA Regimes

CA storage has been used successfully to extend the storage life of pears and to maintain greater capacity for ripening. The optimum and safe CA atmosphere for commercial use is 1 to 2.5 kPa $O_2 + 0$ to 1.5 kPa CO_2 (Table 12.4). Use of short-term high CO_2 treatment improves keeping quality of 'd'Anjou' pears (Wang and Mellenthin, 1975). Treatment with 12% CO_2 for 2 weeks immediately after harvest has a beneficial effect on retention of ripening capacity. Keeping 'd'Anjou' pears in a low O_2 atmosphere (1.0 kPa) with ≤0.1 kPa CO_2 can also maintain higher dessert quality and reduce incidence of superficial scald after long-term cold storage (Chen, 2004).

A stepwise low O_2 CA procedure was introduced by Chen and Varga (1997) in which 'd'Anjou' pears are stored in 0.8% O_2 + <0.1% CO_2 at −1°C for 3–4 months before transfer to standard CA. Low O_2 effectively controls superficial scald without inducing "black speck" and "pithy brown core." For superficial scald control, the primary method is a postharvest ethoxyquin treatment with a dosage of 2700 ppm. Chen and Varga (1997) combined CA and ethoxyquin for superficial scald control. They showed 'd'Anjou' fruit destined for mid term (5–6 months) or long term (7–8 months) can be stored in CA (1.5 kPa O_2 and <0.5 kPa CO_2 throughout the entire storage period) plus a reduced ethoxyquin dosage (1000 ppm). Ethoxyquin needs to be applied within 2 days.

CA storage regimes for pears are similar to apples, and dynamic CA has a potential to maintain a better quality in extended storage (Prange et al., 2002, 2003, 2005a,b).

12.2.2.3 Pear Ripening

Most pear cultivars require a chilling period after harvest to gain ripening ability. Storage duration at −1°C, which is required to induce normal ripening of pear fruit, is 2–4 weeks for 'Bartlett' (Agar et al., 1999), 2–3 weeks for 'Bosc' (Chen et al., 1982), and 7–8 weeks for 'd'Anjou' (Chen et al., 1982). After receiving enough chilling units, fruit readily ripen at ambient temperatures. The best ripening temperature range is about 15°C–21°C. Higher or lower temperatures result in poorer quality or excessive decay (Prange et al., 1988). Most cultivars fail to ripen at 30°C or above.

Freshly harvested winter pears do not produce ethylene (Wang et al., 1985). Exposure of pears to cold storage stimulates synthesis of 1-aminocyclopropane-1-carboxylic acid (ACC) because low temperature induces biosynthesis of ACC oxidase and ACC synthase (Wang et al., 1985; Lelièvre at al., 1997). Exogenous ethylene, at 100 μL L^{-1}, has been used commercially to precondition underchilled 'Bartlett' and 'd'Anjou' pears at 20°C for 2–3 days before shipment. Preconditioned pear fruits are capable of ripening normally and uniformly upon reaching the retail markets or for the canning process (Chen et al., 1996; Agar et al., 1999).

1-MCP has been considered a potential tool to extend storage life of pears. However, Chen and Spotts (2006) report that when 1-MCP dosage is 30 μL L^{-1} or higher, pear fruit loses its normal ripening ability; on the other hand, when 1-MCP dosages are 20 μL L^{-1} or lower, superficial scald cannot be controlled after 4 months of cold storage. 'Bartlett' pears are less sensitive to 1-MCP in comparison with 'd'Anjou' (Bai et al., 2006). Bai et al. (2006) reinitiated the ripening ability of 1-MCP-treated 'Bartlett' pears by holding fruit at 10°C–20°C for 10–20 days. However, it is difficult to ripen 1-MCP-treated 'd'Anjou' pears (Bai et al., 2006).

12.2.3 MAP

Because pears are sensitive to high CO_2, and because CO_2 concentration in MAP is generally high and difficult to control, there is very limited commercial use for this technology. In CA storage, 'd'Anjou' pears should be stored with the oxygen at least 1 kPa higher than the CO_2 at all times. However, there has been some research indicating fruit stored at >1°C can tolerate higher CO_2 levels. Sugar (2001) reported that 'Comice' in MAP (~4 kPa O_2 + 4 kPa CO_2) maintains storage life for 4–5 months with a comparable quality to those held in standard CA (2 kPa O_2 + <1 kPa CO_2). Similarly, 'Bosc' pears can be stored for 5 months successfully (MAP: 3 kPa O_2 + 4 kPa CO_2), but low O_2 and/or high CO_2 injury is observed after 6 months. Low O_2 and/or high CO_2 injury is also observed in MAP-stored 'Starkrimson' (1–2 kPa O_2 + 4–5 kPa CO_2), 'Bartlett' (1–2 kPa O_2 + 5 kPa CO_2), and 'd'Anjou' (4 kPa O_2 + 4 kPa CO_2) pears while fruit stored in standard CA (2 kPa O_2 + <1 kPa CO_2) maintain good quality (Sugar, 2001).

12.2.4 Edible Coating

According to the Kupferman (1998) survey in Washington, Oregon, and California, which produce more than 90% of U.S. pears, edible coatings are applied on 71% 'd'Anjou,' 25% of 'Bartlett,' and 26% of 'Bosc' fruits. Carnauba-based waxes are the major coatings for pears, but sucrose fatty acid ester and shellac are also used by some packers in some cultivars. Carnauba waxes used in pears are specially formulated for pears or diluted apple waxes. Pear wax has less total solids than apple wax, because a high gas barrier may cause anaerobic metabolism in pears, and disrupt the ripening.

Amarante (1998) and Amarante and Banks (2001) showed that differences in skin structure of pears affect coating performance significantly. For cultivars without lignified cells in the skin ('Bartlett,' 'Comice,' and 'Packham's Triumph'), the diffusion of water vapor, CO_2 and O_2 decreases markedly with a small increase in the total solids in a carnauba wax, because the coating covers the entire skin surface, including lenticels (Figure 12.3). However, for 'Bosc' pears, which characteristically have lignified cells in the

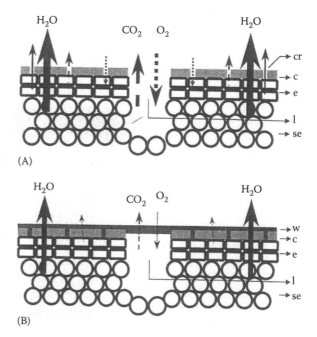

FIGURE 12.3
Hypothetical model for the differences in skin permeance to O_2 (arrow with dotted line), CO_2 (arrow with dashed line), and water vapor (arrow with solid line) in a non-coated (A) and a coated (B) pear without lignified cells in the skin. Arrow size is proportional to the differences in permeance values (cr = crack in the cuticle; c = cuticle; e = epidermis; l = lenticel; se = sub-epidermis; w = wax coating layer). (From Amarante, C., Gas Exchange, Ripening Behaviour and Postharvest Quality of Coated Pears, PhD dissertation, Massey University, Palmerston North, New Zealand, 1998. With permission.)

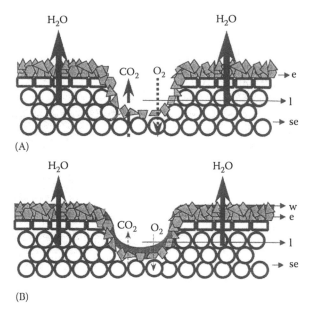

FIGURE 12.4

Hypothetical model for the differences in skin permeance to O_2 (arrow with dotted line), CO_2 (arrow with dashed line), and water vapor (arrow with solid line) in a non-coated (A) and a coated (B) pear with lignified cells in the skin. Arrow size is proportional to the differences in permeance values (e = epidermis; 1 = lenticel; se = sub-epidermis; w = wax coating layer). (From Amarante, C., Gas Exchange, Ripening Behaviour and Postharvest Quality of Coated Pears, PhD dissertation, Massey University, Palmerston North, New Zealand, 1998. With permission.)

skin, only the diffusion of CO_2 and O_2 markedly decreases, whereas the water vapor diffusion decreases little by increasing the total solids content. This was because almost all O_2 and CO_2 exchange occurs through the lenticels, and the coating effectively covers them; on the other hand, increasing the coating concentration is not effective in covering the lignified cells in the epidermis which is the major pathway for water vapor exchange (Figure 12.4).

12.3 Asian Pear

Asian pears were developed from *Pyrus ussuriensis* (cold tolerance, Northeast China and North Korea), *P. bretschneideri* (temperate-zone, Central China), and *P. pyrifolia* (warm rainy regions, South China, South Korea, and Japan). Asian pears are eaten at a firm and crisp stage right after harvest or after cold storage. The texture is like a very crisp and juicy apple. They are also known in the United States as Chinese pears, Japanese pears, and sand or apple pears. There are three types of Asian pears in appearance. They are (1) round or flat fruit with green-to-yellow skin ('Nijisseiki' and 'Shinseiki'), (2) round or flat fruit with bronze-colored skin and a light bronze-russet ('Hosui' and 'Shinko'), and (3) pear-shaped fruit with green or russet skin ('Ya Li' and 'Kuerle'). Some cultivars, such as 'Nijisseiki,' 'Kosui,' and 'Niitaka' have a nonclimacteric respiratory pattern and produce little ethylene (<0.1 μL kg^{-1} h^{-1}). Other cultivars, such as 'Yali,' 'Tsu Li,' 'Hosui,' 'Kikusui,' and 'Chojuro' have a climacteric respiratory pattern and produce ethylene up to 9–14 μL kg^{-1} h^{-1} ('Ya Li' and 'Tsu Li') or 1–3 μL kg^{-1} h^{-1} (other cultivars) at 0°C (Crisosto, 2004).

12.3.1 Maturity and Harvest

Most growers determine harvest time by fruit taste and color. Sugar content over 12.5% usually is adequate and fruit pressure of 35.5–49 N (8–11 lb) seems satisfactory. Fruit pressure is not as good a measure of maturity in Asian pears as it is in European pears. The color of russet-type fruit changes from green to brown, and the ground color of green fruit changes from green to yellow. Color and sugar content best determine time to harvest.

Some green Asian pears, such as 'Ya Li' do not change color much at maturity. All Asian pears must be carefully handled to minimize bruising and brown marks and stem punctures. Overmature fruit quickly show roller bruises, fingerprints, and other signs of handling at harvest. Undermature fruits are poor in flavor and ruin the market for Asian pears. Multiple harvests are necessary to get mature, quality fruit from most cultivars. Asian pears are harder to handle than European pears and they are not suited to large, fast-moving packinghouse lines. Fruit is best field-packed from picking containers to packing boxes or trays.

12.3.2 Storage Regimes

12.3.2.1 Temperature

Optimum storage conditions are $0°C \pm 1°C$ with RH of 90%–95% (Crisosto, 2004). The actual freezing temperature is influenced by SSC of the fruit; at soluble solids levels of 10–12%, the freezing points would be approximately $-1.7°C$ to $-2.2°C$. Asian pears are susceptible to water loss. When water loss has been higher than 5%–7%, fruits become dehydrated and have a shriveled appearance, especially the 'Kosui' and 'Hosui' varieties.

After 2–3 months 'Hosui' and 'Shinko' fruits get spongy, show some storage rot, and after 4 months, may show internal breakdown in the core area. Less mature fruits get spongy sooner than the fully mature fruit. At room temperature of 20°C, the fruit begins to soften or get spongy after 14–21 days.

12.3.2.2 CA Consideration

Based on limited studies, it appears that the magnitude of CA benefits for Asian pears is cultivar-specific and is generally less than that of European pears and apples. O_2 levels of 1–3 kPa for 'Nijisseiki' and 3%–5% for 'Ya Li' help retain firmness and delay changes in skin color. Asian peas are sensitive to CO_2 injury (>2 kPa for most cultivars) when stored for more than 1 month.

References

Agar, I.T., W.V. Biasi, and E.J. Mitcham. 1999. Exogenous ethylene accelerates ripening responses in 'Bartlett' pears regardless of maturity or growing region. *Postharvest Biol. Technol.* 17:67–78.

AgroFresh Inc. 2007. *Factors Affecting Apple Crunch—How to Optimize Storage for Fruit of Varied Quality Levels*. AgroFresh Inc., Springhouse, PA.

Amarante, C. 1998. Gas Exchange, Ripening Behaviour and Postharvest Quality of Coated pears. PhD dissertation, Massey University, Palmerston North, New Zealand.

Amarante, C. and N.H. Banks. 2001. Postharvest physiology and quality of coated fruits and vegetables. *Hort. Rev.* 26:161–237.

Bai, J., E.A. Baldwin, and R.H. Hagenmaier. 2002a. Effect of several alternative coatings on the gloss, internal gases and quality of 'Delicious' apple fruit. *HortScience*. 37:559–563.

Bai, J., R.H. Hagenmaier, and E.A. Baldwin. 2002b. Volatile response of four apple varieties with different coatings during marketing at room temperature. *J. Agric. Food Chem.* 50:7660–7668.

Bai, J., R.H. Hagenmaier, and E.A. Baldwin. 2003. Coating selection for 'Delicious' and other apples. *Postharvest Biol. Technol.* 28:381–390.

Bai, J., E.A. Baldwin, K.L. Goodner, J.P. Mattheis, and J.K. Brecht. 2005. Response of four apple cultivars to 1-methylcyclopropene treatment and controlled atmosphere storage. *HortScience*. 40:1534–1538.

Bai, J., J.P. Mattheis, and N. Reed. 2006. Re-initiating softening ability of 1-methylcyclopropene-treated 'Bartlett' and 'd'Anjou' pears after regular air or controlled atmosphere storage. *J. Hort. Sci. Biotechnol.* 81:959–964.

Baldwin, E.A., 1994. Edible coatings for fresh fruits and vegetables: past, present, and future. In: *Edible Coatings and Films to Improve Food Quality.* J.M. Krochta, E.A Baldwin, and M.O. Nisperos-Carriedo (Eds.), Technomic Publishing Co., Lancaster, PA, pp. 25–64.

Banks, N.H., B.K. Dadzie, and D.J. Cleland. 1993. Reducing gas exchange of fruits with surface coatings. *Postharvest Biol. Technol.* 3:269–284.

Blanpied, G.D. 1990. Controlled atmosphere storage of apples and pears. In: M. Calderon and R. Barkai-Golan (Eds.), *Food Preservation by Modified Atmospheres.* CRC Press, Boca Raton, FL, pp. 265–299.

Brody, A.L. 2005. What's fresh about fresh-cut? *Food Technol.* 56:124–125.

Byers, R.E. 1997. Effect of aminoethoxyvinylglycine (AVG) on preharvest fruit drop and maturity of 'Delicious' apples. *J. Tree Fruit Prod.* 2:53–75.

Chen, P.M. 2004. Pear. In: Gross, K.C., C.Y. Wang, and M. Saltveit (Eds.), The Commercial Storage of Fruits, Vegetables, and Florist and Nursery Crops. *USDA, ARS, Agriculture Handbook 66*, http://www.ba.ars.usda.gov/hb66/107pear.pdf

Chen, P.M. and R.A. Spotts. 2006. Changes in ripening behaviors of 1-MCP-treated 'd'Anjou' pears after storage. *Int. J. Fruit Sci.* 5:3–17.

Chen, P.M. and D.M. Varga. 1997. CA regimes for control of superficial scald of 'd'Anjou' pears. In: *Proceedings of 13th Annual WA Tree Fruit Postharvest Conference*, E. Kupferman (Ed.), Wenatchee WA, pp. 17–26.

Chen, P.M., D.G. Richardson, and W.M. Mellenthin. 1982. Differences in biochemical composition between 'Beurre d'Anjou' and 'Bosc' pears during fruit development and storage. *J. Am. Soc. Hort. Sci.* 107:807–812.

Chen, P., K. Olsen, and M. Meheriuk. 1985. Effects of low-oxygen atmosphere on storage scald and quality preservation of 'Delicious' apples. *J. Am. Soc. Hort. Sci.* 110:16–20.

Chen, P.M., S.R. Drake, D.M. Varga, and L. Puig. 1996. Precondition of 'd'Anjou' pears for early marketing by ethylene treatment. *J. Food Qual.* 19:375–390.

Crisosto, C.H. 2004. Asian pear. In: Gross, K.C., C.Y. Wang, and M. Saltveit (Eds.), The Commercial Storage of Fruits, Vegetables, and Florist and Nursery Crops. *USDA, ARS, Agriculture Handbook 66*, http://www.ba.ars.usda.gov/hb66/031asianpear.pdf

DeEll, J.R., R.K. Prange, and D.P. Murr. 1995. Chlorophyll fluorescence as a potential indicator of controlled-atmosphere disorders in 'Marshall' McIntosh apples. *HortScience.* 30:1084–1085.

DeEll, J.R., R.K. Prange, and D.P. Murr. 1998. Chlorophyll fluorescence techniques to detect atmospheric stress in stored apples. *Acta Hort.* (R. Bieleski, W. Laing, and C. Clark, Eds.) 464:127–131.

DeLong, J.M., R.K. Prange, J.C. Leyte, and P.A. Harrison. 2004a. A new technology that determines low-oxygen thresholds in controlled-atmosphere-stored apples. *HortTechnology.* 14:262–266.

DeLong, J.M., R.K. Prange, and P.A. Harrison. 2004b. The influence of pre-storage delayed cooling on quality and disorder incidence in 'Honeycrisp' apple fruit. *Postharvest Biol. Technol.* 34:353–358 (Erratum version).

DeLong, J.M., R.K. Prange, P.A. Harrison, C.G. Embree, D.S. Nichols, and A.H. Harrison. 2006. The influence of crop-load, delayed cooling and storage atmosphere on post-storage quality of 'Honeycrisp'™ apples. *J. Hort. Sci. Biotechnol.* 81:391–396.

DeLong, J.M., R.K. Prange, and P.A. Harrison. 2007. Chlorophyll fluorescence-based low-O$_2$ CA storage of organic 'Cortland' and 'Delicious' apples. *Acta Hort.* (D.R. Lynch and R.K. Prange, Eds.) 737: 31–37.

Embree, C.G., M.T.D. Myra, D.S. Nichols, and A.H. Wright. 2007. Effect of blossom density and crop load on growth, fruit quality, and return bloom in 'Honeycrisp' apple. *HortScience.* 42:1622–1625.

Fan, X., S.M. Blankenship, and J.P. Mattheis. 1999. 1-Methylcyclopropene inhibits apple ripening. *J. Am. Soc. Hort. Sci.* 124:690–695.

Gout, E., A.-M. Boisson, S. Aubert, R. Douce, and R. Bligny. 2001. Origin of the cytoplasmic pH changes during anaerobic stress in higher plant cells. carbon-13 and phosphorous-31 nuclear magnetic resonance studies. *Plant Physiol.* 125:912–925.

Greene, D.W. 2005. Time of aminoethoxyvinylglycine (AVG) application influences preharvest drop and fruit quality of 'McIntosh' apples. *HortScience*. 40:2056–2060.

Hagenmaier, R.D. and P.E. Shaw. 1992. Gas permeability of fruit coating waxes. *J. Am. Soc. Hort. Sci.* 117:105–109.

Hansen, E. and W.M. Mellenthin. 1979. Commercial handling and storage practices for Winter pears. *Oreg. Agr. Expt. Sta. Spec. Rpt.* 550: pp. 12.

Kader, A.A. 2007. Controlled atmosphere. In: E.J. Mitcham and R.B. Elkins (Eds.), *Pear Production and Handling Manual.* University of California Agriculture and Natural Resources Publication 3483. pp. 175–177.

Kupferman, E. 1997. Management of Gala, Braeburn, and Fuji for quality. In: *Proceedings of Apple Harvesting, Handling, and Storage Workshop*, NRAES 112. Cornell University, Ithaca, NY, pp. 1–5.

Kupferman, E. 1998. Postharvest applied chemicals to pears: a survey of pear packers in Washington, Oregon and California. *Tree Fruit Postharvest J.* 9:3–24. http://postharvest.tfrec.wsu.edu/pgDisplay.php?article = J911A.

Kupferman, E. 2002. Observations on harvest maturity and storage of apples and pears. *Postharvest Information Network.* http://postharvest.tfrec.wsu.edu/EMK2002A.pdf.

Kupferman, E. 2003. Controlled atmosphere storage of apples and pears. *Acta Hort.* (J. Oosterhaven and H.W. Peppelenbos, Eds.) 600:729–735.

Larrigaudiere, C., I. Lentheric, and M. Vendrell. 1998. Relationship between enzymatic browning and internal disorders in controlled atmosphere stored pears. *J. Sci. Food Agric.* 78:232–236.

Lau, O.L. 1983. Effects of storage procedures and low oxygen and high carbon dioxide atmospheres on storage quality of 'Spartan' apples. *J. Am. Soc. Hort. Sci.* 108:953–957.

Lau, O.L. 1985. Harvest guide for B.C. apples. *British Columbia Orchardist* 7:1A–20A.

Lau, O.L. 1997. The effectiveness of 0.7% O_2 to attenuate scald symptoms in 'Delicious' apples is influenced by harvest maturity and cultivar strain. *J. Am. Soc. Hort. Sci.* 122:691–697.

Lau, O.L. 1999. Ethylene scrubbing offers no firmness and scald benefits to 'Golden Delicious,' 'Delicious,' and 'Spartan' apples in low-oxygen storage. *Tree Fruit Postharvest J.* 10:15–17. http://postharvest.tfrec.wsu.edu/pgDisplay.php?article = J10I1D.

Lau, O.L., C.L. Barden, S.M. Blankenship, P.M. Chen, E.A. Curry, J.R. DeEll, L. Lehman-Salada, E.J. Mitcham, R.K. Prange, and C.B. Watkins. 1998. A North American cooperative survey of 'Starkrimson Delicious' apple responses to 0.7% O_2 storage on superficial scald and other disorders. *Postharvest Biol. Technol.* 13:19–26.

Lelièvre, J.M., L. Tichit, P. Dao, L. Fillion, Y.W. Nam, J.C. Pech, and A. Latchè. 1997. Effects of chilling on the expression of ethylene biosynthetic genes in 'Passe-Crassane' pear (Pyrus communis L.) fruit. *Plant Mol. Biol.* 33:847–855.

Little, C.R., J.D. Faragher, and H.S. Taylor. 1982. Effects of initial low O_2 stress treatments in low oxygen modified atmosphere storage of 'Granny Smith' apples. *J. Am. Soc. Hort. Sci.* 107:320–323.

Marlow, G.C. and W.H. Loescher. 1984. Watercore. *Hort. Rev.* 6:189–251.

Mattheis, J.P. 2008. How 1-methylcyclopropene has altered the Washington state apple industry. *HortScience.* 43:99–101.

Maxwell, K. and G.N. Johnson. 2000. Chlorophyll fluorescence—a practical guide. *J. Expt. Botany.* 51:659–668.

Moyls, A.L., D.-L. McKenzie, R.P. Hocking, P.M.A. Toivonen, P. Delaquis, B. Girard, and G. Mazza. 1998. Variability in O_2, CO_2 and H_2O transmission rates among commercial polyethylene films for modified atmosphere packaging. *Trans. ASAE* 41:1441–1446.

Porritt, S.W. 1964. The effect of temperature on postharvest physiology and storage-life of pears. *Can. J. Plant Sci.* 44:568–579.

Prange, R.K. and J.M. DeLong. 1998. Determination of maturity for long-term storage of apples. 14th Annual Postharvest Conference, Yakima, Washington. http://postharvest.tfrec.wsu.edu/pgDisplay.php?article = PC98L.

Prange, R.K., J.M. Delong, J.C. Leyte, and P.A. Harrison. 2002. Oxygen concentration affects chlorophyll fluorescence in chlorophyll-containing fruit. *Postharvest Biol. Technol.* 24:201–205.

Prange, R.K., J.M. DeLong, and P.A. Harrison. 2003. Oxygen concentration affects chlorophyll fluorescence in chlorophyll containing fruits and vegetables. *J. Am. Soc. Hort. Sci.* 128:603–607.

Prange, R.K., J.M. DeLong, B.J. Daniels-Lake, and P.A. Harrison. 2005a. Innovation in controlled atmosphere technology. *Stewart Postharvest Rev.* 1(3):1–11(11).

Prange, R.K., DeLong, J.M., and Harrison, P.A. 2005b. Quality management through respiration control: Is there a relationship between lowest acceptable respiration, chlorophyll fluorescence and cytoplasmic acidosis? *Acta Hort.* (F. Mencarelli and P. Tonutti, Eds.) 682:823–830.

Prange, R.K., C.G. Embree, and H.-Y. Ju. 1988. Effects of simulated shelf-life conditions on consumer acceptance and weight loss in 'Clapp's Favorite', 'Bartlett', 'Flemish Beauty' and 'Anjou' pears. *Fruit Var. J.* 42:76–79.

Rojas-Graü, M.A., R. Grasa-Guillem, and O. Martín-Belloso. 2007. Quality changes in fresh-cut Fuji apple as affected by ripeness stage, antibrowning agents, and storage atmosphere. *J. Food Sci.* 72:S36–43.

Sisler, E.C. and M. Serek. 1997. Inhibitors of ethylene response in plants at the receptor level: recent developments. *Physiol. Plant.* 100:577–582.

Soliva-Fortuny, R.C., M. Ricart-Coll, and O. Martín-Belloso. 2005. Sensory quality and internal atmosphere of fresh-cut Golden Delicious apples. *Int. J. Food Sci. Technol.* 40:369–375.

Streif, J. 1996. Optimum harvest date for different apple cultivars in the 'Bodensee' area. In: A. de Jager, D. Johnson, and E. Hohn (Eds). 1996. European Commission COST 94: The postharvest treatment of fruit and vegetables—determination and prediction of optimum harvest date of apples and pears. *Proceedings of June, 1994 Workshop*, Loftus, Norway.

Sugar, D. 2001. Modified atmosphere packaging for pears. http://postharvest.tfrec.wsu.edu/proc/PC2001R.pdf.

Sugar, D. 2007. Postharvest handling of winter pears. In E.J. Mitcham and R.B. Elkins (Eds.), *Pear Production and Handling Manual*. University of California Agriculture and Natural Resources Publication 3483. pp. 171–174.

Swindeman, A.M. 2002. Fruit packing and storage loss prevention guidelines. http://postharvest.tfrec.wsu.edu/REP2002D.pdf.

Toivonen, P.M.A. 1997. Non-ethylene, non-respiratory volatiles in harvested fruits and vegetables: their occurrence, biological activity and control. *Postharvest Biol. Technol.* 12:109–125.

Ueda, Y., J. Bai, and H. Yoshioka. 1993. Effect of polyethylene packaging on flavor retention and volatile production of 'Starking Delicious' apple. *J. Jpn. Soc. Hort. Sci.* 62:207–213.

Veltman, R.H., J.A. Verschoor, and J.H. Ruijsch van Dugteren. 2003. Dynamic control system (DCS) for apples (*Malus domestica* Borkh. cv 'Elstar'): Optimal quality through storage based on product response. *Postharvest Biol. Technol.* 27:79–86.

Wang, C.Y. and W.M. Mellenthin. 1975. Effect of short-term high CO_2 treatment on storage response of 'd'Anjou' pears. *J. Am. Soc. Hort. Sci.* 100:492–495.

Wang, C.Y., C.E. Sams, and K.C. Gross. 1985. Ethylene, ACC, soluble polyuronide, and cell wall noncellulosic neutral sugar content in 'Eldorado' pears during cold storage and ripening. *J. Am. Soc. Hort. Sci.* 110:687–691.

Wang, Z. and D.R. Dilley. 2000. Initial low oxygen stress controls superficial scald of apples. *Postharvest Biol. Technol.* 18:201–213.

Watkins, C.B. 2006. The use of 1-methylcyclopropane (1-MCP) on fruits and vegetables. *Biotechnol. Adv.* 24:389–409.

Watkins, C.B., E. Kupferman, and D.A. Rosenberger. 2004. Apple. In: Gross, K.C., C.Y. Wang, and M. Saltveit (Eds.), *The Commercial Storage of Fruits, Vegetables, and Florist and Nursery Crops. USDA, ARS, Agriculture Handbook 66*, http://www.ba.ars.usda.gov/hb66/027apple.pdf

Zagory, D. 1997. Advances in modified atmosphere packaging (MAP) of fresh produce. *Perishable Handling Newsletter* No. 90:2–4.

Zanella, A. 2003. Control of apple superficial scald and ripening—a comparison between 1-methyl-cyclopropene and diphenylamine postharvest treatments, initial low oxygen stress and ultra low oxygen storage. *Postharvest Biol. Technol.* 27:69–78.

Zanella A., P. Cazzanelli, A. Panarese, M. Coser, M. Cecchinel, and O. Rossi. 2005. Fruit fluorescence response to low oxygen stress: Modern storage technologies compared to 1-MCP treatment of apple. *Acta Hort.* (F. Mencarelli and P. Tonutti, Eds.) 682:1535–1542.

13

Stone Fruits

Carlos H. Crisosto, Susan Lurie, and Julio Retamales

CONTENTS

13.1 Introduction

Peaches are characteristically soft fleshed and highly perishable fruit, with a limited postharvest life potential. Botanically, stone fruits are drupes. A drupe is a fleshy fruit with thin, edible outer skin (epicarp), an edible flesh of varying thickness beneath the skin (mesocarp) and a hard inner ovary wall that is highly lignified (endocarp), and is commonly referred as a stone or pit, which encloses a seed. In general, stone fruits contain

87% water with give 43 calories per 100 g fruit. The solid content of stone fruits consists of carbohydrates, organic acids, pigments, phenolics, vitamins, volatiles, antioxidants, and trace amounts of proteins and lipids that make them very attractive to consumers (Kader and Mitchell, 1989; USDA, 2003). Apricot, cherry, nectarine, peach, and plum commercial postharvest losses are mainly due to decay and internal breakdown (IB) or chilling injury (CI) (Ceponis et al., 1987; Mitchell and Kader, 1989). Apricot, nectarine, peach, and plum are climacteric fruits that display a rapid increase in ethylene production and an equally rapid rate of ripening. Contrary to the types of stone fruits, cherry is a nonclimacteric fruit, which implies that cherry ripening occurs during the last weeks before harvesting. As much as 25% of final weight is added in the last week of growth before harvesting accompanied with changes in color flavor and texture (Looney et al., 1996). From the commercial point of view, cherries are ready to eat at harvest time and then deteriorate very fast.

The climacteric stone fruits are picked in a preclimacteric state (mature-firm) in order to be marketed successfully. In some cases, this fruit does not acquire upon ripening the sensory characteristics needed to satisfy consumers. Ripening can be delayed by rapid removal of field heat and storage near to 0°C. Peach, nectarine, and plum deteriorate quickly at ambient temperature, therefore, low temperature during storage is used to slow softening, flavor losses, and decay development. However, low-temperature disorders, CI classified as IB, limit the storage life of these fruit types under refrigeration. The onset of CI symptoms determines the postharvest storage/shipping potential because their development reduces consumer acceptance. CI is genetically influenced and triggered by a combination of storage temperature and storage period. It manifests itself as fruit that are dry and have a mealy or woolly texture (mealiness or woolliness), or hard textured fruit with no juice (leatheriness), fruit with flesh or pit cavity browning (internal browning), or with flesh bleeding (internal reddening) and development of off-flavor. This phenomenon (CI or IB) is triggered by storage temperature. It manifests itself as dry, mealy, woolly, or hard-textured fruit (not juicy), flesh or pit cavity browning, and flesh translucency usually radiating through the flesh from the pit. An intense red color development of the flesh (bleeding) usually radiating from the pit may be associated with this problem in some peach cultivars. Recently released cultivars rich in skin red pigment showed flesh bleeding that did not affect fruit taste. The development of this symptom has been associated with fruit maturity rather than storage temperature. In many cultivars, flavor is lost before visual CI symptoms are evident (Crisosto and Labavitch, 2002). There is large variability in CI susceptibility among peach cultivars (Mitchell and Kader, 1989; Crisosto et al., 1999c). In general, most of the mid-season and late-season peach cultivars are more susceptible to CI than early season cultivars (Mitchell and Kader, 1989), although as new cultivars are being released from a new genetic pool, the susceptibility to CI is becoming random in the new cultivar population (Crisosto et al., 1999a; Crisosto, 2002). It has been widely reported that the expression of CI symptoms develops faster and more intensely when susceptible fruit are stored at temperatures between about 2.2°C and 7.6°C (killing temperature zone) than those stored at 0°C or below but above their freezing point (Harding and Haller, 1934; Smith, 1934; Mitchell and Kader, 1989). Therefore, market life is dramatically reduced when fruit are exposed to the killing temperature zone (Crisosto et al., 1999a).

Increasing CO_2 and decreasing O_2 in the atmosphere around the fruit tissue has profound effects on cellular metabolism and storage potential. These controlled atmosphere (CA) conditions reduce the respiration rate of fruits and vegetables (Kader, 1986). However, low oxygen may cause external symptoms such as skin browning and even black pitting on the skin. The internal damage is associated near the skin and surrounding the stone. In both cases, well-defined grayish brown or brown areas are formed. These areas are not associated with mealy tissues and can occur anytime during cold storage (Kader, 1986). Ke et al. (1994) proposed that elevated CO_2 influences respiration rates by regulating carbon flux through the tricarboxylic acid (TCA) cycle. High CO_2 appears to increase carbon flux and maintain energy levels in the cell and enhance the alternative electron pathway by inducing and/or activating alternative oxidase and inhibiting cytochrome oxidase activity (Watkins, 2000). CA can also affect the activity of the enzymes involved in ethylene synthesis such as 1-aminocyclopropane-1-carboxylic acid (ACC) oxidase (Poneleit and Dilley, 1993). Therefore, ethylene is inhibited during storage in CA, but will recover following removal to air. CA also affects the cell wall-degrading enzymes that are responsible for fruit disassembly and whose imbalance or inhibition are associated with mealiness development. At the end of CA storage, the activities of pectin esterase and polygalacturonase were lower than following regular air storage at 0°C (Zhou et al., 2000). However, both the activity and mRNA abundance of both enzymes increased after storage ripening to a greater extent in fruit that had been stored in CA than in fruit stored in air. This recovery of enzyme activity enabled pectin molecules to be cleaved quickly during ripening and led to normal fruit softening and development of juiciness.

CA or modified atmosphere packaging (MAP) and relative humidity management are used as supplements to proper fruit cold storage to limit water loss, delay ripening, and suppress diseases (Smith et al., 1987; Beaudry, 1999). Numerous researchers have contributed to identification of optimum CA to extend postharvest life of various commodities. Nowadays, produce amenable for long storage, such as apples, can be stored under CA regime up to 12 months. However, quality fruit flavor does not always persist after such prolonged storage. Studies have shown that after certain periods under CA, apples can still present an acceptable appearance but lack their characteristic taste (Brackmann et al., 1993).

MAP is a practical way to modify the gas environment surrounding the fruit, and utilizes polymeric films with different permeabilities to oxygen and carbon dioxide to prolong the shelf life of fruits and vegetables. Atmospheric modification evolves within the package as a result of the respiration rate of the fruit, temperature, and the gas diffusion characteristics of the film. Obviously, film selection is important to the system of MAP, since a proper matching of the commodity respiration with the film results in the passive evolution of an appropriate atmosphere within the sealed package (Zanderighi, 2001). The beneficial effects of the MAP technique for keeping quality and extending shelf life of fruits and vegetables are well known (Kader et al., 1989). However, MAP performance is sensitive to temperature and under improper management it can create off-flavor and decay problems, which negate the potential benefit of reducing weight loss. It has been reported that high CO_2 and/or low O_2 levels may accelerate the production of off-flavor (Kader et al., 1989; Golias and Bottcher, 2003).

13.2 Apricot

13.2.1 Physiological Disorders

Apricots (*Prunus armenia*) suffer from a number of cold storage disorders of the flesh, similarly to peaches, nectarines, and plums. Gel breakdown, which detracts from storage quality of apricots, can develop in the orchard, but is aggravated by cold storage (Ginsburg and Combrink, 1972). It begins at the pit and spreads toward the skin (Ryall and Pentzer, 1982). Gel breakdown (Figure 13.1) differs from other IB symptoms as described by Dodd (1984) in that the mesocarp initially does not discolor, but takes the form of a translucent,

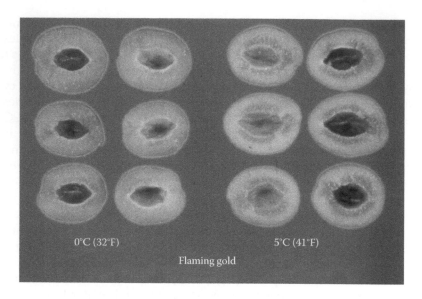

FIGURE 13.1
Apricot gel breakdown symptoms.

gelatinous mass. Another type of IB is called internal browning since the flesh is discolored. Even in advanced stages of IB, apricots may have a normal external appearance (Harvey et al., 1972). The more advanced the stage of fruit maturity at harvest the greater the incidence of gel breakdown after storage (Ginsburg and Combrink, 1972; Taylor and De Kock, 1991). The incidence of these disorders can vary from year to year in the same orchard due to unknown orchard and/or climactic factors. For example, two cultivars of apricots from South Africa, 'Peeka' and 'Royal' had 0.8% and 0.2% gel breakdown in one season and 1.3% and 45% in the next season (Taylor and De Kock, 1991).

CA and MA can delay these flesh disorders, as well as have an accelerating effect if the conditions are unfavorable. IB or gel breakdown occurred in 'Canino' apricots held in air storage or 5 kPa CO_2 and 3 kPa O_2 for 6 weeks, but was prevented when the CO_2 level was 10 or 15 kPa (Kosto et al., 2000). The same was true of 'Perfection' and 'Rival' apricots held for 45 days in 12 kPa or 15 kPa CO_2 and 2 kPa O_2 and another 15 days in regular air storage and then 2 days at 20°C (Drake and Yazdaniha, 1999). Fruit exposed to 15 kPa CO_2 had 7% IB, while fruit held in 9 kPa or less CO_2 had 50% IB. On the other hand, gel breakdown in three apricot cultivars ('Supergold,' 'Imperial,' and 'Peeka') was between 30% and 50% after 4, 5, or 6 weeks storage in 15, 19, or 23 kPa CO_2 and 5 kPa O_2 (Truter et al., 1994a).

13.2.2 Disease Development

Apricots are very susceptible to decay development from a number of fungal species, including *Monolinia fructicola, Botrytis cinerea, Penicillium expansum,* and *Rhizopus stolonifera* (Figure 13.2). The fruit becomes more susceptible as it ripens and softens, but even at harvest small cracks or bruises that develop during or after picking can become sites for decay development. Storing apricots in concentrations of CO_2 above 10 kPa has been found to decrease decay development. For example, storing 'Peeka,' 'Supergold,' and 'Imperial' apricots in 5 kPa O_2 and 15, 19, or 23 kPa CO_2 had an inhibitory effect on decay development (Truter et al., 1994a). The higher the CO_2 level the less was the decay after 4 weeks of cold storage at −0.5°C. Furthermore, the three cultivars differed in their decay

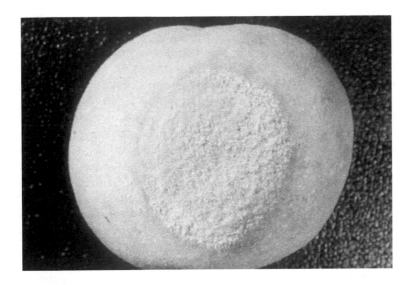

FIGURE 13.2
Monilinia or Botrytis on apricot.

susceptibility. 'Supergold' at 15 kPa CO_2 had 7% decay in shelf life after 4 weeks, 18% after 5 weeks, and 58% after 6 weeks. 'Imperial' had 0.4%, 2.4% and 5% after the same storage times, while 'Peeka' had 1%–2% at all times measured. Drake and Yazdaniha (1999) found that 'Perfection' and 'Rival' apricots stored in 12 or 15 kPa CO_2 and 2 kPa O_2 could be stored for 60 days and after 2 days at 20°C developed 2% decay compared to lower CO_2 concentrations of 3 or 6 kPa where decay development was 39% and 17%, respectively. Apricots in 10 and 15 kPa CO_2 with 3 kPa O_2 decreased decay that developed after 6 weeks of storage (Kosto et al., 2000). Fruits stored at a CO_2 concentration of 5 kPa had decay similar to air stored fruit while 20 kPa CO_2 damaged the fruit, as did concentrations over 16 kPa in another study (McLaren et al., 1997). There is, however, one report of low CO_2 levels retarding decay. Brecht et al. (1982) found that storing immature and overmature 'Patterson' and 'Tilton' apricots in 5 kPa CO_2 and 2 kPa O_2 retarded decay more than in air storage.

One MAP study, which compared different plastic films, found that two treatments that produced 13 kPa to 15 kPa CO_2 and either 3 kPa or 10 kPa O_2 prevented decay development on 'Canino' apricots after 35 days storage and 4 days at 20°C, while control fruits had 30% decay (Kosto et al., 2002). In another MAP study, decay was higher after 5 weeks at 0°C (McLaren et al., 1997). However, the CO_2 concentration in this study was about 5 kPa. Another method to increase fruit resistance to pathogen infection was to give them a prestorage treatment with high CO_2 (10–30 kPa for 24 or 48 h). This was tried on 'Rouge du Rousillon' apricots and the best treatment found to be 20 kPa CO_2 for 48 h. The treated fruit both softened more slowly and developed less decay than untreated fruit (Chambroy et al., 1991).

13.2.3 Fruit Biochemistry and Quality Attributes

Apricot fruit quality is associated with attributes such as appearance, texture, taste, color, and aroma, and all of these are influenced by the ripening conditions. The fruit is harvested commercially before it reaches full color and it colors and softens after harvest. Aroma compounds develop, titratable acidity decreases, and sugar composition does not change during shelf life of the fruit. All these changes occur rapidly and in a few days after harvest the ripened fruit reaches an over ripe stage that makes it unsuitable for consumption. Cold storage can slow but not prevent these changes, and cold storage together with CA or MAP can further slow ripening. A study of three CA conditions (20 kPa CO_2 and O_2, 20 kPa CO_2 and 1 kPa O_2, and 0.03 kPa CO_2 and 1 kPa O_2) found that all three prevented ethylene evolution at 2°C, while air stored 'Bulida' apricots had an ethylene peak in storage after 2 weeks (Pretel et al., 1999). A study of MAP storage of 'Beliana,' 'Rouge de Rousillon,' and 'Polonais' apricots held at 10°C also found inhibition of ethylene production (Pretel et al., 2000). The atmosphere inside the bags ranged from 23 kPa CO_2, 4 kPa O_2 to 6 kPa CO_2, 16 kPa O_2. In the low-permeability films (high CO_2, low O_2) ethanol accumulated, although sixfold less in 'Polonais' fruit than in bags with 'Beliana' fruit. The conclusion was that optimal storage conditions may differ among cultivars. Because of the inhibitory effect on ethylene found in CA and MA conditions, ripening processes such as peel and pulp color changes, firmness loss, titratable acidity loss, and aroma development are slowed (Andrich and Fiorentini, 1986; Drake and Yazdaniha, 1999; Kosto et al., 2000).

These CA/MAP storage conditions also decrease the weight loss that occurs in regular air storage (McLaren et al., 1997). Therefore, for most of the cultivars examined a CO_2 level of between 10 and 15 kPa and an O_2 concentration of 2–5 kPa will extend the storage period of apricots in CA, while the same CO_2 levels and with O_2 levels between 2 and 15 kPa will give good results in MAP. However, the extension of storage is not great. Most studies do not go beyond 5 or 6 weeks, and some look at storage for 3 weeks.

13.2.4 Effects of MAP

Modified atmosphere has been found to be both beneficial and detrimental to apricots, depending on the study. In one study using microperforated and unperforated film the CO_2 levels were from 4 to 6 kPa and O_2 from 10 to 13 kPa. After 5 weeks at 0°C and 5 days at 20°C, the fruit from three cultivars 'Sundrop,' 'Valleygold,' and 'CluthaGold' all had high levels of internal browning (McLaren et al., 1997). Although the external appearance was better from the MAP stored fruit, the internal browning was as high as or higher than in regular air storage. In another MAP trial, CO_2 levels were 15 kPa with O_2 of either 3 or 10 kPa. In this trial with 'Canino' the apricots from MAP had no internal browning after 2 days shelf life and about 20% after 4 days, while control fruits had 40% internal browning (Kosto et al., 2002).

13.2.5 Commercial Use

The use of CA or MAP will slow down the ripening process and maintain for longer time the fruit sensory and quality characteristics. Studies that examined more than one storage time found that the shorter storage time of 3 or 4 weeks give higher quality fruit than the longer time of 5 or 6 weeks. Therefore, the use of this CA/MAP technology in commercial practice is uncommon.

13.3 Cherries

13.3.1 Physiological Disorders

Extending storage/shipping life and assuring good arrival of cherries (*Prunus avium*) is an essential requirement for the world cherry industry. In addition to good temperature management practices, new technology is needed to maintain green stems, flavor, decay

(a)

(b)

FIGURE 13.3
CA container views.

losses, and to avoid skin color darkness. The use of CA technology during storage and shipment (Figure 13.3a and b) is being used successfully by the cherry industry. The benefits of CA/MAP on protecting quality and extending storage life of sweet cherries have been well demonstrated. Because cherries are more tolerant of high CO_2 levels than other stone fruits, CA benefits on slowing down deterioration and decay development during storage are commercially important (Porritt and Mason, 1965; Kader and Morris, 1977; Patterson and Melstad, 1977). Levels as high as 40 kPa are tolerated, and up to 20 kPa CO_2 has been used in commercial practices. When combined with good temperature management, high levels of CO_2 help to reduce decay and retain firmness, acidity, soluble solids, and fruit color (Patterson, 1982). In stored 'Bing' cherries at CA atmospheres featuring either high CO_2 or low O_2, those stored at 20 kPa CO_2 and 10 kPa O_2 maintained the highest percentage of very green stems, brighter fruit color, and higher levels of titratable acids than other atmospheres (Chen et al., 1981). In general, oxygen levels between 3 and 10 kPa delay softening and carbon dioxide levels between 10 and 20 kPa limit decay development and maintain flesh appearance. However, high carbon dioxide concentrations and low oxygen concentrations can cause injury. Oxygen levels below 1 kPa may induce skin pitting and off-flavor while carbon dioxide higher than 30 kPa has been associated with brown skin discoloration (Kader, 1997). Elevated temperatures and prolonged exposure increase the probability of carbon dioxide injury to the fruit. Gas tolerances change with changes in other gas components and with changes in temperature and time. Stem browning precedes the appearance of flesh injury symptoms. Stems are more sensitive to atmospheric extremes, and browning occurs sooner under toxic carbon dioxide levels than in low-oxygen atmospheres. Stems initially develop a red brown color in response to toxic levels of carbon dioxide, whereas anoxic stems tend to develop more of a black-brown color. An accelerated appearance of fruit darkening occurs, which is at least partly due to cell membrane rupture and subsequent leakage. In addition, expressed juice from red cherry cultivars becomes progressively purple and less red with progressive toxicity caused by atmospheres of high carbon dioxide or low oxygen, or both. High carbon dioxide causes droplets of exudates to form on the skin surface before any surface browning appears. Ultimately, brown areas appear on the fruit surface, and injury may be expressed by a blend of brown with red or mahogany and a dull water-soaked appearance. 'Rainier' is more sensitive to surface damage due to high concentrations of carbon dioxide than either 'Bing' or 'Lambert.' Sour and fermented odors and flavors develop before visual symptoms appear. The development of a flavor that is similar to synthetic almond oil is characteristic of fruit injured by carbon dioxide or low oxygen. Taste tests conducted immediately upon the removal of fruit from extremely high levels of carbon dioxide reveal that juice in the fruit may be detectably carbonated. Carbonation of the juice and off-odors of the fruit may be reduced with aeration at 0°C at ambient temperature. Overall, it appears that a storage life based on visual appearance and decay expression of about 8 weeks should be routinely achievable with CA storage technology but flavor may be reduced earlier than visual expression. Modified atmosphere has been mainly targeted to reduce symptoms of softening, decay, and color changes and to maintain quality and extend market life.

13.3.2 Disease Development

Brown rot, caused by *M. fructicola* G. Wint. Honey, is a major disease of all commercial grown Prunus species and causes severe crop losses in the United States and many other countries (Ogawa and English, 1991). For example, approximately half of the market losses in Western sweet cherry fruit have resulted from diseases including brown rot (Ceponis and Butterfield, 1981). It has been reported that high CO_2 reduced the onset of lesion development and the percentage of fruit with brown rot (DeVries-Patterson et al., 1991). Successful control was reported by using an integrated approach for control of postharvest brown rot of sweet cherry fruit. This approach included a preharvest application of fungicide (propiconazole), a postharvest application of a yeast (*Cryptococcus infirmo-miniatus*) and storage in MAP at cold storage (Spotts et al., 1998, 2002).

13.3.3 Fruit Biochemistry and Quality Attributes

When volatile constituents of 'Bing' sweet cherry fruit were analyzed from fruit homogenates at harvest and after air or CA storage, acetic acid and aldehydes were found to be the largest volatile compounds present at harvest (Mattheis et al., 1997). Changes during storage in the concentrations of three compounds previously identified as contributors to sweet cherry fruit flavor, benzaldehyde, E-2-hexenal, and hexanal, were independent of storage conditions. Ethanol accumulated in fruit stored in 15 or 20 kPa CO_2 with 5 kPa O_2 after 6 weeks of storage. Qualitative and quantitative changes in ester production, particularly ethyl acetate, were coincidental with the accumulation of ethanol. The 2-propanol concentrations were consistently highest in fruit stored in 5 kPa O_2 with 0.1 kPa CO_2. A number of esters were not detectable after 4 weeks of storage, and several compounds, including butanal, 2-butanone, and pentyl acetate, were only detected after 4 weeks of storage (Mattheis et al., 1997). When 'Bing' cherries were stored in air or in a mixture of 0, 20, 40, 60, 80, and 100 kPa carbon dioxide at 0°C, they developed off-flavors proportionate to storage time and carbon dioxide level. Off-flavors developed within 1 week in fruit held at 80 kPa carbon dioxide or more and slightly later at 60 kPa carbon dioxide; no off-flavors developed at 40 kPa carbon dioxide or less (Patterson and Melstad, 1977). Thus, high levels of CO_2 when combined with good temperature management help to reduce decay and retain firmness, acidity, soluble solids, and fruit volatiles (Patterson, 1982). Cherries stored at 20 kPa CO_2, 10 kPa O_2 maintained a higher percentage of brighter fruit color, higher levels of titratable acids, conserved fruit brightness but these conditions did not prevent stem dehydration (Chen et al., 1981). Freshly harvested 'Burlat' cherries were held at 5°C for 10 days with five different atmospheres: air, 12 kPa CO_2–4 kPa O_2, 12 kPa CO_2–20 kPa O_2, 5 kPa CO_2–4 kPa O_2, and 5 kPa CO_2–20 kPa O_2. The best results were obtained with cherries kept in high CO_2 atmospheres, independent of O_2 concentrations. In these conditions, the cherries presented a higher acidity level (0.65 vs. 0.60 g malic acid per 100 mL) and lower anthocyanin content (0.40 vs. 0.48 mg g^{-1}). Consequently, $h*$ (18 vs. 20.5) and $C*$ (24 vs. 39) values were lower, which made the cherries visually more reddish, less dull and therefore more attractive to consumers. In addition, cherries kept in high CO_2 atmospheres contained lower levels of peroxidase (469 vs. 737 au g^{-1}) and polyphenoloxidase (73 vs. 146 au g^{-1}) activities, which favored postharvest stability of color (Remon et al., 2004.)

13.3.4 Effects of MAP

MAP is an alternative technology for CA storage/shipment. In a MAP system, the new atmosphere is attained by the respiration of the fruit, as determined by the box liner permeability and fruit temperature. At a given temperature, the two gases (carbon dioxide and oxygen) reach steady concentrations (plateau) when the rate of gas permeation through the package film equals the rate of respiration. The beneficial effect of high CO_2 while maintaining a high level of O_2 (3–10 kPa) to avoid injury is the target for MAP on cherries. Several published studied reported MAP benefits such as reducing weight loss, color changes, decay incidence in various cultivars (Lurie and Aharoni, 1997; Petracek et al., 2002). 'Bing' cherries packed using Xtend MAP film stored well up to 2 months at 0°C but quality rapidly deteriorated during the third month of storage (Lurie and Aharoni, 1997). 'Hedelfingen' and 'Lapins' sweet cherries (*P. avium*) were stored in air or in two types of MAP bags (LifeSpan 204 and 208) at 3°C and 90% relative humidity for 4 weeks. For 'Hedelfingen' cherries, there were differences in CO_2 and O_2 composition within the MAP bags, depending on the bag used. This resulted in slightly better cherry quality for the bag with lower O_2 permeability (LifeSpan 204), which equilibrated at 4 kPa to 5 kPa O_2, and 7 kPa to 8 kPa CO_2. For 'Lapins,' the two MAP bags showed concentrations of 9–10 kPa O_2, and 8–9 kPa CO_2, and similar final fruit quality. The commercial MAP application was promoted by service companies (Viewfresh, Freshold) that established the ideal CO_2 and O_2 desired combination at the time of packaging, by adding CO_2 and sealing the MAP (active MAP). Currently, several MAP bags that do not require establishment of the initial gas composition that are manually hermetically sealed are being successfully used for long-term storage or long distance shipment. However, some buyers prefer heat-sealed MAP because fruit is better displayed (Figure 13.4).

FIGURE 13.4
Cherry MAP bags manually sealed.

The commercial use of MAP on cherry has been developed rapidly in the Pacific Northwest in the United States to deliver high-quality cherries to long distance markets such as Hong Kong and Taiwan. In the last decade, different box liners have become available to the cherry industry; however, to our knowledge few detailed evaluations have been done comparing them in similar conditions. In the California industry, the use of MAP is becoming fashionable to keep stems green based upon customer demand.

Market life potential of 'Bing' cherries packed with different MAP box liners was compared with solid (commercial) and perforated box liners after 15, 30, and 45 days of storage at $1.6°C$ followed by a 4 day simulated shelf life period at $20°C$ (Crisosto et al., 2002). Three higher CO_2/O_2 ratios were reached with the use of MAP box liners such as LifeSpan (8 kPa CO_2–5 kPa O_2), Fresh Fruit Cherry Bag 011 (10 kPa CO_2–5 kPa O_2), and Fresh Fruit Cherry Bag 012 (8 kPa CO_2–10 kPa O_2) than with the solid (4 kPa CO_2–16 kPa O_2) or perforated box liner (0 kPa CO_2–20 kPa O_2). The incidence of decay was significantly lower with any of the MAP box liners, even the solid box liner, compared with fruit packed in perforated box liners at any evaluation time. In a short-term shipment (within 15 days), cherries packed in the Fresh Fruit Cherry Bag 011 had the lowest decay (4.9%), followed by cherries packed using the solid (10.8%) and LifeSpan (7.5%) box liners. However, the use of the solid box liner protected cherries from stem browning, skin color darkness, and firmness losses as well as any of the MAP treatments. In a long-term shipment (within 30–45 days), the use of solid box liners did not protect cherries from decay development and other deterioration factors as well as any of the MAP treatments. MAP delayed 'Bing' cherry deterioration such as decay, stem browning, skin color darkness, firmness losses, and titratable acidity (TA) degradation. Most of these beneficial effects were also carried over during the warming period at $20°C$ that simulated shelf life.

13.3.5 Commercial Use

Cherry is very well adapted to commercially benefit from the use of CA or MAP and the commercial use of CA containers and MAP without establishment of initial gas composition are mainly being used for long-term shipments. Also heat sealed is visually preferred by buyers over manually sealed MAP. The use of MAP is less expensive, and more easily applied than CA under the current handling and marketing conditions. Also, the modified atmosphere conditions inside the container can be maintained differently than CA containers during distribution at the retail point. Thus, the use of MAP is widely used but is not targeting the early cherry production market. This MAP technique is recommended to be used in situations where a specific sweet cherry cultivar will develop symptoms of deterioration and decay incidence. The proper use of MAP will allow industries to access long distance markets with high quality fruit and low transportation costs.

13.4 Nectarine and Peach

13.4.1 Physiological Disorders

For both nectarines and peaches (*Prunus persica*), O_2 levels between 1 and 2 kPa have been related to delayed ripening and CO_2 levels between 3 and 5 kPa induced a limited reduction in CI and fruit softening. An oxygen level below 1 kPa may induce failure to ripen, skin browning, and off flavors while carbon dioxide higher than 10 kPa has been associated with flesh browning and off-flavor (Kader, 1992). Thus, specific evaluation per cultivar should be carried out to conclude if any potential CA benefit can be used commercially.

FIGURE 13.5
Peach chilling injury symptoms.

Original recommendations for CA conditions for peaches and nectarines were similar to those for some apple cultivars: 3–5 kPa CO_2 + 1–2 kPa O_2 at 0°C (Anderson et al., 1969; Kader, 1986). However, later studies have found that higher levels of CO_2 will delay appearance of CI symptoms, especially flesh browning (Figure 13.5), better than the original recommendations (Lurie, 1992; Retamales et al., 1992; Streif et al., 1992), although such high levels were also used successfully to prevent CI in previous investigations in 'J.H. Hale' peaches (Wade, 1981). Exposure to 10 kPa CO_2 + 10 kPa O_2 for 6 weeks has been reported to delay CI in 'Fantasia,' 'Flavortop,' and 'Flamekist' nectarine cultivars (Lurie, 1992). The physiological storage disorders of flesh browning and reddening (flesh bleeding) were almost completely absent in nectarines kept in 10 kPa O_2 + 10 kPa CO_2. Although this CA prevented flesh browning and reddening, after 6 weeks cold storage, fruit did not develop the increased extractable juice during poststorage ripening that occurred in nonstored fruit. Therefore, while reducing levels of storage disorders, CA does not reduce the loss of ripening ability occurring during nectarine storage. In other work, it was demonstrated that 'Fantasia' nectarines stored in 10–20 kPa CO_2 were juicy and had good flavor after 4 weeks at 0°C storage plus 5 days of shelf life (Burmeister and Harman, 1998). However, after 6 weeks of cold storage plus 3 days of shelf life, only fruit stored with 10 kPa CO_2 was still of good eating quality while fruit stored with 15 and 20 kPa CO_2 suffered from skin browning and had off-flavors (Burmeister and Harman, 1998).

In a storage experiment to prevent low temperature storage disorders in nectarine fruits of 'July Red' (Figure 13.6) and 'Autumn Grand,' the fruit was either cooled immediately after harvest or kept at 20°C for 48 h, before transfer to CA conditions at 0°C (Retamales et al., 1992). Combinations of 0, 10, 15, and 20 kPa CO_2 with 8 and 16 kPa O_2, were assayed. Holding the fruit at 20°C before cold storage prevented woolliness in the absence of elevated CO_2 levels but did not affect flesh browning and increased reddish discoloration; further, it enhanced water loss and ripening, increasing fruit softening markedly. Conversely, high CO_2 delayed fruit ripening in CA storage, keeping the fruit firmer, and preventing the development of woolliness, flesh browning, and reddish discoloration

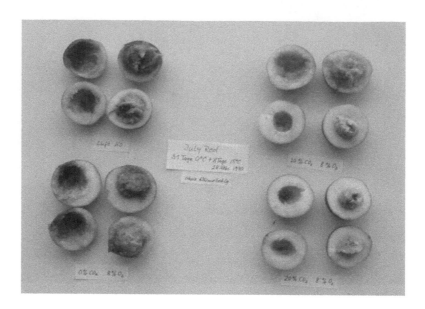

FIGURE 13.6
Nectarine chilling injury symptoms.

during ripening, the best results being mostly obtained with 20 kPa CO_2. O_2 levels assayed did not show clear effects, but decreased O_2 concentrations in the absence of high CO_2 showed some benefit in 'July Red.'

In a recent detailed study testing four CA conditions (air, 2 kPa O_2 + 5 kPa CO_2, 10 kPa CO_2 + 10 kPa, and 17 kPa CO_2 + 6 kPa) for 5 and 6 weeks (simulated shipment) on 11 important nectarine cultivars grown in California, it was concluded that on 9 out of 11 CA was not necessary after 5 weeks storage at 0°C as CI symptoms were not expressed and fruits remained firm (Crisosto and Cantín, unpublished report). In these studies, fruits were held in cold storage for 1 week before CA treatments were established simulating standard shipping practices. 'August Pearl' had high flesh browning incidence that was reduced by any of the three CA treatments. By 5 weeks of simulated shipment, 'August Fire' had high levels of mealiness in all the treatments (Table 13.1). On the other hand, by 6 weeks of simulated shipment at 0°C, only 6 cultivars out of the 11 did not show commercial CI symptoms in any of the storage conditions (Table 13.2). 'Ruby Diamond' had near 80% flesh browning and absence of flesh mealiness on fruit under cold storage and flesh browning was reduced in all the CA treatments. 'Honey Royale,' 'Summer Bright,' 'August Pearl,' and 'August Fire' had mainly high incidence of flesh mealiness that were not reduced by using any of the CA treatments. Despite the lack of good control of flesh mealiness, the CA treatments did not damage fruit except for the 17 kPa CO_2 + 6 kPa O_2 at 1°C. A dull skin browning color (Figure 13.7) and pitting (Figure 13.8) was observed on 'August Pearl' and 'Honey Royale' fruit after 6 weeks cold storage under the 6 kPa O_2 + 17 kPa CO_2.

The CO_2 component appears to be critical for delaying the onset of CI (Anderson et al., 1969; Kajiura, 1975; Wade, 1981). Most studies of CA storage of peaches have found that lowering O_2 and raising CO_2 in the storage atmosphere benefited the fruit and delayed or prevented the appearance of mealiness, internal reddening, and flesh browning (Lurie, 1992; Retamales et al., 1992; Crisosto et al., 1995; Levin et al., 1995; Zhou et al., 2000; Murray et al., 2007). CA conditions have been claimed to delay CI and extend postharvest life. A 17 kPa CO_2 combined with 6 kPa O_2 atmosphere balanced in nitrogen

TABLE 13.1

Influence of CA Conditions on CI Development of Ripe Nectarines after 5 Weeks of CA Cold Storage in Addition to a Previous 1 Week of Cold Storage at 0°C

Nectarine Cultivars	5 Weeks			
	Air	2 kPa + 5 kPa	10 kPa + 10 kPa	6 kPa + 17 kPa
Ruby Diamond	0	0	0	0
Ruby Sweet	0	0	0	0
Spring Bright	0	0	0	0
Honey Royale	0	0	0	0
Fire Pearl	0	0	0	0
Summer Bright	0	0	0	0
Summer Fire	0	0	0	0
Zee Glo	0	0	0	0
August Pearl	100[FB]	10[M]	10[M]	0[M]
August Fire	20[M]	60[M]	30[M]	50[M]
Arctic Snow	0	0	0	0

Source: From Crisosto, C.H., and G. Crisosto, 2008. Effects of controlled atmospheres on nectarines. Central Valley Postharvest Newsletter, 17(2): 7–8.

[M] = Mealiness.
[FB] = Flesh browning.

TABLE 13.2

Influence of CA Conditions on CI Development of Ripe Nectarines after 6 Weeks of CA Cold Storage in Addition to a Previous 1 Week of Cold Storage at 0°C

Nectarine Cultivars	6 Weeks			
	Air	2 kPa + 5 kPa	10 kPa + 10 kPa	6 kPa + 17 kPa
Ruby Diamond	80[FB]	0	0	0
Ruby Sweet	0	0	0	0
Spring Bright	0	0	0	0
Honey Royale	40[M]	40[M]	30[M]	70[M]
Fire Pearl	20[M]	0	0	0
Summer Bright	30[M]	40[M]	50[M]	40[M]
Summer Fire	20[FB]	0	10[M]	10[M]
Zee Glo	10[FB]–10[M]	0	0	0
August Pearl	100[M]–100[FB]	40[FB]–20[M]	20[M]	10[M]
August Fire	30[M]	100[M]	40[M]	80[M]
Arctic Snow	10[FB]–10[M]	0	0	0

Source: From Crisosto, C.H., and G. Crisosto, 2008. Effects of controlled atmospheres on nectarines. Central Valley Postharvest Newsletter, 17(2): 7–8.

[M] = Mealiness.
[FB] = Flesh browning.

is commercially utilized during overseas shipment by the TransFRESH Corp. A 5 kPa CO_2 combined with 2 kPa O_2 atmosphere has also been recommended for long-term shipment/storage (Carrier Corp., 1995). CA conditions of 6 kPa O_2 + 17 kPa CO_2 have been reported to be beneficial for some fresh peaches and nectarines shipped from Chile (Retamales et al., 1992; Streif et al., 1992). In California, the major benefits of CA during storage/shipment are retention of fruit firmness and ground color, and reduction of flesh browning development. CA conditions of 6 kPa O_2 + 17 kPa CO_2, the best combination, at 0°C have shown a limited benefit for reduction of mealiness during shipments for yellow flesh cultivars

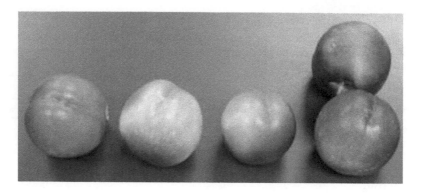

FIGURE 13.7

General view of external fruit damage caused by CA conditions on 'August Pearl'–'Honey Royale' nectarine after 6 weeks cold storage at 0°C. Fruit were stored in air, 5 kPa CO_2 + 2 kPa O_2, 10 kPa CO_2 + 10 kPa O_2, and 17 kPa CO_2 + 6 kPa O_2 (from left to right).

FIGURE 13.8

Detailed view of external damage developed after prolonged storage (6 weeks) under 17 kPa CO_2 + 6 kPa O_2 at 0°C for 'Honey Royale' nectarine.

(Crisosto et al., 1999a,c) and white flesh cultivars (Garner et al., 2001). In California clingstone canning peaches, a large difference in storage potential and canned quality following storage was reported among five clingstone peach cultivars tested (Crisosto et al., 2007). Based on this work, it is recommended that 'Loadel' and 'Carolyn' peaches should be stored for up to 4 weeks under 2 kPa O_2 + 5 kPa CO_2 at 1.1°C while 'Andross,' 'Klamt,' and 'Halford' should be stored for shorter storage period (Brecht et al., 1982). These same benefits of CA on mealiness reduction were reported in Brazil using a local cultivar. There, the overall quality of the 'Chiripa' cultivar was better on CA stored fruit than on air stored (Girardi et al., 2005).

As mealiness is the main CI symptom rather than flesh browning, the use of CA technology in California cultivars has been limited (Lurie and Crisosto, 2005). The CA efficacy is related to cultivar (Mitchell and Kader, 1989; Peace et al., 2005) preharvest factors

(Von Mollendorff, 1987; Crisosto et al., 1997b), temperature, fruit size, marketing period, and shipping time (Crisosto et al., 1999a,b,c). Despite commercial use of CA conditions, the benefits with respect to reducing symptoms of CI have been erratic and unreliable. Prior research (Crisosto et al., 1997a, 1999c) has led us to believe that CA storage performance can be improved once we better understand the relationships between cultivar susceptibility and "orchard factors." Large ($\sim$275 g), medium ($\sim$175 g), and small ($\sim$125 g) 'Elegant Lady' and 'O'Henry' peaches were stored in either air, 5 kPa CO_2 + 2 kPa O_2, or 17 kPa CO_2 + 6 kPa O_2 at 3.3°C (Crisosto et al., 1999a). Large size peach fruit benefited more from the 17 kPa CO_2 + 6 kPa O_2 than from 5 kPa CO_2 + 2 kPa O_2 or air storage atmosphere treatments. Fruit size, storage atmosphere, and temperature all had significant effects on CI development. Small peaches stored in air at 0°C had a longer market life than large fruit. At both storage temperatures, large size 'Elegant Lady' and 'O'Henry' fruits had a longer market life under CA than under air storage. However, at 3.3°C small size 'Elegant Lady' fruit in CA showed browning in the flesh. This suggests that 17 kPa CO_2 + 6 kPa O_2 may induce flesh browning in small size 'Elegant Lady' peaches. In both years, lack of juiciness (i.e., development of mealiness/leatheriness) was observed before the development of flesh browning. Thus, market life was dependent on the incidence of mealiness/leatheriness rather than flesh browning. Market life of 'Elegant Lady' and 'O'Henry' peaches stored at 0°C or 3.3°C in air or CA was predicted using regression equations. Predicted market life was based on incidence of CI for both peach cultivars, but does not include the 4 days storage at 0°C prior to the main storage treatment. Under air storage conditions, the predicted maximum market life (0°C) of large, medium, and small size 'Elegant Lady' peaches was 9, 19, and 21+ days, respectively. Under CA storage conditions, predicted maximum market life at 0°C of large, medium, and small size 'Elegant Lady' fruit was 19, 21+, and 21+ days, respectively. The three similar sizes of 'O'Henry' fruit lasted 13, 16, and 21+ days in air storage at 0°C, respectively. All three sizes of 'O'Henry' fruit lasted 21+ days in CA storage at 0°C. The predicted market life of 'Elegant Lady' in air at 3.3°C (minimum market life) was 4, 8, and 21 days for large, medium, and small sizes, respectively. The predicted minimum market life of 'Elegant Lady' fruit in CA was 9, 15, and 16 days for large, medium, and small sizes, respectively. Large, medium, and small sizes of 'O'Henry' fruit stored in air at 3.3°C had a minimum market life of 5, 6, and 12 days, respectively, while large, medium, and small sizes of 'O'Henry' fruit in CA had a predicted minimum market life of 14, 16, and 18 days, respectively.

13.4.2 Disease Development

Worldwide, the most important pathogen of tree fruits is Botrytis rot, caused by the fungus *B. cinerea*. It can be a serious problem during wet, spring weather. It can occur during storage if fruit have been contaminated through harvest and handling wounds. Avoiding mechanical injuries and good postharvest temperature management are effective controls. Brown Rot is caused by *M. fructicola* with infections beginning during flowering. It is the most important postharvest disease of peaches in California. Rhizopus Rot is caused by *R. stolonifer* and can occur in ripe or near-ripe peaches kept at 20°C–25°C. Cooling and keeping fruit below 5°C is part of an effective control. Good orchard sanitation practices and proper fungicide applications are essential to reduce these problems. It is also common to use a postharvest fungicidal treatment against these diseases. A Food and Drug Administration approved fungicide is often incorporated into a fruit coating/wax for uniformity of application. The regulation on the use of fruit coatings varies according to country. Careful handling to minimize fruit injury, sanitation of packinghouse equipment, and rapid, thorough cooling to 0°C as soon after harvest as possible are also important for effective disease suppression.

It has been demonstrated that reduction on botrytis growth by a 2 kPa O_2 level is not more than 15% below the rate of growth in air. The addition of CO_2 during cold storage affects *B. cinerea* and *M. fructicola* but levels higher than 10 kPa are needed to suppress fungi growth. Thus, the combinations O_2 and CO_2 commonly used have a very modest suppression on fungi (Sommer, 1985).

Effects of short-term exposure to a 15 kPa CO_2 atmosphere on 'Summer Red' nectarine inoculated with brown rot were investigated (Ahmadi et al., 1999). Nectarines were inoculated with spores of *M. fructicola* and incubated at 20°C for 24, 48, or 72 h and then transferred to storage in either air or air enriched with 15 kPa CO_2 at 5°C. The incubation period after inoculation, storage duration, and storage atmosphere had significant effects on fruit decay. 'Summer Red' nectarines tolerated a 15 kPa CO_2 atmosphere for 16 days at 5°C. Development of brown rot decay in fruit inoculated 24 h before 5 or 16 days storage in 15 kPa CO_2 at 5°C was arrested. After 3 days ripening in air at 20°C, the progression of brown rot disease was rapid in all inoculated nectarines, demonstrating the fungistatic rather than fungitoxic effect of 15 kPa CO_2 (Ahmadi et al., 1999). The addition of 11 kPa CO_2 to 4 kPa $O_2 \pm 5$ kPa CO_2 atmospheres completely inhibited growth of *Monilinia*. However, the normal rate of rot development resumed once inoculated plates and fruits were transferred to air at 20°C (Kader et al., 1982).

13.4.3 Fruit Biochemistry and Quality Attributes

CAs have also been assayed for keeping quality of fresh-cut nectarine fruit, given the extreme short life in postharvest (Mencarelli et al., 1998). Thus, the shelf life of slices from 13 cultivars of peaches and 8 cultivars of nectarines, varied between 2 and 12 days at 0°C. CAs of 0.25 kPa O_2 and/or 10 kPa or of 20 kPa CO_2 extended the shelf life at 10°C of 'O'Henry' or 'Elegant Lady' peach slices by 1–2 days beyond the air control. Low (0.25 kPa) O_2 acted synergistically with CO_2 levels of 10 and 20 kPa to induce fermentative metabolism as indicated by ethanol and acetaldehyde production (Gorney et al., 1999).

Other techniques have been assessed as a replacement of or combined with CA. Thus, both delayed storage of 'Flavortop' nectarine fruits held for 48 h at 20°C before storage, and CA storage of 10 kPa CO_2, 3 kPa O_2, alleviated or prevented CI manifested as woolliness in nectarine fruits stored for 4 or 6 weeks at 0°C. Control fruits showed 80% and 100% woolliness during ripening after 4 or 6 weeks at 0°C, respectively (Zhou et al., 2000). Delayed storage and CA were similar in their beneficial effect after 4 weeks and CA was better after 6 weeks storage. According to Zhou et al. (2000), the two storage processes appeared to prevent woolliness by different mechanisms. Delayed storage initiated ripening so that, at removal from storage, polygalacturonase activity was higher and pectin esterase activity was lower than in control fruits. The polygalacturonase activity increased further during ripening, and normal softening occurred in delayed storage fruits. There was no difference in mRNA abundance of polygalacturonase and pectin esterase between delayed storage and control fruits. CA repressed both mRNA levels and activity of polygalacturonase during storage, but allowed recovery of activity during ripening. Endogluconase activity declined during ripening in all fruits, but control fruits retained more activity than delayed storage or CA fruits. The endogluconase mRNA level was high in control fruits during ripening after storage, and almost undetectable in all treatments at all other times. Consequently the authors postulate that the ratio between polygalacturonase/ pectin esterase either at removal or during ripening will determine whether woolliness develops or not (Zhou et al., 2000).

Extremely low oxygen atmospheres have been also studied in relation to nectarine quality. Thus, in a study from Tian et al. (1996) 'Independence' nectarine was kept in 0.3 kPa O_2 atmosphere and in air at 0°C and 6°C for 12, 21, 29, and 36 days to study the effect of low O_2 atmosphere on the postharvest physiology and quality of the fruits. Ethanol and acetaldehyde content increased in low O_2 atmosphere with the prolongation of storage time, particularly at 6°C. The correlation between ethanol and storage life was statistically significant. Low O_2 treatment had no significant effect on methanol content. Flesh firmness and titratable acidity decreased along with storage period. Therefore, low O_2 treatment can delay the decrease in firmness and acidity, but temperature did not affect firmness and acidity. A slow increase in soluble solids content (SSC) occurred during storage. SSC was not influenced by temperature, but fruit stored in 0.3 kPa O_2 atmosphere had less SSC than fruit stored in air. Higher correlations between ethanol and acetaldehyde, SSC, acidity, and firmness were also observed. Nectarine fruits stored in 0.3 kPa O_2 atmosphere at 0°C and 6°C for 36 days did not have any detrimental effects and showed acceptable quality values (Tian et al., 1996).

The effect of postharvest methyl bromide fumigation and CA storage on fruit quality of two nectarine cultivars was investigated in New Zealand (Harman et al., 1999). 'Redgold' and 'Fantasia' nectarines were either not fumigated or fumigated with 64 g/m^3 methyl bromide for 2 h at 12°C, and then cool-stored for up to 6 weeks in air or CA. The main effects of CA storage on nonfumigated 'Redgold' and 'Fantasia' nectarines were to retard softening and loss of green color, to maintain soluble solids concentration (SSC), and to delay onset of storage disorders such as mealiness and IB (Harman et al., 1990). Fumigation retarded fruit softening, lowered SSC, and increased the severity of mealiness and IB in nectarines stored for more than 3 weeks. It did not significantly affect flesh color development. CA storage did not alleviate the effect of fumigation on SSC or the severity of internal disorders. Fumigated 'Redgold' nectarines, stored in either air or CA, showed signs of skin damage after 2 weeks of storage; damage increased in severity with length of time in storage. With 'Fantasia' nectarines, skin damage was found only in fruit that had been forced-air-cooled and CA-stored immediately after fumigation.

CA treatments decreased respiration and ethylene production rates in yellow and white flesh peaches during storage (Kajiura, 1975; Kader et al., 1982). A detailed vitamin evaluation on CA treated peaches was carried out, where CA of 2 kPa O_2, 12 kPa CO_2 in air, and 2 kPa $O_2 + 12$ kPa CO_2, had no effect on provitamin carotenoids in fresh-cut 'Fay Elberta' peaches (*Prunus persica* (L.) Batsch) held for 7 days at 5°C (Wright and Kader, 1997). Under extreme insecticidal low oxygen CA conditions of 0.25 or 0.02 kPa O_2 at 0°C or 5°C for up to 40 days, 'Fairtime' fruit carbon dioxide, ethylene production rates, and internal CO_2 concentrations were reduced but resistance to CO_2 diffusion was slightly increased (Ke et al., 1991a). The low O_2 treatments delayed incidence and reduced severity of CI and decay of peaches stored at 5°C. In this study, low O_2 atmospheres did not significantly influence changes in skin color, flesh firmness, and SSC, but retarded titratable acidity loss and pH rise. Ethanol and acetaldehyde accumulated in peaches kept in 0.02 kPa O_2 at 0°C or 5°C or in 0.25 O_2 at 5°C. The fruits kept in air or 0.25 kPa O_2 at 0°C for up to 40 days and those stored in 0.02 kPa O_2 at 0°C or in air, 0.25 or 0.02 kPa O_2 at 5°C for up to 14 days had good to excellent taste, but the flavor of the fruits stored at 5°C for more than 29 days was unacceptable.

'Elberta' peach fruit stored under varying CO_2 concentrations 5–20 kPa were compared to conventional refrigerated stored fruit (Wankier et al., 1970). This data indicated that firmness, total pectins, titratable acidity, total sugars, and tannins decreased with duration of cold storage time. However, they usually decreased at a slower rate in CA-stored fruit than in refrigerator-stored fruit. Color, pH, and free reducing sugars increased with storage time, but the organic and amino acids content varied erratically with the treatment and length of storage time. The organic acids were generally depleted as storage was extended. Succinic acid increased only under elevated CO_2 concentrations and increasing the CO_2 accelerated accumulation of succinic acid and the depletion of malic acid.

Increased CO_2 also caused alanine to accumulate and aspartic acid to decrease. Peach fruit did not appear to be satisfactorily stored under the conditions of this experiment. In some cases, when evaluations have been made prior to the end of normal market life, CA treatments may not show any benefits. For example, 'Snow King' peaches harvested at commercial maturity were subjected to a large range of CO_2 and O_2 atmosphere combinations for a 2 week simulated transportation under CA (0°C) period after 1 week of cold storage in air (0°C). After 3 weeks of simulated shipment, there were no differences on fruit flesh firmness, SSC, titratable acidity, and CI among the different CA treatments. In studies carried out in Israel, 'Hermoza' peaches were stored at 0°C in air or in CA (10 kPa CO_2, 3 kPa O_2) for 4 weeks and then ripened for 4 days at 20°C. Woolliness developed in the regular air stored fruit while the CA stored fruit ripened normally. In the woolly fruit, symptoms of the disorder were greater in the inner mesocarp than in the outer mesocarp (Zhou et al., 2000). Polygalacturonase and pectin esterase activities differed in the outer and inner mesocarps of the affected fruit. Polygalacturonase activity was low and pectin esterase activity was high in the inner mesocarp of the woolly fruit during ripening relative to the outer mesocarp, while in the healthy fruit, activities were similar in both areas. Cell wall fractions of water-soluble, CDTA-soluble, and carbonate-soluble pectins were prepared from freshly harvested peaches and incubated with pectin esterase and polygalacturonase from ripe peaches at different ratios. Only the CDTA-soluble fraction formed a gel with peach enzymes, and the rate of gelation increased with increasing amounts of pectin esterase relative to polygalacturonase. Both water-soluble and CDTA-soluble pectin fractions formed gels with commercial pectin esterase (extracted from orange peel). The pectin esterase extracted from peaches was stable when stored at 0°C for 9 days, while polygalacturonase activity was stable only for 1 day. The results suggest that pectin esterase, acting on pectins in the cell wall in vivo may cause gel formation and that the CDTA-soluble polymers have the capacity to bind apoplastic water and create the dry appearance observed in woolly fruit. Cold storage alone had a major effect on reducing endopolygalacturonase and exopolygalacturonase, and less impact on pectin esterase. Under CA storage, pectin esterase activity was effectively reduced and the activities of endopolygalacturonase and exopolygalacturonase, which were low during the treatment, dramatically increased 5 days after the end of storage.

13.4.4　Effects of MAP

Another tool to modify atmosphere and reduce CI is the use of MAP. This technique has been intensively tested in several yellow flesh peach cultivars growing in Chile (Zoffoli et al., 2002) without consistent success. Despite high CO_2 levels that were reached during simulated shipping, flesh mealiness and flesh browning development limited the potential benefits of this technology. In some commercial situations when box MAP was used the incidence of decay increased because of lack of proper cooling and condensation during transportation. Also, fruit damage has been observed when MAP-packed fruits (O_2 level <3 kPa and CO_2 levels >13 kPa) were exposed to warm temperatures, which occur during postharvest distribution (Malakou and Nanos, 2005). In the Spanish flat-type peach 'Paraguayo' cultivar, the use of MAP showed reductions in flesh browning (Fernández-Trujillo et al., 1998). In this study, firm-breaker and firm-mature flat white flesh peaches were stored in air for 10 days at 20°C, or precooled and sealed in either one of two unperforated or one macroperforated polypropylene film for 14 or 21 days at 2°C. The atmosphere inside the macroperforated film bags remained close to the composition of air during storage. In unperforated bags, steady state atmospheres were reached within 6 and 9 days: firm-breaker fruit (12 kPa CO_2 and 4 kPa O_2 in standard type polypropylene, 23 kPa CO_2 and 2 kPa O_2 in oriented type polypropylene) and firm-mature fruit

(22 kPa CO_2 and 3 kPa O_2 in standard polypropylene and 21 kPa CO_2 and 2 kPa O_2 in oriented polypropylene). After 14 days storage plus a 3 day shelf-life test, mealiness and slight flesh browning developed in fruit stored in macroperforated polypropylene. Ethanol and acetaldehyde accumulated to higher levels in oriented polypropylene bags for both firm-breaker and firm-mature fruits. Modified atmospheres in both unperforated bags were associated with lower weight loss, less senescence and CI, absence of decay, and delayed ripening changes of the fruit after a shelf-life period. Promising results on the use of MAP and storage potential were also reported on 'Chaoyang' honey peach in China. In this study the quality of honey peach fruit packed using three different thickness (15, 25, and 40 μm) low-density polyethylene (LDPE) films and stored at 2°C was studied (An et al., 2006). MAP treatments inhibited the climacteric peak, reducing color change and softening and retarding the reduction of TTS, TA, and membrane integrity. After 20 days cold storage at 2°C, fruit packed in the LDPE25 bags had a steady-state atmosphere of 5 kPa $CO_2 + 4$ kPa O_2 and had the best quality. A new approach that consists of wrapping an entire pallet following cooling the fruit is currently being tested.

In nectarines, trials were set up with the objective of attaining with MAP high CO_2 levels similar to the ones that provide some control of woolliness (Retamales et al., 1997). Trials were performed using modified atmosphere obtained through use of sealed plastic bags. In the first season, 'Fantasia' and 'Flamekist' nectarines were packed using different films with or without injection of CO_2 at different concentrations (with nitrogen as balance gas) to replace partially the air inside. In the second season, 'July Red,' 'Flamekist,' and 'August Red' nectarines together with 'Calred' peach were packed using two low-permeability films with or without CO_2 injections (with oxygen as balance gas). Fruits were cold-stored for 20 or 30 days and kept afterwards for 5 days at room temperature for ripening (shelf life). In the first season, good control of woolliness was obtained in both nectarine cultivars only in the treatments where high CO_2 concentrations (above 10 kPa) were reached. In the second season, a reduction in woolliness incidence was obtained in nectarines through MAP, but no reduction was apparent in 'Calred' peach. In general, high CO_2 treatments resulted in less pronounced softening after cold storage (Retamales et al., 1997).

13.4.5 Commercial Use of CA and MAP

The use of CA or MAP technologies has some limited benefits on maintaining peach or nectarine postharvest life. Some shipments are conducted using CA technology. Trans-Fresh Corporation uses containers that maintain 17 kPa CO_2 combined with 6 kPa O_2, while Carrier Corporation recommends 5 kPa CO_2 and 2 kPa O_2 atmospheres. TransFresh has also developed a pallet wrap that encloses an entire pallet and then the proper atmosphere is injected into the pallet. All of these systems are for transport to distant ports and not for extended storage. That has yet to see any major utilization.

13.5 Plums

13.5.1 Physiological Disorders

Recently the global production and consumption of plums (*Prunus salicina* Lindell) and prunes (*Prunus domestica* L.) have increased sharply and the need for longer storage periods is also increasing. Maintaining quality for a period of 5 weeks or even longer is needed for orderly overseas marketing. Incorporation of new cultivars has extended the

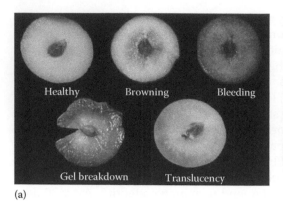

(a)   (b)

FIGURE 13.9
(a) Plum and (b) prune cold storage disorders.

harvest season from late spring through the summer months. Plums are climacteric fruits and undergo rapid deterioration after ripening, including softening, dehydration, and decay. Commercial storage conditions (0°C–5°C and 80%–95% relative humidity) may delay the softening process, but may also lead to the development of storage disorders. Storage disorder symptoms include flesh browning, gel breakdown, mealiness, flesh translucency, red pigment accumulation (bleeding), overripening, and loss of flavor (Figure 13.9a and b) (Dodd, 1984; Taylor et al., 1995; Taylor, 1996; Crisosto et al., 1999a,c). Postharvest life varies among cultivars and is strongly affected by temperature management. Most plum and fresh prune cultivars are susceptible to cold storage disorders when stored at 5°C more than at 0°C. Market life of 'Blackamber,' 'Fortune,' and 'Angeleno' plums growing in California at 0°C was >5 weeks. 'Showtime,' 'Friar,' and 'Howard Sun' plums developed CI symptoms within 4 weeks, even when stored at 0°C. In all plum cultivars, a much longer market life was achieved when stored at 0°C than at 5°C (Table 13.2). Most plum cultivars when handled at temperatures very close to 0°C have a postharvest life that allows them to be marketed within 4 weeks without expressing cold storage disorders (Table 13.3). However, other deterioration factors such as softening, skin color changes, and decay may become important deterioration factors during after storage marketing that limit prolonged storage of plums (Fourie and Holz, 1985).

For some cultivars, as with peaches and nectarines, oxygen levels between 1 and 2 kPa delay ripening and carbon dioxide levels between 0 and 5 kPa suppress fruit softening. Oxygen levels below 1 kPa may induce failure to ripen and off-flavors while carbon dioxide higher than 10 kPa has been associated with flesh browning. Thus, specific evaluation per cultivar must be carried out. The major benefits of CA reported during storage and shipment have been delaying skin and flesh color changes (Eksteen et al., 1986; Ben and Gaweda, 1992a,b). Flesh firmness retention immediately after cold storage removal has not normally been a benefit when fruit have been stored close to 0°C. However, in most of the cases when fruits were stored under cold storage temperatures higher than 0°C, CA-treated fruit had higher firmness than air-stored fruit. The same occurred in studies when fruit were stored at 7.5°C to avoid chilling damage. In this situation, CA-treated fruit had a higher firmness than air-stored fruit (Crisosto and Garner, 2008; Machado and Kader, personal communication).

Currently, CA has a limited use for storage >4 weeks with some cultivars such as 'Angeleno,' 'Casselman,' 'Santa Rosa,' 'Laroda,' and 'Queen Ann' (Couey, 1960, 1965; Eksteen et al., 1986; Kader and Mitchell, 1989; Streif, 1989; Ben and Gaweda, 1992a,b,c;

TABLE 13.3

Postharvest Market Life for Several California Plum Cultivars
Based on CI under 0°C and 5°C Cold Storage

Cultivar	Market Life		
	Weeks 0°C	**Weeks 5°C**	**Type**[a]
Angeleno	5	5	A
Betty Anne	5	5	A
Black Amber	5	2–3	B
Earliqueen	3	2	C
Flavorich	5	5	A
Fortune	5	2	B
Friar	5	3	B
Hiromi Red	4	2–3	B
Howard Sun	4	1	C
Joanna Red	5	5	A
October Sun	5	5	A
Purple Majesty	4	3	B
Showtime	5	3	B

Source: From Crisosto, C.H. and D. Garner. 2008. Effects of controlled atmosphere on plums. Central Valley Postharvest Newsletter, May: 4–6.

[a] Type A = CI nonsusceptible and temperature insensitive (fruit with at least 4 weeks of market life at both temperatures); Type B = CI nonsusceptible (at least 4 weeks of market life) at 0°C but susceptible (<4 weeks of market life) at 5°C (temperature sensitive); and Type C = CI susceptible (<4 weeks of market life) at both storage temperatures.

Truter et al., 1994b). In California, success in plums has also been achieved with the long-term storage temperature of 0.6°C and an atmosphere of 2 kPa O_2 + 5 kPa CO_2 with 8 weeks for 'Friar,' 12 weeks for 'Casselman' and 16 weeks for 'Angeleno' (Mitchell et al., 1981). In South Africa, a period of 8 weeks postharvest life under CA storage at 3 kPa O_2 + 5 kPa CO_2 has been reported for 'Laetitia' and 'Casselman.' For 'Songold' plums, the best CA treatment was also 3 kPa O_2 + 5 kPa CO_2 but the maximum storage time was 7 weeks. During these studies, the incidence of slight overripening was high in all treatments. CA storage resulted in the lowest incidence of over-ripening and the highest firmness. Gel breakdown (Taylor, 1996) was low in fruit of all treatments and no decay developed in any of the fruit. Although the results indicate that air stored fruit also performed well, the taste of these fruit was totally unacceptable after 8 weeks storage and 7 days at 10°C (Truter and Combrink, 1997).

Recently a detailed evaluation on different CO_2 and O_2 combinations (5 kPa CO_2 + 3 kPa O_2, 10 kPa CO_2 + 5 kPa O_2, or 15 kPa CO_2 + 10 kPa O_2) was carried out using well-mature 'Blackamber,' 'Flavorich,' 'Fortune,' and 'Friar' plums (Crisosto and Garner, unpublished report). Fruits were collected from packers, and forced air-cooled overnight to a pulp temperature of approximately 0°C. After cooling, the fruits were stored at 0°C or 7.5°C under a continuous flow of either air or CA combinations. Plum quality was evaluated at receipt, then after 3 and 6 weeks cold storage.

On 'Blackamber' plum CA treatments did not affect fruit firmness changes as softening did not occur during a 3 week period at 0°C. However, CA delayed skin color changes from red to dark fruit during a 6 week period at 0°C. At this time, a higher proportion of red plums (~60%) stored in the 5 kPa CO_2 + 3 kPa O_2 and the 10 kPa CO_2 + 5 kPa O_2 CA treatments turned dark, while in the 15 kPa CO_2 + 10 kPa O_2 treatment only 27% of red plums turned dark. In fruits stored at 7.5°C, CA treatments significantly reduced fruit softening and skin color changes from red to dark. By 6 weeks at 0°C, cold storage disorders such as gel breakdown, flesh browning, and mealiness were not observed in any of the treatments. Flesh bleeding incidence was lower in all of the three CA treatments than in air-stored plums at 0°C.

On 'Flavorich' plum CA treatments did not affect fruit firmness changes as softening did not occur during the 6 week period at 0°C. However, CA delayed skin color changes from red to dark fruit during the 6 week period at 0°C. In fruits stored at 7.5°C, CA treatments significantly reduced fruit softening and skin color changes from red to dark. At 7.5°C, 100% of the red fruits stored in air turned dark after 3 weeks storage, while 40% of CA-stored fruits were still red after 6 weeks storage. Cold storage disorders such as gel breakdown, flesh browning, mealiness, and flesh bleeding were not observed in any of the treatments within this 6 week period at 0°C. However, after 6 weeks storage at 7.5°C, high levels of gel breakdown were measured in CA-stored fruit (both red and black) at 7.5°C, indicating that low oxygen and/or high CO_2 toxicity may have occurred during this long period at 7.5°C and suggesting the importance of keeping fruit close to 0°during storage to avoid any potential injury from the CA treatment.

In 'Fortune' plums, CA treatments did not affect fruit firmness changes as softening did not occur during the 6 week period at 0°C. At 7.5°C, dark red fruit stored in CA remained firmer than air stored fruit after 3 weeks. After 6 weeks storage, red plums stored in 10 or 15 kPa CO_2 were firmer than plums stored in air or 5 kPa CO_2. In this cultivar, CA did not significantly affect skin color changes during this 6 week period at 0°C and 7.5°C. By 3 weeks at 0°C, cold storage disorders appeared and about 20% of plums had mealiness symptoms. Cold storage disorder incidence increased during cold storage at 0°C, reaching only about 20% of sound fruit on air stored compared with none on CA treated after 6 weeks. At 7.5°C, most plums were not marketable because of different cold storage disorder symptoms such as flesh browning and mealiness.

On 'Friar' plum CA treatments did not affect fruit firmness changes as softening did not occur during a 3 week period at 0°C. Fruit remained firm at 0°C under all storage atmospheres for 3 weeks. Some softening occurred after 6 weeks storage, but there was still no difference between storage atmospheres. At 7.5°C, fruit stored in CA remained firmer than air stored fruit after both storage intervals. In this cultivar which is dark at harvest time, CA did not delay skin color changes from dark to dark black fruit during the 6 week period at 0°C. In fact, plums turned dark black in cold storage, regardless of storage atmosphere or temperature. After 3 weeks storage at 0°C or 7.5°C, fruit stored in air had more flesh bleeding than CA stored fruit although onset of flesh browning symptoms were observed in fruit from all the treatments. At this time, fruit from all storage atmospheres remained juicy. At 7.5°C, both dark and dark black fruit from the 10 and 15 kPa CO_2 storage atmosphere treatments started to show some onset symptoms of gel breakdown. After 6 weeks storage, there were high levels of flesh browning and bleeding in all fruit. There was less flesh bleeding incidence in CA-stored fruit than air stored fruit. Fruit still remained juicy and CA treatments did not reduce the flesh browning problems.

13.5.2 Disease Development

It has been demonstrated that reduction on *B. cinerea* growth by a 2 kPa O_2 level is not more than 15 kPa below the rate of growth in air. The addition of CO_2 during cold storage affects *B. cinerea* and *M. fructicola* but levels higher than 10 kPa are needed to suppress fungi growth. However, many plum cultivars will not tolerate low O_2 and high CO_2. Thus, the combinations of O_2 and CO_2 commonly used have a very modest suppression on fungi (Sommer, 1985). Since most cultivars of plums develop increased flesh browning in CO_2 levels above 10 kPa, there is little benefit of CA or MA in reducing decay by fungistatic means. However, in the delay of ripening and softening, which is found under CA and MA conditions, there is an indirect beneficial effect on delaying decay.

13.5.3 Fruit Biochemistry and Quality Attributes

CA storage is known for its ability to reduce the respiration rate, ethylene production, softening, and skin color changes. During 10 weeks 'Wegierka Zwykla' prune (*P. domestica* L.) was stored under different CA storage conditions (3 kPa CO_2 + 3 kPa O_2, 5 kPa CO_2 + 3 kPa O_2, and 15 kPa CO_2 + 3 kPa O_2). The CA conditions had a significant influence on fruit firmness and taste. The possible storage period based on firmness was lengthened with higher concentrations of CO_2 in the chamber with 3 kPa CO_2 + 3 kPa O_2 by 2 weeks, with 5 kPa CO_2 + 3 kPa O_2 by 4 weeks, and in the combination of 15 kPa CO_2 + 3 kPa O_2 by 8 weeks as compared with fruit in air cold storage. High positive coefficients of correlation of fruit firmness with soluble sugars, organic acids, turgor; and high negative coefficients with weight losses and pectins were reported (Ben and Gaweda, 1992c). Taste was better preserved in fruits stored in the combination of 5 kPa CO_2 + 3 kPa O_2 than air stored up to 6 weeks. High positive correlations were obtained between the taste quality of plums and the level of soluble sugars, organic acids, and total phenols in their flesh (Ben and Gaweda, 1992c). On 'Friar' plums stored in air, 2 kPa O_2 + 5 kPa CO_2, or 6 kPa O_2 + 15 kPa CO_2 at 0°C or 10°C up to 8 weeks, ethanol and acetaldehyde accumulation was present in fruit under the 6 kPa O_2 + 15 kPa CO_2 but not in fruit under the 2 kPa O_2 + 5 kPa CO_2 treatment at 0°C or 10°C (Machado and Kader, unpublished report). The presence of high CO_2-induced flesh damage observed by 6 weeks was associated with the high ethanol and acetaldehyde accumulation on plum stored in the 6 kPa O_2 + 15 kPa CO_2 (Machado and Kader, unpublished report). In these studies also fruits stored in 2 kPa O_2 + 5 kPa CO_2, or 6 kPa O_2 + 15 kPa CO_2 at 0°C or 10°C up to 8 weeks had positive influence on fruit firmness and skin and flesh color but CA did not significantly influence soluble solids. CA treatment retarded titratable acidity loss and pH rise (Machado and Kader, unpublished report).

Fruits of 'Angeleno' plum were kept in air and under extreme low oxygen insecticidal CA conditions of 0.25 or 0.02 kPa O_2 at 0°C, 5°C, and 10°C for 3, 7, 14, 25, or 35 days to study the effects of low-O_2 atmospheres on their postharvest physiology and quality attributes (Ke et al., 1991b). SSC, pH, and external appearance were not significantly influenced, but resistance to CO_2 diffusion was increased by the low-O_2 treatments. Exposures to the low-O_2 atmospheres inhibited ripening, including reduction in ethylene production rate, retardation of skin color changes and flesh softening, and maintenance of titratable acidity. On the other hand, the most important detrimental effect of the low-O_2 treatments was development of an alcoholic off-flavor that had a logarithmic relationship with ethanol content of the fruits.

13.5.4 Effects of MAP

Limited testing has been done on the use of MAP as a supplement to refrigeration for maintaining quality of plums (Couey, 1960, 1965). Early work showed that some varieties of plums respond favorably to the atmospheres that occur in polyethylene liners, but others are severely injured (Couey, 1960, 1965). Therefore, each variety must be considered individually before any commercial recommendations can be given. Because 'El Dorado' plums stored well for 6 weeks in an atmosphere of 6.6 kPa CO_2 and 7.2 kPa O_2, the 'Nubiana' variety, which is closely related to the 'El Dorado,' also responded similarly. Natural modification of atmosphere in sealed 1.5 mil polyethylene box liners provided a favorable environment for 'Nubiana' plums during storage. MAs that averaged 7.8 kPa CO_2, and 11.0 kPa O_2 reduced fruit decay, softening, and loss of soluble solids during storage periods up to 10 weeks when compared with fruit in vented liners. The longer the storage period, the greater were the benefits attributable to CA (Couey, 1965). The influence of MAP on quality attributes and shelf life performance of 'Friar' plums was studied in California (Cantín et al., 2008). Plums were stored at 0°C and 85% relative humidity for a 60 day period in five different box liners (LifeSpan L316, FF-602, FF-504, 2.0% vented area perforated, and Hefty liner) and untreated (control). Flesh firmness, SSC, TA, and pH were not affected by the MAP liners. Fruit skin color changes were repressed on plums packed in box liners that modified gas levels and weight loss was reduced by the use of any of the box liners. Plums packed without box liners (bulk packed) had ~6% weight loss. High carbon dioxide and low oxygen levels were measured in boxes with MAP box liners (LifeSpan L316, FF-602, and FF-504). Percentage of healthy fruit was not affected by any of the treatments during the ripening period following 45 days of cold storage. However, after 60 days of cold storage, fruit from the MAP box liners with higher CO_2 and lower O_2 levels had a higher incidence of CI symptoms, evident as flesh translucency, gel breakdown, and off flavor than fruit from the other treatments. Overall, results indicate that the use of MAP box liners is recommended to improve market life of 'Friar' plums up to 45 days cold storage. However, the use of box liners without gas control capability may lead to decay disorders in fruit cold-stored for longer periods. Similar cold storage disorders were reported by García and Farías (personal communication) after 69 days under MAP for 'Larry Ann' plum in Chile.

The storage behavior of the plum variety 'Bühler Frühzwetsche' was tested in modified atmospheres with simultaneous O_2 reductions and CO_2 increase or CO_2 increase alone. After 4 weeks in CA storage with 12 kPa CO_2 and 2 kPa O_2 at 0°C, the plums showed a good appearance and fruit firmness without any unfavorable taste or damages. In CA conditions with higher O_2 and lower CO_2 or in an atmosphere, which was only enriched with CO_2 (12 kPa, 18 kPa, 24 kPa), the delay of ripening was more ineffective and the plums ripened more than in 12 kPa CO_2 – 2 kPa O_2. Although plums are generally considered as a CO_2-sensitive fruit, plums of 'Bühler Frühzwetsche' did not show any CO_2 damage until 16 kPa CO_2; but in higher concentrations the fruits increasingly showed cracked tips (Streif, 1989).

13.5.5 Commercial Use

Some plum cultivars get limited benefit from the use of CA/MAP depending on the storage/shipping period. In Israel, CA storage is used for plums harvested from August through September to allow for export in November and December (Lurie et al., 1999; Ben-Arie et al., 2001). In California, for most current commercial cultivars, using proper temperature management achieves similar results as CA storage. As one of the most consistent benefits of CA/MAP is reduced plum water loss in mid- to long-term storage, it is important to study ways to reduce fruit water loss during storage using less expensive nonmodified atmosphere boxes or pallet liners in the future. Therefore, the MAP use should be evaluated based on the added expenses and risks of the technology.

References

Ahmadi, H., W.V. Biasi, and E.J. Mitcham. 1999. Control of brown rot decay of nectarines with 15 kPa carbon dioxide atmospheres. *J. Am. Soc. Hort. Sci.* 124:708–712.

An, J., M. Zhang, and Z. Zhan. 2006. Effect of packing film on the quality of 'Chaoyang' honey peach fruit in modified atmosphere packages. *Packag. Technol. Sci.* 20:71–76.

Anderson, R.E., C.S. Parsons, and W.L. Smith, Jr. 1969. Controlled atmosphere storage of eastern-grown peaches and nectarines. USDA Marketing Research Report No. 836, 19pp.

Andrich, G. and R. Fiorentini. 1986. Effects of controlled atmosphere on the storage of new apricot cultivars. *J. Sci. Food Agric.* 37:1203–1208.

Beaudry, R.M. 1999. Effect of O_2 and CO_2 partial pressure on selected phenomena affecting fruit and vegetable quality. *Postharvest Biol. Technol.* 15:293–303.

Ben, J. and M. Gaweda. 1992a. Effect of increasing concentrations of CO_2 in controlled atmosphere storage on the development of physiological disorders and fungal diseases in plums (*Prunus domestica* L.). *Folia Hort.* 4:87–100.

Ben, J. and M. Gaweda. 1992b. The effect of increasing concentration of carbon dioxide in controlled atmosphere storage of plums cv. Wegierka Zwykla (*Prunus domestica L.*). I. Firmness of plums. *Acta Physiologiae Plant.* 14(3):143–150.

Ben, J. and M. Gaweda. 1992c. The effect of increasing concentration of carbon dioxide in controlled atmosphere storage of plums cv. Wegierka Zwykla (*Prunus domestica* L.). II. Changes in the content of soluble sugars, organic acids, and total phenolics in the aspect of fruit taste. *Acta Physiologiae Plant.* 14(3):151–158.

Ben-Arie, R., O. Neria, and S. Lurie. 2001. Guidelines for harvest and storage of plums and peaches. *Alon HaNotea* 55:352–355.

Brackmann, A., J. Streif, and F. Bangerth. 1993. Relationship between a reduced aroma production and lipid metabolism of apples after long-term control led-atmosphere storage. *J. Am. Soc. Hort. Sci.* 118:243–247.

Brecht, J.K., A.A. Kader, C.M. Heintz, and R.C. Norona. 1982. Controlled atmosphere and ethylene effects on quality of California canning apricots and clingstone peaches. *J. Food Sci.* 47:432–436.

Burmeister, D.M. and J.E. Harman. 1998. Effect of fruit maturity on the success of controlled atmosphere storage of 'Fantasia' nectarines. *Acta Hort.* 464:363–369.

Cantín, C., C.H. Crisosto, and K.R. Day. 2008. Evaluation of the effect of different modified atmosphere packaging (MAP) box liners on the quality and shelf life of 'Friar' plums (*Prunus salicina* Lindell). *HortTechnology* 18(2): 261–265.

Carrier Corporation. 1995. *Controlled Atmosphere Handbook: A Guide for Shipment of Perishable Cargo in Refrigerated Containers.* Carrier Corporation, Syracuse, NY, pp. 93.

Ceponis, M.J. and J.E. Butterfield. 1981. Cull losses in Western sweet cherries at retail and consumer levels in metropolitan New York. *HortScience* 16:324–326.

Ceponis, M.J., R.A. Cappellini, and G.W. Lightner. 1987. Disorders of sweet cherry and strawberry shipments to the New York market, 1972–1984. *Plant Dis.* 71(5):472–475.

Chambroy, Y., M. Souty, M. Reich, L. Breuils, G. Jacquemin, and J.M. Audergon. 1991. Effects of different CO_2 treatments on post harvest changes of apricot fruit. *Acta Hort.* 293:675–684.

Chen, P.M., W.M. Mellenthin, S.B. Kelly, and T.J. Facteau. 1981. Effects of low oxygen and temperature on quality retention of 'Bing' cherries during prolonged storage. *J. Am. Soc. Hort. Sci.* 106:533–535.

Couey, H.M. 1960. Effect of temperature and modified atmosphere on the storage life, ripening behavior, and dessert quality of 'El Dorado' plums. *Proc. Am. Soc. Hort. Sci.* 75:207–215.

Couey, H.M. 1965. Modified atmosphere storage of Nubiana plums. *Proc. Am. Soc. Hort. Sci.* 86:166–168.

Crisosto, C.H. and D. Garner. 2008. Effects of controlled atmosphere on plums. Central Valley Postharvest Newsletter, 17(2):4–6.

Crisosto, C.H. and G. Crisosto. 2008. Effects of controlled atmospheres on nectarines. Central Valley Postharvest Newsletter, 17(2):7–8.

Crisosto, C.H. 2002. How do we increase peach consumption? *Acta Hort.* 592:601–605.

Crisosto, C.H. and J.M. Labavitch. 2002. Developing a quantitative method to evaluate peach (*Prunus persica*) flesh mealiness. *Postharvest Biol. Technol.* 25:151–158.

Crisosto, C.H., G.M. Crisosto, and M.A. Ritenour. 2002. Testing the reliability of skin color as an indicator of quality for early season 'Brooks' (*Prunus avium* L.) cherry. *Postharvest Biol. Technol.* 24:147–154.

Crisosto, C.H., D. Garner, L. Cid, and K.R. Day. 1999a. Peach size affects storage, market life. *Calif. Agric.* 53:33–36.

Crisosto, C.H., D. Garner, and G. Crisosto. 1997a. Evaluating the relationship between controlled atmosphere storage, peach fruit size and internal breakdown. *Perishables Handling* 90:7–8.

Crisosto, C.H., R.S. Johnson, K.R. Day, B. Beede, and H. Andris. 1999b. Contaminants and injury induce inking on peaches and nectarines. *Calif. Agric.* 53(1):19–23.

Crisosto, C.H., R.S. Johnson, T. DeJong, and K.R. Day. 1997b. Orchard factors affecting postharvest stone fruit quality. *HortScience.* 32:820–823.

Crisosto, C.H., F.G. Mitchell, and R.S. Johnson. 1995. Factors in fresh market stone fruit quality. *Postharvest News Info.* 5:17N–21N.

Crisosto, C.H., F.G. Mitchell, and Z. Ju. 1999c. Susceptibility to chilling injury of peach, nectarine and plum cultivars grown in California. *HortScience* 34(6):1116–1118.

Crisosto, C.H., C. Valero, and D.C. Slaughter. 2007. Predicting pitting damage during processing in Californian clingstone peaches using color and firmness measurements. *Appl. Eng. Agric.* 23(2):189–194.

De Vries-Paterson, R.M., A.L. Jones, and A.C. Cameron. 1991. Fungistatic effects of carbon dioxide in a package environment on the decay of Michigan sweet cherries by *Monilinia fructicola*. *Plant Dis.* 75:943–949.

Dodd, M.C. 1984. Internal breakdown in plums. *Deciduous Fruit Grower* 34:255–256.

Drake, S.R. and A. Yazdaniha. 1999. Short term controlled atmosphere storage for shelf life extension of apricots. *J. Food Process. Preserv.* 23:57–70.

Eksteen, G.J., T.R. Visagie, and J.C. Laszlo. 1986. Controlled atmosphere storage of South African grown nectarines and plums. *Deciduous Fruit Grower* 36(4):128–132.

Fernandez-Trujillo, J.P., A. Cano, and F. Artes. 1998. Physiological changes in peaches related to chilling injury and ripening. *Postharvest Biol. Technol.* 13:109–119.

Fourie, J.F. and G. Holz. 1985. Postharvest fungal decay of stone fruit in the South-Western Cape. *Phytophylactica* 17:175–177.

Garner, D., C.H. Crisosto, and E. Otieza. 2001. Controlled atmosphere storage and aminoethoxyvinyl-glycine postharvest dip delay post cold storage softening of 'Snow King' peach. *HortTechnology* 11(4):598–602.

Ginsburg, L. and J.C. Combrink. 1972. Cold storage of apricots. *Dried Fruit* 4:19–23.

Girardi, C.L., A.R. Corrent, L. Lucchetta, M.R. Zanuzo, T.S. da Costa, A. Brackmann, R.M. Twyman, F.R. Nora, L. Nora, J.A. Silva, and C.V. Rombaldi. 2005. Effect of ethylene, intermittent warming and controlled atmosphere on postharvest quality and the occurrence of woolliness in peach (*Prunus persica* cv. Chiripa) during cold storage. *Postharvest Biol. Technol.* 38(1):25–33.

Golias, J. and H. Bottcher. 2003. Changes in the ethanol content of stored fruits at low oxygen atmosphere. *J. Appl. Bot.* 77:181–184.

Gorney, J.R., B. Hess-Pierce, and A.A. Kader. 1999. Quality changes in fresh-cut peach and nectarine slices as affected by cultivar, storage atmosphere and chemical treatments. *J. Food Sci.* 64(3):429–432.

Harding, P.L. and M.H. Haller. 1934. Peach storage with special reference to breakdown. *Proc. Am. Soc. Hort. Sci.* 32:160–163.

Harman, J.E., M. Layyee, D.P. Billing, C.W. Yearsley, and P.J. Jackson. 1990. Effects of methyl-bromide fumigation, delayed cooling, and controlled atmosphere storage on the quality of Redgold and Fantasia nectarine fruit. *N. Z. J. Crop Hort. Sci.* 18(4):197–203.

Harvey, G.M., L. Wilson, G.R. Smith, and J. Kaufman. 1972. Market diseases of stone fruits. *Agric. Handb.* 414:39.

Kader, A.A. 1986. Biochemical and physiological basis for effects of controlled and modified atmospheres on fruits and vegetables. *Food Technol.* 40(5):99–100, 102–104.

Kader, A.A. 1992. Modified atmospheres during transport and storage. In: A.A. Kader (Ed.), *Postharvest Technology of Horticultural Crops*. University of California Division of Agricultural and Natural Resources, Publication 3311, Oakland, California, pp. 85–92.

Kader, A.A. (Ed.). 1997. Fruits other than apples and pears. *Seventh International Controlled Atmosphere Research Conference*, July 13–18, University of California, Davis, CA, pp. 1–261.

Kader, A.A. and F.G. Mitchell. 1989. Postharvest physiology peaches. In: J.H. LaRue and R.S. Johnson (Eds.), *Peaches, Plums and Nectarines. Growing and Handling for Fresh Market*. University of California Division of Agriculture and Natural Resources, Oakland, CA, Publication 3331, pp. 158–164.

Kader, A.A. and L.L. Morris. 1977. Relative tolerance of fruits and vegetables to elevated CO_2 and reduced O_2 levels. In: D.H. Dewey (Ed.), *Controlled Atmospheres for the Storage and Transport of Perishable Agricultural Commodities. Proceedings of 2nd National Controlled Atmosphere Research Conferences*. Michigan State University Horticulture Report No. 28, East Lansing, Michigan, pp. 260–265.

Kader, A.A., M.A. El-Goorani, and N.F. Sommer. 1982. Postharvest decay, respiration, ethylene production, and quality of peaches held in controlled atmospheres with added carbon monoxide. *J. Am. Soc. Hort. Sci.* 107:856–859.

Kader, A.A., D. Zagory, and E.L. Kerbel. 1989. Modified atmosphere packaging of fruits and vegetables. *Crit. Rev. Food Sci. Nutr.* 28:1–30.

Kajiura, I. 1975. Controlled atmosphere storage and hypobaric storage of white peach 'Okubo'. *Scientia Hort.* 3:179–187.

Ke, D., F. El-Wazir, B. Cole, M. Mateos, and A.A. Kader. 1994. Tolerance of peach and nectarine fruits to insecticidal controlled atmospheres as influenced by cultivar, maturity and size. *Postharvest Biol. Technol.* 4:135–146.

Ke, D., L. Rodriguez-Sinobas, and A.A. Kader. 1991a. Physiological responses and quality attributes of peaches kept in low oxygen atmospheres. *Scientia Hort.* 47:295–303.

Ke, D., L. Rodriguez-Sinobas, and A.A. Kader. 1991b. Physiology and prediction of fruit tolerance to low-oxygen atmospheres. *J. Am. Soc. Hort. Sci.* 116(2):253–260.

Kosto, I., A. Weksler, and S. Lurie. 2000. Extending storage of apricots. *Alon Hanotea* 54:250–254.

Kosto, I., A. Weksler, and S. Lurie. 2002. Modified atmosphere storage of apricots. *Alon Hanotea* 56:173–175.

Levin, A., S. Lurie, Y. Zutkhi, and R. Ben Arie. 1995. Physiological effects of controlled atmosphere storage on 'Fiesta Red' nectarines. *Acta Hort.* 379:121–127.

Looney, N.E., A.D. Webster, and E.M. Kuperman. 1996. Harvest and handling sweet cherries for the fresh market. In: A.D. Webster and N.E. Looney (Eds.), *Cherries: Crop Physiology, Production, and Uses*. CAB International University Press, Cambridge, U.K., pp. 411–441.

Lurie, S. 1992. Controlled atmosphere storage to decrease physiological disorders in nectarines. *Int. J. Food Sci. Technol.* 27:507–415.

Lurie, S. and N. Aharoni. 1997. Storage of cherries. In: A.A. Kader (Ed.), *Proceedings of the International Conference on Controlled Atmosphere*. Davis, CA, Conference 3, pp. 149–156.

Lurie, S. and C.H. Crisosto. 2005. Chilling injury in peach and nectarine. *Postharvest Biol. Technol.* 37:195–208.

Lurie, S., A. Weksler, Z. Lapsker, and Y. Greenblat. 1999. Storage properties of new plum cultivars. *Alon HaNotea* 53:238–241.

Malakou, A. and G.D. Nanos. 2005. A combination of hot water treatment and modified atmosphere packaging maintains quality of advanced maturity 'Caldesi 2000' nectarines and 'Royal Glory' peaches. *Postharvest Biol. Technol.* 38:106–114.

Mattheis, J.P., D.A. Buchanan, and J.K. Fellman. 1997. Volatile constituents of Bing sweet cherry fruit following controlled atmosphere storage. *J. Agric. Food Chem.* 45(1):212–216.

McLaren, G.F., J.A. Fraser, and D.M. Burmeister. 1997. Storage of apricots in modified atmospheres. *The Orchardist* 20:31–33.

Mencarelli, F., F. Garosi, R. Botondi, and P. Tonutti. 1998. Postharvest physiology of peach and nectarine slices. *Acta Hort.* 465:463–470.

Mitchell, F.G. and A.A. Kader. 1989. Factors affecting deterioration rate. In: J.H. LaRue and R.S. Johnson (Eds.), *Peaches, Plums and Nectarines—Growing and Handling for Fresh Market*. Publication 3331, University of California Division of Agriculture and Natural Resources, Oakland, CA, Publication 3331, pp. 165–178.

Murray, R., C. Lucangeli, G. Polenta, and C. Budde. 2007. Combined pre-storage heat treatment and controlled atmosphere storage reduced internal breakdown of 'Flavorcrest' peach. *Postharvest Biol. Technol.* 44(2):116–121.

Ogawa, J.M. and H. English. 1991. Diseases of temperate zone tree fruit and nut crops. University of California, Davis, CA, Publication 3345, pp. 461.

Patterson, M.E. 1982. CA storage of cherries. In: D.G. Richardson and M. Meheriuk (Eds.), *Controlled Atmosphere for Storage and Transport of Perishable Agricultural Commodities*. Timber Press, Beaverton, OR, pp. 149–154.

Patterson, M.E. and J.L. Melstad. 1977. Sweet cherry handling and storage alternatives. In: D.H. Dewey (Ed.), *Controlled Atmospheres for the Storage and Transport of Perishable Agricultural Commodities*. *Proceedings of 2nd National Controlled Atmosphere Research Conferences*. Michigan State University Horticulture Report No. 28, Beaverton, Oregon, pp. 55–59.

Peace, C.P., C.H. Crisosto, D.T. Garner, A.M. Dandekar, T.M. Gradziel, and F.A. Bliss. 2006. Genetic control of internal breakdown in peach. *Acta Hort.* 713:489–496.

Petracek, P.D., D.W. Joles, A. Shirazi, and A.C. Cameron. 2002. Modified atmosphere packaging of sweet cherry (*Prunus avium* L., cv. 'Sams') fruit: Metabolic responses to oxygen, carbon dioxide, and temperature. *Postharvest Biol. Technol.* 24(3):259–270.

Poneleit, L.S. and D. Dilley. 1993. Carbon dioxide activation of 1-aminocyclopropane-1-carboxylate (ACC) oxidase in ethylene synthesis. *Postharvest Biol. Technol.* 3:191–199.

Porritt, S.W. and J.L. Mason. 1965. Controlled atmosphere storage of sweet cherries. *Proc. Am. Soc. Hort. Sci.* 87:128–130.

Pretel, M.T., M. Serrano, A. Amoros, and F. Romojaro. 1999. Ripening and ethylene biosíntesis in controlled atmosphere stored apricots. *Eur. Food Res. Technol.* 209:130–134.

Pretel, M.T., M. Souty, and F. Romojaro. 2000. Use of passive and active modified atmosphere packaging to prolong the postharvest life of three varieties of apricot (*Prunus armeniaca* L.) *Eur. Food Res. Technol.* 211:191–198.

Remón, S., A. Ferrer, P. Lopez-Buesa, and R. Oria. 2004. Atmosphere composition effects on Burlat cherry colour during cold storage. *J. Sci. Food Agric.* 84(2):140–146.

Retamales, J., R. Campos, P. Herrera, and J.M. Camus. 1997. High CO_2 modified atmosphere can be effective in preventing woolliness in nectarines. *Proceedings of 7th International Controlled Atmosphere Research Conference*. UC Davis Postharvest Horticulture Series No. 17, Vol. 3, pp. 46–53.

Retamales, J., T. Cooper, J. Streif, and J.C. Kania. 1992. Preventing cold-storage disorders in nectarines. *J. Hort. Sci.* 67(5):619–626.

Ryall, A.L. and W.T. Pentzer. 1982. Diseases and injuries of deciduous tree fruits during marketing. In: *Handling, Transportation and Storage of Fruits and Vegetables*, 2nd edn. Avi Publishing Company, Westport, CN, pp. 495.

Smith, S., J. Geeson, and J. Stow. 1987. Production of modified atmospheres in deciduous fruits by the use of films and coatings. *HortScience* 22:772–776.

Smith, W.H. 1934. Cold storage of Elberta peaches. *Ice and Cold Storage* 37:54–57.

Sommer, N.F. 1985. Role of controlled environments in suppression of postharvest diseases. *Can. J. Plant Pathol.* 7:331–339.

Spotts, R.A., L.A. Cervantes, and T.J. Facteau. 2002. Integrated control of brown rot of sweet cherry fruit with a preharvest fungicide, a postharvest yeast, modified atmosphere packaging, and cold storage temperature. *Postharvest Biol. Technol.* 24(3):251–257.

Spotts, R.A., L.A. Cervantes, T.J. Facteau, and T. Chand-Goyal. 1998. Control of brown rot and blue mold of sweet cherry with preharvest iprodione, postharvest *Cryptococcus infirmo-miniatus*, and modified atmosphere packaging. *Plant Dis.* 82(10):1158–1160.

Streif, J. 1989. Storage behaviour of plum fruits. *Acta Hort.* 258:177–183.

Streif, J., J. Retamales, T. Cooper, and J.C. Kania. 1992. Preventing physiological disorders in nectarines by CA-storage and high-CO_2 storage. *Gartenbauwissenschaft* 57:166–172.

Taylor, M.A. 1996. Internal disorders in South African plums. *Deciduous Fruit Grower* 46:328–335.

Taylor, M.A. and V.A. De Kock. 1991. Effect of harvest maturity and storage regimes on the storage quality of Peeka apricots. *Deciduous Fruit Grower* 41:139–143.

Taylor, M.A., E. Rabe, G. Jacobs, and M.C. Dodd. 1995. Effect of harvest maturity on pectic substances, internal conductivity, soluble solids and gel breakdown in cold stored 'Songold' plums. *Postharvest Biol. Technol.* 5:285–294.

Tian, S.P., A. Folchi, G.C. Pratella, and P. Bertolini. 1996. The correlations of some physiological properties during ultra low oxygen storage in nectarine. *Acta Hort.* 374:131–140.

Truter, A.B. and J.C. Combrink. 1997. Controlled atmosphere storage of South African plums. *Proceedings of the International Controlled Atmosphere Conference.* University of California, Davis, CA, Vol. 3, pp. 54–61.

Truter, A.B., J.C. Combrink, and L.J. von Mollendorff. 1994a. Controlled atmosphere storage of apricots and nectarines. *Deciduous Fruit Grower* 44:421–427.

Truter, A.B., J.C. Combrink, and L.J. von Mollendorf. 1994b. Controlled-atmosphere storage of plums. *Deciduous Fruit Grower* 44:373–375.

USDA. 2003. http://www.usda.gov

Von Mollendorff, L.J. 1987. Woolliness in peaches and nectarines: A review. 1. Maturity and external factors. *Hort. Sci./Tuinbouwetenskap* 5:1–3.

Wade, N.L. 1981. Effects of storage atmosphere, temperature, and calcium on low temperature injury of peach fruit. *Scientia Hort.* 15:145–154.

Wankier, B.N.M., D.K. Salunkhe, and W.F. Campbell. 1970. Effects of controlled atmosphere storage on biochemical changes in apricots and peach fruit. *J. Am. Soc. Hort. Sci.* 95:604–609.

Watkins, C.B. 2000. Responses of horticultural commodities to high carbon dioxide as related to modified atmosphere packaging. *HortTechnology* 10:501–506.

Wright, K.P. and A.A. Kader. 1997. Effect of controlled-atmosphere storage on the quality and carotenoid content of sliced persimmons and peaches. *Postharvest Biol. Technol.* 10:89–97.

Zanderighi, L. 2001. How to design perforated polymeric films for modified atmosphere packs (MAP). *Packag. Technol. Sci.* 14:253–266.

Zhou, H.W., S. Lurie, A. Lers, A. Khatchitski, L. Sonego, and R. Ben Arie. 2000. Delayed storage and controlled atmosphere storage of nectarines: Two strategies to prevent woolliness. *Postharvest Biol. Technol.* 18:133–141.

Zoffoli, J.P., S. Balbontin, and J. Rodríguez. 2002. Effect of modified atmosphere packaging and maturity on susceptibility to mealiness and flesh browning of peach cultivars. *Acta Hort.* 592:573–579.

14

Subtropical Fruits

Zora Singh, S.P. Singh, and Elhadi M. Yahia

CONTENTS

14.1 Introduction

Subtropical fruits are widely grown all over the world and constitute a significant portion of the global fruit production. Geographical boundaries-free trade has given a massive thrust to the growth of fresh fruit trade worldwide. Consequently large volumes of the fruits are shipped across different continents. As opportunities and challenges are coupled, the present-day challenge is to supply high-quality fresh and safe fruit to the consumer consistently. Like other fruit crops, subtropical fruits are also perishable and need great attention during postharvest handling, storage, transportation, distribution, and marketing. Some of the subtropical fruits like avocado, guava, litchi, longan, rambutan, loquat, and kiwifruit are highly perishable while the others like grapefruit, mandarin, orange, and pomegranate are comparatively less perishable. Ironically, the application of controlled/modified atmosphere (CA/MA) storage technology had been limited to apple and pear in most of the developed countries. But the benefits of CA/MA have neither been fully understood nor exploited in case of subtropical fruits, which are mostly grown in the developing world. With the advancement of postharvest science revealing a multitude of facts about fruit ripening and senescence mechanisms, there is a great scope to develop innovative technologies that could meet the demands of fresh fruit industry and fulfill the consumer indulgences. The Asian countries like India and China, which have major share in the world fruit basket, also contribute greatly to the subtropical fruit production. A paradigm shift in the modernization of the postharvest infrastructure in these developing countries has recently been noticed due to heavy investments in fresh fruit trade by the private sector. Currently, the application of CA/MA for subtropical fruits is commercially negligible. The information in literature on these fruit crops is also scanty and inconclusive. The commercial trends in fruit trade indicate that there will be more demand for sophisticated postharvest handling techniques to provide the consumer with quality fruit. Therefore, the future of CA/MA technology seems very bright in terms of its practical utility in maintaining quality and it can also serve as an alternative approach to encounter fruit specific postharvest problems. The objectives of the present chapter include the following: to review the current status of the literature available on CA/MA storage in subtropical fruits; pragmatic assessment of the CA/MA technology for long-term storage, transportation, and distribution; impact of CA/MA storage on flavor, nutrition, and functional components of subtropical fruits; a critical analysis of potential benefits and harmful effects of CA/MA in these fruit crops; application of CA/MA as an alternative or main approach to combat certain fruit specific postharvest problems, which undermine quality and storage potential of fruit; and future applications and research needs in CA/MA storage of subtropical fruits.

14.2 Individual Fruit Crops

14.2.1 Avocado (*Persea americana* Mill.)

Avocado is a climacteric fruit rich in energy. There are three ecological races including Mexican (subtropical), Guatemalan (semitropical), and West Indian (tropical). The most important cultivars are 'Hass,' a Guatemalan self-fertile cultivar, and 'Fuerte,' a hybrid between Mexican and Guatemalan races. Anthracnose and chilling injury (CI) are the most important causes of postharvest losses. Postharvest life is about 2–4 weeks at 4°C–13°C (depending on cultivar) (Yahia et al., 1997b).

Optimum MA and CA (2–5 kPa O_2 and 3–10 kPa CO_2) delay softening and skin color changes and reduce respiration, ethylene production rates, and CI of avocado fruit (Yahia, 1998). Mature-green 'Hass' avocado can be kept at 5°C–7°C in 2 kPa O_2 and 3–5 kPa CO_2 for 9 weeks, then ripen in air at 20°C to good quality. Exclusion and/or removal of ethylene from CA storage are recommended. Elevated (>10 kPa) CO_2 levels may increase skin and flesh discoloration and off-flavor development, especially when O_2 is <1 kPa (Table 14.1).

TABLE 14.1

Some Reported Effects in Avocado Fruit in Different Conditions of MA and CA

Cultivar	Temperature (°C)	O_2 (kPa)	CO_2 (kPa)	Remarks
Hass	7	2–10	4–10	Storage time of 7–9 weeks
Lula, Booth 8, and Fuchs	7.5	2	10	Increase shelf life two folds
Fuerte, Edranol, and Hass	—	2	10	Reduces internal disorders
Nonspecific	—	—	25	Reduces disorders and increases anthracnose
Fuerte	—	—	25	Delays maturation
Fuerte	5.5	2	10	Less dark spots in the pulp
Fuerte	5.5	—	25	Less dark spots in the pulp
Fuerte	24 h at 17°C	3	0	After this treatment, fruit can be stored at 2°C for 3 weeks
Booth 8 and Lula		2	10	Storage time of 8 weeks
Fuerte and Anaheim		6	10	Storage time of 38 days
Waldin and Fuchs		2	10	Storage of 4 weeks, prevents anthracnose and CI
Hass		2	5	Storage time of up to 60 days

Source: Yahia, E.M., Manejo postcosecha del aguacate (postharvest handling of avocado), 2ª. Parte. Boletín Informativo de APROAM El Aguacatero, year 6, number 32, May 2003, Mexico, 2003.

Very early research by Overholser (1928) reported that the storage life of 'Fuerte' avocados was prolonged 1 month when fruit was held in an atmosphere of 4–5 kPa O_2 and 4–5 kPa CO_2 at 7.5°C compared to air storage. Brooks et al. (1936) reported that fruit could be held in atmospheres containing 20–50 kPa CO_2 at 5°C–7.5°C for 2 days without causing any injury. Atmospheres with CO_2 levels below 3 kPa prolonged the storage life of Florida avocados at all temperatures, and reduced the development of brown discoloration of the skin (Stahl and Cain, 1940). Extensive work was later done also with 'Fuerte' avocado, and concluded that the time for the fruit to reach the climacteric is extended in proportion to the decrease in O_2 concentration from 21 to 2.5 kPa (Biale, 1942, 1946). In later years Young et al. (1962) demonstrated that the delay of the climacteric could also be achieved by 10 kPa CO_2 in air, and the combination of low O_2 and high CO_2 suppresses further the intensity of fruit respiration. Hatton and Reeder (1965, 1969, 1972) and Spalding and Reeder (1972, 1974a) found that a CA of 2 kPa O_2 and 10 kPa CO_2 at 7.5°C doubled the storage life of the cultivars 'Lula,' 'Fuch,' and 'Booth 8.' The percentage of acceptable fruit after storage was increased by absorption of ethylene during CA storage (Hatton and Spalding, 1974). 'Reed' and 'Hass' avocados were reported to be stored for up to 3 and 2 months, respectively, in CA (Sive and Resnizky, 1989). Jordan and Smith (1993) reported that 'Hass' avocados remained firm and unripe for 7–9 weeks in CA of 2–10 kPa O_2 and 4–10 kPa CO_2 at 7°C. Below 4 kPa CO_2 storage life was 5–6 weeks. Truter and Eksteen (1987a,b) reported that a mixture of 2 kPa O_2 and 10 kPa CO_2 extended the shelf life and reduced the gray pulp and virtually eliminated pulp spot of 'Fuerte,' 'Edranol,' and 'Hass,' but an increase in anthracnose was observed. Truter and Eksteen (1987b) found that a 25 kPa CO_2 shock treatment applied 1 day after harvest reduced physiological disorders without any increase in anthracnose. Allwood and Wolstenholme (1995) were able to delay ripening of 'Fuerte' fruit using a 25 kPa CO_2 shock treatment applied in pulses 3 times every 24 h. Marcellin and Chavez (1983) reported that intermittent exposure to 20 kPa CO_2

of 'Hass' avocados stored in air delayed senescence at 12°C, reduced CI at 4°C, and controlled decay at both temperatures.

CA delays the softening process, and thus maintains the resistance of the fruit to fungal development (Spalding and Reeder, 1975). In addition, Prusky et al. (1991, 1993) reported that 30 kPa CO_2 (with 15 kPa O_2) for 24 h increased the levels of the antifungal compound, 1-acetoxy-2-hydroxy-4-oxo-heneicosa-12,15-diene, in the peel and flesh of unripe avocado fruits, and delayed decay development. This diene has been suggested as the basis for decay resistance in unripe avocados (Prusky et al., 1982, 1988, 1991). A 20 kPa CO_2 can be tolerated by thick-skinned avocados such as 'Hass' and 'Lula,' but causes browning of the skin in thin-skinned varieties such as 'Ettinger' (Collin, 1984). High concentration of CO (5–10 kPa) added to CA can reduce decay development (El-Goorani and Sommer, 1981). Moderately high concentrations of CO_2 (up to 10 kPa) were shown to ameliorate CI in 'Taylor' avocados (Vakis et al., 1970). Spalding and Reeder (1972) found less internal and external CI in CA than in air storage of 'Booth 8' and 'Lula' avocados. Intermittent high CO_2 treatment (3 treatments during 21 days) reduced CI symptoms (Marcellin and Chaves, 1983). 'Fuerte' avocados had less pulp spot and blackening of cut vascular bundles after storage in 2 kPa O_2 and 10 kPa CO_2 at 5.5°C for 28 days, or after a "shock" treatment of 25 kPa at 5.5°C for 3 days and an additional 28 days at normal atmosphere at 5.5°C (Bower et al., 1990). Spalding (1977) concluded that the CO_2 must be kept below 15 kPa to prevent fruit injury. Prestorage of 'Fuerte' avocados in 3 kPa O_2 (balance N_2) atmosphere for 24 h at 17°C significantly reduced CI symptoms after storage at 2°C for 3 weeks (Pesis et al., 1993, 1994). Fruits prestored in 97 kPa N_2 had lower respiration and ethylene production, lower ion leakage, higher reducing power (expressed as SH groups, mainly cysteine and glutathione), and longer shelf life than the untreated fruit. 'Booth 8' and 'Lula' avocados were reported to be held successfully for up to 8 weeks in a CA of 2 kPa O_2 with 10 kPa CO_2 at 4°C–7°C and 98%–100% relative humidity (RH), and removal of ethylene further improved the keeping quality of the 'Lula' fruits (Spalding and Reeder, 1972). Fruits of 'Booth 8' had slight CI at 4.5°C. 'Fuerte' and 'Anaheim' fruits were stored in Brazil for up to 38 days in 6 kPa O_2 and 10 kPa CO_2 at 7°C, but only for 12 days in air (Bleinroth et al., 1977a). Storage of 'Waldin' and 'Fuchs' avocados in 2 kPa O_2 and 10 kPa CO_2 for up to 4 weeks at 7°C was also reported to prevent development of anthracnose and CI (Spalding and Reeder, 1974a, 1975). 'Hass' avocado was reported to be stored for up to 60 days in atmospheres of 2 kPa O_2 and 5 kPa CO_2 (Faubion et al., 1992; Jordan and Barker, 1992; McLauchalan et al., 1992). Four commercial CA rooms were constructed in Florida in the season of 1972/73 for storage of 'Lula' avocados in bulk bins (Spalding and Reeder, 1974a). The rooms were run at 2 kPa O_2 and 10 kPa CO_2 at 7.2°C and 95% RH, and fruits were reported to be marketed in excellent conditions after 5 weeks of storage, except for some fruits with rind discoloration (CI) where temperature dropped below 4.4°C. In South Africa, Bower et al. (1989) suggested that even though fruit stored in CA (2 kPa O_2 and 10 kPa CO_2) were superior than in other storage systems, the economic and logistical realities were not significant. Oudit and Scott (1973) reported a considerable extension in the storage life of 'Hass' avocados sealed in polyethylene bags. 'Hass' avocados sealed in polyethylene bags (0.015–0.66 mm) ranging in permeability from 111 to 605 cc O_2/m^2 h atm, and from 0.167 to 0.246 gr H_2O/m^2 h atm and stored at 5°C for up to 4 weeks lost less weight and firmness compared with unsealed fruits (Gonzalez et al., 1990, 1997; Yahia and Gonzalez, 1998).

Low-pressure (LP) atmosphere, especially below 100 mmHg, markedly prolonged the storage life of 'Hass' avocados (Apelbaum et al., 1977). Optimum conditions for LP storage of Florida avocados were suggested to be 20 mmHg at 4.5°C (Spalding and Reeder, 1976a; Spalding, 1977). Fruits stored in these conditions for up to 3 weeks were firmer, and had less decay and CI than fruits stored in 76 or 760 mmHg, however, gases such as CO_2 and

FIGURE 14.1
CA reduces chilling injury in 'Hass' avocados stored for 5 weeks. (Courtesy of Prof. E.M. Yahia.)

CO, cannot be added when LP system is used. CO_2 is considered to be essential for control of decay and to ameliorate CI in avocados (Figure 14.1). 'Hass' avocados maintained in MA (0.1–0.44 kPa O_2, 50–75 kPa CO_2, balance N_2) for up to 5 days at $20°C$ had higher CO_2 production compared to fruit stored in air, most likely reflecting anaerobiosis (Carrillo-Lopez and Yahia, 1990, 1991; Rivera and Yahia, 1994; Rivera et al., 1993; Yahia, 1993; Yahia and Carrillo-Lopez, 1993). Fruit stored in this MA and then ripened in air had mesocarp and exocarp injury after 2 days. On the basis of these results Yahia (1993) and Yahia and Carrillo-Lopez (1993) concluded that 'Hass' avocado fruit is very sensitive to insecticidal atmospheres, tolerating only 1 day at $20°C$. These findings were in agreement with Yahia and Kader (1991) and Ke et al. (1995). 'Hass' avocados kept in 0.25 kPa O_2 alone or in combination with 80 kPa CO_2 for 3 days at $20°C$ had higher concentrations of acetaldehyde and ethanol (Ke et al., 1995).

During CA/MA storage and transportation, O_2 levels of 2–5 kPa and 3–10 kPa CO_2 are commonly used. Low O_2 injury may appear as irregular brown to dark brown patches on the skin and may additionally cause diffuse browning of flesh beneath affected skin (Carrillo-Lopez and Yahia, 1991; El-Mir et al., 2001; Loulakakis et al., 2006; Yahia, 1997a; Yahia and Rivera, 1994). CO_2 atmospheres above 10 kPa can be detrimental by leading to discoloration of the skin and development of off-flavor, particularly when the O_2 concentration is <1 kPa.

The mechanism of low O_2 tolerance in avocado fruit has been explored at the molecular level by Loulakakis et al. (2006). They investigated the steady-state levels of protein and mRNA accumulation of selected hypoxic and ripening genes in response to low O_2 atmospheres. The analysis of mRNA populations in preclimacteric 'Hass' avocado fruit

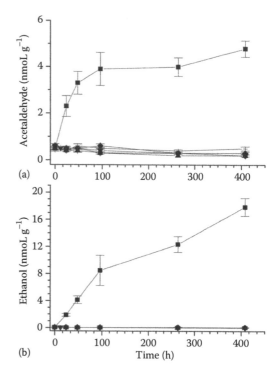

FIGURE 14.2
(a) Acetaldehyde and (b) ethanol accumulation in
'Hass' avocado fruit during storage at 6°C in atmospheres containing 0.1% (■), 1%, 2%, 5%, 10%, and 21%
O_2 (O_2 concentrations >0.1% were not distinguishable).
Error bars are ±SEM. Each data point is the average of
five fruit. (From Burdon, J., Lallu, N., Yearsley, C.,
Burmeister, D., and Billing. D., *Postharvest Biol. Technol.*,
43, 207, 2007. With permission.)

revealed that low O_2 levels induced new mRNA species, which were possibly implicated in the adaptive mechanism under low O_2, suppressed *de novo* synthesized ones, or left unaffected housekeeping and/or preexisting mRNAs, indicating that the low O_2 response is more complex and involves more than a simple adaptation in energy metabolism. New alcohol dehydrogenase (ADH) isoenzymes were present in preclimacteric and ripening-initiated avocado fruits stored in low-oxygen atmospheres and correlated with elevated ADH mRNA levels (Loulakakis et al., 2006).

In order to assess the suitability of avocados for dynamic CA storage, a recent study has shown that preclimacteric 'Hass' avocados are physiologically well suited to dynamic CA storage as the rapid induction of acetaldehyde (AA) and ethanol in response to low O_2 stress (Figure 14.2) and the comparable rapid recovery to basal levels after removal of the stress atmosphere, together with a seemingly high tolerance to O_2 atmospheres (<2 kPa) and the similar but relatively smaller effect of CO_2 was observed as compared with O_2 (Burdon et al., 2007).

MA and CA are commonly used for transporting avocado fruit by sea to distant markets in refrigerated shipping containers (Yahia, 1998). The atmosphere used and technology for controlling the atmosphere vary between shipping companies.

14.2.2 Citrus

Citrus fruits are classified as nonclimacteric and lack any burst in the rate of respiration and ethylene production after harvest. They can be stored for 6–8 weeks at appropriate storage temperatures. However, the susceptibility to pathological decay and physiological disorders are major postharvest problems encountered by the citrus industry. Postharvest rind disorders are very common in citrus fruit stored for longer duration limiting the postharvest storage capability and causing severe economic losses. Rind disorders may result either due to storage of fruit at chilling temperature or even during storage at optimal temperature. The occurrence and symptoms of CI in citrus fruit are species specific. For example, surface pitting is a typical CI symptom in grapefruit, while flavedo and albedo browning are specific to orange and lime, respectively. Rind disorders not related to CI include rind breakdown, stem-end-rind breakdown (SERB), shriveling, and aging indicated by the collapse of the stem-end button (Porat et al., 2004). Nonchilling-related disorders may be enhanced by the increased water loss from the peel tissue. CA/MA technology has proven useful to alleviate these postharvest physiological disorders and decay in citrus fruit, but the commercial application is limited. The sensitivity of citrus fruit to low internal O_2 and high CO_2 conditions is also species-specific; for example, 'Murcott' mandarins are more sensitive to anaerobic stress conditions than 'Star Ruby' grapefruit (Shi et al., 2005). Anaerobic conditions in fruit may result from many postharvest procedures such as application of waxes or coatings, a common practice to improve the shine and reduce water loss, wrapping in plastic films, storage under CA conditions, and also exposure to certain quarantine treatments involving heat and anaerobic conditions. The occurrence of off-flavors in citrus fruit is mainly associated with the accumulation of anaerobic metabolites, ethanol and acetaldehyde. High susceptibility of most of the citrus species to low O_2 and/or high CO_2 atmospheres is a potential hindrance to the commercialization of CA technology in the citrus industry.

14.2.2.1 Grapefruit (Citrus paradisi Macf.)

Grapefruit is a chilling sensitive fruit and should not be stored at <10°C. Early reports suggested that cold storage of grapefruit at 10°C–12°C supplemented with low O_2 (1–16 kPa) with or without elevated CO_2 did not effectively increase the storage life more than 10 weeks (Harding, 1969; Seberry and Hall, 1970). Elevated levels of CO_2 (20 kPa) adversely affected the rind integrity and increased incidence of decay upon removal from CA, while exposure to low O_2 resulted in development of off-flavors due to accumulation of anaerobic metabolites. The short-term exposure of grapefruit to high CO_2 prior to cold storage has been found beneficial to alleviate CI and SERB (Hatton and Cubbedge, 1977; Hatton et al., 1975). Prestorage treatment of 'Marsh' grapefruit with 20 or 40 kPa CO_2 for 3 days at 21°C significantly reduced the SERB during storage at 4.5°C for 8 or 12 weeks compared to fruit either directly stored at 4.5°C or exposed to air at 21°C prior to storage (Hatton and Cubbedge, 1977). The application of CA in combination with hot air may be a potential disinfestation treatment for grapefruit against certain pests. Shellie et al. (1997) showed that establishment of CA containing 1 kPa O_2 with or without 20 kPa CO_2 during heat treatment can reduce the amount of time required to provide quarantine security against the Mexican fruit fly (*Anastrepha ludens* Loew) in 'Rio Red' grapefruit. The potential of ultralow oxygen (ULO) storage in suppressing disease development in grapefruit has also been reported (Shellie, 2002). 'Rio Red' grapefruit stored in ULO containing 0.05–0.10 kPa O_2 at 14°C for 21 days, and additional 14 days in air at 23°C, showed suppressed green mold development during ULO and subsequent air storage at 23°C, but the flavor quality was rated acceptable yet inferior to normal air stored fruit (Shellie, 2002). However, ULO may provide a viable option for nonchemical suppression of green mold during long-distance marine transportation and postharvest insect–pest disinfestations. Grapefruit has been found more tolerant to anaerobic stress conditions as compared to mandarins (Shi et al., 2005). In response to anaerobic stress conditions imposed through N_2 exposure for different durations (6, 12, 24, 48, and 72 h) at 20°C, 'Star Ruby' grapefruit in comparison to 'Murcott' mandarin showed little increase in the rate of respiration and did not show any increase in ethylene production. Shi et al. (2005) reported that the accumulation of fermentation metabolites such as ethanol and acetaldehyde (AA) was also slower and lesser in grapefruit as compared to mandarins (Figure 14.3). The flavor quality of grapefruit exposed to

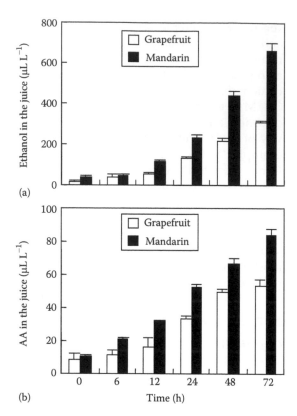

FIGURE 14.3

Effects of exposure to N_2 atmospheres on (a) ethanol and (b) AA concentrations in the juice of 'Murcott' mandarins and 'Star Ruby' grapefruit. Fruit were exposed to N_2 for periods of 0, 6, 12, 24, 48 or 72 h at 20°C. Data are means $\pm$ SE of four measurements and are one of three different experiments with similar results. (From Shi, J.X., Porat, R., Goren, R., and Goldschmidt, E.E., *Postharvest Biol. Technol.*, 38, 99, 2005. With permission.)

N_2 atmospheres for 72 h at 20°C was found acceptable while it was rated unacceptable in mandarins even after 48 h exposure period. The differences in the response patterns of grapefruit and mandarin to anaerobic conditions could be associated with the gas exchange properties of the peel (Table 14.2), as the former is more permeable and thus prevents the build up of ethanol, AA in their internal atmospheres (Shi et al., 2007a). These studies clearly indicate that grapefruit is comparatively more tolerant to anaerobic conditions, and may become a suitable candidate for short-term low O_2 and/or high CO_2 treatments having insecticidal and fungistatic properties.

The application of waxes is a very common practice in postharvest handling of grapefruit. Waxes modify the internal atmosphere of fruit, reduce the weight loss, and also improve the surface shine, leading to more consumer appeal. However, the selection of a wax or coating is very crucial to obtain beneficial results. It may also lead to development of off-flavors and surface pitting. Surface pitting in grapefruit, which was morphologically and causally distinct from CI, has been correlated with the levels of internal O_2 and CO_2 concentrations (Petracek et al., 1998). Postharvest waxing and CA conditions, leading to low internal O_2, stimulated the pitting of grapefruit. The application of more restrictive waxes hinders the gas exchange, and results in surface pitting and off-flavors in fruit. The shellac coating resulted in reduced CI incidence in white 'Marsh' grapefruit stored for 2 months at 4°C compared to nonwaxed fruit or fruit waxed with either carnauba or polyethylene waxes (Dou, 2004). Thus selection of a waxing/coating material for beneficial effects in grapefruit is a prime factor determining the fruit quality. The inherent capability of grapefruit to tolerate low O_2 atmospheres better than mandarins may be exploited for commercial use in CA/MA based quarantine system.

TABLE 14.2

Gas Diffusion Properties of 'Star Ruby' Grapefruit and 'Murcott' Mandarin Peel Disks

Tested Gas	'Star Ruby' Grapefruit	'Murcott' Mandarin
	Gas Diffusion Rate (nmol · min^{-1} · cm^{-2})	
Gas diffusion from the inner to the outer side (albedo → flavedo)		
O_2	148 ± 12	105 ± 13
CO_2	114 ± 8	66 ± 6
C_2H_4	0.052 ± 0.015	0.032 ± 0.003
AA	0.045 ± 0.003	0.033 ± 0.004
Ethanol	0.030 ± 0.003	0.012 ± 0.003
Gas diffusion from the outer to the inner side (flavedo → albedo)		
O_2	108 ± 11	78 ± 8
CO_2	184 ± 17	158 ± 17
C_2H_4	0.076 ± 0.018	0.107 ± 0.020
AA	0.071 ± 0.009	0.048 ± 0.006
Ethanol	0.035 ± 0.005	0.016 ± 0.002

Source: Shi, J.X., Goldschmidt, E.E., Goren, R., and Porat, R., *Postharvest Biol. Technol.*, 46, 242, 2007a.

Note: Peel disks were sealed with silicon and plastic gastight sticky tape between two 10 mL plastic syringes. The narrow ends of both syringes were closed with serum caps and 2 mL of the indicated gases, at known concentrations, were injected through the serum cap into one syringe. After 30 min equilibration, gas samples were withdrawn from the opposite syringe and analyzed by gas chromatography. All measurements were carried out at 25°C. Data represent means ± SE of 10 fresh disks.

14.2.2.2 Lemon (Citrus limon L. Burman f.)

Lemon can be stored at 7°C–12°C, 85%–95% RH for 6 months depending on the maturity-ripeness stage at harvest, season of harvest, storage time, and production area. Lemon is also chilling sensitive, so should not be stored for prolonged periods below 10°C, although 3–4 weeks storage at 3°C–5°C, which is typical for some receivers, is usually tolerated without harm. CAs of 7.5–10 kPa O_2 + up to 10 kPa CO_2 can delay loss of green color and senescence. Not much work has been reported on the response of lemon to CA. Wild et al. (1977a,b) reported the effects of oxygen, carbon dioxide, and ethylene concentrations and their interactions in storage atmospheres on the keeping quality, appearance, and juice characteristics of lemons. Just removal of ethylene from the storage atmosphere reduced the rate of green color loss from lemon rind, and appearance was improved when compared with the fruit stored for 6 months at 10°C under CA containing 10 kPa O_2 without CO_2 (Wild et al., 1977a,b). Oxygen concentration in the atmosphere was also shown to regulate the effect of ethylene on the rate of green color loss from lemon rind. An atmosphere of 11 kPa O_2 and 6 ppm accumulated ethylene caused faster loss of green rind color than that of 5 kPa O_2 and 300 ppm ethylene (Wild et al., 1977a). Thus, reduced levels of O_2 in CA can negate the effects of ethylene present in storage atmosphere on lemon fruit quality. Seal-packaging of individual lemon cv. Eureka fruit with a film of high-density polyethylene (HDPE) having thickness of 0.01 mm reduced weight loss by five folds, maintained fruit freshness, and markedly delayed their deterioration in terms of peel shrinkage, softening, deformation, and loss of flavor (Ben-Yehoshua et al., 1979). HDPE wraps delayed loss of firmness and peel coloration of 'Lisbon' lemons for 8 months in cool storage at 10°C, but developed unacceptable levels of rotting within 4 months of harvest (Sharkey et al., 1985). The available information on the effects of

CA/MA on lemon fruit physiology is limited. The interesting information on the effects of lower O_2 concentrations on the regulation of ethylene action in lemon (Wild et al., 1977a) needs further investigation and may be useful to refine the optimum CA requirements for lemons. The success of individual seal packaging in lemon is a classical example of MAP commercialization.

14.2.2.3   Lime (Tahiti Lime, Citrus latifolia Tan. and Persian Lime, Citrus aurantifolia Swing.)

'Tahiti' lime (*Citrus latifolia* Tanaka) is commercially very important. It can be stored at 10°C–12°C and 85%–95% RH for 4–8 weeks (Kader and Arpaia, 1992). CI symptoms in the form of surface pitting and decay appear when stored below 8°C. Maintaining green skin color of fruit during postharvest handling and marketing is a challenge. Kader (2003) recommended 5 kPa O_2 and 0–10 kPa CO_2 for storage of lime at 10°C–12°C. CA storage can potentially delay the degreening of limes but results in certain unfavorable effects on fruit quality including rind breakdown, peel thickening, juice loss, and enhanced decay (Spalding and Reeder, 1974b, 1976b). Limes stored in CA containing 5 kPa O_2 with 7 kPa CO_2 maintained acceptable green color without any off-flavor (Spalding and Reeder, 1974b). Similarly, fruit stored in CA of 21 kPa O_2 with 7 kPa CO_2 had acceptable juice content and were comparatively greener than those stored in air. Low-pressure storage (LPS) can be beneficial to extend the storage and "green" life of limes without any detrimental effects on fruit quality. Limes stored at low pressure of 20 kPa (150 mmHg) at 10°C could be kept well for 56 days whereas yellowing took place in 14–35 days in fruit held in normal atmosphere (Burg and Burg, 1966). Storage of 'Tahiti' limes at a low atmospheric pressure of 170 mmHg for up to 6 weeks at 10–15.6°C retained green color, juice content, and flavor acceptable for marketing and also had a low incidence of decay compared to fruit in normal air, which turned yellow within 3 weeks (Spalding and Reeder, 1976b). However, low pressure did not ameliorate the CI when fruit were stored for 4 weeks at 170 mmHg at 2.2°C and 98%–100% RH and developed as much CI as comparable to fruit at normal atmospheric pressure (Spalding and Reeder, 1976b). Burg (2004) made a comparative account of the usefulness of CA and LPS in maintaining the quality of limes, and showed that LPS (20 kPa or 150 mmHg) helped in removal of endogenous ethylene, which would otherwise cause degreening and peel thickening effects. Stomata of limes presumably remained open at low pressures, but remained close in CA, and this might have helped to decrease their internal ethylene levels during LPS (Burg, 2004). The use of polyethylene film bags inside cartons reduced the weight loss of limes, but promoted the loss of green color of skin compared to fruit directly kept in cartons without sealing (Thompson et al., 1974). The utility of MAP in extending the storage life can be derived through the use of ethylene absorbents inside the packs as ethylene is known to be responsible for faster degreening and peel thickening in limes. More research is needed to investigate the effects of different atmosphere regimes on lime storage and quality.

14.2.2.4   Mandarins (Citrus reticulata Blanco and Citrus unshiu Marc.)

The moisture loss from the rind is the primary concern in postharvest quality of mandarins. The excessive moisture loss can lead to shriveling and deteriorated external appearance, and thus impairing market quality of fruit. Mandarins, being nonclimacteric in nature, would not be deriving much benefit from control over respiration at postharvest stage. Therefore, prevention or reduction of water loss from the rind constitutes the central core of the postharvest technology in maintaining fruit quality of mandarins. Mandarins have a shorter storage life as compared to oranges, grapefruit, and lemons (Kader and Arpaia, 1992). Like other citrus fruit, CA conditions trigger the fermentative metabolism, and cause physiological stress, leading to the development of off-flavors and poor consumer acceptance of mandarins. The short-term exposure of 'Fortune' mandarins to high CO_2 (95 kPa) at 20°C for 20 h as a complement to standard cold quarantine treatment (14–18 days at 1.1°C–2.2°C) did not have adverse effects on external appearance, texture, and sensory quality of fruit, but concentrations of ethanol and acetaldehyde were higher in treated fruit (Alonso et al., 2005). The combination of high CO_2 pretreatment with standard cold treatment against tephritid pests may be very helpful to shorten the lethality time, and can prove instrumental in expediting product movement in the supply chain to its final destination. The tolerance of different cultivars of mandarin to short-term exposure of CO_2 need to be investigated before such practice is integrated into the quarantine security system. The long-term storage of mandarins under CA conditions is not a commercial practice, and also attracted less attention from the researchers. Recently, Luengwilai et al. (2007) reported that 'Clemenules Clementine' and 'W. Murcott' mandarins could be stored in air at 5°C and 90%–95% RH for up to 5 and 7 weeks, respectively, and there was no additional benefit derived from keeping them in CA containing 3 and 5 kPa O_2. CA conditions reduced the respiration rates in both cultivars and did not influence the soluble solids concentration (SSC), TA content and sensory characteristics, but ethanol and AA concentrations were higher than those kept in air (Luengwilai et al., 2007). The molecular, physiological, and biochemical basis for high susceptibility of mandarins to anaerobic stress conditions has recently been established showing that treatment of mature or immature fruit to anaerobic conditions (N_2 atmosphere) upregulated the expression of pyruvate decarboxylase and ADH genes, and increased the levels of AA and ethanol in fruit (Shi et al., 2007a,b). It appears that the commercial scale application of CA for enhancing the storage life of mandarins is not feasible.

The high sensitivity of mandarins to low O_2 does not warrant the use of low-permeability films for MAP. The restriction over the loss of moisture from peel tissue could be helpful to prevent the shriveling and ageing in fruit, but condensation of water droplets may lead to enhanced decay over prolonged storage. Therefore, the selection of a packaging film with appropriate gas and water vapor permeabilities would be crucial for harnessing the beneficial effects of MAP in mandarins. 'Malvasio' mandarin packed using three different types of films, Goglio-LDPE, Cryovac-MR, and perforated Cryovac PY-8, were stored at 4°C for 12 weeks (D'Aquino et al., 2001). Perforated Cryovac PY-8 film neither modified the atmosphere nor maintained high humidity around the fruit; Cryovac-MR film reduced the weight loss and maintained visual appearance without any adverse effect on sensory characters, while Goglio-LDPE film resulted in severe condensation around the fruit leading to >90% decay incidence after 12 weeks at 4°C plus 1 week at 20°C (D'Aquino et al., 2001). The application of waxes to improve gloss and appearance quality is also a common packing house practice in mandarins. Various studies conducted on the effects of different types of coatings on the flavor and shelf life of mandarins concluded that polyethylene based wax coatings are more suitable than shellac and wood resin-based coatings due to their more permeable nature (Hagenmaier, 2002; Hagenmaier and Baker, 1993; Porat et al., 2005). Hagenmaier (2002) outlined certain critical limits for internal concentration of gases in different mandarin hybrids, and concluded that flavor quality was seriously affected when mean internal O_2 concentration fell below 4 kPa and internal CO_2 >14 kPa, and juice ethanol content >1500 ppm after 7 days at 21°C. The oxygen permeability of coating is determinant of the internal atmosphere of fruit, and eventually the flavor quality. The composite coatings, having constituents from different sources, provide an opportunity to combine several desirable characteristics into an ideal coating type, which matches the product requirements. The proportion of each constituent of the composite coating affects its permeability characteristics leading to drastic changes in the response of the fruit (Perez-Gago et al., 2002). 'Fortune' mandarins coated with hydroxypropyl methylcellulose (HPMC)-lipid coatings containing 60% lipid had lower weight loss and ethanol content whereas low lipid content coatings (20%) induced lower internal O_2 and higher ethanol content than high lipid content coatings (60%), which possibly could be related to the low oxygen permeability of HPMC and final thickness of these coatings from higher viscosity of the emulsions (Perez-Gago et al., 2002). Different types of coatings based on natural and synthetic materials are available in the market, but the success rate will depend upon the selection of an appropriate coating formulation and its method of application keeping in view the gas exchange properties and postharvest handling duration and storage conditions of the product.

14.2.2.5 Oranges (Citrus sinensis (L.) Osbeck)

The response of oranges to modified atmospheres is quite similar to mandarins and grapefruit. The commercial scale application of CA is very limited. The use of waxes and other coatings is much popular in oranges like other citrus fruits. Ke and Kader (1990) reported that 'Valencia' oranges tolerated up to 20 days of exposure to 0.5, 0.25, or 0.02 kPa O_2 at 5°C or 10°C followed by holding in air at 5°C for 7 days without any detrimental effects on external and internal appearance. CA exposure reduced respiration rates, but offered higher resistance to CO_2 diffusion and higher ethanol evolution rates than those stored in air at 10°C. Oranges kept in 60 kPa CO_2 at 5°C for 5–14 days followed by holding in air at 5°C for 7 days developed slight to severe injury that was characterized by skin browning and low appearance quality (Ke and Kader, 1990). Low O_2 or high CO_2 treatments did not influence the juice color, SSC, pH, titratable acidity, and ascorbic acid content of fruit. However, these treatments increased ethanol and AA contents, which correlated with the decrease in flavor score of the fruit (Ke and Kader, 1990). The application of MAP has been reported to maintain freshness, reduce weight loss during storage in oranges apart from reducing the development of various rind disorders including superficial flavedo necrosis (noxan), rind breakdown, SERB, CI, and collapse of button or ageing (Ben-Yehoshua et al., 2001; Porat et al., 2004). Noxan or superficial flavedo necrosis is a serious physiological peel disorder specific to 'Shamouti' oranges, which appears after harvest in the form of superficial pits on the flavedo, and with time, pits enlarge to form a necrotic area. 'Shamouti' oranges stored at 5°C and 90% RH had 14% noxan whereas individual seal-packaged fruit or fruit in polyethylene liners had 2%–3% noxan incidence (Ben Yehoshua et al., 2001). The maintenance of high humidity around the fruit and modification of atmosphere with low O_2 and high CO_2 were considered two modes of action of MAP in reducing the rind disorders during storage of oranges (Porat et al., 2004). Microperforated and macroperforated Xtend® packages reduced rind disorders not related to chilling (rind breakdown, SERB, and ageing) after 5 weeks of storage at 6°C and 5 days of shelf-life conditions by 75% and 50%, respectively, in 'Shamouti' oranges. The elevated levels of CO_2 (2–3 kPa) and slightly reduced levels of O_2 (17–18 kPa) in microperforated Xtend films (0.002% perforated area) than macroperforated Xtend films (0.06% perforated area), which maintained CO_2 and O_2 concentrations of 0.2–0.4 and 19–20 kPa, respectively, might be contributing to the alleviation of these disorders. The use of waxes and coatings is also very helpful to prolong the storage life and maintain quality in oranges (Hagenmaier, 2000; Hagenmaier and Baker, 1993). The general precautions for choosing the coating material for oranges are also the same as for other citrus fruits.

14.2.3 Date Palm (*Phoenix dactylifera* L.)

Date palm is an important fruit crop in Middle East and North African countries. Date fruit is drupe with a single nut. The fruit is oblong in shape, 2.5–7.5 cm long, thick or thin flesh, astringent when premature and become sweet when ripe, skin of specific color at ripening stage, and a hard pit inside grooved down one side (Yahia, 2004). A high-quality fresh date fruit should have adequate size and color, small pit, thick flesh, free from insects and rodents damages, fungi and molds infestation, dirt, sand, and sugar crystals formation. Depending upon the cultivar, skin should have a characteristic color like golden-brown, amber, green, or black and should be smooth and free from shriveling (Yahia, 2004). Due to low moisture content and very low respiratory activity, dates can be stored for longer durations. Darkening of skin and pathological deterioration are major postharvest constraints in dates. These can be stored at 0°C for 1 year without major loss in quality, but may develop some sugar spots (Yahia, 2004). Moisture content is the determinant of the storage potential of dates. Partially dried dates can be stored for a year at 0°C or lower or for a few weeks at ambient temperature whereas dry dates can be held at 20°C for years without significant quality losses. Storage at freezing temperatures (−35°C to −50°C) is recommended for maintaining the dates in good quality (Shomer et al., 1998). Storage of dates at 0°C to −18°C may accelerate the process of ripening due to damage at cellular level resulting into yellow-brown spots, pale color and cell sap leakage (Shomer et al., 1998).

The applications of CA in maintaining date fruit quality and insect pest disinfestation have been reported (Al-Redhaiman, 2005; Baloch et al., 2006; Navarro et al., 2001). Navarro et al. (2001) reported that use of CA during ambient temperature storage is feasible to control insect–pests and maintain quality of dates for 4.5 months. 'Hallawi,' 'Hadrawi,' 'Zahidi,' 'Derei,' and 'Ameri' dates were stored in a CA containing 60–80 kPa CO_2 for 4.5 months with no significant changes in peel sloughing and sugar formation on fruit surface, and quality of CA-stored dates was as good as those stored in normal air at −18°C at the end of storage for 4.5 months (Navarro et al., 2001). Fruit quality of mature 'Barhi' dates stored at 0°C under CA containing 5 or 10 kPa CO_2 (balance air) could be maintained for 17 weeks against 7 weeks in air, while increase in CO_2 to 20 kPa was more effective for maintaining fruit color, firmness, SSC, and total tannins for 26 weeks (Al-Redhaiman, 2005). The degradation of caffeoylshikmic acid, one of the major phenolics undergoing losses during ripening, was also retarded greatly by CA containing 20 kPa CO_2, which suggested that CA was very effective in retarding the ripening process in date. Baloch et al. (2006) investigated the effects of water activity (a_w) and storage atmosphere on the fruit quality of 'Dhakki' date and showed that fruit having lower levels of water activity ($0.52a_w$) were stable for 4 months when stored under nitrogen atmosphere at 40°C. The storage of fruit under air or oxygen atmosphere resulted in an increase in skin darkening and titratable acidity during storage; the same effects were also observed with an increase in the water activity of fruit (Baloch et al., 2006). The studies conducted on the response of dates to CA are not very conclusive. There is a potential to further investigate the effects of different atmosphere regimes on the quality of dates harvested at different maturities and having different moisture levels. The tolerance of dates to high CO_2 has been exploited to develop biologically safe alternatives to fumigation treatments to control storage pests in Israel (Navarro et al., 2001). This technology has the potential to be further extended to other date-producing countries of the developing world.

14.2.4 Fig (*Ficus carica* L.)

Figs are native to Asia Minor and later spread to the Mediterranean and other parts of the world. Figs have been classified as climacteric as well as nonclimacteric fruit, but climacteric nature is more accepted. The harvest maturity depends upon the intended use of the fruit either for fresh purpose or drying. Figs for fresh market are harvested when these are fully ripe and firm to be of good quality. Fruit for drying should be harvested when fully ripe and partially dry on the tree, and is eventually dehydrated to a moisture level of 17%. The storage of fresh figs at $-1°C$ to $0°C$ with 90%–95% RH is recommended and faster cooling to $0°C$ with forced air is strongly recommended to maintain the fruit quality (Crisosto and Kader, 2004). Postharvest life of 'Mission' figs could be extended to 2–3 weeks at $0°C–5°C$ in air (Colelli et al., 1991). CA storage has been reported to extend the storage life and maintain quality of figs (Claypool and Ozbek, 1952; Colelli et al., 1991; Crisosto and Kader, 2004; D'Aquino et al., 1998; Tsantili et al., 2003; Türk et al., 1994). Claypool and Ozbek (1952) reported that storage in atmospheres containing up to 60 kPa CO_2 at $20°C$ was not effective, but pretreatment with 100 kPa CO_2 for 36 h at $5°C$ and $10°C$ delayed the growth of microorganisms. 'Mission' figs could be stored for up to 4 weeks when kept at $0°C$, $2.2°C$, or $5°C$ in atmospheres enriched with 15 or 20 kPa CO_2 (Colelli et al., 1991). CO_2 exposure reduced the ethylene production, softening, decay incidence, and maintained bright external appearance of fruit. Ethanol content of the CO_2-treated fruit increased slightly during the first 3 weeks and moderately during the 4th week, while acetaldehyde concentration increased during the first week, then decreased (Colelli et al., 1991). 'Bursa Siyahi' fig could be stored for 4 weeks at $0°C$ and 90%–95% RH with 3–5 kPa of O_2 and CO_2 (Türk et al., 1994). Ferizli and Emekci (2000) conducted a large-scale trial on dried figs in a gastight flexible storage unit loaded with 2.5 t of fruit in perforated plastic bags. CO_2 treatment resulted in a decrease of O_2 level to 0.8 kPa and increase of CO_2 level to 96 kPa, which remained stable for 5 days and resulted in complete mortality of both insects and mites (Ferizli and Emekci, 2000). Fully ripe 'Mavra Markopoulou' figs stored at $-1°C$ in CA containing 2 kPa O_2 (balance N_2) for 29 days showed decreased softening, weight loss, ethylene production, and retention of bright skin color (Tsantili et al., 2003). CA combinations of 5–10 kPa O_2 + 15–20 kPa CO_2 are effective in decay control, firmness retention, and reduction of respiration and ethylene production for 3 to 4 weeks for California-grown 'Black Mission' and 'Calimyrna' figs (Crisosto and Kader, 2004). The prolonged storage in CA can result in loss of characteristic flavor and fruit exposed to <2 kPa O_2 and/or >25 kPa CO_2 develop off-flavors due to fermentative metabolism (Crisosto and Kader, 2004). Based on various studies conducted so far, it can be concluded that 3–5 kPa O_2 in combination with 5–15 kPa CO_2 supplemented with $0°C$ storage temperature could be beneficial to store and transport fresh figs over longer distances. The short-term treatment of packaged dried figs with high CO_2 could be beneficial to control storage pests.

14.2.5 Guava (*Psidium guajava* L.)

Guava is a native of tropical America. It is a climacteric fruit, but some cultivars may show nonclimacteric or suppressed climacteric behavior. Guava fruit suffers CI if stored below $10°C$. The prolonged storage at suboptimal temperatures may lead to severe CI symptoms in the form of skin and flesh browning, and surface pitting. Short-term exposure of guava fruit to high CO_2 levels (10, 20, and 30 kPa) did not influence the respiration rates, but reduced ethylene evolution during ripening (Pal and Buescher, 1993). Similarly, treating guava with 10 kPa O_2 + 5 kPa CO_2 for 24 h before storage in air at $4°C$ for 2 weeks delayed color development and reduced CI, compared to fruit held in air (Yahia, 1998). Application of continuous CA for long-term storage of guava fruit has been recently reported (Pal et al., 2007; Singh and Pal, 2008a). Singh and Pal (2008a) reported that atmospheres containing O_2 concentrations less than 5 kPa were detrimental for external and flavor quality of two commercial cultivars, 'Lucknow-49' and 'Allahabad Safeda,' while 'Apple Color,' a pink-fleshed cultivar, could not tolerate O_2 concentration below 8 kPa. The best CA conditions for these three cultivars 'Lucknow-49,' 'Allahabad Safeda,' and 'Apple Color' were 5 kPa O_2 + 2.5 kPa CO_2, 5 kPa O_2 + 5 kPa CO_2, and 8 kPa O_2 + 5 kPa CO_2. Under these CA conditions, fruit could be stored for 30 days at 8°C without significant CI symptoms and ripening took place at ambient conditions in 5 days (Figure 14.4). Respiration and ethylene production rates of fruit held in low O_2 ($\leq$ 5 kPa) atmospheres were also found significantly lower than air-stored fruit. CA was also found beneficial in maintaining high ascorbic acid and phenolic content during storage of guava. The benefits of CA may be derived to extend the storage life of guava fruit at temperature slightly below optimal and also to overcome CI problem (Singh and Pal, 2008a). However, the beneficial effects of 1-MCP in extending storability, maintaining quality, reducing decay, and alleviation of CI disorder in guava fruit have also been reported (Singh and Pal, 2008b). Thus, storage of 1-MCP-treated guava fruit in optimum CA conditions at low temperatures (8°C–10°C) could be a good proposition to minimize postharvest storage problems and needs further investigation. Preliminary studies on the short-term exposure of guava fruit to very low O_2 (<1 kPa) and high CO_2 (40 kPa) at 40°C for 12 h have shown that such treatments are beneficial to extend shelf life of fruit by 2–3 days at ambient conditions and may find application in postharvest insect–pest disinfestations for quarantine purposes (Singh and Pal, 2007). The response of guava fruit to CA has been reported quite favorable, but the commercial application is still awaited.

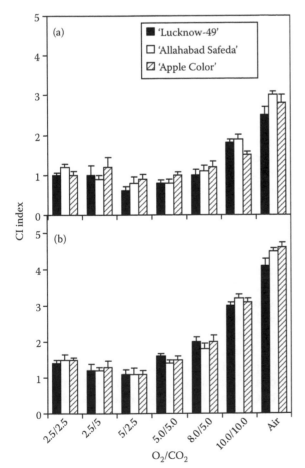

FIGURE 14.4
CI index (1–5 scale) in three guava cultivars after 30 days storage either in controlled atmospheres or air (a) at 8°C and (b) at ambient conditions (25°C–28°C and 60%–70% R.H.) for 5 days. Vertical bars represent ± SE. LSD ($P \leq 0.05$): cultivar = 0.18, atmosphere composition = 0.20, cultivar × atmosphere composition = 0.36. (From Singh, S.P. and Pal R.K., *Postharvest Biol. Technol.*, 47, 304, 2008a. With permission.)

Coating of mature-green guava fruit with cellulose- or carnauba-based emulsions delayed ripening, slowed softening, but caused less coloring and more surface blackening (McGuire and Hallman, 1995). Coating with 2% and 4% hydroxypropylcellulose slowed softening by 35% and 45%, respectively, compared to uncoated fruit. Coating with 5% carnauba formulation slowed softening by 10%–30%, and was effective in reducing weight loss. Pal et al. (2004) reported that 'Lucknow-49' guavas with individual shrink wrapping using 9 μm LLDPE film could be successfully stored for up to 12 and 18 days at ambient and in evaporative cool chamber, respectively, with negligible loss in vitamin C content. The application of Sta-fresh coating was more effective in cool chamber than ambient (Pal et al., 2004). The success of LDPE shrink wrapping in extending the storage life of guava was confirmed by Mohamed et al. (2006). LDPE shrink-wrap was the most effective in reducing weight loss and maintaining skin color and firmness of guava fruit followed by clingwrap packaging and 200 g L^{-1} palm oil emulsion coating during storage at 10°C (Mohamed et al., 2006). Shrink wrapping of a tray containing guava fruit with D-955 film drastically reduced the weight loss, and maintained freshness of fruit without significant retardation of changes in skin color. The display of shrink-wrapped fruit trays at retail outlets appeal more to the consumer and may enhance the market potential of fruit. MAP has not been used commercially to derive multitude of benefits for guava fruit.

14.2.6 Jujube (*Ziziphus mauritiana* Lam. and *Ziziphus jujuba* Mill.)

Jujube is broadly classified into two categories: Indian jujube (*Z. mauritiana* Lam.) and Chinese jujube (*Z. jujube* Mill.). Jujube originated in China and has been cultivated there for more than 4000 years. The fruit is a drupe, varying from round to elongate and from cherry-size to plum-size depending on cultivar. It has a thin, edible skin surrounding whitish sweet flesh and pleasant flavor. Depending upon the cultivar, the changes in skin color progress from green at immature stage to a yellow-green stage with mahogany-colored spots appearing on the skin as the fruit ripens. Chinese and Indian jujubes have been reported to differ in their respiratory and ethylene production behaviors. Indian jujube is considered climacteric in nature (Abbas and Fandi, 2002; QiuPing and WenShui, 2007) and undergoes faster ripening and senescence in response to ethylene. The nonclimacteric nature of Chinese jujube has been reported by Kader et al. (1982). Postharvest decay is a major problem during long-term storage of jujube.

There has been a big research gap on the responses of jujube to CA. Recent studies in China are restricted to publications in local journals, and mostly in Chinese language. Limited commercialization of jujube in the international market might be another factor to the research gaps. Therefore, literature on CA/MA application in jujube has not advanced so far. CA storage at low temperatures is effective for control of various pathological and physiological problems in stored commodities. CA may be capable of retarding decay development in jujube through inhibition of growth, sporulation, and enzyme activity of the pathogens, and also improved physiological condition of the fruit enabling it to resist decay more effectively. Decay problems may be aggravated as the fruit is removed from CA and shifted to ambient conditions (Qin and Tian, 2004). Therefore, CA storage cannot be used as a sole method for decay control in jujube. The integration of CA storage at low temperature (0°C) with low doses of fungicides and a biocontrol agent has been proved successful in controlling storage rots in jujube (Qin and Tian, 2004). Fruit treated with an antagonistic yeast, *Cryptococcus laurentii* plus imazalil at 25 μg a.i. mL^{-1} or kresoxim-methyl at 50 μg a.i. mL^{-1} and stored in CA containing 10 kPa $O_2 + 0$ kPa CO_2 at 0°C showed a lower disease incidence caused by *Alternaria alternata* and *Monilinia fructicola* than fruit stored in air at 0°C. Zhang et al. (2005) reported that 'Dong' jujube could be stored for 140 days at $-2°C$ at atmospheric pressure of 40.5 kPa with regular ozone exposure (300 mg m^3) at 10 days interval. The combination of hypobaric storage with ozone reduced the respiratory rate, inhibited the activities of amylase and ascorbic acid oxidase, slowed down degradation of starch, ascorbic acid, softening, and reduced CI and fruit rot (Zhang et al., 2005). The beneficial effects of superatmospheric O_2 exposure have been studied in jujube (Li et al., 2006; Wang et al., 2006). Exposing 'Dongzao' jujube to 100 kPa O_2 at $-2°C$ for 15 days inhibited the respiration rate and softening and prevented loss of ascorbic acid without any effect on SSC during subsequent shelf life at 20°C. Contrarily, treatment with 70 kPa O_2 promoted the respiration and loss of ascorbic acid (Li et al., 2006). Membrane permeability and activities of oxidative defense enzymes were also maintained in fruit treated with 100 kPa O_2 against no positive effects from treatment with 70 kPa O_2 (Wang et al., 2006).

There is limited information on the use of MAP for extending the storability of jujube. MAP has been shown to reduce weight loss and maintain quality in jujube during storage at 12°C for more than 30 days (Martinez-Madrid et al., 2001). The information on the use of edible coatings in jujube is also scanty. A recent study by Qiuping and Wenshui (2007) has shown that coating of Indian jujube 'Cuimi' with chitosan solution (1.5 g/100 mL) extended the storage life to 12–13 days at ambient conditions (21°C–32°C) against 8 days in control. A combination of chitosan coating with 1-MCP treatment (600 nL L^{-1} for 12 h) extended the storage life of fruit up to 16 days, and was very effective in delaying climacteric ethylene and respiratory peaks, better retention of chlorophyll, SSC, ascorbic acid, and fruit firmness (QiuPing and WenShui, 2007). The development and application of edible coating to prolong the shelf life of jujube is still in infancy and needs more investigations from different perspectives.

14.2.7 Kiwifruit or Chinese Gooseberry (*Actinidia deliciosa* A. Chev. C.F. Liang and A.R. Ferguson)

Kiwifruit, a native of China, is mostly oval in shape with pale greenish-brown skin densely covered with fine hairs and flesh is bright green or creamish yellow in color with numerous small black seeds surrounding the white central core (Cheah and Irving, 1997). It is a climacteric fruit, and undergoes increases in respiration and ethylene production in addition to rapid softening and increase in SSC during ripening. Postharvest softening is a major problem associated with cold storage of kiwifruit. Kiwifruit is highly sensitive to ethylene; even smaller concentrations of ethylene in the storage atmosphere may result in rapid fruit softening, and thus deterioration in fruit quality (Arpaia et al., 1985, 1986; Harman and McDonald, 1989; McDonald and Harman, 1982). Kiwifruit harvested at appropriate maturity can be stored at 0°C and >95% RH for 4–6 months (Cheah and Irving, 1997; Harman and McDonald, 1989; McDonald and Harman, 1982). The main factor limiting the storage life of kiwifruit at 0°C is the rate of softening of the fruit. Kiwifruit softening can be retarded by low-temperature storage (0°C) supplemented with CA (Arpaia et al., 1984, 1985, 1986, 1987; Harman and McDonald, 1989; McDonald and Harman, 1982). McDonald and Harman (1982) recommended that optimum CA conditions containing 5 kPa CO_2 and 2 kPa O_2 at 0°C can increase the storage expectancy by 2–3 months more than normal air and suggested to avoid contamination of storage atmospheres with even trace amounts of ethylene (0.1 µL L^{-1}). Textural abnormality also occurred in fruit stored in 8 kPaCO_2 + 15–20 kPa O_2 after 24 weeks of storage whereas fruit stored in 5 kPa CO_2 + 2 kPa O_2 for 24 weeks retained good texture, appearance, and flavor (Harman and McDonald, 1989). Storage life of 'Bruno' kiwifruit could be extended to 6 months by storing the fruit at $-1°C$ in sealed 0.04–0.05 mm thick polyethylene bags having in-pack atmospheres with 3–4 kPa CO_2 and 15–16 kPa O_2, and C_2H_4 absorbent (Ben-Arie and Sonego, 1985). High CO_2 concentrations (≥ 8 kPa) in CA have been found to promote textual abnormality and acceleration of titratable acidity loss during prolonged storage. Arpaia et al. (1985) tested the storage performance of 'Hayward' kiwifruit in atmospheres containing 2 kPa O_2 and different levels (0, 3, 5, and 7 kPa) of CO_2 either in presence or absence of ethylene at 0°C for 24 weeks. They found that CO_2 concentration has a greater influence than O_2 concentration in retarding flesh softening in kiwifruit and higher CO_2 (5 and 7 kPa) concentrations were necessary in CA to maintain fruit firmness. The presence of ethylene in CA aggravated the softening process, and also resulted in higher incidence of white core inclusions and internal breakdown of fruit in the form of granulation and translucency (Arpaia et al., 1985, 1986). Therefore, it is strongly recommended to keep the storage atmosphere free from ethylene to prevent its detrimental effects on fruit quality. The presence of ethylene under CA conditions in concentrations as low as 50 ppb can negate the beneficial effects of CA though enhanced flesh softening and stimulated internal breakdown of fruit. The delay in establishing CA can possibly affect fruit softening to a greater extent during the storage of fruit and thus determining the overall storage potential of fruit. It is, therefore, essential to avoid any delay between harvesting and CA establishment for better storage life and flesh firmness (Arpaia et al., 1984). Stavroulakis and Sfakiotakis (1997) revealed that the action of propylene, an analogue of ethylene, in kiwifruit was shown to be dependent upon the O_2 concentration in the atmosphere. 'Hayward' kiwifruit treated with 130 $\mu L\ L^{-1}$ propylene under 1, 5, 10, 13, 16, and 21 kPa O_2 at 20°C showed no differences in ripening rate at O_2 levels between 21 and 10 kPa, indicating that ripening at 10 kPa O_2 commenced at the normal rate. The rate of propylene induced ripening was greatly inhibited in the O_2 levels between 1 and 5 kPa (Stavroulakis and Sfakiotakis, 1997). Storage of kiwifruit at very low O_2 atmospheres has been suggested to interfere with the ethylene production capacity of fruit at poststorage stage (Antunes and Sfakiotakis, 2002). Kiwifruit removed from ULO storage (0.7 kPa O_2 + 0.7 kPa CO_2) after 60, 120, and 180 days showed drastically reduced capacity to produce ethylene mainly due to low ACC oxidase activity rather than reduced ACC production or ACC synthase activity. ULO-stored fruit required propylene treatment for normal ripening (Antunes and Sfakiotakis, 2002). Similarly, storage of kiwifruit in atmospheres containing 1 kPa O_2 + 1 kPa CO_2 also resulted in the loss of capacity to produce sufficient ethylene to cause normal ripening. It suggests that kiwifruit can be ideally stored in a narrow range of CA comprising of 2 kPa O_2 and 5 kPa CO_2. Higher levels of CO_2 can also cause potential injuries to the fruit. White core inclusion disorder has been found to be associated with higher levels of CO_2 (5–7 kPa) atmospheres either with or without ethylene (Arpaia et al., 1985, 1986). Fruit stored in atmospheres having CO_2 concentrations of 8 kPa or more for longer than 16 weeks resulted in flesh breakdown and reduced soluble solids content in fruit than air-stored fruit (Harman and McDonald, 1989). A further increase in CO_2 to 10 or 14 kPa CO_2 in CA caused "hard core" condition in which the white core area of fruit remained firm even when the fruit cortex was soft and ripe. As kiwifruit cannot tolerate high CO_2 concentrations in storage environment for longer periods, but a short-term exposure to high CO_2 atmospheres gave some positive results in terms of maintaining fruit firmness during subsequent cold storage (Brigati et al., 1989; Irving, 1992; Nicolas et al., 1989). Brigati et al. (1989) reported that when kiwifruit were pretreated at 20°C for 24 h with 80 kPa CO_2, fruit were firmer after 60 days storage at 0°C than when pretreated with ethylene-free air. However, the differences in fruit firmness had largely disappeared by 135 days of storage at 0°C (Brigati et al., 1989). Nicolas et al. (1989) observed that six intermittent exposures of kiwifruit to 30 kPa CO_2 (3.5 days at 30 kPa CO_2 followed by 3.5 days in air) and subsequent air storage at 0°C reduced firmness loss during 32 weeks storage as compared to air storage. A very high CO_2 (60 or 80 kPa) treatment of 'Hayward' kiwifruit at 0°C for a week resulted in increased rates of respiration and ethylene production, quick fruit softening, and high SSC during subsequent storage at 20°C (Irving, 1992). High CO_2 treatment temporarily retarded fruit softening during subsequent storage at 0°C, but differences in firmness had disappeared after 8 weeks air storage. The increase in duration of CO_2 treatment to 2 or 3 weeks dramatically increased the incidence of fruit rot (Irving, 1992). The application of CA with high CO_2 concentrations and/or very low O_2 can be extended to provide kiwifruit with quarantine security against certain serious pests, which often impede the fruit exports to countries that restrict the possible entry of these pests. Lay-Yee and Whiting (1996) demonstrated that CA treatment of 0.4 kPa O_2, 20 kPa CO_2 at 40°C for 7 h, followed by hydrocooling, controlled nondiapausing type two-spotted spider mite (*Tetranychus urticae* Koch) and did not damage 'Hayward' kiwifruit, although it appeared to result in a decrease in fruit flesh firmness following cold storage at 0°C for 8 weeks. However, the length of treatment required to control the diapausing form of *T. urticae* (10 h) appears to be outside the tolerance limits of the fruit (Lay-Yee and Whiting, 1996). The kiwifruit industry can harness the potential benefits of CA/MA technology to extend storability and prevent fruit softening. The application of CA in kiwifruit to cross over quarantine security barriers could be another promising approach towards more environmentally friendly postharvest treatment eliminating the use of chemicals leaving toxic residues on the fruit.

14.2.8 Litchi (*Litchi chinensis* Sonn.)

Litchi is a nonclimacteric fruit native to southern China. Postharvest life is about 3–5 weeks at 1°C. Loss of bright cherry red color, desiccation, and decay are the major causes of postharvest losses. Litchi fruit begins to loose its bright cherry red color soon after harvest and turn dull brown (Paull and Chen, 1987). CA can be beneficial in prolonging the storage life and maintaining fruit quality in litchi (Chen et al., 1986; Jiang and Fu, 1999; Jiang et al., 2004; Liu et al., 2007; Lonsdale, 1993; Mahajan and Goswami, 2004; Techavuthiporn et al., 2006a,b; Tian et al., 2005; Vilasachandran et al., 1997). Browning of 'Mauritius' fruit was controlled for 28 days at 1°C and 20% CO_2 in N_2 (Lonsdale, 1993). However, Chen et al. (1986) reported that CO_2 levels higher than 5 kPa caused off-flavors. Harvest maturity is one of the important factors affecting the physiological behavior of fruit under CA/MA conditions. Pesis et al. (2002) observed an accumulation of acetaldehyde and ethanol content in litchi fruit in case of delayed harvesting. The levels of these metabolites had direct impact on the storage potential and flavor quality of fruit during subsequent postharvest handling (Pesis et al., 2002). CA containing 3–5 kPa O_2 with 5 kPa CO_2 at 5°C for 22 days helped to maintain the flavor and texture of litchi fruit in good condition with decreased incidence of black spot and stem-end decay, however, higher concentrations of CO_2 (10–15 kPa) led to off-flavor development (Vilasachandran et al., 1997). 'Huaizhi' litchi stored in CA containing 3–5 kPa O_2 and 3–5 kPa CO_2 at 1°C and 90% RH for 30 days significantly reduced the pericarp browning, weight loss, and disease incidence as compared to fruit held in normal air. It also helped to maintain high titratable acidity and ascorbic acid content in litchi fruit (Jiang and Fu, 1999). Mahajan and Goswami (2004) also confirmed the beneficial effects of CA in preventing pericarp browning and maintaining quality in 'Bombay' stored at 2°C and 92%–95% RH under CA containing 3.5 kPa O_2 and 3.5 kPa CO_2 for 56 days. The exposure of litchi to high O_2 (70 kPa) for 1 week prior to CA storage in 5 kPa O_2 with 5 kPa CO_2 at 3°C limited the ethanol production during early period of 42 days storage, and also reduced pericarp browning, anthocyanin decomposition, and decay incidence compared to MAP (Tian et al., 2005).

The exposure of litchi fruit to anoxia also exhibited some beneficial results in maintaining fruit quality (Jiang et al., 2004; Liu et al., 2007). Short-term exposure of litchi fruit to pure N_2 for 3 or 6 h markedly delayed skin browning, reduced disease incidence, and also resulted in higher levels of soluble solids, titratable acidity, and ascorbic acid after 6 days of storage at 20°C and 95%–100% RH (Jiang et al., 2004). Fruits treated under anoxia conditions for 6 h and subsequently kept at 25°C and 95%–100% RH for 6 days exhibited higher concentrations of adenosine triphosphate (ATP), adenosine diphosphate (ADP), adenosine monophosphate (AMP), and adenylate energy charge levels, which led to the lower levels of browning index and membrane permeability compared to nontreated fruit (Liu et al., 2007). The exposure of litchi to superatmospheric oxygen levels also yielded certain benefits in terms of delaying pericarp browning (Duan et al., 2004; Techavuthiporn et al., 2006b). 'Huiazhi' litchi treated with pure O_2 for 6 days at 28°C showed reduced pericarp browning due to inhibition of polyphenol oxidase (PPO) and anthocyanse activities, and also maintained higher soluble solids, titratable acidity levels and with a significant reduction in the ascorbic acid content as compared to nontreated fruit (Duan et al., 2004). 'Hong Huay' litchi treated with superatmospheric O_2 levels of 50–70 kPa at 4°C remained fresh compared to those held in air or 90 kPa O_2. The high O_2 treatment accelerated the rate of respiration, but was helpful to reduce the pericarp browning and weight loss (Techavuthiporn et al., 2006b).

The storage life of litchi was extended very significantly (up to 5 weeks) by storage in polyethylene bags at 2°C (Akamine, 1960; Campbell, 1959; Macfie, 1955a,b; Singh, 1957; Thompson, 1955). Packaging in polyethylene bags and storage at 10°C also extended the postharvest life of the fruit and reduced the incidence of decay (Macfie, 1956). Storage of the cultivars 'Hei Ye' ('Groff' or 'Kak yip') and 'Chen Zi' ('Brewster') in polyethylene bags (0.25 mm thickness) at 2°C or 22°C delayed the onset of pericarp browning (Paull and Chen, 1987). Decay was a problem in fruit packaged in polyethylene bags for 6 days at 22°C, but began to develop after about 20 days in bags held at 2°C. The rate of fruit darkening was reduced by storage in plastic bags (Akamine, 1960; Scott et al., 1982). Pericarp of 'Brewster' browning was delayed significantly by applying polysaccharide coatings, but the effect was reported to be insufficient to be applied commercially (York, 1995). In South Africa, Lonsdale (1993) suggested an alternative treatment to SO_2 which include 'Vitafilm' wrapping, gas mixture of 20 kPa CO_2 in N_2 and irradiation (0.75 or 1.5 kGy). Bleaching of the pericarp was noted due to low levels of O_2 (Akamine, 1960). Pesis et al. (2002) reported that microperforated laminated polyethylene packs containing litchi fruit stored at 2°C for 4 weeks resulted in accumulation of more CO_2, AA, and ethanol in the headspace as compared to macroperforated packs. The late-harvested litchi fruit in MAP had more decay incidence, high AA and ethanol contents than early-harvested ones (Pesis et al., 2002). Tian et al. (2005) compared the effectiveness of MAP using PE film creating in-package atmosphere of 15–19 kPa O_2 + 2–4 kPa CO_2 and CA containing 5 kPa O_2 and 5 kPa CO_2 for storage of litchi at 3°C and 95% RH for 42 days. The quality MA packed litchi in terms of pericarp browning and other quality attributes was never superior to CA-stored fruit (Tian et al., 2005).

14.2.9 Longan (*Dimocarpus longan* Lour.)

Longan is a native of mountainous regions of northern Burma, northeast and southern China. The mature longan fruit is small (1.5–2 cm diameter), conical, heart-shaped or spherical in shape and light brown in color and it has a thin, leathery and indehiscent pericarp surrounding a succulent, edible white aril (Jiang et al., 2002). It has a short postharvest life of 2–3 days at ambient conditions (Jiang, 1999). Postharvest deterioration is mainly associated with desiccation, skin color loss, pericarp browning, and pathogenic decay. Fruit can be stored at low temperature (1°C–5°C) and high humidity (85%–95%) for 30–40 days depending upon the cultivar (Jiang, 1999; Jiang et al., 2002; Zhou et al., 1997).

High humidity in the storage may be conducive for soaking and decay development whereas low humidity promotes desiccation and peel browning. The flesh breakdown and impairment of visual quality limit the long-term storage of fruit. CA and MAP have been found beneficial to maintain fruit quality of longan (Jiang, 1999; Seubrach et al., 2006; Su et al., 2005; Tian et al., 2002a,b; Vangnai et al., 2006; Zhang and Quantick, 1997). Zhang and Quantick (1997) reported that 'Shixia' longan fruit stored in polyethylene bags (0.03 mm film) in MA consisting of 1 kPa O_2 + 5 kPa CO_2 and 3 kPa O_2 + 5 kPa CO_2 for 7 days at room temperature (~25°C) or 35 days at 4°C showed delayed browning of the peel by 2 days and 1 day at room temperature, respectively, and by 7 days and 5 days at 4°C, respectively. MA containing 1 or 3 kPa O_2 with 5 kPa CO_2 also inhibited fruit respiration, maintained total soluble solids, and ascorbic acid levels in the fruit, and partially inhibited PPO activity, but 1 kPa O_2 atmospheres caused slight off-flavor in fruit (Zhang and Quantick, 1997). Jiang (1999) recommended atmospheres with 6–8 kPa CO_2 and 4–6 kPa O_2 for storage of 'Shixia' and 'Wuyuan' cultivars at 1°C and 2.5°C, respectively, for 30 days. These CA conditions provided the best disease control and overall fruit quality in both cultivars. The storage life of longan fruit could be further extended to 60 days when these were stored in atmospheres containing 15 kPa CO_2 and 4 kPa O_2 at 2°C (Tian et al., 2002b). Under these CA conditions, a significant reduction in fruit decay, inhibition of PPO activity, prevention of skin browning, and decreased ethylene productions were observed to favor prolonged storage life of fruit. Tian et al. (2002a) found that CA with high O_2 concentration of 70 kPa without CO_2 significantly decreased ethanol production in the longan fruit, maintained a lower pH value in the peel, and prevented peel browning during 40 days storage at 2°C. Higher CO_2 atmospheres (4 kPa O_2 + 15 kPa CO_2) more effectively reduced decay, but accumulated more ethanol during storage as compared to high O_2 atmospheres (Tian et al., 2002a). MA, which was created by packing 5 kg fruit per PE pack (0.04 mm thick), containing O_2 concentration of 15–19 kPa and CO_2 concentration of 4–5 kPa was not found beneficial to prevent peel browning and also caused accumulation of higher levels of ethanol in fruit during storage at 2°C as compared with other CA treatments comprising either 70 kPa O_2 or 15 kPa CO_2 + 4 kPa O_2 (Tian et al., 2002a).

Another report (Su et al., 2005) on the advantage of high O_2 atmosphere treatment revealed that fruit exposed to pure O_2 for 6 days at 28°C showed lower incidence of peel browning, higher rates of respiration, and higher ATP concentrations, but lower concentrations of AMP as compared with the fruit held in air. However, fruit exposed to pure O_2 exhibited higher activities of PPO and peroxidase (POD), which were not associated with reduced skin browning inhibition supporting the hypothesis that skin browning of longan fruit may be a consequence of membrane injury caused by the lack of maintenance energy (Su et al., 2005). MAP of 'Daw' longan with polyvinyl chloride (PVC) film (15 μm thick) extended the storage life of fruit to 20 days at 4°C against 16 days in nonpacked fruit without substantial weight loss and with good appearance quality (Seubrach et al., 2006). Coating of 'Daw' longan with chitosan (1% or 1.5%) has been reported to slightly inhibit pericarp browning, suppress disease development, and reduce fruit respiration during 20 days of storage at 4°C (Vangnai et al., 2006). There is still research need to explore the possibilities of different CA compositions for better storage life and quality of longan during long-term storage.

14.2.10 Loquat (*Eriobotrya japonica* Lindl.)

Loquat is a nonclimacteric subtropical fruit native to China. It is also chilling-sensitive when stored below 5°C for long duration. CI in loquat is expressed as flesh woodiness, adhesion of peel to flesh, leathery and juiceless pulp, and internal browning. Loquat fruit responds well to MA conditions as shown by various studies. CA containing 12 kPa CO_2 with either air or 2 kPa O_2 has been reported to induce severe internal browning in loquat fruit (Ding et al., 1999). The combination of MAP and low-temperature storage could be ideal to prolong the storage life and avoid the development of disorders in loquat fruit. MAP, using films with low permeance at ambient conditions, may lead to high CO_2 atmospheres surrounding the fruit, leading to internal browning and more decay incidence (Ding et al., 1999, 2002). Ding et al. (2002) found that bagging loquats with 20 μm thick PE at 5°C having in-bag atmosphere of approximately 4 kPa O_2 with 5 kPa CO_2, resulted in the highest scores for appearance and chemical compounds. Under these MA conditions, the fruit had lesser weight loss and retained organic acids without any effect on sugars, and could be stored for 2 months with a higher quality and minimal risk of internal browning (Ding et al., 2002).

Loquat fruit in CA with 10 kPa $O_2 + 1$ kPa CO_2 could be stored for 50 days at 1°C with normal flavor and low decay incidence (Ding et al., 2006). Short-term high O_2 (70 kPa) treatment for 24 h at the beginning of CA storage had little effect on fruit flavor, but stimulated ethanol accumulation in loquat fruit, and reduced activities of endo-PG (polygalacturonase) and exo-PG. CA conditions also reduced PPO and increased POD activities, and had little effect on the phenylalanine ammonialyase activity, leading to reduced flesh browning. Thus, control of tissue browning in loquat fruit through inhibition of oxidases and mitigation of oxidative stress during low-temperature storage are important merits associated with the CA storage in loquat fruit (Ding et al., 2006). Hypobaric storage has also been shown to influence loquat fruit quality positively. Loquat fruit stored for 49 days at 40–50 kPa pressure at 2°C–4°C had lower decay, reduced respiration, and ethylene production rates, inhibited activities of PPO and POD, and resulted in higher titratable acidity and ascorbic acid content (Gao et al., 2006). Therefore, hypobaric storage has also potential to control fruit decay and flesh leatheriness development of loquat fruit stored at low temperature. It is evident that MA supplemented with low-temperature storage helps to maintain loquat fruit quality during long-term storage and also helps to control tissue lignification, internal browning, and decay.

14.2.11 Olive Fruit (*Olea europaea* L.)

Olive fruit is native to the eastern parts of the Mediterranean region. Olive fruit is a drupe with smooth skin (exocarp), fleshy mesocarp, and pit (endocarp). Olives are processed for oil extraction as well as table purpose. There is a need to store olives for many reasons like regulation of fruit flow to the processing industry, nonsynchronization of harvesting with the processing period, and fruit production exceeding the processing capacity of the industry. Preprocessing storage of olive under optimum conditions is very crucial to have high quality oil and table olive products. Olive oil quality is directly related to the physiological condition of the olives from which it is extracted (García et al., 1996). Careless handling and storage of olives accelerates the processes of ripening and senescence increasing sensitivity of fruit to mechanical damage and to the action of pathogenic microorganisms. Due to alteration in physical and chemical structure of olive fruit, the oil quality deteriorates with a characteristic fusty smell, high titratable acidity, low stability, and high values for the indices that measure the level of oxidation (García et al., 1996).

Olive is a nonclimacteric fruit. Table olives can be best stored at 5°C temperature with 90%–95% RH (García and Streif, 1991; Kader et al., 1990; Maestro et al., 1993; Maxie, 1964; Nanos et al., 2002; Woskow and Maxie, 1965). Chilling-sensitive nature of olive does not allow its storage below 5°C and storage at higher temperatures promotes fruit senescence. Typical CI symptoms appear in the form of external and internal browning, loss of flavor, and fruit softening (Kader et al., 1990; Maxie, 1964; Nanos et al., 2002). There is a great variation among different olive cultivars toward chilling sensitivity depending upon the storage temperature (Crisosto and Kader, 2004; Nanos et al., 2002). CI becomes visible on olives stored for >2 weeks at 0°C, 5 weeks at 2°C, or 6 weeks at 3°C (Crisosto and Kader, 2004). The order of susceptibility to CI in decreasing order is 'Sevillano,' 'Ascolano,' 'Manzanillo,' and 'Mission' (Crisosto and Kader, 2004). A comparison of the two green olive cvs. 'Conservolea' and 'Chondrolia' stored at 5°C and 7.5°C in air or various CA showed that 'Chondrolia' olives were very sensitive to CI and lost their capacity to develop skin color and ripen after 2–4 weeks of cold storage with excessive internal browning, resulting in pitting, and external discoloration (Nanos et al., 2002). The effects of CA on postharvest life and quality of table olives have not been very encouraging (Agar et al., 1999; Castellano et al., 1993; García and Streif, 1991; Kader et al., 1990; Maestro et al., 1993; Nanos et al., 2002; Woskow and Maxie, 1965). CA helped to maintain green skin color of fruit, but aggravated the CI symptoms in the form of pitting and external discoloration. Earlier report of Woskow and Maxie (1965) showed that 'Mission' olives could be stored for 10 weeks in green mature stage at 5°C storage temperature supplemented with 2–5 kPa O_2 and 2.5–10 kPa CO_2. Storage life of green 'Manzanillo' olives could be extended to 12 and 9 weeks at 5°C and 7.5°C, respectively, in CA containing 2 kPa O_2 (Kader et al., 1990). The development of 'nailhead' disorder characterized by surface pitting and spotting was delayed by CA conditions, but CA containing CO_2 concentration more than 5 kPa negatively influenced the fruit quality (Kader et al., 1990). Olives appear to be very sensitive to high CO_2 (5 kPa) and low O_2 (<1 kPa) atmospheres (García and Streif, 1991). Quality characteristics of fresh 'Gordal' olive and oil were maintained best during storage at 5°C for 3 months, and storage in CA containing either >5 kPa CO_2 or <1 kPa O_2 affected the fruit quality in terms of firmness loss, skin discoloration, and loss in oil quality (García and Streif, 1991). Another study also showed that storage of 'Picual' olives in atmospheres containing 3 kPa CO_2 + 5 kPa O_2 at 5°C delayed ripening as indicated by retention of green color and flesh firmness (Castellano et al., 1993). However, prolonged storage under these conditions resulted in higher incidence of CI and rot. Maestro et al. (1993) found that the polyphenol content in 'Picual' olive was maintained for 30–45 days at 5°C in normal air and there was no significant influence of reduced O_2 (5 kPa) on polyphenol content during first 30 days of storage, but reduced the decrease in polyphenol content that occurred in air beyond 30 days. CA composition of 3 kPa CO_2 + 5 kPa O_2 resulted in the lowest polyphenol loss during 45 days storage period (Maestro et al., 1993). A study on CA storage of black-ripe 'Manzanillo' olives showed that these could be stored for 4 weeks at 0°C–5°C in air or 2 kPa O_2 without any adverse effect on fruit or oil characteristics (Agar et al., 1999) whereas green 'Manzanillo' olives are susceptible to chilling at <5°C (Kader et al., 1990). Storage of green 'Conservolea' and 'Chondrolia' olives at 5°C and 7.5°C in CA containing 2 kPa O_2 plus 5 kPa CO_2 increased susceptibility to CI, although fruit successfully retained skin green color; 'Conservolea' green olives could be stored up to 37 days at 5°C in air or for up to 22 days at 7.5°C and 2 kPa O_2 + 5 kPa CO_2 (Nanos et al., 2002). Several attempts have been made to derive CA/MA benefits for table or mill olives, but without great success. Cold storage at optimum temperature (5°C) has been found to be sufficient enough to keep olives in good condition for at least 1 month without any additional benefit from atmospheric modification. CA/MA rather aggravated the problem of CI in olives.

14.2.12 Persimmon or Chinese Date Plum or Sharon Fruit (*Diospyros kaki* L.)

Persimmon (*Diospyros kaki* L.) is native to China. Its cultivation first extended to Japan and other parts of East Asia, and was later introduced into California and Southern

Europe. Fruit is very delicious in taste and occasionally fibrous. There are mainly two types of persimmon fruit: astringent and nonastringent. Astringent-type fruit contains very high level of tannins making it unacceptable to consume at mature and firm stage, and is eaten at fully ripe soft stage. Nonastringent fruit contains tannins at very low levels and can be consumed when fruit is still firm and crisp. Persimmon fruit can be stored for 3–5 months at low temperature. The optimum storage conditions for persimmon are 0°C and 90%–95% RH (Crisosto, 2004). Persimmon cultivars can be further classified on the basis of their sensitivity to CI. For example, 'Hachiya' is chilling-insensitive while 'Fuyu' is chilling-sensitive, and experiences CI symptoms at storage temperatures between 2°C and 15°C. CI symptoms appear in the form of flesh browning and softening, and are further aggravated by the presence of ethylene (1 μL L^{-1}) in storage atmosphere (Crisosto, 2004). Ethylene-free atmosphere would be helpful to ameliorate CI and prevent fruit softening during storage. Persimmon is susceptible to physiological damage during postharvest storage. Atmospheres low in O_2 and high in CO_2 have been shown to cause delayed ripening and useful for removal of astringency in persimmon fruit. Postharvest storage problems like skin browning or blackening, CI, and pulp softening can also be addressed to some extent through CA. Storage of 'Fuyu' persimmon in CA containing $CO_2 > 10$ kPa and/or $O_2 < 3\%$ at 0°C for more than 30 days resulted in off-flavor development, brown discoloration, and failure to ripen (Tanaka et al., 1971) as cited by Mitcham et al. (1997). Guelfat-Reich et al. (1975) recommended CA containing 3 kPa CO_2 and 3–5 kPa O_2 for storage at −1°C, but post-CA exposure to 90 kPa CO_2 at 17°C for 12 h was necessary to remove astringency and slow down the rate of fruit softening. High CO_2 levels in the storage atmosphere accelerated softening, increased pulp injuries, and elevated respiration rates during shelf-life (Guelfat-Reich et al., 1975; Tanaka et al., 1971). However, persimmon has been reported to be tolerant to short-term exposure of very low O_2 and/or very high CO_2 (Dentener et al., 1992; Mitcham et al., 1997), and this property can be useful for insecticidal CA treatments to provide quarantine security against certain insect-pests. Treatment of 'Fuyu' persimmon in atmospheres with 0.5 kPa O_2 and 5.3 kPa CO_2 for 4 days at 20°C was beneficial to control light-brown apple moth and long-tailed mealybug (Dentener et al., 1992). Mitcham et al. (1997) found that exposure of 'Fuyu' persimmon to either 0.25 kPa O_2 or 40 kPa CO_2 and a combination of both for 7 days at 20°C did not result into serious injury symptoms and off-flavor except slight external injury symptoms in the form of surface pits. Though ethanol and AA concentrations were higher in treated fruit compared with air controls upon removal from ICA, with the highest levels in persimmons exposed to 0.25 kPa O_2 + 40 kPa CO_2, but did not affect the sensory quality of the fruit (Mitcham et al., 1997). CA alone or in combination with hot water treatment (HWT) has been reported to reduce the CI symptoms in persimmon during long-term cold storage (Burmeister et al., 1997). Following HWT at 47°C–52°C for different durations ranging from 20 to 120 min, 'Fuyu' persimmon could be stored in CA containing either 10 kPa CO_2 and 2 kPa O_2 or 100 kPa N_2 for 6 weeks at 0°C without significant CI symptoms, but N_2 atmospheres promoted the development of external browning to a greater extent (Burmeister et al., 1997). CA containing 3–5 kPa O_2 + 5–8 kPa CO_2 can delay ripening, retain firmness, and can also reduce CI in persimmon (Crisosto, 2004).

MA has a unique and significant role in the postharvest quality improvement of persimmon. Removal of astringency from the fruit can be achieved by exposure of fruit to anaerobic conditions. Matsuo and Ito (1977) proposed a method for astringency removal by application of 80 kPa CO_2 for 1–2 days in which induced AA caused polymerization of the tannin. The rate of astringency removal was positively correlated with the level of endogenous AA attained in the fruit under a CO_2 atmosphere. Application of increasing levels of CO_2 (0, 30, 70, and 100 kPa, balanced with N_2) caused increasing AA formation (Pesis and Ben-Arie, 1984). The application of CO_2 to create anaerobic conditions was comparatively more effective in astringency removal than N_2 exposures (Arnal and Del Río, 2003; Pesis and Ben-Arie, 1986). The increased formation of AA by CO_2-treated fruit was probably because of dark CO_2 fixation that can contribute to higher volatile production via fixation of CO_2 to malic acid by malic enzyme (Pesis and Ben-Arie, 1986). Pesis et al. (1988) reported that MA created either by vacuum or by replacing air with N_2 and CO_2 in polyethylene bags has potential to remove astringency. The rate of deastringency was dependent upon the rate of accumulation of AA. MA-packed fruit stored at 20°C in CO_2 were the first to lose astringency, but showed internal browning after 1 week, whereas fruit under vacuum or N_2, where less volatiles accumulated, maintained their high quality and firmness for 2 weeks at 20°C or 3 months at −1°C (Pesis et al., 1988). Ben-Arie et al. (1991) reported that the maximum storage life of 'Triumph' persimmons in MA packs of LDPE film (0.08 mm) was 8 weeks when packed under vacuum and 20 weeks when packed under nitrogen. The storage life could be extended to 7 months by storing gibberellin-treated (50 ppm) fruit in air for 16 weeks at −1°C prior to MA packaging (Ben-Arie et al., 1991). Prusky et al. (1997) found that accumulation of AA to a level of 80 μg mL^{-1} and ethanol to a level of 900 μg mL^{-1}, and CO_2 up to 30 kPa inside MA packs of persimmon at −1°C for 4 months reduced the incidence of black spot caused by *A. alternata*. This study demonstrated through *in vitro* assays that the increasing concentration of CO_2 during storage was the principal factor in the inhibition of black spot disease development while acetaldehyde concentration inside the packs was not sufficient to cause inhibition of disease development (Prusky et al., 1997). 'Rojo Brillante' persimmon exposed to atmospheres with 98 kPa CO_2 or N_2 at 25°C for 18 or 27 h plus 3 days at 15°C, 80% RH exhibited reduced astringency without any effect on SSC and PPO activity, but firmness reduction was observed in treated fruit (Arnal and Del Río, 2003). According to Cia et al. (2006), atmospheric conditions (1.5–2.1 kPa O_2 and 4.1–4.4 kPa CO_2) generated by the 58 μm multilayer polyolephynic film and 50 μm LDPE film produced satisfactory results to store 'Fuyu' persimmons for up to 84 days at 1°C/90% RH. The 38 μm microperforated PO film did not efficiently modify the atmosphere inside the packages, being ineffective at delaying ripening of the fruit (Cia et al., 2006). Persimmon fruit exposed to N_2 (80 ± 5 kPa) or ethanol vapors (2.5 mL^{-1}) for 2 h showed continuous decline in astringency as indicated by the levels of various phenolics during storage for 5 weeks at ambient conditions (Bibi and Khattak, 2007).

Removal of astringency from persimmon fruit has been the main focus of MA research. A plenty of information has been generated uncovering the mechanism of deastringency in response to anaerobic conditions. More research is required for defining CA conditions for long-term storage. The responses of persimmon to hypobaric conditions need further investigations.

14.2.13 Pomegranate (*Punica granatum* L.)

Pomegranate is native to Mediterranean region and grown world wide. It is a non-climacteric fruit and is harvested when fully ripe. Husk scald, desiccation, and decay are major postharvest problems encountered during long-term storage and marketing. Storage at or below 5°C resulted in severe CI symptoms on the fruit in the form of brown discoloration of the skin, surface pitting, and increased susceptibility to decay organisms (Elyatem and Kader, 1984). Husk scald is another important factor limiting the commercial storage potential of the fruit and its symptoms appear as superficial skin browning initiating from stem end of the fruit and spread toward the blossom end as the severity increases (Defilippi et al., 2006). CA storage has been found very useful to control both of these serious maladies in pomegranate. The initial report on CA storage of pomegranate suggested that storage of 'Wonderful' in low O_2 atmospheres (2 kPa) at 2°C for 6 weeks prevented the development of husk scald, but resulted in accumulation of ethanol leading to off-flavor development (Ben-Arie and Or, 1986). Kupper et al. (1995) found that 'Hicaz' pomegranate could be successfully stored for 6 months in CA containing 6 kPa CO_2 and 3 kPa O_2 at 6°C. CA storage reduced weight loss, decay development, and husk scald in 'Mollar' stored in atmosphere containing 5 kPa O_2 and 0 or 5 kPa CO_2 at 5°C for 2 months without any ethylene scrubber (Artés et al., 1996). The anthocyanin content of arils in 'Wonderful' pomegranate increased during air storage at 10°C for 6 weeks, but the air enriched with 10 or 20 kPa CO_2 suppressed the anthocyanin biosynthesis during storage due to the inhibitory effect on phenylalanine ammonia lyase activity (Holcroft et al., 1998). Hess-Pierce and Kader (2003) recommended 5 kPa $O_2 + 15$ kPa CO_2 as the optimal CA conditions for 'Wonderful' pomegranate at 7°C for 6 months. Further studies on the CA storage of the same cultivar showed that 1 kPa O_2, or 1 kPa $O_2 + 15$ kPa CO_2 resulted in greater accumulation of fermentative volatiles than the CA treatment with 5 kPa $O_2 + 15$ kPa CO_2 (Defilippi et al., 2006). Combination of CA storage with prestorage antifungal treatment has been found very effective in controlling the *Botrytis* decay problem during long-term storage (Palou et al., 2007). Synergistic effects of CA containing 5 kPa $O_2 + 15\%$ CO_2 with antifungal treatment (potassium sorbate 3 min dip in 3% [w/v] solution at 21°C) reduced the decay development, maintained internal and external fruit quality during 15 weeks storage at 7.2°C (Palou et al., 2007). It is evident from the published work that pomegranate is highly sensitive to low-oxygen atmospheres (<5 kPa), CI, and decay. Therefore, the beneficial effects of CA can be harnessed only if the optimal atmosphere conditions are worked out for each cultivar. The combination approach of prestorage antifungal treatments with the optimal storage atmosphere (5 kPa $O_2 + 15$ kPa CO_2), and temperature (~7°C) suggested by Palou et al. (2007) seems quite promising and practical to encounter postharvest storage problems in pomegranate.

MAP can be a cheap and good alternative to control weight loss and CI problems (Artés et al., 2000). MAP of 'Mollar de Elche' pomegranate using perforated polypropylene film of 20 μm thickness reduced weight loss and CI without incidence of decay when stored at 5°C for 12 weeks plus 6 days at 15°C (Artés et al., 2000). The individually shrink film (BDF-2001 and D-955) wrapped fruit of 'Ganesh' could be stored for 12, 9, and 4 weeks at 8°C, 15°C, and 25°C, respectively, whereas nonwrapped fruit could be kept for 7, 5, and 1 week under similar storage conditions (Nanda et al., 2001). There was a tremendous reduction of weight loss in shrink-wrapped fruit in addition to greater firmness retention at the end of 12 weeks storage at 8°C (Figure 14.5). Individual shrink wrapping of pomegranate fruit maintained the harvest freshness for 1 month at ambient conditions and for 3 months at 8°C (Figure 14.6a), and this technology has been transferred to state horticultural development agencies for further adoption by commercial packers (Sudhakar Rao, personal communication). MAP of whole CFB box with D-955 film also resulted in reduced weight loss and maintained pomegranate fruit quality for 3 weeks at ambient and 12 weeks at 8°C (Figure 14.6b). MAP and shrink wrapping also alleviated the development of CI in pomegranate at 8°C (Sudhakar Rao, personal communication). In conclusion, the beneficial effects of CA and MAP in pomegranate include extended storage life, control of husk scald, CI, decay, and weight loss. However, the only limitation is the inhibition of the accumulation of anthocyanins in the aril during long-term storage of fruit in CO_2-enriched atmospheres. Thus, CA/MA technology has great potential to improve the commercialization of this fruit.

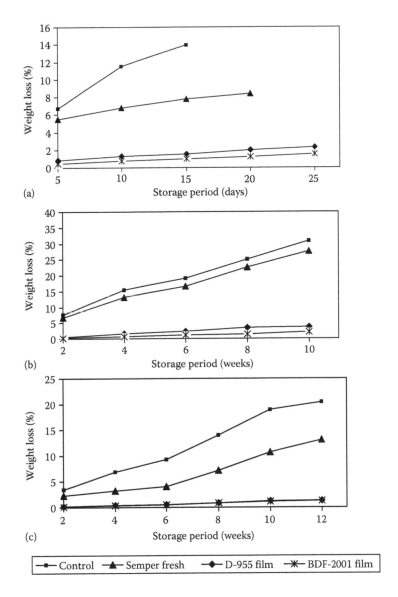

FIGURE 14.5
Effects of film wrapping and skin coating on percent weight loss of pomegranate fruit during storage at (a) 25°C, (b) 15°C, (c) 8°C. (From Nanda, S., Sudhakar Rao, D.V., and Krishnamurthy, S., *Postharvest Biol. Technol.*, 22, 61, 2001. With permission.)

14.2.14 Prickly Pear (*Opuntia* spp.) Fruit and Cladodes

Prickly pear fruits are harvested from various species of the prickly pear cactus, genus *Opuntia* of the cactus family (Cactaceae), and are produced and consumed in several countries. Prickly pear cladodes (called nopal or nopalitos in Mexico) are the rapidly growing succulent stems of the prickly pear cactus (*Opuntia* spp). Cladodes are traditionally consumed in Mexico and some parts of the United States as a vegetable, and also exported to few other countries. The cladodes of many *Opuntia* species are edible but most commercial plantings of nopalitos are from *Opuntia ficus-indica* and *Opuntia inermis*.

FIGURE 14.6
(a) Individual shrink wrapped pomegranate fruit stored either for 3 months at 8°C or for 1 month at ambient conditions. (b) MAP of whole CFB box containing pomegranate fruit with D-955 film and stored either for 12 weeks at 8°C or for 3 weeks at ambient conditions. (Courtesy of Dr. D.V. Sudhakar Rao, IIHR, Bangalore, India.)

The fruit is characterized by high sugar content (12%–17%) and low acidity (0.03%–0.12%), and contain considerable amounts of vitamin C (200–400 µg g^{-1}), among other nutritional components (Yahia, 2008). Nopalitos contain carbohydrates, including fiber (4%–6%) and some protein (1%–2%), minerals, principally calcium (1%), and moderate amounts of vitamin C and vitamin A, and have been suggested to have several health benefits (Yahia, 2008). The fruit is relatively perishable, can be kept for 2–5 weeks at 5°C–8°C with 90% to 95% RH, and several factors can limit storage life such as decay, dehydration, and CI. Cladodes are more perishable than fruit and can be maintained only for few days at 5°C with 95%–99% RH, and major factors limit their storage life including CI, decay and

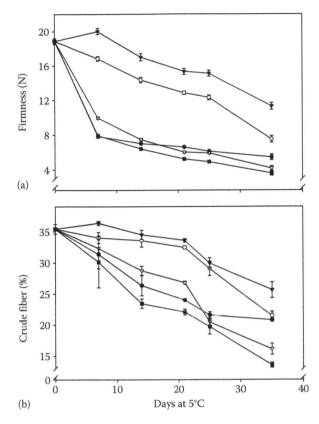

FIGURE 14.7
Changes in (a) firmness (N) and (b) crude fibers (%) in passive and semi-active MAP prickly pear cactus stems stored at 5°C for up to 35 days. Control (●), passive MAP (○), 20% CO_2 (▼), 40% CO_2 (▽), and 80% CO_2 (■). Each point is the average of 30 observations in the case of firmness and five observations in the case of crude fiber. Vertical bars represent standard error of the mean. (From Guevara, J.C., Yahia, E.M., Brito, E., and Biserka, S., *Postharvest Biol. Technol.*, 29, 167, 2003. With permission.)

dehydration (Yahia, 2008; Guevera and Yahia, 2005). MA/CA can delay ripening/senescence and extend storage life of both fruit and cladodes. Very little work has been done on MA and CA of prickly pear fruit and cladodes, but holding at 5°C in 2 kPa O_2 + 2–5 kPa CO_2 can delay ripening and senescence and extend storage life of fruit and cladodes (Guevara et al., 2001, 2003). Packaging nopalitos in Cryovac PD960 films created a MA as a result of the high respiration rate of the cactus stems, whereas O_2 decreased to about 8.6 kPa and CO_2 increased to about 6.9 kPa after 30 days in storage (Guevara et al., 2001, 2003). Cladodes packaged in MA had very low weight loss. The texture and crude fiber contents of cladodes that were maintained in MAP decreased only very slightly during all of the storage period, and there was a close relation between them, indicating that the retention in texture might be due to the positive effects of MA on preventing crude fiber degradation (Figure 14.7). CO_2 concentrations between 7 and 20 kPa decreased the loss in color, firmness, and fiber content, and reduced chlorophyllase activity, and microbial flora load on the stems. These benefits are due to atmospheric modification and not to increased humidity in the atmosphere as shown in Figure 14.8 (Guevara et al., 2001; Yahia et al., 2005). On the other hand, elevated CO_2 levels (≥40 kPa) caused injury in cladodes in comparison with nonpackaged cactus stems. There were no big differences in the quality of prickly pear cactus stems packaged in passive MAP or in semiactive MAP with an initial CO_2 atmosphere of 20 kPa. The relative limit of tolerance of prickly pear cactus stems to CO_2 is 20 kPa. The storage life of the cactus stems can be up to 32 days in passive MAP or in semiactive MAP with an initial CO_2 atmosphere of 20 kPa.

A model has been generated and applied to describe the gas profile of prickly pear cactus stems in passive and semiactive MAP (Yahia et al., 2005; Guevara et al., 2006a,b). The model describes the gas exchange in nonsteady state, taking into consideration the effect

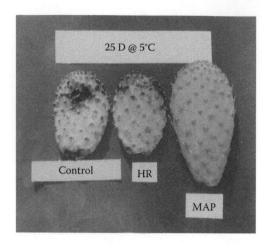

FIGURE 14.8
Prickly pear cactus stems (nopalitos) exposed to MAP, a humid regular air (same humidity as those maintained in MAP), or regular air, all at 5°C for 25 days. (Courtesy of Prof. E.M. Yahia.)

of temperature at 5°C, 14°C, 20°C, and 25°C and the relative humidity from 65% to 90% RH (at intervals of 5%) on film permeability characteristics, respiration rate and tissue permeance of prickly pear cactus stems. The model suitably describes the changes in CO_2 concentration and overestimates the O_2 concentration in passive (no addition of gases) MAP of prickly pear cactus stems. However, when an atmosphere of >20 kPa CO_2 is added to the packages, the model adequately describes the changes in O_2 but underestimates the changes in CO_2. This might be due to high CO_2 concentration alterations in tissue metabolism. The temperature and RH integration and the small number of the measurable parameters in the proposed model may help it to be easily used for a great variety of situations involving different commodities. This approach should facilitate package designing for fresh produce, and therefore maximizing the benefits of MAP for horticultural products. Another major conclusion of this study, however, was that the relative humidity around the packages, for example in the display cabinet of retail stores, can have a major impact on the respiration and gas exchange of MAP packed products. This could well be the missing link in understanding the sometimes erroneous and incomprehensible results obtained with MAP packages in practice.

14.2.15 Rambutan (*Nephelium lappaceum* L.)

A nonclimacteric fruit native to the rain forest of Malaysia. The fruit has a leathery skin covered with fleshy pliable spines. Desiccation, loss of red peel color, browning, and drying of spintern are the principal causes of deterioration. Storage life is about 1–3 weeks at 7°C–12°C (depending on cultivar).

Beneficial atmospheres are 3%–5% O_2 and 7%–12% CO_2 (Kader, 1993; Yahia, 1998). These atmospheres were reported to delay senescence, retard red color loss, and extend postharvest life to about 1 month if water loss is controlled. Storage of 'R162' in 9–12 kPa CO_2 retarded color loss and extended shelf life by 4–5 days, while 3 kPa O_2, 5 ppm C_2H_4, or the inclusion of C_2H_4 absorbents did not significantly affect the rate of color loss (O'Hare, 1995; O'Hare et al., 1994). No further gain in skin color deterioration was observed beyond 12 kPa CO_2. Shelf life was increased from 13 to 17 days at 9 kPa CO_2. Diseases were not observed until at least 22 days of storage. The authors suggested that MA/CA is effective through an elevation of the CO_2 concentration and not through a decrease in O_2. No deleterious CO_2 effects were observed at concentrations of up to 12 kPa.

Storage of rambutan in sealed polyethylene bags (Inpun, 1984; Ketsa and Klaewkasetkorn, 1995; Lam and Ng, 1982; Lee and Leong, 1982; Mendoza et al., 1972) or plastic containers (Mohamed and Othman, 1988) was reported to retard skin color loss, ameliorate CI, and to extend shelf life. The storage life of 'Lebak bulus' packed in 0.03 and 1.0 mm thick polyethylene or polypropylene bags at 10°C was extended to 10 days (Harjadi and Tahitoe, 1992). Wax coatings were reported to be less effective than polyethylene wrapping in extending storage life (Brown and Wilson, 1988; Mendoza et al., 1972). MA/CA containing high concentrations of CO_2 (up to 12 kPa) are beneficial to retard quality loss in rambutan. Research is needed to investigate the beneficial effects of low O_2 atmospheres, and the economic feasibility of MA/CA application.

14.3 Conclusions

The international trade of subtropical fruits is progressively increasing and there are clear signs of further boom in this sector. The increasing gaps between the production and consumption sites further warrant the transportation of produce over large distances, which in turn put forward a daunting task of maintaining fruit quality and safety to provide consumer with a high quality produce at competitive prices consistently. There is a continuous improvement in the existing storage technologies due to increasing understanding of fruit physiology, new product innovations, and also the emerging sophisticated engineering technologies. As a supplement to the use of refrigeration, CA/MA technology has brought multitude of benefits to maintain fresh fruit quality. The atmospheric modification can be achieved by active control of gases as in the case of CA and LPS, while the passive atmospheric modification is possible through the use of polymeric films either in individual packaging or bulk scale packing of whole carton, pallet, or several pallets in a container. Ben-Yehoshua (2005) proposed the "triple bottom line" concept to evaluate the adoption of a postharvest technology based on three aspects: social, economic, and environmental, which are interrelated and interdependent on each other. Environmental issues like climate change are more volatile in the present-day situation, and would have major impact on the technology adoption in near future. If we critically evaluate the CA and LPS technologies for subtropical fruits, these appear to best fit on the social and environmental lines, but may not hold good economy-wise. LPS can outweigh the CA storage technology on environmental front as the former does not involve the addition of CO_2, and appears to be more ecofriendly. But the less expensive alternative like MAP with polymeric films does not seem to be environment friendly. Most of the research on CA/MA issues on subtropical fruit focused on maintaining "freshness" of fruit and the flavor aspects were being almost neglected. The rising consumer awareness for healthy diets and intense competition in the global market has increased consumer expectations in terms of produce freshness, flavor, and toxic residues. Quality, safety, and reliability are three driving factors influencing the acceptance of a produce in the global market. Fresh look and fresh flavor do not necessarily go together when the produce is at the terminal end of the supply chain.

The benefits of CA/MA in subtropical fruits are species specific. The CA/MA requirements of subtropical fruits are shown in Table 14.3. The extension of storage life in avocado, litchi, guava, kiwifruit, loquat, longan, and rambutan has been achieved through CA/MA application. MAP has become popular and commercially used for

TABLE 14.3

Summary of CA Storage Requirements of Subtropical Fruits and Their Sensitivities to O_2, CO_2, and C_2H_4

Fruit	Temperature Range (°C)	CA Range O_2 (kPa)	CA Range CO_2 (kPa)	Sensitivity to O_2, CO_2, and C_2H_4	Outstanding CA/MAP Benefit	References
Avocado	5–13	2–5	3–10	Sensitive to low O_2 (<1 kPa) and high CO_2 (>15 kPa)	Reduction in CI; marine transportation	Yahia (1997a, 1998)
Dates	0	Air	10–20	Tolerant to short-term high CO_2 (60–80 kPa)	Insect disinfestations and transportation of fresh fruit; control of storage pests in dried dates	Al-Redhaiman (2005); Navarro et al. (2001)
Fig	0	5–10	15–20	Sensitive to O_2 < 2 kPa and CO_2 > 25 kPa; dried figs tolerant to very high CO_2 (96 kPa)	Transportation of fresh fruit	Colelli et al. (1991)
Grapefruit	10–15	3–10	5–10	Tolerant to short-term low O_2 (1 kPa) and high CO_2 (20–40 kPa)	Potential application in insect disinfestation	Kader (2003); Shi et al. (2005)
Guava	8–10	5–8	5	Sensitive to low O_2 (<5 kPa)	Reduction in CI and fruit softening	Pal et al. (2007); Singh and Pal (2008a); Singh (2006 unpublished)
Jujube	0–5	10	0	Superatmospheric O_2 (100 kPa) also beneficial	Delayed fruit senescence	Qin and Tian (2004); Wang et al. (2006)
Kiwifruit	0–5	1–2	3–5	Sensitive to high CO_2 (>5 kPa); highly sensitive to ethylene	Retardation of fruit softening, potential in insect disinfestation	Antunes and Sfakiotakis (2005); Arapia et al. (1985, 1986, 1987); McDonald and Harman (1982)
Lemon	10–15	5–10	5–10	Sensitivity to ethylene low in low O_2 atmospheres	MAP is commercial; reduction in water loss	Kader (2003); Wild et al. (1977a,b)

Lime	10–15	5–10	5–10	Tolerate low O$_2$ during low pressure transportation	MAP is commercial; reduction in water loss	Kader (2003)
Litchi	5–12	3–5	3–5	Tolerant to superatmospheric O$_2$ (100 kPa)	Inhibition of pericarp browning	Duan et al. (2004); Jiang and Fu (1999); Vilasachandran et al. (1997)
Longan	1–5	4–6	5–15	Tolerant to superatmospheric O$_2$ (70 kPa)	Reduction in CI symptoms	Jiang (1999); Su et al. (2005); Tian et al. (2002)
Loquat	1–5	4–10	1–5	Tolerant to superatmospheric O$_2$ (70 kPa)	Reduction in CI symptoms and flesh toughening	Ding et al. (1999, 2002, 2006)
Olive	5–10	2–3	0–1	—	No significant advantage	García and Streif (1991); Castellano et al. (1993)
Orange	5–10	5–10	0–5	Sensitive to low O$_2$ and high CO$_2$	MAP recommended; reduction in water loss	Kader (2003); Ke and Kader (1990); Porat et al. (2004)
Persimmon	0	3–5	5–8	Tolerant to low O$_2$ (0.25 kPa) and high CO$_2$ (40 kPa) for short-term; sensitive to ethylene	Removal of astringency and alleviation of CI	Guelfat-Reich et al. (1975); Mitcham et al. (1997); Pesis and Be-Arie (1986); Pesis et al. (1988)
Pomegranate	5–10	3–5	5–10	Sensitive to low O$_2$ (<5 kPa)	Prevention of husk scald and chilling symptoms	Defilippi et al. (2006); Hess-Pierce and Kader (2003); Nanda et al. (2001)
Prickly pear (stems)	5	2	2–5	Tolerate up to 20 kPa CO$_2$	MAP recommended; reduction in weight and color loss	Guevara et al. (2001, 2003)
Rambutan	8–15	3–5	7–12	Sensitive to low O$_2$ (<1 kPa) and high CO$_2$ (>20 kPa)		Yahia (1998); Kader (1993, 2003)

fruit crops such as avocado, citrus fruits, litchi, and pomegranate. The impact of CA/MA technology in encountering product-specific postharvest problems is dramatic. The alleviation of CI in avocado, guava, persimmon, and pomegranate is an advantage associated with the use of CA/MA. The short-term exposure of fruit to insecticidal atmospheres has been demonstrated beneficial to control certain insect–pests impeding the exports of some subtropical fruits. The scope of CA/MA application for dried fruits like date and fig is immense. These fruits treated with gas mixtures high in CO_2 lead to control of stored-products-pests for longer duration at ambient storage conditions. A unique application of CA/MA in quality improvement of a fruit can be best cited as in case of persimmon. The exposure of fruit to atmospheres low in O_2 and high in CO_2 for 1–2 days have been known and commercially adopted for removal of astringency in persimmon fruit. The successful control of the pericarp browning problem in litchi, rambutan, and longan through the use of either low or superatmospheric O_2 treatments is another cutting-edge advantage of the technology. The positive results of superatmospheric oxygen treatments have inspired more research in this area. At this stage, we do not have comprehensive data to conclude something very promising. LPS is a promising technology for atmospheric control and has many advantages over conventional CA systems. But it could not be very popular due to some reasons best outlined by Burg (2004). It appears that there is a tremendous scope of the LPS technology for long-term storage and transportation of subtropical fruits with minimal adverse effects on fruit quality. The high infrastructural costs to set up the research-scale LPS system might have resulted in research gaps. It is not mandatory that every fruit species will respond favorably to the CA/MA conditions. Table or mill olives do not find any favor from CA, rather suffer from CI under these conditions. Similarly, citrus fruits do not respond well to CA due to accumulation of anaerobic metabolites in the fruit tissue. It is evident that there is no additional advantage of CA in either enhancing the storage life or maintaining fruit quality in citrus fruit. However, grapefruit can tolerate CA conditions better than mandarins for a shorter duration. The application of individual seal packaging in enhancing the storage life of lime and lemons is a classical example of the success of MAP. However, the underlying factor involved in this case is the reduction in water loss from the rind tissue and not the atmospheric modification. The same technology has been found quite practical in pomegranate as well. Again the rising problem on the use of plastics creates environment havoc and is opposed by environmentalists. The increasing use of edible coatings and other waxes in some fruits especially citrus is becoming popular as these are more consumer and environmentally friendly.

14.4 Future Research Needs

The literature on CA/MA research has advanced very fast in recent years, but more research is required to bring the level of knowledge in subtropical fruits comparable to the temperate fruits. The effects of different atmosphere regimes on the nutritional quality and antioxidant profiles of subtropical fruits deserve further investigations. The information on the aroma volatiles profiles of subtropical fruits subjected to different CA/MA and other postharvest handling conditions is limited and there is a vast scope for future research in this area. The advancement and simplicity in aroma analytical techniques can be used to ramp up research on changes in flavor components of subtropical fruits in response to CA/MA. There is growing interest in studying the response of fruit to CA/MA at molecular level and identifying the upregulating or downregulating genes controlling vital postharvest processes. More efforts by molecular biologists to unfold the genomic mysteries will certainly be opening new avenues for postharvest technologists along with several other challenges as well. Thus, genetic engineering approaches would be supplementary to encounter some postharvest problems in subtropical fruits, but can never be a "total solution" as there has been a very little success in the history of plant biotechnology era of last century. Till date, more research on CA/MA was carried out at laboratory scale and we must look forward to strengthening linkage with the logistics industry to conduct experiments in a "real-world" situation. A more involvement of researchers with logistics industry would help to exchange knowledge and experience between them, leading to reducing financial risks to the shippers and traders. The exploitation of nonconventional gas mixtures (e.g., nitric oxide, nitrous oxide, and carbon monoxide) as a supplement or complement to the existing CA technologies would be another interesting area of research, but the possible hazards associated with these gas mixtures have to be kept in mind. Innovation is a natural outcome of increased human curiosity and needs. Innovation of 1-MCP has revolutionized the postharvest industry and such innovations will hopefully be coming forward in future also. The continuous maintenance of 1-MCP at very low levels in gas mixtures may be further investigated. The dynamic control system (DCS) technology, which continually adjust the storage atmosphere depending upon the state of the produce, may be conceptualized for subtropical fruits. The biosensors application in regular monitoring of the product response to CA, based on the stress levels and volatile compounds emission, would be helpful to determine critical stress limits, and thus maintaining ideal atmospheres surrounding the fruit in DCS. The research in this area needs to be streamlined so as to impose stress on fruit within acceptable limits. The commercialization of "smart packaging" using biodegradable films and development of low costs sensors/indicators has not been explored much in subtropical fruits. The scientific component involving technology rectification or new technology development is one episode; another equally important aspect is market research to recognize the consumer behavior toward CA-stored produce, where the flavor is often a debatable issue. In a consumer-driven market, our objective is not fulfilled unless we could meet, rather exceed the consumer expectations by providing a flavor- and nutrition-rich fruit.

Acknowledgments

We dedicate this chapter to Prof. Adel A. Kader, who recently retired from UC-Davis, for his lifetime contribution to the CA/MA research. We acknowledge Dr. Ron Porat, ARO Volcani Centre, Israel, Dr. D.V. Sudhakar Rao, IIHR, Bangalore, India, and Dr. J. Burdon, HortResearch, New Zealand for contributing published material/illustrations, which are reproduced in this chapter.

Abbreviations

AA	acetaldehyde
ADH	alcohol dehydrogenase
AMP	adenosine monophosphate
ATP	adenosine triphosphate
CI	chilling injury
CA	controlled atmospheres

DCS   dynamic control system
HDPE   high density polyethylene
HDMC   hydroxypropyl methylcellulose
HWT   hot water treatment
ICA   insecticidal controlled atmospheres
kPa   kilo Pascals
LDPE   low density polyethylene
LPS   low pressure storage
MA   modified atmospheres
MAP   modified atmosphere packaging
POD   peroxidase
PG   polygalacturonase
PPO   polyphenol oxidase
SERB   stem-end-rind-breakdown
SSC   soluble solids concentration
ULO   ultralow oxygen

References

Abbas, M.F. and B.S. Fandi. 2002. Respiration rate, ethylene production and biochemical changes during fruit development and maturation of jujube (*Ziziphus mauritiana* Lamk). *J. Sci. Food Agric.* 82:1472–1476.

Agar, I.T., B. Hess-Pierce, M.M. Sourour, and A.A. Kader. 1999. Identification of optimum preprocessing storage conditions to maintain quality of black ripe 'Manzanillo' olives. *Postharvest Biol.Technol.* 15:53–64.

Akamine, E.K. 1960. Preventing the darkening of fresh lychee prepared for export. Hawaii Agricultural Experiment Station Technical Progress Report No. 127.

Allwood, M.E. and B.N. Wolstenholme. 1995. Modified atmosphere shock treatment and on orchard mulching trial for improving Fuerte fruit quality. *South African Avocado Grower's Association Yearbook*, Vol. 18, pp. 85–88.

Alonso, M., M.A.D. Río, and J.A. Jacas. 2005. Carbon dioxide diminishes cold tolerance of third instar larvae of *Ceratitis capitata* Wiedemann (Diptera: Tephritidae) in 'Fortune' mandarins: Implications for citrus quarantine treatments. *Postharvest Biol. Technol.* 36:103–111.

Al-Redhaiman, K.N. 2005. Chemical changes during storage of 'Barhi' dates under controlled atmosphere conditions. *HortScience* 40:1413–1415.

Antunes, M.D.C. and E.M. Sfakiotakis. 2002. Ethylene biosynthesis and ripening behaviour of 'Hayward' kiwifruit subjected to some controlled atmospheres. *Postharvest Biol. Technol.* 26:167–179.

Apelbaum, A., C. Zauberman, and Y. Fuchs. 1977. Prolonging storage life of avocado fruits by subatmospheric pressure. *HortScience* 12:115–117.

Arnal, L. and M.A. Del Río. 2003. Removing astringency by carbon dioxide and nitrogen-enriched atmospheres in persimmon fruit cv. 'Rojo brillante.' *J. Food Sci.* 68:1516–1518.

Arpaia, M.L., F.G. Mitchell, A.A. Kader, and G. Mayer. 1985. Effects of 2% O_2 and varying concentrations of CO_2 with or without C_2H_4 on the storage performance of kiwifruit. *J. Am. Soc. Hort. Sci.* 110:200–203.

Arpaia, M.L., F.G. Mitchell, A.A. Kader, and G. Mayer. 1986. Ethylene and temperature effects on softening and white core inclusions of kiwifruit stored in air or controlled atmospheres. *J. Am. Soc. Hort. Sci.* 111:149–153.

Arpaia, M.L., F.G. Mitchell, G. Mayer, and A.A. Kader. 1984. Effects of delays in establishing controlled atmospheres on kiwifruit softening during and following storage. *J. Am. Soc. Hort. Sci.* 109:768–770.

Arpaia, M.L., J.M. Labavitch, C. Greve, and A.A. Kader. 1987. Changes in the cell wall components of kiwifruit during storage in air or controlled atmosphere. *J. Am. Soc. Hort. Sci.* 112:474–481.

Artés, F., J. Gines Marin, and J.A. Martínez. 1996. Controlled atmosphere storage of pomegranate. *Z. Lebensm.-Untersuchung Forsch.* 203:33–37.

Artés, F., R. Villaescusa, and A. Tudela. 2000. Modified atmosphere packaging of pomegranate. *J. Food Sci.* 65:1112–1116.

Baloch, M.K., S.A. Saleem, A.K. Baloch, and W.A. Baloch. 2006. Impact of controlled atmosphere on the stability of Dhakki dates. *Lebensm. Wiss. U-Technol.* 39:671–676.

Ben-Arie, R. and E. Or. 1986. The development and control of husk scald on 'Wonderful' pomegranate fruit during storage. *J. Am. Soc. Hort. Sci.* 111:395–399.

Ben-Arie, R. and L. Sonego. 1985. Modified-atmosphere storage of kiwifruit (*Actinidia chinensis* Planch) with ethylene removal. *Scientia Hort.* 27:263–273.

Ben-Arie, R., Y. Zutkhi, L. Sonego, and J. Klein. 1991. Modified atmosphere packaging for long-term storage of astringent persimmons. *Postharvest Biol. Technol.* 1:169–179.

Ben-Yehoshua, S. 2005. Introduction. *Environmentally Friendly Technologies for Agricultural Produce Quality.* CRC Press, Taylor and Francis Group, Boca Raton, FL, pp. 1–10.

Ben-Yehoshua, S., I. Kobiler, and B. Shapiro. 1979. Some physiological effects of delaying deterioration of citrus fruits by individual seal packaging in high density polyethylene film. *J. Amer. Soc. Hort. Sci.* 104:868–872.

Ben-Yehoshua, S., J. Peretz, R. Moran, B. Lavie, and J.J. Kim. 2001. Reducing the incidence of superficial flavedo necrosis (noxan) of 'Shamouti' oranges (*Citrus sinensis*, Osbeck). *Postharvest Biol. Technol.* 22:19–27.

Biale, J.B. 1942. Preliminary studies on modified air storage of the Fuerte avocado fruit. *Proc. Am. Soc. Hort. Sci.* 41:113–118.

Biale, J.B. 1946. Effect of oxygen concentration on respiration of 'Fuerte' avocado fruit. *Am. J. Bot.* 23:363–373.

Bibi, N. and A.B. Khattak. 2007. Effect of modified atmosphere on methanol extractable phenolics of persimmon modified atmosphere effect on persimmon phenolics. *Int. J. Food Sci. Technol.* 42:185–189.

Bleinroth, E.W., J.L.M. Garcia, I. Shirose, and A.M. Carvalho. 1977. Storage of avocado at low temperature and in controlled atmosphere (in Portuguese). *Coletanea Inst. Tecnol. Aliment (Brazil)* 8(2):587–622.

Bower, J.P., J.G.M. Cutting, and A.B. Truter. 1989. Modified atmosphere storage and transport of avocados—what does it mean? *South African Avocado Growers' Association Yearbook*, Vol. 12, pp. 17–20.

Bower, J.P., J.G.M. Cutting, and A.B. Truter. 1990. Container atmosphere as influencing some physiological browning mechanisms in stored 'Fuerte' avocados. *Acta Hort.* 269:315–321.

Brigati, S., G.C. Pratella, and R. Bassi. 1989. CA and low oxygen storage of kiwifruit: Effects on ripening and diseases. *Proceedings of the 5th International Controlled Atmosphere Research Conference, Wenatchee, WA*, Vol. 2: pp. 41–48.

Brooks, C., C.O. Bratley, and L.P. McColloch. 1936. Transit and storage diseases of fruits and vegetables as affected by initial carbon dioxide treatments. USDA Technical Bulletin No. 519.

Brown, B.I. and P.R. Wilson. 1988. Exploratory study of postharvest treatments on rambutan (Nephelium (sic) Iappaceum) 1986/1987 Rare fruit Counc. *Aust. Newsl.* 48:16–18.

Burdon, J., N. Lallu, C. Yearsley, D. Burmeister, and D. Billing. 2007. The kinetics of acetaldehyde and ethanol accumulation in 'Hass' avocado fruit during induction and recovery from low oxygen and high carbon dioxide conditions. *Postharvest Biol. Technol.* 43:207–214.

Burg, S.P. 2004. Horticultural commodities. In: S.P. Burg (Ed.), *Postharvest Physiology and Hypobaric Storage of Fresh Produce.* CABI Publishing, Wallingford, U.K., pp. 371–439.

Burg, S.P. and E.A. Burg. 1966. Fruit storage at subatmospheric pressures. *Science* 153:314–315.

Burmeister, D.M., S. Ball, S. Green, and A.B. Woolf. 1997. Interaction of hot water treatments and controlled atmosphere storage on quality of 'Fuyu' persimmons. *Postharvest Biol. Technol.* 12:71–81.

Campbell, C.W. 1959. Storage behavior of fresh 'Brewster' and 'Bengal' lychees. *Proc. Fla. State Hort. Soc.* 72:356.

Carrillo López, A. and E.M. Yahia. 1990. Tolerancia del aguacate var. 'Hass' a niveles insecticidas de O_2 y CO_2 y el efecto sobre la respiración anaeróbica. *Tecnología de Alimentos (México)* 25 (6):13–18.

Carrillo López, A. and E.M. Yahia. 1991. Avocado fruit tolerance and responses to insecticidal O_2 and CO_2 atmospheres. *HortScience* 26(6):734.

Castellano, J.M., J.M. García, A. Morilla, S. Perdiguero, and F. Gutierrez. 1993. Quality of Picual olive fruits stored under controlled atmospheres. *J. Agric. Food Chem.* 41:537–539.

Cheah, L.H. and D.E. Irving. 1997. Kiwifruit. In: S.K. Mitra (Ed.), *Postharvest Physiology and Storage of Tropical and Subtropical Fruits*. CAB International, Wallingford, U.K., pp. 209–227.

Chen, W.S., M.X. Su, and F.W. Li. 1986. A study on controlled atmosphere storage of lychee. In: K.M. Chau (Ed.), *Selection from the Symposium on Litchi Research Papers (1981–1985) Beijing*. pp. 87–88.

Cia, P., E.A. Benato, J.M.M. Sigrist, C. Sarantopóulos, L.M. Oliveira, and M. Padula. 2006. Modified atmosphere packaging for extending the storage life of 'Fuyu' persimmon. *Postharvest Biol. Technol.* 42:228–234.

Claypool, L.L. and S. Ozbek. 1952. Some influences of temperature and carbon dioxide on the respiration and storage life of the Mission fig. *Proc. Am. Soc. Hort. Sci.* 60:226–230.

Colelli, G., F.G. Mitchell, and A.A. Kader. 1991. Extension of postharvest life of 'Mission' figs by CO_2-enriched atmospheres. *HortScience* 26:1193–1195.

Collin, M. 1984. Conservation de l'avocat par chocs CO_2. *Fruits* 39:561–566.

Crisosto, C.H. 2004. The commercial storage of fruits, vegetables, and florist and nursery stocks. USDA Agricultural Research Services Agricultural Handbook No. 66.

Crisosto, C.H. and A.A. Kader. 2004. The commercial storage of fruits, vegetables, and florist and nursery stocks. USDA Agricultural Research Services Agricultural Handbook No. 66.

D'Aquino, S., M. Angioni, S. Schirru, and M. Agabbio. 2001. Quality and physiological changes of film packaged 'Malvasio' mandarins during long term storage. *Lebensm. Wiss. U-Technol.* 34:206–214.

D'Aquino, S., A. Piga, M.G. Molinu, M. Agabbio, and C.M. Papoff. 1998. Maintaining quality attributes of 'Craxiou de Porcu' fresh fig fruit in simulated marketing conditions by modified atmosphere. *Acta Hort.* 480:289–294.

Defilippi, B.G., B.D. Whitaker, B.M. Hess-Pierce, and A.A. Kader. 2006. Development and control of scald on wonderful pomegranates during long-term storage. *Postharvest Biol. Technol.* 41:234–243.

Dentener, P.R., S.M. Peetz, and D.B. Birtles. 1992. Modified atmospheres for the postharvest disinfestation of New Zealand persimmons. *N. Z. J. Crop Hort. Sci.* 20:203–208.

Ding, C.K., K. Chachin, Y. Ueda, Y. Imahori, and H. Kurooka. 1999. Effects of high CO_2 concentration on browning injury and phenolic metabolism in loquat fruits. *J. Jpn. Soc. Hort. Sci.* 68:275–282.

Ding, C.K., K. Chachin, Y. Ueda, Y. Imahori, and C.Y. Wang. 2002. Modified atmosphere packaging maintains postharvest quality of loquat fruit. *Postharvest Biol. Technol.* 24:341–348.

Ding, Z., S. Tian, Y. Wang, B. Li, Z. Chan, J. Han, and Y. Xu. 2006. Physiological response of loquat fruit to different storage conditions and its storability. *Postharvest Biol. Technol.* 41:143–150.

Dou, H. 2004. Effect of coating application on chilling injury of grapefruit cultivars. *HortScience* 39:558–561.

Duan, X.W., Y.M. Jiang, X.G. Su, and Z.Q. Zhang. 2004. Effects of a pure oxygen atmosphere on enzymatic browning of harvested litchi fruit. *J. Hort. Sci. Biotechnol.* 79:859–862.

El-Goorani, M.A., N.F. Sommer. 1981. Effects of modified atmospheres on postharvest pathogens of fruits and vegetables. *Hort. Rev.* 3:412–461.

El-Mir, M., D. Gerasopoulos, I. Metzidakis, and A.K. Kanellis. 2001. Hypoxic acclimation prevents avocado mesocarp injury caused by subsequent exposure to extreme low oxygen atmospheres. *Postharvest Biol. Technol.* 23:215–226.

Elyatem, S.M. and A.A. Kader. 1984. Post-harvest physiology and storage behaviour of pomegranate fruits. *Scientia Hort.* 24:287–298.

Faubion, D.F., F.G. Mitchell, G. Mayer, and M.L. Arpaia. 1992. Response of 'Hass' avocado to postharvest storage in controlled atmosphere conditions. In: C. Lovatt et al. (Eds.), *Proceedings 2nd World Avocado Congress* April 21–26, 1991. Orange, CA, pp. 467–472.

Ferizli, A.G. and M. Emekci. 2000. Carbon dioxide fumigation as a methyl bromide alternative for the dried fig industry. *Annual International Research Conference on Methyl Bromide Alternatives and Emission Reductions*. Orlando, FL. p. 81.

Gao, H.Y., H.J. Chen, W.X. Chen, J.T. Yang, L.L. Song, Y.H. Zheng, and Y.M. Jiang. 2006. Effect of hypobaric storage on physiological and quality attributes of loquat fruit at low temperature. *Acta Hort*. 712:269–273.

García, J.M., F. Gutierrez, M.J. Barrera, and M.A. Albi. 1996. Storage of mill olives on an industrial scale. *J. Agric. Food Chem*. 44:590–593.

García, J.M. and J. Streif. 1991. The effect of controlled atmosphere storage on fruit quality of 'Gordal' olives. *Gartenbauwissenschaft* 56:233–238.

Gonzalez, G., E.M. Yahia, and I. Higuera. 1990. Modified atmosphere packaging of mango and avocado fruit. *Acta Hort*. 269:355–344.

González-Aguilar, G., E.M. Yahia, and M. Silveira. 1997. Predicción de la atmósfera en aguacate empacado en bolsas de polietileno y evaluación de su calidad durante el almacenamiento. *Horticultura Mexicana* 5(4):351–360 (México).

Guelfat-Reich, S., R. Ben-Arie, and N. Metal. 1975. Effect of CO_2 during and following storage on removal of astringency and keeping quality of 'Triumph' persimmons. *J. Am. Soc. Hort. Sci*. 100:95–98.

Guevara, J.C. and E.M. Yahia. 2005. Pre- and postharvest technology of cactus stems, the nopal. In: R. Dris (Ed.), *Crops: Growth, Quality and Biotechnology*. WFL, Helsinki, Finland, pp. 592–617.

Guevara, J.C., E.M. Yahia, R.M. Beaudry, and Cedeño. 2006b. Modeling the influence of temperature and relative humidity on respiration rate of prickly pear cactus cladodes. *Postharvest Biol. Technol*. 41:260–265.

Guevara, J.C., E.M. Yahia, and E. Brito. 2001. Modified atmosphere packaging of prickly pear cactus stems (*Opuntia* spp.). *Lebensm. Wiss. U-Technol*. 34:445–451.

Guevara, J.C., E.M. Yahia, E. Brito, and S. Biserka. 2003. Effects of elevated concentrations of CO_2 in modified atmosphere packaging on the quality of prickly pear cactus stems (*Opuntia* spp). *Postharvest Biol. Technol*. 29:167–176.

Guevara, J.C., E.M. Yahia, L. Cedeño, and L.M.M. Tijskens. 2006a. Modeling the effects of temperature and relative humidity on gas exchange of prickly pear cactus (*Opuntia* spp.) stems. *Lebensm. Wiss. U-Technol*. 39:796–805.

Hagenmaier, R.D. 2000. Evaluation of a polyethylene–candelilla coating for 'Valencia' oranges. *Postharvest Biol. Technol*. 19:147–154.

Hagenmaier, R.D. 2002. The flavor of mandarin hybrids with different coatings. *Postharvest Biol. Technol*. 24:79–87.

Hagenmaier, R.D. and R.A. Baker. 1993. Reduction in gas exchange of citrus fruit by wax coatings. *J. Agric. Food Chem*. 41:283–287.

Harding, P.R. 1969. Effect of low oxygen and low carbon dioxide combination in controlled atmosphere storage of lemons, grapefruit and oranges. *Plant Dis. Rep*. 53:585–588.

Harjadi, S.S. and D. Tahitoe. 1992. The effects of plastic films bags at low temperature storage on prolonging the shelf life of rambutan (*Nephelium lappaceum*) cv. Levak bulus. *Acta Hort*. 321:778–785.

Harman, J.E. and B. McDonald. 1989. Controlled atmosphere storage of kiwifruit. Effect on fruit quality and composition. *Scientia Hort*. 37:303–315.

Hatton, T.T., Jr. and D.H. Spalding. 1974. Maintenance of market quality in Florida avocados. In: *ASHRAE Transactions*, Part. I, Vol. 80.

Hatton, T.T. and R.H. Cubbedge. 1977. Effects of prestorage carbon dioxide treatments and delayed storage on stem-end rind breakdown of 'Marsh' grapefruit. *HortScience*. 12:120–121.

Hatton, T.T., Jr. and W.F. Reeder. 1965. Controlled atmosphere storage of Lula avocados-1965 tests. *Proc. Carb. Reg. Am. Soc. Hort. Sci*. 9:152–159.

Hatton, T.T., Jr. and W.F. Reeder. 1969. Maintaining market quality of Florida avocado. In: *Proceedings of the Conference Tropical Subtropical Fruits*. Tropical Products Institute Conference, University of London, U.K.

Hatton, T.T., Jr. and W.F. Reeder. 1972. Quality of 'Lula' avocados stored in controlled atmospheres with and without ethylene. *J. Am. Soc. Hort. Sci*. 97:339–341.

Hatton, T.T., R.H. Cubbedge, and W. Grierson. 1975. Effects of prestorage carbon dioxide treatments and delayed storage on chilling injury of 'Marsh' grapefruit. *Proc. Fl. State Hort. Soc.* 88:335–338.

Hess-Pierce, B. and A.A. Kader. 2003. Responses of 'Wonderful' pomegranates to controlled atmospheres. *Acta Hort.* 600:751–757.

Holcroft, D.M., M.I. Gil, and A.A. Kader. 1998. Effect of carbon dioxide on anthocyanins, phenylalanine ammonia lyase and glucosyltransferase in the arils of stored pomegranates. *J. Am. Soc. Hort. Sci.* 123:136–140.

Inpun, A. 1984. Effect of temperature and packaging material (polyethylene bags, plastic baskets) on postharvest quality and storage life of rambutan (*Nephylium lappaceum* L.) Var. Seechompoo. (in Thai). *Monograph.* Department of Horticulture, Kasetsart University, Bangkok, Thailand.

Irving, D.E. 1992. High concentrations of carbon dioxide influence kiwifruit ripening. *Postharvest Biol. Technol.* 2:109–115.

Jiang, Y., X. Su, X. Duan, W. Lin, and Y. Li. 2004. Anoxia treatment for delaying skin browning, inhibiting disease development and maintaining the quality of litchi fruit. *Food Technnol. Biotechnol.* 42:131–134.

Jiang, Y., Z. Zhang, D.C. Joyce, and S. Ketsa. 2002. Postharvest biology and handling of longan fruit (*Dimocarpus longan* Lour.). *Postharvest Biol. Technol.* 26:241–252.

Jiang, Y.M. 1999. Low temperature and controlled atmosphere storage of fruit of longan (*Dimocarpus longan* Lour.). *Trop. Sci.* 39:98–101.

Jiang, Y.M. and J.R. Fu. 1999. Postharvest browning of litchi fruit by water loss and its prevention by controlled atmosphere storage at high relative humidity. *Lebensm.-Wiss. u.-Technol.* 32:278–283.

Jordan, R.A. and L.R. Barker. 1992. Controlled atmosphere storage of 'Hass' avocado. *HPG Biennial Rev. QDPI* 4–5.

Jordan, R.A. and L.G. Smith. 1993. The responses of avocado and mango to storage atmosphere composition. In: *Proceedings of the 6th CA Research Conference, Cornell University*, Ithaca, NY, pp. 629–638.

Kader, A.A. 1993. Modified and controlled atmosphere storage of tropical fruits. In: B.R. Champ, E. Highley, and G.I. Johnson (Eds.), *Postharvest Handling of Tropical Fruits. Proceedings of the International Conference Chiang Mai, Thailand*, July 19–23, 1993. ACIAR Proceeding No. 50, pp. 239–249.

Kader, A.A. 2003. A summary of CA requirements and recommendations for fruits other than apples and pears. *Acta Hort.* 600:737–740.

Kader, A.A. and M.L. Arpaia. 1992. Postharvest handling system: Subtropical fruits. In: A.A. Kader (Ed.), *Postharvest Technology of Horticultural Crops.* ANR Publications University of California, Davis, California, pp. 233–240.

Kader, A.A., Y. Li, and A. Chordas. 1982. Postharvest respiration, ethylene, and compositional changes of Chinese jujube fruits. *HortScience* 17:678–679.

Kader, A.A., G.D. Nanos, and E.L. Kerbel. 1990. Storage potential of fresh 'Manzanillo' olives. *Calif. Agric.* 40:23–24.

Ke, D. and A.A. Kader. 1990. Tolerance of 'Valencia' oranges to controlled atmospheres as determined by physiological responses and quality attributes. *J. Am. Soc. Hort. Sci.* 115:779–783.

Ke, D., E.M. Yahia, B. Hess, L. Zhou, and A. Kader. 1995. Regulation of fermentative metabolism in avocado fruit under oxygen and carbon dioxide stress. *J. Am. Soc. Hort. Sci.* 120:481–490.

Ketsa, S. and O. Klaewkasetkorn. 1995. Effect of modified atmosphere on chilling injury and storage life of rambutan. *Acta Hort.* 398:223–231.

Kupper, W., M. Pekmezci, and J. Henze. 1995. Studies on CA-storage of pomegranate (*Punica granatum* L., cv. Hicaz). *Acta Hort.* 398:101–108.

Lam, P.F. and K.H. Ng. 1982. Storage of waxed and unwaxed rambutan in perforated and sealed polyethylene bags. *Proceedings of the Workshop on Mango and Rambutan.* University of Philippines, Los Baños, pp. 190–210.

Lay-Yee, M. and D.C. Whiting. 1996. Response of 'Hayward' kiwifruit to high-temperature controlled atmosphere treatments for control of two-spotted spider mite (*Tetranychus urticae*). *Postharvest Biol. Technol.* 7:73–81.

Lee, S.K. and P.C. Leong. 1982. Storage studies on the rambutan in Singapore. *Proceedings of the Workshop on Mango and Rambutan.* University of Philippines, Los Baños, Philippines, pp. 172–175.

Li, P.X., G.X. Wang, L.S. Liang, and J.S. Fan. 2006. Effects of high-oxygen treatments on respiration intensity and quality of 'Dongzao' jujube during shelf-life. *Trans. Chin. Soc. Agric. Eng.* 22:180–183.

Liu, H., L. Song, Y. Jiang, D.C. Joyce, M. Zhao, Y. You, and Y. Wang. 2007. Short-term anoxia treatment maintains tissue energy levels and membrane integrity and inhibits browning of harvested litchi fruit. *J. Sci. Food Agric.* 87:1767–1771.

Lonsdale, J.H. 1993. Maintaining market quality of lychee overseas without SO_2. *South African Litchi Growers Association Yearbook*, Vol. 5, pp. 25–28.

Loulakakis, C.A., M. Hassan, D. Gerasopoulos, and A.K. Kanellis. 2006. Effects of low oxygen on in vitro translation products of poly(A) $^+$ RNA, cellulase and alcohol dehydrogenase expression in preclimacteric and ripening-initiated avocado fruit. *Postharvest Biol. Technol.* 39:29–37.

Luengwilai, K., K. Sukjamsai, and A.A. Kader. 2007. Responses of 'Clemenules Clementine' and 'W. Murcott' mandarins to low oxygen atmospheres. *Postharvest Biol. Technol.* 44:48–54.

Macfie, G.R. Jr. 1955a. Wrapping and packging of fresh lychee. *Florida Lychee Growers Association. Yearbook*, Vol. 1, pp. 25.

Macfie, G.R., Jr. 1955b. Packaging and storage of lychee fruits, preliminary experiments. *Florida Lychee Growers Association Yearbook*, Vol. 1, pp. 44.

Macfie, G.R. Jr. 1956. Packaging and handling of fresh lychee fruit. Florida *Lychee Growers Association. Yearbook*, Vol. 2, pp. 15.

Maestro, R., J.M. García, and J.M. Castellano. 1993. Changes in polyphenol content of olives stored in modified atmospheres. *HortScience* 28:749.

Mahajan, P.V. and T.K. Goswami. 2004. Extended storage life of litchi fruit using controlled atmosphere and low temperature. *J. Food Process. Preserv.* 28:388–403.

Marcellin, P. and A. Chavez. 1983. Effects of intermittent high CO2 treatment on storage life of avocado fruits in relation to respiration and ethylene production. *Acta Hort.* 138:155–163.

Martinez-Madrid, M.C., J.C. Martinez-Zamora, and F. Romojaro. 2001. Modified atmosphere packaging preserves sensorial and nutritional quality of *Ziziphus jujuba* Mill. cv. Li. *Acta Hort.* 553:613–614.

Matsuo, T. and S. Ito. 1977. On mechanism of removing astringency in persimmon fruits by carbon dioxide treatment. I. Some properties of the two processes in the de-astringency. *Plant Cell Physiol.* 18:17–25.

Maxie, E.C. 1964. Experiments on cold storage and controlled atmosphere. California Olive Association Annual *Technical Report* No. 43 pp. 12–15.

McDonald, B. and J.E. Harman. 1982. Controlled-atmosphere storage of kiwifruit. I. Effect on fruit firmness and storage life. *Scientia Hort.* 17:113–123.

McGuire, R.G. and G.J. Hallman. 1995. Coating guavas with cellulose- or carnauba-based emulsions interferes with postharvest ripening. *HortScience* 30:294–295.

McLauchalan, R.L., S.N. Ledger, and L.R. Barker. 1992. Simulated export of avocado by sea under controlled atmosphere. *HPG Biennial Rev.* 992.

Mendoza, Jr., D.B., ER.B. Pantastico, and F.B. Javier. 1972. Storage and handling of rambutan (*Nephellium lappaceum* L.). *Philippines Agric.* 55:322–332.

Mitcham, E.J., M.M. Attia, and W. Biasi. 1997. Tolerance of 'Fuyu' persimmons to low oxygen and high carbon dioxide atmospheres for insect disinfestation. *Postharvest Biol. Technol.* 10:155–160.

Mohamed, S. and E. Othman. 1988. Effect of packaging and modified atmosphere on the shelf life of rambutan (*Nephellium lappaceum*). *Pertanika* 11:217–228.

Mohamed, S., K.M.M. Kyi, and S. Yusof. 2006. Effects of various surface treatments on the storage life of guava (*Psidium guajava* L.) at 10°C. *J. Sci. Food Agric.* 66:9–11.

Nanda, S., D.V. Sudhakar Rao, and S. Krishnamurthy. 2001. Effects of shrink film wrapping and storage temperature on the shelf life and quality of pomegranate fruits cv. Ganesh. *Postharvest Biol. Technol.* 22:61–69.

Nanos, G.D., A.K. Kiritsakis, and E.M. Sfakiotakis. 2002. Preprocessing storage conditions for green 'Conservolea' and 'Chondrolia' table olives. *Postharvest Biol. Technol.* 25:109–115.

Navarro, S., J.E. Donahaye, M. Rindner, and A. Azrieli. 2001. Storage of dates under carbon dioxide atmosphere for quality. *Proceedings of the International Conference on Controlled Atmosphere and Fumigation in Stored Products*. Fresno, CA, October 29–November 3, 2000, pp. 231–239.

Nicolas, J., C. Rothan, and F. Duprat. 1989. Softening of kiwifruit in storage. Effects of intermittent high CO_2 treatments. Acta Hort. 258:185–192.

O'Hare, T.J. 1995. Postharvest physiology and storage of rambutan. Postharvest Biol. Technol. 6:189–199.

O'Hare, T.J., A. Prasad, and A.W. Cooke. 1994. Low temperature and controlled atmosphere storage of rambutan. Postharvest Biol. Technol. 4:147–157.

Oudit, D.D. and K.J. Scott. 1973. Storage of Hass avocados in polyethylene bags. Trop. Agric. (Trinidad) 50(3):241–243.

Overholser, E.L. 1928. Some limitations of gas storage of fruits. Ice Refrig. 74:551.

Pal, R.K. and R.W. Buescher. 1993. Respiration and ethylene evolution of certain fruits and vegetables in responses to CO_2 in controlled atmosphere storage. J. Food Sci. Technol. (Mysore) 30:29–32.

Pal, R.K., M.S. Ahmad, S.K. Roy, and M. Singh. 2004. Influence of storage environment, surface coating, and individual shrink wrapping on quality assurance of guava (Psidium guajava) fruits. Plant Foods Human Nutr. 59:67–72.

Pal, R.K., S.P. Singh, C.P. Singh, and R. Asrey. 2007. Response of guava fruit (Psidium guajava L. cv. Lucknow-49) to controlled atmosphere storage. Acta Hort. 735:547–554.

Palou, L., C.H. Crisosto, and D. Garner. 2007. Combination of postharvest antifungal chemical treatments and controlled atmosphere storage to control grey mold and improve storability of 'Wonderful' pomegranates. Postharvest Biol. Technol. 43:133–142.

Paull, R.E. and J.N. Chen. 1987. Effect of storage temperature and wrapping on quality characteristics of litchi fruit. Scientia Hort. 33:223–236.

Perez-Gago, M.B., C. Rojas, and M.A.D. Río. 2002. Effect of lipid type and amount of edible hydroxy-propyl methylcellulose-lipid composite coatings used to protect postharvest quality of mandarins cv. Fortune. J. Food Sci. 67:2903–2910.

Pesis, E. and R. Ben-Arie. 1984. Involvement of acetaldehyde and ethanol accumulation during induced deastringency of persimmon fruits. J. Food Sci. 49:896–899.

Pesis, E. and R. Ben-Arie. 1986. Carbon dioxide assimilation during postharvest removal of astringency from persimmon fruit. Physiol. Plant 67:644–648.

Pesis, E., O. Dvir, O. Feygenberg, R. Ben-Arie, M. Ackerman, and A. Lichter. 2002. Production of acetaldehyde and ethanol during maturation and modified atmosphere storage of litchi fruit. Postharvest Biol. Technol. 26:157–165.

Pesis, E., A. Levi, and R. Ben-Arie. 1988. Role of acetaldehyde production in the removal of astringency from persimmon fruits under various modified atmospheres. J. Food Sci. 53:153–156.

Pesis, E., R. Marinansky, G. Zauberman, and Y. Fuchs. 1993. Reduction of chilling I injury symptoms of stored avocado fruit by prestorage treatment with high nitrogen atmosphere. Acta Hort. 343:252–255.

Pesis, E., R. Marinansky, G. Zauberman, and Y. Fuchs. 1994. Prestorage low-oxygen atmosphere treatment reduces chilling injury symptoms in 'Fuerte' avocado fruit. HortScience 29:1042–1046.

Petracek, P.D., H. Dou, and S. Pao. 1998. The influence of applied waxes on postharvest physiological behavior and pitting of grapefruit. Postharvest Biol. Technol. 14:99–106.

Porat, R., B. Weiss, L. Cohen, A. Daus, and N. Aharoni. 2004. Reduction of postharvest rind disorders in citrus fruit by modified atmosphere packaging. Postharvest Biol. Technol. 33:35–43.

Porat, R., B. Weiss, L. Cohen, A. Daus, and A. Biton. 2005. Effects of polyethylene wax content and composition on taste, quality, and emission of off-flavor volatiles in 'Mor' mandarins. Postharvest Biol. Technol. 38:262–268.

Prusky, D., L. Karni, I. Kobiler, and Plumbey. 1991. Induction of the pre-formed antifungal diene in unripe avocado fruits: Effect of challenge inoculation by Colletotrichum gloesporoides. Physiol. Molecular Plant Pathol. 37:425–435.

Prusky, D., N.T. Keen, J.J. Sims, and S.L. Midland. 1982. Possible involvement of an antifungal diene in the latency of Colletotrichum gloeosporioides on unripe avocados. Phytopathology 72:1578–1582.

Prusky, D., I. Kobiler, R. Ardi, and Y. Fishman. 1993. Induction of resistance of avocado fruit to Colletotrichum gloesporoides attack using CO_2 treatments. Acta Hort. 343:325–330.

Prusky, D., I. Kobiler, and B. Jacoby. 1988. Involvement of epicatechin in cultivar susceptibility of avocado fruits to Colletotrichum gloeosporioides after harvest. J. Phytopathol. 123:140–146.

Prusky, D., A. Perez, Y. Zutkhi, and R. Ben-Arie. 1997. Effect of modified atmosphere for control of black spot, caused by *Alternaria alternata*, on stored persimmon fruits. *Phytopathology* 87:203–208.

Qin, G.Z. and S.P. Tian. 2004. Biocontrol of postharvest diseases of jujube fruit by *Cryptococcus laurentii* combined with a low dosage of fungicides under different storage conditions. *Plant Dis.* 88:497–501.

QiuPing, Z. and X. WenShui. 2007. Effect of 1-methylcyclopropene and/or chitosan coating treatments on storage life and quality maintenance of Indian jujube fruit. *Lebensm. Wiss. u-Technol.* 40:404–411.

Rivera, M. and E.M. Yahia. 1994. Calorimetría diferencial de barrido de frutas de aguacate y mango almacenadas en atmósferas insecticidas. *Tecnología de Alimentos (México)* 29(1):27.

Rivera, M., E.M. Yahia, and L. Vázquez. 1993. Response differences between avocado and mango fruit exposed to insecticidal modified atmospheres. *HortScience* 28:580.

Scott, K.J., B.I. Brown, G.R. Chaplin, M.E. Wilcox, and M. Bain. 1982. The control of rotting and browning of litchi fruit by hot benomyl and plastic film. *Scientia Hort.* 16:253–262.

Seberry, J.A. and E.G. Hall. 1970. C.A. storage of citrus fruit. *Food Preserv. Q.* 30:41–43.

Seubrach, P., S. Photchanachai, V. Srilaong, and S. Kanlayanarat. 2006. Effect of modified atmosphere by PVC and LLDPE film on quality of longan fruits (*Dimocarpus longan* Lour) cv. 'Daw.' *Acta Hort.* 712:605–610.

Sharkey, P.J., C.R. Little, and I.R. Thornton. 1985. Effects of low density polyethylene liners and high-density polyethylene wraps on quality, decay and storage life of lemon and tangor fruits. *Austr. J. Expt. Agric.* 25:718–721.

Shellie, K.C. 2002. Ultra-low oxygen refrigerated storage of 'Rio Red' grapefruit: Fungistatic activity and fruit quality. *Postharvest Biol. Technol.* 25:73–85.

Shellie, K.C., R.L. Mangan, and S.J. Ingle. 1997. Tolerance of grapefruit and Mexican fruit fly larvae to heated controlled atmospheres. *Postharvest Biol. Technol.* 10:179–186.

Shi, J.X., E.E. Goldschmidt, R. Goren, and R. Porat. 2007a. Molecular, biochemical and anatomical factors governing ethanol fermentation metabolism and accumulation of off-flavors in mandarins and grapefruit. *Postharvest Biol. Technol.* 46:242–251.

Shi, J.X., R. Porat, R. Goren, and E.E. Goldschmidt. 2005. Physiological responses of 'Murcott' mandarins and 'Star Ruby' grapefruit to anaerobic stress conditions and their relation to fruit taste, quality and emission of off-flavor volatiles. *Postharvest Biol. Technol.* 38:99–105.

Shi, J.X., J. Riov, R. Goren, and E.E. Goldschmidt. 2007b. Regulatory aspects of ethanol fermentation in immature and mature citrus fruit. *J. Am. Soc. Hort. Sci.* 132:126–133.

Shomer, I., H. Borochov-Neori, B. Luzki, and U. Merin. 1998. Morphological, structural and membrane changes in frozen tissues of Madjhoul date (*Phoenix dactylifera* L.) fruits. *Postharvest Biol. Technol.* 14:207–215.

Singh, R. 1957. Improvement of packaging and storage of litchi at room temperature. *Ind. J. Hort.* 14:205.

Singh, S.P. and R.K. Pal. 2007. Postharvest fruit fly disinfestation strategies in rainy season guava crop. *Acta Hort.* 735:591–596.

Singh, S.P. and R.K. Pal. 2008a. Controlled atmosphere storage of guava (*Psidium guajava* L.) fruit. *Postharvest Biol. Technol.*, 47:296–306.

Singh, S.P. and R.K. Pal. 2008b. Response of climacteric-type guava (*Psidium guajava* L.) to postharvest treatment with 1-MCP. *Postharvest Biol. Technol.*, 47:307–314.

Sive, A. and D. Resinsky. 1989. Storage as a solution to a number of problems in the avocado industry (in Hebrew). *Alon Hanotea* 44(1):39–46.

Spalding, D.H. 1977. Low pressure (hypobaric) storage of avocados, limes, and mangos. In: D.H. Dewy (Ed.), *Controlled Atmospheres for the Storage and Transport of Perishable Agricultural Commodities*. Horticultural. Report No. 28. Michigan State University, East Lansing, MI, pp. 156–164.

Spalding, D.H. and W.F. Reeder. 1972. Quality of 'Booth 8' and 'Lula' avocados stored in a controlled atmosphere. *Proc. Fla. State Hort. Soc.* 85:337–341.

Spalding, D.H. and W.F. Reeder. 1974a. Current status of controlled atmosphere storage of four tropical fruits. *Proc. Fla. State Hort. Soc.* 87:334–337.

Spalding, D.H. and W.F. Reeder. 1974b. Quality of 'Tahiti' limes stored in a controlled atmosphere or under low pressure. *Proc. Trop. Reg. Am. Soc. Hort. Sci.* 18:128–135.

Spalding, D.H. and W.F. Reeder. 1975. Low oxygen and high carbon dioxide controlled atmosphere storage for control of anthracnose and chilling injury avocados. *Phytopathology.* 65:458–468.

Spalding, D.H. and W.F. Reeder. 1976a. Low pressure (hypobaric) storage of avocados. *HortScience* 11:491–492.

Spalding, D.H. and W.F. Reeder. 1976b. Low pressure (hypobaric) storage of limes. *J. Am. Soc. Hort. Sci.* 101:367–370.

Stahl, A.L. and J.C. Cain. 1940. Storage and preservation of miscellaneous fruits and vegetables. Florida Agricultural State Annual Report, p. 88.

Stavroulakis, G. and E. Sfakiotakis. 1997. Regulation of propylene-induced ripening and ethylene biosynthesis by oxygen in 'Hayward' kiwifruit. *Postharvest Biol. Technol.* 10:189–194.

Su, X.G., Y.M. Jiang, X.W. Duan, H. Liu, Y.B. Li, W.B. Lin, and Y.H. Zheng. 2005. Effects of pure oxygen on the rate of skin browning and energy status in longan fruit. *Food Technnol. Biotechnol.* 43:359–365.

Tanaka, Y., N. Takase, and J. Sato. 1971. Studies on the CA-storage of fruits and vegetables. III. Effect of CA-storage on the quality of persimmons. *Res. Bull. Aichi-ken Agric. Res. Cent.* 3:100–106.

Techavuthiporn, C., W. Niyomlao, and S. Kanlayanarat. 2006a. Low oxygen storage of litchi cv. 'Hong Huay.' *Acta Hort.* 712:623–627.

Techavuthiporn, C., W. Niyomlao, and S. Kanlayanarat. 2006b. Superatmospheric oxygen retards pericarp browning of litchi cv. 'Hong Huay.' *Acta Hort.* 712:629–634.

Thompson, B.D. 1955. The effect of pre-packaging and cold storage on the quality of fresh lychees. *Florida Lychee Growers Association Yearbook*, Vol. 1, pp. 22.

Thompson, A.K., Y. Magzoub, and H. Silvis. 1974. Preliminary investigations into desiccation and degreening of limes for export. *Sudan J. Food Sci.Technol.* 6:1–6.

Tian, S.P., B. Li, and Y. Xu. 2005. Effects of O_2 and CO_2 concentrations on physiology and quality of litchi fruit in storage. *Food Chem.* 91:659–663.

Tian, S.P., Y. Xu, A. Jiang, and Q. Gong. 2002a. Physiological and quality responses of longan fruit to high O_2 or high CO_2 atmospheres in storage. *Postharvest Biol. Technol.* 24:335–340.

Tian, S.P., Y. Xu, Q.Q. Gong, A.L. Jiang, Y. Wang, and Q. Fan. 2002b. Effects of controlled atmospheres on physiological properties and storability of longan fruit. *Acta Hort.* 575:659–665.

Truter, A.B. and G.J. Ecksteen. 1987a. Controlled and modified atmospheres to extend storage life of avocados. *South African Avocado Grower's Association Yearbook*, Vol. 10, pp. 151–153.

Truter, A.B. and G.J. Ecksteen. 1987b. Controlled atmosphere storage of avocados and bananas in South Africa. *Frigair '86* 3(20):5.

Tsantili, E., G. Karaiskos, and C. Pontikis. 2003. Storage of fresh figs in low oxygen atmosphere. *J. Hort. Sci. Biotechnol.* 78:56–60.

Türk, R., A. Eris, M.H. Özer, E. Tuncelli, and J. Henze. 1994. Research on the CA storage of fig cv. Bursa Siyahi. *Acta Hort.* 368:830–839.

Vakis, N., W. Grierson, and J. Soule. 1970. Chilling injury in tropical and subtropical fruits III. The role of CO_2 in suppressing chilling injury of grapefruit and avocados. *Proc. Trop. Reg. Am. Soc. Hort. Sci.* 14:89–100.

Vangnai, T., C. Wongs-Aree, H. Nimitkeatkai, and S. Kanlayanarat. 2006. Quality maintaining of 'Daw' longan using chitosan coating. *Acta Hort.* 712:599–604.

Vilasachandran, T., S.A. Sargent, and F. Maul. 1997. Controlled atmosphere storage shows potential for maintaining postharvest quality of fresh litchi fruits. *Proceedings of the 7th International CA Research Conference.* Davis, CA, July, 12–18, 1997.

Wang, G.X., P.X. Li, L.S. Liang, and J.S. Fan. 2006. Effect of postharvest high-oxygen treatments on 'Dongzao' jujube membrane lipids peroxidation and defensive enzyme activities during shelf-life. *Acta Hort. Sinica* 33:609–612.

Wild, B.L., W.B. McGlasson, and T.H. Lee. 1977a. Ethylene in CA long term lemon storage. *Proceedings of the International Society of Citriculture.* Lake Alfred, FL, pp. 259–263.

Wild, B.L., W.B. McGlasson, and T.H. Lee. 1977b. Long term storage of lemon fruit. *Food Technol. Aust.* 29:351–357.

Woskow, M. and E.C. Maxie. 1965. Cold storage studies with olives. California Olive Association Annual Technical Report No. 44, pp. 6–11.

Yahia, E.M. 1993. Responses of some tropical fruits to insecticidal atmospheres. *Acta Hort.* 343:371–376.

Yahia, E.M. 1997a. Avocado and guava fruits are sensitive to insecticidal MA and/or heat. In: J.F. Thompson and E.J. Mitcham (Eds.), *CA Technology and Disinfestations Studies*, Vol. 1 *Proceedings of the CA Research Conference*, University of California, Davis, CA, July, 13–18, 1997, pp. 132–136.

Yahia, E.M. 1997b. Modified/controlled atmospheres for avocado (*Persea americana* Mill). In: A.A. Kader (Ed.), *Fruits Other Than Apples and Pears*, Vol. 3 *Proceedings of the CA Research Conference*, University of California, Davis, CA, July 13–18, 1997, pp. 97–103.

Yahia, E.M. 1998. Modified and controlled atmospheres for tropical fruits. *Hort. Rev.* 22:123–183.

Yahia, E.M. 2003. Manejo postcosecha del aguacate (postharvest handling of avocado), 2ª. Parte. Boletín Informativo de APROAM El Aguacatero, Año 6, Número 32, Mayo de 2003, Mexico.

Yahia, E.M. 2004. Dates. In: *The Commercial Storage of Fruits, Vegetables, and Florist and Nursery Stocks.* USDA Agricultural Research Services, Agricultural Handbook No. 66.

Yahia, E.M. 2008. Prickly pear cactus cladodes (nopal) and fruit. In: D. Rees (Eds.), *Crop Postharvest: Science and Technology. Volume 3: Perishables.* Blackwell Publishing. In press.

Yahia, E.M. and A. Carrillo-López. 1993. Tolerance and responses of avocado fruit to insecticidal O_2 and CO_2 atmospheres. *Lebensm. Wiss. U-Technol.* 26:42–48.

Yahia, E.M. and G. Gonzalez. 1998. The use of passive and semi-active atmospheres to prolong the postharvest life of the fruit of avocado. *Lebensm. Wiss. u Technol.* 31(7 and 8):602–606.

Yahia, E.M. and A.A. Kader. 1991. Physiological and biochemical responses of avocado fruits to O_2 and CO_2 stress. *Plant Physiol.* 96 (Suppl.):96 (Abstract).

Yahia, E.M. and M. Rivera. 1994. Differential scanning colorimetry (DSC) of avocado and mango fruits stored in an insecticidal atmosphere. *HortScience* 29(5):448.

Yahia, E.M., J. Guevara, L.M.M. Tijskens, and L. Cedeño. 2005. The effect of relative humidity on modified atmosphere packaging gas exchange. *Acta Hort.* 674:97–100.

York, G.M. 1995. An evaluation of two experimental polysaccharide nature seal coating in delaying the postharvest browning of the lychee pericarp. *Proc. Fla. State Hort. Soc.* 107:350–351.

Young, R.E., R.J. Romani, and J.B. Biale. 1962. Carbon dioxide effects of fruits respiration. II. Response of avocados, bananas and lemons. *Plant Physiol.* 37:416–422.

Zhang, D. and P.C. Quantick. 1997. Preliminary study on effects of modified atmosphere packaging on postharvest storage of longan fruit. *Postharvest Hort. Ser. Dept. Pomol., University of California, Davis, USA* 17:90–96.

Zhang, Y., J. Han, and R. Zhang. 2005. Study on fresh-keeping physiological activity of Dong jujube using low temperature combined with hypobaric and ozone treatments. *Scientia Agric. Sinica* 38:2102–2110.

Zhou, Y., Z.L. Ji, and W.Z. Lin. 1997. Study on the optimum storage temperature and chilling injury mechanism of longan fruit (Chinese). *Acta Hort. Sinica* 24:13–18.

15

Small Fruit and Berries

Leon A. Terry, Carlos H. Crisosto, and Charles F. Forney

CONTENTS

15.1   Introduction

Small fruits and berries, such as grape (*Vitis* spp.), strawberry (*Fragaria ananassa*), raspberry (*Rubus idaeus*), blueberry and cranberry (*Vaccinium* spp.), and kiwifruit (*Actinidia chinensis*), are not only prized for their unique appearance, flavor, taste and texture, but also for their reported health-promoting properties. Small fruit are, however, characterized by a relatively short postharvest life that limits their marketing. This limitation is related to most small fruit being inherently prone to postharvest disease and general lack of firm texture. In addition, most berries are regarded as nonclimacteric in their postharvest behavior. These attributes alone pose significant and particular difficulties for the postharvest maintenance of quality in terms of handling, transportation, distribution, and shelf life. In addition to maintaining cold storage, proper modifications of oxygen, carbon dioxide, and other gases in storage or inside packaging may extend postharvest life and maintain nutritional value. Thus, the use of controlled atmosphere (CA) or modified atmosphere packaging (MAP) technologies have benefits on extending postharvest life of grapes, strawberries, and other berry fruit. As responses to atmospheric modification vary greatly among species and cultivars, it is very important to understand the specific conditions and limitations of these techniques on the supply chains of each small fruit type. Specific information on the effects of CA conditions on disease incidence, nutritional values, texture, color, taste, and postharvest life will be discussed in this chapter. Recommendations on ideal atmosphere composition for maximum benefits for each commodity will be discussed.

15.2   Research Review on CA and MAP

Due to the difficulties associated with rapid postharvest deterioration of many small fruits and berries, research on the use and applicability of CA has been conducted for over 50 years. Extensive research has really only been conducted over the last 20 years. Most work has concentrated on employing CA and/or MAP to extend postharvest life of strawberry and tables grapes, since these economically important fruits are especially prone to rapid deterioration in postharvest quality, caused, in part, by their susceptibility to infection by *Botrytis cinerea* at flowering and subsequent gray mold disease after a period of quiescence (Terry et al., 2004). CA and/or MAP storage have been shown to significantly suppress decay and maintain quality of berry fruit.

Berry fruit, unlike many fresh produce types, are relatively tolerant of elevated carbon dioxide (CO_2) concentrations and/or low or superatmospheric oxygen (O_2). CA has been shown to increase storage life through decreasing respiration rate, inhibition of gray mold development, and improved retention of fruit firmness. Nonetheless, incorrect control of CO_2 and O_2 concentrations or overextended CA storage can result in deleterious effects on color, appearance, texture, flavor-, taste-, and health-related compounds/parameters.

15.3   Response of Specific Berry Fruits to CA and MAP

15.3.1   Strawberry

15.3.1.1   *Disease Development*

Strawberries (*Fragasia* x *ananassa* Duch.) are highly perishable and especially prone to postharvest disease, principally caused by *B. cinerea*, *Rhizopus stolonifer*, and *Mucor*

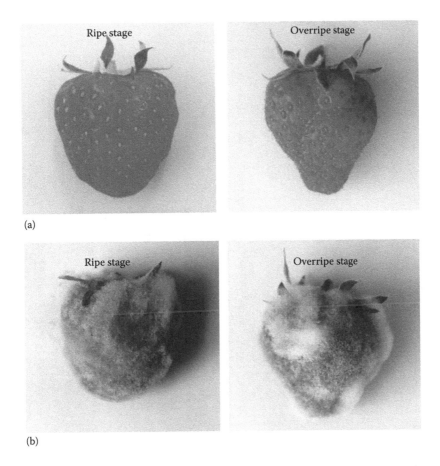

FIGURE 15.1
Example of gray mold disease caused by *B. cinerea* on ripe stage (28 days after anthesis [DAA]) and overripe stage (35 DAA) 'Elsanta' strawberry fruit stored in the dark at 5°C: (a) healthy fruit and (b) fruit showing 100% disease severity where mass of conidia are clearly seen.

spp. *B. cinerea* tends to infect strawberry flowers at anthesis after which it enters a quiescent phase. Gray mold on strawberry fruit caused by *B. cinerea* is almost inevitable when fruit are stored inappropriately (Terry et al., 2004; Figure 15.1). Whilst *Rhizopus* and *Mucor* spp., which can infect strawberry, are generally controlled by maintaining fruit at low temperature (<5°C), *B. cinerea* is able to grow, albeit slowly, at ~0°C, and thus still presents a serious problem for postharvest storage of strawberry fruit. Sulfur dioxide (SO_2) treatment is not appropriate for strawberry fruit. Accordingly, additional treatments to refrigerated storage, such as CA or MAP, have been actively sought for long-distance transport of strawberry to slow down the respiration rate of both the fruit and pathogen, and thus extend postharvest life (Table 15.1). High CO_2 or low or superatmospheric concentrations are fungistatic to *B. cinerea* and hence can inhibit conidial germination, germ-tube elongation, and mycelial growth. Garcia-Gimeno et al. (2002) demonstrated that *B. cinerea* could not grow above 30 kPa CO_2. Wszelaki and Mitcham (2000) showed that in vitro radial growth rate of *B. cinerea* on potato dextrose agar was also reduced at 40 kPa O_2. A combined treatment of 15 kPa CO_2 and 40 kPa O_2 was most effective at suppressing in vitro mycelial growth. Upon removal from CA conditions, *B. cinerea* resumed growth, showing that CA treatment was only fungistatic since no residual inhibition of growth was observed (Wszelaki and Mitcham, 2000).

TABLE 15.1

Examples of Effects of Postharvest CA Storage on Strawberry Fruit Quality

Cultivar	CO₂ Concentration (kPa)	O₂ Concentration (kPa)	Storage Temperature (°C)	Effects on Fruit Quality during Storage	Reference
Chandler	CO₂-enriched atmosphere (ns)	>20	1.7	⇓ Postharvest decay (*B. cinerea*)	Woodward and Topping (1972)
Cambridge Favorite	—	3	0–20	⇓ Respiration rate	Robinson et al. (1975)
Elsanta	20	—	0	⇓ Respiration rate, postharvest decay	Hansen (1986)
Elvira	20	—	24 for 1 day	⇓ Postharvest decay; ⇔ Fruit and calyx color	Baab (1987)
—	15–20	5–10	0–5	⇑ Off-flavors but recovery 6 h after CO₂ treatment; ⇓ Respiration rate, postharvest decay; ⇑ firmness retention	Kader (1989)
Redcoat; 21^A of 25 cvs. studied	<20	<2	0–5	⇑ Off-flavors, ⇑ berry discoloration	Smith (1992); Smith and Skog (1992)
	15	18.5	0 for 42 h	⇑ Firmness, postharvest decay; ⇔ L*, H°, TSS, or pH	
Chandler	50	0.25 or 21	5 for 1–7 days	⇑ Acetylaldehyde, ethanol, ethyl acetate, ethyl butyrate concentrations and pyruvate decarboxylase, alcohol dehydrogenase activities; ⇓ Isopropyl acetate, propyl acetate, butyl acetate concentrations and alcohol acetyl transferase activity,	Ke et al. (1994)
Chandler	20	10	4	⇑ pH, a*, C*, H°; ⇓ L*	Nunes et al. (1995)
—	20	Air	2	⇑ Acetylaldehyde, ethanol and ethyl acetate	Watkins et al. (1999)
Selva	20	16	5	⇑ pH, succinic acid concentration, H°; ⇓ Postharvest decay (*B. cinerea*), loss in firmness, TA (citric and malic acid concentrations), anthocyanin accumulation, internal flesh L*, C*; ⇔ Ascorbic acid, sucrose, fructose, glucose concentration	Holcroft and Kader (1999a,b)
Camarosa	—	100	5 for 14 days	⇓ Postharvest decay (*B. cinerea*), TSS; ⇔ Ethylene production, firmness, TA; ⇑ Acetylaldehyde, ethanol and ethyl acetate, respiration rate, H°, L*	Wszelaki and Mitcham (2000)
Honeoye, Korona	9–12^MAP	11–14^MAP	5	⇓ Postharvest decay	Nielsen and Leufvén (2008)

Note: ⇓, decrease; ⇑, increase; ⇔, no effect; C*, chroma; H° [= arctan b*/a*], hue angle; TA, titratable acidity; TSS, total soluble solids; ns, not specified; MAP, modified atmosphere packaging; ^A = Allstar, Dana, Glooscap, Governor Simcoe, Guardian, GU62e55, Kent, Micmac, Midway, Pajaro, Raritan, Redcoat, Selva, Settler, Sparkle, Tribute, Tristar, Vesper, Vibrant, Selkirk, Scotland.

A large number of publications have shown that CA or MAP can reduce postharvest decay in numerous strawberry cultivars (e.g., Couey and Wells, 1970; Browne et al., 1984; Ke et al., 1991; Smith, 1992; Chambroy et al., 1993; Goto et al., 1995; Nunes et al., 2002; Wszelaki and Mitcham, 2003; Zhang and Watkins, 2005). CA also slows down strawberry respiration and senescence, and thus may maintain natural disease resistance (NDR) of strawberry fruit by retaining endogenous antifungal compounds and/or the capability of fruit to produce these phytoalexins (Terry and Joyce, 2004; Terry et al., 2004), thus delaying the development of disease symptoms. Early work by Woodward and Topping (1972) reported that gray mold could be reduced by storing 'Chandler' strawberry fruit at 1.7°C in a CO_2-enriched atmosphere, however partial pressures >20 kPa CO_2 were injurious. Consensus suggests that a combination of 15–20 kPa CO_2 and 5–10 kPa O_2 is the ideal partial pressure mix for successful reduction of gray mold disease incidence and severity on strawberry fruit when held at 0°C–5°C. However, the wealth of research describing the effects of CA and MAP on strawberry demonstrates that no ubiquitous CA combination has been found that is applicable to all cultivars, growing conditions, and storage scenarios. Slightly different ratios between CO_2 and O_2 have been reported for relatives of the commercial strawberry. Almenar et al. (2006) showed that 10 kPa CO_2 and 11 kPa O_2 were efficacious at prolonging shelf life and quality of 'Reina de los Valles' wild strawberry fruit.

In addition to conventional CA storage (i.e., high CO_2 and low O_2), other gaseous combinations such as superatmospheric oxygen storage (Wszelaki and Mitcham, 2000; Stewart, 2003; Allende et al., 2007; Zheng et al., 2007, 2008), ozone (Peréz et al., 1999), and carbon monoxide (CO; El-Kazzaz et al., 1983) have all been shown to extend postharvest life of strawberry fruit. Recent research by Zheng et al. (2008) demonstrated that 60–100 kPa O_2 in a storage environment could reduce postharvest decay on 'Fengxiang' strawberry fruit. It is, however, unclear how superatmospheric O_2 storage could be used commercially due to concerns over safety, as the potential dangers of using flammable gas at high concentrations are often overlooked. Fumigation in an anaerobic nitrogen atmosphere with nitric oxide (NO; 5–10 μL L^{-1}) for up to 2 h at 20°C doubled the postharvest life of 'Pajaro' strawberry fruit subsequently stored at 5°C (Wills et al., 2000). The effects of this treatment may be transitory in a storage environment as NO is rapidly oxidized to NO_2 in the presence of oxygen.

15.3.1.2 *Fruit Biochemistry*

When used appropriately, CA can preserve strawberry fruit quality by maintaining the concentration and integrity of both flavor- and taste-related compounds (Siro et al., 2006; Zheng et al., 2007). The time period to which CA storage is effective is not only dependent on storage temperature and the correct selection of gas concentrations, but also on genotype, preharvest factors, and management. Nunes et al. (1995) reported that CA (5 kPa O_2 and 15 kPa CO_2) maintained 'Chandler' strawberry fruit quality better than air-stored fruit at both 4°C and 10°C.

Incorrect CA regimes (i.e., >30 kPa CO_2 or <5 kPa O_2) can increase production of off-flavors (e.g., ethanol, acetaldehyde, ethyl acetate), cause berry discoloration (e.g., bleaching; loss of redness), and decrease total titratable acidity (Table 15.1). Holcroft and Kader (1999a) showed that storing 'Selva' strawberry fruit at 16 kPa CO_2 and 16 kPa O_2, not only extended postharvest life, but also maintained nonstructural carbohydrate concentrations (viz. fructose, glucose, and sucrose). However, the main nonvolatile organic acids, citric and malic acid, were reduced. Several authors have demonstrated that an elevated CO_2 atmosphere can help retain the levels of health-related constituents, including anthocyanins, ascorbate (Kader et al., 1989; Holcroft and Kader, 1999a), and phenolic compounds (Pérez and Sanz, 2001). Palmer Wright and Kader (1997) showed that CA had no effect on ascorbate levels in sliced 'Selva' strawberry fruit. In contrast, others have reported that both conventional CO_2-enriched atmospheres and superatmospheric O_2 can reduce the concentration of some phenylpropanoids (Gil et al., 1997) and ascorbate (Agar et al., 1997). 'Camarosa' strawberry fruit stored in 10, 20, or 40 kPa CO_2 at 5°C had lower levels of ellagic acid after 10 days as compared to air-stored fruit even though increased contents of quercetin and kaempferol derivatives were observed for fruit stored in air or CO_2-enriched atmospheres (Gil et al., 1997). 'Camarosa' strawberry fruit stored in 10 kPa CO_2 balanced with N_2 or elevated oxygen (80 kPa O_2) had lower concentrations of total phenolics and ascorbate after 5 and 12 days storage, respectively, when compared to air-stored fruit (Allende et al., 2007). Moreover, concentrations of ellagitannins and procyanidins during the first 9 days of storage were reduced at 80 kPa O_2. In contrast, Zheng et al. (2007) reported the storage of 'Allstar' strawberry fruit in 60–100 kPa O_2 at 5°C increased the accumulation of total phenolics, total anthocyanins, individual phenolics, and antioxidant capacity, as measured using the standard oxygen radical absorbance capacity (ORAC) assay, during the initial 7 days storage. However, this increase was only transitory as no difference in these health-promoting compounds was observed between CA and control fruit after 14 days storage. Pelayo et al. (2003), however, found no effect of storage in 20 kPa CO_2 on the level of total phenolics in 'Aromas,' 'Diamente,' and 'Selva' strawberry fruit. Apart from the reported health-promoting properties of phenylpropanoids, it is possible that the level of phenolic compounds in strawberry fruit as affected by postharvest abiotic treatments may affect natural disease resistance (NDR) of the fruit and, thus the susceptibility of the fruit to infection (Terry et al., 2004).

Storage in superatmospheric O_2 (60–100 kPa) has been shown to extend the storage life of strawberry fruit (Wszelaki and Mitcham, 2000; Zheng et al., 2007, 2008), however, it is questionable whether this treatment will be commercially viable in the near future as it can lead to increased production of fermentation-derived metabolites that negatively affect fruit organoleptic (Wszelaki and Mitcham, 2000) and nutritional quality (Allende et al., 2007). Berna et al. (2007) reported that the production of ethyl acetate, as measured using headspace gas chromatography–mass spectrometry, was reduced in 'Elsanta' strawberry fruit after 4 and 7 days of storage under superatmospheric O_2 (without CO_2).

15.3.1.3 Quality Attributes

The quality of strawberry fruit is fundamentally based on visual appearance (including sepal condition), texture, organoleptic qualities, and nutritional value. The supposed effectiveness of CA is dependent on whether postharvest life is based on appearance or the composition of flavor components and other quality attributes. Pelayo et al. (2003) highlighted this difficulty by showing that postharvest life based on appearance for 'Aromas,' 'Diamente,' and 'Selva' fruits stored in air was extended by 2–4 days when stored at 20 kPa CO_2. However, when postharvest life was based on maintenance of acceptable levels of flavor- and taste-related components (sugars, organic acids, and aroma compounds) and sensory evaluation, postharvest life was reduced (albeit longer in CA-treated fruit) as compared to appearance-based life. Together with genotype and storage temperature, the judgment of when postharvest life is over is critical and may have added to the confusion about what are the ideal or recommended CA conditions required to maintain strawberry fruit quality. Moreover, ripeness at harvest also affects the efficacy of CA since Nunes et al. (1995) reported that three-quarter colored 'Chandler' fruit responded better to CA than fully red fruit, maintaining better color and firmness over 2 weeks storage at 4°C.

Despite strong evidence that CA storage reduces postharvest decay, too high CO_2 and/or too low O_2 atmospheres have been shown to negatively affect organoleptic properties of strawberry fruit by increasing levels of fermative volatiles (viz. ethyl acetate, ethanol, and acetylaldehyde)/off-flavors (Couey and Wells, 1970; Shaw, 1970; Harris and Harvey, 1973; Browne et al., 1984; Ke et al., 1991, 1994; Larsen and Watkins, 1995; Watkins et al., 1999; Wszelaki and Mitcham, 2000; Pérez and Sanz, 2001; Van der Steen et al., 2002, Pelayo et al., 2003). Guichard et al. (1992) showed that <20 kPa CO_2 had no ill effect on aroma of strawberry fruit, however, accumulation of off-flavors is known to be cultivar-dependent (Watkins et al., 1999). Depending on cultivar, storage temperature, and specific gaseous composition of the storage environment, CA has been shown to either increase (Kader, 1989; Smith, 1992; Smith and Skog, 1992; Larsen and Watkins, 1995; Watkins et al., 1999; Wszelaki and Mitcham, 2000; Van der Steen et al., 2002; Pelayo et al., 2003), reduce (Holcroft and Kader, 1999a), or have no effect (Wszelaki and Mitcham, 2000) on strawberry fruit firmness. Zhang and Watkins (2005) demonstrated that 20 kPa CO_2 enhanced 'Jewel' strawberry fruit firmness at 2°C, but not at 20°C. Genotypic differences in the response of strawberry to CA were shown by Pelayo et al. (2003) who reported that flesh firmness was maintained in 'Aroma' and increased in 'Diamente' and 'Selva' strawberries when held at 5°C in both air and 20 kPa CO_2. Since high CO_2 concentrations have been shown to increase apoplastic pH of strawberry fruit tissue during storage (Nunes et al., 1995), it is believed that this may lead to greater cell-to-cell adhesion, precipitation of soluble pectins, and greater firmness retention (Harker et al., 2000). Work by Benitez-Burraco et al. (2003) stated that expression of pectin lyase genes, isolated and characterized from 'Chandler' fruit, and involved in cell wall disassembly and fruit softening were reduced in fruit held at high CO_2 concentration.

CA storage may reduce the rate of accumulation of anthocyanins (Zheng et al., 2007), normally seen after harvest, and has been shown to have variable effects on external and internal fruit coloration (Gil et al., 1997; Holcroft and Kader, 1999b). Smith and Skog (1992) demonstrated that 15 kPa CO_2 had no effect on objective color or pH for 25 strawberry cultivars. Nonetheless, Gil et al. (1997) showed that CO_2-enriched air had minimal effect on anthocyanin content of external tissues, but dramatically reduced internal color of cortex tissue in 'Selva' strawberry fruit. It is likely that increases in cellular pH commonly associated with CA may alter the stability, copigmentation, and spectra of the anthocya-nins found in strawberry fruit. Specifically, pelargonidin 3-glucoside, which is the main anthocyanin found in strawberry fruit, usually increases after harvest, but was reduced in internal tissue from fruit stored under high CO_2 concentrations (Gil et al., 1997). Cyanidin 3-glucoside was not found in internal tissue in either air- or CO_2-stored fruit, but was found in external tissue and appeared to be resistant to CO_2-induced degradation. Straw-berry cultivars are known to differ radically in their degree of both external and internal redness (Sacks and Shaw, 1994), and hence it is likely that CA may have differential effects on anthocyanin profiles according to genotype.

15.3.1.4 *Effects of MAP*

In general, strawberry fruit are still predominately sold in open or vented punnets, thus making effective MAP impossible. However, shrouded pallets are being increasingly used commercially, whereby stacked pallets are wrapped with plastic film, creating a transient modified atmosphere during transit. Research has shown that MAP may be a viable option for extending shelf life of strawberry fruit for short periods (Ferreira et al., 1994; Sanz et al., 1999; Allende et al., 2007; Almenar et al., 2007; Nielsen and Leufvén, 2008). A greater understanding of the underlying interactions between strawberry fruit, gaseous composi-tion, temperature and disease may lead to optimum packaging being selected (Hertog et al., 1999). Incorporation of an antifungal volatile compound naturally found in straw-berry fruit (2-nonanone) with MAP has been reported to prolong shelf life of 'Reina de los

Valles' wild strawberry fruit by suppressing *B. cinerea* (Almenar et al., 2007). A modified atmosphere environment can also be created using edible coatings by controlling respiratory gas exchange and ongoing research continues in this area (Han et al., 2004).

15.3.2 Grape

Several CA conditions including CO_2 and O_2 combinations have been studied on different table grape (*Vitis vinifera* L.) cultivars during cold storage with the main objective to control *B. cinerea* development (Figure 15.2). One of the problems associated with the ubiquitous use of SO_2 fumigation of grapes is the constant potential for injury to the berries and rachis. Injured tissue first shows bleaching of color, followed by sunken areas where accelerated water loss has occurred. These injuries first appear on the berry where some other injury has occurred, such as a harvest wound, transit injury, or breakage at the cap peduncle attachment. Symptoms may also be seen around the cap stem and slowly spread over the berry. Additional problems with SO_2 fumigation of grapes or slow release of SO_2 using metabisulfite generator pads are berry shatter (detachment of berries from the rachis) and the level of sulfite residue remaining at time of final sale. Considerable research has been conducted to find alternatives to SO_2 treatments for gray mold control. This is especially desirable for organic table grape growers who are prohibited from using SO_2. Despite many alternative treatments showing promise in the laboratory, none are being used routinely in a commercial setting. Limitations include phytotoxicity and difficulties in getting good penetration of the treatment through the grape bunches. Among them high CO_2 is one that has showed promise under some conditions. Earlier, CA research was carried out on 'Ribier' and 'Razaki' (Eris et al., 1993), South African cultivars (Laszio, 1985), 'Emperor' (Uota, 1957), and some Italian cultivars (Cimino et al., 1987). Detailed studies have been done on 'Thompson Seedless' (Nelson, 1969; Yahia et al., 1983; Berry and Aked, 1996, 1997, and 1998; Crisosto et al. 2002a,b). These studies concluded that the use

FIGURE 15.2
Example of gray mold disease caused by *B. cinerea* on 'Thompson Seedless' grapes after 8 weeks of commercial storage.

of 10–15 kPa CO_2 yields the best and safest control of *B. cinerea* development. These studies also pointed out that high CO_2 can cause damage in some cultivars under some conditions. Thus, specific CO_2 levels are dependent on cultivar, storage length, and maturity.

15.3.2.1 Disease Development

Gray mold caused by *B. cinerea* is the most destructive of the postharvest diseases of table grapes, primarily because it develops at temperatures as low as $-0.5°C$ and can spread from berry to berry (Figure 15.2). Gray mold first turns berries brown, then loosens the skin of the berry, its white, thread-like hyphal filaments erupt through the berry surface, and finally masses of gray colored conidia develop. Wounds near harvest also provide opportunities for infections. No wound is required for new infection when wet conditions occur. *Botrytis* rot on grapes can be diagnosed by its characteristic "slipskin" that develops on the surface of infected berries. Areas of the berry skin infected with *Botrytis* are brown in color and slip freely when rubbed with the fingers, leaving the firm underlying pulp exposed (Luvisi et al., 1992). Uncontrolled infections result in the development of aerial mycelium that spread to adjacent berries (nests). *Rhizopus* sp. and *Aspergillus niger* can infect table grapes, but they do not develop during storage at low temperatures. *Penicillium* sp. do not tend to infect healthy grapes, but may develop through wounds. *Cladasporium herbarum, Alternaria* sp., and *Stemphylium* sp. infections develop slowly in cold storage and can be distinguished from *Botrytis* by the black lesions they form. Table grapes are treated with SO_2 primarily to control gray mold, which is not inhibited sufficiently by rapid cooling alone. Standard practice is to fumigate with SO_2 immediately after harvesting and/or packing followed by lower dose SO_2 treatments weekly during storage (Luvisi et al., 1992).

High CO_2 levels have been known to inhibit the growth of fungi (Luvisi et al., 1992; Tian et al., 2001). On table grapes, high CO_2 has shown promise for the control of *Botrytis* (Crisosto et al., 2002a,b; Retamales et al., 2003; Artés-Hernández et al., 2004). In fact, CO_2 at 5–10 kPa can provide decay control equally effective to SO_2 (Yahia et al., 1983). However, sensitivity to CO_2 varied dependent on cultivar and maturity (Crisosto et al., 2002a,b). In 'Thompson Seedless' table grapes from Egypt, Berry and Aked (1997) observed an inhibition of *B. cinerea* by exposure to a CA of 15 kPa CO_2 and 5 kPa O_2.

The relationship between CA conditions and maturity for decay control in 'Thompson Seedless' (Crisosto et al., 2002a) and 'Red Globe' (Crisosto et al., 2002b) table grapes was tested (Figure 15.3). Early maturity (16.5% soluble solids concentration [SSC]) and late maturity (19% SSC) grapes were exposed to 5, 10, 15, 20, or 25 kPa CO_2 combined with 3, 6, or 12 kPa O_2. In these evaluations, all fruit were initially fumigated with SO_2 and air-stored grapes were used as controls. Natural quiescent botrytis infection ranged from none to 5.5% in the first month for air-stored early harvested grapes. For late harvested grapes, natural botrytis infection varied from 1.0% to 32.8%. In all cases, decay incidence was not related to O_2 or the O_2–CO_2 interaction. Storage atmospheres did not affect SSC, titratable acidity (TA), or sugar:acid ratio (SSC:TA). The main storage limitations for early harvested grapes were presence of off-flavors and rachis and berry browning, which resulted from exposure to >10 kPa CO_2 (Figures 15.4 and 15.5). However, ≥15 kPa CO_2 was needed to control total decay and nesting development independent of O_2 concentrations under low inoculum conditions (Crisosto et al., 2002a,b). High CO_2 atmospheres (≥15 kPa) were more effective in decay control without detrimental effects on quality when late harvested grapes were tested. Fifteen kPa CO_2 is suggested for up to 12 weeks storage only for late harvested 'Thompson Seedless' table grapes; CA is not recommended for early harvested 'Thompson Seedless' grapes. For 'Red Globe,' a combination of 10 kPa CO_2 is suggested for up to 12 weeks storage for late

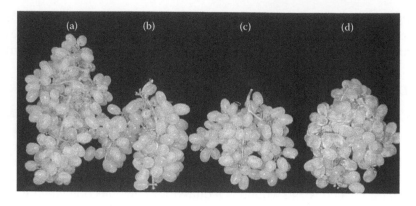

FIGURE 15.3
Effect of CA treatments on visual appearance and gray mold of early 'Thompson Seedless' grapes after 3 months storage at 0°C: (a) 15 kPa CO_2 and 6 kPa O_2, (b) 10 kPa CO_2 and 12 kPa O_2, (c) 15 kPa CO_2 and 12 kPa O_2, and (d) air.

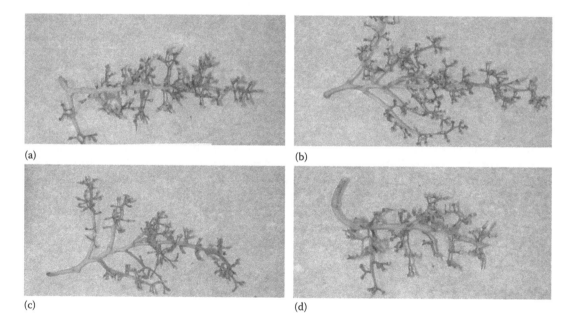

FIGURE 15.4
Effect of CA treatment with or without SO_2 of early 'Flame Seedless' grapes after 2 months storage at 0°C: (a) air, (b) air and SO_2 pad, (c) 10 kPa CO_2 and 6 kPa O_2, (d) 10 kPa CO_2 and 6 kPa O_2 and SO_2 pad.

harvested grapes. For early harvested 'Red Globe' grapes, an atmosphere of 10 kPa CO_2 is suggested for periods shorter than 4 weeks. Other researchers indicated that high CO_2 during small scale cold storage control gray mold better than using 0.7 g kg^{-1} $Na_2S_2O_5$. The use of ≥15 kPa CO_2 to delay *Botrytis* development under low inoculum conditions has been observed (Crisosto et al., 2002a,b). In conditions of high inoculum (simulating bad years), initial SO_2 fumigation was necessary to keep decay development low during postharvest life.

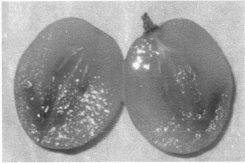

FIGURE 15.5
Acute physiological damage of 'Thompson Seedless' grapes caused by prolonged storage at 25 kPa CO_2 for 12 weeks at 0°C.

15.3.2.2 Fruit Biochemistry

There are no publications describing the effects of CA on respiration, volatiles, and vitamins for table grapes. However, an indirect evaluation of fruit flavor and/or volatiles was done on 'Thompson Seedless' and 'Red Globe' (Crisosto et al., 2002a,b). Using a trained panel approach, judges perceived development of off-flavor in 'Thompson Seedless' table grapes and related this only to CO_2 levels. Off-flavor development on early and late harvested grapes was perceived after one month at 0°C followed by 2 days at 20°C. Early harvested grapes had more off-flavors than late harvested grapes. In the evaluation carried out after 2 months at 0°C followed by 2 days at 20°C, approximately 35% of the judges detected off-flavor in air- and 10 kPa CO_2-stored grapes, while approximately 60%–80% of the judges detected off-flavor in early harvested grapes stored at $\geq$15 kPa CO_2. In grapes stored at 5 kPa CO_2, none of the judges detected off-flavor. In all of the sensory evaluations, grapes from the 5 kPa CO_2 treatments had the same or less off-flavor development than air-stored grapes. Off-flavor was induced by CA treatments when grapes at both maturities were stored in $\geq$15 kPa CO_2. It has been reported that the use of 10–15 kPa CO_2 did not alter SSC, TA, sugar composition, or organic acid composition during cold storage and shelf life for 'Autumn Seedless,' 'Red Globe,' and 'Thompson Seedless' in different countries (Crisosto et al., 2002a,b; Retamales et al., 2003; Artés-Hernández et al., 2004). In 'Kyoho' which is a cross between *Vitis rabruscana* and *V. vinifera*, CO_2 exhibited good decay control without altering fruit detachment force, firmness, and SSC, or TA after 60 days cold storage. However, grapes kept at CO_2 concentrations higher than 9 kPa had an unacceptable alcoholic flavor and browning after 45 days cold storage (Deng et al., 2006, 2007).

15.3.2.3 Quality Attributes

Efficacy of CA conditions for decay control in 'Thompson Seedless' and 'Red Globe' table grapes was carried out during two seasons (Crisosto et al., 2002a,b). Early 'Thompson Seedless' (16.5% SSC) and late harvested (19% SSC) grapes were exposed to 5, 10, 15, 20, and 25 kPa CO_2 combined with 3, 6, and 12 kPa O_2. In these two cultivars, storage atmospheres did not affect SSC, TA, SSC:TA, or berry shatter. Some storage limitations for early harvested 'Thompson Seedless' table grapes were rachis and berry browning development, which resulted from exposure to >10 kPa CO_2. Earlier work also reported that berry internal browning incidence overcame the potential benefits of CA in 'Thompson Seedless' table grapes from Coachella Valley; the earliest production area in California (Nelson, 1969). The same phenomenon was described later on 'Thompson Seedless' from Coachella Valley using 5 kPa CO_2 and 2 kPa O_2, indicating that *B. cinerea* decay (gray mold) and internal berry browning were the main deterioration symptoms beyond 8 weeks storage (Yahia et al., 1983).

15.3.2.4 Effects of MAP

Information on the use of MAP in single containers on grape is limited because of the incompatibility with the safe use of a SO_2 generator and slow and uneven cooling performance. Currently the use of a pallet wrap in combination with SO_2 generator pads is being evaluated under commercial conditions as a way to assure proper cooling prior to the pallet being wrapped. By using a pallet wrap after fast cooling and initial SO_2 fumigation, initial-coding operation is not affected. A short cold storage period (simulating European handling conditions) using MAP as a replacement for SO_2 was studied in Spain. 'Superior Seedless' table grapes were stored for 7 days at 0°C followed by 4 days at 8°C plus 2 days at 20°C under MAP. During this short cold storage period, there were no changes in skin color, firmness, soluble solids content (SSC), pH, TA, maturity index, off-flavor, phenolic composition, and berry browning. After shelf-life MAP-treated clusters showed slight to moderate stem browning, except under SO_2 where practically no browning occurred, while control clusters showed extreme stem browning. Thus, it was concluded that SO_2-free MAP kept the overall quality of clusters close to that at harvest, with few differences when SO_2 was added (Artés-Hernández et al., 2006). Another novel approach has been using MAP in combination with essential oils such as eugenol, thymol, or menthol (Valverde et al., 2005). In this work, 'Crimson Seedless' grapes packed using MAP combined with 0.5 mL of eugenol, thymol, or menthol inside the packages was stored at 1°C for 35 days. During this cold storage period, the inside box gas composition reached about 1.4–2.0 and 10.0–14.5 kPa of CO_2 and O_2, respectively. These results showed that the addition of essential oils improved the beneficial effect of MAP in terms of delaying weight loss and color changes, retarding SSC:TA ratio evolution, and maintaining firmness. In addition, the total viable counts for both mesophilic aerobics, especially yeasts and molds, were significantly reduced in the grapes packaged with essential oils. Further studies on the potential benefit of these SO_2 substitutes to maintain table grape storage quality are being carried out.

Ethanol is approved as a disinfectant or sanitizer for organic crop use by the USDA National Organic Program. Evidence from several independent studies performed on several cultivars suggests that immersion of grape clusters in 50%, 40%, or 33% ethanol, prior to packaging, inhibited berry decay and was equivalent to, or better than, the effectiveness of SO_2 released from metabisulfite generator pads. Decay was acceptably controlled for a cold storage period of 4–5 weeks and sometimes longer and the quality of the grapes was not affected (Lichter et al., 2002; Milkota Gabler et al., 2005). The combination of MAP and ethanol dipping prior to MAP was tested to control decay under conditions of extended storage or high postharvest decay pressure on 'Superior' grapes (Lichter, personal communication; Lichter et al., 2005). Quality after storage was good if the level of CO_2 did not exceed 10 kPa, and any off-flavor that was present at removal from storage dissipated with 24 h after the liners were opened. The rationale behind this combined technology is that the ethanol dip treatment prevents development of *B. cinerea* in the first phase of storage before sufficient CO_2 accumulation occurs (Chervin et al., 2005). The viability of this approach was confirmed on ethanol-dipped grapes stored in water permeable liners using two ethanol delivery methods (Lurie et al., 2006). However, rachis and berry browning occurred on 'Thompson Seedless' grapes when ethanol-impregnated papers were used, but not when ethanol was applied from a paper wick dipped in ethanol. The efficacy of ethanol as a volatile may be associated with its conversion by the berry alcohol dehydrogenase to acetaldehyde, which is very toxic to microorganisms. Due to its cross-linking capacity, a higher concentration of this volatile may promote polymerization of phenolic compounds and browning. Further studies to adapt and evaluate this new technique to commercial operations are being carried out.

15.3.3 Blueberry

15.3.3.1 Disease Development

Decay is the primary cause of postharvest quality loss in fresh blueberries (*Vaccinium* spp.) (Cappellini et al., 1982). The most common postharvest disease of blueberries is *Botrytis* rot caused by *B. cinerea* (Caruso and Ramsdell, 1995; Barkai-Golan, 2001). *Botrytis* rot, also known as gray mold, is able to grow under conditions of high humidity even at low temperatures of 0°C, making it a major problem during cold storage and marketing. Other fungi causing postharvest rots include *Alternaria alternata* and *Alternaria tenuissima* (Alternaria rot) and *Colletotrichum gloeosporioides* (Anthracnose fruit rot, ripe rot; Caruso and Ramsdell, 1995; Barkai-Golan, 2001). Atmosphere modification is effective in inhibiting the growth of these decay-causing fungi (Wells and Uota, 1970; Agar et al., 1990). Elevated concentrations of CO_2 have proven to be most beneficial in inhibiting the development of decay, while the benefits of O_2 reduction are still being debated (Ceponis and Cappellini, 1985; Forney et al., 2003; Harb and Streif, 2004).

For effective decay control, CO_2 concentrations must be ≥ 10 kPa (Sargent et al., 2006). Many studies have demonstrated the effectiveness of high concentrations of CO_2, ranging from 10 to 25 kPa, to reduce decay in highbush (Ceponis and Cappellini, 1985; Forney et al., 1998; Harb and Streif, 2004), rabbiteye (Smittle and Miller, 1988), and lowbush (Prange et al., 1995) blueberries. Concentrations of $CO_2 < 10$ kPa are less effective in controlling decay. Decay was not reduced by 6 kPa CO_2 in 'Duke' blueberry fruit during 7 weeks of storage at 0°C–1°C (Harb and Streif, 2004). In 'Bluecrop' fruit, however, 6 kPa CO_2 reduced decay after 4 weeks but not after 7 weeks when compared to air stored fruit (Harb and Streif, 2006). The minimal benefit of 8 kPa CO_2 compared to air storage observed during shipping of fruit from Chile to the United States was also likely a result of the low CO_2 concentration (Beaudry et al., 1998). While higher concentrations of CO_2 may be more effective in inhibiting decay, they may cause physiological damage to the blueberry fruit.

Injury caused by excessive CO_2 is characterized by fruit softening, flesh discoloration, and the development of off-flavor (Figures 15.6 through 15.8). At CO_2 concentrations ≥ 15 kPa, firmness of 'Burlington' (Forney et al., 2003), 'Duke' (Harb and Steif, 2004), and 'Bluecrop' (Krupa and Tomala, 2007) decreased below initial values following 3 or more weeks of storage and flesh discoloration developed. Associated with this loss of texture was the development of off-flavors. Harb and Steif (2004, 2006) reported that 'Duke' and 'Bluecrop' blueberries stored at 0°C in 18 or 24 kPa CO_2 for 5 weeks developed

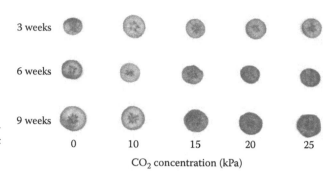

FIGURE 15.6
Internal discoloration of 'Burlington' blueberry fruit caused by prolonged storage at 0°C for 3, 6, or 9 weeks in 0–25 kPa CO_2.

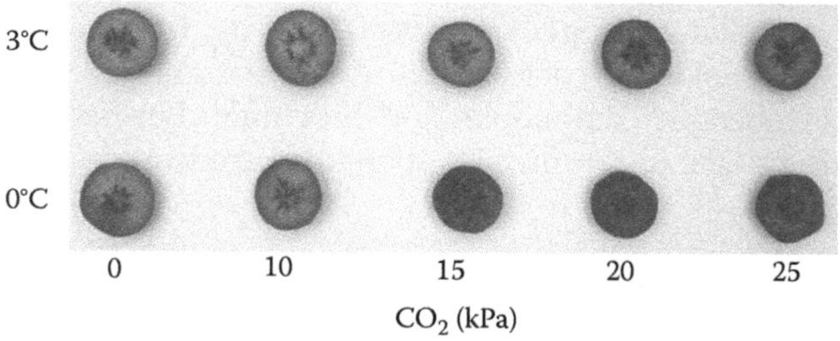

FIGURE 15.7
Internal appearance of 'Burlington' blueberry fruit following storage for 9 weeks in 0–25 kPa CO_2 at 0 or 3°C plus an additional week in air at 7°C.

FIGURE 15.8
Example of acute internal physiological injury of blueberry fruit caused by storage for 9 weeks in 20 kPa CO_2 at 0°C.

off-flavors and mealy texture. Similarly, 'Woodward' rabbiteye fruit stored in 20 kPa CO_2 for 42 days at 5°C developed a slightly fermented flavor (Smittle and Miller, 1988). Concentrations of 25 kPa CO_2 induced an 18-fold increase in ethanol and a 25-fold increase in ethyl acetate after 6 weeks at 0°C, which was associated with off-flavor (Forney et al., 2003).

Factors determining fruit tolerance to CO_2 include CO_2 concentration, storage duration, storage temperature, and cultivar and must be considered in optimizing CA storage conditions. Blueberry fruit can tolerate high concentrations of CO_2 for short durations without injury, but as exposure duration increases tolerable concentrations decrease. For example, 'Burlington' fruit were able to tolerate 25 kPa CO_2 for 3 weeks at 0°C with little injury, but after 6 weeks fruit became soft and after 9 weeks many fruits were discolored (Figure 15.6, Forney et al., 1998). As storage temperature decreases, blueberry fruit become less tolerant to elevated concentrations apparently due to the increased solubility of CO_2 into the fruit tissue. When blueberry fruit were held in 100 kPa CO_2 at 0°C or 10°C, a carbonic acid-like flavor was detected after 1 h, however when fruit were at 20°C or 30°C this flavor was not detected until 3 h (Saltveit and Ballinger, 1983). Concentrations of 15, 20, or 25 kPa caused 'Burlington' fruit to soften after 3 weeks at 0°C, whereas at 3°C, softening of fruit held in 20 or 25 kPa CO_2 did not occur until after 6 weeks (Forney et al., 1998, 2003). Cultivars also appear to vary in their tolerance to elevated concentrations of CO_2. When five cultivars of blueberries were held for 9 weeks in 12.5 kPa CO_2 at 0°C, 'Brigitta,' 'Burlington,' and 'Bluegold' had no significant loss of firmness, while 'Coville' and 'Reka' lost 23% and 41% of their initial firmness, respectively (Forney et al., 1997). 'Woodward' rabbiteye fruit stored in 20 kPa CO_2 for 42 days at 5°C developed a slightly fermented flavor, while 'Climax' fruit stored under the same conditions did not (Smittle and Miller, 1988). Optimum concentrations of CO_2 for maintaining the quality of lowbush fruit also varied with 10 kPa being optimum for 'Blomidon,' while 15 kPa was optimum for 'Fundy' (Prange et al., 1995).

The benefits of reduced O_2 in combination with elevated CO_2 concentrations rarely has shown benefits in reducing decay or increasing marketable fruit following storage. Combining 2 kPa O_2 with high concentrations of CO_2 provided no additional control of decay or maintenance of quality of highbush blueberry fruit than high CO_2 alone when stored at 0°C–2°C (Ceponis and Cappellini, 1985; Harb and Streif, 2004, 2006). When 'Bluecrop' fruit were stored in O_2 concentrations ranging from 0.5 to 2.0 kPa in combination with CO_2 concentrations of 0–15 kPa at 0°C, no benefit of the reduced O_2 concentrations was observed (Fan et al., 1993). Similar results were obtained with lowbush blueberries where O_2 concentrations ranging from 1 to 5 kPa provided no additional reduction in decay beyond that resulting from elevated CO_2 atmospheres (Prange et al., 1995). However, when 'Coville' fruit were held at 15°C for 10 days with elevated partial pressures of CO_2, O_2 partial pressures of 3 and 9 kPa reduced decay more than 15 kPa (Kim et al., 1995). Reduced partial pressures of O_2 without elevated CO_2 were reported to reduce decay, but never to the extent of 10 kPa or greater CO_2 levels (Prange et al., 1995).

Recent research has reported benefits of elevated concentration of oxygen on reducing decay. High O_2 atmospheres ($\geq$60 kPa) decreased decay of 'Duke' blueberry fruit during storage at 5°C. After 35 days, decay of fruit held in 80 and 100 kPa O_2 averaged 4.5% and 6.7%, respectively, compared to 40% for air-stored fruit (Zheng et al., 2003).

15.3.3.2 Fruit Biochemistry

The respiration rate of blueberry fruit varies among cultivars and as a result of storage conditions. Respiration rates among freshly harvested fruit ranged from 62 mg CO_2 kg^{-1} h^{-1} for 'Herbert,' 82 mg CO_2 kg^{-1} h^{-1} for 'Bluecrop,' 105 mg CO_2 kg^{-1} h^{-1} for 'Elliot,' and 106 mg CO_2 kg^{-1} h^{-1} for 'Coville' when measured in air at 20°C (Plestenjak et al., 1998). When respiration rates were modeled for highbush blueberry fruit held at 15°C under various aerobic atmospheres, respiration rates decreased as CO_2 concentration increased to 15 kPa, but O_2 concentrations between 4 and 19 kPa had little effect (Song et al., 1992). Higher concentrations of 40 and 60 kPa CO_2 induced fermentation in 'Bluecrop' and 'Elliot' blueberries held at 15°C for 4 days, resulting in increased rates of CO_2 evolution (Beaudry, 1993). However, these CO_2 concentrations had no effect on rates of O_2 consumption. Short exposures of <2 h of highbush and rabbiteye blueberry fruit to 100 kPa N_2 reduced CO_2 evolution by about 20% (Saltveit and Ballinger, 1983). Following 28 days of storage at 0°C, the respiration rates of 'Duke' fruit increased depending on CO_2 concentration (Harb and

Streif, 2004). Rates of fruit stored in 24 kPa CO_2 was 29 mg CO_2/kg h compared to 22 mg CO_2/kg h in fruit that had been stored in air, when measured 24 h after removal from CA storage and warming to 20°C. Respiration rates of fruit from the elevated CO_2 concentrations also remained higher than air controls 48 h after removal. Respiration rate from fruit held in atmospheres containing 2 kPa O_2 had no effect on respiration rates. Fruit held in air had a respiratory quotient of 2.0, which was reduced by elevated CO_2 and reduced O_2 atmospheres. However, changes in respiration rate of fruit under CA conditions and refrigerated temperatures after extending storage periods have not been measured.

While soluble solids (SS) are commonly measured following storage, there are few reports of changes in sugars. The predominate sugars in blueberries are glucose, fructose, and sucrose. At harvest, the concentrations of these sugars averaged 20–40 mg g^{-1} FW and were found in a 0.8:1:1 ratio, respectively, in 'Bluecrop,' 'Herbert,' and 'Patriot' fruit (Plestenjak et al., 1998). During storage, concentrations of glucose and fructose increased and sucrose declined in both air and 6 kPa CO_2 and 3 kPa O_2 atmospheres at 1°C. The most prominent acid in blueberry fruit is citric followed by malic, quinic, and succinic. During 40 days of storage in air or in 6 kPa CO_2 and 3 kPa O_2 at 1°C, acid concentrations in 'Bluecrop,' 'Herbert,' and 'Patriot' fruit decreased and were not significantly affected by CA conditions (Plestenjak et al., 1998). Similarly, when 'Duke,' 'Bluecrop,' and 'Brigitta' fruit were stored in atmospheres of 0–15 kPa CO_2 in combination with 1–16 kPa O_2 for 9 weeks at 0°C, atmosphere composition had no significant effect on acid composition (Forney et al., 2007). Elevated concentration of CO_2 helps to preserve pectin in blueberry fruit. Total pectin decreased and became more soluble in 'Woodward' and 'Climax' rabbiteye fruit following 21–42 days of storage in air at 5°C (Smittle and Miller, 1988). However, total pectin and its solubility did not change when fruit were stored in atmospheres of 10–20 kPa CO_2.

Total anthocyanin content of blueberry fruit are not strongly affected by atmosphere modification during storage. Total anthocyanin concentrations increased 2 weeks after harvest and then declined during an additional 6 weeks of storage in 'Bluecrop' fruit stored at 0°C in air or combinations of 12 or 18 kPa CO_2 and 1.5 to 12 kPa O_2 (Krupa and Tomala, 2007). Fruit stored in atmospheres of 1.5 and 3 kPa O_2 retained higher concentrations of anthocyanins following 4–8 weeks of storage than air-stored fruit although these effects were only significant in one of two seasons tested. Concentrations of 12 and 18 kPa CO_2 had no significant effects. In another study, total anthocyanins in 'Coville' highbush blueberries did not change significantly during 4 weeks of storage in 10 kPa CO_2 and 15 kPa O_2 at 0°C or during 1 week in air at 10°C (Song et al., 2003). However, when 'Duke' fruit were held in $\geq$60 kPa O_2 for 7 weeks at 5°C, total anthocyanins increased and remained 10%–19% higher than initial values after 7 weeks of storage (Zheng et al., 2003). Total anthocyanins in air-stored fruit increased slightly after 2 weeks and then declined to 88% of initial concentrations after 7 weeks. Total phenolics in blueberry fruit change in a manner similar to total anthocyanins. When 'Duke,' 'Bluecrop,' and 'Brigitta' fruit were stored in atmospheres of 0–15 kPa CO_2 in combination with 1–16 kPa O_2 for 9 weeks at 0°C, total phenolics, measured as gallic acid equivalents, declined slightly during storage, but atmosphere composition had no significant effect on total phenolic concentration (Forney et al., 2008). Total phenolics remained unchanged in 'Coville' highbush blueberries during 4 weeks of storage in 10 kPa CO_2 and 15 kPa O_2 at 0°C or during 1 week in air at 10°C (Song et al., 2003). Total phenolics increased during storage in $\geq$60 kPa O_2 at 5°C, while those in air or 40 kPa O_2 increased initially and then remained constant. After 7 weeks, the fruit from the high O_2 treatments had about 20% higher phenolics than initials, while phenolic concentrations in those stored in air were unchanged (Zheng et al., 2003).

Blueberries are considered to have high antioxidant content when compared to other fruits. Therefore, it is important to understand how the fruit antioxidant capacity changes during storage. During storage in air for 8 days at temperatures ranging from 0°C to 30°C, there was no significant change in antioxidant capacity, determined by the Oxygen Radical Absorbance Capacity (ORAC) assay, of either 'Bluecrop' highbush blueberries or wild lowbush blueberries (Kalt et al., 1999). Similarly, 'Coville' fruit stored for 4 weeks in 10 kPa CO_2 and 15 kPa O_2 at 0°C also had no significant change in antioxidants determined using the ORAC assay (Song et al., 2003). When 'Bluecrop' fruit were stored for up to 8 weeks at 0°C, the antioxidant capacity of the fruit, determined using the 2,2-diphenyl-1-picrylhydrazyl radical assay, did not change significantly regardless of whether fruit were stored in air or in CA ranging from 12 to 18 kPa CO_2 and 1.5 to 12 kPa O_2 (Krupa and Tomala, 2007). In contrast, when five cultivars including 'Bluecrop' were stored 4 weeks in air or 10 kPa CO_2 and 10 kPa O_2 at 0°C or 8°C, the antioxidant capacity, determined using the ferric reducing ability of plasma (FRAP) assay, decreased an average of 24% and 36% following storage in air and CA, respectively (Remberg et al., 2003). Loss of antioxidant capacity of fruit was greater at 8°C than at 1°C. 'Centurion' and 'Maru' rabbiteye fruit had a slight reduction in antioxidant capacity, measured using the FRAP assay, following 4–8 weeks of storage in air or 15 kPa CO_2 and 2.5 kPa O_2 at 1.5°C (Schotsmans et al., 2007). However, when the rabbiteye fruit was held an additional 6 days at 20°C, the antioxidant activity of the fruit increased 10%–20%. In contrast to traditional CA or air storage, when 'Duke' fruit were stored in ≥ 60 kPa O_2 at 5°C, the antioxidant activity, determined using the ORAC assay, increased, while those in air or 40 kPa O_2 increased initially and then declined (Zheng et al., 2003). After 7 weeks, the fruit from the high O_2 treatments had about 50% higher antioxidant capacity than at harvest, while antioxidant activity of fruit stored in air had decreased about 9%.

15.3.3.3 Quality Attributes

In the absence of decay and excessive dehydration, most blueberry fruits tend to increase in firmness when stored in air (Forney et al., 1998; Harb and Streif, 2004; Schotsmans et al., 2007). Elevated concentrations of CO_2 in the storage atmosphere inhibit this firming, and fruit firmness tends to decrease as CO_2 concentrations increase in the storage atmosphere (Fan et al., 1993; Forney et al., 2003). Firming of air-stored fruit has been associated with thickening of parenchyma and sclerenchyma cell walls in the fruit flesh, which is inhibited by CO_2 (Allan-Wojtas et al., 2001). Bünemann et al. (1957) observed a corrugation of cell walls of 'Rubel' fruit stored in air at 4°C or 10°C, while the cell walls of fruit stored in 15 kPa CO_2 remained smooth and similar to the unstored fruit. Loss of water during storage causes a loss in turgidity resulting in fruit softening (Forney et al., 1998). The high humidity that is normally maintained during CA storage minimizes water and turgidity loss and thus helps to maintain fruit firmness.

Injurious concentrations of CO_2 and O_2 can cause the development of off-flavors in fresh blueberries. 'Duke' and 'Bluecrop' fruit stored in ≥ 18 kPa CO_2 for 5 or more weeks and 'Woodward' rabbiteye fruit stored in 20 kPa CO_2 for 6 weeks developed off-flavors (Smittle and Miller, 1988; Harb and Streif, 2004, 2006). Oxygen partial pressures of 2 kPa or less also cause off-flavor development (Saltveit and Ballinger, 1983; Ceponis and Cappellini, 1985; Smittle and Miller, 1988) and can enhance off-flavors induced by CO_2 (Harb and Streif, 2004, 2006). Injurious atmospheres have been reported to induce the production of ethanol and ethyl acetate (Saltveit and Ballinger, 1983; Forney et al., 2003). However, injurious atmospheres of 15 and 25 kPa CO_2 did not alter the flavor volatiles benzaldehyde, γ-butyrolactone, hexyl 2-methylbutanoate, and ethyl-1-hexanol in 'Burlington' blueberries, while concentrations of methyl acetate and methyl butanoate were reduced (Forney et al., 2003).

When blueberries are held in noninjurious atmospheres, no adverse effects on flavor have been reported. The flavor of 'Duke' and 'Bluecrop' fruit held in 6 and 12 kPa CO_2

was not rated significantly different from fruit held in air (Harb and Streif, 2004, 2006). 'Woodward' rabbiteye fruit developed a tart, musty off-flavor when stored in air at 5°C for 14 days, but this off-flavor did not develop when fruit were stored in atmospheres of 15 kPa CO_2 and 5 kPa O_2 (Smittle and Miller, 1988). This preservation of flavor by the CA environment may have been a result of the repression of fungal growth.

Postharvest changes in SSC and titratable acidity (TA) are variable depending on cultivars and storage conditions. Soluble solids declined during storage in air at 0°C–5°C in 'Burlington,' 'Bluecrop,' and 'Duke' highbush blueberries (Forney et al., 2003; Zheng et al., 2003; Krupa and Tomala, 2007) and 'Woodward' and 'Climax' rabbiteye blueberries (Smittle and Miller, 1988) with decreases averaging 3%–15% after 35 or more days of storage. Controlled atmospheres of up to 25 kPa CO_2 with or without reduced O_2 concentrations had no significant effects on this decline in SS. However, $\geq$60 kPa O_2 reduced this decline (Zheng et al., 2003). In other studies with a variety of highbush cultivars, no significant changes in SSC content were reported in fruit stored in either air or CA for 40 or more days (Plestenjak et al., 1998; Harb and Streif, 2004, 2006). Soluble solids in 'Centurion' rabbiteye fruit also did not change during storage in air, but decreased slightly ($\sim$0.5%) when stored in 15 kPa CO_2 and 2.5 kPa O_2 for 42 days at 1.5°C where as SS in 'Maru' fruit decreased slightly in air ($\sim$0.8%) and did not change in CA (Schotsmans et al., 2007). When 'Coville' fruit were held at 15°C for 10 days, SS increased 30% when held in 3 kPa CO_2 but as CO_2 partial pressure increased, the increase in SS was less being only 15% in 18 kPa CO_2 (Kim et al., 1995). Reduced O_2, as low as 3 kPa, had no effect on SS. As with SS, postharvest changes in TA are variable among studies. Titratable acidity has been reported to decrease during storage in air in highbush blueberry fruit (Plestenjak et al., 1998; Forney et al., 2003; Zheng et al., 2003; Harb and Streif, 2004; Krupa and Tomala, 2007). Contrary to this in other studies TA increased in highbush (Kim et al., 1995; Harb and Streif, 2006), lowbush (Prange et al., 1995) and rabbiteye (Smittle and Miller, 1988; Schotsmans et al, 2007) blueberry fruit. However, in either case, both elevated concentrations of CO_2 and reduced concentrations of O_2 reduced postharvest change in TA in all of the studies cited.

15.3.3.4 Effects of MAP

To date, no commercial application of MAP is being used for the marketing of fresh blueberries. This is primarily due to several factors including the limitations in the gas transmission properties of commercial packaging materials, the biological variability in fruit respiration rates and tolerance to CO_2, and the variability of temperature in the market chain. To optimize blueberry market life, target atmospheres should be 10–12 kPa CO_2 with >2 kPa O_2. However, these atmospheres are difficult to achieve using available plastic films due to the higher permeability of CO_2 than O_2. In modeling studies with a biodegradable film, Makino and Hirata (1997) found equilibrium atmospheres for packaged blueberries with the experimental film were 10 kPa O_2 and 3.4 kPa CO_2, which was not effective in extending market life.

The effects of variable temperatures on fruit respiration are also a challenge. Oxygen and CO_2 permeation rates of low-density polyethylene packaging do not increase as rapidly as blueberry fruit respiration rates when subjected to increasing temperatures resulting in greater atmosphere modification (Beaudry et al., 1992). This dilemma is common for most commercial films, making it difficult to match the permeability properties of common packaging films with the respiration rates of blueberry fruit over a range of temperatures. Therefore, packages continually run the risk of developing anaerobic atmospheres or failing to maintain beneficial ones as a result of fluctuation in holding temperature. The use of microperforation is another option to obtain high concentrations of CO_2 while avoiding anaerobic conditions. However, perforations have little response to temperature change (Cameron et al., 1994). Predicting fruit respiration rates is also a challenge due to biological variability. Respiration rates of fruit can vary substantially among and within cultivars due to maturity or growing environment (Cameron et al., 1994). In addition, handling of fruit can affect respiration. When 'Brightwell' and 'Tifblue' rabbiteye blueberry fruit were mechanically harvested, respiration rates were about 30% greater than hand-harvested fruit when measured at 1°C or 22°C (Nunez-Barrios et al., 2005). This elevated rate of respiration was maintained for a week following harvest. Variation in fruit respiration rates makes it very difficult to predict fruit gas exchange and equilibrium package atmosphere composition. Variation in respiration and limitations in package permeation properties have prevented the development of commercial MAP for blueberry fruit.

15.3.4 Raspberry, Other Brambles, and Minor Small Fruits

Compared to strawberry fruit, relatively little CA research has been conducted on raspberry, blackberry, gooseberry, or currants. Whilst increasing in recent years, the fresh market demand for soft fruit, other than strawberry and raspberry, is comparatively small. Blackcurrants (*Ribes nigrum* L.), for example, are almost exclusively produced for processing. This said, recent work by Harb et al. (2008) reported that decreasing the O_2 levels and increasing CO_2 levels retarded the capacity of 3-week-stored 'Titania' blackcurrant fruit at 1°C to synthesize terpenes as compared to air-stored fruit; however partial recovery was evident after 6 weeks.

Of the CA work that has been carried out, the majority has concentrated on extending postharvest life of raspberry and blackberry fruit along similar lines to that used for strawberry fruit (Joles et al., 1994; Agar and Streif, 1996; Haffner et al., 2002; van der Steen et al., 2002). CA storage at both 15 kPa CO_2 and 10 kPa O_2 and 31 kPa CO_2 and 10 kPa O_2 was shown to suppress postharvest disease mainly caused by *B. cinerea* on Norwegian-grown 'Glen Ample' and 'Glen Lyon' 'Malling Admiral,' 'Malling Orion,' and 'Veten' raspberries (Haffner et al., 2002). In contrast to air-stored fruit where berries became darker after 7 days storage at 2°C, no significant change in berry pigmentation was observed under CA conditions. Disease incidence of erect-type 'Navaho' and 'Arapaho' blackberry fruit was reduced under 15 kPa CO_2 and 10 kPa O_2 stored at 2°C for 7 days as compared to fruit stored in air (Perkins Veazie and Collins, 2002). Consumer taste panels did not detect the presence of off-flavors in blackberries treated with CA for 3 and 7 days storage at 2°C. However, CA treatment resulted in a 20% reduction in total anthocyanins. Red currants are stored for up to 5 months in plastic covered pallets in atmospheres of 18–20 kPa CO_2 and 2 kPa O_2 in the Netherlands (van Schaik and Verschoor, 2003). Decay caused by *Botrytis* was significantly reduced in red currant fruit stored in CA (Roelofs and Waart, 1993).

Like strawberry, alternative CA treatments, such as superatmospheric oxygen (van der Stern et al., 2002; Siro et al., 2006) or ozone (Barth et al., 1995) have been investigated to reduce postharvest disease of raspberry and other berries. High oxygen atmospheres improved the keeping quality of raspberry fruit mainly by reducing decay caused by *B. cinerea* (van der Stern et al., 2002). Similarly, high oxygen atmospheres (60–100 kPa) have been found to inhibit decay in 'Wumei' Chinese red bayberry (*Myrica rubra*) fruit when stored at 5°C (Zheng et al., 2008). When fruits were removed from oxygen-enriched atmospheres and then held in air for 2 days at 20°C, fruits still exhibited less decay, suggesting that high oxygen atmospheres had a residual effect on postharvest disease control. Research by Agar et al. (1997) demonstrated that CA has differential effects on ascorbic acid and dehyroascorbic acid in various berry fruit. Ascorbic acid tended to be reduced more than dehydroascorbic acid at high CO_2. High CO_2 levels (10–30 kPa) tended to reduce total vitamin C content in blackcurrant and blackberries, but had almost no effect in raspberries and red currants. This research suggests that CA may not appropriate for some berry fruit types.

15.3.5 Cranberry

15.3.5.1 Disease Development

The main causes of cranberry fruit loss during storage are decay and physiological breakdown (Forney, 2003). Unlike other small fruit, decay development during storage is caused by a complex of fungal organisms mostly unique to cranberry fruit. They include *Allantophomopsis lycopodina* (black rot), *Allantophomopsis cytisporea* (black rot), *Strasseria geniculata* (black rot), *Coleophoma empetri* (ripe rot), *Fusicoccum putrefaciens* (end rot), *Phyllosticta elongata* (berry speckle), *Physalospora vaccinii* (blotch rot), and *Botrytis* spp. (yellow rot) (Boone, 1995a,b; Carris, 1995; Caruso, 1995; Pepin and Boone, 1995; Oudemans et al., 1998). Decay normally is characterized by discoloration in the form of external lesions and softening of the fruit. Physiological breakdown is also a major cause of storage loss and is also known as sterile breakdown because there is no association with a fungal pathogen (Bristow and Patten, 1995). Physiological breakdown that develops during storage has been associated with overmature fruit (Doughty et al., 1967), bruising (Patterson et al., 1967), chilling injury (Hruschka, 1970), freezing (Bristow and Patten, 1995), extended water immersion (Ceponis and Stretch, 1983), and anoxia (Stark et al., 1974). It is characterized by a dull appearance, rubbery texture, and diffusion of red pigment throughout the fruit flesh (Figure 15.9).

Modification of O_2 and/or CO_2 in the storage atmosphere does not appear to have a strong effect in delaying decay or physiological breakdown. Storage of 'Howes' cranberries in combinations of 0, 5, and 10 kPa CO_2 with 3, 10, and 21 kPa O_2 at 0°C and 3°C resulted in more decay and physiological breakdown of fruit than air storage (Anderson et al., 1963). Similar results were found with 'Stevens' fruit stored in 0, 5, 10, or 15 kPa CO_2 and 1 or 15 kPa O_2 at 5°C (Forney, 2009). Stark et al. (1969) found that cranberries stored at 22°C for 3 weeks in atmospheres of 5 or 10 kPa CO_2 with 3 kPa O_2 had the same levels of decay as air-stored fruit. Doughty et al. (1967), citing unpublished work by Patterson in Washington State, also reported that cranberries did not benefit from CA conditions.

Attempts to use more extreme atmospheres have also had limited success. Berries held in 100 kPa N_2 for 3 weeks developed physiological breakdown and had a fermented

Healthy fruit   Physiological breakdown

FIGURE 15.9
Comparison of healthy cranberry fruit with fruit expressing physiological breakdown.

odor (Stark et al., 1969; Lockhart et al., 1971). Concentrations of 70 kPa O_2 also had no beneficial effect in reducing decay or physiological breakdown and induced production of acetaldehyde, ethanol, and ethyl acetate (Gunes et al., 2002). However, Gunes et al. (2002) did find that atmospheres of 30 kPa CO_2 and 21 kPa O_2 reduced decay and breakdown of 'Pilgrim' and 'Stevens' fruit when stored for 2 months at 3°C, but after 4 months these benefits were no longer apparent due to extensive breakdown (>97%) in all atmospheres tested. In addition, the beneficial effect of 30 kPa CO_2 atmospheres was diminished with late-harvested fruit.

High-relative humidity (RH) normally associated with CA storage appears to be detrimental to cranberry fruit storage life. Forney (2008) reported that storage in 75%–81% RH for 5 months resulted in 1.46- and 2.46-fold more marketable fruit than storage in 88% or 98% RH, respectively. Lowering the humidity in CA storage chambers reduced fruit breakdown and decay in some atmospheres giving results similar to the air controls (Anderson et al., 1963). However, lowering the humidity in CA storage rooms creates an additional challenge for the limited and variable potential benefits.

15.3.5.2 Fruit Biochemistry

Cranberry fruit respiration rates are affected by storage atmosphere composition. After removal from storage, fruit held in O_2 concentrations as low as 1 kPa and CO_2 concentrations as high as 30 kPa tended to have lower respiration rates than fruit held in air (Anderson et al., 1963; Gunes et al., 2002). Cranberry fruit produce little ethylene (<0.10 μL kg^{-1} h^{-1}), which was not affected by O_2 or CO_2 concentrations (Gunes et al., 2002). Fermentation is induced in cranberry fruit when held in 100 kPa N_2, which is associated with increased production of acetaldehyde and ethanol (Stark et al., 1974; Gunes et al., 2002). Atmospheres of 70 kPa O_2 with 15 or 30 kPa CO_2 also induced increased concentrations of acetaldehyde and ethanol as well as ethyl acetate (Gunes et al., 2002). Atmosphere modification during storage of cranberry fruit for 2 months had no effect on phenolic content, only a slight effect on flavonoid content, and reduced fruit antioxidant activity (Gunes et al., 2002).

15.3.5.3 Quality Attributes

Atmosphere modification has had variable effects on fruit firmness. When weight loss was minimized (<1%), fruit firmness, measured by force of penetration, increased an average of 9% during 2 months of storage in air and was not affected by O_2 concentration; however, concentrations of 15 and 30 kPa CO_2 caused a 10%–14% reduction in firmness, respectively, compared to fruit stored in 0 kPa CO_2 (Gunes et al., 2002). In another study where weight loss of 'Stevens' cranberry fruit averaged 1.6% after 2 months of storage, fruit firmness, measured as compression, decreased 13% in air stored fruit (Forney, 2002). Concentrations of O_2, ranging from 1 to 15 kPa, and CO_2, ranging from 0 to 15 kPa, had no effect on fruit firmness.

Most CA storage conditions did not appear detrimental to cranberry flavor. Controlled atmospheres of 1, 3, 10, or 21 kPa O_2 combined with 0, 5, or 10 kPa CO_2 caused no off-flavor after cooking in cranberries stored for up to 30 weeks (Anderson et al., 1963). Gunes et al. (2002) reported that atmospheres of 30 kPa CO_2 and 21 kPa O_2 had no detrimental effect on flavor based on informal assessments. Cranberry sauce made from fruit stored in 100 kPa N_2 for 14 months had superior flavor to that made from frozen berries as determined by a taste panel (Stark et al., 1974). However, fruit exposed to 100 kPa N_2 atmospheres were not suitable for use as fresh fruit or juice due to off-odors and flavors (Stark et al., 1969).

15.3.5.4   Effects of MAP

No reported research has been conducted on the use of MAP to extend the storage life of fresh cranberries due to the lack of benefit to atmosphere modification. High RH generally found in MAP would also be expected to be detrimental to fruit storage life. Anderson et al. (1963) stored cranberries in plastic bags as part of their CA storage trials, which resulted in the most fruit loss. They attributed this to the high RH in the bags.

15.3.6   Kiwifruit

Many studies have been conducted to evaluate the potential benefits of CA storage for kiwifruit. The major benefits are a delay in flesh softening and reduction in gray mold (*B. cinerea*) during cold storage (McDonald and Harman, 1982; Arpaia et al., 1984, 1985, 1986, 1994; Scott et al., 1984; Ben-Arie and Sonego, 1985; Mitchell, 1990; Figure 15.10). Atmospheres containing elevated concentrations of CO_2, with 15–20 kPa O_2, caused retardation in the softening of kiwifruit. This delay on fruit softening increased as the CO_2 content of the atmosphere increased from 4 to 10 kPa, but additional CO_2 above 10 kPa had no further effect on fruit firmness (McDonald and Harman, 1982). Although CA cold storage prolongs storage life of kiwifruit, the magnitude of the effect was found to vary from year to year and/or vineyard to vineyard. Contamination of the CA with very low concentrations of ethylene (0.1 μL L^{-1}) severely reduced the effectiveness of CA in maintaining kiwifruit firmness during cold storage at 0°C (Arpaia et al., 1986). The best results on delaying the rate of kiwifruit softening and increasing storage life up to 3–4 months beyond normal air-storage life have been reached using a CA of 5 kPa CO_2 and 2 kPa O_2 (McDonald and Harman, 1982; Mitchell, 1990) (Figure 15.11). However, fruit must be promptly cooled and placed under CA conditions (ethylene-free) within 1 week after harvest (Arpaia et al., 1984) for best results.

FIGURE 15.10

Example of gray mold disease caused by *B. cinerea* on 'Hayward' kiwifruit after 6 weeks of air storage at different storage temperatures −1°C (30°F), −0.5°C (31°F), 0°C (32°F), and 0.5°C (33°F).

2 kPa O$_2$ 5 kPa CO$_2$ 2 kPa O$_2$ and
5 kPa CO$_2$

−C$_2$H$_4$

+C$_2$H$_4$

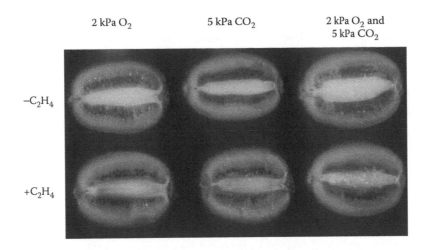

FIGURE 15.11
Effect of controlled atmosphere storage at 0°C with and without 1.0 μL L^{-1} ethylene (C$_2$H$_4$) on 'Hayward' kiwifruit stored for 6 weeks at 0°C.

15.3.6.1 Disease Development

Most kiwifruit decay problems occur as a result of infection by *B. cinerea* (Sommer et al., 1982; Tonini et al., 1989). While the rate of growth and spread of this organism is low at 0°C, it is a major cause of fruit loss because of the relatively long-term storage of kiwifruit. While *Botrytis* can directly invade the fruit, it also enters through wounds, by invasion of dead floral parts or other organic matter on the fruit surface, and by spreading from infected fruit to healthy surrounding fruit (nesting). During long-term storage, some individual kiwifruit become rotten and produce ethylene that can affect flesh softening of healthy fruit (Sommer et al., 1982). Even fruit farthest from the rotten fruit can soften more rapidly than fruit in trays that are rot-free. Because *Botrytis* rot invasion and development is associated with soft kiwifruit, any practice such as the proper use of CA that maintains firmness during storage will decrease the fruit rotting problem and prolong storage life.

15.3.6.2 Fruit Biochemistry

CO$_2$ and O$_2$ levels used during storage may affect internal chemical composition, which in turn may affect the volatile production of ripe fruit. It is suggested that alcohol metabolism contributed significantly to the ripe fruit volatile profile, particularly ester production when kiwifruit are exposed to CO$_2$ concentrations higher than 8–10 kPa for a long cold storage period (Arpaia et al., 1985). In many studies, CO$_2$ production, ethylene production, SSC increase, and TA decrease during storage and after ripening, but were not affected by CA conditions (Sfakiotakis et al., 1989; Arpaia et al., 1994; Ozer et al., 1999). However, excessive levels of CO$_2$ in the atmosphere increased ethanol accumulation without affecting SSC and TA. Ripeness stage influenced ethanol accumulation within kiwifruit tissues since as fruit approached senescence ethanol accumulation increased (Yanez-Lopez et al., 1999).

The effects of ultralow oxygen (ULO) (0.7 kPa O$_2$ + 0.7 kPa CO$_2$ and 1 kPa O$_2$ + 1 kPa CO$_2$) on ethylene biosynthesis and ripening of 'Hayward' kiwifruit during storage at 0°C and post-storage at 20°C was investigated. Kiwifruit removed from ULO-storage had drastically reduced capacity to produce ethylene mainly due to low 1-Aminocyclopropane-1-carboxylic acid (ACC) oxidase activity rather than reduced ACC production or ACC synthase activity (Sfakiotakis et al., 1989; Antunes et al., 2002).

15.3.6.3 Quality Attributes

Extensive studies have been conducted on the potential benefits of CA storage. The major reported benefits for kiwifruit are a substantial slowing in the rate of flesh softening during storage and a reduction in the development of *Botrytis* rot (McDonald and Harman, 1982; Arpaia et al., 1984, 1985, 1986, 1994; Scott et al., 1984; Ben-Arie and Sonego, 1985; Tonini et al., 1989; Mitchell, 1990). In these CA studies, fruit injury symptoms, including surface pitting, accelerated flesh softening, with long-term storage (beyond 4 months), and loss of green flesh color were reported at 10 kPa CO_2, with some evidence that 8 kPa CO_2 may be the upper limit tolerated by kiwifruit. Solubilization of cell wall softening-related components such as uronic acids and neutral sugar residues usually associated with pectic polymers (galactose, arabinose, and rhamnose) was faster in air-stored kiwifruit than CA cold-stored kiwifruit (Arpaia et al., 1994). Thus, the use of $\sim$5 kPa CO_2 and 2 kPa O_2 is recommended (McDonald and Harman, 1982; Mitchell, 1990) to prolong storage without affecting other quality attributes (Figure 15.11). Even in CA, the lowest levels of ethylene tested (0.05 μL L^{-1}) caused rapid flesh softening (Arpaia et al., 1986) (Figure 15.11). Furthermore, a fruit injury problem called white core inclusion (WCI) (Figures 15.11 and 15.12), which was earlier identified in California by Arpaia, et al., 1986 and involves a ethylene–CO_2 interaction, appeared after exposure to C_2H_4 as low as 0.05 μL L^{-1} in CA. A similar disorder (hard columella) was described in northern Italy when CA storage was used, particularly with early harvested fruit (Testoni, 1991). Reduction of this problem has been attained by removing ethylene in air and CA-stored kiwifruit.

It has been reported that the rate of softening bin-stored kiwifruit at 0°C in ethylene-free air or 5 kPa CO_2 + 2 kPa O_2 during 4 months was related to fruit size and storage conditions (Crisosto et al., 1999; Figure 15.11). Under both environmental storage

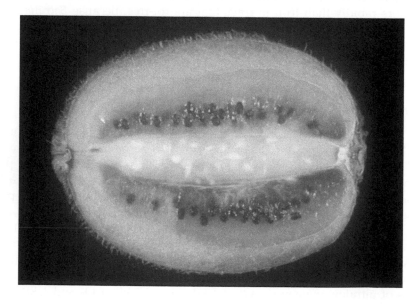

FIGURE 15.12

Example of WCIs in core tissue of 'Hayward' kiwifruit stored under CA conditions (2 kPa O_2 and 5 kPa CO_2) in the presence of 0.5 μL L^{-1} ethylene.

conditions, large size (~101 g) fruit had a slower rate of softening than medium (93 g) and small size (81 g) kiwifruit. Air-stored kiwifruit softened approximately 2.6-fold faster than CA-stored fruit. Because kiwifruit are more susceptible to physical damage during packaging when they soften below 2.3 kg-force, the number of weeks to reach this firmness under each of the different size storage conditions was calculated. Under air conditions large, medium, and small size kiwifruit reached 2.3 kg-force fruit firmness by 12, 10, and 8 weeks, respectively. Large, medium, and small size kiwifruit under CA conditions reached 2.3 kg-force fruit firmness by 49, 30, and 20 weeks, respectively.

15.3.6.4 *Effects of MAP*

Information on the use of MAP on kiwifruit is very limited in spite of several reports on the potential benefits of elevated CO_2 as a modified atmosphere treatment for kiwifruit (McDonald and Harman, 1982; Scott et al., 1984; Ben-Arie and Sonego, 1985; Mitchell, 1990). MAP conditions were generated by using sealed polyethylene liners, containing an ethylene absorbing material, with different thicknesses (0.04–0.05 mm), which results in an average 3–4 kPa CO_2 concentration during the 6 months storage life for kiwifruit stored at −1°C. The gaseous composition of the MAP generated by the respiration of the fruit and benefits of the high CO_2 was related to the liner type (Ben-Arie and Sonego, 1985). The use of plastic tents inside large air storage room to obtain CA cold storage was also tested in kiwifruit where fruit were well cooled prior to tenting (Harman and McDonald, 1983). Unfortunately, this large plastic tent established ideal CA conditions too late during cold storage for best results since the respiration rate of kiwifruit is very low at 0°C.

15.4 Future Research Needs

The storage and market-life of many small fruits can be extended with the use of CA and/or MAP technologies. However, past research has tended to be biased toward judging end of storage-life based on visual appearance rather than on other quality attributes, such as health-promoting properties. Many small fruits and berries are naturally rich in antioxidant products that have been associated with reduction in the incidence of cancer, cardiovascular diseases, and other degenerative diseases. This high concentration of antioxidants is mainly due to the abundance of phenylpropanoids, notably anthocyanins and hydroxycinnamic acids, and ascorbic acid. However, understanding the effects of atmosphere modification and extended storage on these health-promoting compounds is in its infancy. Opportunities for the use of atmosphere modification to preserve and enhance these healthful components need to be identified and utilized.

Often expectations of CA have been overexaggerated, since they have traditionally been centered on extending shelf-life and have somewhat ignored or given less emphasis to flavor and taste. Longer shelf-life and good taste are not mutually exclusive. Greater recognition of the effects of CA and MAP on overall fruit quality (viz. flavor, nutritional value, texture etc.) is required in the future. Concomitant to this, the underlying mechanisms driving spatial and temporal changes in berry fruit biochemistry, and texture should be investigated further. The genetic control and effect of CA on such processes (e.g., cell wall metabolism) need to be better understood and the mechanisms by which physiological injury results and how this may be mediated or elevated should also be considered. Techniques such as chemometric profiling of target analytes with reported flavor, taste, and health-related properties could be used to better model the dynamic and multivariant effects of CA and MAP.

It is well established that CA can reduce postharvest disease incidence and severity. The underlying mechanisms of suppressing disease through using CA have mainly been attributed to CA atmospheres being outside the ideal cardinal atmospheres for normal growth of the pathogen. Appropriate CA storage not only reduces the respiration rate of the product, thereby maintaining product quality and NDR (Terry and Joyce, 2004), but also by reducing the respiration rate of disease-causing fungal pathogens, leading to suppression or inhibition of growth. More research is required to establish whether upholding sufficient levels of NDR of berry fruit using CA will extend the period of postharvest quiescence and thus postharvest life through maintaining concentrations of preformed antifungal compounds (Terry et al., 2004) and/or the ability of fruit to produce de novo phytoalexins and pathogenesis-related proteins.

The vast majority of CA/MAP research has been conducted on North American-derived cultivars. There has understandably been a tendency in the past to believe that optimum CA conditions for one cultivar can be readily transferred to another. Moreover, the influence of preharvest growing conditions and management need also to be considered in more detail. It is for these reasons that more research is required on defining appropriate CA conditions for each cultivar and postharvest scenario. Salveit (2003) argues that due to heterogeneity in fresh produce and its dynamic response to CA it may be impossible to truly define and agree upon optimum storage atmospheres for each fresh produce type and cultivar. In addition, more research is warranted whereby CA conditions are attested under "real world" conditions rather than merely relying on laboratory-based studies when, for example, temperature control is more precise. The effect and impact of cool chain abuse (viz. variable temperature and associated RH%) should be considered in future CA research, such that alterative gaseous environments, and combined postharvest treatments are explored.

15.5 Conclusions

Despite extensive research on elucidating optimum CA concentrations for small fruits and berries, CA and MAP are still relatively underutilized by industry. The comparative slow uptake in CA storage regimes for berry fruit is manifest, in part, by short storage life, the tendency for rapid product degradation after removal from CA conditions and common problems associated with prolonged exposure during long-distance transport. What is clear is that when used appropriately CA and MAP can prolong postharvest life while maintaining product quality. However, the use of CA for small fruit has primarily been for long-distance transport (e.g., United States to EU, Chile to United States) and extending market seasons. Due to the added costs and logistics, it is rarely used commercially for the local marketing.

References

Agar, I.T. and J. Streif. 1996. Effect of high CO_2 and controlled atmosphere (CA) storage on the fruit quality of raspberry. *Gartenbauwiss* 61:261–267.

Agar, I.T., J. Streif, and F. Bangreth, F. 1997. Effect of high CO_2 and controlled atmosphere (CA) on the ascorbic and dehydroascorbic acid content of some berry fruits. *Postharvest Biol. Technol.* 11:47–55.

Agar, T., J.M. Garcia, U. Miedtke, and J. Streif. 1990. Effect of high CO_2 and low O_2 concentrations on the growth of *Botrytis cinerea* at different temperatures. *Gartenbauwiss* 55:219–222.

Allan-Wojtas, P.M., C.F. Forney, S.E. Carbyn, and K.U.K.G. Nicholas. 2001. Microstructural indicators of quality-related characteristics of blueberries—An integrated approach. *Lebensm.-Wiss. u.-Technol.* 34:23–32.

Allende, A., A. Marín, B. Buendía, F. Tomás-Barberán, and M. Gil. 2007. Impact of combined postharvest treatments (UV-C light, gaseous O_3, superatmospheric O_2 and high CO_2) on health promoting compounds and shelf-life in strawberries. *Postharvest Biol. Technol.* 46:201–211.

Almenar, E., O. Hernández-Muñoz, J.M. Lagarón, R. Catalá, and R. Gavara. 2006. Controlled atmosphere of wild strawberry fruit (*Fragaria vesca* L.). *J. Agric. Food Chem.* 54:86–91.

Almenar, E., D.V. Del Valle, R. Catala, and R. Gavara. 2007. Active packaging for wild strawberry fruit. (*Fragaria vesca* L.). *J. Agric. Food Chem.* 55:2240–2245.

Anderson, R.E., R.E. Hardenburg, and H.C. Vaught. 1963. Controlled-atmosphere storage studies with cranberries. *Proc. Am. Soc. Hort. Sci.* 83:416–422.

Antunes, M.D.C. and E.M. Sfakiotakis. 2002. Ethylene biosynthesis and ripening behavior of 'Hayward' kiwifruit subjected to some control atmospheres. *Postharvest Biol. Technol.* 26:167–179.

Arpaia, M.L., F.G. Mitchell, G. Mayer, and A.A. Kader. 1984. Effects of delays in establishing controlled atmospheres on kiwifruit softening during and following storage. *J. Am. Soc. Hort. Sci.* 109:768–770.

Arpaia, M.L., F.G. Mitchell, A.A. Kader, and G. Mayer. 1985. Effects of 2% O_2 and varying concentrations CO_2 with or without C_2H_4 on the storage performance of kiwifruit. *J. Am. Soc. Hort. Sci.* 110:200–203.

Arpaia, M.L., F.G. Mitchell, A.A. Kader, and G. Mayer. 1986. Ethylene and temperature effects on softening and white core inclusions of kiwifruit stored in air or controlled atmospheres. *J. Am. Soc. Hort. Sci.* 111:149–153.

Arpaia, M.L., F.G. Mitchell, and A.A. Kader. 1994. Postharvest physiology and causes of deterioration. In: *Kiwifruit Growing and Handling*, pp. 88–93. UC-DANR, Publication 3344.

Artés-Hernández, F., E. Aguayo, and F. Artés. 2004. Alternative atmosphere treatments for keeping quality of 'Autumn Seedless' table grapes during long-term cold storage. *Postharvest Biol. Technol.* 31:59–67.

Artés-Hernández, F., F.A. Tomas-Barberan, and F. Artés. 2006. Modified atmosphere packaging preserves quality of SO_2-free 'Superior Seedless' table grapes. *Postharvest Biol. Technol.* 39:146–154.

Baab, G. 1987. Improvement of strawberry keeping quality. *Obstbau* 12:265–268.

Barkai-Golan, R. 2001. *Postharvest Diseases of Fruits and Vegetables*. Elsevier, Amsterdam, the Netherlands.

Barth, M.M., C. Zhou, J. Mercier, and F.A. Payne. 1995. Ozone storage effects on anthocyanin content and fungal growth in blackberries. *J. Food Sci.* 60:1286–1288.

Beaudry, R.M. 1993. Effect of carbon dioxide partial pressure on blueberry fruit respiration and respiratory quotient. *Postharvest Biol. Technol.* 3:249–258.

Beaudry, R.M., A.C. Cameron, A. Shirazi, and D.L. Dostal-Lange. 1992. Modified-atmosphere packaging of blueberry fruit: Effect of temperature on package O_2 and CO_2. *J. Am. Soc. Hort. Sci.* 117:436–441.

Beaudry, R.M., C.E. Moggia, J.B. Retamales, and J.F. Hancock. 1998. Quality of 'Ivanhoe' and 'Bluecrop' blueberry fruit transported by air and sea from Chile to North America. *HortScience* 33:313–317.

Ben-Arie, R. and L. Sonego. 1985. Modified-atmosphere storage of kiwifruit (*Actinidia chinensis* Planch.) with ethylene removal. *Scientia Hort.* 27:263–273.

Benitez-Burraco, A., R. Blanco-Portales, J. Redondo-Nevado, E. Moyano, J.L. Caballero, and J. Munoz-Blanco. 2003. Cloning and characterisation of two ripening-related strawberry (*Fragaria* × *ananassa* cv. Chandler) pectate lyase genes. *J. Exp. Bot.* 54:633–645.

Berna, A.Z., S. Geysen, S. Li, B.E. Velinden, J. Lammertyn, and B.M. Nicolai. 2007. Headspace fingerprint mass spectrometry to characterize strawberry aroma at super-atmospheric oxygen conditions. *Postharvest Biol. Technol.* 46:230–236.

Berry, G. and J. Aked. 1996. Packaging for fresh produce—A case study on table grapes. *Postharvest News Info.* 7:40N–44N.

Berry, G. and J. Aked. 1997. Controlled atmosphere alternatives to the postharvest use of sulfur dioxide to inhibit the development of *Botrytis cinerea* in table grapes. CA '97 Proceedings, Vol. 3, Postharvest Horticulture Series No. 17. Postharvest Outreach Program, University of California, Davis, CA, p. 100.

Berry, G. and J. Aked. 1998. The control of *Botrytis cinerea* on table grapes under controlled atmosphere storage. In: *Nonconventional Methods for the Control of Postharvest Disease and Microbiological Spoilage. Proceedings of the COST 914 and COST 915 Joint Workshop*, October 9–11, 1997, Bologna, Italy, pp. 189–194.

Boone, D.M. 1995a. Blotch rot. In: *Compendium of Blueberry and Cranberry Diseases*, F.L. Caruso and D. C. Ramsdell (Eds.). St. Paul, MN: ASP Press, p. 32.

Boone, D.M. 1995b. Yellow rot. In: *Compendium of Blueberry and Cranberry Diseases*, F.L. Caruso and D. C. Ramsdell (Eds.). St. Paul, MN: ASP Press, p. 42.

Bristow, P.R. and K.D. Patten. 1995. In: *Compendium of Blueberry and Cranberry Diseases*, F.L. Caruso and D.C. Ramsdell (Eds.). St. Paul, MN: ASP Press, pp. 68–69.

Browne, K.M., J.D. Geeson, and C. Dennis. 1984. The effects of harvest date and CO_2-enriched storage atmospheres on the storage and shelf-life strawberries. *J. Hort. Sci.* 59:197–204.

Bünemann, G., D.H. Dewey, and D.P. Watson. 1957. Anatomical changes in the fruit of the Rubel blueberry during storage in controlled atmospheres. *Proc. Am. Soc. Hort. Sci.* 70:156–160.

Cameron, A.C., R.M. Beaudry, N.H. Banks, and M.V. Yelanich. 1994. Modified-atmosphere packaging of blueberry fruit: Modeling respiration and package oxygen partial pressures as a function of temperature. *J. Am. Soc. Hort. Sci.* 119:534–539.

Cappellini, R.A., M.J. Ceponis, and G. Koslow. 1982. Nature and extent of losses in consumer-grade samples of blueberries in greater New York. *HortScience* 17:55–56.

Carris, L.M. 1995. Black rot. In: *Compendium of Blueberry and Cranberry Diseases*, F.L. Caruso and D.C. Ramsdell (Eds.). St. Paul, MN: ASP Press, pp. 31–32.

Caruso, F.L. 1995. Botryosphaeria fruit rot and berry speckle. In: *Compendium of Blueberry and Cranberry Diseases*, F.L. Caruso and D.C. Ramsdell (Eds.). St. Paul, MN: ASP Press, pp. 32–33.

Caruso, F.L. and D.C. Ramsdell. 1995. *Compendium of Blueberry and Cranberry Diseases*. St. Paul, MN: APS Press.

Ceponis, M.J. and R.A. Cappellini. 1985. Reducing decay in fresh blueberries with controlled atmospheres. *HortScience* 20:228–229.

Ceponis, M.J. and A.W. Stretch. 1983. Berry color, water-immersion time, rot, and physiological breakdown of cold-stored cranberry fruits. *HortScience* 18:484–485.

Chambroy, Y., M.H. Guinebretiere, G. Jacquemin, M. Riech, L. Breulis, and M. Souty. 1993. Effects of carbon-dioxide on shelf-life and post harvest decay of strawberries [sic.] fruit. *Sci. Aliments* 13:409–423.

Chervin, C., P. Westercamp, and G. Monteils. 2005. Ethanol vapours limit *Botrytis* development over postharvest life of table grapes. *Postharvest Biol. Technol.* 36:319–322.

Cimino, A., M. Mari, and A. Marchi. 1987. U.L.O. storage of table grapes and kiwifruit. *Proceedings of the 17th International Congress on Refrigeration*, Vienna, Austria, pp. 642–646.

Couey, H.M. and J.M. Wells. 1970. Low oxygen or high carbon dioxide atmospheres to control postharvest decay of strawberries. *Phytopathology* 60:47–49.

Crisosto, C.H., D. Garner, and K. Saez. 1999. Kiwifruit size influences softening rate during storage. *Calif. Agric.* 53:29–31.

Crisosto, C.H., D. Garner, and G.M. Crisosto. 2002a. High carbon dioxide atmospheres affect stored 'Thompson Seedless' table grapes. *HortScience* 37:1074–1078.

Crisosto, C.H., D. Garner, and G. Crisosto. 2002b. Carbon dioxide-enriched atmospheres during cold storage limit losses from *Botrytis* but accelerate rachis browning of 'Red Globe' table grapes. *Postharvest Biol. Technol.* 26:181–189.

Deng, Y., Y. Wu, and L. Yunfei. 2006. Physiological responses and quality attributes of 'Kyoho' grapes to controlled atmosphere storage. *Lebensm. Wiss. u. Technol.* 39:584–590.

Deng, Y., Y. Wu, and L. Yunfei. 2007. Effects of high CO_2 and low O_2 atmospheres on the berry drop of 'Kyoho' grapes. *Food Chem.* 100:768–773.

Doughty, C.C., M.E. Patterson, and A.Y. Shawa. 1967. Storage longevity of the 'McFarlin' cranberry as influenced by certain growth retardants and stage of maturity. *Proc. Am. Soc. Hort. Sci.* 91:192–204.

El-Kazzaz, M.K., N.F. Sommer, and R.J. Fortlage. 1983. Effect of different atmospheres on postharvest decay and quality of fresh strawberries. *Phytopathology* 73:282–285.

Eris, A., C. Turkben, M.H. Ozer, and J. Henze. 1993. A research on CA-storage of grape cultivars 'Alphonse Lavallee' and 'Razaki.' *Proceedings of the 6th International CA Research Conference,* NRAES-71, Ithaca, NY: Cornell University, pp. 705–710.

Fan, X., M.E. Patterson, J.A. Robbins, J.K. Fellman, and R.P. Cavalieri. 1993. Controlled atmosphere storage of 'Bluecrop' blueberries (*Vaccinium corymbosum* L.). *Proceedings of the 6th International Controlled Atmosphere Research Conference,* NRAES-71, Vol. 2, pp. 699–704.

Ferreira, M.D., J.K. Brecht, S.A. Sargent, and J.J. Aracena. 1994. Physiological responses of strawberry to film wrapping and precooling methods. *Proc. Fla. State Hort. Soc.* 107:265–269.

Forney, C.F. 2002. Improvement of the storage-life of fresh cranberry fruit. AAFC Technical Report, Kentville, Nova Scotia.

Forney, C.F. 2003. Postharvest handling and storage of fresh cranberries. *HortTechnology* 13:267–272.

Forney, C.F. 2008. Optimizing the storage temperature and humidity for fresh cranberries: A reassessment of chilling sensitivity. *HortScience* 43:439–446.

Forney, C.F. 2009. Postharvest issues in blueberry and cranberry and methods to improve market-life. *Acta Hort.* (in press).

Forney, C.F., M.A. Jordan, M.R. Vinquist-Tymchuk, S.A.E. Fillmore. 2007. Effects of controlled atmosphere storage on fruit quality and composition in three highbush blueberry cultivars. AAFC Technical Report, Kentville, Nova Scotia.

Forney, C.F., W. Kalt, M.A. Jordan, M.R. Vinquist-Tymchuk, S.A.E. Fillmore. 2008. Effects of controlled atmosphere storage on antioxidant capacity and phenolics in three highbush blueberry cultivars. *HortScience* 43:1169.

Forney, C.F., K. Nicholas, and J. Leyte. 1997. The effects of controlled atmosphere on storage life and quality of highbush blueberries. AAFC Technical Report, Kentville, Nova Scotia.

Forney, C.F., K.U.K.G. Nicholas, and M.A. Jordan. 1998. Effects of postharvest storage on firmness of 'Burlington' blueberry fruit. *Proceedings of the 8th North American Blueberry Research Extension Workers Congress,* Wilmington, North Carolina, pp. 227–232.

Forney, C.F., M.A. Jordan, and K.U.K.G. Nicholas. 2003. Effects of CO_2 on physical, chemical, and quality changes in 'Burlington' blueberries. *Acta Hort.* 600:587–593.

Garcia-Gimeno, R.M., C. Sanz-Martinez, J.M. Garcia-Martos, and G. Zurera-Cosano. 2002. Modeling *Botrytis cinerea* spores growth in carbon dioxide enriched atmospheres. *J. Food Sci.* 67:1904–1907.

Gil, M.I., D.M. Holcroft, and A.A. Kader. 1997. Changes in strawberry anthocyanins and other polyphenols in response to carbon dioxide treatments. *J. Agric. Food Chem.* 45:1662–1667.

Goto, T., M. Goto, K. Chachin, and T. Iwata. 1995. Effects of high carbon dioxide with short term treatment on quality of strawberry fruits. *J. Jpn. Soc. Food Sci.* 42:176–182.

Guichard, E., Y. Chambroy, M. Reich, N. Fournier, and M. Souty. 1992. Effect of carbon-dioxide concentration on aroma of strawberries after storage. *Sci. Aliments* 12:83–100.

Gunes, G., R.H. Liu, and C.B. Watkins. 2002. Controlled-atmosphere effects on postharvest quality and antioxidant activity of cranberry fruits. *J. Agric. Food Chem.* 50:5932–5938.

Haffner, K., H.J. Rosenfield, G. Skrede, and L. Wang. 2002. Quality of red raspberry *Rubus idaeus* L. cultivars after storage in controlled and normal atmospheres. *Postharvest Biol. Technol.* 24:279–289.

Han, C., Y. Zhao, S.W. Leonard, and M.G. Traber. 2004. Edible coatings to improve storability and enhance nutritional value of fresh and frozen strawberries (*Fragaria × ananassa*) and raspberries (*Rubus ideaus*). *Postharvest Biol. Technol.* 33:67–78.

Hansen, H. 1986. Use of high CO_2 concentration in transport and storage of soft fruit. *Obstbau* 11:268–271.

Harb, J.Y. and J. Streif. 2004. Controlled atmosphere storage of highbush blueberries cv. 'Duke'. *Eur. J. Hort. Sci.* 69:66–72.

Harb, J.Y. and J. Streif. 2006. Einfluss verschiedener Lagerbedingungen auf Haltbarkeit und Fruchtqualität von Heidelbeeren der Sorte 'Bluecrop.' *Erwerbs-Obstbau* 48:115–120.

Harb, J., R. Bisharat, and J. Streif. 2008. Changes in volatile constituents of blackcurrants (Ribes nigrum L. cv. 'Titania') following controlled atmosphere storage. *Postharvest Biol. Technol.* 47:271–279.

Harker, F.R., H.J. Elgar, C.B. Watkins, P.J. Jackson, and I.C. Hallet. 2000. Physical and mechanical changes in strawberry fruit after high carbon dioxide treatments. *Postharevst Biol. Technol.* 19:139–146.

Harman, J.E. and B. McDonald. 1983. Controlled atmosphere storage of kiwifruit: Effects on storage life and fruit quality. *Acta Hort.* 138:195–201.

Harris, C.M. and J.M. Harvey. 1973. Quality and decay of Californian strawberries stored in CO_2-enriched atmospheres. *Plant Dis. Rep.* 57:44–46.

Hertog, M.L.A.T.M., H.A.M. Boerrigter, G.J.P.M. van den Boogard, L.M.M. Tijskens, and A.C.R. van Schaik. 1999. Predicting keeping quality of strawberries (cv. 'Elsanta') packed under modified atmospheres: an integrated model approach. *Postharvest Biol. Technol.* 15:1–12.

Holcroft, D.M. and A.A. Kader. 1999a. Controlled atmosphere-induced changes in pH and organic acid metabolism may affect color of stored strawberry fruit. *Postharvest Biol. Technol.* 17:19–32.

Holcroft, D.M. and A.A. Kader. 1999b. Carbon-dioxide-induced changes in color and anthocyanin synthesis of stored strawberry fruit. *HortScience* 34:1244–1248.

Hruschka, H.W. 1970. Physiological breakdown in cranberries-inhibition by intermittent warming during cold storage. *Plant Dis. Rep.* 54:219–222.

Joles, D.W., A.G. Cameron, A. Shiraziu, P.D. Petracek, and R.M. Beaudry. 1994. Modified atmosphere packaging of Heritage red raspberry fruit: Respiratory response to reduced oxygen, enhanced carbon dioxide and temperature. *J. Am. Soc. Hortic. Sci.* 119:545–550.

Kader, A.A. 1989. A summary of CA requirements and recommendations for fruit other than pome fruit. *5th International Controlled Atmosphere Conference Proceedings*, June 14–16, Wenatchee, WA.

Kalt, W., C.F. Forney, A. Martin, and R.L. Prior. 1999. Antioxidant capacity, vitamin C, phenolics, and anthocyanins after fresh storage of small fruits. *J. Agric. Food Chem.* 47:4638–4644.

Ke, D., L. Goldstein, M. O'Mahony, and A.A. Kader. 1991. Effects of short-term exposure to low O_2 and high CO_2 atmosphere on quality attributes of strawberries. *J. Food Sci.* 56:50–54.

Ke, D., L. Zhou, and A.A. Kader. 1994. Mode of oxygen and carbon dioxide action on strawberry ester biosynthesis. *J. Am. Soc. Hortic. Sci.* 119:971–975.

Kim, H.K., Y. Song, and K.L. Yam. 1995. Influence of modified atmosphere on quality attributes of blueberry. *Foods Biotechnol.* 4:112–116.

Krupa, T. and K. Tomala. 2007. Antioxidant capacity, anthocyanin content profile in 'Bluecrop' blueberry fruit. *Vegetable Crops Res. Bull.* 66:129–141.

Larsen, M. and C.B. Watkins. 1995. Firmness and concentrations of acetaldehyde, ethyl acetate and ethanol in strawberries stored in controlled and modified atmospheres. *Postharvest Biol. Technol.* 5:39–50.

Laszio, J.C. 1985. The effect of controlled atmosphere on the quality of stored table grapes. *Decid. Fruit Grower* 32:436–438.

Lichter, A., Y. Zutkhy, L. Sonego, O. Dvir, T. Kaplunov, P. Sarig, and R. Ben-Arie. 2002. Ethanol controls postharvest decay of table grapes. *Postharvest Biol. Technol.* 24:301–308.

Lichter, A., Y. Zutahy, T. Kaplunov, N. Aharoni, and S. Lurie. 2005. The effect of ethanol dip and modified atmosphere on prevention of *Botrytis* rot of table grapes. *HortTechnology* 15:284–291.

Lockhart, C.L., F.R. Forsyth, R. Stark, and I.V. Hall. 1971. Nitrogen gas suppresses microorganisms on cranberries in short term storage. *Phytopathology* 61:335.

Lurie, S., E. Pesis, O. Gadiyeva, et al. 2006. Modified ethanol atmosphere to control decay of table grapes during storage. *Postharvest Biol. Technol.* 42:222–227.

Luvisi, D.A., H. Shorey, J. Smilanick, J. Thompson, B. Gump, and J. Knutson. 1992. Sulfur dioxide fumigation of table grapes. University of California, DANR, Bulletin 1932.

Makino, Y. and T. Hirata. 1997. Modified atmosphere packaging of fresh produce with a bio-degradable laminate of chitosan-cellulose and polycaprolactone. *Postharvest Biol. Technol.* 10:247–254.

McDonald, B. and J.E. Harman. 1982. Controlled atmosphere storage of kiwifruit. I. Effect on fruit firmness and storage. *Scientia Hort.* 17:113–123.

Mitchell, F.G. 1990. Postharvest physiology and technology of kiwifruit. *Acta Hort.* 282:291–307.

Mlikota Gabler, F., J.L. Smilanick, J.M. Ghosoph, and D.A. Margosan. 2005. Impact of postharvest hot water or ethanol treatment of table grapes on gray mold incidence, quality, and ethanol content. *Plant Dis.* 89:309–316.

Nelson, K.E. 1969. Controlled atmosphere storage of table grapes. *Proceedings of National CA Research Conference*, Michigan State University, Horticulture Report No. 9, pp. 69–70.

Nielsen, T. and A. Leufvén. 2008. The effect of modified atmosphere packaging on the quality of Honeoye and Korona strawberries. *Food Chem.* 107:1053–1063.

Nunes, M.C.N., A.M.M.B. Morais, J.K. Brecht, and S.A. Sargent. 1995. Quality of strawberries after storage in controlled atmospheres at above optimum storage temperatures. *Proc. Fla. State. Hort. Soc.* 108:2733–278.

Nunes, M.C.N., A.M.M.B. Morais, J.K. Brecht, and S.A. Sargent. 2002. Fruit maturity and storage temperature influence response of strawberries to controlled atmospheres. *J. Am. Soc. Hort. Sci.* 127:836–842.

Nunez-Barrios, A., D.S. NeSmith, M. Chinnan, and S.E. Prussia. 2005. Dynamics of rabbiteye blueberry fruit quality in response to harvest method and postharvest handling temperature. *Small Fruits Rev.* 4:73–81.

Oudemans, P.V., F.L. Caruso, and A.W. Stretch. 1998. Cranberry fruit rot in the Northeast: A complex disease. *Plant Dis.* 82:1176–1184.

Özer, M.H., A. Eris, R. Turk, and N. Sivritepe. 1999. A research on controlled atmosphere storage of kiwifruit. *Acta Hort.* 485:293–300.

Palmer Wright, K. and A.A. Kader. 1997. Effect of slicing and controlled-atmosphere storage on the ascorbate content and quality of strawberries and persimmons. *Postharvest Biol. Technol.* 10:39–48.

Patterson, M.E., C.C. Doughty, S.O. Graham, and B. Allan. 1967. Effect of bruising on postharvest softening, color changes and detection of polygalacturonase enzyme in cranberries. *Proc. Am. Soc. Hort. Sci.* 90:498–505.

Pelayo, C., S.E. Ebeler, and A.A. Kader. 2003. Postharvest life and flavour quality of three strawberry cultivars kept at 5°C in air or air + 20kPa CO_2. *Postharvest Biol. Technol.* 27:171–183.

Pepin, H.S. and D.M. Boone. 1995. End rot. In: *Compendium of Blueberry and Cranberry Diseases*, F.L. Caruso and D.C. Ramsdell (Eds.). St. Paul, MN: ASP Press, pp. 36–37.

Peréz, A.G. and C. Sanz. 2001. Effect of high-oxygen and high-carbon-dioxide atmospheres on strawberry flavor and other quality traits. *J. Agric. Food Chem.* 49:2370–2375.

Peréz, A.G., C. Sanz, J.J. Ríos, R. Olías, and J.M. Olías. 1999. Effects of ozone treatments on postharvest strawberry quality. *J. Agric. Food Chem.* 47:1652–1656.

Perkins Veazie, P. and J.K. Collins. 2002. Quality of erect-type blackberry fruit after short intervals of controlled atmosphere storage. *Postharvest Biol. Technol.* 25:235–239.

Plestenjak, A., U. Brodnik, J. Hribar, and M. Simcic. 1998. Post-harvest behaviour of highbush blueberries (*Vaccinium corymbosum* L.). In: Woltering, E.J. et al. (eds.) *The Post-Harvest Treatment of Fruit and Vegetables*, E.J. Woltering et al. (Eds.). Current status and future prospects, *Proceedings of the 6th International Symposium of the Europeian Concerted Action Program COST 94*, October 19–22, 1994, Oosterbeek, the Netherlands, pp. 23–30.

Prange, R.K., S.K. Asiedu, J.R. DeEll, and A.R. Westgarth. 1995. Quality of Fundy and Blomidon lowbush blueberries: Effects of storage atmosphere, duration and fungal inoculation. *Can. J. Plant Sci.* 75:479–483.

Remberg, S.F., K. Haffner, and R. Blomhoff. 2003. Total antioxidant capacity and other quality criteria in blueberries cvs 'Bluecrop', 'Hardyblue', 'Patriot', 'Putte' and 'Aron' after storage in cold store and controlled atmosphere. *Acta Hort.* 600:595–598.

Retamales, J., B.G. Defilippi, M. Arias, P. Castillo, and D. Manríquez. 2003. High-CO_2 controlled atmospheres reduce decay incidence in Thompson Seedless and Red Globe table grapes. *Postharvest Biol. Technol.* 29:177–182.

Robinson, J.E., K.M. Brown, and W.G. Burton. 1975. Storage characteristics of some vegetables and soft fruits. *Ann. Appl. Biol.* 81:339.

Roelofs, F.P.M.M. and A.J.P.v.d. Waart. 1993. Long-term storage of red currants under controlled atmosphere conditions. *Acta Hort.* 352 217–222.

Sacks, E.J. and D.V. Shaw. 1994. Optimum allocation of objective color measurements for evaluating fresh strawberries. *J. Am. Soc. Hort. Sci.* 119:330–334.

Saltveit, M.E. 2003. Is it possible to find an optimal controlled atmosphere? *Postharvest Biol. Technol.* 27:3–13.

Saltveit, M.E. Jr., and W.E. Ballinger. 1983. Effects of anaerobic nitrogen and carbon dioxide atmospheres on ethanol production and postharvest quality of blueberries. *J. Am. Soc. Hort. Sci.* 108:459–462.

Sanz, C., A.G. Peréz, R. Olías, and J.M. Olías. 1999. Quality of strawberries packed with perforated polypropylene. *J. Food Sci.* 64:748–752.

Sargent, S.A., J.K. Brecht, and C.F. Forney. 2006. Blueberry harvest and postharvest operations: quality maintenance and food safety. In: *Blueberries for Growers, Gardeners, Promoters,* N.F. Childers and P.M. Lyrene (Eds.), DeLeon Springs, FL: E.O. Painter Printing Co.

Schotsmans, W., A. Molan, and B. MacKay. 2007. Controlled atmosphere storage of rabbiteye blueberries enhances postharvest quality aspects. *Postharvest Biol. Technol.* 44:277–285.

Scott; K.J., J. Guigni, and W.M.C. Bailey. 1984. The use of polyethylene bags and ethylene absorbent to extend the life of kiwifruit (*Actinidia chinensis* Planch.) during cool storage. *J. Hort. Sci.* 59:563–566.

Sfakiotakis, E., P. Ververidis, and G. Stavroulakis. 1989. The control of autocatalytic ethylene and ripening in kiwifruit by temperature and controlled atmosphere storage. *Acta Hort.* 258:115–123.

Shaw, G.W. 1970. The effects of controlled atmosphere stoareg on the quality and shelf-life of fresh strawberries with special references to *Botrytis cinerea* and *Rhizopus nigricans. Diss. Abstr. Int. Sect. B Sci. Eng.* 30:1343.

Siro, I., F. Devlieghere, L. Jacxsens, M. Uyttendaele, and J. Debevere. 2006. The microbial safety of strawberry and raspberry fruits packaged in high-oxygen and equilibrium-modified atmospheres compared to air storage. *Int. J. Food Sci. Technol.* 41:93–103.

Smith, R.B. 1992. Controlled atmosphere storage of 'Redcoat' strawberry fruit. *J. Am. Soc. Hort. Sci.* 117:260–264.

Smith, R.B. and L.J. Skog. 1992. Postharvest carbon dioxide treatment enhances firmness of several cultivars of strawberry. *HortScience* 27:420–421.

Smittle, D.A. and W.R. Miller. 1988. Rabbiteye blueberry storage life and fruit quality in controlled atmospheres and air storage. *J. Am. Soc. Hort. Sci.* 113:723–728.

Sommer, N.F., M.K. El-Kazzaz, J.R. Buchanan, R.J. Fortlage, and M.A. El-Goorani. 1982. Carbonmonoxide suppression of postharvest diseases of fruits and vegetables. In: *Proceedings of the 3rd National Controlled Atmosphere Research Conference,* D.G. Richardson, and M. Meheriuk (Eds.), Beaverton, OR: Timber Press, pp. 289–297.

Song, J., L. Fan, C.F. Forney, M.A. Jordan, P.D. Hildebrand, W. Kalt, and D.A.J. Ryan. 2003. Effect of ozone treatment and controlled atmosphere storage on quality and phytochemicals in highbush blueberries. *Acta Hort.* 600:417–423.

Song, Y., H.K. Kim, and K.L. Yam. 1992. Respiration rate of blueberry in modified atmosphere at various temperatures. *J. Am. Soc. Hort. Sci.* 117(6):925–929.

Stark, R., I.V. Hall, F.R. Forsyth, and P.R. Dean. 1969. Cranberries, evaluated for fresh fruit and processing quality, after reduced oxygen storage. *Cranberries* 37(6):14,16 and 37(7):14,16.

Stark, R., F.R. Forsyth, C.L. Lockhart, and I.V. Hall. 1974. Processing quality of cranberries after extended storage in N_2 atmosphere with low and high relative humidities. *Can. Inst. Food Sci. Technol. J.* 7:9–10.

Stewart, D. 2003. Effect of high O_2 and N_2 atmospheres on strawberry quality. *Acta Hort.* 600:567–570.

Terry, L.A. and D.C. Joyce. 2004. Elicitors of induced resistance in postharvest horticultural crops: A brief review. *Postharvest Biol. Technol.* 32:1–13.

Terry, L.A., D.C. Joyce, N.K.B. Adikaram, and B.P.S. Khambay. 2004. Preformed antifungal compounds in strawberry fruit and flowers. *Postharvest Biol. Technol.* 31:201–212.

Testoni, A. 1991. Influence of storage conditions on softening of the columella. *Acta Hort.* 297:587–593.

Tian, S., Q. Fan, Y. Xu, F. Wang, and A. Jiang. 2001. Evaluation of the use of high CO_2 concentrations and cold storage to control *Monilinia fructicola* on sweet cherries. *Postharvest Biol. Technol.* 22:53–60.

Tonini, G., S. Brigati, and D. Caccioni. 1989. CA storage of kiwifruit: Influence on rots and storability. *Proceedings of the 5th International CA Research Conference*, June 14–16, 1989, Wenatchee, WA, Vol. 2, pp. 69–76.

Uota, K. 1957. Preliminary study on storage of 'Emperor' grapes in controlled atmospheres with and without sulfur dioxide fumigation. *Proc. Am. Soc. Hort. Sci.* 69:250–253.

Valverde, J.M., F. Guillen., D. Martinez-Romero, S. Castillo, M. Serrano, and D. Valero. 2005. Improvement of table grapes quality and safety by the combination of modified atmosphere packaging (MAP) and eugenol, menthol and thymol. *J. Agric. Food. Chem.* 53:7458–7464.

Van der Steen, C., L. Jacxsons, F. Devlieghere, and J. Debevere. 2002. Combining high oxygen atmospheres with low oxygen modified atmosphere packaging to improve the keeping quality of strawberries and raspberries. *Postharvest Biol. Technol.* 26:49–58.

Van Schaik, A.C.R. and J.A. Verschoor. 2003. CA-storage: Technology, application and research. State of the art in the Netherlands. *Acta Hort.* 600:181–187.

Watkins, C.B., J.E. Manzano-Mendez, J.F. Nock, J.J. Zhang, and K.E. Maloney. 1999. Cultivar variation in response of strawberry fruit to high carbon dioxide treatments. *J. Sci. Food Agric.* 79:886–890.

Wells, J.M. and M. Uota. 1970. Germination and growth of five fungi in low-oxygen and high-carbon dioxide atmospheres. *Phytopathology* 60:50–53.

Wills, R.B.H., V.V.V. Ku, and Y.Y. Lesham. 2000. Fumigation with nitric oxide to extend the postharvest life of strawberries. *Postharvest Biol. Technol.* 18:75–79.

Wooward, J.R. and A.J. Topping. 1972. The influence of controlled atmospheres on the respiration rates and storage behaviour of strawberry fruits. *J. Hort. Sci.* 47:547–553.

Wszelaki, A.L. and E.J. Mitcham. 2000. Effects of superatmospheric oxygen on strawberry fruit quality and decay. *Postharvest Biol. Technol.* 20:125–133.

Wszelaki, A.L. and E.J. Mitcham. 2003. Effect of combinations of hot water dips, biological control and controlled atmospheres for control of gray mold on harvested strawberries. *Postharvest. Biol. Technol.* 27:255–264.

Yahia, E.M., K.E. Nelson, and A.A. Kader. 1983. Postharvest quality and storage life of grapes as influenced by adding carbon monoxide to air or controlled atmospheres. *J. Am. Soc. Hort. Sci.* 108:1067–1071.

Yanez-Lopez, L., R.M. Galicia, R. Alonso, D. Cabrera, J. Rocha, J.A. Rodriguez, A. Torres, and M. Armelia. 1999. Effect of CO_2 enriched atmospheres on ethanol production accumulation in different ripeness stages of kiwifruit. *Acta Hort.* 498:293–298.

Zhang, J.Z.J. and C.B. Watkins. 2005. Fruit quality, fermentation products, and activities of associated enzymes during elevated CO_2 treatment of strawberry fruit at high and low temperature. *J. Am. Soc. Hort. Sci.* 130:124–130.

Zheng, Y., C.Y. Wang, S.Y. Wang, and W. Zheng. 2003. Effect of high-oxygen atmospheres on blueberry phenolics, anthocyanins, and antioxidant capacity. *J. Agric. Food Chem.* 51:7162–7169.

Zheng, Y., S.Y. Wang, C.Y. Wang, and W. Zhang. 2007. Changes in strawberry phenolics, anthocyanins, and antioxidant capacity in response to high oxygen treatments. *LWT* 40:49–57.

Zheng, Y., Z. Yang, and X. Chen. 2008. Effect of high oxygen atmospheres on fruit decay and quality in Chinese bayberries, strawberries and blueberries. *Food Control* 19:470–474.

16

Tropical Fruits

Elhadi M. Yahia and S.P. Singh

CONTENTS

16.1 Introduction

The increase in the demand, and thus, in the export of tropical crops, have increased the need to investigate and to develop technologies that can maintain the quality and postharvest life of these crops for prolonged periods. Modified atmosphere (MA) and controlled atmosphere (CA) are adequate technologies that can help extend the postharvest life of these crops. MA and CA are not used for storage of tropical crops, but are used for their marine transport (Yahia, 1998a, 2006a).

Until recently many tropical fruits were only grown in home gardens and small farms primarily for local consumption. Currently several tropical fruits are among the most important horticultural crops of international trade. The market for tropical fruits has increased significantly in the last few decades. This is due to several factors including changes in diet habits, demand for exotic articles, improved technologies such as storage and transport, and free under World Trade Organization (WTO) regime. Prices of agricultural commodities have been steadily declining in the last 20 years, however, tropical fruit prices have been rising (Buchanan, 1994). The improved prices, improved technologies, and increased demand are resulting in increased plantations of a diversity of tropical fruits in several regions of the world.

Tropical crops are chilling sensitive (Yahia, 2006b). Some types of avocados, bananas, breadfruit, carambola, cherimoya, jackfruit, mamey, mango, mangosteen, papaya, pineapple, sapota, soursop, and white sapote are very sensitive to chilling injury (CI). Durian, feijoa, sugar apple, and tamarillo are moderately sensitive. Chilling sensitivity of tropical fruits does not permit their long-term maintenance in low temperature, and as a consequence all these crops have a relatively short postharvest life compared to many temperate and subtropical fruits, only few weeks at the most. MA and CA have been shown to ameliorate chilling sensitivity in several crops, including those of tropical origin (Yahia, 1998a, 2006b).

The tropics are characterized by conditions that favor the spread of insects and diseases (high temperature and relative humidity). Some of the most important diseases, which infect tropical fruits and cause major losses, include anthracnose (caused by *Colletotrichum gloeosprioides* Penz.) and stem end rot (caused by *Diplodia natalensis* P. Evans). Anthracnose is the major postharvest problem in bananas, mangoes, papayas, and contributes to most of their losses. MA/CA can control some decay either directly or indirectly by delaying ripening and senescence of the commodity, and thus maintaining the resistance to pathogen attack (Yahia, 1998a, 2006b). Effective control of pathogens is essential for the postharvest maintenance of these crops, and for the successful application of MA/CA.

Many insects infect tropical fruit crops and impede their commercialization due to quarantine restrictions imposed by various countries (Yahia, 1998b, 2004, 2006a). Some of the most important insects include various species of fruit flies such as the *Ceratitis* in several regions of the world, several *Anastrefa* species in South and Central America and the West Indies, the genus *Dacus* in Africa and Asia, etc. (Yahia, 2006a). Quarantine treatments for tropical fruits are needed in order to distribute them around the world. Traditionally, chemical fumigants (mainly ethylene dibromide [EDB]/and methyl bromide [MB]) have been the principal treatments used for this purpose. However, EDB has been banned (Federal Register, 1984) because of health risks. MB is still been used for some crops, but with restrictions due to concerns over the use of toxic chemicals, and the uncertainty of prohibition. Several physical treatments have been tried for quarantine purposes (Yahia and Ariza, 2001; Yahia et al., 2004). Low temperatures (0°C–2.2°C for 10–16 days) can be used for control of the Mediterranean fruit fly. However, these temperatures cannot be used for most tropical fruits because of chilling sensitivity. Hot water treatments (HWTs) are been used in several countries to control fruit flies in mangoes (46.1°C for 65–90 min) and have been used in papaya (two-stage heating process with temperatures of 42 for 30 min and 49°C for 20 min) (Yahia, 2006a). Vapor heat treatments have been developed and used (Fons, 1990). Injury have been reported in both mango and papaya fruits treated with heat (Paull and Armstrong, 1994). Irradiation has proved potentially applicable for insect control in some tropical fruits such as mango and papaya (Paull and Armstrong, 1994; Yahia, 2006a), however, commercial application is still very limited due to several problems including possible injury to the fruit, high costs, and consumer concerns. MA and CA ($\leq 1.0\,kPa$ O_2 and/or $\geq 50\,kPa$ CO_2) have insecticidal and fungistatic effects, and the potential to be developed as quarantine treatments as a safe alternative to the health-hazardous fumigants (Yahia, 1998a,b; 2006a). For more details, see Chapter 11.

Despite all the previously mentioned problems encountered for tropical fruit in postharvest (chilling sensitivity, disease and insect infestations, and short postharvest life), tropical fruits need to be shipped to markets far away from growing areas, usually by air or sea. Shipping by sea can take a very long time, but is economical compared to air-freight. For example, the minimum time required for sea freight from Eastern Australia to South East Asia, Japan and North America, and Europe are 3, 4, and 6 weeks, respectively (McGlasson, 1989). Minimum shipping periods from Mexico to Europe and Japan are about 18 and 21 days, respectively (Yahia, 1993a, 1995c). Therefore, it is essential to assure a sufficiently long postharvest life for these crops to be able to be distributed in distant markets. In addition to better postharvest handling systems, which include optimum harvesting time, control of insects and diseases, and the use of ideal postharvest temperature management, MA and CA can be of major benefits to preserve the quality of these fruits and to prolong their postharvest life.

The rising importance of tropical fruits in world trade and the recent developments in MA and CA technology and applications require the review of published research in order to suggest some valid conclusions and recommendations to improve the handling of tropical fruits, and in particular to increase the use of MA and CA. Several excellent books and reviews have been written on MA and CA (Dewey et al., 1969; Dewey, 1977; Smock, 1979; El-Goorani and Sommer, 1981; Richardson and Meheriuk, 1982; Blankenship, 1985; Kader, 1986; Fellman, 1989; Calderon and Barkai-Golan, 1990; Blanpied, 1993). However, these covered mainly temperate (especially pome) fruits. Very little research has been done on MA/CA of tropical crops as compared to temperate crops such as apples, pears, and kiwifruit. For example, out of 1982 articles published on MA/CA of fruits and vegetables from 1977 to 1993 <6% were on tropical fruits and vegetables (Kader and Morris, 1981, 1997; Kader, 1985; Zagory and Kader, 1989, 1993). Very few reviews have been written on MA/CA of tropical fruits (Hatton and Spalding, 1990; Kader, 1993; Yahia and Paull, 1997; Yahia, 1998a; Yahia, 2006c), and information has been largely limited to mangoes, bananas, pineapples, and papayas. Information on many of the minor tropical crops is either missing or dispersed in nonestablished journals and local reports.

16.2 MA and CA

MA and CA have several potential benefits for tropical fruits. These include retardation of ripening and senescence, alleviation and/or control of CI, control of some pathogenic disorders, physiological disorders, and insects (Shetty et al., 1989; Hatton and Spalding, 1990; Jang, 1990; Ke and Kader, 1992; Yahia, 1998a,b, 2006b).

Optimum gas composition for different products is very variable, and depends on many factors such as type of product, physiological age, temperature, and duration of treatment (Smock, 1979; Yahia, 1998a,b, 2006b). Exposure of tropical fruits to O_2 levels below to, and/or CO_2 levels above to their optimum tolerable range can cause the initiation and/or aggravation of certain physiological disorders, irregular ripening, increased susceptibility to decay, development of off-flavors, and could eventually cause the loss of the product (Kader, 1986). Optimum levels of O_2 and CO_2 for long-term storage of some tropical fruits are listed in Table 16.1. Most tropical fruits can tolerate extreme levels of gases when stored for only short periods.

TABLE 16.1

Summary of Recommendations of MA/CA Conditions for Major Tropical Fruits

Fruit	Intended Use	Atmosphere		Degree of Benefits	Comments	References
		O_2 (kPa)	CO_2 (kPa)			
Banana	Transport, MAP, storage	2–5	2–5	Excellent	Sensitive to low O_2 (<1 kPa) and high CO_2 (>10 kPa). Outstanding response to MA/CA among all tropical fruits. Commercial application for transport (used), and for storage (not used yet).	Gane (1936), Wardlaw (1940), Parsons et al. (1964), Woodruff (1969b)
Cherimoya	Transport, storage	2–5	3–10	Good	MA/CA with ethylene removal has good potential for use during transport.	De La Plaza et al. (1979), Palma et al. (1993), Alique and Oliveira (1994), Zamorano et al. (1999)
Durian	Transport	3–5	10–20	Fair	Research is still needed to further define most appropriate MA/CA conditions and potential use.	Tongdee and Suwanagul (1988a,b), Tongdee and Neamprem (1989), Tongdee et al. (1990a)
Feijoa	Transport	Not well defined		Fair	Short-term MA treatment containing 98 kPa N_2/CO_2 for 24 h at 20°C enhanced flavor quality and storage life to 13 days which also implied its high tolerance to anaerobic conditions. Optimum MA/CA conditions and potential applications are not defined yet.	Pesis et al. (1991)
Lanzones	Transport	5	0	Fair	Further research is needed to establish optimum MA/CA conditions and potential use.	Pantastico et al. (1969)
Mango	Transport Insect control	3–5 ≤0.5	5–10 ≥50	Good Excellent	Sensitive to low O_2 (<2 kPa). Very tolerant to short-term high levels of CO_2 (25–80 kPa). Commercially used for transportation. Huge potential for insect control especially in combination with high temperatures.	Yahia (1993a,b, 1994a, 1995a, 1997a,b), Yahia and Ortega (2000c), Bender et al. (2000a), Nakamura et al. (2004), Yahia (2004), Lalel et al. (2005), Lalel and Singh (2006)

Fruit	Purpose				Comments	References
Mangosteen	Transport	5	5–10	Fair	Further research is needed to determine potential application.	Godfrey Lubulwa and Davis (1994), Rattanachinnakorn et al. (1996)
Papaya	Transport	2–5	5–8	Good	Commercial scale adoption of CA/MA is needed. MAP could be beneficial in CI alleviation and has excellent commercial potential in transport.	Hatton and Reeder (1969b), Nazeeb and Broughton (1978), Yahia et al. (1989), Yahia (1991), Yahia et al. (1992), Yahia (1993b), González-Aguilar et al. (2003), Singh and Sudhakar Rao (2005a,b)
	Insect control	≤ 0.5	≥ 50	Fair		
Passion fruit	Not determined	?	?	?	Research is still needed to determine adequate MA/CA conditions and potential applications.	
Pineapple	Transport	2–5	5–10	Not determined	Reduction of internal browning due to CA effects. Controlled studies are needed to define ideal MA/CA conditions and potential use.	Akamine (1971), Akamine and Goo (1971), Haruenkit and Thompson (1994)
Sapodilla	Not determined	5–10	5–10	Fair	CI alleviation through CA/MA could be beneficial. Research is needed to fully identify ideal MA/CA conditions and potential applications.	Broughton and Wong (1979), Hatton and Spalding (1990)
Sapote mamey	Not determined	5	5	Fair	MA could be beneficial in alleviation of certain physiological disorders. Highly tolerant to elevated levels of CO_2 (50 kPa) for short duration (38 h).	Manzano (2001), Martínez-Morales and Alia-Tejacal (2004), Martínez-Morales et al. (2004, 2005)
Sugar apple	Not determined	5	10	Fair	Research is still needed to determined optimum MA/CA conditions and potential applications.	Broughton and Guat (1979)

16.2.1 MA and CA for Storage

MA and CA are not used for the storage of tropical fruits. This is due to several reasons, including those related to the crop such as availability and quantity, preharvest and postharvest handling, and availability of technology. Some tropical fruits cannot be stored for prolonged periods to justify the use of CA storage. There have been several reports (Hatton and Spalding, 1990) of experimental work indicating that MA/CA storage might be beneficial for tropical crops. However, some of these positive results might be due to factors other than MA/CA, such as humidity control. Many factors, should be considered when evaluating the potential use of MA/CA for storage, such as (1) biological factors, (2) value of the fruit, (3) quantity and quality produced, (4) the specific reason for the use of MA/CA (control of metabolism, control of pathogens, control of insects, etc.), (5) availability of alternative treatments that can be more accessible and/or cheaper, (6) competitions with other production regions, (7) type of market (local, distant, export), (8) distance to market, (9) type of preharvest and postharvest technology available in the region, and (10) research results. Lougheed and Feng (1989) suggested that a fruit to be compatible to the use of CA storage should preferably be characterized by (1) a long postharvest life, (2) resistance to CI, (3) a large range of noninjured atmospheres, (4) resistance to fungal and bacterial attack, (5) adaptation to a humid atmosphere, (6) a climacteric fruit that can be ripened during or after storage, (7) absence of negative MA/CA residual effect, and (8) that MA/CA can reduce the production and effects of ethylene. Other economic, logistic, and marketing factors that can favor the potential use of MA/CA storage include (Lougheed and Feng, 1989) (1) high value of the fruit, (2) adaptability of crop to storage in bulk, (3) less competition with fresh fruits during marketing after storage, (4) storage costs competitive with importation costs of fresh fruits or alternative methods of production, (5) alternative methods not accessible or not recommended for use, (6) fruit quantity sufficient to justify the use of the technology, (7) storage life can be improved with MA/CA, and (8) compatibility of use with other cultivars, temperatures, atmospheres, season, etc. Apple, a fruit very compatible with the use of CA, is characterized by (Lougheed and Feng, 1989): (1) high production, value, and consumption, (2) long postharvest life, (3) climacteric respiration, (4) production and action of ethylene is controlled by CA, (5) big variation in the tolerance levels of O_2 and CO_2, (6) CA permits the use of lower temperatures in some cultivars, (7) relatively less pathological and insects problems than other crops (especially those of tropical origin), (8) some physiological disorders can be alleviated by CA, (9) fruit can be harvested and stored in bulk, and (10) a great deal of research.

16.2.2 MA/CA for Transport

MA and CA are used during marine transport of tropical fruits. For example, MA has been used for more than 30 years during banana transport from Central America to the rest of the world (Yahia, 1998a,b). In the last few years the use of (and interest in using) MA and CA for transport has increased. In recent years many advances have been accomplished. This technology is much more promising for tropical fruits than the storage technology. Since that transport periods of tropical crops can be long (up to several weeks), MA/CA can be very helpful in maintaining the quality of these fruits. The use of MA/CA can encourage the use of sea transport, since it is cheaper than air transport.

Atmospheres for transport can be developed passively, semiactively, or actively. The most common systems for transport in the last 30 years have been developed on a semiactive basis. These systems, which have been (and still) used for transport of banana (Woodruff, 1969a,b), are usually less efficient, but less expensive than active systems.

The use of CA for transport has been contemplated for several years; however, several problems hindered its success including unavailability of adequate gas tight containers, adequate systems for gas control and analysis, and adequate CA generating systems. Existing systems and companies before the late 1980s were unable to deliver on promised applications and benefits of CA transport. In the late 1980s the concept of CA transport became very promising and more practical mainly due to the availability of technologies to build gas tight containers, the use of adequate gas control systems, and the ability to establish controlled gas mixes. The use of air separation technologies in CA in the late 1980s, especially the introduction of membrane technology in 1987, made CA transport practical and feasible for several tropical fruits such as banana and mango (Malcom, 1989; Yahia, 1998a).

16.2.3 Modified Atmosphere Packaging

Modified atmosphere packaging (MAP) has been reported to maintain the quality of some intact and some minimally processed tropical fruits such as durian, jackfruit, mangosteen, papaya, and pineapple (Yahia, 1998a). Reported results are very variable, due to very little experimental control. Some of the beneficial effects noticed might be due to maintaining a humid atmosphere around the commodity, and not related to gas control. Many different types of polymers are usually used, although the most common are different types of polyethylene (PE). Different thicknesses of the same type of film, or different conditions (temperature and RH) surrounding the package result in different permeabilities and different in-package atmospheres. Strict control must be performed in order for the different results to be compared adequately. Research reported on sealed PE packages was done with one or various fruits, and with different size packages. This contributes further to the lack of experimental control and variations of results.

Very active research is being carried out, and commercial interest is rising, especially for minimally processed products. More theoretical research is still needed to study gas exchange characteristics of different fruits, ideal packaging materials for different crops with respect to gases and to water vapor permeability.

MAP has been reported to maintain the quality of several tropical fruits (Scott and Roberts, 1966; Scott et al., 1970; Oudit and Scott, 1973; Scott, 1975; Olorunda, 1976; Salazar and Torres, 1977; Stead and Chithambo, 1980; Brown et al., 1985; Chaplin et al., 1986; Miller et al., 1986; Paull and Chen, 1989; Sornsrivichai et al., 1989; Gonzalez et al., 1990; Ketsa and Leelawatana, 1992; Satyan et al., 1992a,b; Singh and Sudhakar Rao, 2005a). It has also been reported to be advantageous in maintaining some minimally processed tropical fruits such as durian, jackfruit, mangosteen, papaya, and pineapple (Powrie et al., 1990; Siriphanich, 1994).

Some researchers assumed, without very much basis, that packaging in sealed PE bags can substitute for the use of ideal low temperatures (Chaplin et al., 1982). MAP is an inexpensive method compared to CA/MA storage and transport, and thus it was suggested as an alternative to shipping in MA/CA (McGlasson, 1989). MAP may not be very appropriate during transport, especially during long-term sea transport, due to variation in temperature, which would stimulate water condensation inside the packages, change permeability of packaging films, and thus change the adequate atmosphere inside the packages. Many factors must be considered when trying to develop a MAP system including type, thickness and method of fabrication of film, package size, temperature, humidity, length of storage, type, quantity, and physiological stage of crop, and tolerance of each crop to the different gases (O_2, CO_2, C_2H_4). Water saturated atmosphere around the commodity favors the development of decay pathogens. In addition, there is usually an accumulation of ethylene inside packages (Gonzalez et al., 1990; Yahia and Rivera, 1992). Films used for tropical crops should be characterized by a relatively higher permeability to gases and to water vapor. There is usually a need for effective disease control treatments (Yahia and Rivera, 1992), and practical and effective methods for absorbing water and ethylene in the packages (Kader et al., 1989; Yahia and Rivera, 1992). MAP should be used properly and only as a compliment for ideal postharvest handling, including optimum low temperature.

16.2.4 Low-Pressure (Hypobaric) Atmospheres

Low-pressure (LP) atmospheres (hypobaric) refers to holding the commodity in an atmosphere under a reduced pressure, generally, less than 200 mmHg (Burg, 2004). In this system, the O_2 level is reduced depending on the atmospheric pressure. LP has been reported to extend the storage and shelf life of several crops (Gemma et al., 1989; Yahia, 1998a,b; Burg, 2004) including mangoes (Burg and Burg, 1966; Burg, 1975; Apelbaum et al., 1977c; Spalding, 1977a; Spalding and Reeder, 1977; Ilangantileke and Salocke, 1989), bananas (Burg and Burg, 1966; Burg, 1975; Apelbaum et al., 1977a), papayas (Alvarez, 1980; Chau and Alvarez, 1983;), and cherimoyas (Plata et al., 1987). LP was used for a short period in the 1970s during transport of some food products including meats, flowers, and fruits (Burg, 1975; Byers, 1977). However, this system is more expensive than the traditional MA/CA systems. Furthermore, some fruits require other gases that cannot be administered during low pressure storage. Currently this system is not used commercially.

16.2.5 Insecticidal Atmospheres

Atmospheres with very low O_2 (LO) content and/or very high CO_2 levels have insecticidal effects (Yahia, 1998a,b, 2006a; Yahia and Ortega, 2000a). Insect control with MA and CA depends on the O_2 and CO_2 concentration, temperature, RH, insect species and insect development stage, and duration of the treatment. The lower the O_2 concentration, the higher the CO_2, the higher the temperature, and the lower the RH, the shorter the time necessary for insect control. MA and CA have several advantages in comparison with other means for insect control. These are physical treatments that do not leave toxic residues on the fruit, and are competitive in costs with chemical fumigants (Yahia, 1998a,b; Aegerter and Folwell, 2001). MA and CA do not accelerate fruit ripening and senescence as compared to the use of high temperatures, and have better consumer acceptance as compared to the use of irradiation (Yahia, 2006a).

The control of insects by CA has commercially been tried for grains and for dried fruits and vegetables (Banks, 1994a), but not yet fully developed for fresh fruits and vegetables. Atmospheres needed for this application should contain very low levels of oxygen (<1 kPa) or/and very high levels of CO_2 (up to 50–80 kPa). These insecticidal atmospheres (IA) can eliminate insects within a period of 2–4 days at room temperature (Yahia and Carrillo, 1993; Yahia and Tiznado, 1993; Yahia and Vazquez, 1993; Ke et al., 1995; Yahia, 1998a, 2006a), and within a period of 2–4 h at higher temperatures (Ortega and Yahia, 2000a; Yahia and Ortega, 2000a,b). Potential problems that hinder the possibility for developing insecticidal CA for fresh fruits and vegetables include possible fermentation due to the use of anaerobic gas mixtures, and heat injury. The advantages of using this system at room temperature include the low energy input and the avoidance of heat injury. However, the advantage of using it at high temperatures is to accomplish the mortality of insects in a very short period, and thus eliminate the accumulation of products before treatments. In addition, insecticidal CA can be used either in permanent stores or in transport marine containers.

Wrapping of fruits in semipermeable shrink-wrap films for 3–6 days was reported to control some fruit flies (Shetty et al., 1989; Jang, 1990). Shetty et al. (1989) reported that shrink-wrap significantly reduced survivorship of oriental fruit fly (*Dacus dorsalis* Hendel) eggs and first instar larvae in infested papaya in 96 h, and larvae of *Drosophila melanogaster* in mango in 72 h. However, fruit will remain infested if not properly wrapped.

However, fruit tolerance to insecticidal MA and CA is different. For example, mango fruit was found to be tolerant when IA are applied at room temperature or at high temperatures (Yahia, 1997b). 'Manila' and 'Oro' mangoes were treated with a combination of dry-forced heat at 44°C and insecticidal CA (0.5 kPa O_2 and 50 kPa CO_2) for 160 min (Yahia et al., 1997), and the treatment reduced texture loss and caused no injury to the fruit. The basis for sensitivity/tolerance of fruits to IA is still not known (Ke and Kader, 1992; Yahia, 1993b; Kader and Ke, 1994; Ke et al., 1995). Differential scanning calorimetry did not show differences between sensitive (avocado) and tolerant (mango) tissues (Yahia and Rivera-Dominguez, 1995).

There is no current commercial use of IA for horticultural crops. However, treatments with MA/CA alone or in combination with other treatments (such as cold or heat) will most likely be established as a commercial insect control means for tropical crops in the future (Kader and Ke, 1994; Yahia, 1998a, 2006a).

16.2.6 Additional Treatments

The use of carbon monoxide (CO) in addition to MA and CA can provide several advantages such as the control of pathogenic diseases and insects (Woodruff, 1977; El-Goorani and Sommer, 1981), and tissue browning and discoloration (Kader, 1986). There is a potential for the use of CO for tropical fruits, however, development of safe methods of application such as the addition of odors are required before any commercial use can be considered. The use of CO has to be combined with low (≤4 kPa) O_2 levels (Kader, 1986). CO can increase the production of C_2H_4, however, at high concentrations this did not appear to happen (Woodruff, 1977). It has been reported that for fruits that tolerate elevated CO_2, CO appear to be more effective.

Ethylene removal during transport and storage of tropical crops can be beneficial in delaying ripening.

16.2.7 Potential Problems and Hazards of MA and CA

Despite the long list of potential benefits and advantages of MA and CA for tropical fruits, they can have some potential problems and hazards, especially when used inadequately (Smock, 1979; Hatton and Spalding, 1990; Yahia, 1997a,b,c; Bender et al., 2000b; Pesis et al., 2000a,b). Inadequate atmospheres can cause the aggravation and/or the initiation of physiological disorders and fermentation in intact fruits (Kader, 1986; Yahia, 1998a,b). After removal from CA storage to ambient conditions, tropical fruits suffer huge losses due to a sudden increase in the disease incidence, thus limiting shelf life. Other problems associated with the CA-stored tropical fruits include poor skin color development, staggered ripening, faster fruit softening, lack of characteristic aroma, and limited shelf life. Inappropriate atmospheres can increase the hazard of microbial contamination of minimally processed products. MA and CA can be deadly to humans getting inside a CA container without proper security equipments or before the container is properly ventilated. MA and CA can cause structural damage to containers that lack proper pressure relief systems, or due to inadequate use of some gases such as propane.

16.3 Individual Crops

16.3.1 Bananas and Plantains (*Musa* spp.)

Bananas and plantains are climacteric fruits and one of the most important in world trade. Crown rot, anthracnose (caused by *Colletotrichm musae*), "cigar end" stylar rot, and CI are the major causes of postharvest losses of bananas. Postharvest life is about 4 weeks at 12°C–16°C.

MA/CA extends the storage life of green bananas (Mapson and Robinson, 1966; Scott and Roberts, 1966; Smock, 1967; Bardan and Lima, 1969; Woodruff, 1969b; Liu, 1970; Quazi and Freebairn, 1970; Scott et al., 1970; Fuchs and Temkin-Forodeiski, 1971; Scott, 1971; Duan et al., 1973; Scott and Gandanegara, 1974; Burg, 1975; Scott, 1975; Liu, 1976a,b; Scott and Brown, 1981; Banks, 1984a,b; Hesselman and Freebairn, 1986; Kanellis et al., 1989; Satyan et al., 1992a,b). Bananas are very responsive to MA/CA when the fruit is at the preclimacteric stage (Smock, 1979). Optimum atmospheres differ for different cultivars but are about 2–5 kPa O_2 and 2–5 kPa CO_2, and optimum temperature for MA/CA storage is 13°C (Woodruff, 1969b). Cooking bananas have similar CA requirements (Satyan et al., 1992b). Atmospheres containing 5 kPa O_2 and 5 kPa CO_2 were found to be suitable for 'Gros Michel' bananas held for 20 days at 12°C (Wardlaw, 1940). 'Lacatan' and 'Dwarf Cavendish' bananas were kept for 3 weeks in atmospheres containing 6–8 kPa CO_2 and 2 kPa O_2 at 15°C (Smock, 1967). The recommended atmosphere for two Malaysian cultivars of bananas at 20°C and 80% RH was 5–10 kPa CO_2 and a continuous removal of ethylene (Broughton and Wu, 1979). Atmosphere of 1 kPa O_2 inhibited ripening in green bananas and was considered as the lower limit at 15.5°C (Parsons et al., 1964). O_2 atmospheres <1 kPa cause fruit injury, which includes dull yellow to brown skin discoloration, failure to ripen, flaky gray flesh, and off-flavors (Parsons et al., 1964). However, 1 kPa O_2 was reported by other researchers (Mapson and Robinson, 1966; Chiang, 1970) to result in poor quality and more stalk rot. Furthermore, research by Hesselman and Freebairn (1986) has indicated that O_2 levels <2.5 kPa affect the taste of 'Valery' bananas. CO_2 levels higher than 5 kPa are reported to result in undesirable flavor and texture after fruit ripening (Woodruff, 1969b). CO_2 level of 10 kPa was considered to be the upper limit for 'Gros Michel' bananas (Gane, 1936).

Experiments were conducted with storage of bananas in CA sealed rooms (Woodruff, 1969b). Rooms were flushed with N_2 to reduce the O_2 level and supplemental CO_2 was added. Water scrubber was used to control the CO_2 concentration. Purifiers containing brominated, activated carbon were used to absorb volatiles (including ethylene). CA markedly reduced the crown rot. Woodruff (1969b) listed four advantages of CA storage of bananas including (1) fruit can be held for long periods without significant ripening, (2) decreases incidence of rots and molds, (3) maintains a fresher appearance fruit, and (4) more flexibility in coping with glutted markets. However, there has been no commercial CA storage for bananas. Bananas are available all year around, and therefore there is no need for long-term storage. Gastight CA chamber would have to be belt aboard marine ships, since most of the postharvest life of bananas is maintained on transit. In the past, this was not technologically feasible, however, the recent advances in CA technology and marine containers facilitates the application of CA aboard marine ships, which can provide a postharvest life of up to 2 months.

MA has been used commercially for the last three decades during marine shipments of banana (Woodruff, 1969a,b; Yahia, 1998a,b). In this system, green fruits are usually packed in PE bags of about 0.04 mm (1.5 mil) thickness, which are then evacuated (usually using a vacuum cleaner) and sealed (Woodruff, 1969a). High temperatures at the time of evacuation accelerate the establishment of the desirable atmosphere

(Woodruff, 1969b). The atmosphere in these bags usually averages about 2.5 kPa O_2 (1–4.5 kPa) and 5.2 kPa CO_2 (4–6 kPa) after 3–4 weeks. This system has been called 'Banavac' by United Fruit Company (Woodruff, 1969b; Smock, 1979). Bananas can be held for 30 days by this method, and can be maintained green for up to 60 days but rots increase and quality declines after 30 days (Woodruff, 1969b). Fermentation problems have occurred in up to 1% of fruits shipped in this system (Woodruff, 1969a). Only green fruits should be used, and care should be taken not to use punctured bags. Punctured bags will not allow the development of an appropriate atmosphere. Ripe fruits would increase the accumulation of ethylene inside the bags, and would further stimulate fruit ripening. Ethylene concentration of 10 ppm accelerated the ripening of 'Valery' bananas (Woodruff, 1969b). A concentration of 10 ppm or more of ethylene can also stimulate the softening of green fruit (Chiang, 1968, 1970), a condition known as "soft-green" (Woodruff, 1969b) or "green ripeness" (Scott, 1975). High temperature, high CO_2, and LO in the storage atmosphere were suggested to be the main factors causing this disorder; however, the exact mechanism is not fully understood (Zhang et al., 1993). The use of ethylene absorbent agents such as potassium permanganate absorbed on aluminum silicate or vermiculite inside the bags can prevent this disorder and prolong the postharvest life of the fruit (Scott et al., 1968; Liu, 1970; Scott, 1975). Ethylene removal with brominated carbon was found to extend the storage life of 'Lacatan' and 'Cavendish' bananas held in 2–3 kPa O_2 and 8 kPa CO_2 (Smock, 1967), and was found to be more effective than using molecular sieve 5A in a continuous air and C_2H_4 stream (Chiang, 1968). The use of MA for bananas was found to prolong their storage life even at ambient temperatures (Scott and Gandanegara, 1974).

The use of sealed PE (0.1 mm thickness) bags containing 100 g vermiculite impregnated with a saturated solution of $KMnO_4$ allowed a storage life of 'Williams' bananas for up to 6 weeks at 20°C–28°C and 16 weeks at 13°C (Satyan et al., 1992a,b). 'Latundan' banana was stored in 0.08 mm thick PE bags for up to 13 days at 26°C–30°C (Agillon et al., 1987). Storage of green mature 'Cavendish' bananas in low-density polyethylene (LDPE) (0.05 mm thickness) bags for up to 30 days at 8°C, 11°C, and 14°C developed an in-package atmosphere of 3–11 kPa O_2 and 3–5 kPa CO_2 (Hewage et al., 1995). However, these authors reported that these storage conditions did not affect ripening and sensory quality, nor did they alleviate CI symptoms developed at 8°C and 11°C. 'Emas' bananas stored in PE bags (0.04 mm thickness) for 6 days at 24°C generated an atmosphere of up to 3% C_2H_4, up to 14.6 kPa CO_2, and as low as 2.9 kPa O_2 (Tan et al., 1986). Accumulation of 10 kPa CO_2 or more, especially from day 3 to day 6, and an O_2 concentration below 2 kPa in the bags caused abnormal ripening when fruit was ripened later in air. Fruit had skin and pulp darkening, and softening of the inner portion of the pulp, even though the outer portion remained hard. Water insoluble protopectins decreased, and water-soluble pectins and pectates increased in wrapped fruit. The authors suggested that a minimum of 10 kPa CO_2 for few days is required to cause injury in 'Emas' bananas. CO_2 (10 kPa) injury was also reported for 'Mas' bananas (Abdullah et al., 1987). Several cultivars of cooking banana ('Bluggoe,' 'Pacific plantain,' 'Blue Lubin,' and 'Pisang Awak') behaved similarly to the dessert cultivar 'Cavendish' when stored in PE bags (0.1 mm) with or without an ethylene absorbent (potassium permanganate on aluminum oxide) at 7°C, 13°C, 20°C, and 28°C (Satyan et al., 1992b). The storage life increased by a factor of 2 in the absence of an ethylene absorbent and a factor of 3 in the presence of the ethylene absorbent. CO_2 pressure inside the packages increased up to 15 kPa and pressure as high as 32 kPa were reported at the end of storage. Packaging with or without ethylene absorbent had no effect on the incidence of CI neither in the cooking banana cultivars nor in 'Cavendish.' The authors suggested that this method of sealing in PE bags "appears to be an alternative method to refrigeration."

Liu (1976a) suggested pretreatment of the fruit with ethylene at the production or packing site before storage or shipping, to avoid post shipping treatment due to high costs, and to provide even ripening. 'Dwarf Cavendish' pretreated with ethylene and stored for 28 days in 1 kPa O_2 or in 0.1 atmospheric pressure at 14°C remained green and firm until the end of the storage period, and started to ripen almost immediately after being placed in air at 21°C without additional ethylene treatment. However, the period of ethylene pretreatment is critical and should not exceed a threshold length of time (TLT). The TLT is defined by Liu (1976b) as the minimum time required for a fixed concentration of ethylene treatment to induce banana ripening response. Only bananas, which had been pretreated with ethylene for a period equal to TLT, were successfully stored in CA (Liu, 1976a). Neither CA nor LP could prevent the ripening of bananas pretreated with ethylene for a period longer than TLT. Fruits are not uniform in their TLT. Commercially mature bananas may have TLT between 4 and 20 h, and a test for TLT requires 1–2 days (Liu, 1976b). Therefore, from a practical point of view, the author concluded that it would be extremely difficult to select large lots of fruit with uniform TLT, and thus the potential hazard of fruit ripening during storage or shipping after excessive ethylene pretreatment jeopardize the commercial applicability of this method.

Treatment with Pro-long (a mixture of sucrose esters of fatty acids and sodium salt of carboxymethylcellulose) extended the shelf life of bananas (Lowings and Cuts, 1982; Banks, 1984a,c). The commercial wax 'Decco Luster 202' at a 1:2 (wax:water, v/v) delayed ripening of 'Saba' bananas (Pastor and Pantastico, 1984), but other formulations ('Carbowax' and 'Prima Fresh') had no effect. The action of "Pro-long" has been attributed to increased resistance to CO_2 diffusion and to O_2 creating an internal atmosphere with a reduced O_2 and elevated CO_2 (Lizada and Noverio, 1983).

'Gros Michel' bananas held in a LP of 150 mmHg at 15°C were maintained longer in a better quality than those held in normal pressure (Burg and Burg, 1966). Fruits held in LP of 760, 250, and 80 mmHg at 14°C were reported to be maintained for 30, 60, and at least 120 days, respectively. The authors reported that fruit had an acceptable texture, taste, and aroma, and no injury.

The quality of green bananas was not affected when fruits were held for up to 7 days in 100 kPa N_2 at 15.5°C, but had dark-brown to black skin blemishes when held for 10 days (Parsons et al., 1964). After 4 days in 100 kPa N_2 at 15.5°C, fruits ripened to a normal color and flavor in 13 days at 20°C. However, fruit failed to ripen in air, and developed decay, brown skin discoloration and off-flavor after storage in 100 kPa N_2 for 7 days. Fruit were ripened normally in air at 20°C after being held in 99 kPa N_2 and 1 kPa O_2 at 15.5°C for 10 days. LO (2.5 kPa) suppressed the activity of acid phosphatase and the addition of 500 mL of ethylene to the LO atmosphere did not reverse this suppression (Kanellis and Solomos, 1985; Kanellis et al., 1989). However, this atmosphere either alone or in combination with 500 mL ethylene prevented the decline in the activity of pectin methyl esterase. Kanellis et al. (1989) suggested that there were differential effects of LO on metabolic processes since that the accumulation of sugars increased gradually for 4 days in LO, but no increase in acid phosphatase was observed throughout the duration of the LO treatment. LO (3 kPa) limited the operation of the Krebs cycle in fruits of *Musa paradisiaca* L., but high CO_2 showed no rate limiting steps in this cycle (McGlasson and Wills, 1972). Ali Azizan (1988) reported that high CO_2 suppressed the activities of alcohol dehydrogenase (ADH), lactate dehydrogenase (LDH), pyruvate decarboxylase (PDC), and phosphofructokinase, but not malic enzyme and phosphoenol pyruvate carboxylase in 'Pisang Mas' bananas.

A study conducted by Stewart et al. (2005) has concluded that the silicone membrane system offers an inexpensive and easy to use alternative to the traditional methods used for MA storage of bananas. 'Cavendish' bananas were stored for 42 days at 15°C under MA conditions using silicone membrane and diffusion channel systems. The smallest area of silicone membrane achieved gas levels of $3.5\,kPa\ CO_2/3\,kPa\ O_2$ in about 10 days while the shortest diffusion channel achieved $5\,kPa\ CO_2/3\,kPa\ O_2$, in 12–16 days. Fruit in these atmospheres remained unripe for 42 days, had harvest-fresh appearance, good color, minimum mold, and excellent marketability compared with controls and fruit stored in different gas compositions. In general, these authors found that the silicone membrane system was found to be superior; it achieved stability more quickly than the diffusion channel system, maintained more stable gas levels throughout storage and had better physiological and sensory ratings. The diffusion channel system had higher CO_2 levels that may have resulted in peel discoloration in some chambers and may have affected other quality attributes.

N_2O treatments extended the storage life of banana fruit without causing adverse effect on physicochemical quality (Poubol and Izumi, 2005a). The response to N_2O was found to be dose- and time-dependent; the delay of ripening by N_2O was not detectable at $20\,kPa$, but steadily rose at increasing pressure above $40\,kPa$, and its effects on ripening appeared to be saturated at $80\%\ N_2O$. Combinations of N_2O with LO (8 and $12\,kPa$) had a synergic effect on the ripening delay. The capability of N_2O to slow down fruit ripening is thought to be due to its antiethylene activity, as suggested by the delay in the climacteric associated rise in ACC oxidase activity.

Effects of LO atmospheres on quality and ethanol and acetaldehyde formation in stored ethylene-treated bananas ('Cavendish') were investigated by Imahori et al. (1998). Treatment of mature-green (MR) bananas with 200 ppm ethylene for 24 h at 20°C accelerated acetaldehyde and ethanol formation. The skins of bananas stored in 0 and $1\,kPa\ O_2$ remained green in color and CO_2 production was suppressed; however, ethanol formation was accelerated compared with those stored in air, leading to ethanol accumulation and the development of off-flavors. Off-flavor development in fruit stored in $0\,kPa\ O_2$ was intensive. Likewise, in fruit stored in $2\,kPa\ O_2$ skins remained green, CO_2 production was suppressed and ethanol formation was accelerated; however, ethanol levels in fruit flesh were lower than in those stored in 0 or $1\,kPa\ O_2$ and remained constant during storage.

MAP (about $12\,kPa\ O_2$ and $4\,kPa\ CO_2$) resulted in less visible CI in 'Kluai Khai' bananas at 10°C (Nguyen et al., 2004). Total free phenolics in the peel of control bananas decreased more rapidly than in fruit held in the MA package, and phenylalanine ammonia lyase (PAL) and polyphenol oxidase (PPO) activities were considerably higher than in MAP fruit. Pulp softness, sweetness, and flavor of MAP fruit were better than in control fruit. Banana fruit ('Sucrier') packed in polyvinyl chloride film and held at 29°C–30°C prevented the early senescent peel spotting, typical for this cultivar (Choehom et al., 2004). Carbon dioxide and ethylene concentrations within the packages increased, but inclusion of CO_2 scrubbers or ethylene absorbents, which considerably affected gas composition, had no effect on spotting. Experiments with continuous LO concentrations confirmed that the effect of the package was mainly due to LO. Relative humidity was higher in the packages but this had no effect on spotting. The positive effect of MAP on peel spotting was accompanied by reduced PAL and increased PPO activity in the peel. Therefore, senescent spotting of banana peel seems to require rather high oxygen levels. Recently, it has been confirmed that high O_2 ($90 \pm 2\,kPa$) is associated with the enhanced peel spotting in banana (Maneenuam et al., 2007). The *in vitro* activities of PAL and PPO measured both in the whole peel and in peel spots, were lower in high oxygen than in the controls. It has been concluded that peel spotting was not correlated with *in vitro* PAL and PPO activities, but decrease in dopamine levels correlated with peel spotting, indicating that it might be used as a substrate for the browning reaction (Maneenuam et al., 2007).

The application of the antiethylene compound 1-methylcyclopropene (1-MCP) in combination with the use of PE bags was reported to extend the postharvest life of banana fruit (Yueming et al., 1999). 1-MCP treatment delayed peel color change and fruit softening, and extended shelf life in association with suppression of respiration and C_2H_4 evolution. Banana fruit ripening was delayed when exposed to 0.01–1.0 µL 1-MCP/L for 24 h, and increasing concentrations of 1-MCP were generally more effective for longer periods of time. Similar results were obtained with fruit sealed in 0.03 mm PE bags containing 1-MCP at various concentrations, but longer delays in ripening were achieved. The greatest longevity of about 58 days was obtained by packing fruit in sealed PE bags with 1-MCP.

Fresh-cut banana slices have a short shelf life due to fast browning and softening after processing, and therefore the effects of atmospheric modification, exposure to 1-MCP, and chemical dips on their quality were determined by Vilas-Boas and Kader (2006). Low levels of O_2 (2 and 4 kPa) and high levels of CO_2 (5 and 10 kPa), alone or in combination, did not prevent browning and softening, but softening and respiration rates were decreased in response to 1-MCP treatment (1 µL/L for 6 h at 14°C), although ethylene production and browning rates were not influenced. A 2 min dip in a mixture of 1% (w/v) $CaCl_2$ + 1% (w/v) ascorbic acid + 0.5% (w/v) cysteine effectively prevented browning and softening of the slices for 6 days at 5°C, dips in <0.5% cysteine promoted pinking, while concentrations between 0.5% and 1% cysteine delayed browning and softening and extended the post-cutting life to 7 days at 5°C.

A MAP system to extend the shelf life of 'Kolikuttu' bananas at room temperature (approximately 25°C and 85% RH) was developed by Chamara et al. (2000). Effect of various MAP conditions on sensory properties of fruit was evaluated and efficacy of potassium permanganate as an ethylene absorber was also examined. MAP systems were created using LDPE bags with no ethylene absorber, or with wrapped or unwrapped bricks impregnated with permanganate. In-package O_2 and CO_2 levels were determined on days 10, 14, 17, and 20, and physicochemical properties and color of fruit were monitored. MAP with wrapped ethylene absorber produced the optimal results, with reduced in-package ethylene and CO_2, increased O_2, and minimal changes in firmness, total soluble solids, weight, titratable acidity, and pH.

The skins of bananas stored in 0–1 kPa O_2 remained green in color and CO_2 production was suppressed; however, ethanol formation was accelerated compared with those stored in air, leading to ethanol accumulation and the development of off-flavors (Imahori et al., 1998). Off-flavor development in fruits stored in 0 kPa O_2 was intensive. Likewise, in fruit stored in 2 kPa O_2 skins remained green, CO_2 production was suppressed and ethanol formation was accelerated; however, ethanol levels in fruit flesh were lower than in those stored in 0 or 1 kPa O_2 and remained constant during storage.

Blankenship (1996) reported that bananas that have ripened under CA conditions were not as high quality as those ripened in air in terms of visual appearance. Research needs for this fruit include investigation on the cost and technological feasibility of the establishment and use of CA, especially on board of sea ships.

16.3.2 Cherimoya (*Annona cherimola* Mill.)

A subtropical fruit with an active metabolism and a climacteric respiratory behavior. Rapid peel browning, loss of firmness, and CI are the main causes of postharvest losses. Post-harvest life is about 2–4 weeks at 8°C–10°C.

Fruits of 'Fino de Jete' were stored in air and in CA (3 kPa O_2 in combination with 0, 3, 6, and 9 kPa CO_2) at 8°C (Alique and Oliveira, 1994). The combination of high CO_2/LO had an additive effect on reducing ethylene production and fruit softening, but did not significantly affect sugars and citric acid. The authors concluded that 3 kPa O_2 in combination with 3 or 6 kPa CO_2 increased the storage life of 'Fino de Jeta' fruits at

9°C by 2 weeks over that of fruits stored in air. There were no differences between the 3 and 6 kPa CO_2. Earlier studies by De La Plaza et al. (1979) also showed that CA (2 kPa O_2+ 10 kPa CO_2 at 9°C) retarded fruit softening and prolonged the storage life of 'Fino de Jete' and 'Campa' by 1 week compared to storage in air. This study found that CA had no effect on reducing sugars and titratable acidity in both cultivars, and the high CO_2 increased the respiration rate in 'Fino de Jete,' 'Fino de Jete' and 'Campa' fruit had a higher maximum climacteric in CA (2 kPa O_2 and 10 kPa CO_2) than in air (De La Plaza, 1980). Palma et al. (1993) concluded that 'Concha lisa' cherimoyas can be maintained in 5 kPa O_2 and 10°C for up to 43 days and still ripen normally after 4 days at room temperature. In this cultivar 10 kPa and 15 kPa O_2 delayed the climacteric, but at 5 kPa O_2 there was neither detectable climacteric nor ethylene production for up to 43 days. 'Fino de Jete' fruit held in a combination of 10 kPa O_2 and 10, 15, or 20 kPa CO_2 at 8°C and 98% RH for 3, 6, or 9 days, ripened later than those held in air (Alique, 1995). Fruit treated with 20 kPa CO_2 for 9 days showed unacceptable quality due to bitterness. 20 kPa CO_2 for 3 days delayed fruit softening, retarded the accumulation of polygalacturonase-related protein, and maintained chlorophyll content of the peel (Del Cura et al., 1996).

Ethylene absorption using $KMnO_4$-sepiolite extruded round rods (Green Keeper) in fruit of 'Fino de Jete' packed in PE films and stored at 8.5°C and 98% RH was found to be beneficial at a dose of 3.5 g/kg (De La Plaza et al., 1993). This dose was efficient in reducing respiration, ethylene production, and kept fruit in good quality for 18 days.

Several mechanical parameters obtained by compression and penetration tests, and changes in cherimoya (*Annona cherimola* Mill.) fruit quality during storage in air and in 2 CA treatments (3 kPa O_2 + O kPa CO_2 and 3 kPa O_2 + 3 kPa CO_2) were analyzed by Zamorano et al. (1999). A gradient of softening was found among the equatorial and apical areas of the flesh during CA storage, as assessed by localized penetration tests. The combination of LO/elevated CO_2 (3 kPa O_2 + 3 kPa CO_2) increased this gradient and had a greater inhibiting effect on skin softening than LO. The prevention of softening by CA storage was stronger in the less mature tissues (equatorial and outer areas) than in the more mature tissues (apical and inner areas around the longitudinal axis). CA storage delayed or inhibited changes in fruit quality observed during air storage. Effects of short CA treatments, involving exposure to 10 kPa O_2 combined with 10, 15, and 20 kPa CO_2 on a selection of biochemical parameters and related enzymic activities of cherimoya fruit stored for 0–9 days were studied by Sanchez et al. (1998). A marked increase in the activity of soluble and cell wall invertases was found in cherimoyas stored in air, but this was inhibited by CA treatment, which also reduced the accumulation of soluble sugar and malic acid. CA had no effect on citric acid content and enhanced the rise in fumaric acid concentration, which was directly related to CO_2 level. Differences in the softening of fruit stored in air and in CA were found after 9 days storage; however, the role of the cell-wall hydrolases polygalacturonase and carboxymethyl-cellulase in this process was not clear. Assis et al. (2001) reported that exposure of cherimoya fruit to either air or 20 kPa CO_2 for 3 days at 20°C and then transferred to air, showed that total polyphenol levels remained constant while a rapid decline in lignin content was observed in fruit stored in air. Compared to air-stored fruit, CO_2 treatment inhibited ethylene production without affecting PAL activity showing that increase in PAL activity does not relate to ethylene. The CO_2 treatment inhibited flesh softening and maintained lignin at levels found in freshly harvested fruit and also improved internal color of fruit (Assis et al., 2001).

Cherimoya fruit exhibited active nitrogen metabolism mainly under high CO_2 (20 kPa) levels (Merodio et al., 1998), the high nitrogen reassimilation was correlated with a high PAL activity. On the contrary, fruit showing low PAL activity accumulated high levels of endogenous ammonia (Maldonado et al., 2002). Specifically, this metabolic situation occurred in fruit during the first days of storage at chilling temperature (6°C) (Maldonado et al., 2002). High CO_2 (20 kPa) treatment improved tolerance to prolonged storage at chilling temperature and was found closely linked to the maintenance of fruit energy metabolism, pH stability, and the promotion of synthesis of defense compounds that prevented or repaired damage caused by chilling temperature (Maldonado et al., 2004). *Annona* fruit has been considered a good model to analyze the phenylalanine–cinnamate pathway and the regulation of PAL enzyme (Merodio et al., 1998; Assis et al., 2001; Maldonado et al., 2002, 2004, 2007).

Coating cherimoya fruit with wax reduced its respiration and ethylene production and extended the shelf life by 5 days, with less weight loss and minimal browning (Yonemoto et al., 2002). Optimum atmospheres for cherimoya are about 2–5 kPa O_2 and 3–10 kPa CO_2, which can be potentially beneficial during marine transport.

16.3.3 Durian (*Durio zibethinus* Murray)

A climacteric fruit native to Southeast Asia that can weigh up to 5 kg, and has thick fibrous skin and short sharp pines. Thailand is the biggest producer and exporter of durian, and 'Chanee' and 'Monthong' are the two most popular cultivars. The fruit is used as fresh, candy, paste, dehydrated powder, or frozen. Postharvest life is about 3–7 weeks at 4°C–6°C.

Reduction of the O_2 level in the atmosphere to 10 kPa caused a significant reduction in respiration and ethylene production in fruits of 'Chanee,' 'Kan Yao,' and 'Mon Tong,' but did not delay fruit ripening (Tongdee and Suwanagul, 1988a,b; Tongdee and Neamprem, 1989; Tongdee et al., 1990a). The same authors reported that ripening was delayed in fruits stored in 5 or 7.5 kPa O_2 for up to 7 days at 22°C, and fruit ripened normally upon transfer to air. However, fruit stored in 2 kPa O_2 was injured and failed to ripen even when transferred to air. High CO_2 (10 or 20 kPa) caused only a slight reduction in ethylene production and did not effect fruit ripening, but had a greater effect on the aril condition when combined with LO. The effect of LO and/or high CO_2 on durian ripening was reported to be largely influenced by the stage of maturity at harvest (Suwanagul and Tongdee, 1989; Tongdee et al., 1989). Fruit that were allowed to begin their ripening process, before storage in CA, had a slightly longer shelf life (Siriphanich, 1996).

Waxing of 'Chanee' and 'Mon Tong' fruit with different "FMC SF" formulas restricted gas movement through the rind resulting in higher internal CO_2, lower O_2, and C_2H_4 concentrations and delayed fruit ripening (Tongdee et al., 1990b). Sriyook and Siriphanich (1989) reported that sucrose ester coating delayed ripening for 1–3 days at 25°C. Storage of minimally processed fruit in an atmosphere of 10 kPa CO_2 in air was found to inhibit fungal growth for 1 week at room temperature (Siriphanich, 1994).

Optimum atmospheres are about 5 kPa O_2 and up to 20 kPa CO_2. More studies are still needed to further define the most appropriate atmosphere, and the feasibility of MA/CA applications.

16.3.4 Feijoa (*Feijoa sellowiana* Berg.)

A climacteric fruit, originated in Brazil, and grown in several countries. Postharvest life is about 2–4 weeks at 5°C–10°C. Treatment of hand picked feijoa with 98 kPa N_2 or CO_2 for 24 h at 20°C increased the accumulation of acetaldehyde, ethanol, ethyl acetate, and ethyl butyrate, which improved aroma and flavor for up to 13 days of storage (Pesis et al., 1991). Fruit were judged to be sweeter than the control, but there were no differences in total soluble solids or acidity. Treatment with N_2 for 24 h was the most effective in increasing volatile production and in maintaining the best appearance. Work is still needed to identify ideal atmosphere composition, and potential benefits of MA/CA.

16.3.5 Lanzones (*Lansium domesticum* Correa)

The fruit is 2–5 cm ellipsoid to round, native to the southeastern Malay archipelago, consumed mostly fresh but can be candied. CA atmospheres containing 5 kPa O_2 with no CO_2 at 14.4°C increased the storage life of 'Paete type' lanzones to 16 days compared to 9 days in air, and also reduced browning (Pantastico et al., 1969). Increasing the CO_2 from 0 to 5 kPa, regardless of the O_2 concentration, did not increase the acidity of the fruit, and increased surface browning, especially at 10 kPa O_2. Waxing with Johnson's Prima fresh wax emulsion reduced weight loss, but aggravated the browning effect after only 5 days of storage, and waxed fruits were sweeter than the control, which indicated that waxing did not impair the ripening process to offset the normal reduction in acidity. The authors concluded that the optimum atmosphere for storage of lanzones was 5% O_2 and zero CO_2.

Holding fruit in PE bags (27×22 cm, 0.08 mm thick) with 16, 32, and 64 holes for 4 days at 28°C reduced water loss but increased surface browning (Brown and Lizada, 1984). Atmospheres (O_2/CO_2) developed in the bags were 15.2/4.4, 10.9/7.6, and 8.8/8.4 kPa, respectively. Browning was thought to be caused by the increased concentration of CO_2. Further research is still needed to investigate the potential benefits of MA/CA of lanzones.

16.3.6 Mango (*Mangifera indica* L.)

A climacteric fruit and one of the most important tropical fruits in world trade. Anthracnose and CI are the most important causes of losses in postharvest. Postharvest life is about 2–4 weeks at 10°C–15°C.

MA and CA are commercially used during sea transport of mangoes. They have several beneficial effects on mango fruit (Yahia, 1998a,b; Lalel et al., 2005). The history of research on CA storage of mango is as old as for pome fruits, but the commercial success is far below the pome fruits. Mango fruit production was (and still) mainly restricted to the tropical regions of world, which mostly fall in the developing world with poor postharvest handling infrastructure. Consequently, the commercialization and distribution of mango in nonproducing countries was very slow. The first report of Singh et al. (1937) showed that mango can be stored in CA containing 9.2 kPa O_2 to prolong their ripening period. Various researchers have tried to optimize CA conditions for different cultivars of the world (Table 16.2). The CA requirements of mangoes vary among cultivars and also depend upon the harvest maturity. A slight variation in CA from its optimum may result into the development of poor flavor quality of mango. A very LO and/or high CO_2 shift the equilibrium from aerobic to anaerobic metabolism. The tolerance of mango to a LO and/or a high CO_2 level has been evaluated (Yahia et al., 1989; Yahia and Hernandez, 1993; Bender et al., 2000b; Lalel et al., 2005). Mango is tolerant to high CO_2 concentrations for a short period (Yahia et al., 1989; Yahia and Tiznado, 1993), which warrant the use of high CO_2 for insect disinfestation purposes. MR 'New World' mangoes can tolerate 25% CO_2 for 3 weeks at 12°C (Bender et al., 1994). Atmospheres with CO_2 concentrations higher than 25% result in elevated ethanol production and damage to skin color development of the mangoes. LO concentrations below 2% resulted in the accumulation of ethanol and thus impaired the flavor quality of 'Tommy Atkins' (Bender et al., 2000a) and 'Delta R2E2' (Lalel and Singh, 2006) mangoes. The extent of ripeness also determines the tolerance of mango to LO concentration. 'Haden' mangoes when stored for 2 weeks at the onset of the climacteric peak produced 10 times more ethanol than preclimacteric mangoes of the same cultivar (Bender et al., 2000a). Bender et al. (2000a,b) reported that preclimacteric 'Haden' and 'Tommy Atkins' mangoes are able to tolerate 3 kPa O_2 for 2 or 3 weeks at 12°C–15°C and that tolerance to LO decreases as mangoes ripen. 'Haden' and 'Tommy

TABLE 16.2

Summary of CA Research on Mango Cultivars in Different Parts of World

Cultivars	Country	CA O_2 (kPa)	CA CO_2 (kPa)	Temperature (°C)	Storage (days)	References
Alphonso	India	—	7.5	8.3–10	35	Kapur et al. (1962)
Amelie	Senegal	5	5	10–12	28	Kane and Marcellin (1979)
Chok Anan	Malaysia	2 or 5	0	15	28	Shukor et al. (2000)
Carlota	Brazil	6	10	8	35	Bleinroth et al. (1977a,b)
Delta R2E2	Australia	3	6	13	34	Lalel et al. (2005), Lalel and Singh (2006)
Haden	Brazil	6	10	8	30	Bleinroth et al. (1977a,b)
Irwin	Japan	5	5	8–12	28	Maekawa (1990)
Irwin (Tree-ripe)	Japan	5	5–10	5–15	30	Nakamura et al. (2004)
Jasmin	Brazil	6	10	8	35	Bleinroth et al. (1977a,b)
Julie	Senegal	5	5	10–12	28	Kane and Marcellin (1979)
Keitt	United States	5	5	13	—	Spalding and Reeder (1977)
Kensington Pride	Australia	2–4	4–6	13	30–35	Lalel et al. (2004), McLauchlan and Barker (1992)
Rad	Thailand	6	4	13	25	Noomhorn and Tiasuwan (1995)
Raspuri	India	0	7.5	5.5–7.2	49	Kapur et al. (1962)
Sao Quirino	Brazil	6	10	8	35	Bleinroth et al. (1977a,b)
South-East Asia cultivars	United States	5	5	13	—	Kader (1993)
Tommy Atkins	United States	3	20 or 30	13	28	Abdulah and Basiouny (2000)
	United States, Chile	3–5	0–5	12–15	21–31	Bender et al. (2000a,b), Lizana and Ochagavia (1997)

Atkins' mangoes were stored in air, 2, 3, 4, or 5 kPa O_2 plus N_2, or 25 kPa CO_2 plus air for 14 days at 15°C or 21 days at 12°C, respectively, then in air for 5 days at 20°C to determine their tolerance to reduced O_2 levels for storage times encountered in typical marine shipments (Bender et al., 2000a,b). All LO treatments reduced mature ripe mango respiration, however, elevated ethanol production occurred in 2 and 3 kPa O_2 storage, with the levels 2–3 times higher in 'Tommy Atkins' than in 'Haden.' In contrast, 'Haden' fruit at the onset of the climacteric also accumulated ethanol in 4 kPa O_2 and produced 10–20 times more ethanol in 2 and 3 kPa O_2 than preclimacteric fruit. There were no visible injury symptoms, but off-flavor developed in MR fruit at 2 kPa O_2 and in ripening initiated fruit at 2 and 3 kPa O_2. Ethanol production was not affected by storage in 25 kPa CO_2. Ethylene production was reduced slightly by LO; however, 'Haden' fruit also showed a residual inhibitory effect on ethylene production at 2 or 3 kPa O_2 storage, while 'Tommy Atkins' fruit stored in 2 kPa O_2 produced a burst of ethylene upon transfer to air at 20°C. Fruit firmness, total sugars, and starch levels did not differ among treatments, but 2, 3, or 4 kPa O_2 and 25 kPa CO_2 maintained significantly higher acidity than 5 kPa O_2 or air. The epidermal ground color responded differently to LO and high CO_2 in the two mango cultivars. Only 2 kPa O_2 maintained 'Haden' color better than air, while all LO levels maintained 'Tommy Atkins' color equally well and better than air. High CO_2 was more effective than LO in maintaining 'Haden' color, but had about the same effect as LO on 'Tommy Atkins.' Mangoes have high tolerance to short-term elevated CO_2 atmospheres (Yahia, 1998a,b). Bender et al. (2000a,b) found that mangoes can tolerate CO_2 atmospheres of up to 25 kPa for 2 weeks at 12°C. High (25 kPa) CO_2 completely inhibited ethylene production, but increased ethanol production. Aroma volatiles were reduced following 25 kPa CO_2 treatment, while 10 kPa CO_2, LO atmospheres and storage temperature did not significantly influence production of terpene hydrocarbons, the compounds suggested to be characteristic of Florida-type mangoes.

The quality of 'Keitt' mangoes were evaluated during storage for 6 days at 20°C under LO (approximately 0.3 kPa) atmospheres before storage in MAP in three LDPE films of different characteristics (Gonzalez-Aguilar et al., 1997). After LO treatment, fruits were individually packaged and stored for 30 days at 10°C and 20°C. Both LO and MA treatments delayed the losses of color, weight, and firmness. Fruits maintained a good appearance with a significant delay of ripening. It was observed that mangoes were very tolerant to LO treatment. However, some individual MAP fruits developed a fermented taste after 10 and 20 days at 20°C. Short duration (6 days) storage of mangoes at LO did not have any deleterious effect on fruit quality during subsequent storage under MA or normal atmosphere. Properly selected atmospheres, which prolong mango shelf life by slowing ripening processes, seem to allow the fruit to be shipped without sacrificing their superior aroma quality.

MG and tree-ripe (TR) mangoes ('Tommy Atkins') were stored for 21 days in air or in CA (5 kPa O_2 + 10 kPa or 25 kPa CO_2) (Bender et al., 2000a,b). MG fruit were stored at 12°C and TR fruit at either 8°C or 12°C. TR mangoes produced much higher levels of all aroma volatiles accept hexanal than did MG fruit. Both MG and TR mangoes stored in 25 kPa CO_2 tended to have lower terpene (especially *p*-cymene) and hexanal concentration than did those stored in 10 kPa CO_2 and air. Acetaldehyde and ethanol levels tended to be higher in TR mangoes from 25 kPa CO_2 than in those from 10 kPa CO_2 or air storage, especially at 8°C. Inhibition of volatile production by 25 kPa CO_2 was greater in MG than in TR mangoes, and at 8°C compared to 12°C for TR fruit. However, aroma volatile levels in TR mangoes from the 25 kPa CO_2 treatment were in all cases equal to or greater than those in MG fruit treatments.

In previous years, the research on CA storage of mango was aimed at maximizing the storage life of MR fruit with less attention to flavor quality. However, the flavor of mango fruit post-CA is of prime importance for increasing the consumer acceptance. Recently, the research priorities have been refocused more on flavor and aroma quality of mango fruit (Lalel et al., 2004; Lalel and Singh, 2006). CA storage of 'Kensington' mangoes in atmospheres containing >6 kPa CO_2 increased the concentration of total fatty acids as well as palmitic acid, palmitoleic acid, stearic acid, and linoleic acid (Lalel et al., 2004). CA storage comprising 2 kPa O_2 and 3 kPa CO_2 or 3 kPa O_2 in combination with 6 kPa CO_2 at 13°C seems to be promising for extending the shelf life of the 'Kensington Pride' mango while still maintaining a high concentration of the major volatile compounds responsible for the aroma of ripe mangoes. 'Kensington Pride' mangoes stored under CA containing 3 kPa O_2 + 5 kPa CO_2 at 13°C for 4 weeks remained green in color, but the fruit held in normal air at the same temperature turned yellow in color (Figure 16.1). Another approach is to subject the tree ripe fruit to CA storage, which can yield the storage life of 2–3 weeks, which is generally considered enough for sea freight and distribution of fruit at the destination market (Brecht et al., 2003). TR 'Irwin' mango could be stored under CA containing 5 kPa O_2 and 5–10 kPa CO_2 for 30 days at 5°C–15°C, and there were no CI symptoms even at 5°C (Nakamura et al., 2004).

Some fruit coatings can create a MA, and therefore two types of fruit coatings, one was polysaccharide-based while the other had carnauba wax as the main ingredient, were tested for their effect on external and internal mango fruit atmospheres and quality factors during simulated commercial storage at 10°C or 15°C with 90%–99% RH followed by simulated marketing conditions at 20°C and 56% RH (Baldwin et al., 1999). These two coatings exhibited markedly different O_2 permeability characteristics under laboratory conditions. Polysaccharide coatings were less permeable to respiratory gases, such as O_2, and more permeable to water vapor compared to carnauba wax. When applied to fruit under simulated commercial conditions, however, the difference between the coatings in permeance to respiratory gases was much reduced, most likely due to the high humidity during chilled storage. Both coatings created a MA, reduced decay, and improved appearance by imparting a subtle shine; but only the polysaccharide coating delayed ripening and increased concentrations of flavor volatiles. The carnauba wax coating significantly reduced water loss compared to uncoated and polysaccharide-coating treatments.

MAP consisting of 4 kPa O_2, 10 kPa CO_2, and 86 kPa N_2, vacuum packaging or 100 kPa O_2 has been applied to minimally processed mango and pineapple fruits, and gas mixture treatment achieved the longest shelf life (Martínez-Ferrer et al., 2002). A study by Beaulieu and Lea (2003) was performed to assess volatile and quality changes in stored fresh-cut

FIGURE 16.1
'Kensington Pride' mangoes stored under CA (3 kPa O_2 + 5 kPa CO_2) and normal air for 4 weeks at 13°C. (Courtesy of: Maria F. Sumual, Curtin University of Technology, Perth, Australia.)

mangoes prepared from "firm-ripe" (FR) and "soft-ripe" (SR) fruit, and to assess what effect MAP may have on cut fruit physiology, overall quality, and volatile retention or loss. Florida-grown 'Keitt' and 'Palmer' mangoes were used, without heat treatment. Subjective appraisals of fresh-cut mangoes based on aroma and cut edge or tissue damage indicated that most SR cubes were unmarketable by day 7 at 4°C. Both varieties stored in MAP at 4°C had almost identical O_2 consumption, which was independent of ripeness. CO_2 and O_2 data for cubes stored in passive MAP indicated that the system was inadequate to prevent potential anaerobic respiration after 7 days storage.

'Keitt' mango was found to be very tolerant to as low as 0.2 kPa O_2 and as high as 80 kPa CO_2 for up to 5 days at 20°C without any type of injury upon ripening, although fermentative odors could be noted while the fruit is under stress (Yahia, 1993a,b, 1994a, 1995a, 1997a,b,c, 2004; Yahia and Ortega, 2000c). Some other mango cultivars were later evaluated and shown to be also tolerant (Yahia, 1998a,b; Yahia et al., 2000a,b).

Kapur et al. (1962) reported that 'Alfonso' mangoes were kept satisfactorily in 7.5 kPa CO_2 at 8.3°C–10.0°C for 35 days, and 'Raspuri' mango in 7.5 kPa CO_2 at 5.5°C–7.2°C for 49 days. Maekawa (1990) concluded that it is possible to maintain 'Irwin' mangoes for up to 4 weeks in 5 kPa O_2 and 5 kPa CO_2 at 12°C and the use of an ethylene absorbent (activated charcoal/vanadium oxide catalyst). In addition, the author reported that temperature can be safely reduced to 8°C. 'Rad' mangoes was reported to be successfully kept for up to 25 days in 6 kPa O_2 and 4 kPa CO_2 at 13°C and 94% RH (Noomhorn and Tiasuwan, 1995). In Brazil, 'Haden' mangoes were held for 30 days and 'Carlota,' 'Jasmin,' and 'Sao Quirino' were held for 35 days in 6 kPa O_2 and 10 kPa CO_2 at 8°C and 90% RH (Bleinroth et al., 1977b). In France, 'Amelie' mangoes stored for 4 weeks in 5 kPa O_2 and 5 kPa CO_2 at 10°C–12°C had less decay, and fruits were reported to be more acceptable after CA than after air storage (Kane and Marcellin, 1979). Sive and Resinsky (1989b) claimed to maintain 'Tommy Atkins' and 'Keitt' mangoes treated with prochloraz (to control *Alternaria alternata*) in CA for up to 10 weeks at 13°C. The Philippines Council for Agriculture and Resource Research (1978) reported that 'Caraboa' mangoes can be kept for 28 days in 5 kPa O_2 and 5 kPa CO_2 at 10°C, however, this cultivar was reported by other researchers in the Philippines (Gautam and Lizada, 1984; Nuevo et al., 1984a,b) to be very susceptible to MA injury. Gautam and Lizada (1984) reported that storage in MA using PE bags for more than 1 day causes ripening abnormalities. It has been suggested that CA is not, or only slightly beneficial for mango (Hatton and Reeder, 1966, 1967, 1969a; Spalding and Reeder, 1974). The best atmosphere for 'Keitt' mango was reported to be 5 kPa O_2 and 5 kPa CO_2 at 13°C, however, quality was not significantly better than in air storage (Hatton and Reeder, 1966; Spalding and Reeder, 1977). A 10 kPa CO_2 atmosphere alleviated chilling symptoms in fruit of the cultivar 'Kensington,' but higher concentrations were injurious, while LO (5 kPa) had no significant effect (O'Hare and Prasad, 1993). Higher levels of CO_2 (more than 10 kPa) were found by these authors to be ineffective in alleviating CI at 7°C, and tended to cause tissue injury and high levels of ethanol in the pulp. 'Rad' mangoes had internal browning and off-flavor in atmospheres containing 6 and 8 kPa CO_2 (Noomhorn and Tiasuwan, 1995). The presence of starchy mesocarp in 'Carabao' mango, which is characteristic of internal breakdown, increases in this cultivar during storage in MA (Gautam and Lizada, 1984). Fruit stored for 4–5 days exhibited severe symptoms, which included air pockets in the mesocarp resulting in spongy tissue (Nuevo et al., 1984a,b). Parenchyma cells of affected tissues had an average of 18 starch granules per cell, compared to an average of two starch granules in healthy adjacent cells. However, no difference in starch granule shape was detected between the two tissues. The spongy tissue, which usually occurs in the inner mesocarp near the seed and becomes evident during ripening, had almost 10 times of starch content compared to the healthy tissue in the same fruit. External symptoms of the internal browning due to MA were reported to consist in the failure of the peel to develop color beyond the half-yellow stage. 'Carabao' mango stored in PE bags (0.04 mm thickness) had faint fermented odor that disappeared during ripening when fruit was kept for 1 day (Gautam and Lizada, 1984). The fermented odor was stronger the longer the storage duration, and persisted throughout ripening when fruits were kept for 2–5 days in PE bags. The respiratory quotient of this cultivar ranged from 0.59 at 21 kPa O_2 to 6.03 at 2.4 kPa O_2, which indicates a progressively anaerobic metabolism (Sy and Mendoza, 1984). The same authors reported that CO_2 production decreased from 21 to 3 kPa O_2, but increased at concentrations below 3 kPa. Fermented odor was explained as a possible indication of fermentative decarboxylation as was reported in 'Alfonso' mango subjected to elevated concentrations (more than 15 kPa) of CO_2 (Lakshminarayana and Subramanyam, 1970). Injury in 'Kensington' mango caused by higher levels of CO_2 appeared to be more severe at lower temperatures (O'Hare and Prasad, 1993), which could be due either to compounding injury (chilling + CO_2) or to reduced sensitivity of ripe mango to CO_2.

Diseases, especially anthracnose and stem-end rot, are the principal limiting factors for mango storage. Moderate O_2 ($\geq$5 kPa) and CO_2 ($\leq$5 kPa) atmospheres are not sufficient for control of diseases. Pronounced decay incidence appeared after storage of 'Rad' mangoes for 20 days in atmospheres containing 4–6 kPa O_2 and 4–8 kPa CO_2 at 13°C and 94% RH, and severe incidence appeared after 25 days (Noomhorn and Tiasuwan, 1995). Greater incidence of decay (stem-end rot and anthracnose) was observed in 'Carabao' mango stored in MA for 2–5 days at 25°C to 31°C (Gautam and Lizada, 1984). The enrichment of CA with 5–10 kPa CO was suggested for better disease control (Woodruff, 1977). However, CO is potentially toxic and explosive, and should not be used unless safe measures for application are developed.

Mango fruit wrapped in 0.08 mm thick PE bags, with and without perlite-$KMnO_4$, and stored for 3 weeks at 10°C before treatment with ethylene, were ripened to normal color, texture, and flavor (Esguerra et al., 1978). Individually sealed 'Keitt' mangoes in LDPE and high-density polyethylene (HDPE) films for 4 weeks at 20°C delayed ripening, reduced weight loss, and did not result in any off-flavors (Gonzalez et al., 1990). LDPE had a thickness of 0.010 mm and permeabilities of 700 cc O_2/cc.m^2 h atm, and 0.257 g H_2O/m^2 h atm. HDPE film had a thickness of 0.020 mm and permeabilities of 800 cc O_2/m^2 h atm, and 0.166 g H_2O/m^2 h atm. Combined effect of hot benomyl (1000 ppm) solution at 55°C for 5 min, and seal packaging in 0.01 mm PVC extended the storage life of MR 'Nam Dok Mai' mango stored at 13°C (Sosnrivichai et al., 1992). The authors found that fruit quality was not affected by film packaging after 4 weeks, but fruit showed inferior quality after 6 weeks. The inhibition of carotene pigmentation in the peel of this variety was suggested by Yantarasri et al. (1994) to be related to O_2 concentration inside the package and not to CO_2 concentration. The authors suggested that 16 kPa O_2 is essential to develop peel color to the marketable stage (greenish). 'Tommy Atkins' mangoes individually sealed in heat shrinkable films and stored for 2 weeks at 12.8°C and then ripened at 21°C had less weight loss, but did not show differences in firmness, skin color development, decay development, or time to fruit ripening, and had more off-flavors than unwrapped fruit (Miller et al., 1983). Polyethylene films used were Clysar EH-60 film of 0.01 nominal thickness, Clysar EHC-50 copolymer film of 0.013 mm nominal thickness, and Clysar EHC-100 copolymer film of 0.025 mm nominal thickness. Individual mature fruits of the same variety were later sealed in Clysar EHC-50 copolymer film with 0.013 mm thickness and Cryovac D955 with 0.015 thicknesses, and stored at 21°C and 85%–90% RH (Miller et al., 1986). O_2 permeabilities of the films were 620 cm^3/24 h m^2 atm and 9833 cm^3/24 h m^2 atm, respectively. Water permeability was 1.5 g/24 h m^2, and 2.0 g/h m^2 at 23°C, respectively. Fruit had less weight loss, but higher incidence of decay and off-flavor at soft-ripeness than unsealed fruit. The authors concluded that there were no practical benefits by wrapping this cultivar in these films and storing them at 21°C or even at lower temperatures. They have even suggested that "film wrapping mangoes at various stages of ripeness after harvest is not a technique which will improve the maintenance of mango quality during storage for ripening." 'Kensington' mango treated with heated benomyl (0.5 g/L at 51.5 for 5 min) and sealed in PE bags (0.04 mm thickness) for various durations at 20°C, had off-flavor and lacked normal skin color when ripened, but ripened satisfactorily when held in perforated bags (Chaplin et al., 1982). The postharvest life of these fruits was not consistently longer than the control. CO_2 in the bags exceeded 20 kPa and that of O_2 was lower than 5 kPa. The incidence of off-flavors was reduced by the inclusion in the bags of C_2H_4 absorbent blocks ($KMnO_4$ on vermiculite/cement block). The authors concluded that "mangoes cannot be stored satisfactory at ambient temperature by such technique." However, Stead and Chithambo (1980) reported that fruit ripening at 20°C–30°C was delayed 5 days by sealing in PE bags (0.02 mm thickness) containing potassium permanganate, without any abnormal flavor. Gas composition in the bags was not reported by these authors. 'Tommy Atkins' and 'Keitt' mangoes were individually sealed in shrinkable Cryovac polyolefin films (15 or 19 mm thickness), either nonperforated (MD film) or perforated with eight holes of 1.7 mm diam./sq. inch (MPY) or 8 holes of 0.4 mm diameter/in.^{2} (SM60M) (Rodov et al., 1994). After 2–3 weeks storage at 14°C and an additional week at 17°C mango packaged in perforated polyelefin films ripened normally and best results were achieved when film with 0.4 mm perforations was combined with increased free volume inside the package by sealing the fruit within polystyrene trays. After 3 weeks of storage and 1 week of shelf life, sealed 'Keitt' mango had inferior quality than the control because it was less ripe, but beyond 4 weeks (up to 6 weeks) sealed fruits had better quality scores because it was less overripe. Sealing did not reduce decay of fruit stored for long periods. Nonperforated PVC film packaging of 'Nam Dork Mai' mangoes was not sufficiently permeable for O_2 exchange to allow proper ripening (Yantarasri et al., 1995). Therefore a so-called "perforated MA" was used where fruits were wrapped in polystyrene trays (3 fruits/pack) at 20°C with perforation area of ≥ 0.004 cm^2. Fruits were reported to ripen normally and with no production of off-flavors. Color development in the peel was reported to require a higher concentration of O_2 than the flesh, and a film of pore area ≥ 0.008 cm^2 allowed fruit color to develop after 3 weeks while a pore area of ≥ 0.39 cm^2 allowed the fruit to color within 2 weeks. Chaplin et al. (1986) reported that symptoms of CI were reduced in four cultivars of mango stored in sealed PE bags for up to 15 days at 1°C.

'Julie' mangoes treated with 0.75% w/v aqueous solution of Pro-long (a mixture of sucrose esters of fatty acids and sodium salt of caboxy methyl cellulose) and stored at 25°C and 85%–95% RH reduced weight loss, retarded ripening, and increased storage life (6 days longer) without causing any adverse effects on quality (Dhalla and Hanson, 1968). A treatment with 1.0% increased ethanol concentration in the pulp of some fruits. Treatment with "Prolong" (0.8%–2.4%) delayed ripening of 'Haden' mangoes (Carrillo-López et al., 1996).

Mango fruits were artificially infested with larvae of *Drosophila melanogaster*, and individually wrapped with a Cryovac D-955 cross-linked, 60-gauge polyofin shrink film (Shetty et al., 1989). None of the insects survived in fruits wrapped for 72 h or more. Gould and Sharp (1990) reported a 99.95% mortality of the Caribbean fruit fly in film wrapped mangoes for 15 days, but the fruit deteriorated after only 6 days.

Storage of 'Keitt' mangoes in an insecticidal MA (0.03–0.26 kPa O_2, 72–79 kPa CO_2, balance N_2), and CA (0.2 kPa O_2, balance N_2 or 2 kPa O_2 + 50 kPa CO_2, balance N_2) for up to 5 days at 20°C delayed fruit ripening as indicated by respiration, flesh firmness, and color development (Yahia, 1993b; Yahia and Tiznado, 1993; Yahia and Vazquez, 1993; Yahia, 1997; Yahia 1998a). These atmospheres increased the activity of phosphofructokinase, ADH, and PDC, but did not affect the activity of pyruvate kinase, succinate dehydrogenase, and α-ketoglutarate dehydrogenase. Although these atmospheres caused changes in glycolysis and tricarboxylic acid cycle, there was no indication of injury and fruit ripened normally after exposure to air. Sensory evaluation conducted after fruit ripening showed no presence of off-flavors, and there were no differences between fruit maintained in MA/CA and those maintained continually in air. On the basis of these results the authors concluded that 'Keitt' mango is very tolerant to IA. It is assumed that the 5 days tolerated by mango are sufficient to control many insects (Rojas-Villegas et al., 1996; Yahia, 1994b; Yahia, 1995b; Ortega and Yahia, 2000b). Storage of 'Keitt' and 'Tommy Atkins' mangoes for 21 days at 12°C in an atmosphere containing 25, 45, 50, and 70 kPa CO_2 plus either 3 kPa O_2 or air induced the production of 0.18–3.84 mL ethanol/kg h after transfer to air at 20°C for 5 days (Bender et al., 1995). These authors reported that atmospheres containing 50 and 70 kPa CO_2 caused fruit injury, and resulted in highest ethanol production rates. The enclosure of 'Haden' and 'Tommy Atkins' mangoes in sealed 20 L jars with an initial atmosphere of 90 kPa CO_2 in air or 97 kPa N_2 + 3 kPa O_2 for 24 h prior to storage delayed their ripening (Pesis et al., 1994b). 'Tommy Atkins' mangoes held in CA (3 kPa O_2 + 97 kPa N_2 or 3 kPa O_2 + 10 kPa CO_2 + 87 kPa N_2) directly or after a hot water quarantine treatment (46°C for 75 min), for 2 weeks at 10°C, and subsequent ripening at 25°C had extended shelf life without adversely affecting the nutritional profile of the fruit (Kim et al., 2007). Studies supplementing CA storage with a prestorage HWT showed that it is possible to gain dual advantage of retarded ripening in CA containers during transportation and controlling insect-pests and diseases with HWT for overcoming the quarantine barriers. Therefore, a combination of CA and HWT may be integrated into the postharvest supply chain of mango to comply with the phytosanitary regulations and providing consumer with a chemical residue-free fruit.

Burg (1975) reported that 'Haden' mangoes ripened four times slower at 150 mmHg than in air. Several cultivars of mango including 'Irwin,' 'Keitt,' 'Kent,' and 'Tommy Atkins' were found to be firmer after storage for 3 weeks in 76/152 mmHg at 13°C and 98%–100% RH (Spalding and Reeder, 1977). These fruits ripened normally after storage, had less decay, and higher percentage of acceptable fruits. A pressure of 76 mmHg resulted in the greener fruit however and caused splitting. Therefore, a pressure of 152 mmHg is considered as the optimum LP. Mango was shipped experimentally in LP (80 mmHg at 10°C) from Mexico to Japan and arrived in satisfactory condition after 28 days from picking (Spalding, 1977a). LP (152 mmHg) was reported to be suitable for shipping or storage of mango, together with bananas and limes (Spalding, 1977a,b). 'O Krong' mangoes precooled at 15°C and waxed were maintained at 60–100 mmHg at 13°C for up to 4 weeks, and then ripened normally (Ilangantileke and Salocke, 1989). 'Rad' mangoes were kept in 100 mmHg and 15°C for 30 days (Chen, 1987). However, due to cost considerations, no commercial use of LP is reported at the present.

There is no current use of CA for mango storage; however, long-term marine shipping in MA and CA is commercially used in several countries including Mexico and Australia (Yahia, 1993a, 1998a,b).

'Keitt' mangoes were individually vacuum packaged in LDPE film (24.5 μm thick, 25 g/m²) and were stored at 7°C/80%–90% RH, 12°C/75%–85% RH, 17°C/70%–80% RH, 22°C/65%–75% RH or 25°C/65%–75% RH (Yamashita et al., 1997). After mass transfer had reached a steady state, respiration rates, moisture loss, permeability of peel and film to water vapor and composition of atmosphere round the fruits were determined during storage for 33 days. Daily rates of weight loss increased from 4.1 g/kg fruit at 7°C to 10.9 g/kg at 25°C. Respiration rates also increased with storage temperature in both packaged and unpackaged mangoes, and were 21%, 38%, and 43% less in packaged fruits at 12°C, 17°C, and 22°C, respectively. Using mass transfer equations and the experimental results, it was calculated that permeability of peel was 600 times greater than that of the plastics film. CO_2 levels of atmosphere round the mangoes increased and that of O_2 decreased with time at the temperature studied; changes were greatest during the first 10–15 days of storage and were more marked at the higher temperature. Experimental and calculated values for CO_2 levels differed by 29% depending on temperature.

Effect of storing mangoes ('Tommy Atkins' or 'Keitt') in microperforated PE or Xtend® film (XF) bags on fruit quality was investigated by Pesis et al. (2000a,b). XF and PE films were used to create a modified atmosphere of 5 and 10 kPa CO_2 and O_2, respectively. Fruits were stored at 12°C and 85% RH for 21 days before being transferred to 80% RH and 20°C for determination of shelf life. Development of peel color was greatest in control fruit; waxing the fruit reduced development of color during storage at 20°C. Development of peel color was reduced by 50% in fruit packed in PE or XF bags after 21 days of storage at 12°C and 5 days at 20°C. Fruit packed in PE had the lowest color development. CI was very noticeable in control and waxed fruits; packing fruits in PE or XF significantly reduced incidence of CI. Fruit packaged in XF also had reduced sap adhesion levels due to low RH (~90%) within the packages compared to PE packaged fruits (~99%).

Fresh-cut 'Carabao' and 'Nam Dokmai' mango cubes were stored in air or in high CO_2 atmospheres (3, 5, and 10 kPa) at 5°C and 13°C (Poubol and Izumi, 2005a). Freshly sliced 'Carabao' mango cubes had a lower respiration rate and total bacterial count and higher L-ascorbic acid content and firmness than 'Nam Dokmai' mango cubes. The shelf life of fresh-cut mango, based on browning discoloration and water-soaked appearance, was 6 days at 5°C and 4 days at 13°C for 'Carabao' and 2 days at 5°C and <1 day at 13°C for 'Nam Dokmai.' High CO_2 atmospheres retarded the development of water-soaked 'Carabao' cubes at 5°C and 13°C and 'Nam Dokmai' cubes at 5°C. Texture of 'Carabao' cubes was enhanced by high CO_2, but ethanol and L-ascorbic acid contents were not affected at 5°C and 13°C. Total bacterial count was lower in 'Carabao' cubes than in 'Nam Dokmai' cubes during storage at both temperatures, and a 10 kPa CO_2 only reduced the bacterial count on 'Carabao' and 'Nam Dokmai' cubes stored at 13°C. Among 40–100 kPa O_2 atmospheres, 60 kPa O_2 reduced the respiration of fresh-cut 'Carabao' mango cubes the most when held at 5°C or 13°C for 42 h (Poubol and Izumi, 2005b). The high O_2 did not affect texture or ascorbic acid content of 'Carabao' and 'Nam Dokmai' mango cubes at either temperature. Counts of lactic acid bacteria and molds were below the detection level (2.4 log colony-forming units [CFU]/g) during storage at both temperatures. However, 60 kPa O_2 stimulated the growth of mesophilic aerobic bacteria on 'Carabao' cubes and yeasts of 'Nam Dokmai' cubes at 13°C. The increased microbial count may have been due to the higher pH of cubes stored in 60 kPa O_2 at 13°C than at 5°C or in air. However, other results have indicated that 60 kPa O_2 is not desirable for mango cubes when held at 13°C (Poubol and Izumi, 2005b).

Effects of LO atmospheres on quality and ethanol and acetaldehyde formation in stored ethylene-treated bananas ('Cavendish') were investigated by Imahori et al. (1998). Research needs on mango are diverse, including the following:

1. Controlled research is needed to establish ideal atmospheres and potential applications for the different cultivars in different regions. Of immediate importance is the ideal gas composition for the different cultivars, and controlled conditions to establish the potential feasibility of MAP and CA storage. No current information is available on the MA/CA research and its commercial use in mango cultivars of India, which is the world's largest diversity and production base of mango.

2. Studies are still needed to investigate the mortality of different insects in MA and CA, in combination with other treatments such as heat.

3. Research and policy support are required for the development of MA/CA as a quarantine treatment for insect control.

4. Tolerance/sensitivity of different mango cultivars to IA and its basis is needed to be established.

5. The efficacy of biological disease control agents in supplementation to CA needs to be tested for effective control of serious postharvest diseases.

6. The tagging of aroma volatile compounds' emission database to the ripening stages and disease incidence of mango fruit may be used to develop the sensors for product-state monitoring in the supply chain.

7. Development of MA/CA technology for storage and/transportation of "ripe" mangoes can minimize the ripening- and flavor-related problems, which are often encountered with CA-stored mature-green fruit.

16.3.7   Mangosteen (*Garcinia mangostana* L.)

A nonclimacteric fruit native to the Malay Archipelago. The fruit is a 5–10 cm round thick shelled, eaten mostly fresh but can be made into paste and candy. Postharvest life is about 2–4 weeks at 13°C and 85%–90% RH. Fruit was stored at 5°C in 5 kPa O_2 and 5 kPa CO_2 for 1 month (Godfrey Lubulwa and Davis, 1994). Mangosteen at turning stage stored in LO (1, 3, 5, or 10 kPa) and/or high CO_2 (5 or 10 kPa) at 15°C and 85%–90% RH had delayed peel color development and calyx deterioration for up to 7 weeks (Rattanachinnakorn et al., 1996). Atmospheres containing 5 kPa O_2 and 10 kPa CO_2 showed best results in maintaining both external appearance and internal quality for up to 4 weeks, with an extra 5 days of shelf life. Atmospheres with 1 kPa O_2 induced fermentation of the fruit, and increased the incidence of disease development. Packaging in PE bags (not characterized) reduced weight loss and postharvest diseases (Daryono and Sabari, 1986). It is not known whether the effect is due to humidity and/or gas modification. Further research is needed to establish the most ideal atmospheres and to investigate the real potential of MA/CA. Further research is still needed to reveal the potential benefits of MA/CA.

16.3.8   Papaya (*Carica papaya* L.)

A climacteric fruit native to Central America. Anthracnose and CI are the major postharvest problems. Postharvest life of the fruit is about 1–3 weeks at 7°C–13°C.

'Solo' papaya in Hawaii held for 6 days in 10 kPa CO_2 at 18°C developed less decay than fruit stored in air or in higher levels of CO_2 (Akamine, 1959, 1969). In 1986, Chen and Paull reported that ripening of 'Kapaho Solo' papaya was delayed by storage in 1.5–5 kPa O_2 with or without 2 or 10 kPa CO_2, but CI symptoms were not reduced. In Florida, papaya held in 1 kPa O_2 and 3 kPa CO_2 at 13°C for 3 weeks and then ripened at 21°C was 90% acceptable, with fair appearance, slight or no decay, and good flavor (Hatton and Reeder, 1969b). Storage life of 'Bentong' and 'Taiping' papaya in Malaysia was extended by maintaining the fruit in 5 kPa CO_2 at 15°C and removal of ethylene (Nazeeb and Broughton, 1978). CA storage of papaya was reported to be beneficial only when the O_2 concentration is kept under 1 kPa, and when it is used along with low temperature, HWT, and EDB (Akamine and Goo, 1969). EDB was banned in 1984 (Federal Register, 1984). Arriola et al. (1980) concluded that MA raised the cost but did not maintain better quality fruit. Spalding and Reeder (1974) concluded that CA is not beneficial to prolong the storage life of papaya. Akamine and Goo (1969) reported that the shelf life of papaya held in 1 kPa O_2 and at 13°C for 6 days was only 1 day longer than fruit held in air. However, Hatton and

Reeder (1969b) reported that they held papayas for 21 days with an acceptable quality in an atmosphere containing 1 kPa O_2 and 5 kPa CO_2. Optimum maturity for CA stored fruit is mature green or 10% yellow (Akamine and Goo, 1969). CA was suggested to supplement the HWT for a potential storage or shipping period of up to 12 days (Akamine and Goo, 1969). Sankat and Maharaj (1989) and Maharaj and Sankat (1990) reported that 'Known You No.1' and 'Tainung No.1' papayas at the color break stage treated with hot water (48°C for 20 min) and dipped in heated (1.23–1.50 g/L) Benlate (52°C for 2 min) were maintained for up to 29 days in 1.5%–2.0% kPaO_2 and 5 kPa CO_2 at 16°C, compared to 17 days in air. Akamine and Goo (1968) suggested that it is feasible to use CA during the shipment of hot-water treated or irradiated papayas when shipping period is from 6 to 12 days.

'Kapoho' and 'Sunrise' papayas individually sealed in HDPE (0.18 mm thickness) had less CI symptoms than unsealed fruits, but developed off-flavor (Chen and Paull, 1986). Seal packaging of 'Backcross Solo' papayas in 3 layers of LDPE (0.0125 mm thickness) and storage at 24°C–28°C for 18 days retarded development of peel color and fruit softening, and reduced the increase in titratable acidity (Lazan et al., 1990). In addition, seal packaging alleviated water stress and modified internal and external atmospheres. Internal CO_2 increased to 2.2 kPa, and O_2 decreased and was maintained at 1.2 kPa. The retardation in fruit softening was attributed partly to a decrease in polygalacturonase activity, and to polyuronide solubilization. MAP (using 0.05 mm shrinkable PE films) at 15°C retarded the firmness loss in 'Exotica' papaya fruit (Lazan et al., 1993). 'Sunset,' 'Sunrise,' and 'Kapoho Solo' papayas had a double HWT, dipped in 0.65 g/L thiabendazole, and either dipped in various wax solutions or shrink wrapped with various films (Paull and Chen, 1989). Films used were Cryovac MPD-2055, Cryovac D-955, Dupont 75EHC, Dupont 60EHC, and Dupont 50EHC. Type of wax solutions used were Brogdex 505-20 (1:11), FMC-7051 (1:9), FMC 560 (1:4), FMC-219B (1:4), Decco-261 (1:4), Agric Chem 93-8510078 (1:0), Prima Fresh-30 (1:3), Wax-On shellac (1:4), and Wax-On PE (1:4). After holding of fruit for up to 2 weeks at 10°C weight loss was reduced by 14%–40% by waxing, and 90% by shrink wrapping. Some treatments delayed ripening by 1–2 days after fruit was ripened in air; however, some off-flavor was also developed. CO_2 atmosphere that caused off-flavor was found to be 7–8 kPa; no off-flavor was developed at 6 kPa CO_2.

Applying a cellulose-based film to papaya altered the internal gas concentration, retarded ripening, and extended the shelf life of the fruit (Baldwin et al., 1992). 'Kapoho' and 'Sunrise' papayas treated with Sta-Fresh 7051 wax solution (1:10 v/v) and stored at 2°C for 14 days or 10°C for 24 days had less CI symptoms (Chen and Paull, 1986).

The mechanisms by which MAP in heat-shrinkable film and storage at moderately low (15°C) temperature (MLT) retarded firmness loss in 'Exotica' papaya fruit were investigated by Lazan et al. (1993). MAP and MLT treatments delayed as well as retarded firmness decrease, with the former being more effective than the latter in retarding texture change particularly when the fruit was stored at the lower-than-ambient (15°C) temperature. Efficacy of the different treatments in retarding firmness loss was reflected closely by the varying pattern of the increase in the cell wall hydrolases activity, particularly β-galactosidase, during storage ripening. Besides the cell wall enzymes, depolymerization of wall pectins was also affected by the storage treatments. The overall close correlations between the enzymes activity and variations in fruit firmness with ripening, tissue position, and storage treatments suggest that the cell wall hydrolases may collectively play a significant role in the softening of papaya.

Fruits shipped in hypobaric containers from Hawaii to Los Angeles and New York (20 mmHg, 10°C, and 90%–98% RH) for 18–21 days had longer postharvest life, developed less diseases, and mostly were ripened normally after removal from hypobaric containers (Alvarez, 1980). Fruit held in hypobaric storage had 63% less peduncle infection, 55% less stem end rot, and 45% less fruit surface lesions than those held in normal atmospheric pressure. Fruit stored for 21 days at 10 mmHg and 10°C immediately after being inoculated with *Colletotrichum gloesporioides* and then ripened for 5 days at room temperature had less anthracnose than the control fruit (Chau and Alvarez, 1983). However, the authors indicated that LP only retard pathogen and disease development and thus will only be effective if disease control programs are used to reduce fruit infection.

Fruit infested with eggs or first instar larvae of the Oriental fruit fly (*Dacus dorsalis* Hendel), wrapped in a Cryovac D-955 cross-linked, 60-gauge polyolefin shrink film, and stored at 24°C–25°C showed a reduction in the number of insects survived after 96 h (Shetty et al., 1989). Eggs and first instar larvae survived when the wrap was present for <48 h. The authors suggested that shrink wrap may affect the survival of eggs and larvae by creating a modified environment due to the depletion or accumulation of certain gases. However, they did not report any gas analysis in the packages. Jang (1990) infested 'Solo' papayas with eggs or one of three larval stages of *Ceratitis capitata* or *Dacus cucurbitae*, and individually wrapped the fruit in the same film (Cryovac, D-955) used by Shetty et al. (1989). Fruits were held at 22°C–24°C for 72–144 h. Fruit infestation significantly decreased as the storage period increased, especially after 96 h. Infestation with eggs of *Ceratitis* decreased about 80% between 72 and 120 h, and larval infestation was also reduced but some infestation remained even after 6 days of wrapping. *Dacus* larvae were found to be more resistant than *Ceratitis* eggs and larvae. The author reported that more than 90% of the infestations found after 120–144 h were due to loosely wrapped fruits or holes in the wrap. Larvae were observed to exit the fruit onto the fruit surface within 30–60 min, and often die between the fruit surface and the wrap. The author suggested that this might be due to modification of gases or altered metabolism inside the fruit. No gas monitoring was reported.

'Sunrise' papayas stored in an IA (0.17–0.35 kPa O_2, balance is N_2) for up to 5 days at 20°C had less firmness loss than the control, and no apparent external or internal injury (Yahia et al., 1989; Yahia, 1991; Yahia et al., 1992; Yahia, 1993b). However, about 30% of the fruit had very weak fermentative odor after 3 days, and increased in intensity as the exposure period to LO was prolonged. Activity of LDH and PDC increased after 3 and 5 days, respectively, while concentration of pyruvate and lactate did not change. On the basis of off-flavor development, papaya was suggested to tolerate these IA for <3 days. Decay was evident after 1 day storage in LO, indicating that LO alone is not sufficient, and there is a necessity for an antifungal treatment (Yahia et al., 1989).

Powrie et al. (1990) patented a preservation procedure for cut and segmented fruit pieces where they claimed to store papaya pieces in MAP for up to 16 weeks at 1°C with little loss in taste and texture. This was supposedly done in a high gas barrier package (DuPont LP 920^{TM}) consisting of PE/tie/ethylene vinyl alcohol/tie/PE plastic laminated pouches. Papaya was cut into pieces of 10–25 g, dipped in 5% citric acid, and the package was flushed with 15–20 kPa O_2 and 3 kPa helium before sealing. The ratio of gas to fruit volume was 1:4.

Exposure of 'Sunrise' papaya fruit to 10^{-5} or 10^{-4} M methyl jasmonate (MJ) vapors for 16 h at 20°C inhibited fungal decay and reduced CI development and loss of firmness during storage for 14–32 days at 10°C and 4 days shelf life at 20°C (González-Aguilar et al., 2003). MJ-treated fruit also retained higher organic acids than the control fruit. LDPE film packaging prevented water loss and further loss of firmness as well as inhibiting yellowing. The MA (3–5 kPa O_2 and 6–9 kPa CO_2) created inside the package did not induce any off-flavor development during storage at 10°C, and the postharvest quality of papaya was enhanced by combining the MJ treatments and MAP. MR 'Solo' papayas individually packed in LDPE, PP, and Pebax-C® films stored at 13°C and 85%–90% relative humidity prevented the development of CI symptoms during storage for 30 days and also during ripening at ambient conditions (Singh and Sudhakar Rao, 2003, 2004a,b, 2005a). Respiration and ethylene production rates during ripening of fruit were affected by the residual MA effect. Fruit packed in PP film, which created an atmosphere having more than 8 kPa CO_2 (Singh and Sudhakar Rao, 2004a), developed off-flavor when ripened after 20 or 30 days storage, while those packed in LDPE and Pebax-C film did not develop off- flavor upon ripening. LDPE and Pebax-C packed papayas were superior in sensory and nutritional quality with higher levels of dietary antioxidants. Singh and Sudhakar Rao (2003, 2005b) also investigated the potential usefulness of individual shrink wrapping (ISW) in a climacteric fruit like papaya, and found that ISW of 'Solo' papayas with LDPE film resulted in storage life of 2 weeks under ambient conditions (26°C–28°C) and up to 1 month at low temperature (13°C). ISW gave an excellent control of weight loss during storage at both the conditions. To confirm these preliminary findings of ISW in papaya, further studies were conducted on another cultivar 'Taiwan Red Lady' using different types of films such as BDF-2001, D-955, and LDPE, and the results were found satisfactory in terms of storage life extension and overall fruit quality (Singh et al., 2005). ISW is a simple postharvest operation which involves loose sealing of the fruit in a high permeability film, followed by passing loose sealed fruit through a heat tunnel with sufficient stay time to allow film-shrinkage, which further increases its permeability due to stretching effect. Microperforated films for ISW of tropical fruits, which are climacteric in nature, are advantageous to facilitate more gas exchange through the film, thus preventing the accumulation of fermentation metabolites beyond the acceptable limits. ISW technique in papaya is demonstrated pictorially in Figure 16.2. There is a huge commercial potential of MAP for extending the storage and shelf life of papaya fruit. Commercial use of MAP in papaya is now gaining momentum for export markets.

No commercial use of CA for storage and transportation is reported at present. Ideal atmospheres are not yet fully defined, but range between 2–5 kPa O_2 and 5–8 kPa CO_2. It is not known yet if MA/CA have potential application for papaya. Further controlled studies are still needed to establish the beneficial applications and adequate atmospheres.

16.3.9 Passion Fruit (Yellow) (*Passiflora edulis* (Sims) *f. flavicarpa* Deg.)

Passion fruit is a climacteric fruit native to the tropical region of North and South America. Postharvest life is about 2–3 weeks at 7°C–10°C. The yellow passion fruit is the basis of almost all the industry.

Fruits placed in polystyrene trays, overwrapped with plasticized PVC film (VF-60), and stored for 15–30 days at 10°C, had less weight loss and better external appearance, although did not effectively modify O_2 or CO_2 (Arjona et al., 1994). The CO_2 level within the packages never exceeded 0.5 kPa and that of O_2 never dropped below 13 kPa through-out the 30 days of the experiment. Therefore, considering the slight increases in CO_2 and decreases in O_2 atmospheres in the packages, most of the beneficial effect is obviously due to humidity control rather than to gas control. Wrapping did not affect fruit sugar and juice pH. About 80% of the fruits were judged marketable after storage of yellow passion fruit in PE bags (bags were not characterized) for 14 days at 23°C (Salazar and Torres, 1977). The authors recommended that fruit should be treated with fungicides before packaging. Fruit sealed in PE bags and stored at 6°C–10°C had less shriveling for 3–4 weeks (Campbell and Knight, 1983). Fruit stored in PE bags or treated with paraffin wax and stored at 7.2°C and 85%–90% RH remained marketable for up to 30 days (Cerrada et al., 1976). Coating of purple passion fruit with 1.0% and 1.5% Semperfresh reduced weight loss and extended the shelf life by 4 days (Bepete et al., 1994). There were no differences between the effect

FIGURE 16.2

Demonstration of different operations ((a) washing, (b) surface drying at ambient, (c) loose packing, (d) and (e) feeding of fruit through heating tunnel of shrink wrapped machine, (f) shrink wrapped fruit at the exit of the tunnel, (g) papaya fruit shrink wrapped with different types of films) involved in shrink wrapping of papaya fruit at the Division of Postharvest Technology, IIHR, Bangalore, India. (Courtesy of: S.P. Singh, Curtin University of Technology, Perth, Australia and D.V. Sudhkar Rao, IIHR, Bangalore, India.)

of 1% and 1.5%. It is not clear yet from previous studies the real potential benefits and feasibility of use of MA/CA for passion fruit, or the ideal atmosphere for this fruit.

16.3.10 Pineapple (*Ananas comsus* L., Merrill.)

A nonclimacteric fruit native to Brazil and widely distributed throughout the tropics. Fruit flies attack the fruit but do not survive inside. Posthravest life of the fruit is about 1–5 weeks at 7°C–13°C. Internal breakdown caused by low temperature is a major disorder during storage and transport.

Dull (1971) concluded that decreased O_2 and increased CO_2 concentrations had no obvious effect on fruit quality, and therefore no major advantage of quality maintenance was to be gained by manipulation of the concentration of these two gases. However, in a previous report, Dull et al. (1967) found that CA extended the storage life by 1–3 days. Akamine (1971) and Akamine and Goo (1971) also concluded that CA storage (2 kPa O_2 and 98 kPa N_2) at 7.2°C had no effect on the crown of the fruit, weight loss, decay, or on the incidence of indigenous brown spot. However, the same authors reported that CA storage delayed shell color development, improved fruit appearance by reducing superficial mold growth on the butt of the fruit, and extended the shelf life. In addition they indicated the possibility of shipping pineapple in CA. Optimum O_2 atmosphere was suggested to be 2 kPa. Haruenkit and Thompson (1994) stored 'Smooth Cayenne' pineapples imported to England from Mexico in CA of 1–2 kPa O_2 and 0–10 kPa CO_2 at 4°C, 8°C, and 12°C, and then ripened the fruit at 22°C for 3 and 6 days. Fruit stored in 1 or 2 kPa O_2 and 10 kPa CO_2 showed a delay in the development of internal browning. The authors concluded that pineapple could be stored in these conditions for <3 weeks. However, Paull and Rohrback (1985) reported that internal browning of 'Smooth Cayene' pineapple is not reduced by storage in 3 kPa O_2, with or without 5 kPa CO_2 at 8°C. Fruit stored in 3 kPa O_2 at 22°C for 1 week and followed by 1 week at 8°C had reduced symptoms.

Hypobaric storage was reported to extend the storage life by up to 30–40 days (Staby, 1976). Packaging of 'Mauricious' pineapple in PE bags for 2 weeks at 10°C accumulated an atmosphere of 10 kPa O_2 and 7 kPa CO_2 and resulted in black heart development (Hassan et al., 1985). However, Abdullah et al. (1985) reported that the same variety packaged in 0.07 mm thick PE bags and stored at 10°C for up to 4 weeks followed by 1 week at 28°C developed less black heart than the control. Atmosphere in these bags had 5–10 kPa O_2 and 7–13 kPa CO_2. Waxing 'Smooth Cayenne' fruit with PE–paraffin mixture (20%–50% v/v) increased the concentration of CO_2 in the internal atmosphere, reduced the loss in pH and ascorbic acid in the juice, and delayed the appearance and severity of internal browning after storage for up to 4 weeks at 8°C (Paull and Rohrback, 1982).

The influences of storage temperature and modified O_2 and CO_2 concentrations in the atmosphere on the postcutting life and quality of fresh-cut pineapple (*Ananas comosus*) were studied by Marrero and Kader (2006). As expected, temperature was the main factor affecting postcutting life, which ranged from 4 days at 10°C to over 14 days at 2.2°C and 0°C. The end of postcutting life was signaled by a sharp increase in CO_2 production followed by an increase in ethylene production. The main effect of reduced O_2 levels (8 kPa or lower) was better retention of the yellow color of the pulp pieces, as reflected in higher final chroma values, whereas elevated (10 kPa) CO_2 levels led to a reduction in browning. MAP allowed conservation of pulp pieces for over 2 weeks at 5°C or lower without undesirable changes in quality parameters. Powrie et al. (1990) patented a preservation procedure claiming to maintain pineapple pieces for up to 10 weeks in MAP at 1°C, without CI and loss of taste or texture. The fruit is sliced and cut into pieces of 6–15 g and packed in DuPont LP 920 plastic pouches. The packages were flushed with a gas mixture containing 15–20 kPa O_2 and 3 kPa argon, sealed, and immediately cooled to 1°C. The ratio of gas to fruit volume was 1–3.3.

Reports on pineapple are contradictory. Ideal atmospheres are not fully defined, but are about 2 kPa O_2 and 5–10 kPa CO_2. Very little research has been done on LP and MAP. Currently no commercial shipping or storage in CA is conducted. Research is still needed in a controlled manner to better conclude on the potential benefits of MA/CA, and fully define the ideal atmospheres.

16.3.11   Sapodilla (*Manilkara achras* L., syn. *Achras sapota* L.)

A climacteric fruit native to the Yucatan peninsula of Mexico and the province of Peten in Guatemala, and grown in several tropical regions. The fruit is called by several names such as sapodilla, chiku, dilly, nasberry, sapodilla plum, chico zapote, or néspero. It is a spherical, ellipsoidal fruit ranging from 100 to 500 g in nut. Postharvest life is about 2–3 weeks at 12°C–16°C and 85%–90% RH.

Storage life was increased by removing ethylene and adding 5–10 kPa CO_2 to the storage atmosphere (Broughton and Wong, 1979). An atmosphere of 20 kPa CO_2 was deleterious. Storage life was also increased when fruit was maintained in an atmosphere with 5–10 kPa O_2 at 20°C and C_2H_4 is removed (Hatton and Spalding, 1990). 'Kalipatti' fruits treated with 6% Waxol or 250 or 500 ppm Bavistin, or hot water (50°C for 10 min) and wrapped in 150 gauge thick PE film with 1% ventilation ripened later than those of the control, but fungal rot was high (Bojappa and Reddy, 1990). This is most probably due to high humidity rather than to atmosphere modification. Further research is still needed to identify optimum atmospheric composition and potential benefits of MA/CA.

The physicochemical changes in chiku (*Achras sapota* L, sapodillas) during storage at 5°C, 10°C and 15°C and the effect of MAP (fruits sealed in a 0.05 mm LDPE, fruit sealed in double-layered 0.05 mm LDPE, fruit vacuum-packed in 0.05 mm LDPE, and fruit shrink-wrapped in 0.025 mm PVC under a hot air tunnel at 150°C for 20 s), and ambient temperature was examined by monitoring fruit texture, weight loss, soluble solids content, pH, sucrose, fructose, glucose, pectin, tannin and ascorbic acid contents, and microbial infection (Mohamed et al., 1996). Using MAP, chiku could be stored for 4 weeks at 10°C and 3 weeks at 15°C, while without MAP the storage life was 1 week shorter. Packaging in LDPE was highly effective in maintaining texture and weight of cold stored fruit. Fruit stored at 5°C experienced CI, reflecting their inability to ripen properly, even after 3 days at room temperature in the presence of 50 g/kg calcium carbide. The ascorbic acid content of chiku was highest in vacuum-packed fruit followed by fruit in other LDPE packaging. LDPE packaged fruit achieved the highest sensory scores for taste, color, texture, and overall acceptability in cold-stored chiku. The unsealed nature and heating involved in shrink wrapping did not favorably affect the storage life of chiku and MAP alleviated the CI which occurred in chiku stored at 10°C but not at 5°C.

16.3.12   Sapote Mamey (*Pouteria sapota* (Jacq.) H.E. Moore & Stearn)

A climacteric tropical fruit native from Mexico and Central America. Its fruit is a good source of nutrients and is highly appreciated for its pleasant and sweet flavor and the bright deep orange-red color of the pulp. Apart from its good demand in Mexico, Central and Southern American countries, it is gaining popularity in other countries such as Australia, Israel, Philippines, and Spain (Alia-Tejacal et al., 2007). Fruit shelf life varies from 3 to 7 days at ambient conditions depending upon the temperature. It is also a chilling sensitive fruit. Thus, optimum storage temperature ranges from 10°C to 15°C depending upon the harvest maturity and season (Alia-Tejacal et al., 2007). Typical CI symptoms are flesh browning, uneven ripening, and softening, adherence of the flesh to the seed, off odors, and flavors and flesh lignification, without any visual expression of injury on the rind (Alia-Tejacal et al., 2007).

Villanueva-Arce et al. (1999) evaluated the ripening of sapote mamey stored in perforated PE bags at 25°C, 40%–45% RH, and observed that weight loss was reduce by 50% and soluble solids content and color development slowed down in comparison to uncovered fruit. Sapote mamey held at 15°C under a continuous flow of 5.1 kPa $CO_2 + 5.6$ kPa O_2 and balance N_2 maintained good quality for 3 weeks with no incidence of physiological disorders (Manzano, 2001). However, no information was provided about the fruit quality after their transfer to room temperature. Martinez-Morales et al. (2004, 2005) reported that storage at 10 kPa CO_2 and 5 kPa O_2 in a static system, decreased ethylene production, and delayed ripening of sapote mamey and these effects were also observed after the transfer of the fruit to room temperature. According to Martinez-Morales and Alia-Tejacal (2004), short storage periods (0–38 h) in 50 kPa CO_2 and between 29°C and 46°C limited the incidence of rots by *Botryodiplodia theobromae*. They also reported that increasing periods (0–38 h) of storage at this CO_2 levels did not affect ripening and quality of sapote mamey. Individual wrapping of sapote mamey fruit with Peakfresh and Kleen Pack films reduced weight loss by about 8%, delayed fruit ripening for 3 days without any adverse effects on fruit quality at 20.5°C ± 2.0°C and 29.4% RH (Ramos et al., 2005). Ergun et al. (2005) evaluated carnauba wax and 1-MCP treatment on 'Magana' sapote mamey and found that both treatments maintained fruit quality, but fruit treated with 1-MCP had a marketable life longer than waxed fruit.

More research efforts are required to define the tolerance limits of sapote mamey for LO and high CO_2. CA/MA could be of great importance for marine transportation of sapote mamey fruit to various nonproducing countries.

16.3.13 Starfruit (Carambola) (*Averrhoa carambola* L.)

Starfruit is a nonclimacteric fruit native to Southeast Asia. The fruit is 6–15 cm long ellipsoid with five distinct ribs, and a yellow to light orange color. The fruit is usually eaten fresh but can be dried or processed in candied confections. In addition of being popular in Asia, the fruit is been grown in Florida and some Central and South American countries. The fruit is very fragile when ripe and easily injured during handling. Storage life is about 3–4 weeks at 5°C–10°C.

Fresh-cut slices of carambola have great potential in fruit salads and other preparations. Fruit softening is not a major problem associated with it, which renders it more satisfactory for fresh-cut use (Teixeira et al., 2008). However, browning of cut-surface of carambola slices is a great concern. Postcutting dip in enzymatic browning inhibitors has been tested for controlling browning problems. However, the potential of LO atmospheres either alone or in association with these antibrowning agents has been investigated recently (Teixeira et al., 2008). 'Maha' carambola slices treated with 1% ascorbic acid in association with 0.4 kPa O_2 at 4.5°C did not present significant browning or loss of visual quality for up to 12 days, 3 days longer than LO (0.4 kPa) alone, thus, their quality can be significantly improved by combining both treatments (Teixeira et al., 2008). The response of fresh starfruit to different atmospheres requires investigations, and the lower O_2 and/or high CO_2 limits need to be established for this fruit.

16.3.14   Sugar Apple (Custard Apple, Sweet Sop) (*Annona squamosa* L.)

Sugar apple is a climacteric fruit indigenous to South America. Storage life is about 4–6 weeks at 5°C–7°C and 85%–90% RH. Ripening was delayed by addition of 10 and 15 kPa CO_2 or removal of O_2 (Broughton and Guat, 1979). An atmosphere of 5 kPa CO_2 did not cause any effect, while 15 kPa CO_2 caused abnormal ripening. The absence of O_2 inhibited fruit ripening and the climacteric rise in respiration. Babu et al. (1990) reported that fruit dipped in 500 ppm Bavistin and kept in PE bags containing $KMnO_4$ were maintained for up to 9 days. Recommended atmosphere is 10 kPa CO_2 at 15°C–20°C and 85%–90% RH. Further research is still needed to investigate the potential benefits of MA/CA.

16.3.15   Wax Apple (*Syzygium samarangense*)

This fruit is sensitive to CI at 2°C–10°C. Sealed PE packaging reduced CI and fruit rotting (Horng and Peng, 1983). There is no indication of whether the effect is due to gas or to humidity control. Research is still needed to determine optimum gas concentrations and potential use.

16.4   Conclusions

Tropical fruits occupy a distinguished place in world's horticultural trade. They are grown in conditions that favor pathogens and insects infestations, grow far from important markets, and frequently in regions in developing countries that lack basic infrastructure for postharvest handling and storage. In addition, most of these crops are chilling sensitive and are characterized with a short postharvest life. Tropical fruits are grown both in northern and southern hemispheres of the world which ensure the year around availability of these fruits to the nonproducing countries. Therefore, transportation and distribution of these fruit crops is very important to increase the scope of commercialization. MA and CA can provide major benefits for preserving the quality of these crops, especially during long-term sea transport. Major advances have been accomplished lately in the development of better MA and CA technology. The rising importance of tropical crops in international trade, and thus the need for technologies to prolong their postharvest life and maintain their quality, increased the need for the use of MA and CA, especially during sea transport. Very little research has been done on MA and CA of tropical crops and most of it has been done on very few crops, and therefore, basic data required for the application of this technology (such as optimum gas concentrations) are not yet available for many tropical crops. Most research conducted on MA/CA of tropical crops was on bananas, mango, and papaya. Some research has been done on cherimoya, durian, and pineapple, and little on feijoa, lanzones, mangosteen, passion fruit, sapodilla, sugar apples, and wax apples. No research has been reported on atemoya, birba, breadfruit, cacao, cashew, jackfruit, langsat, macadamia, mammee-apple, mountain apple, tomatillo, pulsan, white sapote, soursop, and tamarind. Studies on the mode of action of MA/CA were done almost exclusively on bananas, but very little to none on other tropical crops. Potential benefits of MA and CA for tropical crops depend on the type of the crop, handling methods during pre- and postharvest, and length of shipping period. Adequate handling system, including temperature management, humidity control, avoidance of mechanical damage, sanitation, and ethylene removal treatment (for some crops), is essential for the successful application of MA/CA.

16.5 Future Research Needs

Potential benefits and ideal MA/CA for many lesser-known tropical fruits have not been explored and investigated in detail. More research efforts should be directed on these undercommercialized fruits. Some fruit crops such as jackfruit and breadfruit would be more convenient in fresh-cut form to the consumer, and there is ample scope of commercialization of these fruits with the use of MA/CA technology. IA, especially in combination with other treatments such as heat, seem to be very promising for quarantine purposes and should be further investigated for all tropical fruits. Information needed include tolerance of different crops to these atmospheres, mortality of different species of insects of quarantine importance, ideal gas composition, temperature, and duration of treatment. The mode of action of MA/CA in alleviating some physiological disorders, especially CI, is still not clearly understood. The mechanisms by which some physiological disorders are initiated or augmented by MA/CA are also not yet understood. Research aiming at investigating the cause and developing control methods for these physiological disorders will certainly improve the application of MA/CA for tropical fruits. Variable results reported for MAP are obviously due to use of variable conditions (differences in cultivars used, stages of maturity, types of films, sealing methods, sizes of packages, temperatures, RH, etc.). Therefore, experiments should be controlled to distinguish effects due to atmosphere modification or to other factors. The behavior of fruit after MA/CA is still not fully understood, and therefore the methods of handling MA/CA treated crop are not very established. Further research is needed to investigate the metabolic changes due to MA/CA, and thus to implement adequate methods of handling. Potential use of LP for transport of tropical fruits, especially those that are very sensitive to ethylene and do not require the addition of other gases (such as CO_2 and CO), should be further investigated. Inexpensive LP technology is needed to be developed. The hidden potential of LP needs to be harnessed in tropical fruits. More in-depth studies are needed to investigate the potential use of other nonconventional gases such as carbon monoxide, nitrous oxide, and nitric oxide in combination with MA/CA, especially during transit. Treatments and methods to permit safer use should be developed. More research is needed to study the influence of MA/CA on the dynamics of microbial load on fruits during storage. MA/CA effects on flavor, nutritional, and functional components of tropical fruits need more investigations. There is a lack of information on the effects of various factors such as season of harvest, growing environment, and other preharvest factors on the postharvest quality and storage potential of tropical fruits especially under MA/CA. Further and in-depth research on the mode of action of MA/CA is still needed in order to increase the commercial use of the technology for tropical fruits. These studies should contribute further to our understanding of the mechanism by which LO/high CO_2 control fruit ripening/senescence or cause tissue injury. Very little is known on the protein turnover and gene expression in fruits held in MA/CA. Molecular studies are needed to identify clones for genes, which are switched on or off in response to LO/high CO_2, to identify molecular markers to monitor responses of fruits to MA/CA, and to try to manipulate tissue response. Collaboration of MA/CA researchers with the logistics industry is the need of hour to provide solutions to some practical problems often encountered by the traders and shippers. Research objectives need to be defined based on the feedback from the logistics industry. The contribution of MA/CA technologies, which are currently used worldwide and prediction of future growth trends, to the world's carbon emissions must be assessed to know their impact on the environment, and an action plan should be drawn to make them more environment friendly in future.

References

Abdulah, A. S. and F. M. Basiouny. 2000. Effects of elevated CO_2, liquid coating and ethylene inhibitors on postharvest storage and quality of mango. *Phyton-Int. J. Expt. Bot.* 66:137–144.

Abdullah, H., A. R. Abd Shukor, M. A Rohaya, and P. Mohd Salleh. 1987. Carbon dioxide injury in banana (*Musa sp.* cv Mas) during storage under modified atmosphere. *MARDI Annual Senior Staff Conference*, University of Malaysia, Kuala Lampur, January 14–17, 1987.

Abdullah, H., M. A. Rohaya, and M. Z. Zaipun. 1985. Effect of modified atmosphere on black heart development and ascorbic acid contents in 'Mauritius' pineapple (*Ananas comosus* cv 'Mauritius') during storage at low temperature. *ASEAN Food J.* 1:15–18.

Aegerter, A. F. and R. J. Folwell. 2001. Selected alternatives to methyl bromide in the postharvest and quarantine treatment of almonds and walnuts: An economic perspective. *J. Food Process. Preserv.* 25:(6):389–410.

Agillon, A. B., N. L. Wade, and M. C. C. Lizada. 1987. Wound-induced ethylene production in ripening. *ASEAN Food J.* 3(3 and 4):145–148.

Akamine, E. K. 1959. Effects of carbon dioxide on quality and shelf life in papaya. Hawaii Agricultural Experiment Station Technical Progress Report No. 120.

Akamine, E. K. 1969. Controlled atmosphere storage of papayas. *Hawaii Univ. Ext. Misc. Publ.* 64:23–24.

Akamine, E. K. 1971. Controlled atmosphere storage of fresh pineapple. *Univ. Hawaii Ext. Publ.* Honolulu, p. 8.

Akamine, E. K. and T. Goo. 1968. Controlled atmosphere storage for shelf life extension of irradiated papayas (*Carica papaya* L. var. Solo), Annual Report 1967–68. Division of Isotopes, U.S. Atomic Energy Commission, pp. 63–111.

Akamine, E. K. and T. Goo. 1969. Effects of controlled atmosphere storage of fresh papaya (*Carica papaya* L. var. Solo) with special reference to shelf life extension of fumigated fruits. Research Bulletin, Hawaii Agricultural Experiment Station Technical Progress Report No. 144 (BF).

Akamine, E. K. and T. Goo. 1971. Controlled atmosphere storage of fresh pineapple (*Ananas cosmos* L. 'Smooth cayene'). Research Bulletin, Hawaii Agricultural Experiment Station Technical Progress Report No. 152.

Ali Azizan, M. 1988. Effects of carbon dioxide on the process of ripening and modified atmosphere storage of 'Mas' bananas. PhD thesis, University of Kebangsaan, Bangi, Malaysia.

Alia-Tejacal, I., R. Villanueva-Arce, C. Pelayo-Zaldívar, M. T. Colinas-León, V. López-Martínez, and S. Bautista-Banos. 2007. Postharvest physiology and technology of sapote mamey fruit (*Pouteria sapota* (Jacq.) H.E. Moore & Stearn). *Postharvest Biol. Technol.* 45:285–297.

Alique, R. 1995. Residual effects of short term treatments with high CO_2 on the ripening of cherimoya (*Annona cherimola* Mill) fruit. *J. Hort. Sci.* 70:609–615.

Alique, R. and G. S. Oliveira. 1994. Changes in sugars and organic acids in cherimoya (*Annona cherimola* Mill) fruit under controlled atmosphere storage. *J. Agric. Food Chem.* 42:799–803.

Alvarez, A. M. 1980. Improved marketability of fresh papaya by shipment in hypobaric containers. *HortScience* 15:517–518.

Apelbaum, A., Y. Aharoni, and N. Temkin-Gorodeiski. 1977a. Effects of subatmospheric pressure on the ripening processes of banana fruit. *Trop. Agric. (Trinidad)* 54:39–46.

Apelbaum, A., C. Zauberman, and Y. Fuchs. 1977b. Prolongin storage life of avocado fruits by subatmospheric pressure. *HortScience* 12:115–117.

Arjona, H. E., F. B. Matta, and J. O. Garner, Jr. 1994. Wrapping in polyvinyl chloride film slows quality loss of yellow passion fruit. *HortScience* 29:295–296.

Arriola, M. C., J. F. Calzada, J. F. Menchu, C. Rolz, R. García, and S. de Cabrera. 1980. Papaya. In: S. Nagy and P. E. Shaw (Eds.), *Tropical and Subtropical Fruits*, Avi, Westport, CT, pp. 316–340.

Assis, J. S., R. Maldonado, R. Munoz, M. I. Escribano, and C. Merodio. 2001. Effect of high carbon dioxide concentration on PAL activity and phenolic contents in ripening cherimoya fruit. *Postharvest Biol. Technol.* 23:33–39.

Babu, K. B., Md. Zaheeruddin, and P. K. Prasad. 1990. Studies on postharvest storage of custard apple. *Acta Hort.* 269:299.

Badran, A. M. and L. Lima. 1969. Controlled atmosphere storage of green bananas. U.S. Patent 3:450,542.

Baldwin, E. A., J. K. Burns, W. Kazokas, J. K. Brecht, R. D. Hagenmaier, R. J. Bender, and E. Pesis. 1999. Effect of two edible coatings with different permeability characteristics on mango (*Mangifera indica* L.) ripening during storage. *Postharvest Biol. Technol.* 17(3):215–226.

Baldwin, E., M. Nispero-Carriedo, and C. Cambell. 1992. Extending storage life of papaya with edible coating. *Hort Science* 27:679.

Banks, H. J. 1984a. Current methods and potential systems for production of controlled atmospheres for grain storage. *Dev. Agric. Eng.* 5:523–542.

Banks, N. H. 1984b. Some effects of TAL Pro-long coating on ripening bananas. *J. Exp. Bot.* 35:127–137.

Banks, N. H. 1984c. Studies on the banana fruits surface in relation with the effects of TAL prolong coating on gaseous exchange. *Sci. Hort.* 24:279–286.

Beaulieu, J. C. and J. M. Lea. 2003. Volatile and quality changes in fresh-cut mangoes prepared from firm-ripe and soft-ripe fruit, stored in clamshell containers and passive MAP. *Postharvest Biol. Technol.* 30(1):15–28.

Bender, R. J., J. K. Brecht, E. A. Baldwin, and T. M. M. Malundo. 2000b. Aroma volatiles of mature-green and tree-ripe Tommy Atkins mangoes after controlled atmosphere vs. air storage. *HortScience* 35(4):684–686.

Bender, R. J., J. K. Brecht, and C. A. Campbell. 1995. Responses of Kent and 'Tommy Atkins' mangoes to reduced CO$_2$. *Proc. Fla. State Hort. Soc.* 107:274–277.

Bender, R. J., J. K. Brecht, S. A. Sargent, and D. J. Huber. 2000a. Mango tolerance to reduced oxygen levels in controlled atmosphere storage. *J. Am. Soc. Hort. Sci.* 25(6):707–713.

Bepete, M., N. Nenguwo, and J. E. Jackson. 1994. The effect of sucrose coating on ambient temperature storage of several fruits. In: B. R. Champ, E. Highley, and G. I. Johnson (Eds.), *Postharvest Handling of Tropical Fruits. Proceedings of the International Conference*, Chiang Mai, Thailand, July 19–23, 1993. ACIAR Proc. 50, pp. 427–429.

Blankenship, S. M. (Ed.). 1985. Controlled atmospheres for storage and transport of perishable agricultural commodities. *Proceedings of the 4th National CA Research Conference*, Raleigh, NC, July 23–26, 1985.

Blankenship, S. M. 1996. The effect of ethylene during controlled atmosphere of bananas. *HortScience* 31:638 (Abstr.).

Blanpied, G. D. (Ed.). 1993. CA'93. *Proceedings of the 6th International CA Research Conference*, Northeast Region Cultural Engineering Service, Cornell University, Ithaca, NY, NRAES-71, Vols. 1 and 2.

Bleinroth, E. W., J. L. M. Garcia, I. Shirose, and A. M. Carvalho. 1977a. Storage of avocado at low temperature and in controlled atmosphere (in Portuguese). *Coletanea Inst. Tecnol. Aliment. (Brazil)* 8(2):587–622.

Bleinroth, E. W., J. L. M. Garcia, and Y. Yokomizo. 1977b. Low temperature, controlled atmosphere conservation of four varieties of mango. *Coletanea Inst. Tecnol. Aliment. (Brazil)* 8(1):217–243.

Bojappa, K. K. M. and T. V. Reddy. 1990. Postharvest treatments to extend the shelf life of sapota fruit. *Acta Hort.* 269:391 (Abstr.).

Brecht, J. K., K. V. Chau, S. C. Fonseca, F. A. R. Oliveira, F. M. Silva, M. C. N. Nunes, and R. J. Bender. 2003. Maintaining optimal atmosphere conditions for fruits and vegetables throughout the postharvest handling chain. *Postharvest Biol.Technol.* 27:87–101.

Broughton, W. J. and T. Guat. 1979. Storage conditions and ripening of the custard apple (*Annona squamosa* L.). *Scientia Hort.* 10:73–82.

Broughton, W. J. and H. C. Wong. 1979. Storage conditions and ripening of chiku fruits *Achras sapota* L. *Scientia Hort.* 10:377–385.

Broughton, W. J. and K. F. Wu. 1979. Storage conditions and ripening of two cultivars of banana. *Season. Hort.* 10:83–93.

Brown, D. J. 1981. The effects of low O$_2$ atmospheres and ethylene and CO$_2$ production and 1-aminocyclopropane-1-carboxylic acid concentration in banana fruits. MS thesis, University of Maryland, College Park, MD.

Brown, E. O. and M. C. C. Lizada. 1984. Modified atmospheres and deterioration in lan-zones (*Lansium domesticum Correa*). *Postharvest Res. Notes* 1(2):36.

Brown, B. I., L. S. Wong, and B. I. Watson. 1985. Use of plastic film packaging and low temperature storage for postharvest handling of rambutan, carambola and sapodilla. *Proceedings of the Postharvest Horticulture Workshop*, Melbourne, Australia, pp. 272–286.

Buchanan, A. 1994. Tropical fruits: The social, political and economic issues. In: B. R. Champ, E. Highley, and G. I. Johnson (Eds.), *Postharvest Handling of Tropical Fruits. Proceedings of the International Conference*, Ching Mai, Thailand, July 19–23, ACIAR Proc. 50, pp. 18–26.

Burg, S. P. 1975. Hypobaric storage and transportation of fresh fruits and vegetables. In: N. F. Haard and D. K. Salunke (Eds.), *Postharvest Biology and Handling of Fruits and Vegetables*, Avi, Westport, CT, pp. 172–188.

Burg, S. P. 2004. *Postharvest Physiology and Hypobaric Storage of Fresh Produce*, CABI Publishing, Wallingford, U.K. pp. 654.

Burg, S. P. and E. A. Burg. 1966. Fruit storage at subatmospheric pressures. *Science* 153:314–315.

Byers, B. 1977. The Grumman Dormavac system. In: D. H. Dewey (Ed.), *Controlled atmospheres for the Storage and Transport of Perishable Agricultural Commodities. Proceedings of the 2nd National CA Research Conference*, Horticultural Report No. 28. Department of Horticulture, Michigan State University, East Lansing, MI, pp. 82–88.

Calderon, M. and R. Barkai-Golan (Eds.). 1990. *Food Preservation by Modified Atmospheres*, CRC Press, Boca Raton, FL.

Campbell, C. W. and R. J. Knight, Jr. 1983. Produccion de granadilla. Comunicacion XIII congreso NORCOFEL. Ministerio de agricultura, pesca y alimentación, Canary Islands, Spain, pp. 223–231.

Carrillo-López, A., R. Rojas-Villagas, and E. M. Yahia. 1996. Ripening and quality of mango fruit as affected by coating with 'Seperfresh.' *Acta Hort.* 370:206–216.

Cerrada, E., M. P. Cerrada, and M. A. Brasil. 1976. Conservaçao do maracujá amarelo para utilizaçao in natura. *Acta Hort.* 57:145–151.

Chamara, D., K. Illeperuma, and G. P. Theja. 2000. Effect of modified atmosphere and ethylene absorbers on extension of storage life of Kolikuttu banana at ambient temperature. *Fruits* 55(6):381–388.

Chaplin, G. R., D. Graham, and S. P. Cole. 1986. Reduction of chilli injury in mango fruit by storage in polyethylene bags. *ASEAN Food J.* 2:139–142.

Chaplin, G. R., K. J. Scott, and B. I. Brown. 1982. Effects of storing mangoes in polyethylene bags at ambient temperature. *Singapore J. Pri. Ind.* 10:84–88.

Chau, K. F. and A. M. Alvarez. 1983. Effects of low pressure storage on *Colletotrichum gloeosprioides* and postharvest infection of papaya. *HortScience* 18:953–955.

Chen, R. C. 1987. Effect of precooling and waxing treatment on mango under hypobaric storage. MSc thesis, AIT, Bangkok, Thailand.

Chen, J. N. and R. E. Paull. 1986. Development and prevention of chilling injury in papaya fruit. *J. Am. Soc. Hort. Sci.* 111:639–643.

Chiang, M. N. 1968. Studies on the removal of ethylene from CA storage of bananas. Special Publication of the College of Agriculture, National Taiwanese University 20, Taipei, China.

Chiang, M. N. 1970. The effect of temperature and the concentration of O_2 and CO_2 upon the respiration and ripening of bananas, stored in a controlled atmosphere. Special Publication of the China, College of Agriculture, National Taiwan University, Taipei, Vol. 11, pp. 1–13.

Choehom, R., S. Ketsa, and W. G. van Doorn. 2004. Senescent spotting of banana peel is inhibited by modified atmosphere packaging. *Postharvest Biol. Technol.* 31(2):167–175.

Daryono, M. and S. Sabari. 1986. The practical method of harvest time on mangosteen fruit and its characteristics in storage. *Bull. Penelitian-Hortikultura Indones.* 14(2):38–44.

De La Plaza, J. L. 1980. Controlled atmosphere storage of chirimoya. *Proceedings of the 15th International Congress on Refrigeration*, Venice, Italy, Vol. 3, pp. 701–712.

De La Plaza, J. L., L. Muñoz-Delgado, and C. Iglesias. 1979. Controlled atmosphere storage of chirimoya. *Bull. I'Inst. Int. Froid* 59(4):1154.

De La Plaza, J. L., S. Rossi, and M. L. Calvo. 1993. Inhibitory effects of the ethylene chemisorption on the climacteric of cherimoya fruit in modified atmosphere. *Acta Hort.* 343:181–183.

Del Cura, B., M. I. Escribano, J. P. Zamorano, and C. Merodio. 1996. High carbon dioxide delay postharvest changes in RuBPCase and polygalacturonase-related protein in cherimoya peel. *J. Am. Soc. Sci.* 121:735–739.

Dewey, D. H. (Ed.). 1977. Controlled atmosphere for the storage and transport of perishable agricultural commodities. *Proceedings of the 2nd National CE Research Conference*, Horticultural Report No. 28. Department of Horticulture, Michigan State University, East Lansing, MI.

Dewey, D. H., R. C. Herner, and D. R. Dilley (Eds.). 1969. Controlled atmosphere for the storage and transport of horticultural crops. *Proceedings of the National CA Research Conference*, Horticultural Report No. 9. Department of Horticulture, Michigan State University, East Lansing, MI.

Dhalla, R. and S. W. Hanson. 1968. Effect of permeable coating on the storage life of fruit. II Pro-long treatment of mangoes (*Mangifera indica* L. cv. Julie). *Int. J. Food Sci. Technol.* 23:107–112.

Duan, H., S. G. Gilbert, Y. Ashkenazi, and Y. Hening. 1973. Storage quality of bananas packaged in selected permeability films. *J. Food Sci.* 38:1247–1250.

Durand, B. J., L. Orcan, U. Yanko, G. Zauberman, and Y. Fuchs. 1984. Effect of waxing on moisture loss and ripening of 'Fuerte' avocado. *HortScience* 19:421–422.

El-Goorani, M. A. and N. F. Sommer. 1981. Effects of modified atmospheres on postharvest pathogens of fruits and vegetables. *Hort. Rev.* 3:412–461.

Ergun, M., S. A. Sargent, A. J. Fox, J. A. Crane, and D. J. Huber. 2005. Ripening and quality responses of mamey sapote fruit to postharvest wax and 1-methylcyclopropene treatments. *Postharvest Biol. Technol.* 36:127–134.

Esguerra, E. B., D. B. Mendoza, Jr., and E. R. B. Pantastico. 1978. II. Use of perlite-KMnO^4 insert as an ethylene absorbent. *Philipp. J. Sci.* 107:1–2.

Federal Register 1984. Ethylene dibromide; amendment of notice to cancel registration of pesticide products containing ethylene dibromide. *Federal Register* 49:14182–14185.

Fellman, J. K. (Ed.). 1989. *Proceedings of the 5th International CA Research Conference*, Wenatchee, WA, June 14–16, 1989, Vols. 1 and 2.

Fons, J. F. 1990. Quarantine barriers treatments as a means to facilitate trade. *Acta Hort.* 269:435–439.

Fuchs, Y. and N. Temkin-Forodeiski. 1971. The course of ripening of banana fruits in sealed polyethylene bags. *J. Am. Soc. Hort. Sci.* 96:401–402.

Gane, R. 1936. A study of the respiration of bananas. *New Phytol.* 35:383.

Gautam, D. M. and M. C. C. Lizada. 1984. Internal breakdown in 'Carabao' mango subjected to modified atmospheres. Storage duration and severity symptoms. *Postharvest Res. Notes* 1(2):28. (Postharvest Horticultural Training and Research Center, Department of Horticulture, University of Philippines, Los Baños, Philippines).

Gemma, H., C. Oogaki, M. Fukushima, T. Yamada, and Y. Nose. 1989. Preservation of some tropical fruits with an apparatus of low pressure storage. *J. Jpn. Soc. Food Sci. Technol.* 36:508–518.

Godfrey Lubulwa, A. S. and J. S. Davis. 1994. An economic evaluation of postharvest tropical fruit research: Some preliminary results. In: B. R. Champ, E. Highley, and G. I. Johnson (Eds.), *Postharvest Handling of Tropical Fruits. Proceedings of the International Conference*, Ching Mai, Thailand, July 19–23, 1993. ACIAR Proc. 50, pp. 32–49.

Gonzalez, G., E. M. Yahia, and I. Higuera. 1990. Modified atmosphere packaging of mango and avocado fruit. *Acta Hort.* 269:355–344.

González-Aguilar, G. A., J. G. Buta, and C. Y. Wang. 2003. Methyl jasmonate and modified atmosphere packaging (MAP) reduce decay and maintain postharvest quality of papaya 'Sunrise.' *Postharvest Biol. Technol.* 28(3):361–370.

Gonzalez-Aguilar, G., A. Gardea, M. A. Martinez-Tellez, R. Baez, and L. Felix. 1997. Low oxygen treatment before storage in normal or modified atmosphere packaging of mangoes to extend shelf life. *J. Food Sci. Technol.* 34(5): 399–404.

Gould, W. P. and J. L. Sharp. 1990. Caribbean fruit fly (Diptera: Tephritidae) mortality induced by shrink-wrapping infested mangoes. *J. Econ. Entomol.* 83:2324–2326.

Haruenkit, R. and A. K. Thompson. 1994. Storage of fresh pineapples. In: R. Champ, E. Highley, and G. I. Johnson (Eds.), *Postharvest Handling of Tropical Fruits. Proceedings of the International Conference*, Ching Mai, Thailand, July 19–23, 1993. ACIAR Proc. 50, pp. 422–426.

Hassan, A., R. T. Atan, and Z. M. Zain. 1985. Effect of modified atmosphere on black heart development and ascorbic acid contents in 'Mauritius' pineapples (*Ananas comosus* cv. 'Mauritius') during storage at low temperature. *ASEAN Food J.* 1(1):15–18.

Hatton, T. T. and W. F. Reeder. 1966. Controlled atmosphere of Keitt mangoes, 1965. *Proc. Caribb. Reg. Am. Soc. Hort. Sci.* 10:114–119.

Hatton, T. T., Jr. and W. F. Reeder. 1967. Controlled atmosphere storage of Keitt mangoes, 1965. *Proc. Am. Soc. Hort. Sci. (Caribb. Reg.)* 10:114–119.

Hatton, T. T., Jr. and W. F. Reeder. 1969a Responses of Florida avocados, mangoes and limes in several controlled atmospheres. *Proceedings of the National CA Research Conference*, Michigan State University Horticulture Report No. 9, East Lansing, MI, pp. 72–73.

Hatton, T. T., Jr. and W. F. Reeder. 1969b. Controlled atmosphere storage of papayas (1968). *Proc. Trop. Reg. Am. Soc. Hort. Sci.* 13:251–256.

Hatton, T. T., Jr. and D. H. Spalding.1990. Controlled atmospheres storage of some tropical fruits. In: M. Calderon and R. Barkai-Golan (Eds.), *Food Preservation by Modified Atmospheres*, CRC Press, Boca Raton, FL, pp. 301–313.

Hesselman, G. W. and H. T. Freebairn. 1986. Rate of ripening of initiated bananas as influenced by oxygen and ethylene. *J. Am. Soc. Hort. Sci.* 94:635–637.

Hewage, S. K., H. Wainwright, S. W. Wijerathnam, and T. Swinburne. 1995. The modified atmosphere storage of bananas as affected by different temperatures, In: A. Ait Oubahou and M. El Otmani (Eds.), *Postharvest Physiology and Technologies for Horticultural Commodities: Recent Advances*, Institut Agronomique and Vetérinaire Hassan II, Agadir, Morocco, pp. 172–176.

Horng, D. and C. Peng. 1983. Studies on package, transportation and storage of waxapple fruits (*Syzygium samarangense*) (in Japanese). *NCHU Horti.* 8:31–39.

Ilangantileke, S. and V. Salocke. 1989. Low-pressure storage of Thai mango. In: *Other Commodities and Storage Reconditions, Vol. 2. Proceedings of the 5th CA Research Conference*, Wenatchee, WA, June 14–16, 1989, pp. 103–117.

Imahori, Y., M. Kota, Y. Ueda, and K. Chachin. 1998. Effects of low-oxygen atmospheres on quality and ethanol and acetaldehyde formation of ethylene-treated bananas. *J. Jpn. Soc. Food Sci. Technol. (Nippon-Shokuhin-Kagaku-Kogaku-Kaishi)* 45(9):572–576.

Jang, E. B. 1990. Fruit fly disinfestations of tropical fruits using semipermeable shrink-wrap film. *Acta Hort.* 269:453–458.

Kader, A. A. 1985. Modified atmospheres. An indexed reference list with emphasis on horticultural commodities, Supplement No. 4 (January 1, 1981 to May 31, 1985). Postharvest Horticulture Series 3, University of California, Davis, CA.

Kader, A. A. 1986. Biochemical and physiological basis for effects on controlled and modified atmospheres on fruits and vegetables. *Food Technol.* 40(5):100, 102–104.

Kader, A. A. 1993. Modified and controlled atmosphere storage of tropical fruits. In: B. R. Champ, E. Highley, and G. I. Johnson (Eds.), *Postharvest Handling of Tropical Fruits. Proceedings of the International Conference*, Chiang Mai, Thailand, July 19–23, 1993. ACIAR Proc. 50, pp. 239–249.

Kader, A. A. and D. Ke. 1994. Controlled atmospheres. In: R. E. Paul and J. W. Armstrong (Eds.), *Insect Pest and Fresh Horticultural Products: Treatments and Responses*, CAB International, Wallingford, U.K., pp. 223–236.

Kader, A. A. and L. L. Morris. 1981. Modified atmospheres. An indexed reference list with emphasis on horticultural commodities, Supplement No. 3 (March 1, 1977 to December 31, 1980). Vegetable Crops Series 213, University of California, Davis, CA.

Kader, A. A. and L. L. Morris. 1997. Modified atmospheres. An indexed reference list with emphasis on horticultural commodities, Supplement No. 2 (May 1, 1974 to February 28 1977). Vegetable Crops Series 187, University of California, Davis, CA.

Kader, A. A., D. Zagory, and E. L. Kerbel. 1989. Modified atmosphere packaging of fruits and vegetables. *CRC Crit. Rev. Food Sci. Nutr.* 28(1):1–30.

Kane, O. and P. Marcellin. 1979. Effects of controlled atmosphere on the storage of mangoes (varieties Amelie and Julie) (in French). *Fruits* 34(2):123–129.

Kanellis, A. K. and T. Solomos. 1985. The effect of low oxygen on the activities of pectin-methylester-ase and acid phosphatase during the course of ripening of bananas. In: S. M. Blankenship (Ed.), *Proceedings of the 4th National CA Research Conference*, Horticultural Report No. 126. Department of Horticultural Science, North Carolina State University, Raleigh, NC, pp. 20–26.

Kanellis, A. K., T. Solomos, and A. K. Mattoo. 1989. Changes in sugars, enzymic activities and acid phosphatase profiles of bananas ripened in air or stored in 2.5% O_2 with and without ethylene. *Plant Physiol.* 90:251–258.

Kapur, N. S., K. S. Rao, and H. S. Srivastava. 1962. Refrigerated gas storage of mangoes. *Food Sci.* 11:228–231.

Ke, D. and A. A. Kader. 1992. Potential of controlled atmospheres for postharvest insect disinfestations of fruits and vegetables. *Postharvest News Info.* 3(2):31N–37N.

Ke, D., E. Yahia, B. Hess, L. Zhou, and A. Kader. 1995. Regulation of fermentative metabolism in advocado fruit under oxygen and carbon dioxide stresses. *J. Amer. Soc. Hort. Sci.* 120:481–490.

Ketsa, S. and K. Leelawatana. 1992. Effect of precooling and polyethylene film liners in corrugated boxes on quality of lychee fruits. In: S. Subhadrabandhu (Ed.), *Frontier in Tropical Fruit Research*, International Society for Horticultural Science, Wageningen, the Netherlands, pp. 742–746.

Kim, Y., J. K. Brecht, and S. T. Talcott. 2007. Antioxidant phytochemical and fruit quality changes in mango (*Mangifera indica* L.) following hot water immersion and controlled atmosphere storage. *Food Chem.* 105:1327–1334.

Lakshminarayana, S. and H. Subramanyam. 1970. Carbon dioxide injury and fermentative decarboxylation in mango fruit at low temperature storage. *J. Food Sci. Technol.* 7(3):148–152.

Lalel, H. J. D. and Z. Singh. 2006. Controlled atmosphere storage of 'Delta R2E2' mango fruit affects production of aroma volatile compounds. *J. Hort. Sci. Biotechnol.* 81:449–457.

Lalel, H. J. D., Z. Singh, and S. C. Tan. 2004. Biosynthesis of aroma volatile compounds and fatty acids in 'Kensington Pride' mangoes after storage in a controlled atmosphere storage at different oxygen and carbon dioxide concentrations. *J. Hort. Sci. Biotechnol.* 79:343–353.

Lalel, H. J. D., Z. Singh, and S. C. Tan. 2005. Controlled atmosphere storage affects fruit ripening and quality of 'Delta R2E2' mango. *J. Hort. Sci. Biotechnol.* 80:551–556.

Lazan, H., Z. M. Ali, and W. C. Sim. 1990. Retardation of ripening and development of water stress in papaya fruit seal-package with polyethylene film. *Acta Hort.* 269:345–358.

Lazan, H., Z. M. Alid, and M. K. Selamat. 1993. The underlying biochemistry of the effect of modified atmosphere and storage temperature on firmness decrease in papaya. *Acta Hort.* 343:141–147.

Liu, F. W. 1970. Storage of bananas in polyethylene bags with and ethylene absorbent. *HortScience* 5:25–27.

Liu, F. W. 1976a. Storing ethylene-pretreated bananas in controlled atmosphere and hypobaric air. *J. Am. Soc. Hort. Sci.* 101:198–201.

Liu, F. W. 1976b. Correlation between banana storage life and minimum time required for ethylene response. *J. Am. Soc. Hort. Sci.* 101:63–65.

Lizada, C. C. and V. Noverio. 1983. The effect of pro-long on patterns of physico-chemical and physiological changes in the ripening of bananas, Annual Report, PHTRC, Laguna, Philippines, 1983.

Lizana, L. A. and A. Ochagavia. 1997. Controlled atmosphere storage of mango fruits (*Mangifera indica* L.) cvs. 'Tommy Atkins' and 'Kent.' *Acta Hort.* 455:732–737.

Lougheed, E. C. and S. Q. Feng. 1989. Pragmatic assessment of the potential of CA. In: *Other Commodities and Storage Recommendations, Vol. 2 Proceedings of the 5th CA Research Conference*, Wenatchee, WA, June 14–16, 1989, pp. 217–223.

Lowings, P. H. and D. F. Cuts. 1982. The preservation of fresh fruits and vegetables. *IFST Proc.* 15:52–54.

Maekawa, T. 1990. On the mango CA storage and transportation from subtropical to temperate regions in Japan. *Acta Hort.* 269:367–374.

Maharaj, R. and C. K. Sankat. 1990. Storability of papayas under refrigerated and controlled atmosphere. *Acta Hort.* 269:375–385.

Malcom, G. L. 1989. Controlled atmospheres: Now a reality with membranes. *Proceedings of the Cargo Systems Containers and Intermodal Conference*, Hamburg, Germany, November 28–30, 1989.

Maldonado, R., O. Goni, M. I. Escribano, and C. Merodio. 2007. Regulation of phenylalanine ammonia-lyase enzyme in annona fruit: Kinetic characteristics and inhibitory effect of ammonia. *J. Food Biochem.* 31:161–178.

Maldonado, R., A. D. Molina-Garcia, M. T. Sanchez-Ballesta, M. I. Escribano, and C. Merodio. 2002. High CO_2 atmosphere modulating the phenolic response associated with cell adhesion and hardening of *Annona cherimola* fruit stored at chilling temperature. *J. Agric. Food Chem.* 50:7564–7569.

Maldonado, R., M. T. Sanchez-Ballesta, R. Alique, M. I. Escribano, and C. Merodio. 2004. Malate metabolism and adaptation to chilling temperature storage by pretreatment with high CO_2 levels in *Annona cherimola* fruit. *J. Agric. Food Chem.* 52:4758–4763.

Maneenuam, T., S. Ketsa, and W. G. van Doorn. 2007. High oxygen levels promote peel spotting in banana fruit. *Postharvest Biol.Technol.* 43:128–132.

Manzano, E. J. 2001. Caracterizacion de algunos parametros de calidad en frutos de zapote mamey (*Calocarpum sapota* (Jacq.)) Merr. en diferentes condiciones de almacenamiento. *Proc. Interam. Soc. Trop. Hort.* 43:53–56.

Mapson, L. W. and J. E. Robinson. 1966. Relation between O_2 tension, biosynthesis of ethylene, respiration and ripening changes in banana fruit. *J. Food Technol.* 1:215–225.

Marrero, A. and A. A. Kader. 2006. Optimal temperature and modified atmosphere for keeping quality of fresh-cut pineapples. *Postharvest Biol. Technol.* 39:163–168.

Martínez-Ferrer, M., C. Harper, F. Pérez-Muñoz, and M. Chaparro. 2002. Modified atmosphere packaging of minimally processed mango and pineapple fruits. *J. Food Sci.* 67(9):3365–3371.

Martinez-Morales, A. and I. Alia-Tejacal. 2004. Almacenamiento en atmosferas controlada y temperatura alta de zapote mamey (*Pouteria sapota*), Cong. Nal. Agroindustrial, Universidad Autonoma de Chapingo, Mexico, May 12–14, 2004.

Martinez-Morales, A., I. Alia-Tejacal, M. T. Colinas-Leon, and M. T. Martinez-Damian. 2004. Storage of zapote mamey fruit under controlled atmosphere. *HortScience* 39:806.

Martinez-Morales, A., I. Alia-Tejacal, M. T. Colinas-Leon, and M. T. Martinez-Damian. 2005. Respiraci'on y producci'on de etileno de frutos de zapote mamey (*Pouteria sapota*) almacenados previamente en atmosferas modificadas. *Inv. Agropecuaria* 2004, 2:26–31.

McGlasson, W. B. 1989. MA Packaging. A practical alternative to CA shipping containers. In: *Other Commodities and Storage Recommendations Vol. 2. Proceedings of the 5th International CA Research Conference*, Wenatchee, WA, June 14–16, 1989, pp. 235–240.

McGlasson, W. B. and R. B. H. Wills. 1972. Effects of oxygen and carbon dioxide on respiration storage life, and organic acids of green bananas. *Aus. J. Biol. Sci.* 25:35–42.

McLauchlan, R. L. and L. R. Barker. 1992. Controlled atmospheres for Kensington mango storage: Classical atmosphere. *ACIAR Proc.* 58:41–44.

Merodio, C., M. T. Munoz, B. Del Cura, M. D. Buitrago, and M. I. Escribano. 1998. Effect of high CO_2 levels on the titers of α-aminobutyric acid, total polyamines and some pathogenesis-related proteins in cherimoya fruit stores at low temperature. *J. Expt. Bot.* 49:1339–1347.

Miller, W. R., D. H. Spalding, and P. W. Hale. 1986. Film wrapping mangoes at advanced stages of postharvest ripening. *Trop. Sci.* 26:9–17.

Miller, W. R., P. W. Hale, D. H. Spalding, and P. Davis. 1983. Quality and decay of mango fruit wrapped in heat-shrinkable film. *HortScience* 18:957–958.

Mohamed, S., B. Taufik, and M. N. A. Karim. 1996. Effects of modified atmosphere packaging on the physicochemical characteristics of ciku (*Achras sapota* L) at various storage temperatures. *J. Sci. Food Agric.* 70(2):231–240.

Nakamura, N., D. V. Sudhakar Rao, T. Shiina, and Y. Nawa. 2004. Respiratory properties of tree-ripe mango under CA condition. *Jph. Agric. Res. Q.* 38(4):221–226.

Nazeeb, M. and W. J. Broughton. 1978. Storage conditions and ripening of papaya 'Bentong' and 'Taiping.' *Scientia Hort.* 9(3):265–277.

Nguyen, T. B. T., S. Ketsa, and W. G. van Doorn. 2004. Effect of modified atmosphere packaging on chilling-induced peel browning in banana. *Postharvest Biol. Technol.* 31(3):313–317.

Noomhorn, A. and N. Tiasuwan. 1995. Controlled atmosphere storage of mango fruit, *Mangifera indica* L. cv. Rad. *J. Food Proc. Preserv.* 19:271–281.

Nuevo, P. A., A. U. Cua, and M. C. C. Lizada. 1984a. Internal Breakdown in 'Carabao' mango subjected to modified atmospheres. III. Starch in the spongy tissue. *Postharvest Res. Notes*: 1(2):63 (Postharvest Horticultural Training and Research Center, Department of Horticulture, University of Philippines, Los Baños, Philippines).

Nuevo, P. A., E. R. B. Pantastico, and D. B. Mendosa. 1984b. Gas diffusion factors in fruits. I. Anatomical structure in mango fruit. *Postharvest Res. Notes* 1(1):1 (Postharvest Horticultural Training and Research Center, Department of Horticulture, University of Philippines, Los Baños, philippines).

O'Hare, T. J. and A. Prasad. 1993. The effect of temperature and carbon dioxide on chilling symptoms in mango. *Acta Hort.* 343:244–250.

Olorunda, A. O. 1976. Effect of ethylene absorbent on the storage life of plantain packed in polyethylene bags. *Nig. J. Sci.* 10(1/2):19–26.

Ortega, D. and E. M. Yahia. 2000a. Tolerance and quality of mango fruit exposed to controlled atmospheres at high temperatures. *Postharvest Biol. Technol.* 2000:20:195–201.

Ortega, D. and E. M. Yahia. 2000b. Mortality of eggs and larvae of *Anastrepha obliqua* (MACQUART) and *A. ludens* (LOWE) (Diptera: Tephritidae) with controlled atmospheres at high temperature in mango (*Mangifera indica*) cv Manila. *Folia Entomol. Mex.* (in Spanish) 109:43–53.

Palma, T., D. W. Stanley, J. M. Aguilera, and J. P. Zoffoli. 1993. Respiratory behavior of cherimoya (*Annona cherimola* Mill) under controlled atmospheres. *HortScience* 28:647–649.

Pantastico, E. R. B., B. Mendoza, Jr., and R. M. Abilay. 1969. Some chemical and physiological changes during storage of Inzones (*Lansium domesticum* Correa). *Philipp. Agric.* 52:505–517.

Parsons, C. S., J. E. Gates, and D. H. Spalding. 1964. Quality of some fruits and vegetables after holding in nitrogen atmospheres. *Proc. Am. Soc. Hort. Sci.* 84:549–566.

Pastor, R. L. and E. R. B. Pantastico. 1984. Storage characteristics of waxed cooking bananas. *Postharvest Res. Notes* 1:23–24. (Postharvest Horticultural Training and Research Center. Department of Horticulture, University of Philippines, Los Baños, Philippines).

Paull, R. E. and J. W. Armstrong (Eds.). 1994. *Insect Pest and Fresh Horticultural Products. Treatments and Responses*, CAB International, Wallingford, U.K.

Paull, R. E. and J. N. Chen. 1989. Waxing and plastic wraps influence water loss from papaya fruit during storage and ripening. *J. Am. Soc. Hort. Sci.* 114:937–942.

Paull, R. E. and K. G. Rohrbackh. 1985. Symptoms development of chilling injury in pineapple fruit. *J. Am. Soc. Hort. Sci.* 110:100–105.

Pesis, E., D. Aarón, Z. Aarón, R. Ben-Arie, N. Aharoni, and Y. Fuchs. 2000a. Modified atmosphere and modified humidity packaging alleviates chilling injury symptoms in mango fruit. *Postharvest Biol. Technol.* 19(1):93–101.

Pesis, E., D. Aharoni, Z. Aharon, R. Ben-Arie, N. Aharoni, and Y. Fuchs. 2000b. Modified atmosphere and modified humidity packaging alleviates chilling injury symptoms in mango fruit. *Postharvest Biol. Technol.* 19(1):93–101.

Pesis, E., G. Zauberman, and I. Avissar. 1991. Induction of certain aroma volatiles in feijoa fruit by postharvest application acetaldehyde or anaerobic conditions. *J. Sci. Food Agric.* 54(3):329–337.

Philippines Council for Agriculture and Resource Research. 1978. The Philippines recommendations for mango. PCARRD Technical Bulletin, Series 38, University of Philippines, Los Baños, Philippines.

Plata, M. C., L. S. de Medina, M. Martinez-Cayuela, M. J. Faus, and A. Gil. 1987. Changes in Texture, protein content and polyphenoloxidase and peroxidase activities in chirimoya induced by ripening in hypobaric atmospheres or in presence of sulfite (in Spanish). *Rev. Agroquímica Technol. Aliment.* 27:215–224.

Poubol, J. and H. Izumi. 2005a. Shelf life and microbial quality of fresh-cut mango cubes stored in high CO_2 atmospheres. *J. Food Sci.* 70(1):M69–M74.

Poubol, J. and H. Izumi. 2005b. Physiology and microbiological quality of fresh-cut mango cubes as affected by high-O_2 controlled atmospheres. *J. Food Sci.* 70(6):M286–M291.

Powrie, W. D., R. Chiu, H. Wu, and B. J. Skura. 1990. Preservation of cut and segmented fresh fruit pieces. U.S. Patent No. 4,859,729.

Quazi, M. G. and H. T. Freebairn. 1970. The influence of ethylene, oxygen and carbon dioxide on the ripening of bananas. *Bot. Gaz.* 131:5–14.

Ramos, R. X., T. I. Alia, G. S. Valle, L. M.Víctor, and L. M. T. Colinas. 2005. Efecto del almacenamiento de películas plásticas en la maduración de zapote mamey (*Pouteria sapota*). In: *Resúmenes VIII Congress*, National Agricultural Department Fitotecnia, Chapingo, México, April 27–28, 2005.

Rattanachinnakorn, B., J. Phumhiran, and S. S. Nanthachai. 1996. Controlled atmosphere storage of mangosteen (in Thai), Annual Technical Conference. *Proceedings of the Horticultural Research Institute*, Department of Agriculture, Bangkok, Thailand.

Richardson, D. G. and M. Meheriuk (Eds.). 1982. *Controlled Atmospheres for Storage and Transport of Perishable Agricultural Commodities*, Timber Press, Beaverton, OR.

Rodov, V., S. Ben-Yehoshua, S. Fishman, S. Gotlieb, T. Fierman, and D. Q. Fang. 1994. Reducing decay and extending shelf life of bell-peppers and mangoes by modified atmospheres packaging. In: B. R. Champ, E. Highley, and G. I. Johnson (Eds.), *Postharvest Handling of Tropical Fruits. Proceedings of the International Conference*, Chiang Mai, Thailand, July 19–23, 1993. ACIAR Proc. 50, pp. 416–418.

Rojas-Villegas, A., A. Carrillo-Lopez, M. Silveira, R. Avena-Bustillos, and E. Yahia. 1996. Effects of insecticidal atmospheres on the mortality of fruits flies in mango. *Acta Hort*. 370:89–92.

Salazar, R. and R. Torres. 1977. Almacenamiento de frutos de maracuya en bolsas de polyetileno. *ICA-Bogotá, Colombia* 12(1):1–11.

Sanchez, J. A., J. P. Zamorano, T. Hernandez, and R. Alique. 1998. Enzymatic activities related to cherimoya fruit softening and sugar metabolism during short-term controlled-atmosphere treatments. *Food Sci. Technol. (Lebens. Wiss. u-Technol.)* 207(3):244–248.

Sankat, C. K. and R. Maharaj. 1989. Controlled atmosphere storage of papayas. *Proceedings of the 5th International CA Research Conference*, Wenatchee, WA, June 14–16, 1989, pp. 161–170.

Satyan, S. H., K. J. Scott, and D. J. Best. 1992b. Effects of storage temperature and modified atmospheres on cooking bananas grown in New South Wales. *Trop. Agric*. 69:263–267.

Satyan, S. H., K. J. Scott, and D. Graham. 1992a. Storage of banana bunches in sealed polyethylene tubes. *J. Hort. Sci*. 67:283–287.

Scott, K. J. 1971. Polyethylene bags and ethylene absorbent for transporting bananas. *Agric. Gaz. N. S. W*. 82:267–269.

Scott, K. J. 1975. The use of polyethylene bags to extend the life of bananas after harvest. *Food Technol. Aust*. 27:481–482.

Scott, K. J. and S. Gandanegara. 1974. Effect of temperature on the storage life of bananas held in polyethylene bags with ethylene absorbent. *Trop. Agric. (Trinidad)* 51:23–26.

Scott, K. J. and E. A. Roberts. 1966. Polyethylene bags to delay ripening of bananas during transport and storage. *Aust. J. Expt. Agric. Anim. Husb*. 6:197–199.

Scott, K. J., W. B. McGlasson, and E. A. Roberts. 1968. Ethylene absorbent increase age life of bananas packed in polyethylene absorbent. *Agric. Gaz. N. S. W*. 79:52.

Scott, K. J., W. B. McGlasson, and E. A. Roberts. 1970. Potassium permanganate as an ethylene absorbent in polyethylene bags to delay ripening of bananas during storage. *Aust. J. Expt. Agric. Anim. Husb*. 10:237–240.

Shetty, K. K., M. J. Klowden, E. B. Jang, and W. Koshan. 1989. Individual shrink wrapping technique for fruit fly desinfestation in tropical fruits. *HortScience* 24:317–319.

Shukor, A. R. A., M. Razali, and D. Omar. 2000. Poststorage respiratory suppression and changes in chemical compositions of mango fruit after storage in low-oxygen atmosphere. *Acta Hort*. 509:467–470.

Singh, B. N., P. V. V. Seshagiri, and S. S. Gupta. 1937. The response of the respiratory system in mango and guava to alteration in the concentrations of oxygen and nitrogen. *Ann. Bot. (London)* 1(2):311–323.

Singh, S. P. and D. V. Sudhakar Rao. 2003. Shrink wrapping of 'Solo' papaya for extension of storage life and quality assurance. *International Food Convention (IFCON-2003)*, CFTRI, Mysore, India, December 13–18, 2003 (Abstr.).

Singh, S. P. and D. V. Sudhakar Rao. 2004a. Effect of storage temperature and packaging film on the in-package atmosphere of modified atmosphere packed fresh papaya. *Food and Bioprocess Engineering*, Anamaya Publishers, New Delhi, India, pp. 182–185.

Singh, S. P. and D. V. Sudhakar Rao, 2004b. Modified atmosphere packaging of individual papaya (*Carica papaya* L. cv. Solo) fruit for extension of storage life and quality maintenance. *1st Indian Horticulture Congress (IHC-2004)*, New Delhi, India, November 6–9, 2004 (Abstr.).

Singh, S. P. and D. V. Sudhakar Rao. 2005a. Effect of modified atmosphere packaging on the alleviation of chilling injury and dietary antioxidants levels in 'Solo' papaya during cold storage. *Eur. J. Hort. Sci*. 70(5):246–252.

Singh, S. P. and D. V. Sudhakar Rao. 2005b. Quality assurance of papaya (*Carica papaya* L. cv. 'Solo') by shrink film wrapping during storage and ripening. *J. Food Sci. Technol*. 42(6):523–525.

Singh, S. P., S. S. Baskar, and D. V. Sudhakar Rao. 2005. Individual shrink-wrapping of papaya (cv. Taiwan Red Lady) fruit with plastic film enhances the postharvest life and maintains quality. *International Conference on Plasticulture and Precision Farming (ICPPF-2005)*, New Delhi, India, November 17–21, 2005 (Abstr.).

Siriphanich, J. 1994. Minimal processing of tropical fruits. In: B. R. Champ. E. Highley, and G. I. Johnson (Eds.), *Postharvest Handling of Tropical Fruits*. Proceedings of the International Conference, Ching Mai, Thailand, July 19–23, 1993. ACIAR Proc. 50, pp. 127–137.

Siriphanich, J. 1996. Storage and transportation of tropical fruits: A case of study of Durian. In: S. Vijaysegaran, M. Pauziah, M. S. Mohamed, and S. Ahmed Tarmizi (Eds.), *Proceedings of the International Conference on Tropical Fruits*, Kuala Lumpur, Malaysia, July 23–26, 1996, pp. 439–451.

Sive, A. and D. Resinsky. 1989b. CA storage trials of mangoes. *Rep. 1988 Season. Alon Hanotea* 44(1):53–56 (in Hebrew).

Smock, R. M. 1967. Methods of storing bananas. *Philipp. Agric.* 51:501–517.

Smock, R. 1979. Controlled atmosphere storage of fruits. *Hort. Rev.* 1:301–336.

Sonsrivichai, J., P. Anusadorm, C. Oogaqki, and H. Gemma. 1989. Storage life and quality of mango (*Mangifera indica* L. cv. Keaw Sawoey) fruits stored in seal packaging by plastic films and under low pressure at different temperatures. *Jpn. J. Trop. Agric.* 33(11):6–17.

Sonsrivichai, S., S. Gomolmanee, D. Boonyakiat, J. Uthaibutra, P. Boon-Long, and H. Gemma. 1992. Seal packaging by plastic film as a technique for limiting fungal decay of mangoes. *Acta Hort.* 296:23–32.

Spalding, D. H. 1977a. Low pressure (hypobaric) storage of avocados, limes and mangoes. In: D. H. Dewey (Ed.), *Controlled Atmospheres for the Storage and Transport of Perishable Agricultural Commodities*, Horticulture Report No. 28. Michigan State University, East Lansing, MI, pp. 156–164.

Spalding, D. H. 1977b. Current recommendations of atmospheres for transport and storage of tropical fruits. In: D. H. Dewey (Ed.), *Controlled Atmospheres for the Storage and Transport of Perishable Agricultural Commodities*, Horticulture Report No. 28. Michigan State University, East Lansing, MI, pp. 242–249.

Spalding, D. H. and W. F. Reeder. 1974. Current status of controlled atmosphere storage of four tropical fruits. *Proc. Fla. State Hort. Soc.* 87:334–337.

Spalding, S. H. and W. F. Reeder. 1977. Low pressure (hypobaric) storage of mangoes. *J. Am. Soc. Hort. Sci.* 102:367–369.

Sriyook, S. and J. Siriphanich. 1989. *The Use of Surface Coating to Prolong the Shelf Life for Fruit and Vegetables for Export* (in Thai), Institute of Scientific and Technological Research, Bangkok, Thailand.

Staby, G. L. 1976. Hypobaric storage—An overview. *Combined Proc. Int. Plant Prop. Soc.* 26:211–215.

Stead, D. E. and G. S. G. Chithambo. 1980. Studies on the storage of tropical fruits in polyethylene bags. *Luso. J. Sci. Tech. (Malawi)* 1:3–9.

Stewart, O. J., G. S. V. Raghavan, and Y. Gariépy. 2005. MA storage of Cavendish bananas using silicone membrane and diffusion channel systems. *Postharvest Biol. Technol.* 35(3):309–317.

Suwanagul, A. and S. C. Tongdee. 1989. Effect of controlled atmosphere on internal ethylene content and ripening of durian. *Proceedings of the Durian Workshop* (in Thai), Thailand Institute of Scientific and Technological Research, Bangkok, Thailand, pp. 46–54.

Sy, D. A. and D. B. Mendoza, Jr. 1984. Respiratory responses of 'Caraboa' mango to different levels of oxygen. *Postharvest Res. Notes* 1(1):3 (Postharvest Horticulture Training Research Center, Department of Horticulture, University of Philippines, Los Baños, Philippines).

Tan, S. C., P. F. Lam, and H. Abdullah. 1986. Changes of the pectic substance in the ripening of bananas (*Musa sapientum*, cv. *Emas*) after storage in polyethylene. *ASEAN Food J.* 2:76–77.

Teixeira, G. H. A., J. F. Durigan, R. E. Alves, and T. J. O'Hare. 2008. Response of minimally processed carambola to chemical treatments and low-oxygen atmospheres. *Postharvest Biol. Technol.*, 48:415–421.

Tongdee, S. C. and S. Neamprem. 1989. Effects of low oxygen atmospheres on ripening in Durian (*Durio zibethinus* Murray) (in Thai). *Proceedings of the Durian Workshop*, Thailand Institute Scientific and Technological Research, Bangkok, Thailand, pp. 37–45.

Tongdee, S. C. and A. Suwanagul. 1988a. Effects of low oxygen atmospheres in Durian (*Durio zibethinus* Murray). *Seminar on Durian*, Proceedings of the Thailand Institute of Scientific and Technological Research, Bangkok, Thailand, pp. 37–45.

Tongdee, S. C. and A. Suwanagul. 1988b. Effects of controlled atmosphere storage on internal ethylene content and ripening of durian fruits. *Seminar on Durian Proceedings*, Thailand Institute of Scientific and Technological Research, Bangkok, Thailand, pp. 46–54.

Tongdee, S. C., A. Chayasombat, and S. Neamprem. 1989. Effects of harvesting maturity on respiration, ethylene production, and the composition in the internal atmospheres of Durian (*Durio zibethinus* Murray) (in Thai). *Proceedings of the Durian Workshop*, Thailand Institute of Scientific and Technological Research, Bangkok, Thailand, pp. 31–36.

Tongdee, S. C., A. Suwanagul, S. Meamprem, and U. Bunruengsri. 1990b. Effect of surface coatings on weight loss and internal atmosphere of durian (*Durio zibethinus* Murray) fruit. *ASEAN Food J.* 5:103–107.

Tongdee, S. C., A. Suwanagul, and S. Neamprem. 1990a. Durian fruit ripening and the effect of variety maturity stage at harvest, and atmospheric gases. *Acta Hort.* 269:323–324.

Vilas-Boas, E. V. de B., and A. A. Kader. 2006. Effect of atmospheric modification, 1-MCP and chemicals on quality of fresh-cut banana. *Postharvest Biol. Technol.* 39:155–162.

Villanueva-Arce, R., S. Bautista-Baños, and J. C. Diaz-Perez. 1999. Efecto de cubiertas plasticas en la maduraci'on de frutos de mamey (*Pouteria sapota*), *Res. XLVReuni'on Anual Soc. Interamer. Horticultura Tropical*, Lima, Peru, November 15–19, 1999, p. 49.

Wardlaw, C. W. 1940. Preliminary observations on the refrigerated gas storage of 'Gros Michel' bananas. *Trop. Agric. (Trinidad)* 17:103–105.

Woodruff, R. E. 1969a. Overseas transport of bananas. In: *Controlled Atmospheres for the Storage and Transport of Horticultural Crops*. Proceedings of the National CA Research Conference, Horticultural Report No. 9. Michigan State University, East Lansing, MI, p. 54.

Woodruff, R. E. 1969b. Modified atmosphere storage of bananas. In: *Controlled Atmospheres for the Storage and Transport of Horticultural Crops*. Proceedings of the National CA Research Conference, Horticultural Report No. 9. Michigan State University, East Lansing, MI, pp. 80–94.

Woodruff, R. E. 1977. Use of carbon monoxide in modified atmospheres for fruits and vegetables in transit. In. D. H. Dewey (Ed.), *Controlled Atmospheres for the Storage and Transport of Perishable Agricultural Commodities*. Proceedings of the 2nd National CA Research Conference, Horticultural Report No. 28. Department of Horticulture, Michigan State University, East Lansing, MI, pp. 52–54.

Yahia, E. M. 1991. Responses of papaya to insecticidal atmospheres (in Spanish). *Proc. Interam. Soc. Trop. Hort.* 35:84–100.

Yahia, E. M. 1993a. Modified/controlled atmosphere storage in Mexico. In: *CA'93. Proceedings of the 6th International CA Conference*, Ithaca, NY, June 14–16, 1993, pp. 570–579.

Yahia, E. M. 1993b. Responses of some tropical fruits to insecticidal atmospheres. *Acta Hort.* 343:371–376.

Yahia, E. M. 1994a. The potential use of insecticidal atmospheres for mango, avocado, and papaya fruit. In: B. R. Champ, E. Highley, and G. I. Johnson (Eds.), *Postharvest Handling of Tropical Fruits*, Australian Centre for International Agricultural Research (ACIAR) Proceedings No. 50, Canbera, Australia, pp. 373–374.

Yahia, E. M. 1994b. The use of modified and controlled atmospheres for tropical fruits (in Spanish). In: *Proceedings of the 1st Regional Symposium on Postharvest Handling of Tropical Products*, University of Costa Rica, San José, Costa Rica, July 25–29, 1994.

Yahia, E. M. 1995a. Application of differential scanning calorimetry in the study of avocado and mango fruit responses to hypoxia. In: A. Ait-Oubahou and M. El-Otmani (Eds.), *Postharvest Physiology, Pathology and Technologies for Horticultural Crops: Recent Advances*, Institut Agronomique and Vétérinaire Hassan II, Agadir, Morocco, pp. 206–209.

Yahia, E. M. 1995b. Insecticidal atmospheres for tropical fruits. *Proceedings of the International Symposium: Harvest and Postharvest Technologies for Fresh Fruits and Vegetables*, Guanajuato, Gto, Mexico, February 20–24, 1995, pp. 282–286.

Yahia, E. M. 1995c. Postharvest handling of horticultural crops in Mexico. In: A. Ait Oubahou and M. El-Otmani (Eds.), *Postharvest Physiology Pathology and Technologies for Horticultural Crops: Recent Advances*, Institut Agronomique and Vetérinaire Hassan II, Agadir, Morocco, pp. 1–10.

Yahia, E. M. 1997a. Modified/controlled atmospheres for banana and plantain. In: A. A. Kader (Ed.), *CA'97 Proceedings*, Vol. 3: *Fruits Other than Apples and Pears*, University of California, Davis, CA, pp. 104–109.

Yahia, E. M. 1997b. Modified/controlled atmospheres for mango. In: A. A. Kader (Ed.), *CA'97 Proceedings*, Vol. 3: *Fruits Other than Apples and Pears*, University of California, Davis, CA, pp. 110–116.

Yahia, E. M. 1997c. Modified/controlled atmospheres for papaya. In: A. A. Kader (Ed.), *CA'97 Proceedings*, Vol. 3: *Fruits Other than Apples and Pears*, University of California, Davis, CA, pp. 117–120.

Yahia, E. M. 1998a. Modified and controlled atmospheres for tropical fruits. *Horticul. Rev. 1998* 22:123–183.

Yahia, E. M. 1998b. The use of modified and controlled atmospheres for insect control (in Spanish). *Phytoma* (Spain) 97:18–22.

Yahia, E. M. 2004. The development of quarantine insect control system for tropical fruits using controlled atmospheres at high temperatures. *Proceedings of the Frigair 2004, International Refrigeration and Air Conditioning Conference*, Cape Town, South Africa, June 2–4, pp. 1:7–16.

Yahia, E. M. 2006a. Effects of insect quarantine treatments on the quality of horticultural crops. *Stewart Postharvest Rev. 2006* 1(6):1–18.

Yahia, E. M. 2006b. Handling tropical fruits. *Scientists Speak (World Food Logistics Organization Advisory Council) Proceedings*, Walt Disney World Swan Hotel, Orlando, FL, pp. 5–9.

Yahia, E. M. 2006c. Controlled atmospheres for tropical fruits. In: Yahia, E.M. (Ed.), *The Current Status and Future Application of Modified and Controlled Atmospheres for Horticultural Commodities*. Stewart Postharvest review, 5(6):1–10.

Yahia, E. M. and R. Ariza. 2001. Physical treatments for the postharvest handling of horticultural products (in Spanish). Horticultura Internacional (Spain), Extra, pp. 80–88, 153.

Yahia, E. M. and A. Carrillo. 1993. Responses of avocado fruit to insecticidal O_2 and CO_2 atmospheres. *Lebensm. Wiss. U-Technol.* 26:307–311.

Yahia, E. M. and D. Ortega. 2000a. The use of controlled atmospheres at high temperature to control fruit flies (*Anastrepha ludens* and *A. oblique*) and their effect on mango quality. In: W. J. Florkowski, S. E. Prussia, and R. L. Shewfelt (Eds.), *Proceedings of the International Multidisciplinary Conference: Integrated View of Fruit & Vegetable Quality*, Technomic Publishing, Lancaster, Basel, pp. 143–153.

Yahia, E. M. and D. Ortega. 2000b. Mortality of eggs and third instar larvae of *Anastrepha ludens* and *A. obliqua* with insecticidal controlled atmospheres at high temperatures. *Postharvest Biol. Technol.* 20(3):295–302.

Yahia, E. M. and D. Ortega. 2000c. Responses and quality of mango fruit treated with insecticidal controlled atmospheres at high temperatures. *Acta Hort.* 509:479–486.

Yahia, E. M. and R. Paull. 1997. The future for modified atmosphere (MA) and controlled atmosphere (CA) uses with tropical fruit. *Chron. Hort.* 37(4):18–19.

Yahia, E. M. and M. Rivera. 1992. Modified atmospheres packaging of muskmelon. *Lebensm. Wiss. U-Technol.* 25:38–42.

Yahia, E. M. and M. Rivera-Dominguez. 1995. Application of differential scanning. In: A. Ait Oubahou and M. El-Otmani (Eds.), *Postharvest Physiology, Pathology and Technologies for Horticultural Crops: Recent Advances*, Institut Agronomique Vétérinaire Hassan II, Agadir, Morocco.

Yahia, E. M. and M. Tiznado. 1993. Tolerance and responses of harvested mango to insecticidal oxygen atmospheres. *HortScience* 28(10):1031–1033.

Yahia, E. M. and L. Vazquez. 1993. Tolerance and responses of mango to insecticidal oxygen and carbon dioxide atmospheres. *Food Sci. Technol. (Lebens. Wiss. u-Technol.)* 26(1):42–48.

Yahia, E. M., C. Barry-Ryan, and R. Dris. 2004. Treatments and techniques to minimize the postharvest losses of perishable food crops. In: R. Dris and S. M. Jain (Eds.), *Production Practices and Quality Assessment of Food Crops*, Vol. 4: *Postharvest Treatment and Technology*, Kluwer Academic, Dordrecht, the Netherlands, pp. 95–133.

Yahia, E. M., F. Medina, and M. Rivera. 1989. The tolerance of mango and papaya to atmospheres containing very high levels of CO_2 and/or very low levels of O_2 as a possible insect control treatment. *Proceedings of the 5th International CA Research Conference*, Vol. 2, Wenatchee, WA, June 14–16, 1989, pp. 77–89.

Yahia, E. M., D. Ortega, A. Martinez, and P. Moreno. 2000a. The mortality of artificially infested third instar larvae of *Anastrepha ludens* and *A. obliqua* in mango fruit with insecticidal controlled atmospheres at high temperatures. *Acta Hort.* 509:833–839.

Yahia, E. M., D. Ortega, P. Santiago, and L. Lagunez. 1997. Responses of mango and mortality of *Anastrepha ludens* and *A. obliqua* to modified atmospheres at high temperatures. In: J. F. Thompson and E. J. Mitcham (Eds.), *CA'97 Proceedings*, Vol. 1: *CA Technology and Disinfestation Studies*, University of California, Davis, CA, pp. 105–112.

Yahia, E. M., M. Rivera-Dominguez, and O. Hermandez. 1992. Responses of papaya to short-term insecticidal oxygen atmospheres. *J. Am. Soc. Hort. Sci.* 117:96–99.

Yahia, E. M., G. Villagomez, and A. Juarez. 2000b. Tolerance and responses of different fruits and vegetables to hot air treatments at 43°C or 48°C and 50% RH for 160 minutes. In: F. Artes, M. I. Gil, and M. A. Conesa (Eds.), *Improving Postharvest Technologies of Fruits, Vegetables and Ornamentals*, Vol. 2, International Institute of Refrigeration, Spain, pp. 714–718.

Yamashita, F., J. Telis-Romero, and T. G. Kieckbusch. 1997. Gaseous composition estimation in modified atmosphere packaging of mangoes (*Mangifera indica* L.) cv. Keitt. *Ciencia e Tecnologia de Alimentos* 17(2):172–176.

Yantarasri, T., S. Ben Yehoshua, V. Rodov, W. Kumpuan, J. Uthaibutra, and J. Sornsrivichal. 1995. Development of perforated modified atmosphere package for mango. *Acta Hort.* 398:81–91.

Yantarasri, T., J. Uthaibutra, J. Sornrivichai, W. Kumpuan, V. Sardsud, and N. Kanathum. 1994. Modified atmosphere packaging by perforated polymeric film and its effect on physical properties of mango fruit. In: B. R. Champ, E. Highley, and G. I. Johnson (Eds.), *Postharvest Handling of Tropical Fruits. Proceedings of the International Conference*, Chiang Mai, Thailand, July 19–23, 1993. ACIAR Proc. 50, pp. 438–440.

Yonemoto, Y., H. Higuchi, and Y. Kitano. 2002. Effects of storage temperature and wax coating on ethylene production, respiration and shelf-life in cherimoya fruit. *J. Jpn. Soc. Hort. Sci.* 71:643–650.

Yueming, J. D. C. Joyce, and A. J. Macnish. 1999. Extension of the shelf life of banana fruit by 1-methylcyclopropene in combination with polyethylene bags. *Postharvest Biol. Technol.* 16(2):187–193.

Zagory, D. and A. A. Kader. 1989. Modified atmospheres: An indexed references list with emphasis on horticultural commodities, Supplement No. 5 (June 1, 1985 to March 31, 1989). Postharvest Horticulture Series 6, University of California, Davis, CA.

Zagory, D. and A. A. Kader. 1993. Modified atmospheres. An indexed reference list with emphasis on horticultural commodities, Supplement No. 6 (April 1, 1989 to April 30, 1993). Postharvest Horticultural Series 7, University of California, Davis, CA.

Zamorano, J. P., R. Alique, and W. Canet. 1999. Mechanical Parameters to assess quality changes in cherimoya fruit. *Food Sci. Technol.* (*Lebens. Wiss. u-Technel.*) 208(2):125–129.

Zhang, D., B. Y. Huany, and K. J. Scott. 1993. Some physiological and biochemical changes of "green ripe" bananas at relative high storage temperatures. *Acta Hort.* 343:81–85.

17

Vegetables

Jeffrey K. Brecht

CONTENTS

17.1 Introduction

Increased worldwide trade in vegetables has expanded the market for vegetables in many countries to allow year-round availability. This increased trade has been supported by expanded use of controlled atmosphere (CA) and modified atmosphere (MA) technology to maintain vegetable quality and shelf life for extended transport in marine containers and subsequent marketing periods while reducing the economic incentive for long-term stationary MA or CA storage of most vegetables since they are available with better quality from distant producers. Therefore, most current applications of MA and CA for vegetables are found in long distance transport, as well as in modified atmosphere packaging (MAP), especially of fresh-cut vegetable products, topics which are covered in Chapters 3, 4, and 18, respectively. In recent years, CA, MA, and MAP have been applied to several previously underexploited vegetables and fresh-cut vegetable products, and the potential for using superatmospheric O_2, primarily in MAP, has also been extensively evaluated. Advances have also been made in modeling the effects of CA, MA, and MAP on vegetable shelf life and microbiological quality. This chapter will provide information on recent research advances and applications of MA and CA for vegetables. It will also explore the use of vegetables to elucidate the underlying physiological and biochemical mechanisms related to CA and MA effects on produce.

The amount of research on CA, MA, and MAP for vegetables has historically lagged behind that of fruits, and of the research with vegetables, most of the effort has been expended on a few crops, i.e., lettuce (*Lactuca sativa*), onion (*Allium cepa*), and tomato (*Solanum lycopersicum*). The recommended CA and MA conditions for most of the vegetable crops for which such research has been conducted were reviewed by Thompson (1998), Saltveit (2003; Figure 17.1), in the commodity chapters in Bartz and Brecht (2003), and in *USDA Agriculture Handbook 66* (available online at http://www.ba.ars.usda. gov/hb66/), which contains overviews of CA (Kader, 2004) and MAP (Mir and Beaudry, 2004), as well as recommended CA conditions in most of the individual commodity chapters. Fonseca et al. (2002a) reviewed the modeling of respiration rate for the design of MAP systems for vegetables.

Most research on CA, MA, and MAP involves empirical observations of changes in various shelf life-limiting quality factors over time in experiments in which a product is placed in several combinations of gas atmospheres and sometimes also different temperatures. Reports of MAP experiments in which a vegetable product is packaged in some number of packages and shelf life observed and compared are not particularly useful unless the packaging materials were selected because they were expected to create particular desired gas atmospheres for which there is some expectation of benefit to the product.

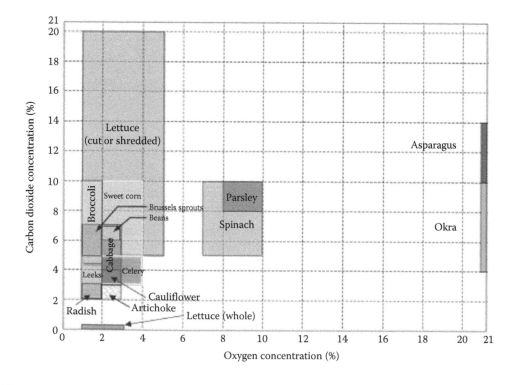

FIGURE 17.1

Recommended oxygen and carbon dioxide ranges for the storage of some harvested vegetable commodities. (From Saltveit, M.E., *Postharvest Biol. Technol.*, 27, 3, 2003. With permission.)

Not included in this chapter are those papers in which MAP was used, but no gas concentrations in the packages were reported.

In 2003, Saltveit asked, "Is it possible to find an optimal controlled atmosphere?" (Saltveit, 2003). His contention was that searching for a single, invariable, and constant set of storage conditions for each inherently variable and dynamic biological system (i.e., fruit or vegetable) has not proven to be terribly successful in many cases. He proposed that CA systems should also be dynamic, that is, they should measure and respond to the commodity's changing metabolic state. One way that has been proposed in which variable environmental conditions during distribution could be addressed is by using a combination CA/MAP system during transportation (Brecht et al., 2003). In this approach, the MAP is designed to produce an optimal gas atmosphere during the higher temperature retail display period, and a surrounding gas atmosphere is selected for the lower temperature transportation period that will interact with the MAP to produce the optimal atmosphere within the packages during that portion of distribution.

17.2 CA and MA in Vegetable Storage and Transport

17.2.1 CA Storage of Vegetables

Commercial use of CA storage is an economic decision based on the benefit to be expected from extending the storage life of a crop in order to extend the marketing period. Cabbage (*Brassica oleracea*, var. *capitata*) and sweet onions are the only two vegetable crops commonly stored in stationary CA or MA storage rooms. In the case of cabbage, the crop value is too low to justify much international marketing, so CA or MA storage is used primarily to extend local supplies through the winter. The situation with sweet onions is different: sweet onions are mostly short-day types that are grown in specific, limited production areas having very low soil sulfur content, which reduces bulb pungency, have limited availability, and are often trademarked (e.g., Vidalia onions, OsoSweets, Walla Walla Sweet onions, Maui Sweets, etc.). The normal marketing season for sweet onions of 1–3 months in the late spring to early summer can be extended to 6 or 7 months by using CA storage (Figure 17.2), which allows marketers to maintain relatively high prices into late fall.

FIGURE 17.2
Vidalia sweet onions from CA storage. (Courtesy of J.K. Brecht.)

17.2.2 MA and CA in Vegetable Transport

Expanding international trade in fresh fruits and vegetables has led to increased attempts to successfully transport vegetables over very long distances. For many vegetables, the transit times involved in international shipping are at or beyond the product shelf life in normal refrigerated air. Thus, there has been increasing interest in using CA or MA marine container transport equipment to allow certain vegetables to be shipped extended distances (Figure 17.3). Some of the vegetables that are more commonly transported in MA or CA include asparagus (*Asparagus officinalis* L.), broccoli (*Brassica oleracea* var. *botrytis*), lettuce, and tomato. However, use is limited because in most markets the price of vegetables cannot absorb the added cost of CA of MA.

17.2.3 Duration of Exposure Influences Responses to MA/CA

The traditional objective of CA storage is to maximally extend the storage time of a product. However, the rise in recent years of international trade in horticultural products has reduced the motivation to extend storage when newer and fresher product from other production areas competes in the market with stored product. However, the consequent extension of transport times inherent in international trade has challenged the ability of postharvest handlers to deliver high-quality products. The new opportunity for application of CA and MA is to products, including many vegetables, for which a 2- or 3-week-long transit time may represent a very significant portion of their potential postharvest life.

The optimum atmosphere conditions for transport of a horticultural product for some limited time period up to a few weeks will be much different than for the same product

FIGURE 17.3

Out-turn appearance of vegetables after transport in MA marine containers: lettuce after 18 days (above left); broccoli after 27 days (below left); asparagus after 15 days (right). (Courtesy of W. Benson, Thermo King Corp.)

intended for long-term storage (Brecht et al., 2003). Traditionally, optimal conditions for CA and MA storage have been selected based on the goal of achieving maximum extension of postharvest life. This usually leads to selection of the product maturity stage that will provide minimally acceptable quality for the longest time in the most extreme O_2 and CO_2 environment that the particular product is able to tolerate without injury, until its storage life is finally ended due to deterioration of one quality parameter or another. However, if the required storage time is shorter, the product can tolerate lower O_2 and higher CO_2 concentrations (Hansen, 1986; Makhlouf et al., 1989; Brecht et al., 2003), a fact that can be useful in the development of transportation CA and MA and MAP systems that are used to maintain product quality only during the limited durations of distribution, domestic marketing, or retail display (Hertog et al., 1998; Saltveit, 2002; Brecht et al., 2003; Mir and Beaudry, 2004).

17.3   CA, MA, or MAP for Vegetables

There are a few recent examples of reports on determination of useful atmospheres for CA storage of vegetables for which little previous work has been done. A CA of 10 kPa O_2 plus 10 kPa CO_2 at 7°C was found to best preserve the visual quality of Belgian endive (*Cichorium intybus*; Van De Velde and Hendrickx, 2001) as determined by a 40- to 45-member consumer panel. Air plus 3–5 kPa CO_2 was beneficial for spinach (*Spinacia oleracea*; Mizukami et al., 2002), 3 kPa O_2 plus 3 kPa CO_2 at 8°C minimized compositional changes in green beans (*Phaseolus vulgaris*; Sanchez-Mata et al., 2003a,b); and 5 kPa O_2 plus 15 kPa CO_2 at 4°C reduced the respiration rate, maintained green color, inhibited decay and internal leaf growth, and decreased pithiness and butt end cut surface browning of green celery (*Apium graveolens*; Gomez and Artes, 2004a). Sweetcorn (*Zea mays* subsp. *mays*) was held in 2 kPa O_2 plus 15 kPa CO_2 for a limited period of 2 weeks to simulate marine transport (Riad and Brecht, 2003). CA storage maintained higher levels of sugars, significantly reduced loss of greenness in the husks, and improved the silk and kernel appearance compared with the air control, but it had no effect on kernel denting. The CA also maintained significantly higher content of dimethyl sulfide, the main characteristic aroma component in sweet corn.

17.3.1   Application of Modified Atmosphere Packaging to Vegetables

Recent development of MAP systems for vegetables includes white asparagus, which Siomos et al. (2000) found to be more tolerant of reduced O_2 than previously reported for green asparagus, reporting that the best quality was maintained in 1 kPa O_2 plus 6 kPa CO_2 at 2.5°C. Guevara et al. (2001, 2003) found that prickly pear cactus stems (*Opuntia* spp.) can be stored for 30–35 days at 5°C in a passive MAP of 8–9 kPa O_2 plus 7–7.5 kPa CO_2 or a semiactive MAP initially flushed with 20 kPa CO_2 that equilibrated at 13 kPa O_2 plus 13.5 kPa CO_2 without significant losses in quality nor any significant increase in microbial counts. A passive MAP of 16–17 kPa O_2 plus 4 kPa CO_2 at 0°C was effective for storage of fennel (*Foeniculum vulgare*; Artes et al., 2002), inhibiting browning of the butt end cut zone more effectively than 1% ascorbic acid or 5% citric acid. Gil-Izquierdo et al. (2002) compared six different films to a perforated polypropylene package control to determine that artichoke (*Cynara scolymus*) vitamin C and phenolic contents were unaffected by O_2 concentrations in the range of 7.7–14.4 kPa, and that an MAP system using low-density polyethylene (LDPE) that equilibrated at

7.7 kPa O_2 plus 9.8 kPa CO_2 was the best choice for maintaining vitamin C content and increasing phenolic content. Van De Velde and Hendrickx (2001) determined that an MAP of 10 kPa O_2 plus 10 kPa CO_2 was optimum for Belgian endive held for 10 days at 5°C; the rate of visual quality degradation was greater at both higher and lower temperatures. Hu et al. (2004) reported that ginseng (*Panax ginseng*) was success- fully stored for 150 days at 0°C in a passive MAP that equilibrated at 14.3 kPa O_2 plus 6.1 kPa CO_2 after 3 months. Gomez and Artes (2004b) stored green celery in MAP using oriented polypropylene (OPP) or LDPE films that equilibrated at about 8–9 kPa O_2 plus 7 kPa CO_2 and 8 kPa O_2 plus 5 kPa CO_2, respectively, after 10 days at 4°C. Both MAP systems inhibited decay and internal leaf growth, decreased pithiness, and reduced chlorophyll degradation and butt end cut surface browning, but the more extreme atmosphere of the OPP film MAP maintained better visual appearance than the LDPE package.

Much more effort has been expended in recent years on developing beneficial MAP systems for fresh-cut products including bulky storage organs such as jicama (*Pachyrhizus erosus*), rutabaga (*Brassica napus* var. *napobrassica*), sweet potatoes (*Ipomoea batatas*), shredded carrot (*Daucus carota*), and pumpkin (*Cucurbita maxima*). Jicama pieces were found to respond well to elevated CO_2 atmospheres (5 to 10 kPa) at 5°C in terms of retarding microbial growth and discoloration for 8–14 days depending on whether the roots were fresh or stored at the time of processing (Aquino-Bolanos et al., 2000). Zhu et al. (2001) modeled the respiration rate of rutabaga cubes and sticks at 5°C as functions of O_2 and CO_2 partial pressures and the degree of cutting, which should be useful for the design of MAP systems for fresh cut rutabaga. Sweet potato slices (Erturk and Picha, 2002) and shreds (McConnell et al., 2005) from cured roots main- tained dry matter, alcohol-insoluble solids, ascorbic acid, glucose, fructose, sucrose, total sugars, and total carotenoids with minimal changes and had fewer total aerobic bacteria and enteric bacteria than air controls over 14 days in MAP systems that equilibrated at 4 kPa O_2 plus 4 kPa CO_2 at 2°C or 6 kPa O_2 plus 4 kPa CO_2 at 4°C. A 5 kPa O_2 plus 5 kPa CO_2 MAP did not extend the shelf life of shredded orange carrots at 5°C, whereas MAP maintained higher anthocyanin and carotenoid contents and delayed browning and off-odor development in purple carrots for 10 days at 5°C (Alasalvar et al., 2005). Habibunnisa et al. (2001) reported that fresh-cut pumpkin could be stored for a 25 days at 5 ± 2°C in a passive LDPE MAP that equilibrated at 2–3 kPa O_2 plus 17–18 kPa CO_2 with minimal weight loss and little change in vitamin C, total soluble solids, carotenoids, and titratable acidity.

MAP systems have also recently been developed for leafy and succulent crops such as green onion (*Allium cepa* x *A. fistulosum*), galega kale (*Brassica oleracea*, var. *sabellica*), kohlrabi (*Brassica oleracea* var. *gongylodes*), salad savoy (*Brassica oleracea* var. *viridis*), cilantro (*Coriandrum sativum*), celery sticks, fennel, wild rocket (*Diplotaxis tenuifolia*), and sweet corn. Green onion stalks were crosscut into 10 cm length pieces and sealed in gas- tight glass containers that had initially been purged with air or 9 kPa O_2 plus 91 kPa N_2 with or without a CO_2 absorbent by Hong and Kim (2001) in order to determine the influence of O_2 and CO_2 and temperature on respiration rate. They found the lower O_2 limit to be about 1.0 kPa O_2 irrespective of temperature (0°C, 10°C, 20°C) on the basis of respiratory quotient increase and modeled the dependence of respiration rate on O_2 level as well as temperature, but found little affect of CO_2 accumulation up to 14 kPa on respiration rate.

Fonseca et al. (2002b) measured shredded galega kale respiration in all combinations of 1, 5 and 10 kPa O_2 plus 0, 10 and 20 kPa CO_2, as well as ambient air at 1°C, 5°C, 10°C, 15°C, and 20°C. Temperature was the variable with the greatest influence on respiration rate and the dependence of respiration rate on gas composition was well described by a Michaelis–Menten type equation with uncompetitive CO_2 inhibition. An atmosphere of 1–2 kPa O_2 plus 15–20 kPa CO_2 extends the shelf life of shredded galega kale in terms of sensory attributes, color alterations, and water, chlorophyll and ascorbic acid contents to 4–5 days at 20°C, compared with 2–3 days in air storage (Fonseca et al., 2005).

Fresh-cut kohlrabi dices stored for 14 days at 0°C in MAP that equilibrated at 6 kPa O_2 plus 14 kPa CO_2 had less browning and lower microbial counts than air-stored product but did not differ in sugar content, soluble solids content, pH, titratable acidity, and sensory attributes (Escalona et al., 2003).

White and violet types of salad savoy were stored at 5°C for 25 days in MAP using several films with different O_2 transmission rates (Kim et al., 2004). Beneficial atmospheres of 1.4–3.8 kPa O_2 plus 3.6–6.3 kPa CO_2 developed within 10 days and contributed to maintenance of color and overall quality, including reduction of cut-end discoloration (browning), off-odor, and decay. White salad savoy had higher respiration rates, exhibited more browning, and had lower quality scores than violet salad savoy.

Luo et al. (2004) reported that fresh-cut cilantro leaves quality was best maintained for 14 days at 0°C in MAP systems that developed atmospheres in the range of 1.5–5.6 kPa O_2 plus 2.7–4.1 kPa CO_2. These MAP systems maintained higher tissue integrity (i.e., low electrolyte leakage) and visual quality scores than product held in air or in MAP with more extreme atmosphere modification.

Holding fresh-cut celery sticks (15 cm length) in MAP that equilibrated at 6 kPa O_2 plus 7 kPa CO_2 in 8 days at 4°C improved the sensory quality, reduced losses of green color, sugars, organic acids, and vitamins B_5 and C, decreased the development of pithiness, and retarded the growth of microorganisms with no occurrence of off-odors or off-flavors for 15 days (Gomez and Artes, 2005).

Escalona et al. (2005) stored fennel bulb slices in passive MAP at 0°C or 5°C for 14 days. They reported no physiological disorders or decay occurred and that MAP did not prevent declines in soluble solids content or titratable acidity. However, an MAP that developed an atmosphere of 4 kPa O_2 plus 14 kPa CO_2 during the 14-day storage period at 0°C without reaching a steady state maintained the sensory quality of the fresh-cut fennel near at-harvest levels.

Wild rocket leaves with trimmed stems were held in CA of 5 kPa O_2 plus 10 kPa CO_2 to ascertain the potential for use of MAP during postharvest handling (Martinez-Sanchez et al., 2006). Compared with air, the CA better maintained visual quality, total flavonoid content, antioxidant capacity, and total vitamin C content for 10 days at 4°C. The CA also controlled aerobic mesophilic and psychrotropic microorganisms as well as coliforms on the fresh-cut wild rocket.

An atmosphere of 2 kPa O_2 plus 10 kPa CO_2 at 1°C best maintained fresh-cut sweet corn kernel quality by reducing sugar loss and free amino acid accumulation from protein degradation (Riad and Brecht, 2001). The same CA also prevented after cooking brown discoloration of sweet corn kernels stored at 5°C (Riad and Brecht, 2001; Riad et al., 2003) and significantly reduced the increase in total aerobic microbial count during storage (Riad et al., 2003).

MAP systems for mixed fresh-cut items have also been explored, including 15 kPa O_2 plus 6 kPa CO_2 for 10 days at 4°C for fresh coleslaw made from cabbage plus shredded carrot (Cliffe-Byrnes and O'Beirne, 2005) and 3 kPa O_2 plus 2 kPa CO_2 for 10 days at 5°C for a ready-to-cook fresh-cut vegetable mixture of parsley (*Petroselinum crispum*), beet (*Beta vulgaris*), spinach, zucchini (*Cucurbita pepo*), pumpkin, carrot, celery, tomato, savoy cabbage (*Brassica oleracea* var. *sabauda*), leek (*Allium porrum*), onion, and rehydrated peas (*Pisum sativum*) and 'Borlotti' beans (*P. vulgaris*; Amodio et al., 2006).

17.4 Underlying Physiological and Biochemical Mechanisms Related to CA and MA Effects on Different Vegetable Tissues

McKenzie et al. (2004) investigated the role of sugar metabolism in the observed inhibition of senescence in CA-stored broccoli and asparagus by measuring glycolytic metabolites, the activities of sucrose-metabolizing and glycolytic enzymes, and intracellular sugar concentrations. Asparagus in CA appeared to engage the ethanolic fermentation pathway and broccoli appeared to engage both the lactic and ethanolic fermentation pathways, but only transiently. The results suggested that asparagus is more prone to irreversibly engage ethanolic fermentation under CA, while broccoli only transiently engages both lactic and ethanolic fermentation. The authors suggested that CA may permit more controlled use of whichever vacuolar sugar the plant tissue normally draws upon (fructose in asparagus and sucrose in broccoli) and delays the senescence-related increase in plasma membrane permeability.

Escalona et al. (2006a) investigated the effect of different CA combinations on the metabolic activity and the quality of kohlrabi stems and slices. The authors found that storage of kohlrabi stems for 28 days or kohlrabi slices for 14 days at 5°C with 95% RH in 5 kPa O_2 plus 5 or 15 kPa CO_2 followed by 3 days at 15°C and 60%–70% RH in air slightly delayed the decline in sugars and organic acids. The most appropriate gas composition for maintaining the respiration rate, ethylene production, sugar, and organic acid contents, and sensory attributes (appearance, taste, and texture) of kohlrabi stems and slices was 5 kPa O_2 plus 5 or 15 kPa CO_2.

Tsouvaltzis et al. (2008) studied the effects of storage atmosphere on the main quality attributes of trimmed leek stalks. Storage in 1 kPa O_2 or 15 kPa CO_2 at 6.5°C for 14 days minimized leaf and root growth as well as color changes at the center of the basal cut surface, but did not prevent peripheral discoloration of the basal cut surface; storage in 1 kPa O_2 plus 15 kPa CO_2, however, resulted in an additional beneficial effect compared with 1% O_2 alone by preventing the appearance of peripheral discoloration on the basal cut surface (Figure 17.4). Exposure to 1 kPa O_2 plus 14 kPa CO_2 at 6.5°C for 12 or 24 h prior to

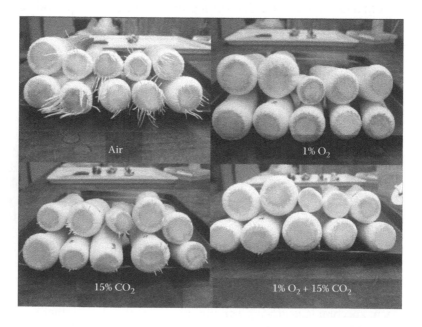

FIGURE 17.4

Root growth and discoloration at the basal cut surface of leeks after storage in air, 1 kPa O_2, 15 kPa CO_2 or the combination of 1 kPa O_2 plus 15 kPa CO_2 at 6.5°C for 14 days. (Courtesy of P. Tsouvaltzis.)

processing did not further contribute to quality maintenance of minimally processed leeks during storage in either air or 1 kPa O_2 plus 14 kPa CO_2 at 6.5°C for 14 days.

Plant polyphenols are able to scavenge free radicals and inactivate other prooxidant chemicals. Plant polyphenols in the diet are considered to be important for human health due to their physiological antioxidant action. Jones (2007) reviewed the effects of post-harvest environmental factors, such as temperature, RH, CA, and MAP on phytochemical content. He concluded that, in most instances, conditions that contribute to maintenance of cellular integrity improve the retention of antioxidant phytochemicals. Thus, CA, MA, and MAP systems that inhibit vegetable senescence, especially systems that include a reduced O_2 component, would be expected to provide nutritionally beneficial retention of antioxidant phytochemical content.

17.5 Innovative MAP Designs

17.5.1 MAP Achieved via Semipermeable Film, Membrane Patch, and Perforation

Almost all MAP systems utilize semipermeable plastic film to regulate the flow of O_2 into and CO_2 out of the package. Fonseca et al. (2000, 2003) described perforation-mediated MAP (PM MAP) and showed how tube dimensions, tube location, and package geometry affect gas transfer while temperature in the range 5°C–20°C does not. The PM MAP method utilizes restricted diffusion through a pore or tube to achieve combinations of reduced O_2 and elevated CO_2 with higher CO_2 concentrations than can be achieved using semipermeable plastic films. Either reducing the diameter of the tube opening or increasing the tube length restricts gas transfer. The ratio of CO_2 to O_2 mass transfer coefficients (0.81) is not affected by temperature, tube dimensions, or tube placement.

Paul and Clarke (2002) described the mathematical modeling of MAP systems using a combination of semipermeable film, membrane patch, and perforations. They showed that a combination of a high flux membrane patch with perforations in a semipermeable film package offers a versatile route to obtaining a wide range of gas atmospheres in MAP—whatever O_2 and CO_2 environment may be needed for a given product can be created (Figure 17.5). An additional benefit of using perforations in MAP that was mentioned is that convective transport of gases through perforations is useful to control the package-free volume. The authors show how mathematical models are developed for describing and designing such systems.

17.5.2 Integrating MAP into Vegetable Distribution Systems with Fluctuating Temperature

For any given product, different atmosphere conditions may be called for depending on the anticipated storage/transport length and temperature. Brecht et al. (2003) described a procedure for designing a combination CA/MAP system that produces optimal atmospheres during transport and for retail display conditions. The proposed MAP/CA system could also be applied to mixed loads of products with different atmosphere requirements.

17.5.3 Combination Treatments

Several papers have been published that report on combining irradiation treatment with MAP to improve whole or fresh-cut vegetable quality (Prakash et al., 2000; Fan and Sokorai, 2002; Han et al., 2004; Ahn et al., 2005; Lafortune et al., 2005) and to control

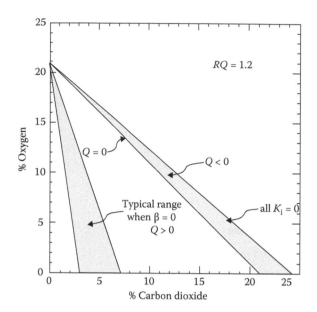

FIGURE 17.5
Steady-state relationships between oxygen and carbon dioxide contents for MAP systems of various designs. When convective flow through the perforations is into the package (i.e., the region where $Q > 0$), the x_{O_2} vs. x_{CO_2} relationship is linear. The shaded region to the left shows the family of lines that can be attained for most polymer membrane materials in the absence of nonselective permeation through holes. The shaded region on the right indicates the small region where the convective flow is out of the package ($Q < 0$) where nonlinear equations describe the relationship between steady-state O_2 and CO_2 levels in the package. (From Paul, D.R. and Clarke, R., *J. Membrane Sci.*, 208, 269, 2002. With permission.)

microbial contamination (Allende and Artes, 2003; Lacroix and Lafortune, 2004; Niemira et al., 2005; Caillet et al., 2006a,b). An attraction of this combination seems to be that, when irradiation is used as a cold pasteurization treatment, the chances of recontamination occurring are minimized if the product is in a sealed MAP system.

McKellar et al. (2004) reported that treating fresh-cut lettuce with chlorine for 30 s at 48°C compared with the typical 4°C chlorine treatment allowed better quality retention using lower CO_2 concentration in MAP. Fan et al. (2003) showed that a 2 min, 47°C water treatment reduced the sensitivity of fresh-cut lettuce in MAP to the undesirable effects of irradiation. However, Delaquis et al. (2002) reported that a 3 min, 47°C water treatment favored the growth of *Listeria monocytogenes* and *Escherichia coli* O157:H7 on lettuce stored at 10°C but not 1°C. Heat treatment and MAP combination treatments have also been tested with fresh-cut green onions (Hong et al., 2000) and leeks (Tsouvaltzis et al., 2007) to control leaf base discoloration, root growth, and leaf growth ("telescoping").

17.6 Superatmospheric Oxygen

Day (1996) described the scientific rationale behind using superatmospheric O_2 in MAP for fresh-cut produce and asserted that this approach is particularly effective at inhibiting enzymatic tissue discoloration, preventing anaerobic fermentation reactions, and inhibiting microbial growth. He suggested that superatmospheric O_2 in MAP is capable of overcoming many of the perceived shortcomings of low O_2 MAP.

Following on the work of Day (1996), a number of investigators have reported on the response of various vegetables to storage in superatmospheric O_2 concentrations, including Amanatidou et al. (2000), Allende et al. (2002, 2004), Escalona et al. (2006b), and Limbo and Piergiovanni (2006).

Jacxsens et al. (2001) validated the concept of superatmospheric O_2 MAP for fresh-cut produce using mixed vegetable salad collected from a commercial processing plant that was stored using the technique for 8 days at 4°C with two films tested, an initial O_2 concentration of 95%, and inoculation with *Listeria monocytogenes* and *Aeromonas caviae*. They concluded that superatmospheric O_2 MAP is most beneficial for fresh-cut vegetables such as celeriac, mushrooms, and endive that are sensitive to enzymatic browning and spoilage by yeasts, but it is questionable how much of the reported benefit was due to elevated O_2 levels and how much was due to the elevated CO_2 (>20 kPa) that they reported quickly developed in the packages that were used.

However, after reviewing the literature, and based on their own experience investigating the use of superatmospheric O_2, Kader and Ben-Yehoshua (2000) found little support for the idea that superatmospheric O_2 is of much practical benefit for most fresh or fresh-cut vegetables. They concluded that the effects of superatmospheric O_2 are highly variable, with numerous examples of completely opposite effects for different products. It is also suggested that the claimed inhibitory effects of superatmospheric O_2 on the growth of microorganisms may be more or less dependent on the presence of elevated concentrations of CO_2, which is a fungistatic gas.

17.7 Modeling the Effects of CA, MA, and MAP on Microbiological Quality

17.7.1 Growth of Microbial Pathogens in MAP

It is well established that CA and MA can reduce the growth of various microorganisms on stored vegetables (Bennik et al., 1998; Jacxsens et al., 2003). Charles et al. (2005), for example, reported that both active and passive MAP with equilibrium atmospheres of 3% O_2 plus 5% CO_2 reduced total aerobic mesophile, yeast, and mold population growth, with active MAP giving better inhibition of *Pseudomonas* spp. and Enterobacteriaceae. Beuchat (2002), however, highlighted the potential for not only survival but also growth of human pathogens on raw, especially fresh-cut, produce. This paper served to focus the attention of researchers, government regulators, and the fresh-cut industry on the potential for survival and growth of human pathogens on fresh-cut produce, especially vegetables, and the possibility of increased risk of food poisoning if treatments that extend the shelf life of fresh-cuts are not coupled with good agricultural practices and good manufacturing practices to minimize the possibility of microbial contamination. It has since been suggested by several authors that MAP, almost universally used with fresh-cut vegetables, could favor the survival or proliferation of human pathogenic microorganisms by limiting competition from the natural microbial population and extending the potential shelf life of fresh-cut products, potentially allowing populations of pathogens to reach levels of significant risk. A number of papers have been published demonstrating the survival and sometimes growth of *L. monocytogenes*, *Salmonella enteritidis*, *E. coli* O157:H7, and other pathogenic species on fresh-cut vegetables in MAP (Francis and O'Beirne, 2001a,b, 2002; Bagamboula et al., 2002; Jacxsens et al., 2002; Sanz et al., 2003; Bourke and O'Beirne, 2004; Gomes and De Martinis, 2004).

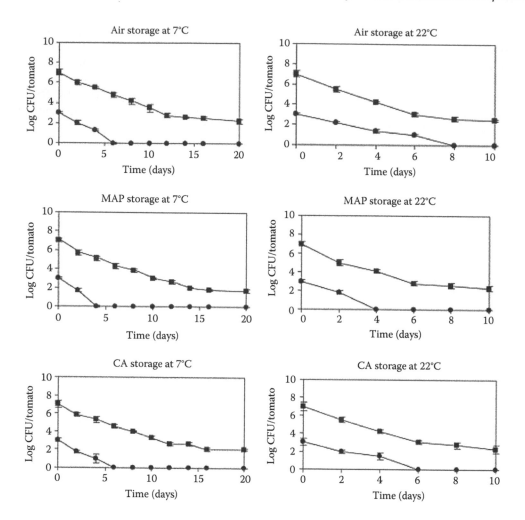

FIGURE 17.6

Survival characteristics of *Salmonella enteritidis* spot inoculated to the surface of cherry tomatoes stored in different atmospheres at 7°C or 22°C for 20 or 10 days, respectively (● low dose inoculum; ■ high dose inoculum). The MAPs were passive systems that equilibrated to approximately 6 kPa O_2 plus 4 kPa CO_2 at both 7°C and 22°C. The CA was 17.3 kPa O_2 plus 5 kPa CO_2 at 7°C and 15 kPa O_2 plus 5 kPa CO_2 at 22°C. (From Das, E., Gurakan, G.C., and Bayindirli, A., *Food Microbiol.*, 23, 430, 2006. With permission.)

However, when Das et al. (2006) recently analyzed the survival and growth of *Salmonella enteritidis* on cherry tomatoes during passive MAP, CA, and air storage at 7°C and 22°C, they found that the death rate of *S. enteritidis* on the fruit surface was faster in MAP than in air or in CA (Figure 17.6). During MAP, the gas composition equilibrated to 6% O_2 plus 4% CO_2 and the CO_2 level in CA was maintained at 5%.

Drosinos et al. (2000) reported that lactic acid bacteria were the predominant microorganisms on a Greek style tomato salad and that the tissue pH dropped and organic acids increased more during storage in MAP than in air, but that there was no effect on survival of *S. enteritidis*. Tassou and Boziaris (2002) reported that increased populations of *Lactobacillus sp.* did not prevent the survival of *S. enteritidis* on grated carrots in MAP irrespective of the amount of *Lactobacillus* spp. inoculum and despite the pH reduction caused by *Lactobacillus*. Geysen et al. (2005) found that superatmospheric O_2 had no effect on the *in vitro* growth of *Listeria innocua* but that elevated CO_2 (12.5% or 25%) did reduce the maximum specific growth rate and prolonged the lag time. Francis and O'Beirne (2001a) suggested that acid adaptation of *Listeria* spp. renders them more resistant to relatively high (25%–30%) CO_2 atmospheres.

17.8 Conclusions and Future Directions

The rapid commercial growth in recent years of fresh-cut products, most of which are vegetables, has stimulated much applied research on MAP with regard to product quality retention and microbiological food safety concerns. The other major application of MA/CA for vegetables occurs during international transport. There is increasing awareness that optimum gas atmospheres can vary for any single commodity based on many factors, but especially the timeframe for distribution of the product. A better understanding is needed of the basic mechanisms that explain differences in vegetable product tolerance to different gas atmospheres and the indicators that product tolerance has been exceeded. This knowledge will help in the development of new CA, MA, and MAP systems that are able to respond to physiological and biochemical cues indicating product stress by continuously adjusting the gas atmosphere to minimize tissue metabolism in order to maintain the highest possible product quality during storage and distribution.

References

Ahn, H.J., J.H. Kim, J.K. Kim, D.H. Kim, H.S. Yook, and M.W. Byun. 2005. Combined effects of irradiation and modified atmosphere packaging on minimally processed Chinese cabbage (*Brassica rapa* L.). *Food Chemistry* 89:589–597.

Alasalvar, C., M. Al-Farsi, P.C. Quantick, F. Shahidi, and R. Wiktorowicz. 2005. Effect of chill storage and modified atmosphere packaging (MAP) on antioxidant activity, anthocyanins, carotenoids, phenolics and sensory quality of ready-to-eat shredded orange and purple carrots. *Food Chemistry* 89:69–76.

Allende, A. and F. Artes. 2003. Combined ultraviolet-C and modified atmosphere packaging treatments for reducing microbial growth of fresh processed lettuce. *Lebensmittel-Wissenschaft Und-Technologie-Food Science and Technology* 36:779–786.

Allende, A., L. Jacxsens, F. Devlieghere, J. Debevere, and F. Artes F. 2002. Effect of super-atmospheric oxygen packaging on sensorial quality, spoilage, and *Listeria monocytogenes* and *Aeromonas caviae* growth in fresh processed mixed salads. *Journal of Food Protection* 65:1565–1573.

Allende, A., Y.G. Luo, J.L. McEvoy, F. Artes, and C.Y. Wang. 2004. Microbial and quality changes in minimally processed baby spinach leaves stored under super atmospheric oxygen and modified atmosphere conditions. *Postharvest Biology and Technology* 33:51–59.

Amanatidou, A., R.A. Slump, L.G.M. Gorris, and E.J. Smid. 2000. High oxygen and high carbon dioxide modified atmospheres for shelf-life extension of minimally processed carrots. *Journal of Food Science* 65:61–66.

Amodio, M.L., R. Rinaldi, and G. Colelli. 2006. Influence of atmosphere composition on quality attributes of ready-to-cook fresh-cut vegetable soup. *Acta Horticulturae* 712:677–684.

Aquino-Bolanos, E.N., M.I. Cantwell, G. Peiser, and E. Mercado-Silva. 2000. Changes in the quality of fresh-cut jicama in relation to storage temperatures and controlled atmospheres. *Journal of Food Science* 65:1238–1243.

Artes, F., V.H. Escalona, and F. Artes-Hdez. 2002. Modified atmosphere packaging of fennel. *Journal of Food Science* 67:1550–1554.

Bagamboula, C.F., M. Uyttendaele, and J. Debevere. 2002. Growth and survival of *Shigella sonnei* and *S. flexneri* in minimal processed vegetables packed under equilibrium modified atmosphere and stored at 7°C and 12°C. *Food Microbiology* 19:529–536.

Bartz, J.A. and J.K. Brecht (Eds.). 2003. *Postharvest Physiology and Pathology of Vegetables*. 2nd ed. Marcel Dekker, New York.

Bennik, M.H.J., W. Vorstman, E.J. Smid, and L.G.M. Gorris. 1998. The influence of oxygen and carbon dioxide on the growth of prevalent Enterobacteriaceae and Pseudomonas species isolated from fresh and controlled-atmosphere-stored vegetables. *Food Microbiology* 15:459–469.

Beuchat, L.R. 2002. Ecological factors influencing survival and growth of human pathogens on raw fruits and vegetables. *Microbes and Infection* 4:413–423.

Bourke, P. and D. O'Beirne. 2004. Effects of packaging type, gas atmosphere and storage temperature on survival and growth of *Listeria* spp. in shredded dry coleslaw and its components. *International Journal of Food Science and Technology* 39:509–523.

Brecht, J.K., K.V. Chau, S.C. Fonseca, F.A.R. Oliveira, F.M. Silva, M.C.N. Nunes, and R.J. Bender. 2003. Maintaining optimal atmosphere conditions for fruits and vegetables throughout the postharvest handling chain. *Postharvest Biology and Technology* 27:87–101.

Caillet, S., M. Millette, S. Salmieri, and M. Lacroix. 2006a. Combined effects of antimicrobial coating, modified atmosphere packaging, and gamma irradiation on *Listeria innocua* present in ready-to-use carrots (*Daucus carota*). *Journal of Food Protection* 69:80–85.

Caillet, S., M. Millette, M. Turgis, S. Salmieri, and M. Lacroix. 2006b. Influence of antimicrobial compounds and modified atmosphere packaging on radiation sensitivity of *Listeria monocytogenes* present in ready-to-use carrots (*Daucus carota*). *Journal of Food Protection* 69:221–227.

Charles, F., J. Sanchez, and N. Gontard. 2005. Modeling of active modified atmosphere packaging of endives exposed to several postharvest temperatures. *Journal of Food Science* 70:E443–E449.

Cliffe-Byrnes, V. and D. O'Beirne. 2005. Effects of chlorine treatment and packaging on the quality and shelf-life of modified atmosphere (MA) packaged coleslaw mix. *Food Control* 16:707–716.

Das, E., G.C. Gurakan, and A. Bayindirli. 2006. Effect of controlled atmosphere storage, modified atmosphere packaging and gaseous ozone treatment on the survival of *Salmonella enteritidis* on cherry tomatoes. *Food Microbiology* 23:430–438.

Day, B.P.F. 1996. High oxygen modified atmosphere packaging for fresh prepared produce. *Postharvest News & Information* 7:31–34.

Delaquis, P., S. Stewart, S. Cazaux, and P. Toivonen. 2002. Survival and growth of *Listeria monocytogenes* and *Escherichia coli* O157:H7 in ready-to-eat Iceberg lettuce washed in warm chlorinated water. *Journal of Food Protection* 65:459–464.

Drosinos, E.H., C. Tassou, K. Kakiomenou, and G.J.E. Nychas. 2000. Microbiological, physicochemical and organoleptic attributes of a country tomato salad and fate of *Salmonella enteritidis* during storage under aerobic or modified atmosphere packaging conditions at 4°C and 10°C. *Food Control* 11:131–135.

Erturk, E. and D.H. Picha. 2002. Modified atmosphere packaging of fresh-cut sweetpotatoes (*Ipomoea batatas* L.). *Acta Horticulturae* 583:223–230.

Escalona, V.H., E. Aguayo, and F. Artes. 2003. Quality and physiological changes of fresh-cut kohlrabi. *HortScience* 38:1148–1152.

Escalona, V.H., E. Aguayo, and F. Artes. 2006a. Metabolic activity and quality changes of whole and fresh-cut kohlrabi (*Brassica oleracea* L. *gongylodes* group) stored under controlled atmospheres. *Postharvest Biology and Technology* 41:181–190.

Escalona, V.H., F. Artes-Hernandez, and F. Artes. 2005. Gas composition and temperature affect quality of fresh-cut fennel. *HortScience* 40:737–739.

Escalona, V.H., B.E. Verlinden, S. Geysen, and B.M. Nicolai. 2006b. Changes in respiration of fresh-cut Butterhead lettuce under controlled atmospheres using low and superatmospheric oxygen conditions with different carbon dioxide levels. *Postharvest Biology and Technology* 39:48–55.

Fan, X.T. and K.J.B. Sokorai. 2002. Sensorial and chemical quality of gamma-irradiated fresh-cut Iceberg lettuce in modified atmosphere packages. *Journal of Food Protection* 65:1760–1765.

Fan, X.T., P.M.A. Toivonen, K.T. Rajkowski, and K.J.B. Sokorai. 2003. Warm water treatment in combination with modified atmosphere packaging reduces undesirable effects of irradiation on the quality of fresh-cut Iceberg lettuce. *Journal of Agricultural and Food Chemistry* 51:1231–1236.

Fonseca, S.C., F.A.R. Oliveira, J.K. Brecht, and K.V. Chau. 2002a. Modelling respiration rate of fresh fruits and vegetables for MAP: A review. *Journal of Food Engineering* 52:99–119.

Fonseca, S.C., F.A.R. Oliveira, J.K. Brecht, and K.V. Chau. 2003. Perforation-mediated modified atmosphere packaging: Influence of package geometry and perforation location on O_2 and CO_2 transfer. *Acta Horticulturae* 600:333–336.

Fonseca, S.C., F.A.R. Oliveira, J.M. Frias, J.K. Brecht, and K.V. Chau. 2002b. Modelling respiration rate of shredded Galega kale for development of modified atmosphere packaging. *Journal of Food Engineering* 54:299–307.

Fonseca, S.C., F.A.R. Oliveira, J.M. Frias, J.K. Brecht, and K.V. Chau. 2005. Influence of low oxygen and high carbon dioxide on shredded Galega kale quality for development of modified atmosphere packages. *Postharvest Biology and Technology* 35:279–292.

Fonseca, S.C., F.A.R. Oliveira, I.B.M. Lino, J.K. Brecht, and K.V. Chau. 2000. Modelling O_2 and CO_2 exchange for development of perforation-mediated modified atmosphere packaging. *Journal of Food Engineering* 43:9–15.

Francis, G.A. and D. O'Beirne. 2001a. Effects of acid adaptation on the survival of *Listeria monocytogenes* on modified atmosphere packaged vegetables. *International Journal of Food Science and Technology* 36:477–487.

Francis, G.A. and D. O'Beirne. 2001b. Effects of vegetable type, package atmosphere and storage temperature on growth and survival of *Escherichia coli* O157:H7 and *Listeria monocytogenes*. *Journal of Industrial Microbiology & Biotechnology* 27:111–116.

Francis, G.A. and D. O'Beirne. 2002. Effects of vegetable type and antimicrobial dipping on survival and growth of *Listeria innocua* and *E. coli*. *International Journal of Food Science and Technology* 37:711–718.

Geysen, S., B.E. Verlinden, A.H. Geeraerd, J.F. Van Impe, C.W. Michiels, and B.M. Nicolai. 2005. Predictive modelling and validation of *Listeria innocua* growth at superatmospheric oxygen and carbon dioxide concentrations. *International Journal of Food Microbiology* 105:333–345.

Gil-Izquierdo, A., M.A. Conesa, F. Ferreres, and M.I. Gil. 2002. Influence of modified atmosphere packaging on quality, vitamin C and phenolic content of artichokes (*Cynara scolymus* L.). *European Food Research and Technology* 215:21–27.

Gomes, B.C. and E.C.P. De Martinis. 2004. Fate of *Helicobacter pylori* artificially inoculated in lettuce and carrot samples. *Brazilian Journal of Microbiology* 2004: 35:145–150.

Gomez, P.A. and F. Artes. 2004a. Controlled atmospheres enhance postharvest green celery quality. *Postharvest Biology and Technology* 34:203–209.

Gomez, P.A. and F. Artes. 2004b. Keeping quality of green celery as affected by modified atmosphere packaging. *European Journal of Horticultural Science* 69:215–219.

Gomez, P.A. and F. Artes. 2005. Improved keeping quality of minimally fresh processed celery sticks by modified atmosphere packaging. *Lebensmittel-Wissenschaft Und-Technologie-Food Science and Technology* 38:323–329.

Guevara, J.C., E.M. Yahia, and E.B. De La Fuente. 2001. Modified atmosphere packaging of prickly pear cactus stems (*Opuntia* spp.). *Lebensmittel-Wissenschaft Und-Technologie-Food Science and Technology* 34:445–451.

Guevara, J.C., E.M. Yahia, E.B. De La Fuente, and S.P. Biserka. 2003. Effects of elevated concentrations of CO_2 in modified atmosphere packaging on the quality of prickly pear cactus stems (*Opuntia* spp.). *Postharvest Biology and Technology* 29:167–176.

Habibunnisa, B.R., R. Prasad, and K.M. Shivaiah. 2001. Storage behaviour of minimally processed pumpkin (*Cucurbita maxima*) under modified atmosphere packaging conditions. *European Food Research and Technology* 212:165–169.

Han, J., C.L. Gomes-Feitosa, E. Castell-Perez, R.G. Moreira, and P.F. Silva. 2004. Quality of packaged Romaine lettuce hearts exposed to low-dose electron beam irradiation. *Lebensmittel-Wissenschaft Und-Technologie-Food Science and Technology* 37:705–715.

Hansen, H. 1986. Use of high CO_2 concentrations in the transport and storage of soft fruit. *Obstbau* 11:268–271.

Hertog, M.L.A.T.M., H.W. Peppelenbos, R.G. Evelo, and L.M.M. Tijskens. 1998. A dynamic and generic model of gas exchange of respiring produce: The effects of oxygen, carbon dioxide and temperature. *Postharvest Biology and Technology* 14:335–349.

Hong, S.I. and D.M. Kim. 2001. Influence of oxygen concentration and temperature on respiratory characteristics of fresh-cut green onion. *International Journal of Food Science and Technology* 36:283–289.

Hong, G., G. Peiser, and M.I. Cantwell. 2000. Use of controlled atmospheres and heat treatment to maintain quality of intact and minimally processed green onions. *Postharvest Biology and Technology* 20:53–61.

Hu, W.Z., S. Tanaka, T. Uchino, D. Hamanaka, and Y. Hori. 2004. Effects of packaging film and storage temperature on the quality of fresh ginseng packaged in modified atmosphere. *Journal of the Faculty of Agriculture Kyushu University* 49:139–147.

Jacxsens, L., F. Devlieghere, and J. Debevere. 2002. Temperature dependence of shelf-life as affected by microbial proliferation and sensory quality of equilibrium modified atmosphere packaged fresh produce. *Postharvest Biology and Technology* 26:59–73.

Jacxsens, L., F. Devlieghere, P. Ragaert, E. Vanneste, and J. Debevere. 2003. Relation between microbiological quality, metabolite production and sensory quality of equilibrium modified atmosphere packaged fresh-cut produce. *International Journal of Food Microbiology* 83:263–280.

Jacxsens, L., F. Devlieghere, C. Van Der Steen, and J. Debevere. 2001. Effect of high oxygen modified atmosphere packaging on microbial growth and sensorial qualities of fresh-cut produce. *International Journal of Food Microbiology* 71:197–210.

Jones, R.B. 2007. Effects of postharvest handling conditions and cooking on anthocyanin lycopene, and glucosinolate content and bioavailability in fruits and vegetables. *New Zealand Journal of Crop and Horticultural Science* 35:219–227.

Kader A.A. 2004. Controlled atmosphere storage. In: K.C. Gross, C.Y. Wang, and M. Saltveit (Eds.). *The Commercial Storage of Fruits, Vegetables, and Florist and Nursery Stocks. Agriculture Handbook Number 66.* USDA, ARS, Washington, D.C.

Kader, A.A. and S. Ben-Yehoshua. 2000. Effects of superatmospheric oxygen levels on postharvest physiology and quality of fresh fruits and vegetables. *Postharvest Biology and Technology* 20:1–13.

Kim, J.G., Y.G. Luo, and K.C. Gross. 2004. Effect of package film on the quality of fresh-cut salad savoy. *Postharvest Biology and Technology* 32:99–107.

Lacroix, M. and R. Lafortune. 2004. Combined effects of gamma irradiation and modified atmosphere packaging on bacterial resistance in grated carrots (*Daucus carota*). *Radiation Physics and Chemistry* 71:79–82.

Lafortune, R., S. Caillet, and M. Lacroix. 2005. Combined effects of coating, modified atmosphere packaging, and gamma irradiation on quality maintenance of ready-to-use carrots (*Daucus carota*). *Journal of Food Protection* 68:353–359.

Limbo, S. and L. Piergiovanni. 2006. Shelf life of minimally processed potatoes Part 1. Effects of high oxygen partial pressures in combination with ascorbic and citric acids on enzymatic browning. *Postharvest Biology and Technology* 39:254–264.

Luo, Y.G., J.L. McEvoy, M.R. Wachtel, J.G. Kim, and Y. Huang. 2004. Package atmosphere affects postharvest biology and quality of fresh-cut cilantro leaves. *HortScience* 39:567–570.

Makhlouf, J., F. Castaign, J. Arul, C. Willemot, and A. Gosselin. 1989. Long-term storage of broccoli under controlled atmosphere. *HortScience* 24:637–639.

Martinez-Sanchez, A., A. Marin, R. Llorach, F. Ferreres, and M.I. Gil. 2006. Controlled atmosphere preserves quality and phytonutrients in wild rocket (*Diplotaxis tenuifolia*). *Postharvest Biology and Technology* 40:26–33.

McConnell, R., V.D. Truong, W.M. Walter, and R.F. McFeeters. 2005. Physical, chemical and microbial changes in shredded sweet potatoes. *Journal of Food Processing and Preservation* 29:246–267.

McKellar, R.C., J. Odumeru, T. Zhou, A. Harrison, D.G. Mercer, J.C. Young, X. Lu, J. Boulter, P. Piyasena, and S. Karr. 2004. Influence of a commercial warm chlorinated water treatment and packaging on the shelf-life of ready-to-use lettuce. *Food Research International* 37:343–354.

McKenzie, M.J., L.A. Greer, J.A. Heyes, and P.L. Hurst. 2004. Sugar metabolism and compartmentation in asparagus and broccoli during controlled atmosphere storage. *Postharvest Biology and Technology* 32:45–56.

Mir, N. and R.M. Beaudry. 2004. Modified atmosphere packaging. In: K.C. Gross, C.Y. Wang, and M. Saltveit (Eds.). *The Commercial Storage of Fruits, Vegetables, and Florist and Nursery Stocks. Agriculture Handbook Number 66.* USDA, ARS, Washington, D.C.

Mizukami, Y., T. Saito, and T. Shiga. 2002. Effects of controlled atmosphere on respiratory characteristics and quality preservation of spinach. *Journal of the Japanese Society for Food Science and Technology-Nippon Shokuhin Kagaku Kogaku Kaishi* 49:794–800.

Niemira, B.A., X.T. Fan, and K.J.B. Sokorai. 2005. Irradiation and modified atmosphere packaging of endive influences survival and regrowth of *Listeria monocytogenes* and product sensory qualities. *Radiation Physics and Chemistry* 72:41–48.

Paul, D.R. and R. Clarke. 2002. Modeling of modified atmosphere packaging based on designs with a membrane and perforations. *Journal of Membrane Science* 208:269–283.

Prakash, A., A.R. Guner, F. Caporaso, and D.M. Foley. 2000. Effects of low-dose gamma irradiation on the shelf life and quality characteristics of cut Romaine lettuce packaged under modified atmosphere. *Journal of Food Science* 65:549–553.

Riad, G.S. and J.K. Brecht. 2001. Fresh-cut sweetcorn kernels. *Proceedings of the Florida State Horticultural Society* 114:160–163.

Riad, G.S. and J.K. Brecht. 2003. Sweetcorn tolerance to reduced O_2 with or without elevated CO_2 and effects of controlled atmosphere storage on quality. *Proceedings of the Florida State Horticultural Society* 116:390–393.

Riad, G.S., J.K. Brecht, and S.T. Talcott. 2003. Browning of fresh-cut sweet corn kernels after cooking is prevented by controlled atmosphere storage. *Acta Horticulturae* 628:387–394.

Saltveit, M.E. 2003. Is it possible to find an optimal controlled atmosphere? *Postharvest Biology and Technology* 27:3–13.

Sanchez-Mata, M.C., M. Camara, and C. Diez-Marques. 2003a. Extending shelf-life and nutritive value of green beans (*Phaseolus vulgaris* L.), by controlled atmosphere storage: Macronutrients. *Food Chemistry* 80:309–315.

Sanchez-Mata, M.C., M. Camara, and C. Diez-Marques. 2003b. Extending shelf-life and nutritive value of green beans (*Phaseolus vulgaris* L.), by controlled atmosphere storage: Micronutrients. *Food Chemistry* 80:317–322.

Sanz, S., M. Gimenez, and C. Olarte. 2003. Survival and growth of *Listeria monocytogenes* and enterohemorrhagic *Escherichia coli* O157:H7 in minimally processed artichokes. *Journal of Food Protection* 66:2203–2209.

Siomos, A.S., E.M. Sfakiotakis, and C.C. Dogras. 2000. Modified atmosphere packaging of white asparagus spears: Composition, color and textural quality responses to temperature and light. *Scientia Horticulturae* 84:1–13.

Tassou, C.C. and J.S. Boziaris. 2002. Survival of *Salmonella enteritidis* and changes in pH and organic acids in grated carrots inoculated or not with *Lactobacillus* sp. and stored under different atmospheres at 4 degrees C. *Journal of the Science of Food and Agriculture* 82:1122–1127.

Thompson, A.K. 1998. *Controlled Atmosphere Storage of Fruits and Vegetables*. CAB International, New York.

Tsouvaltzis, P., J.K. Brecht, A.S. Siomos, and D. Gerasopoulos. 2008. Responses of minimally processed leeks to reduced O_2 and elevated CO_2 applied before processing and during storage. *Postharvest Biology and Technology* 49:287–293.

Tsouvaltzis, P., D. Gerasopoulos, and A.S. Siomos. 2007. Effects of base removal and heat treatment on visual and nutritional quality of minimally processed leeks. *Postharvest Biology and Technology* 43:158–164.

Van De Velde, M.D. and M.E. Hendrickx. 2001. Influence of storage atmosphere and temperature on quality evolution of cut Belgian endives. *Journal of Food Science* 66:1212–1218.

Zhu, M., C.L. Chu, S.L. Wang, and R.W. Lencki. 2001. Influence of oxygen, carbon dioxide, and degree of cutting on the respiration rate of rutabaga. *Journal of Food Science* 66:30–37.

18

Modified Atmosphere Packaging for Fresh-Cut Produce

Peter M.A. Toivonen, Jeffrey S. Brandenburg, and Yaguang Luo

CONTENTS

Modified atmosphere packaging (MAP) is effective in maintaining quality through its effects on modification of the gas composition in the package headspace (Schlimme and Rooney, 1994; Jacxsens et al., 2002; Kim et al., 2003; Luo et al., 2004). The degree of atmospheric modification within a modified atmosphere package is a consequence of the respiratory O_2 uptake and CO_2 evolution of the packaged produce and the rate of gas transfer across the package film (Al-Ati and Hotchkiss, 2002). But several factors impinge on this basic understanding and directly influence the final package atmosphere, including temperature, product weight, and package surface area (Bell, 1996). There have been previous reviews of the effects of modified atmospheres (Mir and Beaudry, 2004; Varoquaux and Ozdemir, 2005) and so the discussion in this chapter will focus on recent findings or those which have not been discussed in detail by other authors.

The intent of this chapter is to provide the reader with a broad appreciation for MAP of fresh-cut processed fruit and vegetable products. To achieve this goal, the discussion will range from packaging design considerations to understanding the effects of modified atmospheres and humidity on the quality and shelf life of the fresh-cut products within the package. Finally, the authors' views on future research and technology needs will be discussed.

18.1 Packaging Technology

18.1.1 Packaging Materials

There is an ever increasing variety of packaging materials available for MAP for fresh-cut produce. The polymers, films, structures, and formats, which make up this type of packaging can include flexible packages, rigid containers, engineered oxygen transmission rate (OTR) polymers, microperforated materials, as well as combinations of all of the above. Within these broad categories, there are also ancillary technologies that have been developed to improve package performance, including easy-peel lid structures, antimicrobial packaging, cook-in packaging, and active and intelligent packaging. Each of these formats has strengths and weaknesses that impact their desirability for a particular fresh-cut produce application. The choice of the correct structure must be determined by matching the strengths of particular structure with the fresh-cut product characteristics (i.e., physiological characteristics), engineering, marketing, and manufacturing requirements of the package, while at the same time minimizing or masking particular weakness.

Specific parameters that have a direct impact on the choice of packaging format include product type, product quantity, market application (food service, retail, or club store), package dimensions, stiffness, graphics, marketing, cost, environmental impact, reusability, easy open feature, cook-in requirement, etc. In order to understand and then effectively match the key parameters to the optimal packaging format, it is essential that communication between grower, processor, and packaging supplier exists. In addition, the MAP design process must be seamlessly incorporated into the overall new product development process.

18.1.1.1 Polymers

There are a variety of polymers used in fresh-cut produce MAP. A portion of these polymers are used in primarily flexible packaging structures, a portion are used in primarily rigid packaging structures, and a portion are used in both types of applications. Each specific polymer has physical, chemical, and gas transmission rate properties that are unique to that polymer. Design of a packaging structure entails matching the specific polymer properties to the requirements of the MAP application. See Table 4.1 for a listing of commonly used polymers and their corresponding attributes. For fresh-cut modified atmosphere applications a polymer's gas transmission rates, specifically, OTR and carbon transmission rate (CO_2TR) are key attributes. The transmission of gases across packaging structures governed by several factors that are related through Fick's law:

$$J_{gas} = \frac{A \times \Delta C_{gas}}{R} \qquad (18.1)$$

where
  J_{gas} is the total flux of gas (cm^3/s)
  A is the surface area of the film (cm^2)
  ΔC_{gas} is the concentration gradient across the film
  R is the resistance of the film to gas diffusion (s/cm)

The gas flow across a film increases with increasing surface area and increasing concentration gradient across the film. In contrast, the gas flow across the film decreases with increasing film resistance to gas diffusion. Designing and controlling gas transmission rate is one of the fundamental underlying principles of successful MAP application in commercial practice. Blends, coextrusions, and films comprised of specific polymers combine to create a package with a specific OTR. The optimal OTR selection for a MAP film for a specific fresh-cut product is dependent upon the respiration rate(s) of the produce, product weight, the internal package dimensions, the targeted atmosphere composition, and product handling temperature.

The goal of MAP of fresh produce is to create an equilibrium package modified atmosphere with an O_2 partial pressure low enough and a CO_2 partial pressure high enough to result in beneficial effects to the produce and not be too low or too high, respectively, such that they become injurious (Zagory, 1998). Therefore both O_2 and CO_2 gas transmission selection is necessary to create an optimal modified atmosphere. However, gases diffuse through polymeric films at different rates: carbon dioxide generally diffuses through most polymer films at rate from between two to five times faster than oxygen (Al-Ati and Hotchkiss, 2002). The ratio of CO_2 transmission rate to O_2 transmission rate of a film is termed the beta value of a particular film. Polymeric films therefore generally have beta values ranging from 2 to 5, with an average of 3. The beta value of a modified atmosphere package will have a direct bearing on the final modified atmosphere achieved within the package (Al-Ati and Hotchkiss, 2002). For example, in standard polymer films, it is generally possible to achieve a low O_2 level (e.g., 2 kPa) in combination with a moderate CO_2 level (e.g., 7 kPa).

The respiration of specific fresh-cut produce items is determined through routine testing procedures. A number of universities specialize in product respiration calculations (e.g., University of California, Davis, California). In addition, there are produce respiration testing protocols available through private consultancy companies (e.g., The JSB Group, LLC) and these can be performed in-house at the processing plant. It must be kept in mind that the respiration of a piece produce is temperature dependent (Varoquaux and Ozdemir, 2005). Product temperature is therefore critical in determining the product respiration rate. Since package OTR is respiration rate dependent, then package OTR determination is temperature dependent as well. Therefore if temperature varies throughout the product life cycle the MAP cannot be optimized or relied upon. Temperature is so critical to MAP performance and if temperature variations are too significant it can actually do more harm than good (Tano et al., 1999; Varoquaux and Ozdemir, 2005).

18.1.1.2 *High OTR Materials and Technologies*

The relatively high respiration rates of many fresh-cut produce items (Gorny, 1997) require that they be packaged in MAP films that have OTRs above 750 cc/100 in.^{2}/atm/day ($\sim$12,000 cc/m^2/atm/day). Generally, there are two approaches to achieving the required high OTRs: either high OTR polymers or microperforations can be used. High OTR polymers include ultralinear low-density polyethylene (ULDPE), plastomer metallocenes, and high percent ethylene vinyl acetate (EVA). Refer to Table 4.1 for typical values of specific high OTR polymers. The density and crystallinity of the polymer dictate the

OTR: the lower the density the higher the OTR. However, OTR is not the only important physical parameter that is altered as polymer density changes. Stiffness, clarity, hot tack, and ultimate seal strength are also impacted. As the density decreases, stiffness decreases, hot tack increases, clarity decreases, and ultimate seal strength decreases. Hence, high OTR resins are very tacky and require additional slip and antiblock additives to optimize behavior during manufacture and fresh-cut product processing. The additional additives can also have a significant impact on final OTR, often reducing the OTR to a lower than desirable level. The specific choice of combination and processing protocol of high OTR polymers is dependent on what attributes, in addition to high OTR, are necessary in the finished film. There is also an upper limit to the OTR for these polymers. Currently without any additives, the upper limit to OTR is 1000–1200 cc/100 in.^{2}/mil/atm/day (15,500–18,600 cc/m^2/mil/atm/day). In a slip and antiblock modified format, this upper limit is nearer to 900–1000 cc/100 in.^{2}/mil/atm/day (13,950–15,500 cc/m^2/mil/atm/day). These OTR levels are for the specific polymer alone. Combined with other polymers in either a coextrusion or blend will further reduce the effective OTR. See Chapter 4 for equations used to calculate OTR in both polymer blends and film coextrusions.

18.1.1.3 Microperforated Films

An alternative method for achieving high OTR packaging structures is through the use of microperforation technology (Lougheed, 1992; Aharoni and Richardson, 1997). This technology involves creating, through various means, holes of several micrometers in the packaging structure. The hole size and configuration can vary with the specific perforation technology but microperforations are not visible to the naked eye and can range from 40–200 μm in diameter. In addition, the spatial placement of the holes within the package itself is critical. Ensuring that the holes are not blocked or obstructed in any way is critical to the success and desired control of gas transmission. Microperforated films are similar to high OTR polymers in that there are OTR limitations. However, the limitations are at the lower end of the range and not critical for MAP applications in fresh-cut produce. Typical OTRs for microperforated films begin at 250 cc/100 in.^{2}/mil/atm/day (i.e., 3875 cc/m^2/mil/atm/day) at the lower end. Gas transmission rates of microperforated film are determined by the size of the individual micro holes and their corresponding placement and frequency. The number of microperforations required to be in a package are directly proportional to the desired OTR (i.e., as the OTR requirements increase, so do the number of holes). Since microperforated structures cannot have less than one hole, the gas transmission rate through a single hole dictates the lowest possible transmission rate. It must be cautioned that microperforated MAP with only one hole can be problematic due to the high potential risk of blockage for that single perforation. With two or more holes, the risk of blockage of all holes declines and also there is more uniform gas transmission throughout the package. In order to avoid a package design with only one microperforation, consideration should be given to altering other key OTR control parameters, such as product weight or package dimension.

The diffusion rates of atmospheric gases through microperforations are very similar. In effect, the diffusion rates of CO_2 and O_2 are virtually identical, i.e., the films have a beta value of 1. Therefore, in these films targeted atmospheres with low O_2 partial pressures (e.g., 2 kPa) and high CO_2 partial pressures (e.g., 19 kPa) can be achieved. The beta value difference between polymeric and microperforated structures provide for significantly different ranges in final modified atmosphere composition. This fact needs to be accounted for in the packaging design process.

Macroperforations, which are visible to the naked eye, should not be considered to be equivalent to microperforations. The gas movement through these larger, visible holes utilized in macroperforated films is too great to consistently modify and control the gas composition within a package. Therefore a low O_2 modified atmosphere is not feasible with macroperforated film. This does not mean, however, there is not a need for macroperforated films. Since the gas transmission is so high, a macroperforated structure will virtually never become anaerobic (i.e., O_2 levels falling to 0 kPa) even under temperature abuse situations. If it is a priority to have a fresh-cut produce package designed such that it cannot become anaerobic under any circumstance and exposure atmospheric levels of O_2 do not significantly reduce shelf life of that product, then macroperforation technology may be applicable.

18.1.2 Package Formats

Choosing the correct packaging format is an integral step in modified atmosphere package design. The process of determining the correct packaging format must include input from all stake holders, including the fresh-cut processor, packaging manufacturer, raw material suppliers, and the consumer. In general there are three types of formats that are most commonly used in fresh-cut produce modified atmosphere packaging: flexible packaging, rigid packaging, and active/intelligent packaging. Within each of these broader categories, there are numerous variations and permeations.

18.1.2.1 *Flexible Packaging*

Flexible packaging is the most common format for fresh-cut produce MAP in North America. Typically this format is available as preformed bags, roll stock, and stand-up pouches. A variety of additional features, such as easy opening and resealability, can be designed into the package. Each format type has its own challenges, strengths, and weaknesses.

Preformed bags are ideal for initial testing and start up operations since order quantities can often be minimized allowing for smaller initial packaging costs. As the name implies preformed bags are preformed with three sides closed and one side remaining open for filling with product. Although not always the case, most preformed bag applications coincide with a hand filling operation in the processing line. Generally, since the final seal on the bags are sealed by hand, it is important that they are sealed consistently ensuring the package dimensions are constant. Failure to situate the seal at a consistent location in a bag will result in variation in internal package surface area and ultimately variation in the total bag OTR. Preformed bags come in a variety of structures and graphics, but are often sold as an off-the-shelf item versus custom designed for each specific application. Ultimately the decision process involves the reconciliation of trade-offs between ideal OTR, off-the-shelf convenience, and cost.

Roll stock is the most common form of modified atmosphere flexible packaging for retail applications. Roll stock packaging configurations can range from the very basic monolayer films to complex reverse printed multilayer laminations. Unlike preformed bags, the packaging structure is not supplied sealed at all but rather delivered cut to the final running width and wound on rolls. In order to obtain the final filled package, the roll stock is run on a form-fill-seal packaging machine. Most common for fresh-cut modified atmosphere packaging applications is the vertical form, fill, and seal (VFFS) machine (Figure 18.1). Form, fill, and seal machines can be configured to run either a lap or a fin-and-crimp seal (see Chapter 4 for a detailed explanation of lap versus fin-and-crimp sealing). No matter what the sealing configuration, the width of the seals cannot be included in the surface area calculations for total package OTR determination. Depending upon the specific machine configuration the seal width can be a significant percentage of

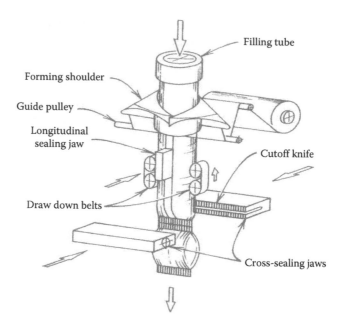

FIGURE 18.1

Schematic representation of a VFFS packaging machine. A precise weight of fresh-cut product is introduced through the filling tube at the top of the drawing. Simultaneously, a bag is being formed around that tube from packaging roll stock which has a width dimension specified for this particular machine. The longitudinal sealing jaw welds the side seam of the newly formed bag. Cross-sealing jaws simultaneously seal the top of the previously filled bag and also the bottom of the currently forming bag in one operation. A cutoff knife is interposed between the two sealing bars of the cross-sealing jaws to separate the two bags as the seals are being formed. Once the bottom seal of the current formed bag is completed, the weighed fresh-cut product is dropped into it and the draw down belts pull the film down one-length of a bag, allowing the top to be sealed with the cross-sealing jaws, while also making the bottom of the next bag. (Brandenburg, unpublished figure.)

the package area. With roll stock structures, which are converted into finished packaging on VFFS machines, there are physical structural properties that become more important compared with preformed bags. These parameters include coefficient of friction (COF), seal initiation temperature, and hot tack seal strength. Although flexible modified atmosphere roll stock packaging is sometimes offered off-the-shelf it is most commonly offered and effective as a customized structure for a specific fresh-cut application to be run of a specific VFFS machine.

Stand-up pouches can be obtained either in a preformed or roll stock format. The distinguishing feature is that at least one side of the bag is gusseted, allowing the pouch to stand on its own. This stand-up feature allows for product differentiation on the store shelf. Stand-up pouches can be designed and manufactured with either solid polymer or microperforated film technology. Due to the stand-up feature, effective package surface area is increased, which allows for a lower OTR specification as compared with a conventional bag design. However the stand-up feature also requires stiffer polymers. Since polymer OTR decreases as polymer stiffness increases the polymers of choice do not have a high enough transmission rate for most fresh-cut applications. Therefore the majority of stand-up pouches for fresh-cut fruits and vegetables utilize microperforated film technology.

18.1.2.2 Rigid Packaging

In rigid packaging, a rigid tray or container with a removable lid is utilized instead of a flexible film structure. There are a number of rigid packaging formats commonly used in fresh-cut produce applications. The key distinguishing parameter from a modified atmosphere perspective is the method utilized to close or seal the tray or container. Three of the more common rigid tray formats include clamshell, snap on lid, and sealable, easy-peel lidding film. The first two formats clamshell and snap on lid are not true modified atmosphere applications since package gas transmissions are not engineered or controlled. Frequently these styles of rigid packages are refered to as natural aspiration packages, meaning that the final atmosphere in the package can vary according to how tight the lid is attached or "snapped" onto the tray. The third format, sealable, easy-peel lidding film, is a true MAP application. The lidding film is designed to a specific gas transmission rate and hermetically sealed onto upper lip of the tray. In all current rigid packaging formats, the tray itself is exempt from the package OTR calculations since it has a negligible OTR, often less than 5 cc/100 in.^{2}/mil/atm/day (77 mL/m^2/mil/atm/day). Therefore when calculating the target OTRs of the package the surface area portion of the calculation is limited to the surface area of the lidding. This generally is a small portion of the total tray surface area and hence the lidding film structures need to have a very high OTR. This high OTR requirement dictates that, for most rigid tray lidding applications, microperforation technology is the only acceptable choice. This requirement is often exacerbated by a printed label placed on the lid further reducing the effective surface transmission area. These gas transmission surface area limitations can lead to microenvironments within the tray. Produce that is in close proximity to the high OTR microperforated lid may have close to the targeted modified atmosphere but produce that is located in the lower corners of the very low OTR rigid tray can experience a significantly different modified atmosphere. Anaerobic conditions are not uncommon in this latter location of a rigid tray packaging system especially if the package is designed to have a low O$_2$ target atmosphere. Micro-atmospheres within the tray can lead to shortened shelf life and there is currently research underway to design trays which can participate in the overall OTR of such packages.

The sealing properties of lidding films to rigid trays are significantly different than desired flexible packaging properties. Unlike flexible packaging films, which generally seal similar polymers to each other, rigid packaging design have to produce and effect seal between very dissimilar polymers. Rigid trays are commonly made from polymers such as amorphous polyethyleneterapthalate (APET), polyvinyl chloride (PVC), polypropylene (PP), and polystyrene (PS). Sealing polymers, especially if they are easy-peel, are frequently made from high percentage ethylene vinyl acetate (EVA) and acid-modified vinyl acetates. The sealing parameters required to achieve an effective seal with dissimilar polymers is much narrower than when sealing a like polymer and hence there is a higher risk for improper sealing with such lidding film. Packaging, which has not been properly sealed, is prone to leaks and since the purpose of MAP is to design and control gas diffusion through a package, leaks in the seal area will cause the system to become out of control. When the packaging system is not a controlled process, optimal shelf life is not attainable.

Rigid packaging provides a number of distinct offerings over flexible packaging including; greater physical protection for sensitive produce, provision of a rigid bowl or tray may be used as the serving vessel by the consumer and the rigid packaging can be made stackable allowing for ease of retail display. However, with the last feature caution must be exercised when rigid trays are stacked the effective OTR surface area of the lidding may be significantly decreased.

18.1.2.3 Active and Intelligent Packaging

Packaging engineers and designers have been working on ways to engineer packaging such that MAP takes an even more active role in protecting and maintaining the quality of fresh-cut produce (Ozdemir and Floros, 2004). Common examples of active packaging include antimicrobial packaging, temperature sensitive switches, absorbent packaging, and cook-in packaging.

The concept behind antimicrobial packaging is to have the package actively participate in the safety of the fresh-cut produce by providing a kill step for any pathogenic bacteria that may be present. There have been and continue to be numerous attempts to develop effective antimicrobial packaging. The focus of the research has been to incorporate a biocide into the package that kills the potentially harmful bacteria. There are a number of challenging technical and regulatory hurdles that must be overcome in order to achieve a working solution including identification of effective antimicrobials, which can be in direct contact with food, regulations, and labeling requirements. Additives incorporated or placed within the package will need to meet all applicable FDA direct food contact regulations (Shanklin and Sánchez, 2005), as well as EPA regulations for pesticides or biocides. As the issue of food safety continues to be priority in the fresh-cut industry so will the interest to support research to finding effective and safe antimicrobial packaging film technologies.

Landec Corporation has been the leading source for temperature switch polymers (Landec, 2007): their Intelimer® polymers are unique materials that respond to temperature changes in a controllable, predictable way. These polymers can abruptly change their permeability, adhesion, viscosity or volume when heated or cooled by just a few degrees above or below a preset temperature switch. The changes are triggered by a built-in temperature switch, which can be set within temperature ranges compatible with most biological applications. Moreover, because the process of change involves a physical and not a chemical change, it can be reversed. Temperature switch polymers can be an effective packaging tool to avoid anaerobic conditions as a result of temperature variation and abuse during the product life cycle. However, temperature switch polymers should never be used in lieu of proper temperature control procedures.

Absorbent packaging describes a packages ability to absorb liquids or gases produced by fresh-cut produce (Ozdemir and Floros, 2004; Brody, 2005). Build up of gases such as ethylene, the ripening hormone, can significantly reduce shelf life. Gas absorption technologies include sachets placed inside the package, and additives within the inner polymer layer. Unlike gas transmission technology which allows gases to leave the package entirely, gas absorption technology absorbs and traps it within the package. The net effect of the two technologies can be similar. Additives incorporated or placed within the package will need to meet all applicable FDA direct food contact regulations. Gas absorption technology is applicable to all types of packaging formats.

Liquid absorption or liquid control has grown with the introduction and growth of the fresh-cut fruit market. Liquid absorption technologies work by trapping or venting liquid into another portion of the package and holding it there. Technologies can include absorbent pads, absorbent gels, and valve films, in combination with compartmentalized rigid packaging. Removal of the liquid from the fresh-cut produce can extend shelf life and trap microorganisms as well as preventing recontamination from the liquid. Although most commonly found in combination with rigid packaging this technology can be adapted to flexible packaging.

Cook-in packaging within the fresh-cut market has seen significant growth within the last 5 years (Forney, 2007). Although most leafy greens are not appealing when cooked, spinach being the exception, the fresh-cooked vegetables market is enjoying significant growth. Cook-in packaging for fresh-cut vegetables is designed for microwave cooking, directly utilizing the heat generated by the microwaves and also the resultant steam generated within the package by the heat (Forney, 2007). The packaging is designed to allow steam pressure build up to a predetermined level and then automatically vent.

A variety of steam venting technologies and formats are available to the package designer. In addition to the standard fresh-cut MAP requirements, functional and regulatory applicability must be considered. Not all polymers commonly used on fresh-cut produce packaging can functionally withstand the temperatures created by steam cooking. Temperatures are significantly higher when oil-based sauces are in used combination with the produce. In addition, not all polymers maintain their direct food contact regulatory status at elevated temperatures. Therefore when designing packaging for cook-in applications, a thorough and complete understanding of cooking temperature and duration are vital in determining polymer functional and regulatory compliance. In addition any additives added to the inner polymer layer such as antifog agents need to be both functional and regulatory compliant.

18.1.3 Packaging Misconceptions

MAP can extend the shelf life of fresh-cut produce items but only under specified ranges of environmental parameters and conditions. There are a number of misconceptions regarding what MAP can and cannot do. First and foremost MAP is only effective if there is consistent temperature management throughout the entire life cycle of the product. This includes during processing right through the entire distribution channel. Lack of temperature control will result in produce respiration variations, which will prevent the packaging system from consistently achieving its targeted optimal modified atmosphere. As previously noted, temperature switch polymers can, to an extent, negate this problem.

MAP will never improve the quality of the incoming raw material product. Under ideal circumstances, the best that can be achieved is to maintain the existing quality level throughout the product life cycle, including shelf life at the consumer level. In real-world applications, MAP will maintain quality for the majority of the targeted shelf life, but due to factors such as loss of temperature control, quality will only visibly suffer at the very end of the desired shelf life.

Fresh-cut MAP relies on the relationship between produce respiration and package transmission rate to alter the atmosphere within the package. This process generally takes a number of days to reach the target atmosphere and equilibrium. For produce items which are prone to enzymatic browning reactions or "pinking," which can be exacerbated by O_2 partial pressures above 3 kPa, this gradual descent may be too long to be helpful for quality maintenance. Hence, gas flushing of fresh-cut MAP has been developed as an approach to establish an initial low oxygen atmosphere within the package, which can be beneficial in reducing enzymatic browning reactions and "pinking." However, gas flushing is not a substitute for proper package design or a compensation for packages that leak after sealing. Effects of gas flushing are only realized when it is employed in combination with proper package design and a leak-free package seal. Since the primary goal of gas flushing is to reduce the initial O_2 level within the package N_2 has proven to be the most effective and economical gas to use for this purpose.

18.2 Atmosphere Effects

18.2.1 Physiology

It has been generally established that prolonged shelf-life of fresh-cut produce is achieved through the reduction of respiration rate, decrease in ethylene biosynthesis and action, and a delaying of senescence (Mir and Beaudry, 2004; Varoquaux and Ozdemir, 2005). This

TABLE 18.1

Summary of Modified Atmosphere Recommendation for Fresh-Cut Vegetables

Fresh-Cut Vegetable	Temperature (°C)	Atmosphere (kPa O_2)	Atmosphere (kPa CO_2)	Efficacy
Beets (red), grated, cubed, or peeled	0–5	5	5	Moderate
Broccoli, florets	0–5	2–3	6–7	Good
Cabbage, shredded	0–5	5–7.5	15	Good
Cabbage (Chinese), shredded	0–5	5	5	Moderate
Carrots, shredded, sticks, or sliced	0–5	2–5	15–20	Good
Jicama, sticks	0–5	5	5–10	Good
Leek, sliced	0–5	5	5	Moderate
Lettuce (butterhead), chopped	0–5	1–3	5–10	Moderate
Lettuce (green leaf), chopped	0–5	0.5–3	5–10	Good
Lettuce (iceberg), chopped or shredded	0–5	0.5–3	10–15	Good
Lettuce (red leaf), chopped	0–5	0.5–3	5–10	Good
Lettuce (Romaine), chopped	0–5	0.5–3	5–10	Good
Mushroom, sliced	0–5	3	10	Not recommended
Onion, sliced or diced	0–5	2–5	10–15	Good
Peppers, diced	0–5	3	5–10	Moderate
Potato, sliced or whole peeled	0–5	1–3	6–9	Good
Pumpkin, cubed	0–5	2	15	Moderate
Rutabaga, sliced	0–5	5	5	Moderate
Spinach, cleaned	0–5	0.8–3	8–10	Moderate
Tomato, sliced	0–5	3	3	Moderate
Zucchini, sliced	5	0.25–1	—	Moderate

Source: Reproduced from Gorny, J.R., *Proc. 7th Intl. Controlled Atmosphere Res. Conf.*, 5, 1997, pp. 30–66. With permission.

directly mediated by low O_2 and/or high CO_2 partial pressures (Varoquaux and Ozdemir, 2005). However, there must be caution raised in regards to package overmodification (i.e., anaerobic levels of O_2 and excessive levels of CO_2), which leads to significantly altered respiratory metabolism and results in development of off-odors and off-flavors (Al-Ati and Hotchkiss, 2002; Watada et al., 2005).

The effect of MAP on the physiology of fresh-cut produce is well documented (Toivonen and DeEll, 2002; Mir and Beaudry, 2004; Varoquaux and Ozdemir, 2005). Tables 18.1 and 18.2 list the currently accepted optimal atmospheres and handling temperatures for selected fresh-cut fruits and vegetables. However, caution must be taken in the application of the recommendations for the optimum atmospheres for given commodity since there may be varying responses by a product as a consequence of differences in physiological maturity, growing conditions, postharvest handling conditions prior to cutting and the expected storage/distribution temperature (Toivonen and DeEll, 2002). It is prudent to verify the efficacy and safety of a given recommendation for a specific situation before applying in commercial practice.

Membrane stability, or retention of membrane function, has not been widely discussed in reviews on MAP effects on fresh-cut product. Loss of membrane integrity is often considered a consequence of oxidative injury in the tissues but the underlying cause may be one of numerous stress factors (Hodges and Toivonen, 2007). Luo et al. (2004) have been able to demonstrate that one of the effects of inadequate permeability is the loss of membrane integrity in packaged cilantro leaves. Cilantro leaves packaged in moderate and fast OTR and microperforated film reached atmospheric equilibrium

TABLE 18.2

Summary of MA Recommendation for Fresh-Cut Fruits

Fresh-Cut Fruit	Temperature (°C)	Atmosphere (kPa O_2)	Atmosphere (kPa CO_2)	Efficacy
Apple, sliced	0–5	<1	4–12	Moderate
Cantaloupe, cubed	0–5	3–5	6–15	Good
Grapefruit, sliced	0–5	14–21	7–10	Moderate
Honeydew, cubed	0–5	2	10	Good
Kiwifruit, sliced	0–5	2–4	5–10	Good
Mango cubes	0–5	2–4	10	Good
Orange, sliced	0–5	14–21	7–10	Moderate
Peach, sliced	0	1–2	5–12	Poor
Pear, sliced	0–5	0.5	<10	Poor
Persimmon, sliced	0–5	2	12	Poor
Pomegranate, arils	0–5	—	15–20	Good
Strawberry, sliced	0–5	1–2	5–10	Good
Watermelon, cubes	0–5	3–5	10	Good

Source: Reproduced from Gorny, J.R., *Proc. 7th Intl. Controlled Atmosphere Res. Conf.*, 5, 1997, pp. 30–66. With permission.

ranging from 1.5 to 21 kPa O_2 and 0 to 3.6 kPa of CO_2. The cilantro in these three package types maintained a low tissue electrolyte leakage throughout the 14 days storage at 0°C. However, those held in packages with low OTR developed an overly modified atmosphere of 0.02 kPa O_2 and 9.0 kPa CO_2. The tissue electrolyte leakage for cilantro in the low OTR packages was significantly higher within 6 days storage and continued to increase over the remainder of the 14 days storage. This data suggests that tissue membranes are plastic and can readily adapt to a wide range of atmospheres in MAP. However, the membrane plasticity has a limit and when anaerobic conditions develop, they rapidly lose their function and cell membrane breakdown occurs and consequently quality declines.

The duration of exposure to anaerobic atmospheres can influence the tissue electrolyte leakage also. Kim et al. (2005b) studied the effect of initial oxygen level on the physiological response of fresh-cut romaine lettuce. Increasing the initial headspace O_2 concentration at the time of sealing the bag, delayed O_2 depletion within the packages but the final steady-state atmospheres were similar in all cases (Figure 18.2). This was associated with a reduction in the levels of tissue electrolyte leakage (Figure 18.3) and also reduced levels of anaerobic metabolite accumulation (Kim et al., 2005b). Therefore, initial gas composition can have a strong influence on membrane breakdown, particularly in packages made from low OTR films which result in low oxygen atmospheres.

Further studies by Kim et al. (2005a) have suggested that a delay after cutting and before packaging results in an altered initial package atmosphere composition at equilibrium. The reason for this result is relatively straightforward. It is well known that wound-induced respiration results in a rapid utilization of oxygen in freshly cut produce (Toivonen and DeEll, 2002). Delay in packaging allows the produce to undergo this period of high respiration rate and consequently when it is finally packaged, there is reduced initial oxygen consumption. This results in higher equilibrium O_2 partial pressures in the package and consequently less fermentative volatile production and tissue electrolyte leakage (Luo, 2007a,b). However, on the down side, there is an increased potential for enzymatic browning due to prolonged exposure to the ambient air after cutting and wounding.

In general in industry practice, some fresh-cut vegetables and especially lettuce, a rapid establishment of a low O_2 and/or elevated CO_2 environment (otherwise known as active

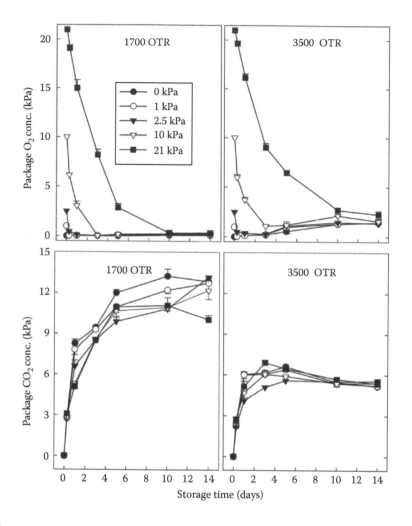

FIGURE 18.2

Effect of packaging film OTR and initial O_2 concentration on aerobic mesophillic bacterial, yeast, and mold populations of packaged fresh-cut romaine lettuce stored at 5°C for 14 days. Packages were flushed with oxygen–nitrogen gas mixtures, sealed in polypropylene film (OTR of 1700 and 3500 mL/days/m² at 5°C) packages and stored at 5°C for up to 14 days. Each symbol represents the mean of three replicate measurements and error bars represent standard errors of the mean. (Reproduced from Kim, J., Luo, Y., Saftner, R.A., Tao, Y., and Gross. K.C., *J. Sci. Food Agric.*, 85, 1622, 2005b. With permission.)

MAP) is considered critical for the prevention of cut surface browning (Kim et al., 2005b). This can be attained by flushing the package with N_2, to create an initial low O_2 atmosphere immediately prior to sealing. While gas flushing a package does not alter the equilibrium O_2 and CO_2 concentrations in the headspace, it merely accelerates attainment of the equilibrium concentrations (Kim et al., 2005b). The result of such treatment is not intended to control membrane deterioration, rather to control the activity of polyphenol oxidase, which is largely inactivated at oxygen concentrations below 1 kPa (Smyth et al., 1998). Many researchers have observed that once such packages are opened, they brown extremely rapidly and this suggests that tissue and membrane degradation have occurred, but since oxygen levels were low the interaction between O_2-polyphenol oxidase-phenolic substrates was inhibited until such time as the lettuce was reintroduced to a higher oxygen atmosphere (Smyth et al., 1998).

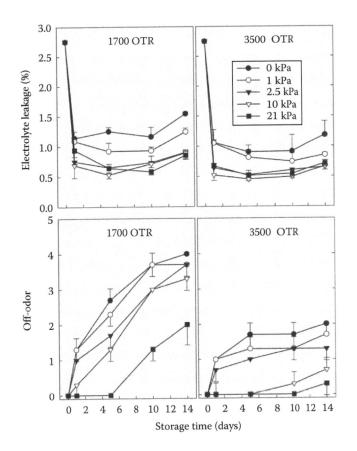

FIGURE 18.3

Effect of packaging film OTR and initial O_2 concentration on electrolyte leakage and off-odor scores of packaged fresh-cut romaine lettuce stored at 5°C for 14 days. Packages were flushed with oxygen–nitrogen gas mixtures, sealed in polypropylene film (OTR of 1700 and 3500 mL/days/m^2 at 5°C) packages and stored at 5°C for up to 14 days. Each symbol represents the mean of three replicate measurements and error bars represent standard errors of the mean. (Reproduced from Kim, J., Luo, Y., Saftner, R.A., Tao, Y., and Gross. K.C., *J. Sci. Food Agric.*, 85, 1622, 2005b. With permission.)

18.2.2 Microbiology

The composition of microbial populations found on commercial fresh-cut products is highly variable. The surface microflora of vegetables and fruits largely comprises *Pseudomonas* spp., *Erwinia herbicola, Flavobacterium, Xanthomonas,* and *Enterobacter agglomerans,* as well as various yeasts and molds. *Pseudomonas,* particularly *P. fluorescens,* generally dominates the microbial population on many vegetables (Nguyen-The and Prunier, 1989; Babic et al., 1996; Zagory, 1999). Lactic acid bacteria, such as *Lactobacillus* and *Lactococcus,* are also commonly found on fresh-cut fruits and vegetables such as carrots (Zagory, 1999; Allende et al., 2004; Ruiz-Cruz et al., 2006; Luo, 2007a,b). Figure 18.4 shows the typical population counts of aerobic mesospheric bacteria, yeasts, and molds found for various commercially packaged fresh-cut fruits.

The effect of modified atmosphere package conditions on microbial ecology is diverse and is dependent on the microbial species, fresh-cut produce type, and storage conditions (Wang et al., 2004; Watada et al., 2005). The gas composition within packages of fresh-cut produce affects the microbial ecology on the produce. The modified package atmospheres commonly used for fresh-cut produce do not exert biocidal effects on microorganisms,

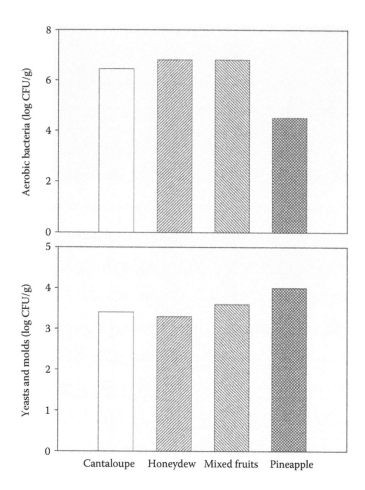

FIGURE 18.4

Aerobic mesophillic bacterial, and yeast and mold populations on various commercially packaged fresh-cut fruits held at 5°C. (Luo, unpublished data.)

but may have a differential influence on the rate of the growth of specific species and therefore the make up of the microbial populations on the packaged products. Elevated CO_2, reduced O_2, or a combination of both can favor the growth of certain classes of microorganisms. For example, a low O_2 plus high CO_2 condition is likely to favor the growth of lactic acid bacteria (Brackett, 1994; Allende et al., 2004).

In general, MAP itself has a limited effect on the growth of aerobic bacteria and any differences in effect may actually be an indirect consequence of the response of the host produce tissue to physiological stress and/or the specific background microbial populations among different fresh-cut products. Luo et al. (2004) examined the microbial growth on packaged fresh-cut cilantro under different atmospheres and reported that there was a gradual increase in aerobic bacterial populations over time regardless of the package atmosphere. Beuchat and Brackett (1990) compared the growth of mesophilic bacteria on shredded lettuce packaged with 3 kPa O_2 and ambient air and found no significant differences between the treatments. Additional studies conducted by Beuchat and Brackett (Brackett, 1994) further concluded that growth of both mesophilic and psychrotrophic aerobic bacteria, yeasts, and molds on sliced tomatoes were essentially unaffected by MAP treatments. In contrast, Babic and Watada (1996) indicated a significant difference of aerobic bacterial growth at different package atmospheres on fresh-cut spinach stored at 5°C in 0.8% O_2 plus 10% CO_2. They reported a 1–2 log reduction in aerobic bacterial populations compared with spinach stored in air at the same temperature. Low O_2, as opposed to high CO_2, seemed to be the limiting factor on the growth of aerobic bacteria on spinach leaves at 5°C, but this relationship did not hold when the spinach was stored at 10°C (Babic and Watada, 1996). *Pseudomonas* sp. counts were lower in a 0 kPa O_2 atmosphere as opposed to a 21 kPa O_2 atmosphere, irrespective of the CO_2 level in the atmosphere. *Pseudomonas* sp. were found to be the predominant spoilage microorganism species in packages in a 21 kPa O_2 atmosphere, whereas *Enterobacter* sp. predominated at a 0 kPa O_2 atmosphere (Babic and Watada, 1996).

On the other hand, growth of anaerobic bacteria can be significantly affected by package atmospheres. Luo et al. (2004) noticed a significant growth in anaerobic bacteria on fresh-cut cilantro leaves packaged in low OTR film when the O_2 was depleted and the CO_2 level was elevated at the end of storage. The effect of MAP on lactic acid bacteria can vary depending on the type of produce packaged. Growth of lactic acid bacteria in response to the elevated CO_2 and decreased O_2 concentrations used in MAP can expedite the spoilage and off-odor development of produce sensitive to lactic acid bacteria, for example lettuce and carrots (Nguyen-The and Carlin, 1994; Ruiz et al., 2006; Luo, 2007a,b).

The growth of yeasts is not generally affected by the package atmosphere composition. However, the extent of yeast growth on different products seems to be extremely variable (Babic et al., 1992, 1996). While high yeast populations were noted in one study on packaged salads toward the end of the MAP storage period (Allende et al., 2006), other researchers have found that yeast populations remain at low levels (10^3–10^4 CFU/g) during an entire storage period in air or controlled atmospheres, at 5°C and 10°C (Babic and Watada, 1996). Fungal growth, on the other hand, can be inhibited by elevated level of CO_2 in the packages (Wells and Uota, 1970) and the population of molds in fresh-cut vegetables is often reported to be very low. The concentration of CO_2 commonly found in packaged salads is usually not considered to be fungicidal.

Atmospheres with O_2 levels higher than 70%, or superatmospheric O_2, have been shown to inhibit microbial growth and enzymatic discoloration and prevent anaerobic fermentation (Day 1996, 2000, 2001). However, the results on packaged vegetable salads are variable (Heimdal et al., 1995; Amanatidou et al., 1999; Day 2001; Allende et al., 2002). Lactic acid bacteria and enterobacteria are inhibited, yeast and *Aeromonas caviae* are stimulated, but psychrotrophic bacteria are unaffected. In general, exposure to high O_2 alone does not have a strong inhibition on microbial growth, while elevated CO_2 generally reduces microbial growth to some extent. The combination of superatmospheric O_2 and elevated CO_2 often exhibits a strong inhibition on microbial growth. As reported by Amanatidou et al. (2000), growth of enterobacteria on fresh-cut carrots was inhibited under 50 kPa O_2 and 30 kPa CO_2, but stimulated under 80 or 90 kPa O_2. With minimally processed vegetables, where CO_2 levels of around 20 kPa or above cannot be used because of physiological damage to the produce, the combined treatment of high O_2 and 10–20 kPa CO_2 may provide significant suppression of microbial growth. However, recent studies by Kader and Ben-Yehoshua (2000) showed that only O_2 atmospheres close to 100 kPa or lower pressures (40 kPa) in combination with CO_2 (15 kPa) are truly effective. These conditions may be difficult to achieve in industry since working with such high O_2 levels can be hazardous due to flammability issues. As with most MAP gases, superatmospheric O_2 has varied effects depending on the commodity, and further research is required in this area to elucidate the utility of this technique in the fresh-cut produce industry.

18.2.3 Quality and Shelf Life

MAP has been shown to increase shelf life of many fresh-cut products (Barriga et al., 1991; Bennik et al., 1996). This phenomenon may or may not be associated with spoilage microorganism growth on the cut surface of the product. Bennik et al. (1996) found that modified atmosphere conditions that were generally favorable for product quality maintenance also retarded growth of spoilage microorganisms at low storage temperatures. However, Barry-Ryan et al. (2000) found that an atmosphere of 3 kPa O_2 and 10 kPa CO_2 maintained acceptable visual quality of lettuce, without appreciably affecting microbial growth (Barriga et al., 1991).

Characteristic quality factors that can shorten shelf life are numerous, including dehydration, discoloration, microbial growth and decay, and off-odor development. While an appropriately developed MAP can assist in increasing shelf life by reducing enzymatic browning, respiration rate, moisture loss, and some microbial growth, it must be accompanied by appropriate storage temperature, minimal physiological damage, and other microbial reduction methods, i.e., produce wash, in order to obtain maximum shelf-life extension. A combination treatment including chlorine prewash, mild heat treatment, MAP, and 5°C storage was able to extend shelf-life significantly for fresh-cut grapes, while none of these processing steps by themselves was able to provide satisfactory quality retention (Kou et al., 2007). There are other examples of the success that can be attained with the use of combined treatment to enhance quality retention (Toivonen and Lu, 2007; Toivonen, 2008) and this trend to combined treatment approaches is likely to increase in the future.

Atmosphere composition may interact synergistically with other protective factors such as storage temperature. While storage of fresh-cut spinach at 5°C in 0.8 kPa O_2 and 10 kPa CO_2 atmospheres reduces aerobic bacterial population by 1–2 log CFU/g as compared with air, bacterial growth increases significantly at 10°C regardless of package atmospheres (Babic et al., 1996).

Although, under conditions that support the physiology of the host plant tissues, some reduction in microbial growth may be attributed to MAP, under conditions of temperature abuse or physiological deterioration the reduction due to MAP is overcome by enhancement of microbial growth on the compromised tissues. As a result of the delicate nature of living plant tissue, extension of shelf-life of fresh-cut produce is best achieved by controlling numerous factors. Shelf-life can be maximized by starting with physiologically healthy, fresh produce; controlling temperature and atmosphere conditions optimally at every stage to minimize microbial growth, dehydration, and senescence. Care must be given to minimize tissue damage during cutting and washing. Packaging film OTR should be selected to meet product respiratory requirements and initial gas mixture should be selected carefully taking into consideration of the unique characteristics of fresh-cut fruits and vegetables.

18.3 Elevated Modified Humidity Effects

While the application of MAP has been targeted to impacting the respiratory metabolism of whole and fresh-cut produce, a side effect (usually considered a side-benefit) has been the maintenance of high humidity around the produce. It has been stated that one of the greatest impacts MAP has in fresh-cut produce is in controlling water loss (Gorny, 1997). Nevertheless, there is only sparse information on the impact of humidity and its control in modified atmosphere package systems. There are some indications that more research in this area may provide new opportunities for MAP technology.

Humidity is generally considered a driving force for water loss, however, it is actually the water vapor deficit (WVPD) between the product and its surrounding atmosphere, which truly defines rate of water loss (van den Berg, 1987). The WVPD is not only affected by relative humidity, but is also influenced by the ambient air temperature and product temperature. The relationship between these factors is clearly shown in the equation which describes the calculation of the WVPD:

$$\text{WVPD} = \text{VP}_{\text{sat-prod}} - (\text{VP}_{\text{sat-air}} \times \text{RH}_{\text{air}}) \tag{18.2}$$

where
   $\text{VP}_{\text{sat-prod}}$ is the saturated vapor pressure at the temperature of the fruit or vegetable
   $\text{VP}_{\text{sat-air}}$ is the saturated vapor pressure at the temperature of the air surrounding the product
   RH_{air} is the relative humidity of the air surrounding the product

Equation 18.2 has several implications in the understanding of the interaction of product with its surrounding humidity in a package. Probably the most important issue relates to the product temperature. Proper postharvest precooling has been the single most cited issue for maintaining quality in fresh fruits and vegetables (Gillies and Toivonen, 1995). One of the reasons for this is exemplified in an experiment where broccoli was properly hydrocooled or not precooled and then placed into the same relative humidity conditions (Gilles and Toivonen, 1995). The result was the broccoli that was not precooled had a much higher calculated WVPD than did properly precooled broccoli. Resultant quality retention was reduced and weight loss was significantly increased by for the broccoli that was not precooled. This same principle holds true for any product that would be placed into a modified atmosphere package. Hence a caveat for using relative humidity to maintain quality is that the product must also be properly temperature managed. Attempts to manipulate humidity in a package will risk failure if good product precooling is not implemented in practice.

Another good example of the importance of temperature control is exemplified by work in packaged broccoli (Tano et al., 2007; Figure 18.5). Temperature fluctuations can significantly increase weight loss in modified atmosphere package produce. Similar increases in moisture loss with temperature fluctuations have been reported previously (Patel and Sastry, 1988; Tano et al., 1999). There are likely a number of factors involved in this increase, including the fact that temperature fluctuations can lead to significant fluctuations in vapor pressure deficit in a fruit or vegetable and potential condensation can occur when temperatures decline during a fluctuation cycle (Patel and Sastry, 1988; Patel et al., 1988; Tano et al., 2007).

Packaging materials can differ in their water transmission properties, and water transmission rates (H_2OTR) do not always parallel gas transmission rates (Table 18.3). A high H_2OTR film would result in a greater flux of water vapor from the package, hence creating a larger WVPD between the fruit or vegetable and the package atmosphere. However, it has been suggested that the humidity in solid film packaging is relatively uniform and hence would not vary largely between packages having large differences in gas transmission rates (Mir and Beaudry, 2004). This is verified by work showing that broccoli weight loss is not significantly different in packages made from films widely varying in both their gas and in H_2OTRs (Table 18.3). Hence, an attempt to control humidity in a solid film package requires the addition of a moisture absorbent.

There have been numerous reports in the literature in regards to modifying moisture in solid film packages using various absorbents, including desiccants, salts, sugar, alcohols, and clay materials (Shirazi and Cameron, 1992; Roy et al., 1995, 1996; Toivonen, 1997a,b; Song et al., 2001; Toivonen et al., 2002; Villaescusa and Gil, 2003; DeEll et al., 2006).

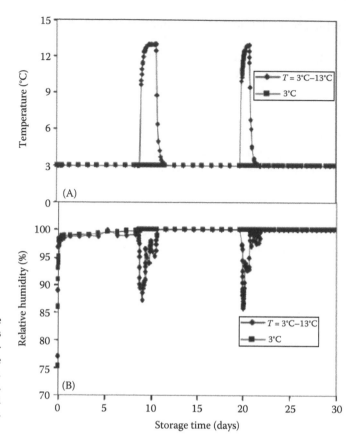

FIGURE 18.5

Changes in temperature (A) and relative humidity (B) levels inside MA packages of broccoli stored at constant temperature: (■) 3°C; under temperature fluctuating conditions: (♦) 3°C–13°C. (Reproduced from Tano, K., Oulé, M.K., Doyon, G., Lencki, R.W., and Arul. J., *Postharvest Biol. Technol.*, 46, 212, 2007. With permission.)

TABLE 18.3

Carbon Dioxide and Water Vapor Transmission Rates (CO_2TR and WVTR, respectively) and Their Ratio for Two Modified Atmosphere Packaging Films and Measured Steady-State CO_2 Levels within Packages Made with These Films Contained Broccoli Heads and the Resultant Weight Loss over 9 Days of Storage at 1°C

Relative Gas Transmission	CO_2TR (mL/m²/day)	Ratio[a] of CO_2TR	Steady-State CO_2 Concentration (kPa)	WVTR (g/m²/day)	Ratio[a] of WVTR	Weight Loss (%)
Fast	71424	2.93	10	13.61	3.67	0.92
Moderate	24389		5	3.71		0.94
Significance	**[b]	**	**	Ns		

Sources: Data extracted from Moyls, A.L., McKenzie, D.-L., Hocking, R.P., Toivonen, P.M.A., Delaquis, P., Girard, B., and Mazza, G., *Trans. ASAE.*, 41, 1441, 1998; DeEll, J.R. and Toivonen, P.M.A., *HortScience*, 35, 256, 2000.

[a] The ratio of transmission rates: the rate for the Fast film divided by the rate for the Moderate film.

[b] Significant at p < 0.01.

Another approach to controlling relative humidity inside a package is to modify the film with microperforations. This approach results in a significant modification of the water transmission characteristics of film (Aharoni and Richardson, 1997) and also significant modification of O_2 and CO_2 transmission (Lougheed, 1992). The mode of action of the microperforated film relates to the effect of the microperforations allowing more rapid movement of water vapor into the atmosphere surrounding the package, thereby lowering the internal relative humidity and consequently a slight increase in water loss from product can be measured (Lougheed, 1992).

18.3.1 Physiology

Once a fruit or vegetable is harvested, the main determinants of its water status are the relative humidity surrounding the product and the time since it has been harvested. Washing and hydrocooling steps can "recharge" the product, but that effect is only transient (Shibairo et al., 1998). Therefore a major impact for relative humidity is the determination of water status of the fruit or vegetable tissue. Water status effects on physiology are well studied and so the implications of not maintaining high water status has been discussed in detail elsewhere (Shamaila, 2005).

Water is essential for all metabolic activity in a living cell and hence decline in levels will lead to some form of metabolic impairment or even permanent injury if the level of water loss is great enough (Shamaila, 2005). The level of water loss that leads to significant changes has not been delineated in any of the literature with regard to postharvest handling of fruits and vegetables. It has been demonstrated that peroxidase activity in diced, modified atmosphere packaged onions was sensitive to minor levels of water loss, with a range of approximately 0.2%–0.8% in weight loss coinciding with lowest levels of activity after 9 days storage (Figure 18.6). The initial level of peroxidase in the diced onions at cutting was 1.9 units/mg protein, therefore the peroxidase activity rose significantly over the 9 days in the package, with those packages experiencing moderate water loss showing the lowest levels. These findings suggest that some water loss can be beneficial to modulating wound associated peroxidase activity. Peroxidase activity is important since it is implicated in the development of off-flavors in improperly preserved fruits and vegetables (Burnette, 1977).

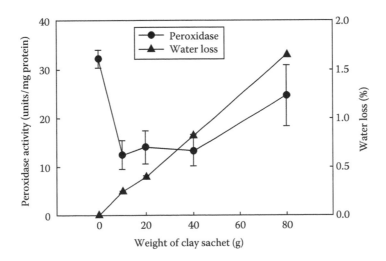

FIGURE 18.6
Changes in peroxidase activity in 'Yellow Colossal' packaged, diced onions associated with differences in water loss within the package due presence of differing weights of sachets containing an adsorbent clay. Data represent means of three replicates and error bars, where not obscured by data points, represent the standard error of those means. A unit of peroxidase activity is defined as a 0.01 unit change in absorbance at 470 nm/min in a mixture containing guaiacol as the substrate. Water loss of the diced onions within a package was inferred by measuring the weight increase of the clay sachet and normalizing this weight against the original total weight of diced onion in a package. (Toivonen, unpublished data.)

TABLE 18.4

Effect of Relative Humidity and Seal-Packaging in High-Density Polyethylene Film on Weight Loss, Firmness, Membrane Integrity, Water Saturation Deficit of Green Bell Pepper Fruit Kept for 4 Weeks at 17°C

	Treatment			
Parameter Examined	WSA[a]	Sealed in HDPE	Nonsealed	Sealed in HDPE + CaCl$_2$ [b]
Weight loss, %	1.76b[c]	1.2a	15.9d	10.5c
Firmness, mm deformation	4.6b	3.3a	12.5d	9.9c
Amino acid leakage, %	14.4ab	11.3a	21.5b	17.3ab
Water saturation deficit, %	11.5a	12.7a	24.4b	20.9b

Source: Data extracted from Ben-Yehoshua, S., Shapiro, B., Chen, Z.E., and Lurie, S., *Plant Physiol.*, 73, 87, 1983. With permission.

[a] WSA refers to a treatment where unpackaged peppers were placed into controlled chamber where relative humidity was maintained at 85% throughout the 4 weeks of holding at 17°C.

[b] Each fruit sealed in a plastic bag containing 5 g of CaCl$_2$ crystals.

[c] Mean separation by Duncan's multiple range test, 1% level.

Ben-Yehoshua et al. (1983) demonstrated that protection against water loss, afforded by plastic film packaging, had a direct effect on the softening process in bell peppers. They found that firmness retention in MA-packaged peppers and those held in a high humidity chamber was significantly better than unpackaged fruit held in normal ambient humidity and those packaged with a desiccant (Table 18.4). The firmness retention in the MA-packaged peppers was partially attributable to better retention of less soluble and insoluble pectin fractions over the 4 weeks storage at 17°C. This retention of less soluble and insoluble pectin fractions was linked to lower levels of polygalacturonase (PG) activity in peppers held in MA compared with those held unpackaged in ambient humidity conditions (Ben-Yehoshua et al., 1983). The direct relationship between water loss and the PG activity was not established, but it may have been associated with the disruption of membrane integrity in the unpackaged peppers as indicated by elevated amino acid leakage values (Table 18.4).

It has been suggested that small amounts of water loss will prevent accumulation of detrimental wound- or ripening-related volatiles (Toivonen, 1997a,b; Toivonen et al., 2002) however, direct proof of this hypothesis is lacking.

18.3.2 Microbiology

Research on the use of humidity control strategies has focused mainly for the control of spoilage microorganisms. It is a well-cited postulate that too high of a humidity leads to condensation within a package and this results in growth of spoilage microorganisms and hence slight lowering of humidity will relieve this problem (Shirazi and Cameron, 1992; Roy et al., 1995; Roy et al., 1996; Fallik et al., 2002; Rodov et al., 2000, 2002; Villaescusa and Gil, 2003). This postulate has been demonstrated to be true in numerous experiments with a large variety of fruits and vegetables, using either absorbents or microperforated films.

Surface browning and/or yellowing and bacterial blotch are symptoms of a *Pseudomonas tolasii* infection in mushrooms and its development is encouraged in sealed modified atmosphere packages that have no humidity control (Roy et al., 1995). Addition of sufficient sorbitol to result in approximately 15% moisture loss at 9 days of 12°C storage provided a significant reduction in discoloration and bacterial surface growth in *Agaricus*

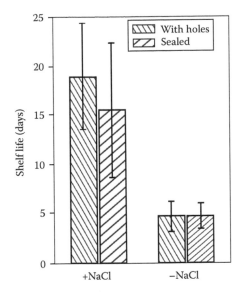

FIGURE 18.7
Effects of controlled humidity on storage life of red-ripe 'Summer Flavor 6000' tomato fruit in modified atmosphere packages. Each bar represents mean shelf life for 18 fruit $\pm$ SD. Dry NaCl (10 g/225 g fruit) packed in spun-bonded polyethylene was used to control in-package relative humidity at $\sim$80%. Average O_2 and CO_2 were $\sim$3 kPa and 6 kPa in sealed packages and 17 kPa and 4 kPa in packages with holes. Note: End of shelf life was determined by the growth of *Fusarium* sp. and *Alternaria* sp. on the stem scar of the fruits. (Redrawn from Shirazi, A. and Cameron. A.C., *HortScience*, 27, 336, 1992.)

bisporus mushrooms (Roy et al., 1995). However, if greater amounts of sorbitol were added, greater water loss occurred and quality suffered due to the direct effect of dehydration. The appropriate amount of sorbitol (in regards to controlling bacterial discoloration) resulted in an internal package relative humidity of approximately 85% (Roy et al., 1996).

The fungal decay organisms *Alternaria* sp. and *Fusarium* sp. infecting the stem scar tissue of whole tomatoes can limit shelf life when they are packaged in modified atmosphere packages, which are either solid or microperforated (Shirazi and Cameron, 1992). Application of a desiccant such as sodium chloride will reduce package humidity from near saturation down to approximately 80%, significantly reducing decay-associated loss of shelf life in either solid or microperforated film packages (Figure 18.7). The experience with snap beans has been similar. Fallik et al. (2002) found that maintaining a relative humidity of approximately 90%, using a specialized microperforated film lead to significant reductions in fungal decay of pods when stored at 5°C. In sweet cherries, lowering of relative humidity to approximately 95%, using the same microperforated film, lowered fungal decay incidence from 45% in conventional packages down to 7% after 2 weeks storage at 0°C (Lurie and Aharoni, 1997). Rodov et al. (2002) found that a similar package could essential control rots caused by primarily *Aspergillus* sp. and *Penicillium* sp. in Chanterais-type melons stored at 7°C for 12 days. These results suggest that even moderate reductions in relative humidity can significantly reduce the growth of fungal decay organisms in fruit and vegetable products.

18.3.3 Shelf Life

While the main focus of this chapter is on fresh-cut and modified humidity systems in modified atmosphere packages, there is a significant review on the effects of humidity and water loss existing (Shamaila, 2005). The reader is asked to refer to those reviews for background effects of water loss on quality retention.

The effects on quality and shelf life are the best documented aspect (as opposed to physiological and microbiological aspects) for modified humidity technologies. The most significant effort, on this aspect, has been expended to develop the use of modified humidity systems to preserve the quality of mushrooms (Roy et al., 1995, 1996; Villaescusa and Gil, 2003). That work has focused on the use of various moisture absorbers and applied a varying product to absorber weight ratios. However, it can be difficult to work with such salts. A good example is that provided by the work of Ben-Yehoshua et al. (1983). They applied a treatment where humidity was maintained at approximately 85% in a controlled humidity chamber and a treatment which consisted of a $CaCl_2$ sachet to maintain relative humidity at about 85%. Less than 2% weight loss was experienced by peppers in the controlled humidity treatment, whereas over 10% weight loss was experienced in the sealed packaging containing 5 g $CaCl_2$. Quite clearly, while the $CaCl_2$ was maintaining relative humidity, it was doing it at the expense of the water available from the fruit. This problem is supported by what is known about how saturated salts and other hygroscopic materials maintain relative humidity in experimental systems (Shirazi and Cameron, 1992): if too much adsorbent is used, significant weight loss causing quality loss will occur (Song et al., 2001).

It is clear that the level of water loss experienced is the most important dictate of the resultant quality of fresh-cut products within a humidity-controlled package. There are a few examples which show this principle quite clearly. While much of the data indicates that there is a maximum water loss after which quality declines, there is some data that suggests that no water loss may also be deleterious, especially in regard to flavor change (Figure 18.8). It has been suggested that water loss may enable the removal of wound-induced compounds from tissues, minimizing changes caused by the cutting process in fresh-cut products (Toivonen, 1997a). Hence, in the future when the more subtle issues of flavor change become more important to the fresh-cut industry there may be reason to reexamine the benefits of mild induced water loss and its affect on quality. Meanwhile the greatest impact of humidity control in modified atmosphere-packaged product is the reduction of several postharvest defects including, decay or mold growth in peppers, lemons, and mushrooms (Ben-Yehoshua et al., 1983; Shirazi and Cameron, 1992; Roy et al., 1995, 1996), chilling injury alleviation in mango (Pesis et al., 2000), and prevention of the development of rind disorders in citrus (Porat et al., 2004). The result of humidity control is generally an extended shelf life and quality.

18.4 Challenges Facing the Industry and Future Research Directions

The fresh-cut produce industry faces many challenges, primarily due to the fact that it is a rapidly expanding industry with many new products introduced every year (Brody, 2005; Forney, 2007). As mentioned in the first part of this chapter there is ongoing research into new and improved packaging materials and formats, which may allow the industry to continue to expand, if these new technologies can overcome the inherent limitations of current packaging. New approaches to modifying package atmospheres, such as superatmospheric oxygen combined with high carbon dioxide, may provide new opportunities especially if they can provide a flavor and/or nutritional advantage over existing packaging systems. Certainly greater exploration of combined treatments should be encouraged as the current data clearly show synergistic effects on quality of fresh-cut produce. However, the fundamental shift in research will likely relate to greater focus on flavor, nutritional, and functional quality of fresh-cut products and how they can be better managed with MAP technologies. The reason for this shift is that the category title (i.e., fresh-cut) brings with it an implicit expectation that the product be in a condition which approximates the initial quality at the time of cutting. While at this time visual quality retention is feasible, fresh taste and nutritional and functional qualities of the fresh-cut

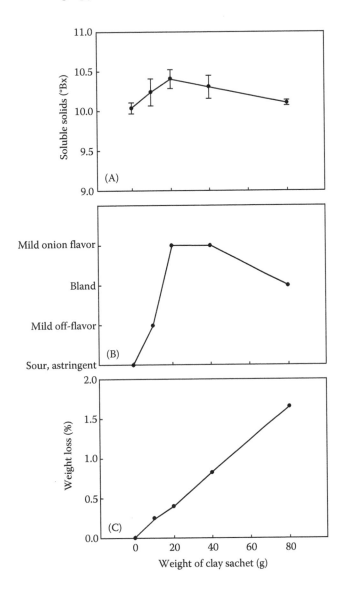

FIGURE 18.8
Effect varying the weight of clay absorbent on soluble solids content (A), flavor (B), and weight loss (C) of modified atmosphere packaged, diced onions (cultivar 'Yellow Colossal') after 9 days of storage at 5°C. Data represent means of three replicates and error bars, where not obscured by data points, represent the standard error of those means. Water loss of the diced onions within a package was inferred by measuring the weight increase of the clay sachet and normalizing this weight against the original total weight of diced onion in a package. (Toivonen, unpublished data.)

product may suffer. There is limited data in regard to the nutritional/functional quality of fresh-cut products suggesting that this does not decline to a large degree over a short shelf life for some products (Gil et al., 2006). However, questions remain as to whether similar results would be obtained under all MAP conditions and under more abusive (realistic) temperature handling conditions. Flavor retention is probably the biggest challenge and perhaps the most important emerging issue for fresh-cut fruit products (Beaulieu and Baldwin, 2002).

References

Aharoni, Y. and D.G. Richardson. 1997. New, higher water permeable films for modified atmosphere packaging of fruits and vegetables. Prolonged MAP storage of sweet corn. *Proc. 7th Int. Controlled Atmosphere Res. Conf.* 4, pp. 73–77.

Al-Ati, T. and J.H. Hotchkiss. 2002. Application of packaging and modified atmosphere to fresh-cut fruits and vegetables, In: O. Lamikanra (Ed.). *Fresh-Cut Fruits and Vegetables: Science, Technology, and Market.* CRC Press, Boca Raton, FL, pp. 305–338 (chap. 10).

Allende, A., L. Jacxsens, F. Devlighere, F. Debevere, and F. Artes. 2002. Effect of superatmospheric oxygen packaging on sensorial quality, spoilage, and *Listeria monocytogenes* and *Aeromonas caviae* growth in fresh processed mixed salads. *J. Food Prot.* 65: 1565–1573.

Allende, A., Y. Luo, J.L. McEvoy, F. Artés, and C.Y. Wang. 2004. Microbial and quality changes in fresh-cut baby spinach stored under MAP and super atmospheric oxygen conditions. *Postharvest Biol. Technol.* 33: 51–59.

Allende, A., J.L. McEvoy, Y. Luo, and C.Y. Wang. 2006. Effectiveness of two-sided UV-C treatments in inhibiting natural microflora and extending the shelf-life of minimally processed 'Red Oak Leaf' lettuce. *J. Food Microbiol.* 23: 241–249.

Amanatidou, A., E.J. Smid, and L.G.M. Gorris. 1999. Effect of elevated oxygen and carbon dioxide on the surface growth of vegetable-associated microorganisms. *J. Appl. Microbiol.* 86: 429–438.

Amanatidou, A., R.A. Slump, L.G.M. Gorris, and E.J. Smid. 2000. High oxygen and high carbon dioxide modified atmospheres for shelf-life extension of minimally processed carrots. *J. Food Sci.* 65: 61–66.

Babic, I., G. Hilbert, C. Nguyen-The, and J. Guiraud. 1992. The yeast flora of stored ready-to-use carrots and their role in spoilage. *Intl. J. Food Sci. Technol.* 27: 473–484.

Babic, I., S. Roy, A.E. Watada, and W.P. Wergin. 1996. Changes in microbial populations on fresh cut spinach. *Intl. J. Food Microbiol.* 31: 107–119.

Babic, I. and A.E. Watada. 1996. Microbial populations of fresh-cut spinach leaves affected by controlled atmospheres. *Postharvest Biol. Technol.* 9: 187–193.

Barriga, M.I., G. Trachy, C. Willemot, and R.E. Simard. 1991. Microbial changes in shredded iceberg lettuce stored under controlled atmospheres. *J. Food Sci.* 56: 1586–1588, 1599.

Barry-Ryan, C., J.M. Pacussi, and D. O'Beirne. 2000. Quality of shredded carrots as affected by packaging film and storage temperature. *J. Food Sci.* 65: 726–730.

Beaulieu, J.C. and E.A. Baldwin. 2002. Flavor and aroma of fresh-cut fruits and vegetables. In: O. Lamikanra (Ed.). *Fresh-Cut Fruits and Vegetables: Science, Technology, and Market.* CRC Press, Boca Raton, FL, pp. 391–425.

Bell, L. 1996. Sealed package containing respiring perishable produce. U.S. Patent # 430,123.

Bennik, M.H.J., H.W. Peppelenbos, C. Nguyen-The, F. Carlin, E.J. Smid, and L.G.M. Gorris. 1996. Microbiology of minimally processed, modified-atmosphere packaged chicory endive. *Postharvest Biol. Technol.* 9: 209–221.

Ben-Yehoshua, S., B. Shapiro, Z.E. Chen, and S. Lurie. 1983. Mode of action of plastic film in extending life of lemon and bell pepper fruits by alleviation of water stress. *Plant Physiol.* 73: 87–93.

Beuchat, L.R. and R.E. Brackett. 1990. Survival and growth of *Listeria monocytogenes* on lettuce as influenced by shredding, chlorine treatment, modified atmosphere packaging and temperature. *J. Food Sci.* 55: 755–758, 870.

Brackett, R.E. 1994. Microbiological spoilage and pathogens in minimally processed refrigerated fruits and vegetables. In: R.C. Wiley (Ed.). *Minimally Processed Refrigerated Fruits and Vegetables.* Chapman & Hall, New York, pp. 269–312.

Brody, A.L. 2005. What's fresh about fresh-cut? *Food Technol.* 56: 124–125.

Burnette, F.S. 1977. Peroxidase and its relationship to food flavor and quality: A review. *J. Food Sci.* 42: 1–6.

Day, B. 1996. High oxygen modified atmosphere packaging for fresh prepared produce. *Postharvest News Inform.* 7: 31–34.

Day, B. 2000. Novel MAP for freshly prepared fruit and vegetable products. *Postharvest News Inform.* 11: 27–31.

Day, B. 2001. *Fresh Prepared Produce: GMP for High Oxygen MAP and Nonsulphite Dipping. Guideline No. 31*, Campden and Chorleywood Food Research Association Group, Chipping Campden, Gloucestershire, U.K., 76 p.

DeEll, J.R. and P.M.A. Toivonen. 2000. Chlorophyll fluorescence as a non-destructive indicator of broccoli quality during storage in modified atmosphere packaging. *HortScience.* 35: 256–259.

DeEll, J.R., P.M.A. Toivonen, F. Cornut, C. Roger, and C. Vigneault. 2006. Addition of sorbitol with $KMnO_4$ improves broccoli quality retention in modified atmosphere packages. *J. Food Qual.* 29: 65–75.

Fallik, E., D. Chalupowicz, Z. Aharon, and N. Aharoni. 2002. Modified atmosphere in a water vapour-permeable film maintains snap bean quality after harvest. *Folia Hort.* 14: 85–94.

Forney, C.F. 2007. New innovations in the packaging of fresh-cut produce. *Acta Hort.* 746: 53–60.

Gil, M.I., E. Aguayo, and A.A. Kader. 2006. Quality changes and nutrient retention in fresh-cut versus whole fruits during storage. *J. Agric. Food. Chem.* 54: 4284–4296.

Gillies, S.L. and P.M.A. Toivonen. 1995. Cooling method influences the postharvest quality of broccoli. *HortScience.* 30: 313–315.

Gorny, J.R. 1997. A summary of CA and MA requirements and recommendations for fresh-cut (minimally processed) fruits and vegetables. *Proc. 7th Intl. Controlled Atmosphere Res. Conf.* 5, pp. 30–66.

Heimdal, H., B.F. Kuhn, L. Poll, and L.M. Larsen. 1995. Biochemical changes and sensory quality of shredded and MA-packaged iceberg lettuce. *J. Food Sci.* 60: 1265–1268.

Hodges, D.M. and P.M.A Toivonen. 2007. Quality of fresh-cut fruits and vegetables as affected by exposure to abiotic stress. *Postharvest Biol. Technol.* doi:10.1016/j.postharvbio.2007.10.016

Jacxsens, L., F. Develieghere, and J. Debevere. 2002. Temperature dependence of shelf-life as affected by microbial proliferation and sensory quality of equilibrium modified atmosphere packaged fresh produce. *Postharvest Biol. Technol.* 26: 59–73.

Kader, A.A. and S. Ben-Yehoshua. 2000. Effects of superatmospheric oxygen levels on postharvest physiology and quality of fresh fruits and vegetables. *Postharvest Biol. Technol.* 20: 1–13.

Kim, J., Y. Luo, and K.C. Gross. 2003. Effect of packaging film on the quality of fresh-cut salad savoy. *Postharvest Biol. Technol.* 32: 99–107.

Kim, J., Y. Luo, R.A. Saftner, and K.C. Gross. 2005a. Delayed modified atmosphere packaging of fresh-cut romaine lettuce: Effects on quality maintenance and shelf-life. *J. Am. Soc. Hort. Sci.* 130: 116–123.

Kim, J., Y. Luo, R.A. Saftner, Y. Tao, and K.C. Gross. 2005b. Effect of initial oxygen concentration and film oxygen transmission rate on the quality of fresh-cut romaine lettuce. *J. Sci. Food Agric.* 85: 1622–1630.

Kou, L., Y. Luo, D. Wu, and X. Liu. 2007. Effects of mild heat treatment on microbial populations and product quality of packaged fresh-cut table grapes. *J. Food Sci.* 72: S567–S573.

Landec. 2007. *Technology.* Landec Corporation, Menlo Park, CA. http://www.landec.com/technology. html

Lougheed, E.C. 1992. Microperforated plastic packages for fruits and vegetables. Final Report to the Agricultural Research Institute of Ontario, Project #AG2041—Agriculture and Food Research Fund.

Luo, Y. 2007a. Wash operation affect water quality and packaged fresh-cut romaine lettuce quality and microbial growth. *HortScience.* 42: 1413–1419.

Luo, Y. 2007b. Challenges facing the industry and scientific community in maintaining quality and safety of fresh-cut produce. *Acta Hort.* 746: 131–138.

Luo, Y., J.L. McEvoy, M.R. Wachtel, J.G. Kim, and Y. Huang. 2004. Package film oxygen transmission rate affects postharvest biology and quality of fresh-cut cilantro leaves. *HortScience.* 39(3): 567–570.

Lurie, S. and N. Aharoni. 1997. Modified atmosphere storage of cherries. *Proc. 7th Int. Controlled Atmosphere Res. Conf.* 3, pp. 149–152.

Mir, N. and R.M. Beaudry. 2004. Modified atmosphere packaging. In: K.C. Gross, C.Y. Wang, and M.E. Saltveit (Eds.). *The Commercial Storage of Fruits, Vegetables, and Florist and Nursery Stocks,*

USDA Handbook 66. (accessed on November 6, 2007). http://www.ba.ars.usda.gov/hb66/015map.pdf

Moyls, A.L., D.-L. McKenzie, R.P. Hocking, P.M.A Toivonen, P. Delaquis, B. Girard, and G. Mazza. 1998. Variability in O_2, CO_2 and H_2O transmission rates among commercial polyethylene films for modified atmosphere packaging. *Trans. ASAE.* 41: 1441–1446.

Nguyen-The, C. and F. Carlin. 1994. The microbiology of minimally processed fresh fruits and vegetables. *Crit. Rev. Food Sci. Nutr.* 34: 371–401.

Nguyen-The, C. and J.P. Prunier. 1989. Involvement of pseudomonads in deterioration of "ready-to-use" salads. *Intl. J. Food Sci. Technol.* 24: 47–58.

Ozdemir, M. and J.D. Floros. 2004. Active food packaging technologies. *Crit. Rev. Food Sci. Nutr.* 44: 185–193.

Patel, P.N. and S.K. Sastry. 1988. Effects of temperature fluctuation on transpiration of selected perishables: Mathematical models and experimental studies. *ASHRAE Trans.* 94: 1588–1601.

Patel, P.N., T.K. Pai, and S.K. Sastry. 1988. Effects of temperature, relative humidity and storage time on the transpiration coefficients of selected perishables. *ASHRAE Trans.* 94: 1563–1587.

Pesis, E., D. Aharoni, Z. Aharon, R. Ben-Arie, N. Aharoni, and Y. Fuchs. 2000. Modified atmosphere and modified humidity packaging alleviates chilling injury symptoms in mango fruit. *Postharvest Biol. Technol.* 19: 93–101.

Porat, R., B. Weiss, L. Cohen, A. Daus, and N. Aharoni. 2004. Reduction of postharvest rind disorders in citrus fruits by modified atmosphere packaging. *Postharvest Biol. Technol.* 33: 35–43.

Rodov, V., A. Copel, N. Aharoni, Y. Aharoni, A. Wiseblum, B. Horev, and Y. Vinokur. 2000. Nested modified-atmosphere packaging maintain quality of trimmed sweet corn during cold storage and the shelf life period. *Postharvest Biol. Technol.* 18: 259–266.

Rodov, V., B. Horev, Y. Vinokur, A. Copel, Y. Aharoni, and N. Aharoni. 2002. Modified-atmosphere packaging improves keeping quality of Charentais-type melons. *HortScience.* 37: 950–953.

Roy, S., R.C. Anantheswaran, and R.B. Beelman. 1995. Sorbitol increases shelf life of fresh mushrooms stored in conventional packages. *J. Food Sci.* 60: 1254–1259.

Roy, S., R.C. Anantheswaran, and R.B. Beelman. 1996. Modified atmosphere and modified humidity packaging of fresh mushrooms packaging. *J. Food Sci.* 61: 391–397.

Ruiz-Cruz, S., Y. Luo, R.J. Gonzalez, Y. Tao, and G. Gonzalez. 2006. Effect of acidified sodium chlorite applications on microbial growth and the quality of shredded carrots. *J. Sci. Food Agric.* 86: 1887–1893.

Schlimme D.V. and M.L. Rooney. 1994. Packaging of minimally processed fruits and vegetables. In: R.C. Wiley (Ed.). *Minimally Processed Refrigerated Fruits and Vegetables.* Chapman & Hall, New York, pp. 156–157.

Shamaila, M. 2005. Water and its relation to fresh produce. In: O. Lamikanra, S. Imam, and D. Ukuku (Eds.). *Produce Degradation. Pathways and Prevention.* Taylor & Francis, Boca Raton, FL, pp. 267–291.

Shanklin, A.P. and E.R. Sánchez. 2005. Regulatory Report: FDA's Food Contact Substance Notification Program. October/November 2005, Reprinted from *Food Safety Magazine.* http://www.cfsan.fda.gov/~dms/fcnrpt.html#authors

Shibairo, S.I., M.K. Upadhyaya, and P.M.A. Toivonen. 1998. Replacement of postharvest moisture loss by recharging and its effect on subsequent moisture loss during short-term storage of carrots. *J. Am. Soc. Hort. Sci.* 123: 141–145.

Shirazi, A. and A.C. Cameron. 1992. Controlling relative humidity in modified atmosphere packages of tomato fruit. *HortScience.* 27: 336–339.

Smyth, A., J. Song, and A. Cameron. 1998. Modified atmosphere package of packaged cut iceberg lettuce: Effect of temperature and O_2 partial pressure on respiration and quality. *J. Agric. Food Chem.* 46: 4556–4562.

Song, Y., D.S. Lee, and K.L. Yam. 2001. Predicting relative humidity in modified atmosphere packaging system containing blueberry and moisture absorbent. *J. Food Process. Preserv.* 25: 49–70.

Tano, K., J. Arul, G. Doyon, and F. Castaigne. 1999. Atmospheric composition and quality of fresh mushrooms in modified atmosphere packages as affected by storage temperature abuse. *J. Food Sci.* 64: 1073–1077.

Tano, K., M.K. Oulé, G. Doyon, R.W. Lencki, and J. Arul. 2007. Comparative evaluation of the effect of storage temperature fluctuation on modified atmosphere packages of selected fruit and vegetables. *Postharvest Biol. Technol.* 46: 212–221.

Toivonen, P.M.A. 1997a. Non-ethylene, non-respiratory volatiles in harvested fruits and vegetables: Their occurrence, biological activity and control. *Postharvest Biol. Technol.* 12: 109–125.

Toivonen, P.M.A. 1997b. Quality changes in packaged, diced onions (*Allium cepa* L.) containing two different absorbent materials. *Proc. 7th Int. Controlled Atmosphere Res. Conf.* 5, pp. 1–6.

Toivonen, P.M.A. 2008. Application of 1-MCP in fresh-cut/minimal processing systems. *HortScience.* 43: 102–105.

Toivonen, P.M.A. and J.R. DeEll. 2002. Physiology of fresh-cut fruits and vegetables. In: O. Lamikanra (Ed.). *Fresh-Cut Fruits and Vegetables: Science, Technology, and Market.* CRC Press, Boca Raton, FL, pp. 91–123.

Toivonen, P.M.A., C. Kempler, and S. Stan. 2002. The use of natural clay adsorbent improves quality retention in three cultivars of raspberries stored in modified atmosphere packages. *J. Food Qual.* 25: 385–393.

Toivonen, P.M.A. and C. Lu. 2007. An integrated technology including 1-MCP to ensure quality retention and control of microbiology in fresh and fresh-cut fruit products at non-ideal storage temperatures. *Acta Hort.* 746: 223–229.

Toivonen, P.M.A. and M. Sweeney. 1998. Differences in chlorophyll loss at 13°C for two broccoli (*Brassica oleracea* L.) cultivars associated with antioxidant enzyme activities. *J. Agric. Food Chem.* 46: 20–24.

van den Berg, L. 1987. Water vapour pressure. In: J. Weichmann (Ed.). *Postharvest Physiology of Vegetables*, Marcel Dekker, Inc., New York, pp. 203–230.

Varoquaux, P. and I.S. Ozdemir. 2005. Packaging and produce degradation, In: O. Lamikanra, S. Imam, and D. Ukuku (Eds.). *Produce Degradation: Pathways and Prevention.* Taylor & Francis Group, Boca Raton, FL, pp. 117–153 (chap. 5).

Villaescusa, R. and M.I. Gil. 2003. Quality improvement of *Pleurotus* mushrooms by modified atmosphere packaging and moisture absorbers. *Postharvest Biol. Technol.* 28: 169–179.

Wang, H., H. Feng, and Y. Luo. 2004. Microbial reduction and storage quality of fresh-cut cilantro washed with acidic electrolyzed water and aqueous ozone. *Food Res. Int.* 37: 949–956.

Watada, A.E., H. Izumi, Y. Luo, and V. Rodov. 2005. Fresh-cut produce, In: S. Ben-Yehoshua (Ed.). *Environmentally Friendly Technologies for Agricultural Produce Quality.* CRC Press, Boca Raton, FL, pp. 149–203 (chap. 7).

Wells, J.M. and M. Uota. 1970. Germination and growth of five fungi in low-oxygen and high carbon dioxide atmospheres. *Phytopathology.* 60: 50–53.

Zagory, D. 1998. *An Update on Modified Atmosphere Packaging of Fresh Produce.* Davis Fresh Technologies, Davis, CA. http://www.davisfreshtech.com/articles_map.html (accessed in April 1998).

Zagory, D. 1999. Effects of post-processing handling and packaging on microbial populations. *Postharvest Biol. Technol.* 15: 313–321.

19

Ornamentals and Cut Flowers

Andrew J. Macnish, Michael S. Reid, and Daryl C. Joyce

CONTENTS

19.1 Introduction

Ornamental plants and their detached organs (e.g., bulbs, cut flowers) are among the most highly perishable horticultural commodities traded internationally (Reid, 2002). High rates of respiration and transpiration, sensitivity to ethylene, and/or low natural resistance to disease-causing organisms (e.g., *Botrytis cinerea*) often limit the postharvest life of ornamentals (Elad, 1988; Reid, 2002). Extending the time that these products can be stored would provide an opportunity to use less expensive surface (e.g., truck, boat) transport than the current costly reliance on airplanes to supply international markets. Developing appropriate methods of storing ornamentals, particularly cut flowers, could also facilitate product accumulation and distribution for more profitable marketing on special days (e.g., Valentine's Day, Mother's Day) when demand for these commodities often exceeds supply.

Storing ornamental plants at low but nonchilling and nonfreezing temperatures is an effective method for extending postharvest life (Halevy and Mayak, 1981). Low

TABLE 19.1

Published Studies that Have Evaluated Responses of Ornamental Plants and Their Detached Organs to CA and MA Treatments

Plant Species	References
Cut flowers and greens	
Anigozanthos rufus	Seaton and Joyce (1993)
Anthurium andraeanum	Akamine and Goo (1981)
Aralia japonica	Philosoph-Hadas et al. (2007)
Asparagus virgatus	Philosoph-Hadas et al. (2007)
Chamelaucium uncinatum	Seaton and Joyce (1993), Shelton et al. (1996)
Delphinium sp.	Shelton et al. (1996)
Dendrathema grandiflorum	Shelton et al. (1996), Yamashita et al. (1999)
Dianthus caryophyllus	Longley (1933), Smith and Parker (1966), Smith et al. (1966), Hanan (1967), Uota (1969), Mayak and Dilley (1976), Menguc et al. (1993), Irving and Honnor (1994), Shelton et al. (1996), Johnston et al. (1998), Zeltzer et al. (2001)
Gerbera jamesonii	De Pascale et al. (2005)
Gladiolus 'Adi'	Meir et al. (1995b)
Gypsophila elegans (preserved)	Sauer and Shelton (2002), Zeltzer et al. (2001)
Hypericum sp.	Zeltzer et al. (2001)
Iris sp.	Shelton et al. (1996)
Leucadendron 'Safari Sunset'	Philosoph-Hadas et al. (2007)
Lilium sp.	De Pascale et al. (2005)
Lilium asiaticum	Shelton et al. (1996)
Limonium sinuatum (preserved)	Sauer and Shelton (2002)
Limonium sinuatum	Shelton et al. (1996)
Lisianthus grandiflorum	Akbudak et al. (2005)
Matthiola incana	Shelton et al. (1996)
Narcissus pseudonarcissus	Parsons et al. (1967)
Pittosporum sp.	Zeltzer et al. (2001)
Pittosporum tobira variegate	Philosoph-Hadas et al. (2007)
Protea sp.	Shelton et al. (1996)
Rosa hybrida	Thornton (1930), Longley (1933), Hauge et al. (1947), Shelton et al. (1996), Zeltzer et al. (2001), De Pascale et al. (2005)
Ruscus hypoglossum	Philosoph-Hadas et al. (2007)
Solidago sp.	Zeltzer et al. (2001)
Potted plants	
Begonia semperflorens-cultorum	Held et al. (2001)
Chrysanthemum × morifolium	Konjoian et al. (1983), Held et al. (2001)
Impatiens wallerana	Held et al. (2001)
Pelargonium × hortorum	Held et al. (2001)
Vegetative cuttings	
Begonia × hiemalis	Erstad and Gislerod (1994)
Begonia × tuberhybrida	Erstad and Gislerod (1994)
Euphorbia pulcherrima	Erstad and Gislerod (1994)
Pelargonium × hortorum	Erstad and Gislerod (1994)
Pelargonium × peltatum	Erstad and Gislerod (1994)
Picea glauca	Behrens (1986)
Bulbs and tubers	
Begonia × tuberhybrida tubers	Prince and Cunningham (1987)
Iris sp. bulbs	Stuart et al. (1970)
Lilium asiaticum bulbs	Legnani et al. (2004a), Legnani et al. (2004b)
Lilium longiflorum bulbs	Thornton and Imle (1941), Stuart et al. (1970), Prince and Cunningham (1991)
Tulipa gesneriana bulbs	Prince et al. (1981), Prince et al. (1986)

temperatures reduce rates of respiration, ethylene production and sensitivity, and metabolic reactions associated with tissue senescence (Barden and Hanan, 1972; Maxie et al., 1973). Maintaining ornamental commodities at constant optimal storage temperatures also prevents condensation from developing on tissues and thereby reduces the related development of fungal pathogens (Reid, 2002). A range of chemical treatments including antiethylene agents (e.g., silver thiosulfate [STS]), hormones (e.g., cytokinins) and fungicides (e.g., iprodione) are also used to maximize the postharvest life of ornamentals (Halevy and Mayak, 1981).

Exposure to atmospheres of reduced O_2 and/or elevated CO_2 concentrations is an additional strategy with potential to maintain the postproduction longevity of ornamentals. Storage under such an appropriate controlled atmosphere (CA) or within packaging in which a beneficial modified atmosphere (MA) develops is an effective means for extending the postharvest longevity of various fruits (e.g., apple, pear) and vegetables (e.g., broccoli) (Kader et al., 1989). The benefits of CA principally arise from associated reductions in rates of respiration, ethylene production and response, and oxidative processes (Beaudry, 1999). CA treatments can also aid in reducing pest infestations (e.g., arthropods) and disease infections (e.g., *Botrytis*) of harvested commodities (El-Goorani and Sommer, 1981; El-Kazzaz et al., 1983). In contrast to fruits and vegetables, effects of CA storage and MA packaging (MAP) on ornamentals have been less extensively and intensively studied. Thus, relatively few ornamental plant species have been tested for beneficial effects of CA and MA treatments (Table 19.1). Nonetheless, early studies on CA storage of cut flowers date back to 1930 (Thornton, 1930). Some interesting studies have also evaluated hypobaric storage, whereby a reduction in atmospheric pressure reduces the partial pressure of O_2 and other gases (e.g., ethylene) and can extend the longevity of ornamentals (Dilley et al., 1975; Staby et al., 1982).

The principles and effects of CA and MA treatments on ornamentals have previously been well documented by Halevy and Mayak (1981), Eisenberg (1985), Goszczynska and Rudnicki (1988), and Zagory and Reid (1989). This chapter aims to highlight specific examples of ornamental species that have responded favorably to CA and MA in terms of extended storage life, including reduced insect pest and fungal pathogen problems. Key biotic and abiotic factors affecting the efficacy of CA treatments and opportunities for future study and commercial development are also discussed.

19.2 CA and MA Storage of Cut Flowers and Greens

Despite occasional reports of success and over 75 years of experimentation, there is still no general commercial application of CA and MA for cut flowers and greens. Nonetheless, CA and MA treatments have been shown to consistently extend the storage life of certain cut flowers and greens.

19.2.1 *Anthurium*

Anthurium is a tropical ornamental plant, the flowers of which are traded internationally. Akamine and Goo (1981) evaluated effects of CA on *Anthurium andraeanum* 'Ozaki' cut flowers held in vases of tap water. They reported that storage of flowers at 13°C for 7 days in an atmosphere of 2.03 kPa O_2 (balanced with N_2) elicited a modest increase in subsequent vase life from 5.6 days (21.18 kPa O_2 control) to 7.4 days (Figure 19.1). When the storage temperature was 24°C–25°C, flowers exposed to the 2.03 kPa O_2 treatment for 7 days displayed a much longer vase life of 6.8 days than that of air-stored control stems

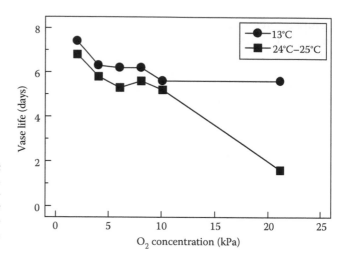

FIGURE 19.1
Vase life of *A. andraeanum* 'Ozaki' cut flowers following treatment with 2.03, 4.05, 6.08, 8.11, 10.13, and 21.18 (air control) kPa O_2 for 7 days at either 13°C or 24°C–25°C. (Data are redrawn from Akamine, E.K. and Goo, T., *HortScience*, 16, 206, 1981.)

(1.6 days). Exposure to higher O_2 concentrations of 4.05, 6.08, 8.11, and 10.13 kPa were not effective during storage at 13°C in increasing poststorage vase life, but were moderately beneficial at 24°C–25°C. Interestingly, the vase life of flowers stored in 2.03 kPa O_2 at 24°C–25°C was only slightly longer (1.2 days) than that of stems held in air at 13°C.

19.2.2   Carnation

Most of the research into the responses of cut flowers to CA and MA treatments has been on carnation (*Dianthus caryophyllus*). Knowledge that cut carnation flowers store particularly well at low temperature (e.g., 0.5°C, 9 weeks; Hanan, 1967) has presumably underpinned this relatively high level of interest. However, results of CA and MA studies with carnation have been highly variable. While several authors have reported that CA storage can extend the longevity of cut carnation flowers (e.g., Longley, 1933; Goszczynska and Rudnicki, 1983; Chen and Solomos, 1996), others have found that CA treatment provides no benefit (e.g., Hanan, 1967; Irving and Honnor, 1994).

Goszczynska and Rudnicki (1983) reported that the cut carnation flowers 'White Sim,' 'Keefer's Cheri Sim,' and 'Scania 3C' could be stored dry at the tight green bud stage within polyethylene bags for up to 6 months at 0°C–1°C provided that the stems were pretreated with 550 mg L^{-1} STS, 100 g L^{-1} sucrose and 0.1% Rovral (a.i. iprodione) fungicide. STS and Rovral treatment prevented ethylene action and growth of *B. cinerea*, respectively, during storage. The encouraging results were attributed to the MA conditions of 8.61–19.50 kPa O_2 and 0.91–6.28 kPa CO_2 that established within the polyethylene bags. These MA stored flowers maintained an acceptable vase life of 6.5–7.4 days following storage.

Two subsequent studies examined shorter-term CA storage treatments for carnation flowers. Joyce and Reid (1985) found that 'White Sim' carnation flowers could be stored in air at 1°C for only 2 weeks before vase life upon removal became unacceptably short (i.e., <4 days) (Figure 19.2). However, when flowers were maintained in 20.27 kPa CO_2 they could be stored for 3 or 4 weeks at 1°C and still possess a vase life of >8 days. Chen and Solomos (1996) reported that 'Elliott's White' cut carnation flowers remained attractive in a vase solution containing 1% sucrose and 150 mg L^{-1} hydroxyquinoline citrate (antibacterial agent) for up to 58 days at 18°C when held at 1.22 kPa O_2. In contrast, matching sets of control flowers in ambient air could be held for only 7–10 days. The hypoxic treatment prevented the typical climacteric rise in respiration and ethylene

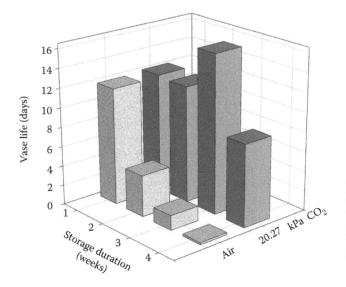

FIGURE 19.2

Vase life of carnation 'White Sim' cut flowers following exposure to air (0.04 kPa CO_2) or 20.27 kPa CO_2 for 1, 2, 3, or 4 weeks at 1°C. (Data are redrawn from Joyce, D.C., and Reid, M.S., *Proceedings of the 4th National Controlled Atmosphere Research Conference*, Raleigh, NC, 1985, pp. 185–198. With permission.)

evolution by flowers. In general, CA treatments utilizing elevated concentrations of CO_2 have been found to reduce rates of ethylene production (Mayak and Dilley, 1976) and response (Smith et al., 1966; Smith and Parker, 1966) by carnation flowers. Likewise, storage of 'Red Sim' and 'White Sim' carnation cut flowers at 0°C for 9 weeks under reduced pressure of 0.066 atm decreased rates of ethylene production and response and thereby extended subsequent vase life over freshly harvested flowers (Dilley et al., 1975).

19.2.3 Daffodil

Daffodil (*Narcissus pseudonarcissus*) flowers can generally be held for only about 2 weeks at low temperature (i.e., 1°C) before vase life is significantly reduced (Parsons et al., 1967). Storage for longer periods is possible via CA treatment. Parsons et al. (1967) showed that cut daffodil 'King Alfred' flowers could be stored for 14, 21, and 25 days at 4.5°C in an atmosphere of pure N_2 without reducing vase life as compared to freshly harvested flowers. Moreover, the vase life of N_2-treated flowers was two- to three-fold greater than control flowers stored in ambient air. Relative to air-stored flowers, N_2 treatment reduced rates of flower respiration by 14% and 39% during storage at 0°C and 15°C, respectively. Parsons et al. (1967) also found that the addition of just 1.01 kPa O_2 to the storage atmosphere reduced the beneficial effects of N_2 treatment.

19.2.4 *Gladiolus*

The postharvest longevity of minigladiolus flower spikes can benefit from storage within MAP. Meir et al. (1995b) reported that storing cut *Gladiolus* 'Adi' flowers inside 40 μm thick polyethylene film packages at 2°C for 2 weeks improved subsequent floret opening and longevity relative to spikes held in unsealed packages or in ambient air. The MA (10.13–14.19 kPa O_2, 4.05–7.09 kPa CO_2) treatment was also associated with significantly retarded bract and leaf senescence. The authors suggested that by delaying bract and leaf senescence, a larger pool of assimilates (e.g., carbohydrates) may be available for export into florets to enhance their opening. Pulse-treating spikes with a solution of 10% sucrose and 0.4 mM STS for 20 h (4 h at 22°C + 16 h at 4°C) prior to placement in MA packages further improved subsequent floral opening and quality.

19.2.5 Rose

There is great demand for cut rose (*Rosa hybrida*) flowers on particular days (e.g., Valentine's Day) and an associated need for effective storage methods. However, CA treatment has not been found to be consistently beneficial for rose flowers. In one of the earliest studies into CA effects on flowers, Thornton (1930) reported that storing cut 'Briarcliff' and 'Mrs F.R. Pierson' rose flowers wrapped in moistened paper in atmospheres containing 5.07–15.20 kPa CO_2 for 3–7 days at 3.3°C and 10°C extended the subsequent vase life by 0.5–2 days over matching control flowers held in ambient air. Longer-term storage (i.e., 17 days) reduced the vase life of flowers in all storage conditions due to retarded opening, excessive petal bleaching, and petal abscission following storage. Joyce and Reid (1985) found that 'Sonia' rose flowers could be stored in 10.13 kPa CO_2 at 1°C for 14 days without a significant reduction in vase life relative to freshly cut flowers. In contrast, air-stored flowers could not be held beyond 7 days at 1°C without loss in vase life.

Hauge et al. (1947) reported that storing cut rose 'Better Times' flowers in gas impervious cellophane bags at 4.5°C–7°C for 5 days delayed rates of flower opening and extended vase life by 1–2 days relative to flowers held in ambient air. The authors attributed the beneficial effects of MA treatment to the accumulation of 5.47 kPa CO_2 inside the packages during the storage period. However, it is also possible that maintenance of higher levels of relative humidity within sealed packages also contributed to improved water relations and flower quality. Similarly, Zeltzer et al. (2001) enclosed several unspecified cultivars of cut rose flowers in a MA box lined with a differentially permeable plastic polymer for 10 days at 2°C. Flowers held in the MA box (7.09 kPa CO_2 and 13.17 kPa O_2) lost ~3.5-fold less fresh weight during storage than control flowers in a conventional shipping box.

19.2.6 Foliage Plant Species

There is interest in expanding sea shipment of decorative cut foliage species to international markets versus the more expensive airfreight option. Philosoph-Hadas et al. (2007) reported that cut branches of *Aralia japonica*, *Asparagus virgatus*, *Leucadendron* 'Safari Sunset,' *Pittosporum tobira variegata*, and *Ruscus hypoglossum* foliage could be stored successfully at 2°C for 4 weeks in 15.20 kPa O_2 and 5.07 kPa CO_2. This CA treatment significantly reduced rates of decay as compared to stems held in ambient air. Moreover, the CA storage treatment maintained the dark green color of *Asparagus* branches, reduced blackening, and desiccation of *Leucadendron* leaves and prevented leaf abscission from and wilting of *Pittosporum* branches. These positive results suggest that further studies are warranted into the response to CA and MA of the many different types of cut greens.

19.3 CA and MA Storage of Bulbs

19.3.1 Lily

Asiatic lily (*Lilium asiaticum*) bulbs can generally only be stored at ambient temperatures for 3–4 weeks before excessive shoot and flower development impairs the ability of the bulb to produce a commercially acceptable flowering plant. Legnani et al. (2004a) showed that holding Asiatic lily 'Marseille,' 'Vermeer,' and 'Vivaldi' bulbs in a CA containing 1.01 kPa O_2 (balanced with N_2) at 22°C–24°C for 30 days can reduce the length of developing shoots and prevent flower bud development during storage (Figure 19.3). Storage in 1.01 kPa O_2 also increased the number of days to anthesis by 4–6 weeks, reduced

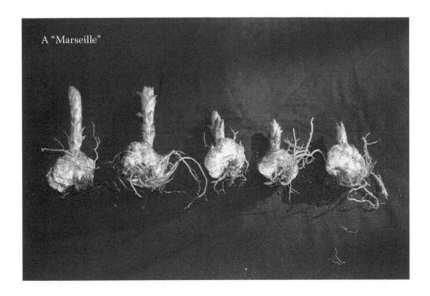

FIGURE 19.3
Photograph of Asiatic lily (*Lilium asiaticum*) 'Marseille' bulbs following treatment in (left to right) 21.18 (air control), 8.11, 4.05, 2.03, and 1.01 kPa O_2 for 30 days at 24°C. Note: Reduced shoot elongation is evident in the bulb exposed to 1.01 and 2.03 kPa O_2. Photograph is a courtesy of W.B. Miller, Cornell University. (From Legnani, G., Watkins, C.B., and Miller. W.B., *Postharvest Biol. Technol.*, 34, 93, 2004b. With permission.)

the number of distorted pistils, and increased the height of plants at flowering as compared to bulbs held in ambient air. Exposure to the higher O_2 concentrations of 2.03, 4.05, and 8.11 kPa were generally not as effective as 1.01 kPa O_2 in reducing shoot elongation, flower bud abortion, and in improving plant height. Nevertheless, these results demonstrate potential for using reduced O_2 atmospheres to extend the retail shelf life of lily bulbs and to improve the postproduction quality of flowering plants. Low O_2 atmospheres are currently being used on a commercial scale in The Netherlands to prolong the storage life of lily bulbs at low temperature (W.B. Miller, Cornell University, personal communication, 2007).

19.3.2 Tulip

The storage life of tulip bulbs showing signs of floral bud differentiation is typically limited to about 1–2 weeks at 15°C (Moe and Hagness, 1975; Prince et al., 1986). Longer-term storage can lead to the development of deformed, blasted, and/or poor quality flowers upon forcing. Prince et al. (1981) demonstrated that placing tulip (*Tulipa gesneriana*) 'Kees Nelis' bulbs into a CA containing 3.04 or 5.07 kPa O_2 at 17°C could extend the storage period to 4 weeks without eliciting detrimental effects on subsequent flowering. Maintenance of bulbs in these low O_2 concentrations also reduced the number of flowers that aborted in response to exposure to 5 μL L^{-1} ethylene during the storage period. The beneficial effects of low O_2 storage were, however, less pronounced for bulbs of 'Prominence.'

In a subsequent study, Prince et al. (1986) reported on the efficacy of a simple MA package composed of low-density polyethylene film to extend the storage life of tulip bulbs. Five 'Kees Nelis' tulip bulbs were placed into each sealed package and held at 15°C, 20°C, or 25°C. The concentration of O_2 and CO_2 inside packages ranged from ~4.05 to 6.08 kPa and 4.05 to 5.07 kPa, respectively, at 15°C, 20°C, or 25°C over a 4 week period.

Bulbs could be held inside MAP for either 3 weeks at 20°C and 25°C or for 4 weeks at 15°C without reducing the number and quality of subsequent flowers. In contrast, bulbs maintained in ambient air for the same periods at each temperature produced poor quality flowers upon forcing. Despite these promising data, there has been no large-scale adoption of CA and MA technology for extending the postharvest life of tulip bulbs.

19.4   CA and MA Storage of Vegetative Cuttings

The potential of CA treatments for extending the storage and/or shipping duration for vegetative cuttings has not been extensively studied. Behrens (1986) examined the response of cuttings from 10 different coniferous plant species to a 4 month exposure to 3.04 kPa O_2 and 3.04 kPa CO_2. However, only one species, *Picea glauca* 'Conica,' showed clear benefits from the CA treatment. Cuttings of this variety stored in CA had 50% higher rooting than those held in air.

19.5   CA and MA Storage of Potted Ornamental Plants

There have been few studies into effects of CA on potted plants. This prospect represents an opportunity for further investigation because such plants are typically high value items that are often transported and/or stored for long periods. CA treatment may offer protection against ethylene-induced flower and leaf abscission and/or senescence, problems commonly afflicting potted plants (Zagory and Reid, 1989). Atmosphere modifications along the lines of CA treatment may also improve production characteristics of ornamental plants. Konjoian et al. (1983) reported that *Chrysanthemum × morifolium* 'Bright Golden Anne plants grown in chambers containing 2.03–5.17 kPa O_2 and 0.07–0.15 kPa CO_2 flowered 10–12 days later than plants held in ambient air. Plants grown in these low O_2 atmospheres produced twice as many inflorescences per plant, 31% more flowers per inflorescence, and accumulated 48% greater dry weight in vegetative tissues than plants grown in air. Plants exposed to the low O_2 atmospheres also produced twice as many viable cuttings through three flushes of growth as plants kept in ambient O_2.

19.6   CA and MA Treatments for Pest and Disease Control

The international trade of ornamental plant species is often limited by infestation of plant tissues with insect pests and fungal diseases (van Gorsel, 1994; Vrind, 2005). Chemical insecticides and fungicides have long been relied upon to control these organisms. However, their continued use is problematical owing to development of resistance and increasing social and environmental concerns over chemical residues (Arthur, 1996; Stehmann and De Waard, 1996). Alternatively, CA and MA treatments have shown promise in controlling arthropod pests and plant pathogens during storage of fruit, vegetables, and grains (El-Goorani and Sommer, 1981; El-Kazzaz et al., 1983). Several studies have evaluated the capacity of CA to reduce insect pest and disease problems on cut flowers and ornamental plants (Table 19.2).

Storing and transporting ornamental commodities at low temperature (e.g., 0.5°C) in atmospheres low in O_2 (e.g., <0.10 kPa) and/or enriched in CO_2 (e.g., >10.13 kPa) has been

TABLE 19.2

Published Studies that Have Evaluated the Capacity of CA Treatments to Reduce Pest and Disease Infestation on Ornamental Plants and Their Detached Organs

Plant Species	Pest Organism	Reference
Anigozanthos rufus cut flowers	Arthropod	Seaton and Joyce (1993)
Begonia semperflorens-cultorum potted plants	Arthropod	Held et al. (2001)
Chamelaucium uncinatum cut flowers	Arthropod	Seaton and Joyce (1993), Shelton et al. (1996)
Delphinium sp. cut flowers	Arthropod	Shelton et al. (1996)
Dendrathema grandiflorum cut flowers	Arthropod	Shelton et al. (1996)
Dendrathema grandiflora potted plants	Arthropod	Held et al. (2001)
Dianthus caryophyllus cut flowers	Arthropod	Shelton et al. (1996)
Gypsophila elegans preserved cut flowers	Arthropod	Sauer and Shelton (2002)
Impatiens wallerana seedlings	Arthropod	Held et al. (2001)
Iris sp. cut flowers	Arthropod	Shelton et al. (1996)
Lilium asiaticum cut flowers	Arthropod	Shelton et al. (1996)
Limonium sinuatum preserved cut flowers	Arthropod	Sauer and Shelton (2002)
Limonium sinuatum cut flowers	Arthropod	Shelton et al. (1996)
Matthiola incana cut flowers	Arthropod	Shelton et al. (1996)
Pelargonium × *hortorum* potted plants	Arthropod	Held et al. (2001)
Protea sp. cut flowers	Arthropod	Shelton et al. (1996)
Rosa hybrida cut flowers	Arthropod	Shelton et al. (1996)
Rosa hybrida cut flowers	Fungal pathogen	Joyce and Reid (1985), Hammer et al. (1990)

suggested as a potential treatment for insect disinfestation (Shelton et al., 1996). Held et al. (2001) reported that exposure of green peach aphids (*Myzus persicae*), two spotted spider mites (*Tetranychus urticae*), western flower thrips (*Frankliniella occidentalis*), and sweet potato whiteflies (*Bemisia* sp.) to pure N_2 or CO_2 for 12–18 h at 20°C caused 100% mortality. In this context, treatment with pure N_2 for either 12 or 18 h did not reduce the quality of *Begonia semperflorens-cultorum* 'Cocktail series' seedlings, potted geranium (*Pelargonium* × *hortorum*) 'Melody Red,' and 'Everglow' plants and potted chrysanthemum (*Dendrathema grandiflora*) 'Pomona,' 'Charm,' and 'Red Remarkable' plants. However, exposure of these species to pure CO_2 caused extensive damage (e.g., discoloration of foliage) to plants. In contrast, exposure of glycerol-preserved cut *Limonium sinuatum* and *Gypsophila elegans* flowers to elevated CO_2 (81.06 kPa CO_2 in N_2) at 32.2°C for 12 h did not affect their quality and vase life (Sauer and Shelton, 2002). This treatment caused 100% mortality of Indian meal moth (*Plodia interpunctella*) pupae.

The postharvest longevity of many cut flowers and potted flowering plants is greatly reduced by infection with *B. cinerea*, the casual agent of gray mold (Elad, 1988). Treatment with >15.20 kPa CO_2 is strongly fungistatic against *B. cinerea* on strawberry fruit (Couey and Wells, 1970; El-Kazzaz et al., 1983). However, there are relatively few published accounts of CA treatments for the control of *B. cinerea* on ornamentals. Joyce and Reid (1985) reported that storing cut carnation 'White Sim' flowers in 20.27 kPa CO_2 at 1°C for 4 weeks prevented *B. cinerea* from developing on flowers. In contrast, control flowers stored in air at 1°C were extensively colonized by *B. cinerea* within 3–4 weeks. Hammer et al. (1990) reported that storing cut rose 'Sonia,' 'Royalty,' and 'Gold Rush' flowers in 10.13 kPa CO_2 for 5 days at 2.5°C reduced the number of *B. cinerea* lesions on petals by 82% relative to control flowers stored in air. Treatment with the higher CO_2 concentration of 20.27 kPa for 7 days at 2.5°C reduced disease severity further but elicited severe leaf discoloration (bronzing) on 'Sonia' and 'Royalty' roses.

Further development, testing, and optimization of treatments is required before CA and MA technologies can be implemented to control pests and diseases on a wide range of ornamental plant species. With regard to both product quality and efficacy of control, such research should consider the complex interactions between the key variables of CO_2 and O_2 concentrations, temperature, and relative humidity. Nevertheless, reports published to date show clear promise for the use of CA as means of managing, if not controlling, pests and diseases on harvested ornamentals.

19.7 Biotic and Abiotic Factors Affecting the Efficacy of CA and MA Treatments

Investigations into the effects of CA and MA treatments on ornamentals have tended to be piecemeal. While some studies have focused on the one plant species (e.g., rose), different investigators have seldom worked with the same cultivars. Considerable variation in response to CA treatments exists among different genotypes. For instance, while treatment with 3.04 or 5.07 kPa CO_2 safely extended the storage life of tulip 'Kees Nelis' bulbs, beneficial effects were not observed for 'Prominence' (Prince et al., 1981). Similarly, Hammer et al. (1990) reported variation in the degree of tolerance of rose flowers 'Sonia,' 'Royalty,' and 'Gold Rush' to elevated CO_2 atmospheres.

The carbohydrate status, physiological age, and/or maturity stage of vegetative and reproductive organs may also affect the efficacy of CA and MA treatments (Joyce and Reid, 1985). Menguc et al. (1993) reported that optimal O_2 and CO_2 concentrations for maintaining the general appearance of cut carnation 'Astor' flowers during storage at 0°C for 60 days varied according to the development stage at which flowers were harvested. For fully opened flowers, exposure to 10.13 kPa CO_2 and 3.04 kPa O_2 was considered optimal. However, treatment with 10.13 kPa CO_2 and 5.07 kPa O_2 was most beneficial for flowers cut at the earlier paintbrush stage. Variation in flower development stages across studies with the one species could well account for some of the diverse findings reported in the literature.

Fungal pathogens such as *B. cinerea* limit the long-term storage of many ornamentals. Zagory and Reid (1989) considered that development of *B. cinerea* on ornamental commodities during storage could contribute variation in responses of ornamentals to CA and MA treatments. Goszczynska and Rudnicki (1983) demonstrated that pretreating cut carnation 'White Sim,' 'Keefer's Cheri Sim,' and 'Scania 3C' flowers with iprodione to control *B. cinerea* was essential for successful long-term (i.e., 6 months) storage within MAP. As *B. cinerea* is a ubiquitous fungal pathogen that thrives in humid, low temperature conditions commonly used for the storage of ornamentals (Elad, 1988), the efficacy of CA and MA protocols should be improved if susceptible commodities are pre-treated with effective fungicides.

Storage temperature is the single most important environmental factor influencing the postharvest longevity of ornamental plants (Maxie et al., 1973; Halevy and Mayak, 1981). Lower temperatures extend the storage life of ornamentals primarily by reducing rates of respiration (Maxie et al., 1973), sensitivity to ethylene (Barden and Hanan, 1972; Macnish et al., 2004b), and metabolic reactions associated with carbohydrate utilization and tissue senescence (Halevy and Mayak, 1981). CA treatments on ornamentals have generally been evaluated at low (e.g., 0°C–5°C) temperatures. Since CA and MA treatments also extend life by reducing rates of respiration, ethylene sensitivity and oxidation reactions, it might be anticipated that relative benefits of CA are greater at higher temperatures (Kader, 2002), which are commonly encountered during commercial handling of ornamentals.

Akamine and Goo (1981) found that treatment of *A. andraeanum* 'Ozaki' flowers with 2.03 kPa O_2 better maintained vase life over air-stored control stems at 24°C–25°C (by 5.2 days) than at 13°C (by 1.8 days) (Figure 19.1). Thornton (1930) also reported that treatment with 5.07–15.20 kPa CO_2 enhanced rates of flower opening and extended the vase life of 'Briarcliff' and 'Mrs F.R. Pierson' rose flowers when applied at 10°C compared to at 3.3°C. Similarly, Parsons et al. (1967) reported greater maintenance of vase life for 'King Alfred' daffodil flowers treated with pure N_2 at 3.5°C, 13°C, and 21°C than at 0°C relative to matching sets of control flowers stored in air at each temperature. Observations that rates of cut 'White Sim' carnation and 'Kardinal' rose flower respiration are more profoundly reduced by exposure to lower temperatures than by CA treatments alone (Figure 19.4; Reid and Granello, unpublished data) are consistent with these findings. The reduction in rates of respiration by 'White Sim' carnation and 'Kardinal' rose flowers in response to a lowering of O_2 concentration from 21.18 to 0.01 kPa was also relatively low as compared to other products such as banana fruit (Wade, 1974). In addition, no 'Pasteur effect' (fermentative CO_2 production at low O_2 partial pressure) was observed at the low O_2 tensions, suggesting tissues were relatively nonresponsive to the changing atmosphere (Figure 19.4). In contrast, an increase in CO_2 production at low O_2 concentration has been reported for other commodities such as apple, asparagus, carrots, pear, and blueberry (Platenius, 1943; Boersig et al., 1988; Beaudry et al., 1992; Peppelenbos et al., 1996). Collectively, these results may help explain why CA treatments have been of limited value in extending the longevity of many ornamental species.

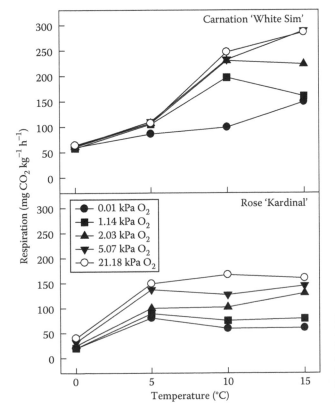

FIGURE 19.4

Rates of respiration by cut carnation (*Dianthus caryophyllus*) 'White Sim' and rose (*Rosa hybrida*) 'Kardinal' flowers during storage at 0°C, 5°C, 10°C, and 15°C in atmospheres of 0.01, 1.14, 2.03, 5.07, and 21.18 kPa O_2. (From Reid and Granello, unpublished data.)

19.8   Future Research Needs and Commercial Applications

Studies into the effects of CA storage on fruit and vegetables have focused increasingly upon understanding the molecular basis of responses at the tissue level. For example, Eason et al. (2007) reported that broccoli floret tissues stored in 10.13 kPa CO_2 and 5.07 kPa O_2 for 4 days at 20°C exhibited upregulation (mRNA expression) of many stress-responsive genes. It follows that analysis of gene regulation and expression for ornamentals combined with a developing understanding of the roles and regulation of their products should facilitate elucidation of their roles in responses to CA and MAP. In turn, this knowledge would contribute to the optimization of CA storage and MAP treatments to suppress senescence in tissues of ornamental plant species. Studies on ornamentals to level of detail may help to explain somewhat curious results, such as why daffodil flowers store well in pure N_2 (Parsons et al., 1967) and why carnations tolerate exposure to high CO_2 levels (e.g., Goszczynska and Rudnicki, 1983; Joyce and Reid, 1985).

It is probable that the international marketing of ornamental plant genotypes would be markedly enhanced with active selection by plant breeders for superior postharvest characteristics, including exceptional performance under CA and MAP. Considerable variation in postharvest longevity has been documented among cultivars of various ornamental crops, including carnation, rose, kalanchoe (*Kalanchoe blossfeldiana*), and Geraldton waxflower (*Chamelaucium uncinatum*), particularly in the presence of ethylene gas (Wu et al., 1991; Muller et al., 1998; Serek and Reid, 2000; Macnish et al., 2004a). Quantification of differential responses to CA and MA treatments across species and cultivars represents a seemingly excellent opportunity to obtain cultivars well suited to long-term transport and/or storage. Similarly, screening genotypes for resistance to infection by *B. cinerea* should help to deliver cultivars better suited for CA storage and MA packaging.

The generally variable response of ornamentals to CA and MA treatments is in apparent contrast to the consistently beneficial effects that these treatments have in delaying the senescence of broccoli. Atmospheres containing elevated CO_2 (e.g., 5.07–10.13 kPa) and reduced O_2 (e.g., 1.01–2.03 kPa) concentrations at temperatures between 0°C and 5°C markedly reduce rates of respiration, chlorophyll degradation, and water loss and also lessen disease development associated with the earlier onset of senescence in air-stored tissues (Makhlouf et al., 1989; Bastrash et al., 1993; Izumi et al., 1996). It follows that comparing and contrasting CA responses in broccoli florets versus the floral tissues of cut flower species represents a potentially lucrative research strategy.

Relative to transport by airplane, the use of surface transport (i.e., truck, boat) is attractive due to reduced shipping costs and better cool chain continuity. Coupled with an increasing availability of refrigerated containers equipped with CA storage capability, applied research, and commercial opportunities exist for attaining increased movement of ornamentals in surface vessels that utilize mobile CA technologies.

Also in this technological context, a range of polymer films with widely varying gas permeability characteristics have been developed for MAP of fresh-cut fruits and vegetables (Kader et al., 1989). To date, there has been limited commercial development and evaluation of such films for use on ornamentals. MA technology might even be applied to value-added ornamentals, such as floral bouquets.

For *Anthurium* flowers, CA and MA treatments applied at ambient temperature can provide a similar extension in storage life as exposure to low temperatures. Akamine and Goo (1981) demonstrated that exposure of *A. andraeanum* 'Ozaki' flowers to 2.03 kPa O_2 at 24°C–25°C for 7 days extended vase life compared to that of flowers stored in air at 13°C. It follows from preceding discussion of the interaction of temperature and gas atmosphere that CA treatments may find increased practical use in situations where the cool chain is either limited or nonexistent. This scenario is probably particularly relevant to tropical ornamentals, such as cut *Anthurium* flowers. Moreover, elevated CO_2 concentrations of 3.04–8.11 kPa have been shown to reduce the severity of chilling injury in fruits, including avocado (Meir et al., 1995a). It would be scientifically interesting and commercially relevant to determine if these findings extend to chilling-sensitive ornamentals. Thus, CA treatments that reduce chilling injury may enhance the storage, transport and/or marketing life of tropical ornamentals, including at temperatures below their usual chilling injury threshold (e.g., 10°C–13°C).

19.9 Conclusions

CA storage and MAP for ornamentals is not widely practiced. However, various CA and/or MA treatments have demonstrated potential to extend the postharvest life of certain cut flower (e.g., *Anthurium*, carnation, daffodil) and bulb (e.g., lily, tulip) crops. Thus, while the CA and MA research and development to date shows promise for selected ornamentals, these technologies have not yet been proven effective across the broad range of diverse species that characterize the ornamentals industry. Accordingly, commercial interest in CA and MA treatments for ornamentals and cut flowers is likely to remain low until the benefits can be demonstrated for a wider selection of high-value species.

References

Akamine, E.K. and T. Goo. 1981. Controlled atmosphere storage of *Anthurium* flowers. *HortScience* 16:206–207.

Akbudak, B., A. Eris, and O. Kucukahmetler. 2005. Normal and modified atmosphere packaging storage of lisianthus (*Lisianthus grandiflorum*) grown in saline conditions. *N. Z. J. Crop Hort. Sci.* 33:185–191.

Arthur, F.H. 1996. Grain protectants: Current status and prospects for the future. *J. Stored Prod. Res.* 32:293–302.

Barden, L.E. and J.J. Hanan. 1972. Effect of ethylene on carnation keeping life. *J. Am. Soc. Hort. Sci.* 97:785–788.

Bastrash, S., J. Makhlouf, F. Castaigne, and C. Willemot. 1993. Optimal controlled-atmosphere conditions for storage of broccoli florets. *J. Food Sci.* 58:338–341.

Beaudry, R.M. 1999. Effect of O_2 and CO_2 partial pressure on selected phenomena affecting fruit and vegetable quality. *Postharvest Biol. Tech.* 15:293–303.

Beaudry, R.M., A.C. Cameron, A. Shirazi, and D.L. Dostal-Lange. 1992. Modified atmosphere packaging of blueberry fruit: Effect of temperature on package O_2 and CO_2. *J. Am. Soc. Hort. Sci.* 117:436–441.

Behrens, V. 1986. Kuhlagerung von unbewurzelten Koniferensteckling. II. Lagertemperatur und–atmosphare (Cool storage of unrooted conifer cuttings. II. Storage temperature and atmosphere). *Gartenbauwissenschaft* 51:118–125.

Boersig, M.R., A.A. Kader, and R.J. Romani. 1988. Aerobic-anaerobic respiratory transition in pear fruit and cultured pear fruit cells. *J. Am. Soc. Hort. Sci.* 113:869–873.

Chen, X.H. and T. Solomos. 1996. Effects of hypoxia on cut carnation flowers (*Dianthus caryophyllus* L): Longevity, ability to survive under anoxia, and activities of alcohol dehydrogenase and pyruvate kinase. *Postharvest Biol. Tech.* 7:317–329.

Couey, H.M. and J.M. Wells. 1970. Low oxygen or high carbon dioxide atmospheres to control postharvest decay of strawberries. *Phytopathology* 60:47–49.

De Pascale, S., T. Maturi, and V. Nicolais. 2005. Modified atmosphere packaging (MAP) for preserving *Gerbera, Lilium* and *Rosa* cut flowers. *Acta Hort.* 682:1145–1152.

Dilley, D.R., W.J. Carpenter, and S.P. Burg. 1975. Principles and applications of hypobaric storage of cut flowers. *Acta Hort.* 41:249–267.

Eason, J.R., D. Ryan, B. Page, L. Watson, and S.A. Coupe. 2007. Harvested broccoli (*Brassica oleracea*) responds to high carbon dioxide and low oxygen atmosphere by inducing stress-response genes. *Postharvest Biol. Tech.* 43:358–365.

Eisenberg, B.A. 1985. A summary of CA requirements, limitations and recommendations for ornamentals. In: S.M. Blankenship (Ed.), *Proceedings of the 4th National Controlled Atmosphere Research Conference*, Raleigh, NC, pp. 493–505.

Elad, Y. 1988. Latent infection of *Botrytis cinerea* in rose flowers and combined chemical and physiological control of the disease. *Crop Prot.* 7:631–633.

El-Goorani, M.A. and N.F. Sommer. 1981. Effects of modified atmospheres on postharvest pathogens of fruits and vegetables. *Hort. Rev.* 3:412–461.

El-Kazzaz, M.K., N.F. Sommer, and R.J. Fortlage. 1983. Effect of different atmospheres on postharvest decay and quality of fresh strawberries. *Phytopathology* 73:282–285.

Erstad, J.L.F. and H.R. Gislerod. 1994. Water uptake of cuttings and stem pieces as affected by different anaerobic conditions in the rooting medium. *Sci. Hort.* 58:151–160.

Goszczynska, D. and R.M. Rudnicki. 1983. Long-term storage of bud-cut carnations. *Acta Hort.* 141:203–212.

Goszczynska, D. and R.M. Rudnicki. 1988. Storage of cut flowers. *Hort. Rev.* 10:35–62.

Halevy, A.H. and S. Mayak. 1981. Senescence and postharvest physiology of cut flowers—Part 2. *Hort. Rev.* 3:59–143.

Hammer, P.E., S.F. Yang, M.S. Reid, and J.J. Marois. 1990. Postharvest control of *Botrytis cinerea* infections on cut roses using fungistatic storage atmospheres. *J. Am. Soc. Hort. Sci.* 115:102–107.

Hanan, J.J. 1967. Experiments with controlled atmosphere storage of carnations. *Proc. Am. Soc. Hort. Sci.* 90:370–376.

Hauge, A., W. Bryant, and A. Laurie. 1947. Prepackaging of cut flowers. *Proc. Am. Soc. Hort. Sci.* 49:427–432.

Held, D.W., D.A. Potter, R.S. Gates, and R.G. Anderson. 2001. Modified atmosphere treatments as a potential disinfestations technique for arthropod pests in greenhouses. *J. Econ. Entomol.* 94:430–438.

Irving, D.E. and L. Honnor. 1994. Carnations: Effects of high concentrations of carbon dioxide on flower physiology and longevity. *Postharvest Biol. Technol.* 4:281–287.

Izumi, H., A.E. Watada, and W. Douglas. 1996. Optimum O_2 or CO_2 atmosphere for storing broccoli florets at various temperatures. *J. Am. Soc. Hort. Sci.* 121:127–131.

Johnston, J.W., A. Carpenter, and R.E. Lill. 1998. Effects of anoxic and hypercarbic controlled atmospheres on anaerobic volatile emission and ethylene production from 'Klemaxi' and 'Doranja' carnation (*Dianthus caryophyllus* L.). *N. Z. J. Crop Hort. Sci.* 26:153–160.

Joyce, D.C. and M.S. Reid. 1985. Effect of pathogen-suppressing modified atmospheres on stored cut flowers. In: S.M. Blankenship (Ed.), *Proceedings of the 4th National Controlled Atmosphere Research Conference*, Raleigh, NC, pp. 185–198.

Kader, A.A. 2002. Modified atmospheres during transport and storage. In: A.A. Kader (Ed.). *Postharvest Technology of Horticultural Crops*. University of California, Oakland, CA, pp. 135–144.

Kader, A.A., D. Zagory, and E.L. Kerbel. 1989. Modified atmosphere packaging of fruits and vegetables. *Crit. Rev. Food Sci. Nutr.* 28:1–30.

Konjoian, P.S., G.L. Staby, and H.K. Tayama. 1983. The growth and development of *Chrysanthemum* × *morifolium* 'Bright Golden Anne' in low oxygen environments. *J. Am. Soc. Hort. Sci.* 108:582–585.

Legnani, G., C.B. Watkins, and W.B. Miller. 2004a. Low oxygen affects the quality of Asiatic hybrid lily bulbs during simulated dry-sale storage and subsequent forcing. *Postharvest Biol. Technol.* 32:223–233.

Legnani, G., C.B. Watkins, and W.B. Miller. 2004b. Light, moisture, and atmosphere interact to affect the quality of dry-sale lily bulbs. *Postharvest Biol. Technol.* 34:93–103.

Longley, L.E. 1933. Some effects of storage of flowers in various gases at low temperatures on their keeping qualities. *Proc. Am. Soc. Hort. Sci.* 30:607–609.

Macnish, A.J., D.E. Irving, D.C. Joyce, V. Vithanage, A.H. Wearing, and A.T. Lisle. 2004a. Variation in ethylene-induced postharvest flower abscission responses among *Chamelaucium* Desf. (*Myrtaceae*) genotypes. *Sci. Hort.* 102:415–432.

Macnish, A.J., D.E. Irving, D.C. Joyce, A.H. Wearing, and V. Vithanage. 2004b. Sensitivity of Geraldton waxflower to ethylene-induced flower abscission is reduced at low temperature. *J. Hort. Sci. Biotechnol.* 79:293–297.

Makhlouf, J., F. Castaigne, J. Arul, C. Willemot, and A. Gosselin. 1989. Long-term storage of broccoli under controlled-atmosphere. *HortScience* 24:637–639.

Maxie, E.C., D.S. Farnham, F.G. Mitchell, N.F. Sommer, R.A. Parsons, R.G. Snyder, and H.L. Rae. 1973. Temperature and ethylene effects on cut flowers of carnation (*Dianthus caryophyllus* L.). *J. Am. Soc. Hort. Sci.* 98:568–572.

Mayak, S. and D.R. Dilley. 1976. Regulation of senescence in carnation (*Dianthus caryophyllus*). *Plant Physiol.* 58:663–665.

Meir, S., M. Akerman, Y. Fuchs, and G. Zauberman. 1995a. Further studies on the controlled atmosphere storage of avocados. *Postharvest Biol. Technol.* 5:323–330.

Meir, S., S. Philosoph-Hadas, R. Michaeli, H. Davidson, M. Fogelman, and A. Schaffer. 1995b. Improvement of the keeping quality of mini-gladiolus spikes during prolonged storage by sucrose pulsing and modified atmosphere packaging. *Acta Hort.* 405:335–342.

Menguc, A., A. Eris, M.H. Ozer, and M. Zencirkiran. 1993. A research on the CA storage of Astor carnation flowers harvested at different growth stages. In: *Proceedings of the 6th International Controlled Atmosphere Research Conference*, Cornell University, Ithaca, NY, pp. 620–628.

Moe, R. and A.K. Hagness, 1975. The influence of storage temperature and 2-chloroethylphosphonic acid (ethephon) on shoot elongation and flowering in tulips. *Acta Hort.* 47:307–318.

Muller, R., A.S. Andersen, and M. Serek. 1998. Differences in display life of miniature potted roses (*Rosa hybrida* L.). *Sci. Hort.* 76:59–71.

Parsons, C.S., S. Asen, and N.W. Stuart. 1967. Controlled atmosphere storage of daffodil flowers. *Proc. Am. Soc. Hort. Sci.* 90:506–514.

Peppelenbos, H.W., L.M.M. Tijskens, J. van't Leven, and E.C. Wilkinson. 1996. Modelling oxidative and fermentative carbon dioxide production of fruits and vegetables. *Postharvest Biol. Technol.* 9:283–295.

Philosoph-Hadas, S., S. Droby, I. Rosenberger, Y. Perzelan, S. Salim, I. Shtein, and S. Meir. 2007. Sea transport of ornamental branches: Problems and solutions. *Acta Hort.* 755:267–276.

Platenius, H. 1943. Effect of oxygen concentration on the respiration of some vegetables. *Plant Physiol.* 18:671–684.

Prince, T.A. and M.S. Cunningham. 1987. Response of tubers of *Begonia* × *tuberhybrida* to cold temperatures, ethylene, and low oxygen storage. *HortScience* 22:252–254.

Prince, T.A. and M.S. Cunningham. 1991. Forcing characteristics of Easter lily bulbs exposed to elevated ethylene and carbon dioxide and low oxygen atmospheres. *J. Am. Soc. Hort. Sci.* 116:63–67.

Prince, T.A., R.C. Herner, and A.A. DeHertogh. 1981. Low oxygen storage of special precooled 'Kees Nelis' and 'Prominence' tulip bulbs. *J. Am. Soc. Hort. Sci.* 106:747–751.

Prince, T.A., R.C. Herner, and J. Lee. 1986. Bulb organ changes and influences of temperature on gaseous levels in a modified atmosphere package of precooled tulip bulbs. *J. Am. Soc. Hort. Sci.* 111:900–904.

Reid, M.S. 2002. Postharvest handling systems: Ornamental crops. In: A.A. Kader (Ed.), *Postharvest Technology of Horticultural Crops.* University of California, Oakland, CA, pp. 315–325.

Sauer, J.A. and M.D. Shelton. 2002. High temperature controlled atmosphere for post-harvest control of Indian meal moth (Lepidoptera: Pyralidae) on preserved flowers. *J. Econ. Entomol.* 95:1074–1078.

Seaton, K.A. and D.C. Joyce. 1993. Effects of low temperature and elevated CO_2 treatments and of heat treatments for insect disinfestations on some native Australian cut flowers. *Sci. Hort.* 56:119–133.

Serek, M. and M.S. Reid. 2000. Ethylene and postharvest performance of potted kalanchoe. *Postharvest Biol. Technol.* 18:43–48.

Shelton, M.D., V.R. Walter, D. Brandl, and V. Mendez. 1996. The effects of refrigerated, controlled atmosphere storage during marine shipment on insect mortality and cut flower vase life. *HortTechnology* 6:247–250.

Smith, W.H. and J.C. Parker. 1966. Prevention of ethylene injury to carnations by low concentrations of carbon dioxide. *Nature* 211:100–101.

Smith, W.H., J.C. Parker, and W.W. Freeman. 1966. Exposure of cut flowers to ethylene in the presence and absence of carbon dioxide. *Nature* 211:99–100.

Staby, G.L., M.P. Bridgen, B.A. Eisenberg, M.S. Cunningham, J.W. Kelly, P.S. Konjoian, and C.L. Holstead. 1982. Hypobaric storage of floral crops. Department of Horticulture, Ohio State University, Columbus, The Ohio Agricultural Research and Development Center, Wooster, OH. Report HO-82–18, 8 pp.

Staby, G.L., M.S. Cunningham, C.L. Holstead, J.W. Kelly, P.S. Konjoian, B.A. Eisenberg, and B.S. Dressler. 1984. Storage of rose and carnation flowers. *J. Am. Soc. Hort. Sci.* 109:193–197.

Stehmann, C. and M.A. De Waard. 1996. Factors influencing activity of triazole fungicides towards *Botrytis cinerea. Crop Prot.* 15:39–47.

Stuart, N.W., C.S. Parsons, and C.J. Gould. 1970. The influence of controlled atmospheres during cool storage on the subsequent flowering of Easter lilies and bulbous iris. *HortScience* 5:356 (Abstract).

Thornton, N.C. 1930. The use of carbon dioxide for prolonging the life of cut flowers, with special reference to roses. *Am. J. Bot.* 17:614–626.

Thornton, N.C. and E.P. Imle. 1941. Effect of mixtures of oxygen and carbon dioxide on the development of dormancy in Easter lilies. *Proc. Am. Soc. Hort. Sci.* 38:708. (Abstract).

Uota, M. 1969. Carbon dioxide suppression of ethylene-induced sleepiness of carnation blooms. *J. Am. Soc. Hort. Sci.* 94:598–601.

van Gorsel, R. 1994. Postharvest technology of imported and trans-shipped tropical floricultural commodities. *HortScience* 29:979–981.

Vrind, T.A. 2005. The *Botrytis* problem in figures. *Acta Hort.* 669:99–102.

Wade, N.L. 1974. Effects of oxygen concentration and ethephon upon the respiration and ripening of banana fruits. *J. Exp. Bot.* 25:955–964.

Wu, M.J., L. Zacarias, and M.S. Reid. 1991. Variation in the senescence of carnation (*Dianthus caryophyllus* L.) cultivars. II. Comparison of sensitivity to exogenous ethylene and of ethylene binding. *Sci. Hort.* 48:109–116.

Yamashita, I., K. Dan, and H. Ikeda. 1999. Storage of cut spray type chrysanthemum (*Dendrathema grandiflorum* Tzvelev) by active modified atmosphere packaging. *J. Jpn. Soc. Hort. Sci.* 68:622–627.

Zagory, D. and M.S. Reid. 1989. Controlled atmosphere storage of ornamentals. In: *Proceedings of the 5th International Controlled Atmosphere Research Conference*, Wenatchee, WA, pp. 353–358.

Zeltzer, S., S. Meir, and S. Mayak. 2001. Modified atmosphere packaging (MAP) for long-term shipment of cut flowers. *Acta Hort.* 553:631–634.

20

Dried Fruits and Tree Nuts

Judy A. Johnson, Elhadi M. Yahia, and David G. Brandl

CONTENTS

20.1 Introduction

Dried fruits and tree nuts are relatively high-value products used primarily for snack foods or as confectionary ingredients, and their successful marketing requires strict attention to quality control. The United States alone produces nearly 1.5 million metric tons each year of almonds, hazelnuts, macadamias, pecans, pistachios, walnuts, dates, figs, prunes, raisins, and dried apricots, worth more than $3 billion (USDA, 2007). These are also valuable products for the foreign export market, important to the economies of such major producers as the United States and Turkey.

Dried fruit and tree nuts typically have one or more preharvest insect pests that feed directly on the product and are capable of causing considerable damage and quality loss (Simmons and Nelson, 1975). Although many of these may be present at the time of harvest and are often brought into storage, they generally do not reproduce under storage conditions (Johnson et al., 2002). However, because they may continue to feed and cause additional damage, and often present phytosanitary problems for processors, they are considered postharvest pests. Feeding damage by these insects also may provide entry to aflatoxin-producing molds (*Aspergillus* spp.) (Campbell et al., 2003). Initial disinfestation of an incoming product is sufficient to control these pests and reduce their damage. These commodities are also susceptible to attack by a number of common stored product moths and beetles, the most serious being the Indianmeal moth, *Plodia interpunctella* (Simmons and Nelson, 1975). Because stored product pests are capable of repeated infestation during storage, long-term protective treatments or repeated disinfestation treatments are necessary for their control.

Current insect control measures for dried fruit and nuts depend largely on fumigation to disinfect large volumes of incoming product during harvest, as well as to control storage infestation (Johnson, 2004). Methyl bromide, a fumigant used in a wide range of postharvest applications, is scheduled for worldwide withdrawal from routine use as a fumigant in 2015 under the Montreal Protocol on ozone-depleting substances (UNEP, 2006), and its use has already been severely restricted in developed countries. Resistance to phosphine, often used as an alternative to methyl bromide for dried fruit and nut crops in developed countries, has been documented in many insect populations (Benhalima et al., 2004), and some regulatory agencies have expressed concerns over worker safety with this compound (Bell, 2000). Sulfuryl fluoride, long used for structural fumigation, has recently been registered for commodity fumigation in several countries including the United States (Prabhakaran and Williams, 2007), but reduced toxicity of this compound against insect eggs and at lower temperatures (Bell and Savvidou, 1999) may limit its applicability. Moreover, there is a mounting pressure against the general use of chemical fumigants due to atmospheric emissions, safety, or health concerns, and an increased interest in organic food production, resulting in efforts to develop nonchemical technologies as alternative control methods for insects. Among these technologies are low and high temperatures, irradiation, and modified atmosphere (MA) and controlled atmosphere (CA).

Dried fruits and nuts, like most low-moisture, durable commodities, often tolerate extreme MA and CA (very high CO_2 and/or very low O_2) at levels used to control insects. More than 25 years ago, the U.S. Environmental Protection Agency approved carbon dioxide, nitrogen, and combustion product gases as a means to manage insects infesting raw and processed agricultural products, including dried fruits and tree nuts (Johnson, 1980, 1981). While current insect control measures for dried fruits and nuts still rely on fumigation, there is some limited commercial use of MA and CA for these products, primarily for organic product lines. This chapter will discuss the potential applications and effect on product quality of MA and CA treatments for insect control in dried fruits and tree nuts.

20.2 Applications for Insect Control

A variety of insects, infesting the product either in the field or during storage, may be found in postharvest dried fruits and nuts (Table 20.1). Although infestations cause direct damage through feeding and by contaminating product with excreta, webbing, and exuvia, these loses are relatively low. Increased losses occur when infestations cause the implementation of regulatory actions or, worse yet, loss of consumer confidence. The presence of any live insects may impact trade, particularly to foreign markets. Some states and countries consider certain field pests, such as codling moth, filbertworm, and

TABLE 20.1

Postharvest Insect Pests of Dried Fruits and Nuts

Common Name	Scientific Name	Pest Status
Field pests (do not normally reproduce in storage)		
Carob moth	*Ectomyelois ceratoniae*	Pest of dried fruits and nuts
Codling moth	*Cydia pomonella*	Quarantine pest in walnuts
Dried fruit beetles	*Carpophilus* spp.	Field pests in dates, figs, and on drying fruits, normally reproduces in storage only under high moisture conditions
Filbertworm	*Cydia latiferreana*	Pest of hazelnuts
Navel orangeworm	*Amyelois transitella*	Pest in California tree nuts and figs
Raisin moth	*Ephestia figulilella*	Found on drying fruits, occasionally nuts
Tropical nut borer	*Hypothenemus obscurus*	Pest of macadamias
Various nut weevils	*Curculio* spp.	Pests of pecans, chestnuts, hazelnuts, and acorns
Postharvest Storage Pests (primarily found in storage)		
Almond moth	*Cadra cautella*	Common in nut products
Indianmeal moth	*Plodia interpunctella*	Most serious storage pest, common on all dried fruits and nuts
Grain and flour beetles	*Oryzaephilus* spp. and *Tribolium* spp.	Common in stored dried fruits and nuts
Various beetles	*Cryptolestes* spp., *Cryptophagus* spp., and *Trogoderma* spp.	Occasionally pests of dried fruits and nuts
Vinegar flies	*Drosophila* spp.	Common in dried fruits

Curculio spp., to be quarantine pests, and as such require specific disinfestation treatments of product to prevent pest introductions. While most of the storage insects found in these products are cosmopolitan in distribution (Simmons and Nelson, 1975), and are not usually subject to quarantine restrictions, they can cause the rejection of product when found by inspectors.

Correctly applied preharvest pest management practices result in very low infestation rates for field pests, but disinfestation treatments shortly after harvest are often necessary to prevent further damage and meet phytosanitary standards. Because most field pests rarely reproduce in storage, a single treatment is sufficient. In contrast, populations of common stored product insects such as the Indianmeal moth may increase rapidly under storage conditions. Dried fruits and tree nuts are considered durable products, and may be held in storage for months or even years, during which the product is always at risk of reinfestation by storage pests. Strategies emphasizing exclusion, sanitation, and use of protective treatments are effective in controlling storage pests, but dried fruit and nut processors commonly rely on one or more disinfestation treatments to ensure insect-free products.

Disinfestation treatments must be efficacious, economical, applicable to the existing processing and storage system, and must not harm product quality or leave chemical residues on marketed product. Also, for treatments to be completely accepted, environmental and worker safety concerns should be manageable. Because of the diversity of dried fruit and nut products, their production, storage, and marketing methods, and volumes handled, no single disinfestation method is suitable for all situations (Johnson, 2004). Fumigants are commonly used because of their efficacy, economy, and suitability to a wide range of applications (Hagstrum and Subramanyam, 2006). To be widely adopted, any alternative treatment must be as effective and practical as fumigants, and must not be too costly. For some applications within the dried fruit and nut industry, modified and controlled atmospheres may be useful.

The volumes of dried fruits and nuts that must be treated within relatively short periods of time to meet the demands of various markets are sometimes quite large when compared to fresh horticultural commodities. MA and CA treatments are successful in replacing chemical fumigants only when they can be applied in a practical, timely, and economical manner to these large volumes. Consequently, research has concentrated on the application of MA or CA to existing processing methods and storage systems (Figures 20.1 and 20.2), in order to avoid drastic changes to facilities and to keep costs low. The methods used to generate treatment atmospheres include gas from cylinders (Figure 20.3), exothermically generated low oxygen atmospheres (GLOA) using combustion to reduce oxygen levels (Storey, 1975), and gas separation systems (Johnson et al., 1998, 2002; Figure 20.4). Another form of modified atmosphere treatment is the use of low pressures to obtain reduced oxygen tensions (Navarro et al., 2003; Figure 20.5).

FIGURE 20.1
Some examples of dried fruit and nut storage found in the central valley of California. Clockwise from top: walnut silos, Fibreen covered raisin stacks; and plastic covered almond piles.

FIGURE 20.2
Common method of sun drying raisins on paper trays.

FIGURE 20.3
Almond silos being treated with CO_2.

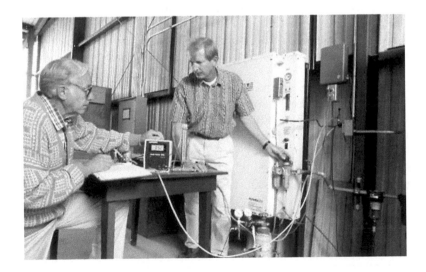

FIGURE 20.4
Hollow fiber membrane gas separation system.

20.2.1 Dried Fruits

Sun-dried fruits such as raisins or figs are susceptible to attack as the product dries by a variety of insects such as raisin moth (*Cadra figulilella*), nitidulid beetles (*Carpophilus* spp.), and vinegar flies (*Drosophila* spp.) (Simmons and Nelson, 1975). Some, in particular, nitidulid beetles and vinegar flies, may continue to reproduce in storage under conditions of excessive moisture. The most common pests infesting product in storage are Indianmeal moth, almond moth (*Cadra cautella*), grain beetles (*Oryzaephilus* spp.), and red flour beetle (*Tribolium castaneum*), although a variety of additional common stored product insects may also be found (Simmons and Nelson, 1975).

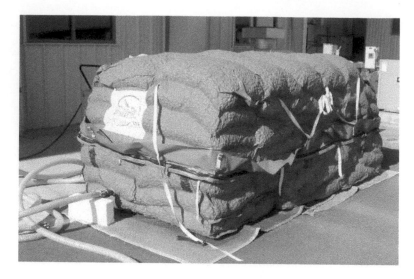

FIGURE 20.5
Flexible, plastic treatment container under vacuum.

Soderstrom and Brandl (1984a) treated raisins using GLOA, with a combustion atmosphere of 0.5 kPa O_2, 12–14 kPa CO_2, 1.0 kPa argon, and the balance nitrogen. Stacks of raisin bins were treated under either polyethylene sheeting or Fibreen, a weatherproof paper laminated with tar and reinforced with fiberglass. The latter is commonly used for commercial raisin storage, and was secured to a wooden framework built around stacked raisin bins and sealed to the ground with oiled dirt. Using a continuous purge rate of 12.7 m^3/h, oxygen levels in small raisin stacks (34 m^3) covered with polyethylene dropped to 0.5 kPa in about 15 h. When adult *Oryzaephilus mercator*, adult red flour beetle and Indianmeal moth pupae were treated in these stacks, complete mortality was reached at 48, 72, and 96 h after purging began, respectively. Larger-scale tests used two Fibreen-covered raisin stacks, 1869 and 3308 m^3, treated at an initial purge rate of 283 m^3/h to reduce oxygen to the target level of <0.5 kPa (Table 20.2). Of the insects treated, *Drosophila*

TABLE 20.2

Efficacy of GLOA Treatments in Two Different Sizes of Fibreen-Covered Raisin Stacks

Parameters	3308 m^3 Stack	1869 m^3 Stack
Initial purge rate	283 m^3/h	283 m^3/h
Time to 0.5 kPa O_2	48 h	24 h
Maintenance purge rate	17 m^3/h	14–17 m^3/h
Average temperature	27°C	16°C
Exposure needed for 100% mortality (time is from the beginning of initial purge)		
Cadra figulilella pupae	72 h	120 h
Plodia interpunctella pupae	48 h	72 h
Drosophila melanogaster pupae	120 h	>336 h
Carpophilus hemipterus mixed cultures	60 h	120 h
Oryzaephilus surinamensis adults	48 h	—
O. surinamensis larvae	48 h	—
Tribolium castaneum adults	48 h	—
T. castaneum larvae	60 h	—

Source: From Soderstrom, E.L. and Brandl, D.G., *J. Econ. Entomol.*, 77, 440, 1984a.

melanogaster pupae proved to be the most tolerant, requiring 5 days (120 h) at around 27°C, and more than 2 weeks at around 16°C. Although these treatment times may seem lengthy, raisins may be stored in Fibreen-covered stacks for several months, and such treatments times are not unacceptable. Phosphine is the fumigant of choice for such stacks and treatment requires a minimum of 3 days at 27°C (Soderstrom et al., 1984).

An economic analysis was made of CA treatment of raisins, comparing costs for methyl bromide, phosphine, GLOA, and nitrogen (Gardner et al., 1982). At the time of the analysis, the GLOA system was found to be competitive with both methyl bromide and phosphine, and further cost savings were seen when heat from the generators was recovered for use in plant heating systems. Use of liquefied nitrogen was found to be the most costly. A more refined analysis (Soderstrom et al., 1984) found the cost of GLOA ($9.66–$10.64/MT) to be slightly lower than phosphine ($10.76/MT). At the time of this analysis, methyl bromide was the least costly ($8.39/MT). Since these studies were done, the cost of methyl bromide has increased considerably, largely due to restrictions imposed by the Montreal Protocol. Methyl bromide prices in the west coast of the United States increased by about 400% from 1995 to 2001 (UNEP/TEAP, 2001), and may have made CA more competitive with the fumigant.

While early GLOA systems were shown to be efficacious and cost effective, concerns about safety, rising natural gas prices, environmental issues, and the continued availability of relatively inexpensive fumigants prevented adoption of the method by processors. As the demand for organic raisins increased, some processors revisited CA treatments as an alternative, and began using liquid nitrogen to disinfest raisins stored in Fibreen-covered yard stacks.

Researchers in the Middle East have used a variety of methods to generate modified atmospheres to treat durable commodities such as dried fruits, including liquid CO_2, hermetic storage, and vacuum treatments. Treatments were often applied within flexible, plastic containers developed for low cost, portable grain storage, known as Volcani Cubes or GrainPro Cocoons (Donahaye et al., 1991). Navarro et al., (1998a) used such a container system to apply CA treatments to disinfest dates of nitidulid beetles. An atmosphere of 60–85 kPa CO_2 and 4–7 kPa O_2 generated from cylinderized liquid CO_2 was applied within a 151 m³ chamber partially filled with 30 t of dates stacked in crates on pallets at temperatures of 22°C–28°C. During the initial purge, a CO_2 concentration of about 85 kPa CO_2 was obtained in the chamber within 1 h by introducing the gas under high pressure, and was maintained for 10 days. After that, CO_2 levels were maintained at about 60 kPa CO_2 for 4.5 months using approximately 0.8 kg CO_2/day. Insect populations were effectively controlled using this regimen, and no significant difference in quality was found between the treated dates and controls stored at −18°C.

Related laboratory studies showed that CO_2 levels >20 kPa, low O_2 levels (<7 kPa) and low pressures (<170 mmHg) caused *Carpophilus hemipterus* to leave artificial refuges at a rate comparable to methyl bromide treatments (Navarro et al., 1989). The low pressure treatments, which provided O_2 levels of <4.7 kPa, were the most effective means of disinfestation, with more than 80% of the beetles leaving refuges after just 10 min of exposure. However, exposure periods for these treatments in excess of 24 h would be needed to cause complete mortality (Navarro et al., 1998b)

Another study used storage under low pressure to control insect infestation in stored dates (Hussain, 1974). When infested dates are packed in polyethylene bags at pressures of 10–20 kPa, insect mortality was 100% after 2 days.

The effect of GLOA on *Cadra cautella* infestation in dried figs was studied by Damarli et al. (1998). An exothermic generator that maintained 1 kPa O_2 and 10–15 kPa CO_2 in a 6 m shipping container was used at 25°C–35°C for 24–72 h. Dried fig samples of 3–4 kg to 8.5 t were infested with *C. cautella* larvae (16–22 days old) and eggs (24, 48, and 72 h old).

There was 100% mortality of all stages in small-scale experiments at 35°C after 24 h exposure. Complete mortality of all stages infesting 8.5 t of dried figs was obtained at 35°C after 30 h exposure. This exposure was comparable to the 24 h methyl bromide treatment recommended to disinfest figs of *C. cautella*.

Laboratory studies (Meyvaci et al., 2003a) investigating methyl bromide alternatives showed that 5 days of exposure to high CO_2 levels (90–96 kPa) caused complete mortality of *C. cautella* eggs, larvae, pupae, and adults at 25°C–35°C. Mixed-stage cultures of the fig mite (*Carpoglyphus lactis*) required 6 days at these temperatures. Results for low-pressure (4.3 kPa) treatments were less consistent, but found that 3 day exposures at 25°C and 2 day exposures at 30°C and 35°C was sufficient to control *C. cautella*, but 3 day exposures at 30°C and 35°C were needed to control fig mites. In a related field study, Emekci et al. (2003) applied cylinderized CO_2 to a 36 m^3 flexible plastic storage container loaded with about 15 MT of dried figs. CO_2 and O_2 levels were >95 kPa and <1 kPa, respectively, and remained stable throughout the 5 day treatment. Test insects included *C. cautella* eggs, *P. interpunctella* larvae, *O. surinamensis* pupae and adults, and mixed stages of *C. hemipterus* and *C. lactis*. The treatment was done at an average temperature of 28°C, and was sufficient to kill all test insects.

Mills et al. (2003) considered CO_2 ($\geq$40 kPa) as an alternative to methyl bromide fumigation for disinfesting dried dates of carob moth (*Ectomyelois ceratoniae*). Use of CO_2 under vacuum did not improve insect control and was not considered to be practical. CO_2 treatment of dates in heated freight containers was more successful, providing adequate insect control after 4 days of exposure. The application of CO_2 to dates under plastic cover was also considered successful and recommended when container capacity is inadequate.

Increasing CO_2 concentrations (60–98 kPa) or decreasing O_2 concentrations (0.5–5 kPa) caused progressively greater mortality of red flour beetle larvae (Soderstrom et al., 1992). Mortality was also increased by increasing temperature (38°C–42°C). However, pretreatment conditioning at 38°C for 24 or 48 h significantly reduced mortalities for the highest CO_2 and lowest O_2 treatments. These results suggest that temperatures above 38°C could be useful in reducing treatment times for CA, but warn that prior exposure to elevated temperatures should be avoided. Donahaye et al. (1994) also showed the effect of increasing temperature on efficacy of low O_2 treatments for nitidulid beetles (*Carpophilus hemipterus* and *Urophorus humeralis*). For both species LT_{95}s for 1 kPa and 2 kPa O_2 at 35°C were usually less than half those at 26°C.

Tarr and Clingeleffer (2005) treated eggs, pupae, and adults of the red flour beetle in hermetically heat-sealed gas barrier bags containing 500 g of golden-colored, sultana raisins, with or without the addition of a commercial O_2 absorber. The bags were held at 30°C for 9 days, 22.5°C for 20 days, or 15°C for 45 days. These times exceeded the minimum time recommended by Annis (1987) for population extinction of red flour beetle under low O_2 conditions. O_2 levels in the bags with or without the O_2 absorber were $\leq$1.7 kPa, and 2.9–16.3 kPa, respectively. All fruit samples treated with O_2 absorbers produced 100% mortality, with no occurrence of a second-generation population, when incubated post-treatment at 27°C for 60 days. Fruit stored in hermetically sealed bags without O_2 absorbers at 22.5°C and 30°C produced 100% adult mortality but eggs and pupae survived. Results at 15°C were confounded by complete control mortality of eggs and pupae, suggesting that the low temperature alone had a great effect. The low O_2 atmosphere was also found to minimize raisin color change during storage.

The psocid *Liposcelis bostrychophila* is a common insect pest, which infests a variety of processed and unprocessed stored foods in households, granaries, and warehouses. Psocids are highly resistant to most forms of pest control, and their populations in stored foods have increased alarmingly in some regions of Asia. Ding et al. (2002) examined the effects of repeated exposures to low O_2 CA (1 kPa O_2, 35 kPa CO_2, and 64 kPa N_2 for 4–14 h) and the insecticide dichlorvos (dimethyl dichloro-vinyl phosphate, DDVP 80%, 0.3 mg/mL for 24 h) on population growth and development of resistance of *L. bostrychophila*. Psocid populations were treated and evaluated every 2 weeks over an 11 week period with CA, with DDVP, or with alternating treatments of CA and DDVP. An untreated population increased 48.1-fold over the 11 week period. Neither CA nor DDVP alone controlled psocid population growth. However, alternating CA and DDVP treatments resulted in a significant increase in mortality compared with either treatment alone. After six exposures, resistance in populations exposed to either CA or DDVP alone increased 1.8- and 2-fold, respectively, and probit analysis of data suggested that further increased resistance was possible. Results of this study have indicated that alternating CA with insecticide applications could be a more effective management measure for control of psocids in stored foods than use of these treatments alone.

20.2.2 Tree Nuts

As with dried fruits, tree nuts may be infested with field pests during harvest. For nut crops such as almonds, that are most often sun dried, or walnuts, which are dehydrated at relatively low temperatures, these field pests are capable of continued damage in storage, and may be a phytosanitary concern in exported product (Johnson, 2004). In addition, nearly all tree nuts are susceptible to infestation by common stored product insects such as Indianmeal moth, grain beetles, and red flour beetles.

In California, the navel orangeworm, *Amyelois transitella*, is a serious pest of the three major tree nut crops: almonds, pistachios, and walnuts. Although primarily a field pest, navel orangeworm may be brought into storage where younger larvae may continue to damage by moving to uninfested nuts (Curtis et al., 1984; Soderstrom and Brandl, 1984b). Storey and Soderstrom (1977) began investigations on the use of CA to control navel orangeworm using an exothermically generated low O_2 atmosphere (<1 kPa O_2, 9–9.5 kPa CO_2, with the balance mostly N_2). Pupae and mature larvae were found to be the most tolerant to low oxygen treatments, with LT_{95}s of 145 and 138 h at 18°C, respectively, and 38 and 37 h at 27°C, respectively. Sublethal exposures of low oxygen to adult navel orangeworm were found to reduce their progeny number; no progeny were produced after adults were exposed for 8 h, while 24 h exposures were required for complete adult mortality.

Brandl et al. (1983) examined the effect of various levels of O_2 and CO_2 on navel orangeworm mortality and determined that low O_2 was more important in reducing treatment times than CO_2. Further study (Soderstrom et al., 1986) looked at the interaction of O_2 concentration, temperature, and relative humidity (RH) on the time needed to control navel orangeworm and Indianmeal moth. The two test species differed in their response; Indianmeal moth was more affected by changes in O_2 concentrations while navel orangeworm responded more strongly to changes in RH. It was suggested that navel orangeworm, as a pest that feeds mainly on product with a higher moisture content, is less tolerant to changes in relative humidities. Indianmeal moth, found in drier storage environments, is better able to survive low relative humidities.

The effect of low O_2 or elevated CO_2 atmospheres on insect feeding was examined by Sodertrom and Brandl (1982). Feeding of navel orangeworm and Indianmeal moth larvae treated for 20 h with various levels of O_2 and CO_2 was determined through a red dye contained in the rearing medium. Both the lowest level of O_2 (1 kPa) and highest level of CO_2 (40 kPa) significantly reduced larval feeding of both species. Navel orangeworm was more susceptible, particularly to high CO_2. Therefore, these results indicate that purging the storage at the beginning of nut filling operation would be beneficial in reducing insect damage to the nuts. CO_2 treatments were considered more practical because it was thought that 40 kPa CO_2 would be easier and cheaper to obtain in a rapid purge of treatment containers.

Soderstrom and Brandl (1984b) conducted a series of successful tests using GLOA to treat bulk stored almonds under plastic covers, in a 3000 ft^3 steel silo, and in 30 m by 7.3 m diameter concrete silos containing 450 MT of inshell almonds. The initial purging of the concrete silo took 8 h, and 100% mortality of Indianmeal moth and navel orangeworm pupae was reached at 36 and 60 h after purging, respectively.

Codling moth (*Cydia pomonella* L.), is a common field pest of California walnuts. Although it does not reproduce in storage, it is considered a quarantine pest by some countries, and processors have relied on methyl bromide to disinfest incoming product (Johnson, 2004). Diapausing codling moth larvae may be present in harvested walnuts, and were found to be the most tolerant life stage to both low O_2 and high CO_2 atmospheres at 25°C, with the lowest LT_{95} (13.6 d) obtained at 60 kPa CO_2 and 60% relative humidity (Soderstrom et al., 1990). Such extended treatment times are not acceptable to processors, who must quickly disinfest incoming product to meet market demands. To reduce treatment times, high-temperature CA treatments were investigated as a methyl bromide alternative for codling moth in inshell walnuts (Soderstrom et al., 1996a). Temperatures (39°C, 41°C, 43°C, and 45°C) selected were those suitable for walnut dehydration without adversely affecting quality. Treatment atmospheres were 98 kPa CO_2 in air, 0.5 kPa O_2 in N_2, and a simulated combustion atmosphere of 0.5 kPa O_2, 10 kPa CO_2, and the balance N_2. Elevated temperatures dramatically reduced treatment times for all atmospheres, including normal air. The most effective treatment tested was 98 kPa CO_2 and 60% RH, with estimated probit 9 values of 3.6 and 6.6 h for 45°C and 43°C, respectively.

Based on the results found in Soderstrom et al. (1996a), a prototype treatment chamber for rapid application of high temperature alone or with CA was designed and tested for disinfesting walnuts of diapausing codling moth larvae (Soderstrom et al., 1996b). The chamber was made of fiberglass insulated polyvinyl chloride pipe, and was designed to allow quick heating and controlled atmosphere recirculation under air tight conditions. The chamber held a mass of about 30 kg of inshell walnuts that was 1.2 m deep and 0.3 m in diameter. Air flow during the initial heating phase was 5.7 m^3/min, which was reduced to 1.4 m^3/min during temperature maintenance. Target temperatures were reached in about 1 h. Treatments evaluated were normal air, 0.5 kPa O_2 (in nitrogen), or 98 kPa CO_2 (balance air) at 43°C, and treatment exposures were based on estimated LT_{95} levels derived from Soderstrom et al. (1996a) (45.2 h for air, 14.6 h for 0.5 kPa O_2, and 5.2 h for 98 kPa CO_2). All treatments exceeded 95% mortality, and any surviving larvae were moribund and subsequently died.

An early economic analysis of the use of ionizing radiation as an alternative to chemical fumigants for dried fruit and nut disinfestation used scenarios that included GLOA treatments (Rhodes and Baritelle, 1986). Although considerably more expensive than fumigants, GLOA treatments had a clear advantage over irradiation for almonds, walnuts, raisins, and prunes during long-term storage. A more current economic analysis was carried out for commercially available methyl bromide alternative treatments for disinfestation of almonds and walnuts, including phosphine fumigation, ionizing irradiation, and CA storage (Aegerter and Folwell, 2001). Costs for each alternative were estimated and compared to benchmark methyl bromide fumigation costs. Irradiation costs ranged from 2× to 14× the benchmark, while CA storage ranged from 1.7× to 2.5× the benchmark costs. PH$_3$ was used only to treat almonds, and its costs were only slightly higher than the benchmark. In all scenarios, CA was less expensive than irradiation. It was concluded that while these alternative treatments could effectively control insect populations in walnuts and almonds at a level comparable to methyl bromide, further research is needed to obtain approval for their use as quarantine treatments. Again, rising fumigant prices since this analysis was done have improved the competitiveness of CA treatments relative to methyl bromide.

Navarro et al. (2003) showed the effectiveness of increased temperatures at reducing treatment times for both CO_2 and vacuum, and demonstrated the utility of flexible plastic treatment containers for the application of vacuum treatments for stored product insect control in durable commodities. Johnson and Valero (2005) showed that relatively short exposures to low pressures of 6.66 kPa could control tree nut pests. Navel orangeworm eggs, diapausing Indianmeal moth, and diapausing codling moth larvae were found to be the most tolerant at temperatures of 25°C and 30°C. Field trials treating almonds in flexible plastic containers were done under both winter (nutmeats in wooden bins) and summer (inshell almonds in 50 kg poly bags) conditions. Winter treatments at 6.3°C–10.5°C required extended treatment times of more than 13 days to get complete control. Summer treatments at 25°C–29.5°C provided complete control with a 48 h exposure.

Pecans are often infested with pecan weevil (*Curculio caryae*), which damages nutmeats and aids invasion of mold (Wells and Payne, 1983). Wells and Payne (1980) used high CO_2 atmospheres to disinfest pecans of this pest, and reduce the level of storage fungi. An atmosphere of 30 kPa CO_2 and 21 kPa O_2 in N_2 was effective at reducing fungal populations and completely killed weevils, but low O_2 atmospheres (1 kPa) with or without elevated CO_2 levels was not.

Macadamia nuts produced on the island of Hawaii are often infested with the tropical nut borer, *Hypothenemus obscurus* (F.) (Delate et al., 1994). The beetle prefers to attack low moisture nuts on the ground or "sticktight" nuts remaining on the tree, and populations may continue to increase in nuts waiting processing. Postharvest control techniques, including CA treatments, were evaluated for control of this pest (Delate et al., 1994). Treatment chambers containing infested nuts and held at 24°C–30°C were flushed daily with either N_2 or CO_2 to maintain gas concentrations of $\geq$95 kPa. Tropical nut borer mortality was lower in nuts with the husk than without, possibly due to absorption of the gas by the husk. Unhusked nuts treated with $\geq$95 kPa CO_2 for 6 days resulted in 97.3% mortality of adult beetles, while all adult insects were killed at this exposure time and concentration when nuts were husked. Although CO_2 treatments provided slightly higher pest mortality, there were concerns over reduced nut quality after extended exposures. A 14 day treatment of $\geq$95 kPa N_2 was required for 100% mortality in unhusked nuts. Holding nuts under CA storage was suggested as a means of alleviating processing loads at the peak of the season without increasing damage due to the beetle.

20.2.3 Combinations with Other Treatments

Combining an initial, short-term CA disinfestation treatment with long-term protective storage treatments in an integrated control program was shown to be promising as an alternative insect control strategy for walnuts (Johnson et al., 1998) and for almonds and raisins (Johnson et al., 2002). CA atmospheres were obtained with a hollow fiber membrane gas separation system in treatment rooms equipped with standard air expansion bags and sealed to a pressure half-life of 1 min after being pressurized to 0.245 kPa. For the initial CA disinfestation treatment, 8 bins of product (3.6 MT of field-run raisins, or 1.8 MT of in-shell nuts) were treated at 0.4 kPa O_2 for 6 days at 25°C after a purge period of 2 days. Test insects for the initial disinfestation treatment were navel orangeworm and raisin moth (*Cadra figulilella*), both common field pests of these products. After initial disinfestation, four bins of product were moved to protected storage environments, either treatment with a preparation containing Indianmeal moth granulosis virus, low temperature ($\leq$10°C) storage, or maintenance CA (5 kPa O_2). Mated female Indianmeal moths, the most serious storage pest of dried fruits and nuts, were added to the storage rooms each week.

TABLE 20.3

Survival of Target Insects after Disinfestation Treatment of 0.4 kPa O_2 for 6 Days at 25°C

Treatment	Number of Insects Treated	Adults Emerged	Survival (%)
Almonds (navel orangeworm as test insect)[a]			
Untreated control	403	393.0	97.5
CA	800	40.0	5.0
Walnuts (navel orangeworm as test insect)[b]			
Untreated control	604	487.0	80.6
CA	1194	4.0	0.3
Raisins (raisin moth as test insect)[a]			
Untreated control	400	251.0	62.7
CA	800	0	0

[a] Johnson et al. (2002).
[b] Johnson et al. (1998).

The initial disinfestation treatment was effective in controlling both test insects (Table 20.3), although there was some evidence that longer exposures might be necessary against raisin moth, or against diapausing navel orangeworm (Johnson et al., 2002). All protective treatments were effective in controlling Indianmeal moth (Table 20.4) and overall product quality was unaffected by any storage treatment. Storage under CA was the most efficacious in preventing infestation by Indianmeal moth, but the sealed storage prevented ready access to the product and the low O_2 atmosphere conditions present worker safety considerations that do not exist for the other methods. Extensive sealing of facilities and equipment for generating CA would be required to provide the needed storage conditions, and would result in considerable expense to processors. However, use of new technologies such as flexible plastic storage containers may help to bring these costs down.

TABLE 20.4

Indianmeal Moth Incidence and Related Damage in Almonds, Walnuts, and Raisins after Extended Storage at 5 kPa O_2

	Live Indianmeal	Indianmeal Moth Damage	
Treatment	Moth	Minor	Serious
Almonds (samples taken after 16 weeks)[a]			
Untreated control	482.5	4.4	28.0
CA	0	0	0
Walnuts (samples taken after 16 weeks)[b]			
Untreated control	81.0	12.6	35.1
CA	0	0	0
Raisins (samples taken after 40 weeks)[a]			
Untreated control	3.2	13.2	
CA	0	0	

Note: Each week mated females (5 for almonds and walnuts, and 15 for raisins) were added to the stored product.
[a] Johnson et al. (2002).
[b] Johnson et al. (1998).

20.3 Effects on Quality

Although most of the uses of controlled atmospheres on bulk-stored dried fruits and nuts are for short-term insect disinfestation, modified or controlled atmospheres are often suggested as means to improve shelf life in these products, particularly for tree nuts. Because tree nuts are high in polyunsaturated lipids they are susceptible to oxidative rancidity (Watkins, 2005). Storage under low O_2 conditions may greatly slow the development of rancidity. As such, vacuum packaging or N_2 flushed packaging is often suggested to improve the shelf life of tree nuts. Consequently, much of the research on the quality effects of controlled atmospheres on tree nuts deals with exposures much longer than those necessary for insect disinfestation treatments. Although dried fruit quality is less affected by rancidity, MAs including elevated CO_2 may reduce fruit darkening and improve shelf life.

20.3.1 Almonds (*Prunus dulcis*)

Kader (1996) states that O_2 is the most important atmospheric component in almond storage, and that high O_2 concentrations result in an increase in rancidity and mold growth. The effect of insecticidal GLOA treatments (>1 kPa O_2, 9–9.5 kPa CO_2, 86–89 kPa N_2, and 1 kPa Ar) on almond flavor quality was determined by exposing shelled and inshell 'Nonpareil' almonds continuously to the atmosphere for 1–12 months (Guadagni et al., 1978a). At normal atmospheres, inshell almonds were found to be more stable than almond meats at both 18.5°C and 27°C, but no differences were detected under low O_2 atmospheres. Low O_2 was as effective as low temperature (18°C) in maintaining stability during storage, and caused less off-flavor development than normal atmosphere for both meats and inshell almonds.

The storage behavior of four varieties of almond ('Marcona,' 'Planeta,' 'Desmayo,' and 'Nonpareil') was investigated at two temperatures (8°C and 36°C), two packaging atmospheres (air and N_2) and two treatments (raw and roasted) for up to 9 months (Garcia-Pascual, 2003). N_2 packaging resulted in O_2 levels <0.25 kPa. No significant differences were observed between air and N_2 packaging for moisture, fat content, peroxide value, α-tocopherol content, and level of aflatoxins. A significant relationship was found between the increase of the peroxide value and the decrease of α-tocopherol content. The aflatoxin contents were always lower than 0.5 μg/kg.

20.3.2 Apricots (*Prunus armeniaca*)

The effects of drying, dehydration, and storage under CA on cell wall components of fresh apricot tissues were studied by Femenia et al. (1998). The degree of browning was also monitored during storage by measuring color and SO_2 content of dehydrated apricots. Apricots were treated with $Na_2S_2O_5$ and acetic acid solutions before dehydration. Dehydrated apricots were then stored at 25°C in air or in five different CA treatments (100 kPa N_2, 20 kPa CO_2 and 80 kPa N_2, 40 kPa CO_2 and 60 kPa N_2, 60 kPa CO_2 and 40 kPa N_2, 100 kPa CO_2). The yield of apricot cell wall material decreased substantially during drying pretreatments and also during dehydration, by 9.5% and 4.7%, respectively. In particular, acetic acid solubilized a large amount of pectic polysaccharides. Further degradation of pectic substances occurred during drying, probably due to the high temperature used. CA storage retarded browning in comparison to the sample stored under air. The content of SO_2 decreased markedly for the sample stored in air, whereas gradual losses were observed for CA-stored samples. In general, samples stored in CA containing low CO_2 levels (20 and 40 kPa) showed minor disruption of cell wall components and better initial characteristics of dehydrated apricots.

20.3.3   Chestnuts (*Castanea sativa*)

Anelli et al. (1982) reported that chestnuts stored in 2 kPa O_2 plus 20 kPa CO_2 at 0.5°C had an increased rate of sugar metabolism. These chestnuts have also been reported to have lower glutamine content as the CO_2 level increased from 5 to 20 kPa. A prestorage treatment of 80 kPa CO_2 for 10 days followed by 80–90 days of storage at 0°C was the best method of reducing Sclerotinia rot in chestnuts (Bertolini and Tonini, 1983). Rouves and Prunet (2002) examined various storage techniques for chestnuts, including low tempera-ture CA (2 kPa O_2 plus 5 kPa CO_2 at −1°C or 1°C). For the varieties 'Marigoule' and 'Bouche de Betizac' water loss was prevented, mold reduced, and taste maintained. The varieties 'Comballe' and 'Marron de Goujonac' did not benefit from CA storage. Chestnuts kept under 20 kPa CO_2 and 2 kPa O_2 had lower rot and insect development after 150 days of storage, while treatment with 2.5 kPa CO_2 and 1.5 kPa O_2 was less effective (Mignani and Vercesi, 2003).

20.3.4   Dates (*Phoenix dactylifera*)

Rygg (1975) suggested inert gas or vacuum packing for storage of high-moisture dates. Mohsen et al. (2003) noted that vacuum packaging is a useful technique for reducing darkening of the date for long-term storage. Mutlak and Mann (1984) reported that browning can be inhibited at low oxygen potentials. CA (5, 10, or 20 kPa CO_2) at 0°C extended storage period and maintained quality of fully mature 'Bahri' date fruit (Al-Redhaiman, 2005). The quality of fruit stored under 20 kPa CO_2 was maintained for up to 26 weeks compared to 17 weeks for fruit held under 5 and 10 kPa CO_2 and only 7 weeks for fruit kept in normal air. A 20 kPa CO_2 maintained acceptable levels of fruit soluble solids, total sugar, total tannins, and caffeoylshikimic acid.

20.3.5   Figs (*Ficus carica*)

The effect of insecticidal GLOA treatments on the quality of dried figs was studied by Damarli et al. (1998). An exothermic generator that maintained 1 kPa O_2 and 10–15 kPa CO_2 atmosphere composition at 35°C for 30 h was used to treat 8.6 MT of product. Under these conditions, considered to be optimal for controlling insect populations, there were no observed negative effects on fruit weight, moisture, color, total sugar content, or acidity. Similarly, the effect of CO_2 proposed for insecticidal treatments on dried fig quality was examined by Meyvaci et al. (2003a). Exposure to 100 kPa CO_2 for 45 days caused color changes as measured on the Hunter color scale with a colorimeter. In the Hunter scale, L measures lightness and varies from 100 for perfect white to zero for black, a measures redness when positive, gray when zero, and greenness when negative, and b measures yellowness when positive, gray when zero, and blueness when negative. In CA-treated figs, mean L values decreased from 57.9 to 51.7, a values increased from 9.8 to 11.7, and b values decreased from 30.6 to 27.7. In studies to improve shelf life of dried figs, Meyvaci et al. (2003b), packaging in 100 kPa N_2, 20 kPa CO_2 in N_2, or under vacuum was examined for up to 7.5 months of storage. Figs stored under N_2 were darker than figs stored in normal air, while CO_2 storage reduced darkening slightly. Figs stored under N_2 also had increased sugar formation. Vacuum packaging resulted in the darkest fruit, but reduced sugar formation. Storage under either N_2 or CO_2 controlled mold growth in rehydrated figs for 3 months more than storage at ambient conditions.

20.3.6 Hazelnuts, Filberts (*Corylus* spp.)

Hazelnut is characterized by high oil content (about 65% of its kernel weight) and is rich in unsaturated fatty acids, and therefore highly sensitive to rancidity (San Martin et al., 2001). Storage conditions that are capable of retarding lipid oxidation and hydrolysis will help preserve quality. Reduced temperature may be effective in combination with other protective measures such as vacuum packaging in extending roasted kernel shelf life to 1 year or more (Ebraheim et al., 1994). Keme et al. (1983) reported that it is possible to store 'Piemonteses,' 'Roman,' and 'Àkcakoca' hazelnuts at ambient temperature under nitrogen ($\geq$99.5 kPa) for prolonged periods of time with a loss of quality that is comparable to that resulting from storage conditions at low temperatures and controlled RH (3°C–6°C, 50%–60% RH). Hazelnut ('Negret') quality was studied after storage in selected CA conditions by San Martin et al. (2001). The shelled and unshelled hazelnuts were stored with different oxygen levels (1, 5, 10, and 20 kPa O_2) at two different temperatures (7°C and 25°C), and quality of the hazelnuts during storage was monitored by determining the peroxide value, acid value, K_{232} and K_{270} indices, percentage of unsaturated fatty acids and sensory analysis. After 1 year in storage, none of the storage conditions tested caused significant rancidity. Storage in atmospheres with O_2 levels lower than 10 kPa significantly reduced autoxidation and the low temperature delayed lipid rancidity.

20.3.7 Macadamia (*Macadamia integrifolia*)

As with other nuts, prolonged exposure to O_2 results in rancidity, and vacuum packaging or nitrogen flush offers protection from O_2 (Cavaletto, 2004).

20.3.8 Pecan (*Carya illinoinsis*)

Shelf life of pecans may be increased by storage in 2–3 kPa O_2 in N_2, and less frequently using CO_2 as the balance gas (Maness, 2004). Storage in <2 kPa O_2 for 52 days can cause the development of a "fruity" flavor (Santerre et al., 1990). O_2 transmission rates for packaging materials should be >0.08 mL/100 cm per 24 h (Dull and Kays, 1988). Vacuum packaging can offer a further benefit of protection from breakage.

20.3.9 Pine Nuts (*Araucaria* spp.)

Pine nuts (piñones), *Araucaria araucana*, stored in different MA (polyethylene bags lined with volcanic dust) and CA (10–5 kPa CO_2–O_2, and 20–5 kPa CO_2–O_2) conditions maintained their moisture and starch content, and 20–5 kPa CO_2–O_2 was the optimum CA condition (Estevez and Galletti, 1997).

20.3.10 Pistachio (*Pistacia vera*)

Although pistachios are fairly stable stored in normal air at 20°C, storage under reduced O_2 (<0.5 kPa), vacuum packaging or N_2 flushed packaging further improves flavor stability (Labavitch, 2004a). Changes in fatty acid composition and peroxide value of pistachio nuts were studied during storage at 10°C–30°C under either air or 98 kPa CO_2 at the monolayer moisture content of pistachios (considered to be the most stable moisture content for storage of dehydrated products), and were compared to those stored under ambient conditions (Maskan and Karatas, 1998). Use of the Arrhenius model to predict fatty acid loss at 10°C, 20°C, and 30°C under both atmospheric conditions, together with prediction of the rate and extent of oxidation during storage, was also investigated. Storage of pistachio nuts under CO_2 provided the greatest stability in relation to fatty acid loss, peroxide values, and free fatty acid formation, especially at 10°C. The Arrhenius model revealed that storage of nuts under CO_2 was more temperature sensitive, and more linolenic acid was lost by oxidation than linoleic acid. Comparison of samples stored at ambient conditions with those stored at or near the monolayer moisture content and under CO_2 storage showed the latter samples to be more stable.

Storage stability, oil characteristics, and chemical composition of whole-split pistachio nuts (*Pistachia vera* L.) were determined on samples stored in CA (2 kPa air, 98 kPa CO_2), air at the monolayer value at 10°C, 20°C, 30°C, and at ambient conditions (Medeni and Sukru, 1999). Most oxidation was observed under ambient storage conditions. CO_2 especially improved the storage stability at low temperatures. Using the reaction rate constants of peroxide formation, it was revealed that as temperature increased, the ratio of rate constants (air/CO_2) approached 1, which means that no significant difference existed between air and CO_2 storages at 30°C. The oxidation activation energies were 8.33 and 13.39 kcal/mol under air and CO_2 storage, respectively.

20.3.11   Raisins (*Vitus vinifera*)

Guadagni et al. (1978b) examined the effect of long-term exposure to insecticidal GLOA treatments on flavor stability of field-run Thompson seedless raisins. After 12 month storage under GLOA (>1 kPa O_2, 9–9.5 kPa CO_2, 86–89 kPa N_2, and 1 kPa Ar) at 18.5°C and 27°C, raisin flavor, based on taste panel evaluations, was equal or superior to that of raisins stored under normal air.

The effect of a low O_2 atmosphere generated by the addition of a commercial O_2 absorber on fruit color of golden-colored, sultana raisins stored in hermetically heat-sealed gas barrier bags at 15°C, 22.5°C, and 30°C was assessed over a 28, 56, or 84 day storage period (Tarr and Clingeleffer, 2005). Fruit color changes were measured using the CIE tristimulus *L*, *a*, *b* measuring system. Low storage temperature (15°C) and low O_2 atmosphere both maintained fruit color during the experimental storage regimes. There were significant negative effects on fruit color at the higher temperature (30°C) and longer storage period without a low O_2 atmosphere.

20.3.12   Walnuts (*Juglans regia*)

Shelf life can be extended by storage in <1 kPa O_2. O_2 <0.5 kPa (balance N_2) or CO_2 levels above 80 kPa in air can be effective in insect control (Labavitch, 2004b).

20.4   Conclusions

Low moisture dried fruits and tree nuts are relatively tolerant of many kinds of MA or CA, making their use for insecticidal treatments or for shelf life extension attractive. Although CA treatment times for bulk-stored product are usually much longer than those for methyl bromide, at temperatures ≥30°C, they are often comparable to phosphine fumigation. For applications that do not require rapid turnaround of product, such as yard stacks of raisins, CA treatments may be appropriate. Currently, the biggest barrier to widescale adoption of CA for disinfestation of bulk dried fruits and nuts is the increased costs, although for high-value organic product, the added cost is more acceptable. MA in packaging (N_2 flushed or vacuum packed) as a means of improving shelf life is currently used, most often by processors of tree nut products.

20.5 Future Research Needs

Much of the cost of CA treatments for bulk-stored dried fruits and nuts is associated with either building new gastight storage structures or retrofitting old storage facilities. Gas leaks in poorly sealed storages also drive up the cost. The recent development of inexpensive, flexible plastic containers capable of maintaining treatment gas levels should make the use of CA treatments more affordable. Costs for bulk CA treatment may be further reduced by more fully exploring the use of vacuum under these flexible plastic containers. Further research is needed to integrate this technology with existing processing methods. An important disadvantage to CA bulk treatments is the lengthy treatment times needed for insect control. Treatment at elevated temperatures is well known to decrease treatment times, but may add considerably to the cost or result in product quality degradation. Methods to cheaply apply CA at high temperatures and protect product quality are also needed.

References

Aegerter, A.F. and R.J. Folwell. 2001. Selected alternatives to methyl bromide in the postharvest and quarantine treatment of almonds and walnuts: An economic perspective. *Journal of Food Processing and Preservation* 25: 389–410.

Al-Redhaiman, K.N. 2005. Chemical changes during storage of 'Barhi' dates under controlled atmosphere conditions. *HortScience* 40: 1413–1415.

Anelli, G., F. Mencarelli, F. Nardin, and C. Stingop. 1982. La conservazione delle castagne mediante l'impiego delle atmosfere controllate. *Industrie Alimentari* 21: 217–220.

Annis, P.C., 1987. Towards rational controlled atmosphere dosage schedules: A review of current knowledge. In: Donahaye, E., Navarro, S. (Eds.), *Proceedings of the Fourth International Working Conference on Stored-Product Protection*, Tel Aviv, Israel, September 21–26, 1986. Bet Dagen, pp. 128–148.

Bell, C.H. 2000. Fumigation in the 21st century. *Crop Protection* 19: 563–569.

Bell, C.H. and N. Savvidou. 1999. The toxicity of Vikane (sulfuryl fluoride) to age groups of eggs of the Mediterranean flour moth (*Ephestia kuehniella*). *Journal of Stored Products Research* 35: 233–247.

Benhalima, H., M.Q. Chaudhry, K.A. Mills, and N.R. Price. 2004. Phosphine resistance in stored-product insects collected from various grain storage facilities in Morocco. *Journal of Stored Products Research* 40: 241–249.

Bertolini, P. and G. Tonini. 1983. Prestorage, high carbon dioxide treatment for the control of *Sclorotinia pseudotuberosa* (Rehm) rot in chestnuts. *Proceedings of the 16th International Congress of Refrigeration*, Paris, France, pp. 231–236.

Brandl, D.G., E.L. Soderstrom, and F.E. Schreiber. 1983. Effects of low-oxygen atmospheres containing different concentrations of carbon dioxide on mortality of the navel orangeworm, *Amyelois transitella* (Walker) (Lepidoptera: Pyralidae). *Journal of Economic Entomology* 76: 828–830.

Campbell, B.C., R.J. Molyneux, and T.F. Schatzki. 2003. Current research on reducing pre- and post-harvest aflatoxin contamination of US almond, pistachio, and walnut. *Journal of Toxicology: Toxin Reviews* 22: 225–266.

Cavaletto, C.G. 2004. Macadamia nut. In: *The Commercial Storage of Fruits, Vegetables, and Florist and Nursery Stocks*, K.C. Gross, C.Y. Wang, and M. Saltveit (Eds.), *USDA Agricultural Handbook No. 66*, http://www.ba.ars.usda.gov/hb66/contents.html

Curtis, C.E., R.K. Curtis, and K.L. Andrews. 1984. Progression of navel orangeworm (Lepidoptera: Pyralidae) infestation and damage of almonds on the ground and on the tree during harvest. *Environmental Entomology* 13: 146–149.

Damarli, E., G. Gun, G. Ozay, S. Bulbul, and P. Oechsle. 1998. An alternative method instead of methyl bromide for insect disinfestation of dried figs: Controlled atmosphere. _Acta Horticulturae_ 480: 209–214.

Delate, K.M., J.W. Armstrong, and V.P. Jones. 1994. Postharvest control treatments for _Hypothenemus obscurus_ (F.) (Coleoptera: Scolytidae) in macadamia nuts. _Journal of Economic Entomology_ 8: 120–126.

Ding, W., J.-J. Wang, Z.-M. Zhao, and J.H. Tsai. 2002. Effects of controlled atmosphere and DDVP on population growth and resistance development by the psocid, _Liposcelis bostrychophila_ Badonnel (Psocoptera: Liposcelididae). _Journal of Stored Products Research_ 38: 229–237.

Donahaye, E., S. Navarro, and M. Rinder. 1994. The influence of temperature on the sensitivity of two nitidulid beetles to low oxygen concentrations. _Proceedings of the 6th International Working Conference on Stored Product Protection_, Canberra, Australia, April 17–23, 1994, pp. 88–90.

Donahaye, E., S. Navarro, A. Ziv, Y. Blauschild, and D. Weerasinghe. 1991. Storage of paddy in hermetically sealed plastic liners in Sri Lanka. _Tropical Science_ 31: 109–21.

Dull, G.G. and S.J. Kays. 1988. Quality and mechanical stability of pecan kernels with different packaging protocols. _Journal of Food Science_ 53: 565–567.

Ebraheim, K.S., D.G. Richardson, and R.M. Tetley. 1994. Effects of storage temperature, kernel intactness and roasting temperature on vitamin E, fatty acids and peroxide value of hazelnuts. _Acta Horticulturae_ 351: 677–684.

Emekci, M., A.G. Ferizli, S. Tütüncü, and S. Navarro. 2003. Modified atmosphere as an alternative to MBr in the dried fig sector in Turkey. _Proceedings of the International Research Conference on Methyl Bromide Alternatives and Emissions Reductions_, San Diego, CA, November 3–6, 2003, pp. 79.1–79.2.

Estevez, A.M. and L. Galletti. 1997. Postharvest storage of "piñones" from _Araucaria araucana_ (Mol.) K. Koch under controlled atmosphere conditions. _Proceedings of the International Controlled Atmosphere Research Conference_, Vol. 3: Fruits Other than Apples and Pears, University of California, Davis, CA, July 13–18, 1997, pp. 185–189.

Femenia, A., E.S. Sanchez, S. Simal, and C. Rosello. 1998. Modification of cell wall composition of apricots (_Prunus armeniaca_) during drying and storage under modified atmospheres. _Journal of Agricultural and Food Chemistry_ 46: 5248–5253.

García-Pascual, P., M. Mateos, V. Carbonell, and D.M. Salazar. 2003. Influence of storage conditions on the quality of shelled and roasted almonds. _Biosystems Engineering_ 84: 201–209.

Gardner, P.D., E.L. Soderstrom, J.L. Beritelle, and K.N. de Lozano. 1982. Assessing alternative methods of pest control in raisin storage, University of California Bulletin No. 1906.

Guadagni, D.G., E.L. Soderstrom, and C.L. Storey. 1978a. Effects of controlled atmosphere on flavor stability of almonds. _Journal of Food Science_ 43: 1077–1081.

Guadagni, D.G., C.L. Storey, and E.L. Soderstrom. 1978b. Effects of controlled atmosphere on flavor stability of raisins. _Journal of Food Science_ 43: 1726–1728.

Hagstrum, D.W. and B. Subramanyam. 2006. _Fundamentals of Stored-Product Entomology_. St. Paul, MN: AACC International.

Hussain, A.A. 1974. Date palms and dates with their pests in Iraq, Baghdad, University of Baghdad, Ministry of Higher Education and Scientific Research, Baghdad, Iraq.

Johnson, E.H. 1980. Tolerance and exemption from tolerances for pesticide chemicals in or on raw agricultural commodities; carbon dioxide, nitrogen, and combustion product gas. _Federal Register_ 45: 75663–75664.

Johnson, E.H. 1981. Carbon dioxide, nitrogen, and combustion product gases: tolerance for pesticides in food administered by the environmental Protection Agency. _Federal Register_ 46: 32865–32866.

Johnson, J.A. 2004. Dried fruit and nuts: United States of America. In: _Crop Post-Harvest Science and Technology, Vol. 2: Durables_, R. Hodges and G. Farrell (Eds.), Oxford, U.K.: Blackwell Science, pp. 226–235.

Johnson, J.A. and K.A. Valero. 2005. Vacuum treatments for tree nuts. _Proceedings of the International Research Conference on Methyl Bromide Alternatives and Emissions Reductions_, San Diego, CA, October 31–November 3, 2005, pp. 81.1–81.4.

Johnson, J.A., P.V. Vail, D.G. Brandl, J.S. Tebbets, and K.A. Valero. 2002. Integration of nonchemical treatments for control of postharvest pyralid moths (Lepidoptera: Pyralidae) in almonds and raisins. _Journal of Economic Entomology_ 95: 190–199.

Johnson, J.A., P.V. Vail, E.L. Soderstrom, C.E. Curtis, D.G. Brandl, J.S. Tebbets, and K.A. Valero. 1998. Integration of nonchemical, postharvest treatments for control of navel orangeworm (Lepidoptera: Pyralidae) and Indianmeal moth (Lepidoptera: Pyralidae) in walnuts. *Journal of Economic Entomology* 91: 1437–44.

Kader, A.A. 1996. In-plant storage. In: *Almond Production Manual*, W.C. Micke (Ed.), University of California, Division of Agriculture and Natural Resources, Publication No. 3364, Oakland, CA, pp. 274–277.

Keme, T., M. Messerli, J. Shejbal, and F. Vital. 1983. The storage of hazelnut at room temperatures under nitrogen (II). *Reviews in Chocolate, Confectionery and Bakery* 8: 15–20.

Labavitch, J. 2004a. Pistachios. In: *The Commercial Storage of Fruits, Vegetables, and Florist and Nursery Stocks*, K.C. Gross, C.Y. Wang, and M. Saltveit (Eds.), *USDA Agricultural Handbook No. 66*, http://www.ba.ars.usda.gov/hb66/contents.html

Labavitch, J. 2004b. Walnut. In: *The Commercial Storage of Fruits, Vegetables, and Florist and Nursery Stocks*, K.C. Gross, C.Y. Wang, and M. Saltveit (Eds.), *USDA Agricultural Handbook No. 66*, http://www.ba.ars.usda.gov/hb66/contents.html

Maness, N. 2004. Pecan. In: *The Commercial Storage of Fruits, Vegetables, and Florist and Nursery Stocks*, K.C. Gross, C.Y. Wang, and M. Saltveit (Eds.), *USDA Agricultural Handbook No. 66*, http://www.ba.ars.usda.gov/hb66/contents.html

Maskan, M. and S. Karataş. 1998. Fatty acid oxidation of pistachio nuts stored under various atmospheric conditions and different temperatures. *Journal of the Science of Food and Agriculture* 77: 334–340.

Meyvaci, K.B., U. Aksoy, F. Sen, A. Altindisli, F. Turanli, M. Emekçi, and A.G. Ferizli. 2003a. Project to phase-out methyl bromide in the dried fig sector in Turkey. *Acta Horticulturae* 628: 73–81.

Meyvaci, K.B., F. Sen, U. Aksoy, F. Ozdamar, and M. Cakr. 2003b. Research on prolonging the marketing period of dried and ready-to-eat type figs (*Ficus carica*). *Acta Horticulturae* 628: 439–445.

Mignani, I. and A. Vercesi. 2003. Effects of postharvest treatments and storage conditions on chestnut quality. *Acta Horticulturae* 600: 781–85.

Mills, K.A., T.J. Wontner-Smith, S.C. Cardwell, and C.H. Bell. 2003. The use of carbon dioxide as an alternative to methyl bromide for the disinfestation of palm dates. *Proceedings of the 8th International Working Conference on Stored Product Protection*, York, England, July 22–26, 2002, pp. 729–735.

Mohsen, A., S.B. Amara, N.B. Salem, A. Jebali, and M. Hamdi. 2003. Effect of vacuum and modified atmosphere packaging on Deglet Nour date storage in Tunisia. *Fruits (Paris)* 58: 205–212.

Mutlak, H.H. and J. Mann. 1984. Darkening of dates: Control by microwave heating. *Date Palm Journal* 3: 303–316.

Navarro, S., E. Donahaye, R. Dias, and E. Jay. 1989. Integration of modified atmospheres for disinfestation of dried fruits, BARD Final Report Project No. I-1095–86, 86 pp.

Navarro, S., E. Donahaye, M. Rindner, and A. Azrieli. 1998a. Storage of dried fruits under controlled atmospheres for quality preservation and control of nitidulid beetles. *Acta Horticulturae* 480: 221–226.

Navarro, S., E. Donahaye, M. Rindner, and A. Azrieli. 1998b. Control of nitidulid beetles in dried fruits by modified atmospheres. *Bulletin-OILB/SROP* 21: 159–63.

Navarro, S., S. Finkelman, G. Sabio, A. Isikber, R. Dias, M. Rindner, and A. Azrieli. 2003. Enhanced effectiveness of vacuum or CO_2 in combination with increased temperatures for control of storage insects. *Proceedings of the 8th International Working Conference on Stored Product Protection*, York, England, July 22–26, 2002, pp. 818–822.

Prabhakaran, S. and R.L. Williams. 2007. Global status and adoption of Profume® gas fumigant. *Proceedings of the International Research Conference on Methyl Bromide Alternatives and Emissions Reductions*, San Diego, CA, October 29–November 1, 2007, pp. 90.1.

Rhodes, A.A. and J.L. Baritelle. 1986. Economic engineering feasibility of irradiation as a postharvest disinfestation treatment for California dried fruits and nuts. In: *Irradiation Disinfestation of Dried Fruits and Nuts: A Final Report from the United States Department of Agricultural Research Service and Economic Research Service*, A.A. Rhodes (Ed.), United States Department of Energy, Energy Technologies Division, pp. 171–229.

Rouves, M. and J.P. Prunet. 2002. Nouvelle technologie pour la conservation des chataignes: l'atmosphere controlee et ses effets. *Infos CTIFL* 186: 33–35.

Rygg, G.L. 1975. Date development, handling, and packing in United States, *USDA Agricultural Handbook No. 482*, United States Department of Agriculture, Agricultural Research Service, Washington DC, 56 pp.

San Martín, M.B., T. Fernández-García, A. Romero, and A. López. 2001. Effect of modified atmosphere storage on hazelnut quality. *Journal of Food Processing and Preservation* 25: 309–321.

Santerre, C.R., A.J. Scouten, and M.S. Chinnan. 1990. Room temperature of shelled pecans: Control of oxygen. *Proceedings Southeastern Pecan Growers' Association* 83: 113–121.

Simmons, P. and H.D. Nelson. 1975. Insects on Dried Fruits, *USDA Agricultural Handbook No. 464*, United States Department of Agriculture, Agricultural Research Service, 26 pp.

Soderstrom, E.L. and D.G. Brandl. 1982. Antifeeding effect of modified atmospheres on larvae of the navel orangeworm and Indianmeal moth (Lepidoptera: Pyralidae). *Journal of Economic Entomology* 75: 704–705.

Soderstrom, E.L. and D.G. Brandl. 1984a. Low-oxygen atmosphere for postharvest insect control in bulk-stored raisins. *Journal of Economic Entomology* 77: 440–445.

Soderstrom, E.L. and D.G. Brandl. 1984b. Modified atmospheres for postharvest insect control in tree nuts and dried fruits. *Proceedings of the 3rd International Working Conference on Stored-Product Entomology*, Manhattan, KS, October 23–28, 1983, pp. 487–497.

Soderstrom, E.L., D.G. Brandl, and B. Mackey. 1990. Responses of codling moth (Lepidoptera: Tortricidae) life stages to high carbon dioxide or low oxygen atmospheres. *Journal of Economic Entomology* 83: 472–475.

Soderstrom, E.L., D.G. Brandl, and B. Mackey. 1992. High temperature combined with carbon dioxide enriched or reduced oxygen atmospheres for control of *Tribolium castaneum* (Herbst). *Journal of Stored Products Research* 28: 235–238.

Soderstrom, E.L., D.G. Brandl, and B.E. Mackey. 1996a. High temperature alone and combined with controlled atmospheres for control of diapausing codling moth (Lepidoptera: Tortricidae) in walnuts. *Journal of Economic Entomology* 89: 144–147.

Soderstrom, E.L., D.G. Brandl, and B.E. Mackey. 1996b. High temperature and controlled atmosphere treatment of codling moth (Lepidoptera: Tortricidae) infested walnuts using a gas-tight treatment chamber. *Journal of Economic Entomology* 89: 712–714.

Soderstrom, E.L., P.D. Garner, J.L. Baritelle, N.K. de Lozano, and D. Brandl. 1984. Economic cost evaluation of a generated low-oxygen atmosphere as an alternative fumigant in the bulk storage of raisins. *Journal of Economic Entomology* 77: 457–461.

Soderstrom, E.L., B.E. Mackey, and D.G. Brandl. 1986. Interactive effects of low-oxygen atmospheres, relative humidity, and temperature on mortality of two stored-product moths (Lepidoptera: Pyralidae). *Journal of Economic Entomology* 79: 1303–1306.

Storey, C.L. 1975. Mortality of three stored product moths in atmospheres produced by an exothermic inert atmosphere generator. *Journal of Economic Entomology* 68: 736–738.

Storey, C.L. and E.L. Soderstrom. 1977. Mortality of navel orangeworm in a low-oxygen atmosphere. *Journal of Economic Entomology* 70: 95–97.

Tarr, C.R. and P.R. Clingeleffer. 2005. Use of an oxygen absorber for disinfestation of consumer packages of dried vine fruit and its effect on fruit color. *Journal of Stored Products Research* 41: 77–89.

UNEP (United Nations Environmental Programme). 2006. *Handbook for the Montreal Protocol on Substances that Deplete the Ozone Layer*, 7th ed., UNEP Ozone Secretariate, Nairobi, Kenya. http://ozone.unep.org/Publications/Handbooks/MP_Handbook_2006.pdf.

UNEP/TEAP (United Nations Environmental Programme/Technology and Economic Assessment Panel). 2001. Report of the Technology and Economic Assessment Report, Section 9.3: Methyl Bromide Technical Options Committee (MBTOC).

USDA (United States Department of Agriculture). 2007. Agricultural statistics, United States Department of Agriculture, National Agricultural Statistics Service, http://www.nass.usda.gov/Publications/Ag_Statistics/2007/index.asp

Watkins, C. 2005. The world of nuts. *International News on Fats, Oils and Related Materials* 16: 200–201.

Wells, J.M. and J.A. Payne. 1980. Reduction of microflora and control of in-shell weevils in pecans stored under high carbon dioxide atmospheres. *Plant Disease* 64: 997–999.

Wells, J.M. and J.A. Payne. 1983. Weather-related incidence of aflatoxin contamination in late-harvested pecans. *Plant Disease* 67: 751–753.

21

Economic Benefits of Controlled Atmosphere Storage and Modified Atmosphere Packaging

Edmund K. Mupondwa

CONTENTS

21.1 Background

This chapter discusses the economic and commercial impact of controlled atmosphere (CA) and modified atmosphere (MA) as increasingly popular food preservation techniques. Their uses span a wide range of food products, including (a) fresh whole and fresh-cut fruit and vegetable products; (b) filled pasta, breaded poultry, meat, or fish; (c) ready meals and other cook-chill products; (d) dairy products; (e) cooked, cured, and processed meat products; (f) cooked and dressed vegetable products. The use of MA and CA technologies has economic implications given global trends over the last 25 years showing significant socioeconomic and business changes and a shift to a foodservice economy driven by demand for more convenient food. The stimulated international trade arising from globalization of markets means that food products have to be transported over greater distances and through a wider variety of climatic zones. These changes have led to a highly competitive food trading environment and demand for postharvest technologies driven by enhanced product quality, extension of shelf life, minimization of waste, and reduced operational costs. It is clear that new MA and CA technologies will have a major economic impact on the food processing industry, especially given the high investment costs associated with developing and marketing these convenience products.

21.2 Introduction

This chapter discusses economic benefits of storage technologies as they relate to horticulture. Storage itself can be viewed in terms of its broadest economic sense. Economists define storage to encompass any activity that transfers a commodity available at any given point in time into a similar commodity available later (Newbery and Stiglitz, 1981; Wright, 2001). All storage activities have a common feature: the inventory in storage is constrained to be nonnegative, otherwise it would be impossible at the margin to draw from the future. The horticulture sector is characterized by the high perishability of its produce. The greatest utility and value in producing fresh perishable commodities comes from the ability to extend their supply over and beyond the harvest season.

There are two types of technologies that have evolved: CA storage and modified atmosphere packaging (MAP). This chapter focuses mainly on CA storage, with a comparison with regular atmosphere (RA) storage, which involves refrigeration or cold storage. Additional discussion is provided on MAP in general. The commercial development of CA storage technology has been fostered by the economic need for postharvest maintenance of fresh produce for ultimate consumption in later periods. In other words, CA is used to extend the utility and useful marketing period of fresh produce during storage, transportation, and distribution so as to maintain quality, nutritional, and market value of the produce, in comparison to that achieved via the use of refrigeration or cold storage alone. CA works by altering oxygen and carbon dioxide concentrations during storage, i.e., reducing oxygen and increasing carbon dioxide in storage facility (Prange et al., 2005). These conditions are narrowly defined depending on the type of fresh produce. In general, the conditions may include setting the lowest temperature to prevent freezing or chilling injury, the lowest level of oxygen to attenuate the rate of respiration and hold the development of senescence; ensuring the highest and safest level of carbon dioxide to minimize the rate of respiration and slow the senescence and ripening process; ensuring the lowest level of ethylene, which is a hormone responsible for triggering the development of ripening and senescence; and finally the maintenance of high levels of humidity to reduce moisture loss from the stored product. The ability to radically suppress the initiation or development of senescence gives significant utility to CA technology in enabling fresh produce to be stored for up to 12 months while retaining high quality, nutrition, and flavor. The ultimate aim of CA from the beginning has been the maintenance of an optimal storage atmosphere to achieve a measurable and desirable response in the stored product (Prange et al., 2005; Schotsmans and Prange, 2006; Yahia, 2006).

It should of course be pointed out that there is a relationship between RA storage and CA storage. The industry uses these two types of storage for short-term and long-term storages, respectively. RA does not suppress or stop the ripening process, it merely slows it down. Therefore, in temperate climates, growers will typically sell RA produce by late January or early February. Because it involves minimal cost relative to CA storage, RA storage is less expensive. On the other hand, CA storage costs more per bin, mainly due to the additional construction and capital equipment needed to monitor and regulate the environmental conditions. Hence, only the best produce is placed in CA storage. It is widely believed that CA should be regarded as supplemental to RA storage. In fact, from a regulatory stand point, once a CA storage facility is opened (due to demand and supply factors), the entire inventory is converted to RA storage.

CA innovation lies at the heart of the development in the horticultural sector. However, it should be noted that CA technology has not been universally adopted commercially to cover all crops. It is principally adopted for fresh produce such as fruits, vegetables, cut flowers, seeds, and nuts. Within this grouping, temperate fruits such as apples and pears are by far the leading crops for which CA technology has been adopted (Yahia, 2006). Commercial use of CA storage is less on cabbages, sweet onions, kiwifruit, avocados, persimmons, pomegranates, and nuts (Kader, 2005a). However, although most MAP and CA technology is not used for storage of tropical crops, it is used for their marine transportation of tropical crops. Developments in CA and humidity control technologies that reduce spoilage and allow the substitution of cheaper ocean shipping for air transport for some of the more perishable items represents a very significant enhancement in the value of low-valued commodities. It is difficult to isolate the contribution of CA technology since it is also connected to the broader development of infrastructure linkages that have made ocean shipment of perishable products both technologically feasible and profitable for all value chain participants.

In terms of its role in improved trade, economists would typically relate the contribution of CA storage technologies in terms of three types of costs in shipping agricultural products: physical shipping costs, time-related costs, and the costs of unfamiliarity (Linnemann, 1966). These costs, collectively, represent a natural tariff barrier to trade and their elimination or reduction would have an impact on trade similar to the impact of the elimination or reduction of an actual trade tariff. Similar analysis can be presented for packaging innovations such as MAP and other techniques such as fruit and vegetable coatings that reduce deterioration of food products and help shippers extend the marketing reach of perishable products.

In temperate climates such as those of Canada and the northern United States (e.g., Washington state), CA storage extends the marketing season for fresh fruits such as apples and pears until May, June, or later. Typically, CA apples can be sold at a premium in January or February due to their better quality and the higher costs attributed to CA storage. In order to maximize these CA benefits, it is important that the fruit is of the best quality. To ensure consumer protection, all CA storage facilities are supervised by government, closed, and sealed. The facility cannot be opened for a minimum of 3 months after the seal is affixed in order for the stored fruit to be designated as CA storage apples.

21.3 The Fruit Sector—Apples

In order to assess the role of CA, it is useful to describe the fresh apple market in terms of the value chain and the degree to which CA has contributed to the sector and its competitiveness. In this section, we examine the apple sector in the temperate regions of Canada and the United States as it relates to CA storage.

21.3.1 U.S. Apple Production

Apples are the third most important fruit crop in the United States, behind grapes and oranges (USDA, 2001; Dimitri et al., 2003) (Figure 21.1). Thirty-five out of 50 states produce apples on approximately 200,000 hectares, yielding approximately 4.2 million metric tons annually, and generating over $1.3 billion in revenue for U.S. apple producers (USDA, 2001). U.S. production accounts for 11% of world production, and is second only to

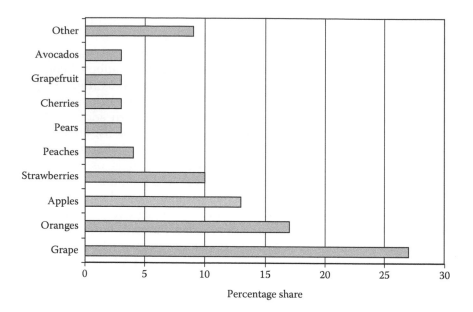

FIGURE 21.1
Percentage share value of U.S. fruit production 1997–2000 (total value: $10.6 billion). The "other" category includes blueberries, nectarines, pineapples, prunes, and tangerines. (From USDA National Agricultural Statistics Service.)

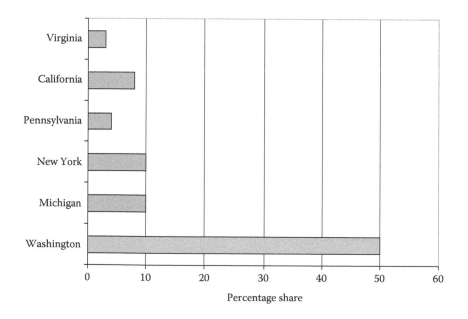

FIGURE 21.2
U.S. apple production by top states. (From USDA National Agricultural Statistics Service.)

China. Within the United States, six states account for more than 85% of U.S. apple production (USDA, 2001) (Figure 21.2).

Washington is by far the largest apple producing state, accounting for more than 50% of all U.S. apple production, with its production expected to continue to increase

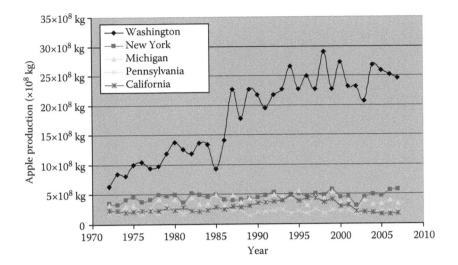

FIGURE 21.3
U.S. apple production by top states—historical trend. (From USDA National Agricultural Statistics Service.)

TABLE 21.1

Top Varieties in the United States and Europe

United States	Europe
1. Red Delicious	1. Golden Delicious
2. Golden Delicious	2. Red Delicious
3. Fuji	3. Jonagold
4. Granny Smith	4. Granny Smith
5. Gala and Royal Gala	5. Gala

(USDA, 2005). Since 1960, apple production in Washington has increased by nearly 400% while production in the rest of the United States declined. This production has come at the expense of other major apple-producing states as shown in Figure 21.3. Washington has a sawtooth shape due to the natural tendency by apple producers for alternating between a large crop in 1 year followed by a smaller crop in the next year. In addition, Washington supplies 85% of U.S.-exported apples, contributing to $249 million U.S. exports of apples. Washington has captured nearly all of the growth in U.S. apple production.

There are over 100 varieties of apples grown commercially in the United States, but 15 varieties accounted for over 90% of the production in 1999. The top five varieties for the United States and Europe are similar (Table 21.1). Apple production is dominated by two varieties: Red Delicious and Golden Delicious. However, newer varieties such as Gala and Fuji are gaining wide market adoption by growers (Figure 21.4). Washington apple growers have traditionally focused on the fresh market.

21.3.2 Canada

The Canadian apple industry is a major component of the horticulture industry. Canada produces approximately 500,000 t of apples annually (Figure 21.5). Although Canada is only one-tenth the size of the U.S. industry, Canadian apples are the number one selling

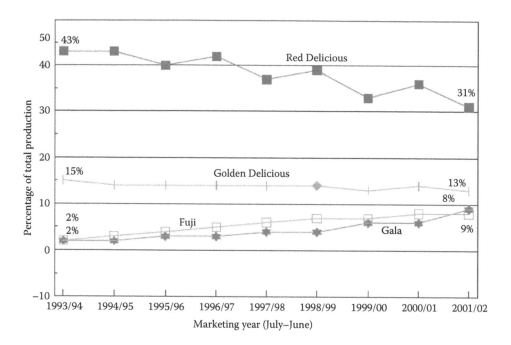

FIGURE 21.4
Varietal shifts in U.S. apple production from traditional varieties. (From USDA National Agricultural Statistics Service.)

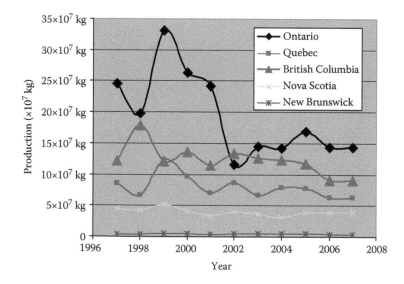

FIGURE 21.5
Canada apple production 1996–2007. (From Statistics Canada.)

fresh fruit in Canadian retail stores, with annual sales of $264 million (*The Okanagan*, 2003). Apple production has generally been on a downward trend, as Figure 21.6 shows. Ontario is the largest producer of apples (56%) followed by British Columbia which accounts for 25.3%. About two-thirds of Canadian apples are retailed in the fresh fruit market, with the remaining one-third directed into the processing market.

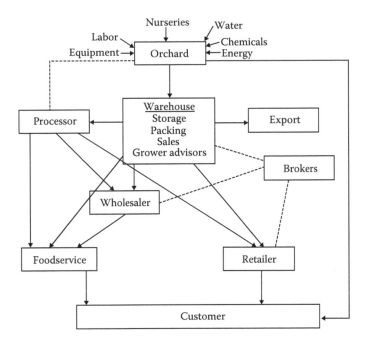

FIGURE 21.6
U.S. apple distribution channel. (From Schotzko, R.T. and Granatstein, D., A Brief Look at the Washington Apple Industry: Past and Present. School of Economic Sciences, College of Human and Natural Sciences, Washington State University and Washington State University Tree Fruit Research, 2005.)

There has been significant decline in production mainly due to intense competition from other countries, including the importation of juice concentrates from China at very low prices, forcing local growers to find alternatives for processing their apples. Substantial global competition continues from Washington state (British Columbia's main competitor), the European Union, Australia, and New Zealand. New competitors include emerging growers in Chile, Argentina, Brazil, India, Iran, Pakistan, Uruguay, and South Africa. Competition between British Columbia and Washington is especially significant and exerts downward pressure on British Columbia. These two regions of the Pacific Northwest are major competitors for both the North American market and the Asian export market. Washington is ranked in the top four most competitive apple producers in the world, along side China, Chile, and New Zealand. Washington has a much larger scale of apple production and marketing institutions compared to British Columbia, giving Washington a cost advantage in the presence of economies of size, CA storage technology, and greater access to migrant labor (Carew et al., 2006). There are confounding factors such as climate conditions, which limit yields compared to warmer areas in Washington. In general, the Canadian apple industry faces a cost-price squeeze through a decline in wholesale prices relative to input prices and an increase in North American and world production.

21.3.3 Distribution Channels

In order to further understand the role of CA storage, it is useful to briefly describe the apple supply chain used in the United States (Washington state) and Canada. In general, fresh produce such as apples go through a lengthy marketing and distribution chain from the orchard, through intermediaries, until they reach retail chains and other stores. In a generic sense, the apple produce system involves four basic activities: production;

harvesting; warehousing, CA and RA storage, packing (grading, sizing, and placing fruit in cartons); and shipment to retailers. Figure 21.6 illustrates a simple apple supply chain and the services provided. Although this pertains to the United States, it is typical of the Canadian apple distribution chain. In both the United States and Canada, the apple industry warehouses sell a set of services to growers and do not typically buy the fruit, but merely supply this set of services. Apple industry packing houses provide a set of services to apple producers, including CA and RA storage.

The apple growing season in Canada starts from May to September. In general, climate and growing season determine the type of apple varieties that are available to Canadian consumers. The development of CA storage technology allowed apples to be marketed over extended periods of time instead of forcing producers to market all their fruit within the 2 month window following harvest in September. RA storage allowed producers to market fresh apples until December. The adoption of CA storage innovations enabled the marketing season to be extended until the next growing season in March or April. As is shown later, benefits accrued both to growers and consumers, as well as the distribution channel. In terms of demand, it is clear that there will be differences in consumer demand for apples close to the harvest season compared to the rest of the year. In fact, in a simple linear regression of price of apples on quantity of apples based on time of storage period (not shown here), the storage season equation had a more inelastic slope (steeper) compared to that for the harvest equation. *A priori*, CA technology should generate an outward shift in the demand apples, given that apples of a higher quality are now available much later in the year. Our regression shows this to be the case, lending further empirical support for the role of CA storage.

CA also had an impact on price stability, which was achieved because of the ability to sell the produce over a longer time period. If CA storage was not available, the price of apples would increase both due to the fact that the quantity of apples sold during the season would be less, and the mere fact that CA storage would be zero. For the most part, apple varieties available to United States and Canadian consumers from mid-winter to summer are mainly supplied from controlled atmosphere. Table 21.2 shows Canadian apple storage holdings by type of storage, region, and apple variety. The quantity of apples held in CA storage has increased gradually over the years, with a corresponding decrease in the volume of apples placed in RA storage. In fact, this period also marked investments by packing houses in CA storage infrastructure leading to increasingly more apples being stored for extended periods. In British Columbia, most CA storage is operated by the BC Tree Fruit Association as a consortium. In Ontario, the organization is more privately held. For instance, Martins Family Fruit Farm Ltd. operates a 340,000 bushel CA building, making it one of the largest and most advanced of its kind in Canada, with atmospheric conditions in each room controlled from a complex automated system imported from Holland. Bins tower to the ceilings. When full, each of the 15 rooms holds 1000 bins (*Ontario Farmer*, 2000).

21.3.4 Apple Storage in the United States

Like Canada, the U.S. apple industry uses CA and RA storage to ensure year-round marketing. Operators of CA storage in the United States must be licensed and certified by the individual states. With proper growing and harvesting techniques, many apple varieties can be stored in CA for 12 months or more. In Washington state, most of the apples are shipped between January and September. RA storage is used for much of the fruit marketed in the fall and early winter months.

Figure 21.7 shows the amount of CA and RA storage apples between 1990 and 2007. Prior to this date, USDA did not disaggregate this data by CA and RA type. The

TABLE 21.2

Canadian Apple Storage Holdings by Type of Storage and Region

Year	CA Storage					Regular Cold Storage				
	British Columbia	Maritimes	Ontario	Quebec	CA Total Canada	British Columbia	Maritimes	Ontario	Quebec	RA Total Canada
1998	59,386	9,775	48,116	20,606	137,883	37,405	3,666	14,231	3,618	58,920
1999	47,889	11,964	58,560	29,487	147,900	22,654	5,794	23,087	12,522	64,057
2000	44,220	12,024	56,401	27,561	140,206	21,064	5,427	13,802	10,413	50,706
2001	45,621	9,499	49,075	27,566	131,761	19,243	3,967	12,091	8,603	43,904
2002	45,964	11,150	30,283	24,176	111,573	14,745	7,288	10,697	7,563	40,293
2003	42,914	10,355	43,349	28,156	124,774	18,755	5,636	15,163	5,010	44,564
2004	48,221	22,666	38,702	29,630	139,219	27,412	8,299	12,463	8,316	56,490
2005	45,517	9,854	56,181	24,914	136,466	23,707	3,351	13,749	7,418	48,225
2006	32,632	12,554	24,874	24,029	94,089	8,873	3,962	5,993	4,784	23,612
2007	54,095	14,029	49,672	29,189	146,985	9,968	4,063	17,309	15,817	47,157

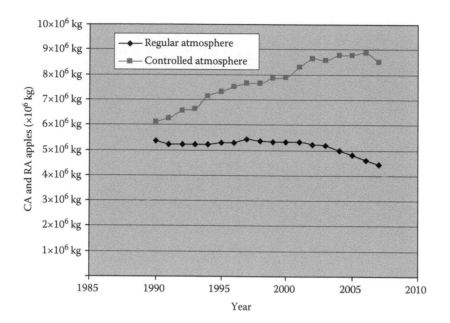

FIGURE 21.7
U.S. controlled atmosphere and regular atmosphere apples: 1990–2007. (Computed from USDA Data.)

quantity of apples held in U.S. storage has increased gradually over the years. In contrast, RA apples decreased, suggesting technology substitution of CA storage for RA storage. From a substitution perspective, it is possible to quantify the degree of RA displacement resulting from CA technology. For instance, using U.S. data, we can run a simple regression using RA as the dependent variable and CA and total apple production as independent variables. The results of the regression are not shown here. The coefficient on CA was negative, indicating an inverse relationship between CA and RA, such that increased adoption of CA technology increasingly led to the substitution of conventional methods of storing apples. Notice that the displacement is only partial and not a greater than total displacement: CA storage displaces RA storage by 30% for every kilogram increase in apple amounts placed in CA storage. This is of course a simple measure of substitution but illustrates the importance and degree of technology substitution that has occurred. One way of relating this to the orchard apple producer level is to determine the impact of CA technology on total apple supply, i.e., the extent to which CA technology shifted the apple supply curve to the right. On a conceptual level, one can introduce the notion of the impact on apple producers in terms of the increase in revenue resulting from CA. This can be calculated easily by multiplying the quantity of CA apples sold in each month by the prevailing average price per month.

When CA technology adoption is broken down by state, the dominance of Washington is evident. In 2007, general refrigeration capacity totaled 3.32 billion gross cubic feet. Apple and pear storage totaled 692 million gross cubic feet. CA capacity amounted to 168 million bushels. The top five states with the largest gross general warehouse capacity (million cubic feet) are depicted in Table 21.3. Washington state, the largest apple producer, accounts for 83% of CA and RA storage capacity in the United States. In 1995, Washington state had over 100 million 42 lb capacity CA storage. The commercial adoption of this CA technology has enabled Washington state to triple its production and effective marketing of apples from 30 million to over 90 million of fresh apples since 1975. Almost all the CA storage in Washington is privately held. This growth in CA

TABLE 21.3

U.S. Gross General Warehouse Capacity (Million Cubic Feet)

California	412
Florida	239
Texas	194
Washington	191
Wisconsin	190

storage and its effect on prices (discussed earlier) shows the rapid industry adoption of this technological innovation. In particular, apple growers and the distribution chain adopted CA storage technology in order to capture benefits associated with higher prices for high-quality fresh late in the marketing season. The figures above show that Washington is by far the most significant adopter of CA storage technology in the United States.

In general, warehouse economics has significantly affected the financial operations of packing houses both in Canada and the United States. Table 21.4 shows reduction in the number of packing houses over the years following warehouse mergers with other firms, liquidation, or disappearance of cooperatives (Schotzko and Granatstein, 2005).

Two primary factors account for this reduction: (1) significant changes in storage technology over the past 25 years, including the innovation of electronic sensing with a (2) significant cost advantage for industry players who made capital investment and volume required to utilize storage technology efficiently.

There are a number of factors that accounted for differentiated adoption in CA technology and hence differences in levels of capital investment among apple growers. This was reflected in our earlier analysis in which we showed that Washington is the top adopter of CA storage (with New York and Michigan following) while other states (e.g., in the Appalachian region) have generally fallen behind. In 2007, CA storage accounted for 82% of total storage in Washington. The key drivers include the level of risk aversion among those who are not willing to carry the additional capital by holding apples later in the season; competitive threats from imports; obsolete CA and unwillingness to invest in retrofits.

The determination of economic returns from investment in CA technology would be made easier with data on the cost of R&D. In order to estimate the economic profitability of CA, one could first examine the costs associated with the use of CA and determine the real economic benefit (financial profitability) to those paying for the technology. We looked at typical CA storage facilities as they exist today in terms of operating costs and capital investment. This implies looking at two components: the revenue component and the cost component of operating a CA storage facility. On the cost side, we would be interested in analyzing differences between CA storage costs and RA costs. This approach is reasonable based on extensive consultations with a several packing houses in Canada and the United States. It can be plausibly assumed that the cost of CA and RA storage does not change by

TABLE 21.4

U.S. Reduction in the Number of Packing Houses

Year	Yakima	Wanatcheo	Total
1985	70	84	154
1995	60	39	99
2002	51	31	84

Source: From Yakima Valley Growers Shippers Association and Warehouse Valley Traffic Association.

any significant amount from year to year, given the large capital cost and relatively small operating costs. Packing houses can store apples in a low price month close to harvest and realize a higher price premium later in the season.

CA plant requirements and costs are a function of processing capacity of the plant. Each of these levels has associated standard costs. The costs include annual operating and maintenance costs, capital costs, and capital cost allowances (CCA). We have used sizes based on United States and Canada, with complete simulation based on altering capacity from 50,000 to 500,000 to assess the impact on average unit costs of operating. Basically, the aim is to assess and compare the effect of building, equipment costs, and annual operating costs on the economic viability of CA versus RA storage.

The second component relates to revenue. This can be calculated using adjusted monthly market prices for apples during the storage period. Prices are then multiplied by storage capacity or volume packed out of CA storage, adjusted for storage losses. Revenue is derived using the simple formula: $r = p(s - c)$ where r is the revenue, s the storage capacity, c the storage loss, and p the price. Here we use monthly apple market prices to derive net differences in revenue from the use of CA, based on subtracting storage cost from total revenue for units sold. Storage decisions involve profit maximization considerations, such that if packing houses maximize profit, their expected marginal revenue would be equal to the marginal storage cost at different storage levels. If CA storage technology offers a marginal cost that is lower than expected marginal revenue at a certain storage level, then packing houses can increase storage such that expected marginal revenue will decline until it is equal to marginal cost. Marginal costs at different storage levels (derived as a percentage of total output) were evaluated and approximated by the average cost given the large fixed capital costs and relatively small operating costs. Furthermore, CA storage in general showed economies of scale in which average cost decreased with increased volume of apples stored. In order to establish the comparison between CA and RA, it is of course important to assume similar quality for apples in CA and RA.

In establishing the basis for assessing storage costs, we consulted extensively with industry to determine standard operating costs. Technologies may not be at the same state of the art. In fact, in regions where investments were made 15 years ago, it is difficult to come up with more current operating costs that reflect efficiencies from new innovations. Besides, most packing houses are reluctant to provide this information. Therefore, although variations may exist between CA facilities in terms of operating cost, it is possible to develop comparative standards on which storage decisions can be made. We have presented a summary of operating and fixed costs associated with CA storage. This is based on 2,000–10,000 t capacity CA storage. A majority of CA surveyed fell within the cluster 2600 t. The results clearly show significant positive return on CA assets relative to RA.

21.3.5 Storage and Price Behavior

A key to understanding some of the benefits of storage is through the behavior of prices. In general, fresh produce such as apples go through a lengthy marketing and distribution chain from the orchard, through intermediaries, until they reach retail chains and other stores. We would therefore observe significant variation between farm gate prices and those paid by consumers at the retail store.

The role of CA storage in pricing behavior needs to be understood in terms of the distribution channel. Within western Canada, for instance, the distribution channel that was described above reflects the dominant position of fruit wholesalers, a position that generally gives them significant market power to extract lower prices from packing houses when purchasing large fruit volumes (British Columbia Ministry of Agriculture

Food and Fisheries, 1999). There are also quality differences among grades and varieties of apples (based on size, time of year, variety, etc.). For instance, most wholesalers pay a price premium for large size fruit of the same grade and variety. A 19 kg tray of medium Royal Gala Canadian Extra Fancy Grade had an average premium of $4.40 over the small-sized grade while the large size had a premium of $6.21 over the small size. Similarly, the popular large Golden Delicious had a premium of $4.55 over the small size.

A related point is the fact that CA storage generates price premiums since apple prices are affected by seasonality and the flexibility that CA storage provides apple producers to sell fresh fruit closer to the following harvest. The significance of storage premiums has been quantified. For instance, Carew and Smith (2004) showed that the premium for CA storage varied depending on the sale month. CA-stored apples are discounted relative to regular stored apples during the fall by $0.52 per tray. During the period when both RA and CA apples are marketed (i.e., from January to June), the net CA storage premium ranged from about $0.96 to $1.89 per tray. As to be expected, the lowest net premium ($0.48) was observed in November when CA apples are first marketed to wholesalers. One of the prices reported by Carew and Smith relates to an earlier point made in the discussion of the role of apples imports, especially from the southern hemisphere during the winter season in Canada. These imports intensify price competition for CA storage apples. Carew and Smith showed that the price of RA storage apples peaked in April and declined through June, while prices in April are viewed as an irregularity since sales volume is relatively low in this month. Their analysis also showed that prices for RA storage apples were higher in September than in any other sale month, possibly due to the start of early-season apple cultivars arriving on the market. Regular storage apple prices in November, December, April, and May were above those in October. A different pattern emerged when apples were stored in CA. The premium paid for CA stored apples during January through June resulted in a price similar to the October regular storage apple price.

21.4 Modified Atmosphere Packing

Modified atmosphere packaging (MAP) represents yet another posharvest innovation that enhances product quality. In general, MAP technologies are based on the use of 1-methylcyclopropene (MCP) with bioactive volatile compounds. MCP is an inhibitor of ethylene action and effectively delays ripening of fruits. Ethylene is a plant growth regulator (hormone) that accelerates ripening of fruits. It a major factor in postharvest losses through undesirable acceleration of ripening. By neutralizing the effect of ethylene, MCP provides many benefits to the fruit sector during transportation and storage. Both U.S. and Canadian apple packing houses reported use of MCP by applying it to harvested fruits in CA or RA storage rooms. Other uses include applications in shipping containers, enclosed truck trailers, and greenhouses. Retailers of fresh-cut fruits and vegetables use MCP in refrigerated or CA food storage facilities.

MAP represents a significant commercial invention given the commercial value of the growing fresh-cut fruit and vegetable market segment. Minimally processed fruits and vegetables constitute one of the major growing segments in food retail establishments. The entire fresh fruit and vegetable marketing system is rapidly evolving to one with an increased focus on value addition and cost minimization by streamlining distribution through direct sales from shippers to final buyers in the foodservice and retail outlets. This new trend is driven by growing consumer demand for convenience in food preparation and consumption, including product form, packaging, and quality preservation, all of which are critical elements in the competitive strategies of retailers. In order to maintain their competitive edge, both foodservice and retail buyers must ensure year-round availability of fresh-cut fruit and vegetable.

From a commercial perspective, marketers of fruit and fresh vegetables want to overcome the first and most obvious changes in produce postharvest, namely, visual and textual changes, such as enzymatic browning (Salunkhe et al., 1991; Verlinden and Nicolai, 2000). Fresh fruit, like most fresh produce, continues the respiration process long after harvest by taking in oxygen and releasing carbon dioxide, water, and heat. The respiration process involves the conversion of carbohydrates into organic acids and then into more simple carbon compounds, resulting in undesirable changes in the color, texture, and flavor of the fruit. Fruits that ripen too early are easily damaged during transport and produce ethylene, which can adversely affect other commodities. It is a chain reaction: ethylene production causes faster ripening and decay of other fruits (Ritenour et al., 1997). Different fruits and vegetables and even different varieties of a given fruit and vegetable have different respiration rates, which has been an impediment to past MAP technologies (Zagory, 1998), and hence the target of MAP innovations for various types of produce. For instance, produce with higher respiration rates, such as asparagus, mushrooms, and broccoli, are more perishable than those with low respiration rates, such as nuts, onions, and potatoes. Also, fruits and vegetables that are cut, sliced, shredded, or otherwise processed have higher respiration rates (Zagory, 1998). The development of technologies to address this problem represents a valuable addition to the fruit value chain.

MAP technologies have also addressed food safety issues related to microbial growth. The harvesting and fruit processing system (such as fresh-cut) results in tissue breakdown via physical bruising or enzymatic browning) and provides a good medium for microbial growth, similar to elevated temperatures. It is important to minimize the microbial load of products by controlling operations at every point of the distribution chain from harvest, handling, processing, storage, and distribution. One cannot place a realistic economic value on the cost of food borne diseases. The are numerous examples from the meat and poultry sector, including recent outbreaks of illness traced to *Escherichia coli* in beef products, events which have spurred endeavors to improve the meat and poultry inspection system. According to the U.S. Centers for Disease Control and Prevention and the Food and Drug Administration, between 6 and 33 million people become ill each year from microbial pathogens in food, including meat and poultry, resulting in as many as 9000 deaths (CAST, 1994). The benefits of reducing pathogens, which include lower medical costs of illness, lower productivity losses, and fewer premature deaths, range from $1.9 to $171.8 billion over 20 years, depending upon the level of pathogen control. These benefits will exceed the costs of a prevention program such as hazard analysis and critical control points (HACCP), which are estimated at between $1.1 and $1.3 billion over 20 years (Crutchfield et al., 1997).

As foods are transported ever greater distances for processing and distribution, the potential for food contamination rises. In the past, microbial growth was largely controlled through a combination of refrigeration and other treatment methods. New MAP technologies provide for much greater control throughout the distribution chain by eliminating the necessity to combine various methods: it controls microbial growth without the necessity of refrigeration. An example of new innovations includes Toivonen and Lu (2006) in a patent application Canada's Ministry of Agriculture and Agri-Food. The invention relates to a composition of matter and process comprising a cyclopropene such as 1-MCP and provides, under controlled conditions, the corelease of the cyclopropene, alcohols, aldehydes, and carbon dioxide, which work together to improve the quality of plants during storage.

Our survey of industry revealed that deficiencies in current methods revolve around space, limited temperature, ethylene zones, and ethylene incompatibility between fruits and vegetables, and attempting to balance this. A number of methods have been used commercially in the fresh-cut market, including refrigeration, ethanol, ethylene oxide, irradiation, treatment procedures and dips, packaging, and MAP. Technologies such as irradiation are used in vegetables to stop enzymatic activity and provide stability. It is very effective in curbing mold and bacteria growth. However, the industry has to deal with considerable negative consumer perception. For this reason, this technology is not widely adopted and used in jurisdictions like Canada although it is more prevalent in other countries.

An increasingly popular broad-spectrum antimicrobial agent used on foods, especially fresh produce, is chlorine dioxide, including other antioxidants (Gunes and Lee, 1997; Agar et al., 1999; Moline et al., 1999; Dong et al., 2000; Gorny et al., 2002). Chlorine dioxide is efficacious against bacteria on a variety of vegetables and fruits including carrots, mushrooms, asparagus, tomatoes, lettuce, cabbage, cherries, strawberries, and apples. Reducing bacterial concentrations on produce increases the shelf lives of these foods. Chlorine dioxide also controls the fungal disease known as late blight and other secondary infections such as soft rot on stored potatoes (Khanna, 2002). In terms of treatment dips, calcium chloride is the compound of choice for maintaining fruit and vegetable processing and subsequent storage since calcium helps in maintaining plant cell wall structure, including cell permeability. It is involved in reducing respiration rates. The commercial disadvantage is that calcium chloride can transfer a taste to the product if too much of the compound is absorbed. Other treatments used include waxing (used to control respiration and improve appearance), ascorbic acid, citric acid, or other whitening agents. Treatment procedures are not universally effective and retailers have to incur significant cost in adapting treatment procedures to a very heterogeneous set of produce (Salunkhe et al., 1991).

Refrigeration has been identified as the best way to reduce respiratory metabolism (Zagory, 1998). Traditionally, tissue respiration has been controlled with rapid cooling which prolongs shelf life. In harvesting situations, when products such as green beans are picked and stored overnight without refrigeration, they can lose up to 60% of their vitamin C content (Salunkhe et al., 1991). The removal of field heat has been identified by industry as a crucial step in retaining and maintaining an acceptable quality of produce. Globally, postharvest losses of agricultural commodities are quite significant, and the losses are compounded in developing countries with higher constraints on the availability of cold storage or temperature controlled transportation. In the United States, fresh fruits and vegetable losses are estimated at 2%–23%, depending on the commodity, with an average of 12% in value chain losses between production and final market (Harvey, 1978; Cappellini and Ceponis, 1984; Kader, 2005b). Kantor et al. (1997) estimated 1995 losses in the U.S. retail, consumer, and foodservice at 23% for fruits and 25% for vegetables. Fresh fruits and vegetables accounted for almost 20% of consumer and foodservice losses. Yaptenco et al. (2001) examined handling practices for highland vegetables transported from La Trinidad Benguet to Manila, comparing them in terms of losses associated with refrigerated versus nonrefrigerated transport systems and loss due to trimmings, physical damage, and disease. Postharvest interventions were introduced at the loading site, including MAP and use of plastic crates as transport containers. Their results showed that postharvest interventions extended shelf life, improved quality, and reduced losses. Previous losses at supermarkets were 40% for broccoli, 10% for sweet peas, and 10% for snap beans. Implementation of their interventions reduced losses to 13%, 3.3%, and 5% respectively. Their financial analysis of transportation by refrigerated truck had an internal rate of return (IRR) of 291% and a payback period of 0.4 years. Nonrefrigerated transport had an IRR of 20% and a payback period of 4.2 years.

The role of temperature is relevant in assessing MAP technologies. Most produce will maintain its best quality at temperatures near 0°. However, some produce of tropical origin require storage temperature of 10°C–13°C in order to prevent chilling injury (Zagory, 1998). For example, tomatoes and avocadoes are dramatically affected by cold storage, which causes loss of taste and sugar, and affects ripening. Toivonen (2006) addresses chilling-sensitive fruits for niche market applications such as tropical fruits where refrigeration is unfavorable or unavailable. Industry surveys suggested that although chilling injury was not reported to produce large losses, industry still spends significant financial resources to constantly monitor and prevent chilling injury.

In terms of commercial technological adoption, 1-MCP is the platform for most MAP technologies. Its safety, toxicity, and environmental profile make it a favorable application (Serek et al., 1995; Sisler and Serek, 1997; Golding et al., 1998; Kim et al., 2001; Blankenship and Dole, 2003; Lu and Toivonen, 2005; Villas-Boas and Kader, 2006). 1-MCP uses a nontoxic mode of action and is chemically similar to naturally occurring substances. It also diffuses out of plant material rapidly. Effective concentrations of 1-MCP vary widely with commodity, time, temperature, and method of application. Furthermore, plant developmental stage need to be taken into account when applying 1-MCP as its effects vary with plant maturity (Blankenship, 2003). The importance of time from harvest to 1-MCP treatment varies with crop species. Generally, the more perishable the crop, the sooner after harvest 1-MCP should be applied. In some crops, such as bananas, certain inconsistencies occur which may limit commercial application unless protocols are devised. Blankenship (2003) found that bananas treated with 1-MCP did not have acceptable color development. The variable effects of 1-MCP across species can also be shown in the way it delays softening in most fruits, but has no effect on others. Temperature conditions are also a variable factor in 1-MCP treatment. A relationship exists between concentration, time, and temperature and applications of 1-MCP. While low temperatures have been used in 1-MCP application, most studies have used temperatures of 20°C–25°C in applications and low temperatures are not effective on some crops (Blankenship, 2003). In terms of industry wide adoption, commercial application of 1-MCP for edible crops has been undertaken by AgroFresh Inc. under the trade name SmartFresh.

A key commercial advantage of new storage technology such as MCP-MAP relates to energy cost implications. Industry sources interviewed confirmed the high cost of refrigerating produce. This cost occurs in the entire distribution chain, on refrigerated trucks from the point of harvest to retail outlets. Refrigerated trucks, which incorporate air conditioning units in order to keep produce cool, also provide less space for product due to the amount of space required for the units. Toivonen (personal communication) estimated that year-round refrigeration in Canada for fresh-cut products at 0°C–1°C consumes 2,766.39 TJ. Studies have shown that application of 1-MCP controlled apple deterioration and heat production at elevated temperatures, resulting in 30–80% reduction in energy use for short-term storage. New technologies such as Toivonen (2006) have been shown to generate energy savings in excess of 10%.

21.5 Packaging and MAP

Packing material innovations is an integral part of MAP. The goal in present packaging systems has been to provide packaging film that is compatible with the respiratory activity of the product (Toivonen, 1997). Since each produce type has differing and often unique packaging requirements, the ability to customize the package to the produce has been the aim of fresh produce package development. The flexible packaging industry has become increasingly responsive to specific gas requirements for fresh produce and is providing packaging films that are product specific, thereby enabling processors of fresh-cut produce provide a wide array of fresh-cut (Zagory, 1998).

Previously, MAP technologies needed to work in synergy with cooling in order to extend shelf life up to 20%. In isolation, MAP technologies could not improve quality; contribute to product safety; improve flavor; enhance product nutritious; substitute for temperature control; or stop microbial growth (Zagory, 2000). More recent technologies (Toivonen and Lau, 2006) are designed to control microbial growth and preserve produce in elevated temperatures. An industry survey by AAFC showed that industry has expressed some dissatisfaction with the deficiencies in a number of MAP technologies, including the deterioration that occurs at elevated temperatures such as MAP packaged produce stored in a consumer's warm vehicle.

Interest in newer technologies that could address these deficiencies was expressed by most of the industry contacts. Some current packaging technologies include FreshHold, Flexible Film, and Side-Chain Polymer. FreshHold is marketed as a high-permeability label that can be attached to a bag or film overwrap and has the ability to admit sufficient gas through a very limited surface area. Flexible Film was developed by Dow Chemical Co. and Exxon Chemical Co. with very high tailored oxygen transmission rate (OTR), low-moisture vapor transmission rate, enhanced clarity, superior strength, low seal initiation temperature, and very rapid bonding of seal. The side-chain polymer technology was developed by Landec Corporation Inc. and allows the film OTR to increase as temperature increases, avoiding anaerobic conditions subsequent to loss of temperature control (Zagory, 1998).

As noted by Hewett (2006), selectively permeable polymeric films have the capability to alter permeability to gases with temperature change. This is a very important development as it lessens the risk of product fermentation occurring. Previous films did not have this capability and oxygen depletion took place inside packages as respiration increased with elevation in temperature. Because of the variation of respiration rate among fruits and vegetables, careful selection of films and package volumes is necessary to ensure that the generated modification of atmosphere inside packs remains within predetermined non-damaging limits (Beaudry, 1999).

MAP is becoming more widely adopted by industry and accepted by consumers as the films now being produced reduce the risk of off-flavors developing. As pointed earlier, MAP technology is especially relevant with the growth of fresh-cut or minimal processing industry where there is a large increase in the number and variety of products now produced in this "ready to eat" format. Hewett (2006) also identifies active packaging being developed that enables some type of interaction between the product and the package. An example is recent development by Jenkins Labels and Hort Research of a small consumer pack with an attached label that indicates to the consumer when the fruit is ready to eat. This is based on technology that induces changes in the especially sensitive label according to the concentration of volatile components produced by the fruit as it begins to ripen in the pack. This has been named "ripeSense™" and is being trailed in several countries for pears at present with the intention to develop other packs for avocados and kiwifruit.*

---

* ripeSense is considered the world's first intelligent sensor label that changes color to indicate the ripeness of fruit. The ripeSense sensor works by reacting to aromas released by the fruit as it ripens. Initially, the sensor is red; it graduates to orange, and finally yellow. A consumer can match the color of the sensor with their eating preferences, enabling the consumer to accurately choose fruit on the basis of degree of ripeness. The technology was voted as one of *Time* magazine's 36 greatest inventions in 2004. It was developed in New Zealand by Jenkins Group, a major supplier to New Zealand's horticultural labeling industry, in partnership with HortResearch, a New Zealand Crown Research Institute.

Another type of active packaging is a consumer pack that holds two to four fruits. The pack can be induced to ripen upon deliberate activation and release of an ethylene following solubilization of a special pellet within the pack. The release of ethylene causes the fruit to ripen. The consumer is able to see the fruit changing color through the clear polymeric material that constitutes the pack. According to Hewett (2006) it remains to be seen how popular these developments are with consumers in this age of convenience foods.

It is clear from this discussion that the economic impact of these storage innovations is significant. The impact on the food distribution chain is considerable especially when one considers the increased demand by supermarkets for safe foods and for traceability of each lot of fruits and vegetables. This places additional demands on growers and packing-houses. The economic value of symmetric information is also clear from the labeling information provided in new packaging technologies as consumers increasingly seek clear and unambiguous information to enable them to make informed decisions that have implications for health. In the past, nutritional information was never obligatory for fresh fruits and vegetables in all markets. The development of smart labels like ripeSense opens a new era in marketing with respect to meeting consumer demand for more information about the nature of the food that they consume.

Overall, a survey of industry stakeholders involved in the development of MAP identified several main impediments and hurdles that have continued to be addressed by new MAP innovations. The first is cost. The produce industry is very sensitive to cost, with a difference in a few cents having the ability to inhibit companies from purchasing the technology. Some of the most sophisticated packaging technologies with tremendous capabilities have not been widely adopted due to cost. Second, industry prefers the ability of a new technology to have compatibility with existing equipment. Packaging companies have invested large amounts of money into machinery and are reticent about investing in technologies that are incompatible with existing plant and equipment. Third, generic conditions of use are important. MAP technologies should not just work efficiently under ideal conditions (0°C–3°C). Some industry observers noted how certain MAP technologies fail under other conditions in which respiration rates increased. More recent technologies have addressed the deficiency by developing technologies that withstand nonideal temperatures. Fourth, the produce market is fragmented and diverse, with each product having its own requirements and physiology such that MAP technologies marketed for one product may have no application in other products. A technology may be able to target newer niche markets, which may have more value (consumer pays more) but the market for packaging these products is extremely miniscule. The produce industry is valued at $75 billion annually, but dispersed into 300 different half-a-billion dollar markets of which 5% is packaging. In order to be economically successful and reach a large enough market, MAP technologies needs to be versatile and capture many different commodities. It is of course recognized that it is not always feasible for one technology to capture the entire breadth of the produce market. Such gaps can be filled in areas of the industry where MAP has not been successful such as stone fruits and chilling sensitive fruits and vegetables. Although MAP is growing rapidly, an industry stakeholder involved in MAP development identified potential for MAP growth in two specific areas. The first involves high respiration rate commodities such as broccoli and asparagus, which require a higher rate of gas exchange. These commodities have not been addressed by major players as a majority of market share is in microperforated films, which do not have precise control over ratio of gas exchange rates. The second area where demand exists is in fresh-cut fruits and vegetables which are increasingly stored in trays (lidstock). Due to limited space for gas exchange in trays (surface only) and the inability of polymers with high exchange rates to seal to trays, there is a demand within the industry for MAP that addresses this constraint.

21.6 Regulatory Issues

An important regulatory aspect involving both CA and MAP is the use of MCP. Most of the technologies around MCP involve proprietary technologies covered by several patents. Further technology development in the incorporation of MCP has generally required negotiations and licenses for freedom to operate. In terms of regulation, MCP is regulated in North America by the U.S. Environmental Protection Agency (EPA) and the Canadian Pest Management Regulatory Agency (PMRA). MCP already has EPA approval (EPA Reg. No. 71297 2). Rohm and Haas through its subsidiary AgroFresh of Spring House, PA received EPA registration to market MCP for use in apple storage. MCP, branded SmartFresh, qualified for review in EPA's biopesticide division as a reduced-risk product, due to its low use rate and favorable safety profile. Since the product does not leave detectable residues on fruit, and EPA granted the product an exemption from tolerance limits. MCP is not intended for use outdoors or in other nonenclosed areas. Rohm and Haas also has registration approval for other uses, including apricots, avocados, broccoli, kiwifruit, papayas, pears, plums, mangoes, melons, nectarines, persimmons, and tomatoes. The EPA stamped "ACCEPTED" label for Smart-Fresh dated July 17, 2002 contains directions for apples, melons, tomatoes, pears, avocados, mangos, papayas, kiwifruit, peaches, nectarines, plums, apricots, and persimmons. The final product label submitted for registration in New York State contains use directions for apples only.

SmartFresh has had a wide adoption rate since 2002 following regulatory approval in a number of countries. It is approved for use on apples in South Africa, United Kingdom, New Zealand, Mexico, Chile, Costa Rica, Argentina, and Israel. In November 2007, Health Canada's PMRA proposed full registration for the sale and use of technical grade active ingredient 1-MCP and the end-use product SmartFresh Technology on harvested apples to maintain fruit firmness and reduce incidence and severity of superficial scald. An evaluation of current scientific data from the applicant, scientific reports, and information from other regulatory agencies resulted in the determination that, under the proposed conditions of use, the end-use product possessed value and did not present an unacceptable risk to human health or the environment.

Prior to full registration in Canada, Canadian apple growers were in a significant competitive disadvantage in export markets, especially since export buyers in countries such as the United Kingdom had requested that SmartFresh be used on their consignments to prevent loss of market share to competing companies.

The expedited management of the regulatory and risk assessments required for final approval is a costly process, with costs ranging from $1 million to $5 million. Registration of a new formulation containing an active ingredient already registered or a new use for a currently registered active ingredient usually takes less time. The EPA has facilitated the development of biopesticides by establishing a tier approval system in which, under some circumstances, several tests are waived. These reduced regulation costs have helped lower the development costs of biopesticides, which are estimated at around $5 million per product. Biological pesticides include microbial pesticides, plant pesticides, and biochemical pesticides. In 1997, the average time to register a biological pesticide was 11 months, compared to 38 months for a conventional pesticide.

Interestingly, other uses for 1-MCP are being advanced. Recently, Rohm and Haas, and Syngenta have entered into a collaborative agreement to develop and commercialize Invinsa crop protection technology, which is currently under development by Rohm and Haas' AgroFresh unit. The alliance will focus on major field crops, including canola, corn, cotton, rice, soybean, and wheat. The companies applied for regulatory approval in the United States for the current formulation of Invinsa, which has already been approved in both Argentina and Chile. Invinsa is a sprayable formulation of 1-MCP that works by preventing ethylene, a natural plant hormone, from triggering stress responses in field crops during extended periods of high temperature, mild-to-moderate drought, and other crop stresses. According to a corporate statement from Rohm and Haas, Invinsa can increase crop yields by 5%–15% and has market potential of more than $500 million.

21.7 Industry Acceptance

CA and MAP technologies are generally perceived well by industry, mainly due to the industry's familiarity with the technologies. As noted earlier, industry has come to appreciate the limitations of MAP and the fact that benefits of a number of current technologies can only be realized within a specific temperature range (Zagory, 1998). This experience shapes the expectations within the industry for MAP and influences how new technologies are received by industry stakeholders. Experience with packaging technologies has also generated a consensus within industry as to which films are appropriate for standard size packages of produce such as garden salad, broccoli, and peeled carrots. The packaged produce industry in general is still fairly well established, therefore a new technologies such as MAP for the fresh-cut segment need to position themselves in relation to meeting current industry expectations in order to secure market access. Industry commented that demand for a new MAP technology will change according to the size of the package, consumer size package, or for mass distribution (i.e., whole truck). One area where market demand is growing is for handling technologies such as pallet-level modified atmospheres (Cook, 2004). The globalization of the fresh fruit and vegetable marketing system is also creating a demand for solutions that can add value and decrease cost by streamlining distribution and understanding customer need. Distribution centers require investment ranging from $500,000 to $1 million, just for the shell. An industry stakeholder estimated that cost for entire distribution operation, refrigerator rooms, trucks, etc. could cost upward to $2.5 million. They also stated that industry is reluctant to adopt new technology unless they are sure that the additional cost can be recovered.

An additional point relates to the growth of the fresh-cut industry, which according to industry represents the fastest growth in retail, next to bottled water. The demand for innovative technologies in this market is also on the rise. With 60% of the fresh-cut market procured by the foodservice industry (Produce Marketing Association, June 2004) and restaurants such as McDonalds adding fresh-cut apple slices to their menu (Cook, 2003), the fresh-cut market has already experienced increasing market acceptance and demand within foodservice. Due to exponential growth in the fresh-cut industry, some industry stakeholders stated that there are increasing concerns regarding food safety because (1) the public is more aware, (2) the industry faces pressure to be highly competitive. This competition contributes to the need to push more number of products, which in turn means more mistakes are made and more health and safety concerns arise.

The issue of technology acceptance was also assessed at the wholesaler–retailer–distributor level. At the grower-shipper level, there are not many concerns with spoilage as this tends to be more of an issue at the retail level. Produce is moved rapidly in order to maintain control over the maturation process. Due to the competitive nature of this industry segment, a grower-shipper must have the ability to guarantee fresh produce, necessitating strict quality standards. At present, grower-shippers seem to be satisfied with current preservation methods. When asked about openness to new technologies, grower-shippers identified cost as the most important factor due to extremely tight margins in the industry. MAP is not currently used extensively at this point in the distribution chain. Industry players identified an opportunity to use MAP to extend shelf life such that specific fruits and vegetables could be offered in the off-season. Demand for a new technology depends heavily on the commodity. Stone fruits were identified as a particularly problematic commodity with respect to deterioration.

Foodservice retailers were also surveyed. The information was communicated via in-person interviews and provided some insight into the concerns of this industry group. In general, foodservice retailers expressed interest in methods that would add convenience and save on product spoilage. Cost and compatibility with existing methods were the two main factors identified that would affect adoption of a new fruit storage technology.

Family restaurants represent a significant component of the away-from-home household food budget. Hence a survey of this segment was important in determining the degree of market adoption of MAP technologies, considering the volume of fresh fruit and vegetables distributed via restaurants. Shrinkage was viewed as the most significant management issue. Restaurants are often very constrained by the lack of storage space, thus requiring managers to manage produce by controlling amounts of produce on hand. Prepackaged produce can create problems with shrinkage due to the lack of quantity control associated with these products. As a result, prepackaged products that are used less frequently on the menu often contribute substantially to produce spoilage. Chilling injury was not identified as a concern. Interest in newer technologies is limited by concerns with respect to cost and storage space. Convenience may weigh heavily with respect to interest in new fresh-cut technologies, especially if significant cost could be saved. There is a role for researchers to innovate for this segment of the food chain. Related to this segment are fast-food restaurants. This industry segment indicated that it had established methods for delivering all supplies on a rotational basis, including produce. Produce spoilage was identified as significant, but this is considered a management issue that relates to proper inventory control. Improper handling of prepackaged produce was identified as the largest problem with respect to spoilage. Chilling injury was only identified in relation to freezing occurring during the winter months.

Grocery retailer is one of the most significant components of the supply chain, as clearly evident from their sheer size and the fact that in many countries, food supermarkets and related food chains are displacing traditional channels to selling fresh fruit and vegetables. A majority of supermarkets surveyed identified spoilage as a significant issue. One retailer estimated shrinkage costs in the $500–$1000 range per day. The problem was not as severe at larger supermarkets that have high turnover rates of produce. However, refrigeration costs are quite substantial and in open top refrigeration units, frequent repairs are required due to high energy needs of motors. Space was not identified as an issue of concern for grocery retailers.

21.8 Consumer Acceptance

Consumer acceptance of technologies such as CA and MAP is inextricably linked to consumers perception of nutrition, safety, quality, taste and flavor, ripeness, and appearance. First, as a result of increased health awareness, consumers are increasingly demanding year-round supplies of a wide variety of fresh produce (Dimitri et al., 2003). Consumption of fruits and vegetables has increased in response to promotional campaigns extolling their nutritional benefits together with the growing array of fresh-cut products, prepackaged salads, processed products, and imported produce available in the market place. The additional demand placed on the fruit distribution chain is evident from increases in per capita fresh fruit consumption. In 2001, each Canadian ate 125 kg of fruit, 13% higher than the early 1990s. Orange juice, bananas, apple juice, apples, oranges, and melons topped the list representing 62% of all fruit consumed in 2006 (Statistics Canada). In fact, Canada imports approximately 86% of the fruit and 39% of the vegetables it consumes, with the United States accounting for approximately 53% and 80% of these imports, respectively. In the United States, per capita consumption of fresh produce increased 12% between 1987 and 1997 (Kaufman et al., 2000). Rising incomes have also allowed consumers to purchase higher quality and greater variety of produce. These growing trends are repeated in Mexico, Brazil, European Union countries, and Latin America, Asia, and Southern Africa.

There is evidence to suggest that consumers acceptance of packaged produce is on the rise. In its 2004 report, the Produce Marketing Association states that while consumers purchase loose fruits and vegetables because they believe these items to be fresher and of better quality, they also feel that prepackaged produce is more sanitary. Less handling of prepackaged produce contributes to consumer perceptions that this produce is more sanitary and safer. These consumer perceptions were confirmed by retailers who noted that consumers associate quality with fresh, and convenience and cleanliness with packaged. Supermarkets and other grocers meet preference of consumers for fresh produce, positioning this produce at the front of the store and fresh-cut nearer the back. However, a few grocers who were surveyed were concerned about potential perceptions about technologies that offer consumers products with longer shelf life, especially prepackaged products that include packing and best before dates, stating that such products may lead consumers to question the freshness of the product. There were also concerns about consumer confusion in understanding the utility of technologies that they perceive to be preservatives. This is a technology positioning problem for marketers of new storage technologies and clearly represents a hurdle that needs to be overcome in the development, commercial transfer, and wide market adoption of new packaging technologies.

Other factors that have driven increased consumer acceptance of fresh-cut products include rising incomes and time demands on households (Kaufman et al., 2000). Certain factors such as package design have also been shown as important in influencing this trend. Over 80% of consumers in a U.S. study expressed preference for a transparent design that allowed them to see what they are purchasing (PMA June 2004). Smaller, more flexible packages were also identified as preferable over larger, rigid packages both because they allow ease of storage and give the consumer the ability to see and feel the produce (PMA, June 2004). Demographics played an interesting role in market acceptance of technologies. An industry survey of associated of the foods industry suggested that higher income shoppers are the largest purchasers of prepackaged produce because of its increased cost, with marketing efforts significantly gender-targeted to the female shopper.

21.9 Summary

This chapter reviewed literature related to the economic and commercial impact of controlled atmosphere (CA) and modified atmosphere (MA) in terms of their increasing popularity as food preservation techniques. Because of the greater industrial adoption of the technologies for temperate fruits, this chapter focused its discussion on North America, specifically Canada and the United States. The main focus was on apples for which CA and MA have had the largest commercial adoption by volume. However, there has also been adoption in a wide range of food products, including (a) fresh whole and fresh-cut fruit and vegetable products; (b) filled pasta, breaded poultry, meat, or fish; (c) ready meals and other cook-chill products; (d) dairy products; (e) cooked, cured, and processed meat products; (f) cooked and dressed vegetable products. The use of MA and CA technologies has economic implications given global trends over the last 25 years, showing significant socioeconomic and business changes and a shift to a foodservice economy driven by demand for convenience.

21.10 Future Research Needs

This analysis was significantly constrained by data availability, in particular primary data and highly disaggregated secondary data, to facilitate quantification of return on investment and value chain economic impacts. This aspect required significant research resources to generate industry surveys required to achieve this aim. Future research in this direction would contribute to further understanding of the magnitude of the impact of technological innovations in storage techniques such as CA and MA.

The rapid pace of globalization coupled with the increasing dominance of international supermarket chains requires additional research effort in understanding the economic impact of CA and MA vis-à-vis their role in enhanced supply chain efficiency from the producer to consumer. There are clear global benefits that arise from a truly global fresh produce international trading chain in which CA and MA innovations enable fresh products to be transported efficiently and safely over wide distances and through an array of climatic zones. Future research can quantify these trends in terms of their commercial and socioeconomic impact, including an assessment of industry conduct, structure, and performance, in particular, the degree to which CA and MA have facilitated the creation of a competitive food trading environment while extending the produce supply window and shelf life, enhancing quality, minimizing product loss, ensuring affordability by consumers, and greatly reducing business operating costs, especially in light of high investment costs that are entailed by such a distribution system. Such future research must include distribution systems in both temperate and nontemperate regions.

References

Agar, I.T., R. Massantini, B. Hess-Pierce, and A.A. Kader. 1999. Postharvest CO_2 and ethylene production and quality maintenance of fresh-cut kiwifruit slices. *Journal of Food Science* 64: 433–440.

Beaudry, R.M. 1999. Effect of O_2 and CO_2 partial pressure on selected phenomena affecting fruit and vegetable quality. *Postharvest Biology and Technology* 15: 293–303.

Blankenship, S.M. and J.M. Dole. 2003. 1-Methylcyclopropene: A review. *Postharvest Biology and Technology* 28: 1–25.

British Columbia Ministry of Agriculture Food and Fisheries. 1999. Industry Overview: Apples and Pears in British Columbia. http://www.agf.gov.bc.ca/treefrt/profile/pome.pdf

Cappellini, R.A. and M.J. Ceponis. 1984. Postharvest losses in fresh fruits and vegetables. In: H.E. Moline (Ed.), *Postharvest Pathology of Fruits and Vegetables: Postharvest Losses in Perishable Crops*. University of California Bulletin 1914, pp. 24–30.

Carew, R., W. Florkowski, and E.G. Smith. 2006. Apple industry performance, intellectual property rights, and innovation: A Canada–US comparison. *International Journal of Fruit Science* 6: 93–116.

Carew, R. and E.G. Smith. 2004. The value of apple characteristics to wholesalers in western Canada: A hedonic approach. *Canadian Journal of Plant Science* 84: 829–835.

CAST. Council for Agricultural Science and Technology. 1994. Foodborne Pathogens: Risks and Consequences. Task 20 An Economic Assessment of Food Safety Regulations Economic Research Service/USDA Force Report, ISSN 01944088, No. 122, Washington, DC, September 1994.

Cook, R.L. 2003. *Globalization and Fresh Produce Marketing*. U.S. Department of Agriculture and Resource Economics. University of California, Davis, CA, July 2003.

Cook, R.L. 2004. The U.S. fresh produce industry: An industry in transition, In A.A. Kader (Ed.), *Postharvest Technology of Horticultural Crops*, University of California Division of Agricultural and Natural Resources, Publication 3311, pp. 5–30, Chapter 2.

Crutchfield, S., J.C. Buzby, T. Roberts, M. Ollinger, and C.-T.J. Lin. 1997. An Economic Assessment of Food Safety Regulations: The New Approach to Meat and Poultry Inspection. U.S. Department of Agriculture Economic Research Service.

Dimitri, C., A. Tegene, and P. Kaufman. 2003. *U.S. Fresh Produce Markets: Marketing Channels, Trade Practices, and Retail Pricing Behaviour*. U.S. Department of Agriculture, Washington DC, Agricultural Economic Report No. 825, September.

Dong, X., R.E. Wrolstad, and D. Sugar. 2000. Extending shelf-life of fresh-cut pears, *Journal of Food Science*, 65: 181–186.

Golding, J.B., D. Shearer, S.G. Wyllie, and W.B. McGlasson, 1998. Application of 1-MCP and propylene to identify ethylene-dependent ripening processes in banana fruit. *Postharvest Biology and Technology* 14: 87–98.

Gorny, J.R., B. Hess-Pierce, R.A. Cifuentes, and A.A. Kader. 2002. Quality changes in fresh-cut pear slices as affected by controlled atmospheres and chemical preservatives. *Postharvest Biology and Technology* 24: 271–278.

Gunes, G. and C. Lee. 1997. Color of minimally processed potatoes as affected by modified atmosphere packaging and antibrowning agents. *Journal of Food Science*. 62: 572–582.

Harvey, J.M. 1978. Reduction of losses in fresh market fruits and vegetables. *Annual Review of Phytopathology* 16: 321–341.

Hewett, E.W. 2006. Postharvest challenges for crops grown under protected cultivation. *Acta Horticulture (ISHS)* 710: 107–112. http://www.actahort.org/books/710/710_8.htm

Kader, A.A. 2005a. Controlled atmosphere. Department of Pomology, University of California, Davis, CA. http://una:usda.gov/hb66/013ca.pdf.

Kader, A.A. 2005b. Increasing food availability by reducing postharvest losses of fresh produce. In: F. Mencarelli and P. Tonutti (Eds.), *Proceedings of the 5th International Postharvest Symposium*. *Acta Horticulture (ISHS)* Verona, Italy, 682:2169–2175.

Kantor, L.S., K. Lipton, A. Manchester, and V. Oliveira. 1997. Estimating and addressing America's food losses. *Food Review* 20: 3–11.

Kaufman, P., C.R. Handy, E.W. McLaughlin, K. Park, and G.M. Green. 2000. *Understanding the Dynamics of Produce Markets*. Agriculture Information Bulletin 758, Washington DC, August 2000.

Khanna, N. 2002. Chlorine dioxide in food applications. In: *Proceedings of the 4th International Symposium*, Chlorine dioxide: The state of science, regulatory, environmental issues, and case histories. AWWA Research Foundation and the American Water Works Association, Las Vegas, NV, February 15–16, 2001.

Kim, H.O., E.W. Hewett, and N. Lallu. 2001. Softening and ethylene production of kiwifruit reduced with 1-methylcyclopropene. *Acta Horticulture* 553: 167–170.

Lau, O.L. 1989. Responses of British Columbia-grown apples to low-oxygen and low-ethylene controlled atmosphere storage. *Acta Horticulture (ISHS)* 258: 107–114.

Linnemann, H. 1966. *An Econometric Study of International Trade Flows*. North-Holland, Amsterdam, the Netherlands. http://www.actahort.org/books/258/258_10.htm

Lu, C. and P.M.A. Toivonen, 2005. Increased maturity enhances 1-MCP efficacy in maintaining quality of summer apples 'Sunrise' and 'Silken' stored at 0°C., *9th International Controlled Atmosphere Research Conference*, East Lansing, MI, July 5–8, 2005.

Moline, H.E., J.G. Buta, and I.M. Newman. 1999. Prevention of browning of banana slices using natural products and their derivatives. *Journal of Food Quality* 22: 499–511.

Newbery, D.M.G. and J.E. Stiglitz. 1981. *The Theory of Commodity Price Stabilization: A Study in the Economics of Risk*. Clarendon, Oxford.

Ontario Farmer. 2000. http://www.ontariofarmer.com/pages/publications/ontariofarmer.html

Prange, R.K., J.M. DeLong, B. Daniels-Lake, and P.A. Harrison. 2005. Innovation in controlled atmosphere technology. *Stewart Postharvest Review* 1(3): 1–11.

Ritenour, M.A., M.E. Mangrich, J.C. Beaulieu, A. Rab, and M.E. Saltveit. 1997. Ethanol effects on the ripening of climacteric fruit. *Postharvest Biology and Technology* 12: 35–42.

Salunkhe, D.K., H.R. Bolin, and N.R. Reddy. 1991. *Sensory and Objective Quality Evaluation. Storage, Processing. And Nutritional Quality of Fruits and Vegetables*, Vol. 1, 2nd ed. CRC Press, Boca Raton, FL.

Schotsmans, W.C. and R.K. Prange. 2006. Controlled atmosphere storage and aroma volatile production. *Stewart Postharvest Review* 2(5): 1–8.

Schotzko, R.T. and D. Granatstein. 2005. *A Brief Look at the Washington Apple Industry: Past and Present*. School of Economic Sciences, College of Human and Natural Sciences. Washington State University; and Washington State University Tree Fruit Research.

Serek, M., E.C. Sisler, and M.S. Reid. 1995. 1-Methylcyclopropene, a novel gaseous inhibitor of ethylene action, improves the life of fruits, cut flowers and potted plants. *Acta Horticulture* 394: 337–345.

Sisler, E.C. and M. Serek. 1997. Inhibitors of ethylene responses in plants at the receptor levels. *Physiologia Plantarium* 100: 577–582.

Statistics Canada. Fruit Situation. http://www.statscan.ca

The Okanagan. 2003. Apples are No. 1 with a bullet. May 24, p. 26.

Toivonen, P. and C. Lu. 2006. Compositions and methods to improve the storage quality of packaged plants. United States Patent Application 20060154822. Assignee: Her Majesty in Right of Canada as Represented by the Minister of Agriculture and Agri-Food. Ottawa, Canada.

Toivonen, P. 1997. Non-ethylene, non-respiratory volatiles in harvested fruits and vegetables: Their occurrence, biological activity and control. *Postharvest Biology and Technology* 12: 109–125.

USDA. 2001. Apple Industry: Situation. Foreign Agricultural Service, Horticultural and Tropical Products Division, September.

USDA. 2005. World Apple Situation. Foreign Agricultural Service, Horticultural and Tropical Products Division, Washington DC. http://www.fas.usda.gov

Verlinden, B.E. and B.M. Nicolaï. 2000. Fresh-cut fruits and vegetables. *Acta Horticulture* 518: 223–230.

Villas-Boas, E.V.B. and A. A. Kader. 2006. Effect of atmospheric modification, 1-MCP and chemicals on quality of fresh-cut banana. *Postharvest Biology and Technology* 39(2):1 55–162.

Wright, B. 2001. Storage and price stabilization. In: *Handbook of Agricultural Economics*, Vol. 1, B. Gardner and G. Rausser (Eds.) Elsevier Science B.V., the Netherlands, Amsterdam.

Yahia, E.M. 2006. Modified and controlled atmosphere for tropical fruits. *Stewart Postharvest Review* 5: 6, 1–10.

Yaptenco, K.F., J.U. Agravante, and E.B. Esquerra. 2001. A case study on refrigerated and non-refrigerated transport of highland vegetables from Benguet to Manila. Quality assurance in marketing of fresh horticultural produce. *Proceedings of the 1st International Postharvest Horticulture Conference*, pp. 118–129.

Zagory, D. 1998. An update on modified atmosphere packaging of fresh produce. *Packaging International* 117, April.

Zagory, D. 2000. What modified atmosphere packaging can and can't do for you, *Washington State University 16th Annual Postharvest Conference and Trade Show*, Yakima Convention Center, Yakima, WA. March 14–15, 2000.

22

Biochemical and Molecular Aspects of Modified and Controlled Atmospheres

Angelos K. Kanellis, Pietro Tonutti, and Pierdomenico Perata

CONTENTS

22.1 Introduction

The quality of fresh fruit and vegetables during storage is greatly influenced by temperature, relative humidity, and atmospheric composition (oxygen, carbon dioxide, and ethylene) of their environment. Application of controlled and/or modified atmosphere (CA–MA) storage (low oxygen alone or coupled with carbon dioxide) prevents and/or delays the rate of ripening of fruit, resulting in the maintenance of their commercial life (Kader, 1986). In addition, ultra low oxygen or anaerobic conditions have been recently applied as new postharvest quarantine treatments to confine insect infestations (Shellie, 2002; Shellie et al., 1997). Furthermore, there are a number of standard commercial practices such as wax coatings, packing in plastic liners, etc. in which the supply of oxygen to fruit in general may be limited. In spite of these commercial practices of CA–MA, the precise mode of action of low O_2 and/or high CO_2 in fresh produce and in ripening is not well understood. In general, plants have developed strategies to manage with low oxygen levels. For example, a rapid decline in respiration, a fall in the adenylate energy charge, and a synchronized downregulation of the Krebs cycle and glycolysis are the first metabolic responses of tissues to low O_2 regimes (Geigenberger, 2003; Solomos, 1982). However, the entire course, beginning with the sensing of oxygen levels and including the acclimation to oxygen deficiency, irrespective of plant tissue (root or plant organs) subjected to stress, includes a series of physiological, biochemical, and molecular mechanisms, which have not been unequivocally defined (Drew, 1997; Fukao and Bailey-Serres, 2004; Geigenberger, 2003; Kanellis, 1994; Solomos and Kanellis, 1997).

In general, the common perception of fruit metabolism under low O_2 and/or high CO_2 regimes was often limited to the repression of respiration and to the stimulation of alcoholic fermentation in combination with the suppressive effects of low O_2 and/or high CO_2 on ethylene biosynthesis and action (Kanellis, 1994; Solomos and Kanellis, 1997). However, research on hypoxia/anoxia is primarily focused on elucidating the regulation of expression of hypoxic/anaerobic genes and the molecular basis for the adaptation to the low oxygen stress in root or vegetative tissues (Bailey-Serres and Chang, 2005; Branco-Price et al., 2005; Chang et al., 2000; Fukao and Bailey-Serres, 2004; Geigenberger, 2003; Subbaiah and Sachs, 2003).

22.2   Low O_2 Effects in CA/MA Applications

22.2.1   Physiological Aspects

Plants are aerobic organisms and rely on oxygen for their life. Occasionally, plants experience a lower availability of oxygen (hypoxia) and, less frequently, the total absence of oxygen (anoxia) due to environmental factors (flooding of the soil) or the anatomical structure of some tissues whose characteristics severely limit the permeability of oxygen (see Perata and Alpi, 1993 for a review). In 2003, Vigeolas et al. measured the internal content of oxygen in developing seeds of oilseed rape. The results showed that even when the external oxygen concentration was 21% (v/v), oxygen fell to 17% (v/v) between and 0.8% (v/v) within seeds (Vigeolas et al., 2003). These results clearly highlight that plant organs such as seeds and fruits can experience hypoxia as a physiological status. A further decline in ambient external oxygen (lower than 21%) will inevitably lead to a dramatic switch from aerobic to anaerobic metabolism. Clearly, the metabolic pathway that is mostly affected by lack of oxygen is mitochondrial respiration. Oxygen acts as the final electron acceptor in the mitochondrial respiration. Without oxygen, the reduced pyridine nucleotides nicotinamide adenine dinucleotide (NADH) produced by the glycolytic pathway, as well as in the Krebs cycle, cannot be reoxidized through the mitochondrial electrons transport chain. This results in the arrest of the glycolysis. There are thus dramatic consequences for cell survival, unless an alternative mechanism for the reoxidation of NADH is activated. Plant cells under low oxygen avoid this otherwise deadly metabolic arrest by switching on the fermentative metabolism, which allows the reoxidation of the NADH produced by the glycolytic pathway so that ATP production can continue. The amount of ATP produced through glycolysis is clearly much lower than the aerobic, mitochondrial one.

Several pathways have been proposed as operating in the reoxidization of NADH (reviewed by Perata and Alpi, 1993). However, the fermentative pathways leading to the production of lactate and, above all, ethanol are believed to play a major role in the anaerobic recycling of pyrimidine nucleotides (Figure 22.1). Lactate is produced from pyruvate by the action of lactate dehydrogenase, while ethanol production arises from the decarboxylation of pyruvate to acetaldehyde (catalyzed by pyruvate decarboxylase) followed by the reduction of the aldehyde to ethanol catalyzed by alcohol dehydrogenase.

Other anaerobic metabolites that are reported to accumulate under low oxygen availability conditions are alanine, succinate, malate, and shikimate (for a complete list of anaerobic metabolites see Crawford, 1982). The regulation of fermentation takes place through the following steps (Davies et al., 1974). Under normoxic conditions (around 21% oxygen), the cytoplasmic pH is above neutrality, and hampers the activity of pyruvate decarboxylase, an enzyme with an acidic pH optimum (Davies, 1980). There is a buildup of

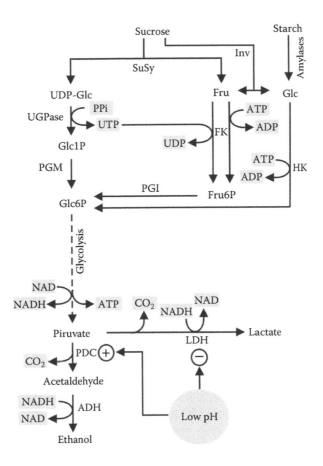

FIGURE 22.1

Fermentative metabolism in plant tissues subjected to low oxygen conditions. Under hypoxia (low oxygen) or anoxia (absent oxygen) the mitochondrial respiration is switched off and most of the adenosine-5′-triphosphate (ATP) is produced through fermentation. The availability of carbohydrates (sucrose, starch) is a prerequisite for an efficient fermentative metabolism. Under low oxygen, starch is degraded by amylases (α-amylases and β-amylases) and the resulting glucose is phosphorylated by a hexokinase (HK), producing glucose-6-phosphate (Glc6P), ready to be used in glycolysis. The metabolism of sucrose is more complex. Two pathways can degrade sucrose, namely the invertase (INV) pathway and the sucrose synthase (SuSy) pathway. It is generally accepted when oxygen is only available in low amounts the SuSy pathway prevails, since it is more energetically advantageous. Indeed, the hexoses (glucose and fructose), resulting from INV action, require phosphorylation by HK and fructokinase (FK), at the cost of two molecules of ATP, to be channeled into glycolysis. The SuSy pathway, instead, produces UDP-glucose (uridine diphosphate glucose-UDP-Glc), which is directly converted into Glc6P by the combined action of UDP-glucose-pyrophosphorylase (UGPase) and phosphoglucomutase (PGM), at the cost of a pyrophosphate (PPi) molecule. Since PPi is a by-product of several biochemical reactions, its use to produce Glc6P is energetically advantageous. The action of UGPase also produces UTP, which can replace ATP in the phosphorylation of fructose by FK. The biochemical steps represented in red indicate ATP-consuming reactions, while green lines indicate ATP-producing steps. Pyruvate cannot be used in the Krebs cycle in the absence of oxygen. It is therefore used by the fermentative pathways (bright blue background in the figure). Initially lactate is produced by the action of lactate dehydrogenase (LDH). However, the production of lactate acidifies the cytoplasm, leading to LDH inhibition with the concomitant activation of pyruvate decarboxylase (PDC). The production of acetaldehyde by the action of PDC allows alcohol dehydrogenase (ADH) to produce ethanol. NAD availability is essential for glycolysis (and thus for ATP production), and both LDH and ADH play a crucial role in recycling NADH, produced in glycolysis, into NAD.

NADH as a result of the inactivity of the mitochondrial respiration, and the reduced pyridine nucleotides are recycled, initially by the lactic fermentation pathway. In the first phases of anaerobic metabolism lactate is produced, leading to a decrease in the cytoplasmic pH. A lower cytoplasmic pH activates pyruvate decarboxylase, thus producing acetaldehyde, with subsequent ethanol production from the action of alcohol dehydrogenase. Therefore, as suggested by Davies et al. (1974), lactate dehydrogenase and pyruvate dehydrogenase act as a metabolic pH-stat regulating the fermentative metabolism.

Whether the initial lactic acid production is the factor causing the fall in cytoplasmic pH is still an open issue, as the time course of cytoplasm acidification and lactic acid production does not correlate well, with a drop in the pH before the lactate concentration reaches a steady state (Saint-Ges et al., 1991). Despite some debate on the regulation of the fermentative metabolism, there are no doubts about the occurrence of alcoholic fermentation in plant tissues exposed to anoxia. Ethanol is almost always found in plant tissues under anoxia or hypoxia (Kimmerer and Mac Donald, 1987), with concentrations of the alcohol usually ranging from 2 to about 50 mM. Ethanol is detected in apple fruits under a controlled atmosphere (CA), with lactate it is only produced in minute amounts (Saquet and Streif, 2008). Treatments under hypoxia cause accumulations of acetaldehyde and ethanol in avocado under CA conditions (0.25% oxygen) while, interestingly, increasing the CO_2 concentration reduces the amount of lactate produced (Ke et al., 1995). In guava (*Psidium guajava* L.), high amounts of fermentative metabolites, ethanol, and acetaldehyde, accumulate in fruit held in atmospheres containing 2.5 kPa O_2 (Singh et al., 2008). Interestingly, in some fruits, it has been found that the exogenous application of ethanol or acetaldehyde alone can affect fruit ripening on the tree, for example, in figs (to induce maturity), banana, and persimmon (to remove astringency), and grape (to increase anthocyanins) (Pesis, 2005).

22.2.2 Gene Expression under Low O_2 Availability

Gene expression is strongly altered by anoxia, and protein synthesis is redirected to the production of a new set of polypeptides (anaerobic polypeptides [ANPs] or anaerobic stress proteins [ASPs]; reviewed by Fukao and Bailey-Serres, 2004; Geigenberger, 2003; Subbaiah and Sachs, 2003). Several ANPs have been identified, and have been shown to be enzymes engaged with the glycolytic and fermentative pathway. Alcohol dehydrogenase, lactate dehydrogenase, cytosolic glyceraldehyde dehydrogenase, aldolase, glucose phosphate isomerase, pyruvate decarboxylase, sucrose synthase, and alanine aminotransferase are induced by anaerobiosis. The induction of alcohol dehydrogenase (ADH; certainly the most studied ANP) by anaerobiosis was first reported by Hageman and Flesher (1960), and it is now considered as quite a common response of plants to anaerobic conditions. Maize *Adh1* gene is one of the few examples of genes whose function has been demonstrated to be essential for tolerance to anaerobic conditions (Schwartz, 1969), while the effect of *Adh2* have been shown to be of minor importance (Roberts et al., 1984). However, the physiological role of the induction of alcohol dehydrogenase activity under hypoxic or anoxic conditions is not clear, as the activity of alcohol dehydrogenase present in plant tissues growing under aerobic conditions often appear to be sufficiently high to explain the rates of ethanol production. Only such alcohol dehydrogenase levels as those found in the *Adh1* null mutants appear to limit both viability and alcoholic fermentation (Roberts et al., 1989). Although the essential role of alcohol dehydrogenase in anoxia tolerance has been demonstrated (Harberd and Edwards, 1982; Schwartz, 1969), the significance of the enhancement of alcohol dehydrogenase activity under conditions of low oxygen availability is still unclear (Roberts et al., 1989).

In addition to enzymes involved in the glycolytic and fermentative pathway, gene expression in roots or plant cell cultures was altered under low oxygen concentrations, and *trans* elements and *cis* factors have been identified (Dolferus et al., 1994; Hoeren et al., 1998; Klok et al., 2002; Olive et al., 1991; Paul et al., 2004). In addition to known ANPs, other hypoxia- or anoxia-induced genes identified consist of transcription factors (de Vetten and Ferl, 1995; Hoeren et al., 1998), signal transduction elements (Baxter-Burrell et al., 2002; Dordas et al., 2003), nonsymbiotic haemoglobin (Dordas et al., 2004), ethylene biosynthetic genes (Olson et al., 1995; Vriezen et al., 1999), nitrogen metabolism (Mattana et al., 1994), and cell wall loosening (Saab and Sachs, 1996). Yet little has been done to clarify these aspects in fruit or other detached plant organs subjected to controlled or modified atmosphere treatments. Lately, a number of reports have attempted to give new insight in the molecular mechanism governing the adaptation of fruit tissues to low oxygen (Eason et al., 2007; Loulakakis et al., 2006; Pasentsis et al., 2007).

Thus, it was shown that transferring preclimacteric or ripening-initiated avocado fruit to different oxygen levels for periods from 2 to 6 days suppressed polypeptides and mRNA species that are de novo synthesized during ripening and also the activity of the immunoreactive protein and mRNA transcripts of known ripening genes such as cellualse and polygacturonase (Kanellis et al., 1989a,b,c, 1993; Loulakakis et al., 2006). Analysis of mRNA populations in preclimacteric avocado fruit revealed that low O_2 levels induced new mRNA species possibly implicated in the adaptive mechanism under low O_2, or kept the housekeeping and/or preexisting mRNAs unaffected, indicating that the low oxygen response is complex and involves more than a simple adaptation in energy metabolism (Loulakakis et al., 2006). Most importantly, it was observed that the range of O_2 levels (0%–5%), which suppressed the induction of ripening enzymes, at the protein and mRNA levels, was similar to those oxygen levels that induced the synthesis of new hypoxic polypeptides and mRNA species as well as isoenzymes of alcohol dehydrogenase, a known low oxygen-regulated gene (Kanellis et al., 1991; Loulakakis et al., 2006), implicating a common mechanism of action. This observation has practical implications in CA–MA storage in the sense that the suppressive effects of low oxygen and the induction of the adaptation of fruit tissues to this environment are expected to be operational only in concentration under 5% O_2.

Recently, a functional genomic approach through the implementation of reverse transcriptase-polymerase chain reaction differential display (DD) laid to the isolation and characterization of novel low O_2-responsive transcripts from orange fruit flavedo tissues and broccoli heads that are up- or downregulated and/or unaffected (Aggelis et al., 1997; Eason et al., 2007; Pasentsis et al., 2007). In terms of *Citrus* flavedo tissues, this approach led to the isolation of 98 transcripts that were differentially expressed in differential display gels in response to hypoxia and anoxia. Northern blot analysis on 25 DD clones showed that 11 genes were stimulated under hypoxia and/or anoxia, 11 displayed constitutive expression and 3 transcripts were repressed by low O_2 regimes. Only two genes, pyruvate decarboxylase, and glutamare decarboxylase, participating in fermentation and γ-aminobutyric acid (GABA) biosynthesis, respectively, overlapped with known hypoxic/anoxic responsive genes shown to be stimulated in root or vegetative tissues of *Arabidopsis* or other plant species (Agarwal and Grover, 2005; Branco-Price et al., 2005; Fenglong et al., 2005; Klok et al., 2002; Liu et al., 2005; Loreti et al., 2005), suggesting common hypoxia/anoxia regulatory mechanisms well conserved among different plant tissues and species, at least for these two genes. However, the rest of the genes were for the first time implicated in low oxygen stress. It is evident that half of the isolated low oxygen-induced cDNAs are either of unknown function in terms of their involvement in hypoxia/anoxia or they share no apparent homology to any expressed sequences in the GenBank/EMBL databases. The other six genes are similar to molecules of the following functions: C-compound and carbohydrate utilization, plant development, amino acid metabolism, and biosynthesis of brasinosteroids. Of the novel genes identified i.e., an auxin-induced protein-like, a putative phosphatase, a putative serine esterase, a hypoxia-responsive family protein, a xylose isomerase family protein, a SPX [SYG1/Pho81/XPR1] domain-containing protein, and a P450, member of the CYP90A family participating in the brassinolide biosynthesis were shown for the first time to be induced by hypoxia/anoxia. It is not known whether the induction of these genes is specific to *Citrus* flavedo tissues, or common to other plant tissues, since bioinformatic analysis did not reveal any similarity with the recently identified low oxygen regulated genes in roots or vegetative tissues (Agarwal and Grover, 2005; Branco-Price et al., 2005; Fenglong et al., 2005; Klok et al., 2002; Liu et al., 2005; Loreti et al., 2005; Pasentsis et al., 2007). The differences observed in transcript profiles between the genes identified in *Citrus* and the above reports might be ascribed to the difference in tissues used (*Citrus* flavedo versus roots and vegetative tissues) and in techniques applied (DD versus DNA microarrays).

It is interesting to note that among all genes identified only two genes (one of unknown function, and the other showing similarities with a cytochrome P450 monooxygenase implicated in brassinolide biosynthesis) seem to be regulated mainly by anoxia as demonstrated by the absence of any apparent alteration in their gene expression profile in response to low and high temperature, wounding and ethylene (Pasentsis et al., 2007). Further, *in silico* analysis revealed that established low oxygen regulatory elements of known promoters (Liu et al., 2005; Mohanty et al., 2005) are present in the 5′ motifs of most of the corresponding homologs of the *Arabidopsis* hypoxic/anaerobic genes identified in the *Citrus* study, indicating control by the same set of transcription factors. However, these elements should be isolated and functionally characterized from the *Citrus* anaerobic genes in order to identify common anaerobic regulatory elements within plant species. Collectively, these data have shed new insight into functions and processes that were not previously connected with low O_2 environment and consequently CA storage and support the notion that the effects of low oxygen of fruit tissues cannot be explained only by the repression of respiration and/or by the concomitant inhibition of ethylene biosynthesis and action.

22.2.3 Acclimation to Low O_2

Plant tissues which are usually sensitive to severe environmental stresses become more tolerant after a period of acclimation under less severe stress. Such strict conditions may exist for example during implementation of ultra low oxygen or anaerobic conditions applied as postharvest quarantine treatments to restrict insect infestations (Shellie, 2002; Shellie et al., 1997). Most fresh fruit and vegetables do not tolerate these low oxygen atmospheres for prolonged periods, though some can tolerate them for short periods (Delate and Bretch, 1989; Ke and Kader, 1992). For example, 'Hass' avocado fruits are very sensitive to insecticidal low oxygen atmospheres and cannot tolerate 0.25% O_2 for longer than 24 h at 20°C (Ke et al., 1995; Yahia and Carrillo-Lopez, 1993; Yahia and Kader, 1991); mesocarp injury observed after exposure to low oxygen for 2 days, usually started as discoloration of the vascular tissue and then extended to the rest of the tissue (Yahia and Carrillo-Lopez, 1993). However, acclimation of avocado fruit to hypoxia (3% O_2) for 24 h resulted in a beneficial increase in tolerance to subsequent ultralow O_2 treatments (El-Mir et al., 2001). In contrast, fruit such as 'Valencia' oranges and mango can tolerate 0.25% O_2 for more than 5 days at 20°C (Ke and Kader, 1992; Yahia and Hernandez, 1993). Further, hypoxic conditioning prior to application of anaerobic conditions was shown to enhance the survival of cut carnation flowers after anoxia (Chen and Solomos, 1996). The potentiating effects of hypoxic conditioning are attended by an increase in the activity of a number of glycolytic enzymes and, in the case of corn roots, the formation of aerenchyma (Chang et al., 2000; Drew, 1997; He et al., 1996b). Within these lines, it is interesting to note a number of anoxic/hypoxic proteins like PDC and ADH showed usually a higher expression in low oxygen level i.e., 1%–3%, than in anoxia, suggesting their participation in the acclimation of the tissue to subsequent anoxia (Eason et al., 2007; El-Mir et al., 2001; Imahori et al., 2002; Pasentsis et al., 2007). In the case of avocados, it was demonstrated that the induction of ADH isoenzymes was saturable with respect to O_2 concentrations in that for the induction to occur, the level of O_2 must drop below 5% (El-Mir et al., 2001; Kanellis et al., 1991; Loulakakis et al., 2006). Thus, the induction of anoxic proteins is initiated at levels of O_2 where the rate of ATP synthesis is expected to be higher than under partial or total anoxia. Since protein synthesis is considered a strong sink for ATP utilization, it is reasonable to assume that the rate of anoxic protein synthesis is faster under O_2 concentrations, which do not create partial anaerobiosis, than it is under partial or total anoxia.

22.2.4 Postanoxic and Posthypoxic Injury

Oxygen toxicity has been proposed as the principle mechanism of postanoxic and post-hypoxic injury (Blokhina et al., 2003; Crawford and Braendle, 1996). When aerobic conditions are restored, a burst in superoxide radicals occurs, resulting in posthypoxic or postanoxic injury to the tissues (Babior, 1987). The natural defense enzymatic systems against such stress are peroxidase, catalase, superoxide dismutase, and low molecular mass antioxidants, i.e., ascorbate, glutathione, and tocopherols, which may contribute significantly to postanoxic survival (Blokhina et al., 2003; Crawford and Braendle, 1996; VanToai and Bolles, 1991). In fact, transferring mature green tomato fruit to air after a 48 h anoxia treatment caused a sharp increase in ascorbate levels accompanied with the concomitant induction of all genes participating in ascorbate biosynthesis and recycling, along with elevated genes expression of catalase and superoxide dismutase (unpublished data, Kanellis' group). It will be interesting to investigate whether the difference observed in the response of various fresh produce to CA–MA storage can be attributed to their ability to stimulate their natural defense systems against postanoxic and posthypoxic injury.

22.2.5 Model of Low O_2 Sensing and Action in Plant Tissues

Direct sensing of molecular oxygen is thought to be mediated by a copper or heme protein that induces an increase in cytosolic Ca^{2+}, which then binds to calmodulin and the produced Ca^{2+}–calmodulin complex activates glutamate decarboxylase (GAD) and production of y-aminobutynic acid (GABA) (Kinnersley and Turano, 2000); in addition, the oxygen sensor mediates a first round of gene expression leading to altered production among other GABA; the latter in turn mediates a second round of gene expression (Figure 22.2). In addition, similar to what happens in yeast, heme may regulate the anaerobic and aerobic metabolism in plants, since it was shown that heme activated protein 5c (HAP5c) exists in *Arabidopsis* (Edwards et al., 1998; Gusmaroli et al., 2001) and tomato (Aggelis et al., 1997; Gherrabi et al., 1999). In addition, NO and reactive oxygen species (similar to animal) may participate in the above mechanism. However, experimental data for plant tissues are lacking.

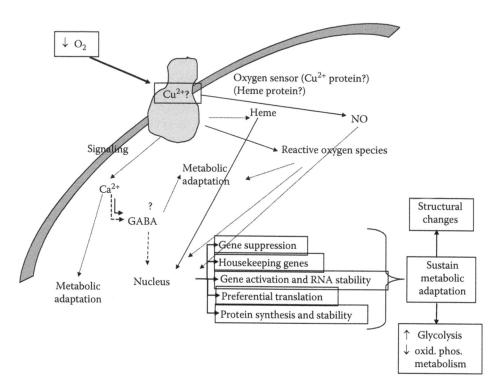

FIGURE 22.2
Schematic diagram of oxygen sensing and signal transduction components and the regulation of metabolism under anoxia and hypoxia in plants.

22.3 Responses to High CO_2

Elevated CO_2 concentrations are highly effective in altering primary and secondary metabolism of harvested products and may have marked impact on many biochemical processes thus affecting quality parameters. The response to high CO_2 concentrations may be beneficial or detrimental depending upon the nature of the product, the concentration of carbon dioxide outside and inside the tissue, the duration of exposure and the concentration of oxygen (Kays and Paull, 2004). Increased CO_2 concentrations are one of the component of the long-term controlled atmosphere storage (and, in this case, it is not easy to separate the effects of elevated carbon dioxide from those induced by low oxygen concentrations) but high CO_2 levels may be used for short-term postharvest treatments that, in general, are effective in prolonging shelf life and maintaining quality of different horticultural produce. The knowledge of the effects of high CO_2 postharvest treatments at biochemical and molecular level is, however, still limited but some basic information is starting to accumulate also due to large-scale analysis of transcriptome using genomic tools as microarrays. In mature green tomato fruit exposed to high CO_2 (20% applied for three days) levels, the delay of ripening is accompanied by an induction of stress-related proteins (heat-shock protein and glutamate decarboxylase) and the block of mRNA accumulation of genes involved in ripening-related changes (Rothan et al., 1997). The effects of elevated concentrations of carbon dioxide on altering postharvest fruit metabolism has been clearly demonstrated by large-scale transcript analysis carried out on strawberry using a cDNA tomato microarray containing 8700 unigenes. In one variety

('Jewel') characterized by a limited tolerance to high CO_2 levels, 178 ESTs were differentially expressed when comparing fruit stored for 48 h in air with those maintained at 20% CO_2 (Ponce-Valadez et al., 2005). Since CO_2 is toxic, many tissues undergo damages following exposure to high CO_2 and high variability is present among and within species in terms of susceptibility to carbon dioxide injury, resulting in internal and external physiological disorders and the appearance of off-flavors often related to the products of fermentation.

Even though this effect is not present in all plant tissues, high carbon dioxide concentrations are well known to decrease respiration and this is due to the inhibition of succinate dehydrogenase that catalyzes the conversion of succinate to fumarate at the tricarboxylic acid. As observed in strawberry (Fernandez-Trujillo et al., 1999), lettuce (Ke et al., 1993), and banana (Liu et al., 2004) succinate level increases following exposure to high carbon dioxide often reaching toxic levels and causing damage to the tissues as observed in apples (Knee, 1973). Carbon dioxide may also inhibit other components of the tricarboxylic acid-cycle (TCA) cycle: isocitrate dehydrogenase appears to be markedly suppressed in ripening banana following exposure to carbon dioxide (Liu et al., 2004) and an inhibition of pyruvate dehydrogenase has been observed in avocado (Ke et al., 1995). Probably due to the acidification of mitocondria and the effects on TCA cycle, stress level of carbon dioxide has the potential to induce the fermentation pathway (Mir and Beaudry, 2002). Accumulation of fermentation products as acetaldehyde, ethanol, and ethyl acetate have been reported in strawberry exposed to high carbon dioxide concentrations (Fernandez-Trujllo et al., 2007; Ke et al., 1994) and this indicates that these treatments represent significant stresses to tissues. ADH induction is regarded as one of the reasons for anaerobic fermentation and accumulation of ethanol. Indeed, Bonghi et al. (1999) showed that anaerobic metabolism is rapidly activated following exposure to high CO_2 (30%) concentrations in peaches and that ADH activity is strongly stimulated following short-term (24 h) treatments that resulted also effective in inducing a slight accumulation of ADH mRNA. The extent of fermentation under elevated CO_2 is, however, strongly influenced by several factors including genotype as observed in strawberry (Fernandez-Trujillo et al., 1999; Pelayo et al., 2002; Watkins et al., 1999). Using a heterologous genomic approach (tomato microarray), Ponce-Valadez et al. (2005) pointed out the presence of different transcript profiles between a tolerant ('Cavendish') and nontolerant ('Jewel') strawberry cultivar to high CO_2 concentrations during storage. Considering primary metabolism, one target corresponding to pyruvate decarboxylase and one to malate dehydrogenase appeared differentially expressed in the two cultivars in response to carbon dioxide (20% for 48 h) treatment.

Other effects of elevated carbon dioxide concentrations in harvested products are the decrease of ethylene biosynthesis and action, a reduced softening rate, altered pigmentation, inhibition of storage pathogens. The inhibition of ethylene and ethylene-dependent processes (including autocatalytic synthesis) by CO_2 has been imputed to the competition of the gas with ethylene at the receptor-binding site but no clear evidences have been, up to now, produced to support this hypothesis. In ripening tomato, inhibition of 1-aminocyclo propane-l-carboxylate (ACC) synthase (ACS) and ACC oxidase (ACO) mRNA accumulation has been observed following CO_2 exposure (Rothan et al., 1997), whereas in peaches a marked effect has been detected in particular on ACS activity and gene expression (Mathooko et al., 2001). Tomato ACC oxidase members are affected by CO_2 differently: in fact, only *LE-ACO1* and *LE-ACO3* but not *LE-ACO4* appear to be downregulated by carbon dioxide (de Wild et al., 2005). Considering that both ethylene-dependent and ethylene-independent (development-dependent) gene expression is affected by high CO_2 levels at ripening (Rothan et al., 1997), it is hypothesized that CO_2 exerts its regulatory effects through different molecular mechanisms including the reduction of the synthesis and/or action of ethylene and some stress-related responses.

One of the most marked effect of short-term postharvest treatments with high CO_2 concentration is the delay in flesh softening of perishable fruit as observed in strawberry and peaches (Bonghi et al., 1999; Harker et al., 2000; Mathooko et al., 2001; Watkins et al., 1999). The maintenance of high firmness values in peaches following 24 h carbon dioxide (30%) treatment is accompanied by a marked reduction of endo-β-1,4-glucanase activity and a decrease in expression of a target corresponding to tomato polygalacturonase gene has been detected in two strawberry cultivars treated with 20% CO_2 for 48 h (Ponce-Valadez et al., 2005). The accumulation of PG transcripts resulted severely delayed in MG tomato kept for 3 days at 20% CO_2, a treatment also effective in downregulating phytoene synthase, a gene involved in carotenoid synthesis and ripening-related color development (Rothan et al., 1997). Considering anthocyanins, pigments responsible for color in some fruit species, a reduction of their concentration in strawberry has been observed following high (10%–40%) carbon dioxide postharvest treatments and this has been associated to both a reduction of the activities of two enzymes involved in the biosynthetic pathway (phenylalanine ammonia lyase [PAL] and UDP glucose:flavonoid glucosyltransferase [UFGT]) and a reduction of anthocyanin stability probably due to changes in pH (Gil et al., 1997; Holcroft and Kader, 1999). Besides color, polyphenol compounds are implicated in modulating fruit taste and flavor and resistance to different biotic and abiotic stresses: little information is available on the effects of high CO_2 levels during storage on the regulation of phenylpropanoid pathway. Table grapes ('Cardinal') maintained for 3 days under 20% CO_2 at 0°C showed, compared to control (air treatment), a reduced accumulation of total anthocyanins and a more pronounced decrease of *trans*-resveratrol and this has been related to a negative effect of carbon dioxide to the expression of PAL, chalcone synthase (CHS), and stilbene synthase (STS) genes (Sanchez-Ballesta et al., 2007). A reduced induction of *PAL*, *CHS*, and *STS* gene expression (and a lower accumulation of *trans*-resevratrol) has also been observed in table grapes stored up to 33 days at 0°C following a pretreatment with 20% CO_2 for 3 days, suggesting that the reduction of fungal decay observed in treated grapes is not mediated by the activation of the stilbene pathway (Romero et al., 2008; Sanchez-Ballesta et al., 2006). The efficacy of high CO_2 pretreatment to control fungal decay during storage at 0°C is associated to a marked reduction of β-1,3-glucanase and, to a lesser extent, of class I chitinase mRNAs, thus suggesting that mechanisms involving other pathogenesis-related proteins might be operating in the control of postharvest fungal infection activated by high CO_2 treatments. Marked changes of metabolism are also observed in whole wine grapes exposed for 10 days to carbon dioxide prior to fermentation (carbonic maceration): 114 volatiles compounds were identified in treated grapes against 60 of the control sample (Dourtoglou et al., 1994). In preliminary large-scale transcriptome analysis using microarrays, altered patterns of gene expression have been observed in skins and, to a lesser extent, in pulp of Sangiovese (red-skinned) and Trebbiano (white-skinned) wine grape berries maintained for 3 days after harvest in a CO_2-enriched atmosphere (Tonutti's group, unpublished data).

An indirect effect of high CO_2 treatment is the removal of astringency in persimmons: this effects if supposed to be the result of acetaldehyde production that would play a role in the insolubilization (polymerization) of condensed tannins, also known as proanthocyanidins (PA), responsible for astringency when present in soluble form (Arnal and Del Rio, 2003; Yamada et al., 2002). Astringency removal has been associated to reduced expression of PAL, CHS, and dehydroflavonol reductase genes in an astringent-type persimmon (Ikegami et al., 2007). A subtraction cDNA library synthesized from astringent and astringent-removed persimmon fruit has been successfully constructed to identify differentially expressed clones, some of them (including anthocyanidin reductase) involved in flavonoid biosynthesis and others as glucose acyltransferase, more expressed in astringent fruit and probably playing a crucial role in PA accumulation in persimmon fruit.

Acknowledgments

Part of the work reviewed by A.K. Kanellis was supported by grants from EU-FAIR-CT95-0225, FAIR-CT-4096, and the Greek General Secretariat for Research and Development (GR-GRST).

References

Agarwal, A. and A. Grover. 2005. Isolation and transcription profiling of low-O_2 stress-associated cDNA clones from the flooding-stress-tolerant FR13A rice genotype. *Annals of Botany* 96: 831–844.

Aggelis, A., A. Makris, A. Christodoulidou, A. Carrillo López, and A.K. Kanellis. 1997. Isolation of low oxygen-regulated genes in tomato fruit applying RNA differential display and yeast complementation approaches. *International Controlled Atmosphere Research Conference*, University of California, Davis, CA, July 13–18, 1997.

Arnal, L. and M.A. Del Rio. 2003. Removing astringency by carbon dioxide and nitrogen-enriched atmospheres in persimmon fruit cv 'Rojo brillante.' *Journal of Food Science* 68: 1516–1518.

Babior, B.M. 1987. The respiratory burst oxidase. *Trends in Biochemical Sciences* 12: 241–243.

Bailey-Serres, J. and R. Chang. 2005. Sensing and signaling in response to oxygen deprivation in plants and other organisms. *Annals of Botany* 96: 507–518.

Basse, C.W. 2005. Dissecting defense-related and developmental transcriptional responses of maize during *Ustilago maydis* infection and subsequent tumor formation. *Plant Physiology* 138: 1774–1784.

Baxter-Burrell, A., Z. Yang, P.S. Springer, and J. Bailey-Serres. 2002. RopGAP4-dependent Rop GTPase rheostat control of *Arabidopsis* oxygen deprivation tolerance. *Science* 296: 2026–2028.

Blokhina, O., E. Virolainen, and K.V. Fagerstedt. 2003. Antioxidants, oxidative damage and oxygen deprivation stress: A review. *Annals of Botany* 91: 179–194

Bonghi, C., A. Ramina, B. Ruperti, R. Vidrih, and P. Tonutti. 1999. Peach fruit ripening and quality in relation to picking time, and hypoxic and high CO_2 short-term postharvest treatments. *Postharvets Biology and Technology* 16: 213–222.

Branco-Price, C., R. Kawaguchi, R.B. Ferreira, and J. Bailey-Serres. 2005. Genome-wide analysis of transcript abundance and translation in *Arabidopsis* seedlings subjected to oxygen deprivation. *Annals of Botany* 96: 647–660.

Carginale, V., G. Maria, C. Capasso, E. Ionata, F. La Cara, M. Pastore, A. Bertaccini, and A. Capasso. 2004. Identification of genes expressed phytoplasma infection in leaves of *Prunus armeniaca* RNA differential display. *Gene* 332: 29–34.

Chang, W.W.P., L. Huang, M. Shen, C. Webster, A.L. Burlingame, and J.K.M. Roberts. 2000. Patterns of protein synthesis and tolerance of anoxia in root tips of maize seedlings acclimated to a low-oxygen environment, and identification of proteins by mass spectrometry. *Plant Physiology* 122: 295–318.

Chen, X. and T. Solomos. 1996. Effects of hypoxia on cut carnation flowers (*Dianthus caryophyllus* L.): Longevity, ability to survive under anoxia, and activities of alcohol dehydrogenase and pyruvate kinase. *Postharvest Biology and Technology* 7: 317–329.

Crawford, R.M.M. 1982. Physiological responses to flooding. In: O.L. Lange, P.S. Nobel, C.B. Osmond, and H. Ziegler (Eds.), *Encyclopedia of Plant Physiology, Physiological Plant Ecology II Water Relations and Carbon Assimilation*, Springer, Berlin, Germany, pp. 453–477.

Crawford, R.M.M. and R. Braendle. 1996. Oxygen deprivation stress in a changing environment. *Journal of Experimental Botany* 47: 145–159.

Davies, D.D. (Ed.), 1980. Anaerobic metabolism and the production of organic acids. In: *The Biochemistry of Plants*, Vol. 2, Academic Press, London, pp. 581–611.

Davies, D.D., S. Grego, and P. Kenworth. 1974. The control of the production of lactate and ethanol in higher plants. *Planta* 118: 297–310.

de Vetten, N.C. and R.J. Ferl. 1995. Characterization of a maize G-box binding factor that is induced by hypoxia. *The Plant Journal* 7: 589–601.

de Wild, H.P.J., P.A. Balk, E.C.A. Fernandes, and H.W. Peppelenbos. 2005. The action site of carbon dioxide in relation to inhibition of ethylene production in tomato fruit. *Postharvest Biology and Technology* 36: 273–280.

Delate, K.M. and J.K. Bretch. 1989. Quality of tropical sweet potatoes exposed to controlled atmosphere treatments for postharvest insect control. *Journal of American Society of Horticultual Science* 114: 963–968.

Dolferus, R., M. Jacobs, W.J. Peacock, and E.S. Dennis. 1994. Differential interactions of promoter elements in stress responses of the *Arabidopsis Adh* gene. *Plant Physiology* 105: 1075–1087.

Dordas, C., B.B. Hasinoff, A.U. Igamberdiev, N. Manac'h, J. Rivoal, and R.D. Hill. 2003. Expression of a stress-induced hemoglobin affects NO levels produced by alfalfa root cultures under hypoxic stress. *The Plant Journal* 35: 763–770.

Dordas, C., B.B. Hasinoff, J. Rivoal, and R.D. Hill. 2004. Class-1 hemoglobins, nitrate and NO levels in anoxic maize cell-suspension cultures. *Planta* 219: 66–72.

Dourtoglou, V.G., N.G. Yannovitis, V.G. Tychopoulos, and M.M. Vamvakias. 1994. Effects of storage under CO_2 atmosphere on the volatile, amino acid and pigment constituents in red grape (*Virtis vinifera* L. var. Agiorgitiko). *Journal of Agricultural and Food Chemistry* 42: 338–344.

Drew, M.C. 1997. Oxygen deficiency and root metabolism: Injury and acclimation under hypoxia and anoxia. *Annual Review of Plant Physiology and Plant Molecular Biology* 48: 223–250.

Eason, J.R., D. Ryan, B. Page, L. Watson, and S.A. Coupe. 2007. Harvested broccoli (*Brassica oleracea*) responds to high carbon dioxide and low oxygen atmosphere by inducing stress-response genes. *Postharvest Biology and Technology* 43: 358–365.

Edwards, D., J.A.H. Murray, and A.G. Smith. 1998. Multiple genes encoding conserved CCAAT-box transcription factor complex are expressed in *Arabidopsis*. *Plant Physiology* 117: 1015–1022.

El-Mir, M., D. Gerasopoulos, I. Metzidakis, and A.K. Kanellis. 2001. Hypoxic acclimation prevents avocado mesocarp injury caused by subsequent exposure to extreme low oxygen atmospheres. *Postharvest Biology and Technology* 23: 215–226.

Fenglong, L., T. VanToai, L.P. Moy, G. Bock, L.D. Linford, and J. Quackenbush. 2005. Global transcription profiling reveals comprehensive insights into hypoxic response in *Arabidopsis*. *Plant Physiology* 137: 1115–1129.

Fernandez-Trujillo, J.P., J.F. Nock, and C.B. Watkins. 1999. Fermentative metabolism and organic acid concentration in fruit of selected strawberry cultivars with different tolerances to carbon dioxide. *Journal of American Society of Horticultural Science* 124: 696–701.

Fukao, T. and J. Bailey-Serres. 2004. Plant responses to hypoxia—is survival a balancing act? *Trends in Plant Science* 9: 449–456.

Geigenberger, P. 2003. Response of plant metabolism to too little oxygen. *Current Opinion in Plant Biology* 6: 247–256.

Gherraby, W., A. Makris, I. Pateraki, M. Sanmartin, P. Chatzopoulos, and A.K. Kanellis. 1999. Manipulation of the expression of heme activated protein HAP5c gene in transgenic plants. In: A.K. Kanellis, C. Chang, H. Klee, A.B. Bleecker, J.P. Pech, and D. Grierson (Eds.), *Biology and Biotechnology of the Plant Hormone Ethylene II*, Kluwer Academic, Dordrecht, the Netherlands, pp. 321–326.

Gil, M.I., D.M. Holcroft, and A.A. Kader. 1997. Changes in strawberry anthocyanins and other polyphenols in response to carbon dioxide treatments. *Journal of Agricultural and Food Chemistry* 45: 1662–1667.

Gusmaroli, G., C. Tonelli, and R. Mantovani. 2001. Regulation of the CCAAT-Binding NF-Y subunits in *Arabidopsis thaliana*. *Gene* 264: 173–185.

Hageman, R.H. and D. Flesher. 1960. The effect of an anaerobic environment on the activity of alcohol dehydrogenase and other enzymes of corn seedling. *Archives of Biochemistry and Biophysics* 87: 203–209.

Harberd, N.P. and K.J.R. Edwards. 1982. The effect of a mutation causing alcohol dehydrogenase deficiency on flooding tolerance in barley. *New Phytology* 90: 631–644.

Harker, F.R., H.J. Elgar, C.B. Watkins, P.J. Jackson, and I.C. Hallett. 2000. Physical and mechanical changes in strawberry fruit after high carbon dioxide treatments. *Postharvest Biology and Technology* 19: 139–146.

He, C., S.A. Finglayson, M.C. Drew, W.R. Jordan, and P.W. Morgan. 1996b. Ethylene biosynthesis during aerenchyma formation in roots of maize subjected to mechanical impedance and hypoxia. *Plant Physiology* 112: 1679–1685.

Hoeren, F., R. Dolferus, W.J. Peacock, and E.S. Dennis. 1998. Evidence for a role for AtMYB2 in the induction of the *Arabidopsis* alcohol dehydrogenase (*ADH1*) gene by low oxygen. *Genetics* 149: 479–490.

Holcroft, D.M. and A.A. Kader. 1999. Carbon dioxide-induced changes in color and anthocyanin synthesis of stored strawberry fruit. *HortScience* 34(7): 1244–1248.

Ikegami, A., S. Eguchi, A. Kitajiama, K. Inoue, and K. Yonemor. 2007. Identification of genes involved in proanthocyanidin biosynthesis of persimmon (*Diospyros kaki*) fruit. *Plant Science* 172: 1037–1042.

Imahori, Y., M. Kota, Y. Ueda, M. Ishimaru, and K. Cachin. 2002. Regulation of ethanolic fermentation in bell pepper fruit under low oxygen stress. *Postharvest Biology and Technology* 25: 159–167.

Kader, A.A. 1986. Biochemical and physiological basis for effects of controlled and modified atmospheres on fruits and vegetables. *Food Technology* 40: 99–104.

Kanellis, A.K. 1994. Oxygen regulation of protein synthesis and gene expression in ripening fruits: Future outlook. In: E. Woltering (Ed.), *Proceedings of the International Symposium on Postharvest Treatment of Fruits and Vegetables*, Oosterbeek, the Netherlands, October 19–22, 1994.

Kanellis, A.K., K.A. Loulakakis, M. Hassan, and K.A. Roubelakis-Angelakis. 1993. Biochemical and molecular aspects of the low oxygen action on fruit ripening. In: C.J. Pech, A. Latche, and C. Balague (Eds.), *Cellular and Molecular Aspects of Biosynthesis and Action of the Plant Hormone Ethylene*, Kluwer Academic, Dordrecht, the Netherlands, pp. 117–122.

Kanellis, A.K., T. Solomos, and A.K. Mattoo. 1989b. Hydrolytic enzyme activities and protein pattern of avocado fruit ripened in air and in low oxygen with and without ethylene. *Plant Physiology* 90: 259–266.

Kanellis, A.K., T. Solomos, and A.K. Mattoo. 1989a. Changes in sugars, enzymatic activities and phosphatase isoenzyme profiles of bananas ripened in air or stored in 2.5% O_2 with and without ethylene. *Plant Physiology* 40: 251–258.

Kanellis, A.K., T. Solomos, A.M. Mehta, and A.K. Mattoo. 1989c. Decreased cellulase activity in avocado fruit subjected to 2.5% O_2 correlates with lower cellulase protein and gene transcript levels. *Plant Cell Physiology* 30: 829–834.

Kanellis, A.K., T. Solomos, and K.A. Roubelakis-Angelakis. 1991. Suppression of cellulase and polygalactorunase and induction of alcohol dehydrogenase isoenzymes in avocado fruit mesocarp subjected to low oxygen stress. *Plant Physiology* 96: 269–274.

Kays, S.J. and R.E. Paull. 2004. *Postharvest Biology*, Exon Press, Athens, GA.

Ke, D. and A.A. Kader. 1992. External and internal factors influence fruit tolerance to low-oxygen atmospheres. *Journal of American Society of Horticultural Science* 117: 913–918.

Ke, D., M. Mateos, J. Siripanich, C. Li, and A.A. Kader. 1993. Carbon dioxide action on metabolism of organic and amino acids in crisphead lettuce. *Postharvest Biology and Technology* 3: 235–247.

Ke, D., E.M. Yahia, B. Hess, L. Zhou, and A.A. Kader. 1995. Regulation of fermentation metabolism in avocado fruit under oxygen and carbon dioxide stresses. *Journal of American Society of Horticultural Science* 120: 481–490.

Ke, D., L. Zhou, and A.A. Kade. 1994. Mode of oxygen and carbon dioxide action on strawberry ester biosynthesis. *Journal of American Society of Horticultural Science* 119: 971–975.

Kimmerer, T.W. and R.C. MacDonald. 1987. Acetaldehyde and ethanol biosynthesis in leaves of plants. *Plant Physiology* 84: 1204–1209.

Kinnersley, A.M. and F.J. Turano. 2000. Gamma aminobutyric acid (GABA) and plant responses to stress. *Critical Review in Plant Science* 19: 479–509.

Klok, E.J., I.W. Wilson, D. Wilson, S.C. Chapman, R.M. Ewing, S.C. Somerville, W.J. Peacock, R. Dolferus, and E.S. Dennis. 2002. Expression profile analysis of the low-oxygen response in *Arabidopsis* root cultures. *The Plant Cell* 14: 2481–2494.

Knee, M. 1973. Effects of controlled atmosphere storage on respiratory metabolism of apple fruit tissue. *Journal of Science, Food and Agriculture* 24: 1289–1298.

Liu, F., T. Vantoai, L. Moy, G. Bock, L.D. Linford, and J. Quackenbush. 2005. Global transcription profiling reveals novel insights into hypoxic response in *Arabidopsis*. *Plant Physiology* 137: 1115–1129.

Liu, S., Y. Yang, H. Murayama, S. Taira, and T. Fukushima. 2004. Effects of CO_2 on respiratory metabolism in ripening banana fruit. *Postharvest Biology and Technology* 33: 27–34.

Loreti, E., A. Poggi, G. Novi, A. Alpi, and P. Perata. 2005. A genome-wide analysis of the effects of sucrose on gene expression in *Arabidopsis* seedlings under anoxia. *Plant Physiology* 137: 1130–38.

Loulakakis, K., M. Hassan, G. Gerasopoulos, and A.K. Kanellis. 2006. Effects of low oxygen on in vitro translation products of poly(A) + RNA, cellulase and alcohol dehydrogenase expression in preclimacteric and ripening-initiated avocado fruit. *Postharvest Biology and Technology* 39: 29–37.

Mathooko, F.M., Y. Tsunashima, E.A.O. Owino, Y. Kubo, and A. Inaba. 2001. Regulation of genes encoding ethylene biosynthetic enzymes in peach (*Prunus persica* L.) fruit by carbon dioxide and 1-methylcyclopropene. *Postharvest Biology and Technology* 21: 265–281.

Mattana, M., I. Coraggio, A. Bertani, and R. Reggiani. 1994. Expression of the enzymes of nitrate reduction during the anaerobic germination of rice. *Plant Physiology* 106: 1605–1608.

Mir, N. and R. Beaudry. 2002. Atmosphere control using oxygen and carbon dioxide. In: M. Knee (Ed.), *Fruit Quality and Its Biological Basis*, Sheffield Academic Press, Sheffield, U.K., pp. 123–156.

Mohanty, B.S., P.T. Krishnan, S. Swarup, and V.B. Bajic. 2005. Detection and preliminary analysis of motifs in promoters of anaerobically induced genes of different plant species. *Annals of Botany* 96: 669–681.

Olive, M.R., W.J. Peacock, and E.S. Dennis. 1991. The anaerobic responsive element contains two GC-rich sequences essential for binding a nuclear protein and hypoxic activation of the maize *Adh1* promoter. *Nucleic Acids Research* 19: 7053–7060.

Olson, D.C., J.H. Oetiker, and S.F. Yang. 1995. Analysis of LE-ACS3, a 1-aminocyclopropane-1-carboxylic acid synthase gene expressed during flooding in the roots of tomato plants. *Journal of Biological Chemistry* 270: 14056–14061.

Pasentsis, K., V. Falara, I. Pateraki, D. Gerasopoulos, and A.K. Kanellis. 2007. Identification and expression profiling of low-oxygen regulated genes from citrus flavedo tissues using RT-PCR differential display. *Journal of Experimental Botany* 58: 2203–2216.

Paul, A.L., A.C. Schuerger, M.P. Popp, J.T. Richards, M.S. Manak, and R.J. Ferl. 2004. Hypobaric biology: *Arabidopsis* gene expression at low atmospheric pressure. *Plant Physiology* 134: 215–223.

Pelayo, C., S. Ebeler, and A.A. Kader. 2002. Postharvest life and flavour quality of three strawberry cultivars kept at 5°C in air or air + 20 kPa CO_2. *Postharvest Biology and Technology* 27: 171–183.

Perata, P. and A. Alpi. 1993. Plant-responses to anaerobiosis. *Plant Science* 93: 1–17.

Pesis, E. 2005. The role of the anaerobic metabolites, acetaldehyde and ethanol, in fruit ripening, enhancement of fruit quality and fruit deterioration. *Postharvest Biology and Technology* 37: 1–19.

Ponce-Valadez, M., C.B. Watkins, S. Moore, and J.J. Giovannoni. 2005. Differential gene expression analysis of strawberry cultivar responses to elevated CO_2 concentrations during storage using a tomato cDNA microarray. *Acta Horticulturae* 682: 255–261.

Roberts, J.K.M., J. Callis, O. Jardetzky, V. Walbot, and M. Freeling. 1984. Cytoplasmic acidosis as a determinant of flooding intolerance in plants. *Proceedings of the National Academy of Science USA* 81: 6029–6033.

Roberts, J.K.M., K. Chang, C. Webster, J. Callis, and V. Walbot. 1989. Dependence of ethanolic fermentation, cytoplasmic pH regulation, and viability on the activity of alcohol dehydrogenase in hypoxic maize root tips. *Plant Physiology* 89: 1275–1278.

Romero, I., M.T. Sanchez-Ballesta, R. Maldonado, M.I. Escribano, and C. Merodio. 2008. Anthocyanin, antioxidant activity and stress-induced gene expression in high CO_2-treated table grapes stored at low temperature. *Journal of Plant Physiology* 165: 522–530.

Rothan, C., S. Duret, C. Chevalier, and P. Raymond. 1997. Suppression of ripening-associated gene expression in tomato fruits subjected to a high CO_2 concentration. *Plant Physiology* 114: 255–263.

Saab, I.N. and M.M. Sachs. 1996. A flooding-induced xyloglucan endotransglycosylase homolog in maize is responsive to ethylene and associated with aerenchyma. *Plant Physiology* 112: 385–391.

Saint-Ges, V., C. Roby, R. Bligny, A. Pradet, and R. Douce, 1991. Kinetic studies of the variations of cytoplasmic pH, nucleotide triphosphates (31P-NMR) and lactate during normoxic and anoxic transition in maize root tips. *European Journal of Biochemistry* 200: 477–482.

Sanchez-Ballesta, M.T., J.B. Jimenez, I. Romero, J.M. Orea, R. Maldonado, A. Gonzales-Urena, M.I. Escribano, and C. Merodio. 2006. Effects of high CO2 pretreatment on quality, fungal decay and molecular regulation of stilbene phytoalexin biosynthesis in stored table grapes. *Postharvest Biology and Technology* 42: 209–216.

Sanchez-Ballesta, M.T., I. Romero, J.B. Jimenez, J.M. Orea, A. Gonzales-Urena, M.I. Escribano, and C. Merodio. 2007. Involvement of the phenylpropanoid pathway in the response of table grapes to low temperature and high CO_2 levels. *Postharvest Biology and Technology* 46: 29–35.

Saquet, A.A. and J. Streif. 2008. Fermentative metabolism in 'Jonagold' apples under controlled atmosphere storage. *European Journal of Horticultural Science* 73: 43–46

Schwartz, D. 1969. An example of gene fixation resulting from selective advantage in suboptimal conditions. *American Nature* 103: 479–481.

Shellie, K.C. 2002. Ultra-low oxygen refrigerated storage of 'Rio Red' grapefruit: Fungistatic activity and fruit quality. *Postharvest Biology and Technology* 25: 73–85.

Shellie, K.C., R.L. Mangan, and S.J. Ingle. 1997. Tolerance of grapefruit to Mexican fruit fly larvae to heated controlled atmospheres. *Postharvest Biology and Technology* 10: 179–186.

Singh, S.P. and R.K. Pal. 2008. Controlled atmospheres storage of guava (*Psidium guajava* L.) fruit, *Postharvest Biology and Technology* 47: 296–306.

Solomos, T. 1982. Effects of low O_2 concentration on fruit respiration: Nature of respiratory diminution. In: D.G. Richardson and M. Meheriuk (Eds.), *Controlled Atmospheres for Storage and Transport of Perishable Agricultural Commodities*, Timber Press, Beaverton, OR, pp. 161–170.

Solomos, T. and A.K. Kanellis. 1997. Hypoxia and Fruit Ripening. In: A.K. Kanellis, C. Chang, H. Kende, and D. Grierson (Eds.), *Biology and Biotechnology of the Plant Hormone Ethylene*, Kluwer Academic, Dordrecht, the Netherlands, pp. 239–252.

Subbaiah, C.C. and M.M. Sachs. 2003. Molecular and cellular adaptations of maize to flooding stress. *Annals of Botany* 91: 119–127.

VanToai, T.T. and C.S. Bolles. 1991. Postanoxic injury in soybean (*Glycine max*) seedlings. *Plant Physiology* 97: 588–592.

Vigeolas, H., J.T. van Dongen, P. Waldeck, D. Huhn, and P. Geigenberger. 2003. Lipid storage metabolism is limited by the prevailing low oxygen concentrations within developing seeds of oilseed rape. *Plant Physiology* 133: 2048–2060.

Vriezen, W.H., R. Hulzink, C. Mariani, and L.A.C.J. Voesenek. 1999. 1-Aminocyclopropane-1-carboxylate oxidase activity limits ethylene biosynthesis in *Rumex palustris* during submergence. *Plant Physiology* 121: 189–195.

Watkins, C.B., J.E. Manzano-Mendez, J.F. Nock, J.J. Zhang, and K.E. Maloney. 1999. Cultivar variation in response of strawberry fruit to high carbon dioxide treatments. *Journal of Science, Food, and Agriculture* 79: 886–890.

Yahia, E.M. and A.A. Kader. 1991. Physiological and biochemical responses of avocado fruits to O_2 and CO_2 stress. *Plant Physiol* 96(suppl.): 96.

Yahia, E.M. and A. Carrillo-Lopez. 1993. Responses of avocado fruit to insecticidal O_2 and CO_2 atmospheres. *Food Science and Technology (Lebenn-Smittel-Wissenschatund Technologied)* 26: 307–311.

Yahia, E.M. and M. Hernandez. 1993. Tolerance and responses of harvested mango to insecticidal low-oxygen atmospheres. *HortScience* 28: 1031–1033.

Yamada, M., S. Taira, M. Ohtsuki, A. Sato, H. Iwanami, H. Yakushji, R. Wan, Y. Yang, and G. Li. 2002. Varietal differences in the ease of astringency removal by carbon dioxide gas and ethanol vapour treatments among oriental astringent persimmons of Japanese and Chinese origin. *Scientia Horticulturae* 94: 63–72.

23

Future Research and Application Needs

Adel A. Kader

CONTENTS

23.1 Introduction

This chapter is an update of my presentation on future research needs at the Ninth International Controlled Atmosphere Research Conference held at Michigan State University in July 2005.

The use of polymeric films for packaging produce and their application in modified atmosphere packaging (MAP) systems at the pallet, shipping container (plastic liner), and consumer package levels continues to increase (Al-Ati and Hotchkiss, 2003; Beaudry, 2000; Fonseca et al., 2000; Kader and Watkins, 2000; Lang, 2000; Watkins, 2000). MAP is widely used in extending the shelf life of fresh-cut vegetable and fruit products (Soliva-Fortuny and Martin-Belloso, 2003). Use of absorbers of ethylene, carbon dioxide, oxygen, and/or water vapor as part of MAP is increasing. Saltveit (2003) concluded that truly significant advances in the use of controlled atmosphere (CA) and modified atmosphere (MA) may require the development of mathematical models that incorporate some measure of the commodity's dynamic response to the storage environment. These measurements should reflect the commodity's changing response to various storage parameters (e.g., a shifting anaerobic compensation point), and should be useful in predicting future changes in quality.

Although much research has been done on use of surface coatings to modify the internal atmosphere within the commodity, commercial applications are still very limited due to the variability of the commodity's gas diffusion characteristics and the stability and thickness of the coating (Amarante and Banks, 2001; Hagenmaier, 2005; Schotsmans et al., 2003).

Several refinements in CA storage technology have been made in recent years (Kader, 2003b). These include the creation of nitrogen on demand by separation from compressed air using molecular sieve beds or membrane systems, use of low (0.7–1.5 kPa) O_2 levels that can be accurately monitored and controlled, rapid establishment of CA, ethylene-free CA, programmed (or sequential) CA (such as storage in 1 kPa O_2 for 2–6 weeks followed by storage in 2–3 kPa O_2 for the remainder of the storage period), and dynamic CA where levels of O_2 and CO_2 are modified as needed based on monitoring some attributes of produce quality, such as ethanol concentration and chlorophyll fluorescence (Prange et al., 2005; Veltman et al., 2003). There is renewed interest in hypobaric storage (Burg, 2004).

The use of CA in refrigerated marine containers continues to benefit from technological and scientific developments. CA transport is used to continue the CA chain for some commodities (such as apples, pears, and kiwifruits) that had been stored in CA since harvest. CA transport of bananas permits their harvest at a more fully mature stage, resulting in higher yield. CA transport of avocados facilitates use of a lower temperature (5°C) than if shipped in air because CA ameliorates chilling injury symptoms. CA combined with precision temperature management may allow nonchemical insect control in some commodities for markets that have restrictions against pests endemic to exporting countries (Mitcham, 2003) and for markets that prefer organic produce.

At the commercial level, CA is most widely applied during the storage and transport of apples and pears (Gross et al., 2004; Kader, 2001). It is also applied to a lesser extent on asparagus, broccoli, cantaloupes, kiwifruits, avocados, persimmons, pomegranates, and nuts and dried fruits. Atmospheric modification during transport is used on apples, avocados, bananas, blueberries, cherries, figs, kiwifruits, mangoes, nectarines, peaches, pears, plums, raspberries, and strawberries. Continued technological developments in the future to provide CA during transport and storage at reasonable cost (positive benefit/cost ratio) are essential to expanding its application on fresh fruits and vegetables.

Current trends that are expected to continue in the future include globalization of produce marketing, consolidation or formation of alliances among producers and marketers from various production areas, consolidation of retail marketing organizations, and increased demand for year-round supply of many produce items with better flavor. Maintaining the cold chain and the MA chain when needed are very important to globalization of produce marketing. For some commodities (such as apples, pears, and kiwifruits), a year-round supply from northern and southern hemisphere countries eliminates price incentives to domestic producers for "out-of-season" produce. This reduces the need for CA storage beyond 6 or 7 months in either hemisphere and has the potential of providing the consumers with better flavor-quality fruits. Also, there will be opportunities for using available CA storage facilities to store fruits from the other hemisphere (that are transported under optimal CA conditions) after the end of storage of locally produced fruits.

23.2 Flavor Quality of Fruits and Vegetables

It is not possible to discuss the future of MA research without considering the broader aspects of research aimed at maintaining quality of fresh horticultural perishables between harvest and consumption. Flavor attributes and associated constituents include sweetness (sugars), sourness or acidity (acids), astringency (tannins), bitterness (isocoumarins), aroma (odor-active volatile compounds), off-flavors due to acetaldehyde, ethanol, and/or ethyl acetate above certain concentrations (Kader, 2003a; Pesis, 2005), and off-odors (sulfurous compounds above certain concentrations). Mattheis and Fellman (2000) reviewed the impact of MAP and CA on aroma, flavor, and quality of horticultural commodities.

Nutritional quality of fruits and vegetables is determined by their contents of vitamins, minerals, dietary fiber, and antioxidant phytochemicals, such as carotenoids and flavonoids. It is important to determine the effects of MA during postharvest handling on these constituents as indicators of flavor and nutritional quality of intact and fresh-cut fruits and vegetables.

Providing better flavored fruits and vegetables is likely to increase their consumption, which would be good for the producers and marketers (making more money or at least staying in business) as well as for the consumers (increased consumption of healthy foods). To achieve this goal, we and all those involved in producing and marketing fruits and vegetables need to do the following:

1. Replace poor flavor cultivars with good flavor cultivars from among those that already exist and/or by selecting new cultivars with superior flavor and good textural quality.

2. Identify optimal cultural practices that maximize flavor quality, such as optimizing crop load and avoiding excess nitrogen and water, which along with low calcium shorten the postharvest life of the fruits and vegetables due to increased susceptibility to physical damage, physiological disorders, and decay.

3. Encourage producers to harvest fruits at partially ripe to fully ripe stages and harvest vegetables at their optimal maturity stages by developing handling methods that protect these commodities from physical damage.

4. Identify optimal postharvest handling conditions (time, temperature, relative humidity, atmospheric composition) that maintain flavor quality of fruits and vegetables and their value-added products. Postharvest life should be determined on the basis of flavor rather than appearance. Most of the published estimates of postharvest life under MA or CA are based on appearance (visual) quality and, in some cases, textural quality (Oosterhaven and Peppelenbos., 2003). These estimates should be revised to reflect the end of flavor-life when the product looks good but does not taste good. The end of flavor-life results from losses in sugars, acids, and aroma volatiles (especially esters) and/or development of off-flavors (due to fermentative metabolism or odor transfer from fungi or other sources). The possible role of MA in delaying these undesirable changes should be investigated.

5. Develop ready-to-eat, value-added products with good flavor. A very important research area is to find the optimal atmospheres for delaying browning and softening of fresh-cut products during distribution within the optimal ranges of temperature and relative humidity.

6. Optimize maturity/ripeness stage at the time of processing and select processing methods to retain good flavor of the processed products.

Future MA research can be part of research on strategies 4–6 in the list.

23.3 Maintaining the MA Chain

Continued improvements in polymeric films and other packaging materials will facilitate expanded use of MAP to extend postharvest life of fresh-cut fruits and vegetables and permit their distribution via vending machines and quick-service restaurants. MAP is an effective way to maintain the desired atmospheric composition between shipping point and the consumer's home. When evaluating polymeric films, it is important to place the control product in perforated plastic bags to separate the effect of the film on reducing water loss from its effect as a barrier to carbon dioxide and oxygen diffusion. Instead of developing more models for MAP, researchers are encouraged to build upon existing models and improve their accuracy.

Although much research has been done on the use of surface coatings to modify the atmosphere within many commodities, this technology has not been used to any extent because of the variability in composition among batches of the coating material. When combined with the natural variation in the gas diffusion characteristics among individual commodity units, a portion of each lot is lost due to off-flavors caused by fermentative metabolites. Further research is needed to overcome these constraints to use of surface coatings for modification of internal atmospheres of fruits and vegetables.

More cost-effective methods for establishing and maintaining MA will facilitate their use during storage at shipping points, transportation, and storage at destination points. Maintaining the MA chain is the second most important factor after the cold chain in keeping quality and safety of fresh produce between harvest and consumption. Brecht et al. (2003) described a procedure for designing a combination CA/MAP system that involves first designing the MAP for a particular commodity that will produce an optimal atmosphere for retail display conditions, then selecting a CA that will interact with the MAP to produce optimal atmosphere within the package during transportation at a lower temperature.

Dohring (2006) concluded that with the variety of atmosphere service choices available, it would help if researchers could quantify the value of each system to commodities being shipped. Since even the most advanced CA system available to shippers has limitations over that of laboratory- or land-based systems, comparative studies using the "actual equipment" available to the market would benefit all.

23.4 Combined Effects of MA and 1-MCP

Further evaluations are needed of the synergistic effects of MA and the ethylene-action-inhibitor,1-methylcyclopropene (1-MCP) (Blankenship and Dole, 2003; Watkins and Miller, 2005), on delaying ripening of partially ripe climacteric fruits and senescence of vegetables and deterioration (browning and softening) of fresh-cut products. Prange et al. (2005) concluded that combining CA with complementary strategies, such as decay or humidity control, delayed CA, high temperature CA, 1-MCP, and other 1-cyclopropenes, is a promising research direction.

23.5 MA for Decay and Insect Control

More research is needed to evaluate the efficacy of MA as a component of postharvest integrated pest management (decay and insect control) in fresh horticultural perishables. The fungistatic and insecticidal effects of low oxygen, elevated carbon dioxide, and super-atmospheric oxygen MA alone or in combination with other treatments merit further investigation (Kader and Ben-Yehoshua, 2000; Lagunas-Solar et al., 2006).

Yahia (2006) concluded that insecticidal atmospheres, especially in combination with other treatments such as heat, seem to be very promising and should be further investigated for all tropical crops. Information needed include the tolerance of different crops to these atmospheres, mortality of different species of insects, ideal gas composition, temperature, and duration of treatment.

23.6 MA and Food Safety Considerations

More research is needed to quantify the effects of MA on growth of pathogenic bacteria on fresh produce and on production of mycotoxins by fungi. Also, we need to understand how superatmospheric oxygen levels alone or in combination with elevated carbon dioxide levels influence growth of decay-causing bacteria and fungi and of human pathogens.

Erkan and Wang (2006) concluded that more research is needed on the microbiological safety of fresh-cut subtropical fruits and the effect of antimicrobial films (Suppakul et al., 2003) on various intact and fresh-cut subtropical commodities. Interaction between various gas concentrations, both MA and CA, the development of pathogenic microorganisms, and finding better film types for different species and cultivars of subtropical fruits also merits further research.

23.7 Biological Bases of MA Effects

Studies of the biological bases of MA effects on fresh horticultural perishables (Kader, 2003a) should be expanded to include superatmospheric oxygen concentrations (Kader and Ben-Yehoshua, 2000) and their interactions with elevated carbon dioxide levels. Brecht (2006) concluded that understanding the basic mechanisms that explain vegetable product tolerance to different gas atmospheres could help to streamline the optimization of MA systems for new products. Feedback systems that allow atmospheres to be adjusted in response to indications of product stress could replace current static MA systems and address the problem of fluctuating temperatures.

Wang (2006) concluded that specific biochemical and physiological reactions to MA and CA in each commodity require further investigation. The great diversity in shape, size, physical structure, and chemical composition of various fruits and vegetables creates variations in the response to MA and CA. Understanding more about these specific responses will be helpful in refining storage technology for each individual commodity.

Prange and DeLong (2006) concluded that recent research on several specific CA-related disorders has resulted in a number of models being proposed to explain how these disorders are affected by CA. To date, there is no consensus on any particular model and future research should focus on validation or improvement of these models. Yahia (2006) concluded that molecular studies are needed to identify clones for genes that are switched on or off in response to low O_2/high CO_2, in order to identify molecular markers to monitor responses of fruits to MA/CA, and try to manipulate tissue response.

23.8 Return on Investment of MA

Future expansion in the commercial use of MAP, MA during transport, and CA storage will depend on demonstrating a positive return on investment (ROI). Thus, it is important to estimate the ROI of every application before recommending its use. Dilley (2006) concluded that economic principles of supply and demand for the commodity in relation to energy use, cost, benefit, and practicality will ultimately determine the use of CA storage technology for postharvest maintenance of the product for ultimate consumption or use.

23.9 Summary

Future research and application needs of MA and CA are discussed in relation to maintaining quality of fresh horticultural perishables between harvest and consumption. Providing better flavored fruits and vegetables is likely to increase their consumption, which would be good for the producers and marketers (making more money or at least staying in business) as well as for the consumers (increased consumption of healthy foods). To achieve this goal, all those involved in producing and marketing fruits and vegetables need to identify and provide optimal postharvest handling conditions (time, temperature, relative humidity, atmospheric composition) that maintain flavor quality of fruits and vegetables and their value-added products. Postharvest life should be determined on the basis of flavor rather than appearance. The end of flavor-life results from losses in sugars, acids, and aroma volatiles (especially esters) and/or development of off-flavors (due to fermentative metabolism or odor transfer from fungi or other sources). Continued improvements in polymeric films and other packaging materials will facilitate expanded use of MAP to extend postharvest life of fresh-cut fruits and vegetables and permit their distribution via vending machines. More cost-effective methods for establishing and maintaining MA will facilitate expanded use during storage at shipping points, transportation, and storage at destination points. Maintaining the MA chain is the second most important factor after the cold chain in keeping quality and safety of fresh produce between harvest and consumption. Further evaluations are needed of (1) the synergistic effects of MA and the ethylene-action-inhibitor,1-MCP, on delaying ripening of partially-ripe climacteric fruits and senescence of vegetables; (2) MA as a component of postharvest integrated pest management (decay and insect control); (3) MA in relation to food safety considerations; and (4) the biological bases of MA effects on fresh horticultural perishables.

References

Al-Ati, T. and J.H. Hotchkiss. 2003. The role of packaging film permselectivity in modified atmosphere packaging. *J. Agric. Food Chem.* 51:4133–4138.

Amarante, C. and N.H. Banks. 2001. Postharvest physiology and quality of coated fruits and vegetables. *Hort. Rev.* 26:161–238.

Beaudry, R.M. 2000. Responses of horticultural commodities to low oxygen: Limits to the expanded use of modified atmosphere packaging. *HortTechnology* 10:491–500.

Blankenship, S.M. and J.M. Dole. 2003. 1-Methylcyclopropene: A review. *Postharvest Biol. Technol.* 28:1–25.

Brecht, J.K. 2006. Controlled atmosphere, modified atmosphere and modified atmosphere packaging for vegetables. *Stewart Postharvest Rev. 2006* 5:5. http://www.stewartpostharvest.com

Brecht, J.K., K.V. Chau, S.C. Fonseca, F.A.R. Oliveira, F.M. Silva, M.C.N. Nunes, and R.J. Bender. 2003. Maintaining optimal atmosphere conditions for fruits and vegetables throughout the postharvest handling chain. *Postharvest Biol. Technol.* 27:87–101.

Burg, S.P. 2004. *Postharvest Physiology and Hypobaric Storage of Fresh Produce.* CABI Publishing, Wallingford, U.K., 654 pp.

Dilley, D.R. 2006. Development of controlled atmosphere storage technologies. *Stewart Postharvest Review 2006* 6:5. www.stewartpostharvest.com

Dohring, S. 2006. Modified and controlled atmosphere reefer container transport technologies. *Stewart Postharvest Rev. 2006* 5:3. http://www.stewartpostharvest.com

Erkan, M. and C.Y. Wang. 2006. Modified and controlled atmosphere storage of subtropical crops. *Stewart Postharvest Rev. 2006* 5:4. http://www.stewartpostharvest.com

Fonseca, S.C., F.A.R. Oliveira, I.B.M. Lino, J.K. Brecht, and K.V. Chau. 2000. Modeling O_2 and CO_2 exchange for development of perforation-mediated modified atmosphere packaging. *J. Food Eng.* 43:9–16.

Gross, K., C.Y. Wang, and M.E. Saltveit (Eds.). 2004. *The Commercial Storage of Fruit, Vegetables, and Florist and Nursery Stocks. USDA Agricultural Handbook 66* (includes chapters on controlled atmosphere storage and modified atmosphere packaging). http://www.ba.ars.usda.gov/hb66/

Hagenmaier, R.D. 2005. A comparison of ethane, ethylene, and CO_2 peel permeance for fruit with different coatings. *Postharvest Biol. Technol.* 37:56–64.

Kader, A.A. (Ed.). 2001. CA Bibliography (1981–2000) and CA Recommendations (2001), CD. Davis: University of California, Postharvest Technology Center, *Postharvest Horticulture Series No. 22* (The CA Recommendations, 2001 portion is also available in printed format as *Postharvest Horticulture Series No. 22A*). http://postharvest.ucdavis.edu/Pubs/Pub_Desc_22A.pdf

Kader, A.A. 2003a. Physiology of CA treated produce. *Acta Hort.* 600:349–354.

Kader, A.A. 2003b. A perspective on postharvest horticulture (1978–2003). *HortScience* 38:1004–1008.

Kader, A.A. and S. Ben-Yehoshua. 2000. Effects of superatmospheric oxygen levels on postharvest physiology and quality of fresh fruits and vegetables. *Postharvest Biol. Technol.* 20:1–13.

Kader, A.A. and C.B. Watkins. 2000. Modified atmosphere packaging-toward 2000 and beyond. *HortTechnology* 10:483–486.

Lagunas-Solar, M.C., T.K. Essert, C. Pina, U.N.X. Zeng, and T.D. Troung. 2006. Metabolic stress disinfection and disinfestation (MSDD): A new non-thermal, residue-free process for fresh agricultural products. *J. Sci. Food Agric.* 86:1814–1825.

Lange, D.L. 2000. New film technologies for horticultural products. *HortTechnology* 10:487–490.

Mattheis, J.P. and J.K. Fellman. 2000. Impacts of modified atmosphere packaging and controlled atmosphere on aroma, flavor, and quality of horticultural commodities. *HortTechnology* 10:507–510.

Mitcham, E.J. 2003. Controlled atmospheres for insect and mite control in perishable commodities. *Acta Hort.* 600:137–142.

Oosterhaven, J. and H.W. Peppelenbos (Eds.). 2003. Proceedings of the 8th International Controlled Atmosphere Research Conference. *Acta Hort.* 600:1–838 (2 volumes).

Pesis, E. 2005. The role of the anaerobic metabolites, acetaldehyde and ethanol, in fruit ripening, enhancement of fruit quality and fruit deterioration. *Postharvest Biol. Technol.* 37:1–19.

Prange, R.K. and J.M. DeLong. 2006. Controlled-atmosphere related disorders of fruits and vegetables. *Stewart Postharvest Rev. 2006* 5:7. www.stewartpostharvest.com

Prange, R.K., J.M. DeLong, B.J. Daniels-Lake, and P.A. Harrison. 2005. Innovations in controlled atmosphere technology. *Stewart Postharvest Rev. 2005* 3:9. http://www.stewartpostharvest.com

Saltveit, M.E. 2003. Is it possible to find an optimal controlled atmosphere? *Postharvest Biol. Technol.* 27:3–13.

Schotsmans, W., B.E. Verlinden, J. Lammertyn, and B.M. Nicolai. 2003. Simultaneous measurement of oxygen and carbon dioxide diffusivity in pear fruit tissue. *Postharvest Biol. Technol.* 29:155–166.

Soliva-Fortuny, R.C. and O. Martin-Belloso. 2003. New advances in extending the shelf-life of fresh-cut fruits: A review. *Trends Food Sci. Technol.* 14:341–353.

Suppakul, P., J. Miltz, K. Sonneveld, and S.W. Bigger. 2003. Active packaging technologies with an emphasis on antimicrobial packaging and its applications. *J. Food Sci.* 68:408–420.

Veltman, R.H., J.A. Verschoor, and J.H. Ruijsch-van Dugteren. 2003. Dynamic control system (DCS) for apples (*Malus domestzca* Borkh cv 'Elstar'): Optimal quality through storage based on product response. *Postharvest Biol. Technol.* 27:79–86.

Wang, C.Y. 2006. Biochemical basis of the effects of modified and controlled atmospheres. *Stewart Postharvest Rev. 2006* 5:8. www.stewartpostharvest.com

Watkins, C.B. 2000. Responses of horticultural commodities to high carbon dioxide as related to modified atmosphere packaging. *HortTechnology* 10:501–506.

Watkins, C.B. and W.B. Miller. 2005. A summary of physiological processes or disorders in fruits, vegetables and ornamental products that are delayed or decreased, increased, or unaffected by application of 1-methylcyclopropene (I -MCP). http://www.hort.cornell.edu/mcp/

Yahia, E.M. 2006. Modified and controlled atmospheres for tropical fruits. *Stewart Postharvest Rev. 2006* 5:6. http://www.stewartpostharvest.com

Index

Milton Keynes UK
Ingram Content Group UK Ltd.
UKHW052028071024
449327UK00027B/2482